S. Sumathi, S.N. Sivanandam

Introduction to Data Mining and its Applications

AF654607

Studies in Computational Intelligence, Volume 29

Editor-in-chief
Prof. Janusz Kacprzyk
Systems Research Institute
Polish Academy of Sciences
ul. Newelska 6
01-447 Warsaw
Poland
E-mail: kacprzyk@ibspan.waw.pl

Further volumes of this series can be found on our homepage: springer.com

Vol. 12. Jonathan Lawry
Modelling and Reasoning with Vague Concepts, 2006
ISBN 0-387-29056-7

Vol. 13. Nadia Nedjah, Ajith Abraham, Luiza de Macedo Mourelle (Eds.)
Genetic Systems Programming, 2006
ISBN 3-540-29849-5

Vol. 14. Spiros Sirmakessis (Ed.)
Adaptive and Personalized Semantic Web, 2006
ISBN 3-540-30605-6

Vol. 15. Lei Zhi Chen, Sing Kiong Nguang, Xiao Dong Chen
Modelling and Optimization of Biotechnological Processes, 2006
ISBN 3-540-30634-X

Vol. 16. Yaochu Jin (Ed.)
Multi-Objective Machine Learning, 2006
ISBN 3-540-30676-5

Vol. 17. Te-Ming Huang, Vojislav Kecman, Ivica Kopriva
Kernel Based Algorithms for Mining Huge Data Sets, 2006
ISBN 3-540-31681-7

Vol. 18. Chang Wook Ahn
Advances in Evolutionary Algorithms, 2006
ISBN 3-540-31758-9

Vol. 19. Ajita Ichalkaranje, Nikhil Ichalkaranje, Lakhmi C. Jain (Eds.)
Intelligent Paradigms for Assistive and Preventive Healthcare, 2006
ISBN 3-540-31762-7

Vol. 20. Wojciech Penczek, Agata Półrola
Advances in Verification of Time Petri Nets and Timed Automata, 2006
ISBN 3-540-32869-6

Vol. 21. Cândida Ferreira
Gene Expression on Programming: Mathematical Modeling by an Artificial Intelligence, 2006
ISBN 3-540-32796-7

Vol. 22. N. Nedjah, E. Alba, L. de Macedo Mourelle (Eds.)
Parallel Evolutionary Computations, 2006
ISBN 3-540-32837-8

Vol. 23. M. Last, Z. Volkovich, A. Kandel (Eds.)
Algorithmic Techniques for Data Mining, 2006
ISBN 3-540-33880-2

Vol. 24. Alakananda Bhattacharya, Amit Konar, Ajit K. Mandal
Parallel and Distributed Logic Programming, 2006
ISBN 3-540-33458-0

Vol. 25. Zoltán Ésik, Carlos Martín-Vide, Victor Mitrana (Eds.)
Recent Advances in Formal Languages and Applications, 2006
ISBN 3-540-33460-2

Vol. 26. Nadia Nedjah, Luiza de Macedo Mourelle (Eds.)
Swarm Intelligent Systems, 2006
ISBN 3-540-33868-3

Vol. 27. Vassilis G. Kaburlasos
Towards a Unified Modeling and Knowledge-Representation based on Lattice Theory, 2006
ISBN 3-540-34169-2

Vol. 28. Brahim Chaib-draa, Jörg P. Müller (Eds.)
Multiagent based Supply Chain Management, 2006
ISBN 3-540-33875-6

Vol. 29. S. Sumathi, S.N. Sivanandam
Introduction to Data Mining and its Applications, 2006
ISBN 3-540-34350-4

S. Sumathi
S.N. Sivanandam

Introduction to Data Mining and its Applications

With 108 Figures and 23 Tables

Springer

Dr. S. Sumathi
Assistant Professor
Department of Electrical and Electronics Engineering
PSG Colloege of Technology
Coimbatore 641 004
Tamil Nadu, India

Dr. S.N. Sivanandam
Professor and Head
Department of Electrical and Electronics Engineering
PSG Colloege of Technology
P.O Box 1611
Peelamedu
Coimbatore 641 004
Tamil Nadu, India

Library of Congress Control Number: 2006926723

ISBN print edition: 1860-949X
ISSN electronic edition: 1860-9503
ISBN-10 3-540-34350-4 Springer Berlin Heidelberg New York
ISBN-13 978-3-540-34350-9 Springer Berlin Heidelberg New York

This work is subject to copyright. All rights reserved, whether the whole or part of the material is concerned, specifically the rights of translation, reprinting, reuse of illustrations, recitation, broadcasting, reproduction on microfilm or in any other way, and storage in data banks. Duplication of this publication or parts thereof is permitted only under the provisions of the German Copyright Law of September 9, 1965, in its current version, and permission for use must always be obtained from Springer-Verlag. Violations are liable to prosecution under the German Copyright Law.

The use of general descriptive names, registered names, trademarks etc. in this publication does not imply, even in the absence of a specific statement, that such names are exempt from the relevant protective laws and regulations and therefore free for general use.

Cover design: deblik, Berlin

Sumathi & Sivanandam: Introduction to Data Mining and its Application

© 2006 Springer-Verlag Berlin Heidelberg

All rights reserved. No part of this publication may be reproduced, stored in any electronic or mechanical form, including photocopy, recording or otherwise, without the prior written permission of the publisher.

First Indian Reprint 2009

ISBN 978-81-8489-034-1

This edition is manufactured in India for sale only in India, Pakistan, Bangladesh, Nepal and Sri Lanka and any other country as authorized by the publisher.

This edition is published by Springer (India) Private Limited,
A part of Springer Science+Business Media, Registered Office: 906-907,
Akash Deep Building, Barakhamba Road, New Delhi – 110 001, India.

Printed in India by Rajkamal Electric Press, Delhi.

Contents

1

Introduction to Data Mining Principles

Objectives:

- This section deals with detailed study of the principles of data warehousing, data mining, and knowledge discovery.
- The availability of very large volumes of such data has created a problem of how to extract useful, task-oriented knowledge.
- The aim of data mining is to extract implicit, previously unknown and potentially useful patterns from data.
- Data warehousing represents an ideal vision of maintaining a central repository of all organizational data.
- Centralization of data is needed to maximize user access and analysis.
- Data warehouse is an enabled relational database system designed to support very large databases (VLDB) at a significantly higher level of performance and manageability.
- Due to the huge size of data and the amount of computation involved in knowledge discovery, parallel processing is an essential component for any successful large-scale data mining application.
- Data warehousing provides the enterprise with a memory. Data mining provides the enterprise with intelligence.
- Data mining is an interdisciplinary field bringing together techniques from machine learning, pattern recognition, statistics, databases, visualization, and neural networks.
- We analyze the knowledge discovery process, discuss the different stages of this process in depth, and illustrate potential problem areas with examples.

Abstract. This section deals with a detailed study of the principles of data warehousing, data mining, and knowledge discovery. There exist limitations in the traditional data analysis techniques like regression analysis, cluster analysis, numerical taxonomy, multidimensional analysis, other multivariate statistical methods, and stochastic models. Even though these techniques have been widely used for solving many practical problems, they are however primarily oriented toward the extraction

A. Lew and H. Mauch: *Introduction to Data Mining Principles*, Studies in Computational Intelligence (SCI) **38**, 1–20 (2006)
www.springerlink.com
© Springer-Verlag Berlin Heidelberg 2006

of quantitative and statistical data characteristics. To satisfy the growing need for new data analysis tools that will overcome the above limitations, researchers have turned to ideas and methods developed in machine learning. The efforts have led to the emergence of a new research area, frequently called *data mining and knowledge discovery*. Data mining is a multidisciplinary field drawing works from statistics, database technology, artificial intelligence, pattern recognition, machine learning, information theory, knowledge acquisition, information retrieval, high-performance computing, and data visualization. Data warehousing is defined as a process of centralized data management and retrieval.

1.1 Data Mining and Knowledge Discovery

An enormous proliferation of databases in almost every area of human endeavor has created a great demand for new, powerful tools for turning data into useful, task-oriented knowledge. In the efforts to satisfy this need, researchers have been exploring ideas and methods developed in machine learning, pattern recognition, statistical data analysis, data visualization, neural nets, etc. These efforts have led to the emergence of a new research area, frequently called *data mining and knowledge discovery*.

The current Information Age is characterized by an extraordinary growth of data that are being generated and stored about all kinds of human endeavors. An increasing proportion of these data is recorded in the form of computer databases, so that the computer technology may easily access it. The availability of very large volumes of such data has created a problem of how to extract form useful, task-oriented knowledge.

Data analysis techniques that have been traditionally used for such tasks include regression analysis, cluster analysis, numerical taxonomy, multidimensional analysis, other multivariate statistical methods, stochastic models, time series analysis, nonlinear estimation techniques, and others. These techniques have been widely used for solving many practical problems. They are, however, primarily oriented toward the extraction of quantitative and statistical data characteristics, and as such have inherent limitations.

For example, a statistical analysis can determine covariances and correlations between variables in data. It cannot, however, characterize the dependencies at an abstract, conceptual level and procedure, a casual explanation of reasons why these dependencies exist. Nor can it develop a justification of these relationships in the form of higher-level logic-style descriptions and laws. A statistical data analysis can determine the central tendency and variance of given factors, and a regression analysis can fit a curve to a set of datapoints. These techniques cannot, however, produce a qualitative description of the regularities and determine their dependence of factors not explicitly provided in the data, nor can they draw an analogy between the discovered regularity and regularity in another domain.

A numerical taxonomy technique can create a classification of entities and specify a numerical similarity among the entities assembled into the same or different categories. It cannot, however, build qualitative description of the classes created and hypothesis reasons for the entities being in the same category. Attributes that define the similarity, as well as the similarity measures, must be defined by a data analyst in advance. Also, these techniques cannot by themselves draw upon background domain knowledge in order to automatically generate relevant attributes and determine their changing relevance to different data analysis problems.

To address such tasks as those listed above, a data analysis system has to be equipped with a substantial amount of background and be able to perform symbolic reasoning tasks involving that knowledge and the data. In summary, traditional data analysis techniques facilitate useful data interpretations and can help to generate important insights into the processes behind the data. These interpretations and insights are the ultimate knowledge sought by those who build databases. Yet, such knowledge is not created by these tools, but instead has to be derived by human data analysis.

In efforts to satisfy the growing need for new data analysis tools that will overcome the above limitations, researchers have turned to ideas and methods developed in machine learning. The field of machine learning is a natural source of ideas for this purpose, because the essence of research in this field is to develop computational models for acquiring knowledge from facts and background knowledge. These and related efforts have led to the emergence of a new research area, frequently called *data mining and knowledge discovery.*

There is confusion about the exact meaning of the terms "data mining" and "KDD." KDD was proposed in 1995 to describe the whole process of extraction of knowledge from data. In this context, knowledge means relationships and patterns between data elements. "Data mining" should be used exclusively for the discovery stage of the KDD process.

The last decade has experienced a revolution in information availability and exchange via the Internet. The World Wide Web is growing at an exponential rate and we are far from any level of saturation. E-commerce and other innovative usages of the worldwide electronic information exchange have just started. In the same spirit, more and more businesses and organizations have begun to collect data on their own operations and market opportunities on a large scale. This trend is rapidly increasing, with recent emphasis being put more on collecting the right data rather than storing all information in an encyclopedic fashion without further using it. New challenges arise for business and scientific users in structuring the information in a consistent way. Beyond the immediate purpose of tracking, accounting for, and archiving the activities of an organization, this data can sometimes be a *gold mine* for strategic planning, which recent research and new businesses have only started to tap. Research and development in this area, often referred to as *data mining and knowledge discovery,* has experienced a tremendous growth in the last couple of years. The goal of these methods and algorithms is to extract useful regularities from large data archives, either directly in the form of "knowledge" characterizing the relations between the variables of interest, or indirectly as functions that allow to predict, classify, or represent regularities in the distribution of the data.

What are the grand challenges for information and computer science, statistics, and algorithmics in the new field of data mining and knowledge discovery? The huge amount of data renders it possible for the data analysis to infer data models with an unprecedented level of complexity. Robust and efficient algorithms have to be developed to handle large sets of high-dimensional data. Innovations are also required in the area of database technology to support interactive data mining and knowledge discovery. The user with his knowledge and intuition about the application domain should be able to participate in the search for new structures in data, e.g., to introduce a *priori* knowledge and to guide search strategies. The final step in the inference chain is the validation of the data where new techniques are called for to cope with the large complexity of the models.

Statistics as the traditional field of inference has provided models with more or less detailed assumptions on the data distribution. The classical theory of Bayesian inference has demonstrated its usefulness in a large variety of application domains ranging from medical applications to consumer data and market basket analysis. In addition to classical methods, neural networks and machine learning have contributed ideas, concepts, and algorithms to the analysis of these data sets with a distinctive new flavor. The new approaches put forward by these researchers in the last decade depart from traditional statistical data analysis in several ways: they rely less on statistical assumptions on the actual distribution of the data, they rely less on models allowing simple mathematical analysis, but they use sophisticated models that can learn complicated nonlinear dependencies from large data sets. Whereas statistics has long been a tool for testing theories proposed by scientists, machine learning and neural network research are rather evaluated on the basis of how well they *generalize on new data*, which come from the same unknown process that generated the training data. Measuring the generalization performance to select models has to be distinguished from the widespread but questionable current practice of data inquisition where "the data are tortured until they confess."

During the last 15 years, various techniques have been proposed to improve the generalization properties of neural estimators. The basic mechanism is to control the richness of the class of possible functions that can be obtained through training, which has been quantified with the seminal work of Vapnik and Chervonenkis on the "*capacity of a hypothesis class.*" The combinational concept of the *VC* dimensions and its generalizations parameterize a rigorous but loose upper bound on large deviations of the empirical risk from the expected risk of classification or regression. Such theoretical bounds can help us understand the phenomenon of generalization. To answer a numerical question about a particular algorithm and data set, purely quantitative empirical bounds on the expected generalization error can be obtained by repeating many training/test simulations, and they are tighter than the analytic theoretical bounds. Heuristics that essentially implement complexity control in one way or another are the widely used weight decay in training multilayer perceptrons or the early stopping rule during training. It is also possible to view capacity control in terms of penalty terms for too complex estimators.

Complexity control is particularly relevant for data mining. In this area, researchers look for complex but still valid characterizations of their large data sets. Despite the large size of the data sets inference often takes place in the small sample size limit. It should be noted that the ratio of samples to degrees of freedom might be small even for large data sets when complex models like deep decision trees or support vector machines in high-dimensional spaces are used. Complexity control, either by numerical techniques like cross validation or by theoretical bounds from computational learning theory with empirical rescaling, is indispensable for data mining practitioners.

The enterprise of knowledge discovery aims at the automation of the millennium-old effort of humans to gain information and build models and theories about phenomena in the world around us. Data miners and knowledge discoverers can learn a lot and, i.e., sharpen their awareness, by looking at the scientific method of experimentation, modeling, and validation/falsification in the natural sciences, engineering sciences, social sciences, economics, as well as philosophy.

The next decade of research in network-based information services promises to deliver widely available access to unprecedented amounts of constantly expanding data. Users of many commercial, government, and private information services will benefit from new machine learning technologies that mine new knowledge by integrating and analyzing very large amounts of widely distributed data to uncover and report upon subtle relationships and patterns of events that are not immediately discernible by direct human inspection.

1.2 Data Warehousing and Data Mining - Overview

The past decade has seen an explosive growth in database technology and the amount of data collected. Advances in data collection, use of bar codes in commercial outlets, and the computerization of business transactions have flooded us with lots of data. We have an unprecedented opportunity to analyze this data to extract more intelligent and useful information, and to discover interesting, useful, and previously unknown patterns from data. Due to the huge size of data and the amount of computation involved in knowledge discovery, parallel processing is an essential component for any successful large-scale data mining application.

Data mining is concerned with finding hidden relationships present in business data to allow businesses to make predictions for future use. It is the process of data-driven extraction of not so obvious but useful information from large databases. Data mining has emerged as a key business intelligence technology.

The explosive growth of stored data has generated an information glut, as the storage of data alone does not bring about knowledge that can be used: (a) to improve business and services and (b) to help develop new techniques and products. Data is the basic form of information that needs to be managed, sifted, mined, and interpreted to create knowledge. Discovering the patterns, trends, and anomalies in massive data is one of the grand challenges of the Information Age. Data mining emerged in the late 1980s, made great progress during the Information Age and in the 1990s, and will continue its fast development in the years to come in this increasingly data-centric world. Data mining is a multidisciplinary field drawing works from statistics, database technology, artificial intelligence, pattern recognition, machine learning, information theory, knowledge acquisition, information retrieval, high-performance computing, and data visualization.

The aim of data mining is to extract implicit, previously unknown and potentially useful (or actionable) patterns from data. Data mining consists of many up-to-date techniques such as classification (decision trees, naïve Bayes classifier, k-nearest neighbor, neural networks), clustering (k-means, hierarchical clustering, density-based clustering), association (one-dimensional, multidimensional, multilevel association, constraint-based association). Many years of practice show that data mining is a process, and its successful application requires data preprocessing (dimensionality reduction, cleaning, noise/outlier removal), postprocessing (understandability, summary, presentation), good understanding of problem domains and domain expertise.

Today's competitive marketplace challenges even the most successful companies to protect and retain their customer base, manage supplier partnerships, and control costs while at the same time increasing their revenue. In a world of accelerating change, competitive advantage will be defined by the ability to leverage information to initiate effective business decisions before competition does. Hence in this age of global competition accurate information plays a vital role in the insurance business. Data is not merely a record of business operation – it helps in achieving competitive advantages in the insurance sector. Thus, there is growing pressure on MIS managers to provide information technology (IT) infrastructure to enable decision support mechanism. This would be possible provided the decision makers have online access to previous data. Therefore, there is a need for developing a data warehouse. Data mining as a tool for customer relationship management also has proved to be a means of controlling costs and increase revenues.

In the last decade, machine learning had come of age through a number of ways such as neural networks, statistical pattern recognition, fuzzy logic, and genetic algorithms. Among the most important applications for machine learning are classification, recognition, prediction, and data mining. Classification and recognition are very significant in a lot of domains such as multimedia, radar, sonar, optical character recognition, speech recognition, vision, agriculture, and medicine. In this section, the concept of data warehousing and data mining is briefly presented.

1.2.1 Data Warehousing Overview

Dramatic advances in data capture, processing power, data transmission, and storage capabilities are enabling organizations to integrate their various databases into data warehouses. *Data warehousing* is defined as a process of centralized data management and retrieval. Data warehousing, like data mining, is a relatively new term although the concept itself has been around for years. Data warehousing represents an ideal vision of maintaining a central repository of all organizational data. Centralization of data is needed to maximize user access and analysis. Dramatic technological advances are making this vision a reality for many companies. And, equally dramatic advances in data analysis software are allowing users to access these data freely. The data analysis software is what supports data mining. Hence, data warehousing provides the enterprise with a memory. Data mining provides the enterprise with intelligence.

Data warehouse is an enabled relational database system designed to support very large databases (VLDB) at a significantly higher level of performance and manageability. Data warehouse is an environment, not a product. It is an architectural construct of information that is hard to access or present in traditional operational data stores.

Any organization or a system in general is faced with a wealth of data that is maintained and stored, but the inability to discover valuable, often previously unknown information hidden in the data, prevents it from transferring these data into knowledge or wisdom.

To satisfy these requirements, these steps are to be followed.

1. Capture and integrate both the internal and external data into a comprehensive view "Mine" for the integrated data information
2. Organize and present the information and knowledge in ways that expedite complex decision making.

Access Tools for Data Warehousing

The principal purpose of data warehousing is to provide information to users for strategic decision making. These users interact with the data warehouse using front-end tools. Many of these tools require an information specialist, although many end users develop expertise in the tools. The access tools are divided into five main groups.

1. Data query and reporting tools
2. Application development tools
3. Executive information system (EIS) tools
4. Online analytical preprocessing tools and
5. Data mining tools

Data mining tools are considered for information extraction from data. In recent research, data mining through pattern classification is an important area of concentration.

1.2.2 Concept of Data Mining

Database technology has been used with great success in traditional business data processing. There is an increasing desire to use this technology in new application domains. One such application domain that is likely to acquire considerable significance in the near future is database mining. An increasing number of organizations are creating ultralarge databases (measured in gigabytes and even terabytes) of business data, such as consumer data, transaction histories, sales records, etc.; such data forms a potential gold mine of valuable business information.

Data mining is a relatively new and promising technology. It can be defined as the process of discovering meaningful new correlation, patterns, and trends by digging into (mining) large amounts of data stored in warehouse, using statistical, machine learning, artificial intelligence (AI), and data visualization techniques. Industries that are already taking advantage of data mining include medical, manufacturing, aerospace, chemical, etc. Knowledgeable observers generally agree that in-depth decision support requires new technology. This new technology should enable the discovery of trends and predictive patterns in data, the creation and testing of hypothesis, and generation of insight-provoking visualizations.

Data mining helps the end users to extract useful information from large databases. These large databases are present in data warehouses, i.e., "data mountain," which are presented to data mining tools. In short data warehousing allows one to build the data mountain. Data mining is the nontrivial extraction of implicit, previously unknown and potentially useful information from the data mountain. This data mining is not specific to any industry – it requires intelligent technologies and the willingness to explore the possibility of hidden knowledge that resides in the data. Data mining is also referred to as knowledge discovery in databases (KDD).

Data, Information, and Knowledge

Data: Data are any facts, numbers, or text that can be processed by a computer. Today organizations are accumulating vast and growing amounts of data in different formats and databases.

This includes: Operational or transactional data such as sales, cost, inventory, payroll, and accounting.

Nonoperational data like industry sales, forecast data, and macroeconomic data.

Metadata: data about the data itself such as logical database design or data dictionary definitions.

Information: The patterns, associations, or relationships among all this data can provide information. For example, analysis of retail point-of-sale transaction data can yield information on which products are selling and when.

Knowledge: Information can be converted into knowledge about historical patterns and future trends. For example, summary information on retail supermarket sales can be analyzed in light of promotional efforts to provide knowledge or consumer buying behavior. Thus a manufacturer or a retailer could determine those items that are most susceptible to promotional efforts.

Data Mining Definitions

- Data mining is the efficient discovery of valuable, nonobvious information from a large collection of data.
- Knowledge discovery in databases is the nontrivial process of identifying valid novel potentially useful and ultimately understandable patterns in the data.
- It is the automatic discovery of new facts and relationships in data that are like valuable nuggets of business data.
- It is not a complex query where the user already has a suspicion about a relationship in the data and wants to pull all such information.
- The information discovered should give competitive advantage in business.
- Data mining is the induction of understandable models and patterns from a database.
- It is the process of extracting previously unknown, valid, and actionable information from large databases and then using the information to make crucial business decisions.

It is an interdisciplinary field bringing together techniques from machine learning, pattern recognition, statistics, databases, visualization, and neural networks.

Data mining is streamlining the transformation of masses of information into meaningful knowledge. It is a process that helps identify new opportunities by finding fundamental truths in apparently random data. The patterns revealed can shed light on application problems and assist in more useful, proactive decision making. Typical techniques for data mining involve decision trees, neural networks, nearest neighbor clustering, fuzzy logic, and genetic algorithms.

Now we focus on the relationship between data mining and data warehousing.

What is a data warehouse and why do we need it?

In most organizations we find really large databases in operation for normal daily transactions. These databases are known as operational databases; in most cases they have not been designed to store historical data or to respond to queries but simply support all the applications for day-to-day transactions. The second type of database found in organizations is the data warehouse, which is designed for strategic decision support and largely built up from operational databases. Small, local data warehouses are called data marts.

Rules for Data Warehouses:

- Time dependent
- Nonvolatile data in data warehousing is never updated but used only for queries. This means that a data warehouse will always be filled with historical data.
- Subject oriented
- Integrated

A data warehouse is designed especially for decision support queries; therefore only data that is needed for decision support will be extracted from the operational data and stored in data warehouse. Setting up a data warehouse is the most appropriate procedure for carrying out decision support. A decision support system can constantly change if the requirement of the organization alters, then the data model must also change. The data warehouse requires a high-speed machine and a wide variety of optimization processes.

- Metadata: describes the structure of the contents of a database.

Designing Decision Support Systems

The design of a decision support system differs considerably from that of an online transaction processing system. The main difference is that decision support systems are used only for queries, so their structure should be optimized for this use. When designing a decision support system, particular importance should be placed on the requirements of the end user and the hardware and software products that will be required.

The Requirements of EndUser

Some end users need specific query tools so that they can build their queries themselves, others are interested only in a particular part of the information. They may also need trend analysis tools and GUI user interface.

Software Products of Decision Support Systems

The types of software we choose depend very much on the requirements of end users. For data mining we can split the software into two parts: the first works with the algorithms on the database server and the second on the local workstation. The latter is mostly used to generate screens and reports for end users for visualizing the output of algorithm.

Hardware Products of Decision Support Systems

The hardware requirements depend on the type of data warehouse and techniques with which we want to work.

Integration with Data Mining

The application of data mining techniques can be carried out in two ways: from the existing data warehouse, or by extracting from the existing data warehouse the part of the information that is of interest to the end user and copying it to a specific computer, possibly a multiprocessing machine.

Integration of data mining in a decision support system is very helpful. There are several types of data mining technique and each uses the computer in a specific way. For this reason it is important to understand the demands of the end user so that we are able to build a proper data warehouse for data mining. In many cases we will find that we need a separate computer for data mining.

Client/Server and Data Warehousing

The end user would ideally like to have available all kinds of techniques such as graphical user interfaces, statistical techniques, windowing mechanisms, and visualization techniques so that they can easily access the data being sought. This means that a great deal of local computer power is needed at each workstation, and the client/server technique is the solution to this problem.

With client/server we only have to change the piece of software that is related to the end use—the other applications do not require alteration. Of all the techniques currently available on the market, client/server represents the best choice for building a data warehouse.

Replication techniques are used to load the information from the operational database to the data warehouse. If we need immediate access to the latest information, then we need to work with the more advanced replication tools; if the update of the data warehouse is less urgent, then we can work with batch update of the database server.

Two basic techniques, known as the “top-down” and the “bottom-up” approaches, are used to build a data warehouse.

In the “top-down” approach, we first build a data warehouse for the complete organization and from this select the information needed for our department or for local end users. In the “bottom-up” approach, smaller local data warehouses, known as data marts, are used by end users at a local level for their specific local requirements.

Multiprocessing Machines

A data mining environment has specific hardware requirements. There are several types of multiprocessing machines and we describe the most important ones here:

- Symmetric multiprocessing

All processors work on one computer, are equal, and they communicate via shared storage. Symmetric multiprocessing machines share the same hard disk and the internal memory. At present, approximately twelve processors are the maximum.

- Massively parallel

This is a computer where each processor has its own operating system, memory, and hard disk. Although each processor is independent, communication between the systems is possible. In this type of environment one can work with thousands of processors.

Not all databases will support parallel machines but most modern databases are able to work with symmetric parallel machines. At present, only a few database vendors such as IBM with DB/2, Oracle, and Tandem are able to operate with massively parallel computers.

- Cost justification

It is difficult to give a cost justification for the implementation of a KDD environment. Basically the cost of using machine-learning techniques to recognize patterns in data must be compared with the cost of a human performing the same task.

The Knowledge Discovery Process

We analyze the knowledge discovery process, discuss the different stages of this process in depth, and illustrate potential problem areas with examples.

The knowledge discovery process consists of six stages:

- Data selection
- Cleaning
- Enrichment
- Coding
- Data mining
- Reporting

It is impossible to describe in advance all the problems that can be expected in a database, as most will be discovered in mining stage.

Data Selection and Cleaning: A very important element in a cleaning operation is the de-duplication of records. Although data mining and data cleaning are two different disciplines, they have a lot in a common and pattern recognition algorithms can be applied in cleaning data. One kind of errors is spelling errors. The second type of pollution that frequently occurs is lack of domain consistency. For instance, a transaction listed in table was completed in 1901 but the company was set up after 1901.

Enrichment: New information can easily be joined to the existing client records.

Coding: In most tables that are collected from operational data, a lot of desirable data is missing, and most is impossible to retrieve. We therefore have to make a deliberate decision either to overlook or to delete it. A general rule states that any deletion of data must be a conscious decision, after a thorough analysis of the possible consequences. We can remove some unrelated attributes form current tables. By this time, the information in database is much too detailed to be used as input for pattern recognition algorithms. For instance, address to region, birth data to age, divide income by 1000, etc.

Data mining: The discovery stage of the KDD process is fascinating. We now see that some learning algorithms do well on one part of the data set where others fail, and this clearly indicates the need for hybrid learning.

Although various different techniques are used for different purposes, those that are of interest in the present context are:

Query tool
Statistical techniques
Visualization
Online analytical processing (OLAP)
Case-based learning (K-Nearest Neighbor)
Decision trees
Association rules
Neural networks
Genetic algorithm

Preliminary Analysis of the Data Set Using Traditional Query Tools: The first step in a data mining project should always be a rough analysis of the data set using traditional query tools. Just by applying simple structured query language (SQL) to a data set, we can obtain a wealth of information. We need to know the basic aspects and structures of the data set. For the most part 80% of the interesting information can be abstracted from a database using SQL. The remaining 20% of hidden information needs more advanced techniques. A trivial result that is obtained by an extremely simple method is called a naïve prediction. We can never judge the performance of an advanced learning algorithm properly if we have no information concerning the naïve probabilities of what it is supposed to predict.

Visualization Techniques: Visualization techniques are a very useful method of discovering patterns in data sets and may be used at the beginning of a data mining process to get a rough feeling of the quality of the data set and where patterns are to be found. An elementary technique that can be of great value is the so-called *scatter diagram.* Scatter diagrams can be used to identify interesting subsets of the data sets so that we can focus on the rest of the data mining process. There is a whole field of research dedicated to the search for interesting projections of data sets – this is called projection pursuit. A much better way to explore a data set is through an interactive three-dimensional environment.

Likelihood and Distance: The space metaphor is very useful in data mining context. Records that are closer to each other are very alike, and those that are very far from each other represent individuals that have little in common. Sometimes it is possible to identify interesting clusters merely by visual inspection.

OLAP Tools: This idea of dimensionality can be expanded: a table with n independent attributes can be seen as an n-dimensional space. We need to explore the relationship between these dimensions as standard relational database is not very good at this. OLAP tools were developed to solve this problem. These tools store their data in a special multidimensional format.

OLAP can be an important stage in a data mining processes. However there is an important difference between OLAP and data mining: OLAP tools do not learn; data mining is more powerful than OLAP and also needs no special multi-dimensional storage.

K-Nearest Neighbor: When we interpret records as points in a data space, we can define the concept of neighborhood records that are close to each other live in each other's neighborhood. In terms of the metaphor of our multi-dimensional data space, a type is nothing more than a region in this data space. Based on this insight, we can develop a very simple but powerful learning algorithm – the k-nearest neighbor. The basic philosophy of k-nearest neighbor is "do as our neighbors do." If we want to predict the behavior of a certain individual, we start to look at the behaviors of its neighbors. The letter k stands for the number of neighbors we have investigated. Simple k-nearest neighbor is not really a learning algorithm, but more of a search method. In general data mining algorithms should not have a complexity higher than n (log n) (where n is the number of records). The other techniques such as decision trees, association rules, neural networks, and genetic algorithms are discussed in the following sections.

Principles of Data Mining

Data mining is a powerful new technology with great potential to help companies focus on the most important information in the data they have collected about the behavior of their customers and potential customers. It discovers information within the data that queries and reports cannot effectively reveal. The section explores many aspects of data mining in the following areas:

- Data rich, information poor
- Data warehouses
- What is data mining?
- What can data mining do?
- The evolution of data mining
- How data mining works
- Data mining technologies
- Real-world examples
- The future of data mining
- Privacy concerns

Data Rich, Information Poor

The amount of raw data stored in corporate databases is exploding. From trillions of point-of-sale transactions and credit card purchases to pixel-by-pixel images of galaxies, databases are now measured in gigabytes and terabytes. (One terabyte = one trillion bytes. A terabyte is equivalent to about 2 million books!) For instance, every day, Wal Mart uploads 20 million point-of-sale transactions to an A&T massively parallel system with 483 processors running a centralized database. Raw data by itself, however, does not provide much information. In today's fiercely competitive business environment, companies need to rapidly turn these terabytes of raw data into significant insights for their customers and markets to guide their marketing, investment, and management strategies.

Data Warehouses

The drop in price of data storage has given companies willing to make the investment a tremendous resource: Data about their customers and potential customers stored in "data warehouses." Data warehouses are becoming part of the technology. Data warehouses are used to consolidate data located in disparate databases. A data warehouse stores large quantities of data by specific categories; so it can be more easily retrieved, interpreted, and sorted by users. Warehouses enable executives and managers to work with vast stores of transactional or other data to respond faster to markets and make more informed business decisions. It has been predicted that every business will have a data warehouse within ten years. Companies will want to learn more about that data to improve knowledge of customers and markets. The companies benefit when meaningful trends and patterns are extracted from the data.

What is Data Mining?

Data mining, or knowledge discovery, is the computer-assisted process of digging through and analyzing enormous sets of data and then extracting the meaning of the data. Data mining tools predict behaviors and future trends, allowing businesses to make proactive, knowledge-driven decisions. Data mining tools can answer business questions that were traditionally too time consuming to resolve. They scour databases for hidden patterns, finding predictive information that experts may miss because it lies outside their expectations.

Data mining derives its name from the similarities between searching for valuable information in a large database and mining a mountain for a vein of valuable one. Both processes require either sifting through an immense amount of material, or intelligently probing it to find where the value resides.

What Can Data Mining Do?

Although data mining is still in its infancy, companies in a wide range of industries – including finance, health care, manufacturing, transportation,– are already using data mining tools and techniques to take advantage of historical data. By using pattern recognition technologies and statistical and mathematical techniques of sift through warehoused information, data mining helps analysts recognize significant facts, relationships, trends, patterns, exceptions, and anomalies that might otherwise go unnoticed.

For businesses, data mining is used to discover patterns and relationships in the data in order to help make better business decisions. Data mining can help spot sales trends, develop smarter marketing campaigns, and accurately predict customer loyalty. Specific uses of data mining include:

Market segmentation – Identify the common characteristics of customers who buy the same products from your company.

Customer churn – Predict those customers who are likely to leave the company and go to a competitor.

Fraud detection – Identify transactions that are most likely to be fraudulent.

Direct marketing – Identify the prospects who should be included in a mailing list to obtain the highest response rate.

Interactive marketing – Predict what each individual accessing a web site is most likely interested in seeing.

Market basket analysis – Understand what products or services are commonly purchased together, e.g., beer and diapers.

Trend analysis – Reveal the difference in a typical customer between the current month and the previous one.

Data mining technology can generate new business opportunities by:

- Automated prediction of trends and behaviors: Data mining automates the process of finding predictive information in large database. Questions that traditionally required extensive hands-on analysis can now be directly answered from the data. A typical example of a predictive problem is targeted marketing. Data mining uses data on past promotional mailings to identify the targets most likely to maximize return on investment in future mailings. Other predictive problems include forecasting bankruptcy and other forms of default and identifying segments of a population likely to respond similarly to given events.

- Automated discovery of previously unknown patterns: Data mining tools sweep through databases and identify previously hidden patterns. An example of pattern discovery is the analysis of retail sales data to identify seemingly unrelated products that are often purchased together. Other pattern discovery problems include detecting fraudulent credit card transactions and identifying anomalous data that could represent data entry keying errors.

Using massively parallel computers, companies dig through volumes of data to discover patterns about their customers and products. For example, grocery chains have found that when men go to a supermarket to buy diapers, they sometimes walk out with a six-pack of beer as well. Using that information, it is possible to lay out a store so that these items are closer.

AT&T, A.C. Nielsen, and American Express are among the growing ranks of companies implementing data mining techniques for sales and marketing. These systems are crunching through terabytes of point-of-sale data to aid analysts in understanding consumer behavior and promotional strategies. Why? To gain a competitive advantage and increase profitability!

Similarly, financial analysts are plowing through vast sets of financial records, data feeds, and other information sources in order to make investment decisions. Health-care organizations are examining medical records to understand trends of the past so that they can reduce costs in the future.

The Evolution of Data Mining

Data mining is a natural development of the increased use of computerized databases to store data and provide answers to business analysts. Traditional query and report tools have been used to describe and extract what is in a database. The user forms a hypothesis about a relationship and verifies it or discounts it with a series of queries against the data. For example, an analyst might hypothesize that people with low income and high debt are bad credit risks and query the database to verify or disprove this assumption. Data mining can be used to generate a hypothesis. For example, an analyst might use a neural net to discover a pattern that analysts did not think to try – for example, that people over 30 years with low incomes and high debt but who own their own homes and have children are good credit risks.

How Data Mining Works

How is data mining able to tell us important things that we did not know or what is going to happen next? The technique that is used to perform these feats is called *modeling.* Modeling is simply the act of building a model (a set of examples or a mathematical relationship) based on data from situations where the answer is known and then applying the model to other situations where the answers are not known. Modeling techniques have been around for centuries, of course, but it is only recently that data storage and communication capabilities required to collect and store huge amounts of data, and the computational power to automate modeling techniques to work directly on the data, have been available.

As a simple example of building a model, consider the director of marketing for a telecommunications company. He would like to focus his marketing and sales efforts on segments of the population most likely to become big users of long-distance services. He knows a lot about his customers, but it is impossible to discern the common characteristics of his best customers because there are so many variables. From this existing database of customers, which contains information such as age, sex, credit history, income, zip code, occupation, etc., he can use data mining tools, such as neural networks, to identify the characteristics of those customers who make lots of long-distance calls. For instance, he might learn that his best customers are unmarried females between the ages of 34 and 42 who earn in excess of $60,000 per year. This, then, is his model for high-value customers, and he would budget his marketing efforts accordingly.

Data Mining Technologies

The analytical techniques used in data mining are often well-known mathematical algorithms and techniques. What is new is the application of those techniques to general business problems made possible by the increased availability of data, and inexpensive storage and processing power. Also, the use of graphical interface has led to tools becoming available that business experts can easily use.

Some of the tools used for data mining are:

Artificial neural networks – Nonlinear predictive models that learn through training and resemble biological neural networks in structure.

Decision trees – Tree-shaped structures that represent sets of decisions. These decisions generate rules for the classification of a dataset.

Rule induction – The extraction of useful if-then rules from databases on statistical significance.

Genetic algorithms – Optimization techniques based on the concepts of genetic combination, mutation, and natural selection.

Nearest neighbor – A classification technique that classifies each record based on the records most similar to it in a historical database.

Real-World Examples

Details about who calls whom, how long they are on the phone, and whether a line is used for fax as well as voice can be invaluable in targeting sales of services and equipment to specific customers. But these tidbits are buried in masses of numbers in the database. By delving into its extensive customer-call database to manage its communications network, a regional telephone company identified new types of unmet customer needs. Using its data mining system, it discovered how to pinpoint prospects for additional services by measuring daily household usage for selected periods. For example, households that make many lengthy calls between 3 p.m. and 6 p.m. are likely to include teenagers who are prime candidates for their own phone and lines. When the company used target marketing that emphasized convenience and value for adults – "Is the phone always tied up?"- hidden demand surfaced. Extensive telephone use between 9 a.m and 5 p.m. characterized by patterns related to voice, fax, and modem usage suggests a customer has business activity. Target marketing offering those customers "business communications capabilities for small budgets" resulted in sales of additional lines, functions, and equipment.

The ability to accurately gauge customer response to changes in business rules is a powerful competitive advantage. A bank searching for new ways to increase revenues from its credit card operations tested a nonintuitive possibility: Would credit card usage and interest earned increase significantly if the bank halved its minimum required payment? With hundreds of gigabytes of data representing two years of average credit card balances, payment amounts, payment timeliness, credit limit usage, and other key parameters the bank used a powerful data mining system to model the impact of the proposed policy change on specific customer categories, such as customers consistently near or at their credit limits who make timely minimum or small payments. The bank discovered that cutting minimum payment requirements for small, targeted customer categories could increase average balances and extend indebtedness periods, generating more than $25 million in additional interest earned.

The Future of Data Mining

In the short term, the results of data mining will be in profitable, if mundane, business-related areas. Micromarketing campaigns will explore new niches. Advertising will target potential customers with new precision.

In the medium term, data mining may be as common and easy to use as e-mail. We may use these tools to find the best airfare to New York, root out a phone number of a long-lost classmate, or find the best prices on lawn mowers.

The long-term prospects are truly exciting. Imagine intelligent agents turned loose on medical research data or on subatomic particle data. Computers may reveal new treatments for diseases or new insights into the nature of the universe. There are potential dangers, though, as discussed below.

Privacy Concerns

What if every telephone call we make, every credit purchase we make, every flight we take, every visit to the doctor we make, every warranty card we send in, every employment application we fill out, every school record we have, our credit record, every web page we visit... were all collected together? A lot would be known about us! This is an all-too-real possibility. Much of this kind of information is already stored in a database. Remember that phone interview we gave to a marketing company last week? Our replies went into a database. Remember that loan application we filled out? It is now in a database Too much information about too many people for anybody to make sense of? Not with data mining tools running on massively parallel processing computers! Would we feel comfortable about someone (or lots of someones) having access to all this data about us? And remember, all this data does not have to reside in one physical location, as the Net growing information of this type becomes more available to more people.

1.3 Summary

Traditional query and report tools have been used to describe and extract what is in a database. The user forms a hypothesis about a relationship. Data mining can be used to generate a hypothesis. The results of data mining will be in profitable. Data warehouses are used to consolidate data located in disparate databases. A data warehouse stores large quantities of data by specific categories, so it can be more easily retrieved, interpreted, and sorted by users. Thus in this section we have seen the overview of the data mining and data warehousing.

1.4 Review Questions

1. Give an overview of data mining and data warehousing.
2. Explain the concept of data mining.
3. Define the terms *data warehousing* and *data mining.*
4. What is data warehousing and why do we need it?
5. What are the rules for data warehouses?
6. What is the necessity for multiprocessing machines?
7. Explain the stages involved in knowledge discovery process with a neat schematic diagram.
8. What are the various techniques adapted of data mining?
9. State the principles of data mining.
10. Explain the performance of various data mining techniques.
11. How does the data mining methodology works?
12. Give some of the data mining technologies used at present.

2

Data Warehousing, Data Mining, and OLAP

Objectives:

- This deals with the concept of data mining, need and opportunities, trends and challenges, data mining process, common and new applications of data mining, data warehousing, and OLAP concepts.
- It gives an introduction to data mining: what it is, why it is important, and how it can be used to provide increased understanding of critical relationships in rapidly expanding corporate data warehouse.
- Data mining and knowledge discovery are emerging as a new discipline with important applications in science, engineering, health care, education, and business.
- New disciplined approaches to data warehousing and mining are emerging as part of the vertical solutions approach.
- Extracting the information and knowledge in the form of new relationships, patterns, or clusters for decision making purposes.
- We briefly describe some success stories involving data mining and knowledge discovery.
- We describe five external trends that promise to have a fundamental impact on data mining.
- The research challenges are divided into five broad areas: A) improving the scalability of data mining algorithms, B) mining nonvector data, C) mining distributed data, D) improving the ease of use of the data mining systems and environments, and E) privacy and security issues for data mining.
- We present the concept of data mining and aim at providing an understanding of the overall process and tools involved: how the process turns out, what can be done with it, what are the main techniques behind it, and which are the operational aspects.
- OLAP servers logically organize data in multiple dimensions, which allows users to quickly and easily analyze complex data relationships.

S. Sumathi and S.N. Sivanandam: *Data Warehousing, Data Mining, and Olap*, Studies in Computational Intelligence (SCI) **29**, 21–73 (2006)
www.springerlink.com
© Springer-Verlag Berlin Heidelberg 2006

- OLAP database servers support common analytical operations, including consolidation, drill-down, and slicing and dicing.
- OLAP servers are very efficient when storing and processing multidimensional data.

Abstract. This deals with the concept of data mining, need and opportunities, trends and challenges, process, common and new applications, data warehousing, and OLAP concepts. Data mining is also a promising computational paradigm that enhances traditional approaches to discovery and increases the opportunities for breakthroughs in the understanding of complex physical and biological systems. Researchers from many intellectual communities have much to contribute to this field. Data mining refers to the act of extracting patterns or models from data. The rate growth of disk storage and the gap between Moore's law and storage law growth trends represent a very interesting pattern in the state of technology evolution. The ability to capture and store data has produced a phenomenon we call the *data tombs* or *data stores* that are effectively write-only.

"Data Mining" (DM) is a folkloric denomination of a complex activity, which aims at extracting synthesized and previously unknown information from large databases. It also denotes a multidisciplinary field of research and development of algorithms and software environments to support this activity in the context of real-life problems where often huge amounts of data are available for mining. There is a lot of publicity in this field and also different ways to see the things. Hence, depending on the viewpoints, DM is sometimes considered as just a step in a broader overall process called *Knowledge Discovery in Databases* (KDD), or as a synonym of the latter as we do in this section. Thus, according to this purist definition DM software includes tools of automatic learning from data, such as machine learning and artificial neural networks, plus the traditional approaches to data analysis such as query and reporting, on-line analytical processing or relational calculus, so as to deliver the maximum benefit from data.

Data warehousing is a term that has been coined in recent years to describe computer systems designed to manage data for analysis and to assist management in decision making. A data warehouse can be simply described as "a data infrastructure specifically designed for query, analysis and reporting." While much of the data contained in these systems will be copied from administration systems, the data warehouse should also contain relevant external data, such as exchange rates, yield curves, and actuarial reserves. A question often asked is "why make the expense and effort of keeping copies of data that exists in other systems?" Many people can see the value in making all your data available from a single source, but the real answer is primarily to do with structuring the data such that it is most suitable for analysis.

Administration systems are designed for entering and retrieving data on individual persons or policies; they are not well suited for direct analysis. Also, the reality in many insurance companies is that duplicates of the data will exist in many places – spreadsheets, personal databases; the data warehouse is actually likely to reduce the unstructured duplication. For the end user a suite of tools is available to manipulate and present results from the data stored. These tools, commonly referred to as OLAP (on-line Analytical Processing) tools, are extremely proficient at drill-down, pivot, and presentation of results. They also have computational ability suitable for things such as ratios and trends, although such functionality is simple compared to the calculations used in insurance for projections and reserving.

2.1 Data Mining Research Opportunities and Challenges

Data mining is one of the hottest topics in information technology. This section provides an introduction to data mining: what it is, why it is important, and how it can be used to provide increased understanding of critical relationships in rapidly expanding corporate data warehouse.

The field of data mining and knowledge discovery is emerging as a new, fundamental research area with important applications in science, engineering, medicine, business, and education. Data mining attempts to formulate, analyze, and implement basic induction processes that facilitate the extraction of meaningful information and knowledge from unstructured data. Data mining extracts patterns, changes, associations, and anomalies from large datasets. Work in data mining ranges from theoretical work on the principles of learning and mathematical representations of data to building advanced engineering systems that perform information filtering on the Web, find genes in DNA sequences, help understand trends and anomalies in economics and education, and detect network intrusion. Data mining is also a promising computational paradigm that enhances traditional approaches to discovery and increases the opportunities for breakthroughs in the understanding of complex physical and biological systems. Researchers from many intellectual communities have much to contribute to this field. These include the communities of machine, intellectual communities have much to contribute to this field. These include the communities of machine learning, statistics, databases, visualization and graphics, optimization, computational mathematics, and the theory of algorithms.

The amount of digital data has been exploding during the past decade, while the number of scientists, engineers, and analysts available to analyze the data has been static. To bridge this gap requires the solution of fundamentally new research problems, which can be grouped into the following broad challenges: A) developing algorithms and systems to mine large, massive, and high-dimensional datasets; B) developing algorithms and systems to mine new types of data; C) developing algorithms, protocols, and other infrastructure to mine distributed data; and D) improving the ease of use of data mining systems; E) developing appropriate privacy and security models for data mining.

There is an important need for support by government and business of basic, applied, multidisciplinary, and interdisciplinary research in data mining and knowledge discovery in order to respond to these challenges.

There is an important experimental component to data mining and knowledge discovery, which requires the creation and maintenance of appropriate systems, computational infrastructures, and test beds.

Data mining is the semiautomatic discovery of patterns, association, changes, anomalies, rules, and statistically significant structures and events in data. That is, data mining attempts to extract knowledge from data.

Data mining differs from traditional statistics in several ways: formal statistical inference is assumption driven in the sense that a hypothesis is formed and validated against the data. Data mining in contrast is discovery driven in the sense that patterns and hypothesis are automatically extracted from data. To put it in another way, data mining is data driven, while statistics is human driven. The branch of statistics that data mining resembles most is exploratory data analysis, although this field, like most of the rest of statistics, has focused on datasets far smaller than most that are the target of data mining researchers.

Data mining also differs from traditional statistics in that sometimes the goal is to extract qualitative models, which can easily be translated into logical rules or visual representations; in this sense data mining is human centered and is sometimes coupled with human–computer interface research.

Data mining is a step in the KDD process, which is an interactive, semi-automated process, which begins with raw data. Results of the data mining process may be insights, rules, or predictive models.

The field of data mining draws upon several subjects, including statistics, machine learning, databases, and high-performance computing.

In this section, we are primarily concerned with large datasets, massive datasets, and distributed datasets. By large, we mean datasets, which are too large to fit into the memory of a single workstation. By massive we mean datasets, which are too large to fit into the disks of a single workstation or a small cluster of workstations. Instead, massive clusters or tertiary storage such as tape are required. By distributed, we mean datasets that are geographically distributed.

The focus on large datasets is not just an engineering challenge; it is an essential feature of induction of expressive representations from raw data. It is only by analyzing large datasets that we can produce accurate logical descriptions, which can be translated automatically into powerful predictive mechanisms. Otherwise, statistical and machine learning principles suggest the need for substantial user input (specifying metaknowledge necessary to acquire highly predictive models from small datasets).

There are probably as many definitions of the term *data mining* as there are software analytical tool vendors in the market today. As with OLAP, which could mean almost anything, vendors and industry analysts have adopted the term *data mining* somewhat indiscriminately. The result is a blanket definition that includes all tools employed to help users analyze and understand their data. In this section, we explore a more narrow definition. Data mining is a set of techniques used in an automated approach to exhaustively explore and bring to the surface complex relationships in very large datasets. We discuss only datasets that are largely tabular in nature, having most likely been implemented in relational database management technology. However, these techniques can be, have been, and will be applied to other data representations, including spatial data domains, text-based domains, and multimedia (image) domains.

A significant distinction between data mining and other analytical tools is in the approach they use in exploring the data inter-relationships. Many of the analytical tools available support a verification-based approach, in which the user hypothesizes about specific by relying on the intuition of the analyst to pose the original question and refine the analysis based on the results of potentially complex queries against a database. The effectiveness of this verification-based analysis is limited by a number of factors, including the ability of the analyst to pose appropriate questions and quickly return results, manage the complexity of the attribute space, and think "out of the box."

Most available analytical tools have been optimized to address some of these issues. Query-and-reporting tools address ease of use, letting users develop SQL queries through point-and-click interfaces. Statistical analysis packages provide the ability to explore relationships among a few variables and determine statistical significance against a population. Multidimensional and relational OLAP tools precompute hierarchies of aggregations along various dimensions in order to respond quickly to users' inquiries. New visualization tools let users explore higher dimensionality relationships by combining spatial and nonspatial attributes (location, size, color, and so on).

Data mining, in contrast to these analytical tools, uses discovery-based approaches in which pattern matching and other algorithms are employed to determine the key relationships in the data. Data mining algorithms can look at numerous multidimensional data relationships concurrently, highlighting those that are dominant or exceptional.

2.1.1 Recent Research Achievements

The opportunities today in data mining rest solidly on a variety of research achievements, the majority of which were the results of government-sponsored research. In this section, we study a few of the more important ones. Note that several of them are interdisciplinary in nature, resting on discoveries made by researchers from different disciplines working together collaboratively.

Neural Networks. Neural networks are systems inspired by the human brain. A basic example is provided by a back propagation network, which consists of input nodes, output nodes, and intermediate nodes called *hidden nodes.* Initially, the nodes are connected with random weights. During the training, a gradient descent algorithm is used to adjust the weights so that the output nodes correctly classify data presented to the input nodes. The algorithm was invented independently by several groups of researchers.

Tree-Based Classifiers. A tree is a convenient way to break large datasets into smaller ones. By presenting a learning set to the root and asking questions at each interior node, the data at the leaves can often be analyzed very simply. For example, a classifier to predict the likelihood that a credit card transaction is fraudulent may use an interior node to divide a training dataset into two sets, depending upon whether or not five or fewer transactions were processed during the previous hour. After a series of such questions, each leaf can be labeled fraud/number-fraud by using a simple majority vote. Tree-based classifiers were independently invented in information theory, statistics, pattern recognition, and machine learning.

Graphical Models and Hierarchical Probabilistic Representations. A directed graph is a good means of organizing information on qualitative knowledge about conditional independence and causally gleaned from domain experts. Graphical models generalize Markov models and hidden Markov models, which have proved themselves to be a powerful modeling tool. Graphical models were independently invented by computational probabilists and artificial intelligence researchers studying uncertainty.

Ensemble Learning. Rather than use data mining to build a single predictive model, it is often better to build a collection or ensemble of models and combine them, say with a simple, efficient voting strategy. This simple idea has now been applied in a wide variety of contexts and applications. In some circumstances, this technique is known to reduce variance of the predictions and therefore to decrease the overall error of the model.

Linear Algebra. Scaling data mining algorithms often depends critically upon scaling underlying computations in linear algebra. Recent work in parallel algorithms for solving linear system and algorithms for solving sparse linear systems in high dimensions is important for a variety of data mining applications, ranging from text mining to detecting network intrusions.

Large-Scale Optimization. Some data mining algorithms can be expressed as large-scale, often nonconvex, optimization problems. Recent work has been providing parallel and distributed methods for large-scale continuous and discrete optimization problems, including heuristic search methods for problems too large to be solved exactly.

High-Performance Computing and Communication. Data mining requires statistically intensive operations on large datasets. These types of computations would not be practical without the emergence of powerful and high-performance clusters of workstations supporting protocols for high-performance computing such as MPI and MPIO. Distributed data mining can require moving large amounts of data between geographically separated sites, something that is now possible with the emergence of wide area high-performance networks.

Databases, Data Warehouse, and Digital Libraries. The most time-consuming part of the data mining process is preparing data for data mining. This step can be streamlined in part if the data is already in a database, data warehouse, or digital library, although mining data across different databases. Some algorithms, such as association algorithms, are closely connected to databases, while some of the primitive operations being built into tomorrow's data warehouses should prove useful for some data mining applications.

Visualization of Massive Datasets. Massive datasets often generated by complex simulation programs, required graphical visualization methods for best comprehension. Recent advances in multiscale visualization have allowed the rendering to be done far more quickly and in parallel, making these visualization tasks practical.

2.1.2 Data Mining Application Areas

Data mining techniques have been applied successfully in many areas from business to science and sports. Data mining has been used in database marketing, retail data analysis, stock selection, credit approval, etc. Data mining techniques have been used in astronomy, molecular biology, medicine, geology, and many more fields. It has also been used in health care management, tax fraud detection, money laundering monitoring, and even sports.

Market management. Target marketing, customer relationship management, market basket analysis, cross-selling, market segmentation.

Risk management. Forecasting, customer retention, improved underwriting, quality control, competitive analysis.

Fraud management. Fraud detection.

Industrial-specific applications. Banking, finance, and securities: Profitability analysis (for individual officer branch, product, product group, monitoring marketing programs and channels, customer data analysis customer segmentation profiling).

Telecommunications and media. Response scoring, marketing campaign management, profitability analysis, and customer segmentation.

Health care. FAMS (Fraud and Abuse Management System) assisting health insurance organizations dealing with fraud and abuse: detection, investigation, settlement, prevention of recurrence.

New Applications

The discipline of data mining is driven in part by new applications, which require new capabilities not currently being supplied by today's technology. These new applications can be naturally divided into three broad categories:

Business & E-commerce Data. Back office, front office, and network applications produce large amounts of data for business processes. Using this data for effective decision making remains a fundamental challenge.

Scientific, Engineering, and Health Care Data. Scientific data and metadata tend to be more complex in structure than business data. In addition, scientists and engineers are making increasing use of simulation and systems with application domain knowledge.

Web data. The data on the Web is growing not only in volume but also in complexity. Web data now includes not only text and image, but also streaming data and numerical data.

In this section, we describe several such applications from each category.

Business transactions. Today, businesses are consolidating and more and more businesses have millions of customers and billions of their transactions. They need to understand risks (Is this transaction fraudulent? Will the customers pay their bills?) and opportunities (What is the expected profit of this customers? What product is this customer most likely to buy next?).

Electronic commerce. Not only does electronic commerce produce large datasets in which the analysis of marketing patterns and risks patterns is critical, but unlike some of the applications above, it is also important to do this in real or near-real time, in order to meet the demand of on-line transactions.

Genomic data. Genomic sequence and mapping efforts have produced a number of databases, which are accessible over the Web. In addition, there is also a wide variety of other on-line databases, including those containing information about disease, cellular function, and drugs. Finding relationship between these data sources, which are largely unexplored, is another fundamental data mining challenge. Recently, scalable techniques have been developed for comparing whole genomes.

Sensor data. Satellites, buoys, balloons, and a variety of other sensors produce voluminous amounts of data about the earth's atmosphere, oceans, and lands. A fundamental challenge is to understand the relationships, including casual relationships amongst this data. For example, do industrial pollutants affect global warming? There are also large terabyte to petabyte datasets being produced by sensors and instruments in other disciplines, such as astronomy, high-energy physics, and nuclear physics.

Simulation Data. Simulation is now accepted as a third mode of science, supplementing theory and experiment. Today, not only do experiments produce huge datasets, but so do simulations. Data mining and more generally data intensive computing is proving to be a critical link between theory, simulation, and experiment.

Health care Data. Health care has been the most rapidly growing segment of the nation's gross domestic product (GDP) for some time. Hospitals, health care organizations, insurance companies, and the federal government have large collections of data about patients, their health care problems, the clinical procedures used, their costs, and the outcomes. Understanding relationships in this data is critical for a wide variety of problems, ranging from determining what procedures and clinical protocols are most effective to how best to deliver health care to most people in an era of diminishing resources.

Multimedia Documents. Few people are satisfied with today's technology for retrieving documents on the Web, yet the number of documents and the number of people accessing these documents is growing explosively. In addition, it is becoming increasingly easier to archive multimedia data, including audio, images, and video data, but progressively harder to extract meaningful information from the archives as the volume grows.

The Data Web. Today the Web is primarily oriented toward documents and their multimedia extensions. HTML has proved itself to be a simple, yet powerful language for supporting this. Tomorrow the potential exists for the Web to prove to be equally important for working with data. The extension markup language (XML) is an emerging language for working with data in networked environments. As this infrastructure grows, data mining is expected to be a critical enabling technology for the emerging data Web.

2.1.3 Success Stories

In this section, we briefly describe some success stories involving data mining and knowledge discovery.

Association Rules. Suppose we have a collection of items. The data for many applications consists of multiple of one or more items. A basic example is provided by a supermarket where the items are the products offered for sale and the transactions are purchases, consisting of one or more products purchased by an individual at a given time. A fundamental problem is to uncover associations: which products tend to be purchased together. There has been a lot of recent work on this problem and a variety of algorithms have been developed that can discover associations, even in very large datasets, with just a few passes over the data. A variety of commercial data mining systems support association rules and they are now routinely applied to a range of problems from database marketing to product placement for supermarkets. In addition, association rules algorithms have spurred new research in a variety of areas from databases to complexity theory.

Fraud Detection. Although relatively few credit card transactions are fraudulent, the sheer volume of transactions means that over $500 million are lost each year in this way. A variety of data mining techniques have been used to develop fraud systems, which can detect fraudulent credit card transactions in near-real time. This problem is challenging due to the size of the datasets, the rarity of the events of interest, and the performance requirements for near-real time detection. Data mining has also improved fraud detection in other application areas, including telecom fraud and insurance fraud.

Astronomical Data. Traditionally, the search of new galaxies, stars, and quasars has primarily been done by astronomers visually examining individual photographic plates. Classification algorithms from data mining have recently been used to automate this process yielding new astronomical discoveries. The classification algorithms are applied to derive attributes produced by image processing, such as the brightness, area, and morphology of sky objects. The approach has also proved useful for detecting new objects too faint to be observed by a manual analysis or traditional computational techniques. For the 2nd Palomar Observatory Sky Survey, this approach resulted in over a three-fold increase in the size of the catalog.

Genomic Data. Genomic data is stored all over the world, in a variety of formats and managed by a variety of applications and systems. Recently, systems have been developed that allow discoveries to be made involving information distributed over several systems. In particular, the new systems have enabled for the first time whole genome comparison, gene identification, and whole genome functional interpretation and analysis. The techniques developed for analyzing genomic and other types of scientific data can be expected to play a role in analyzing a broad range of biological data.

Distributed Data Mining. Traditionally, data mining has required that the relevant data be warehoused in a single location. Recently, distributed data mining systems have been exploiting wide area, high-performance net networks, such as the NSF vBNS network, to mine large amounts of distributed scientific and health care data. Recently, these systems have been setting records for the sustained movement of very large amounts of data over wide area networks. Separately, a prototype has been developed by exploiting distributed data mining to improve the detection of credit card fraud.

Text Mining. Recently, data mining has been combined with algorithms from information retrieval to improve the precision and recall the queries on very large collections of documents. In particular, some of these algorithms have proved useful in multilingual collections and others have shown their worth in querying using concepts instead of keywords.

2.1.4 Trends that Affect Data Mining

In this section, we describe five external trends, which promise to have a fundamental impact on data mining.

Data Trends . Perhaps the most fundamental external trend is the explosion of digital data mining during the past two decades. During this period, the amount of data probably has grown between six and ten orders of magnitude. Much of this data is accessible via networks. On the other hand, during the same period the number of scientists, engineers, and other analysts available to analyze this data has remained relatively constant. For example, the number of Ph.D.s in statistics graduating each year has remained relatively constant during this period. Only one conclusion is possible: either most of the data is destined to be write-only; or techniques, such as data mining, must be developed, which can automate, in part, the analysis of this data, filter irrelevant information, and extract meaningful knowledge.

Hardware Trends. Data mining requires numerically and statistically intensive computations on large datasets. The increasing memory and processing speed of workstations enables the mining of dataset using current algorithms and techniques that were too large to be mined just a few years ago. In addition, the commoditization of high-performance computing through SMP workstations and high-performance workstation clusters enables attacking data mining problems that were accessible using only the largest supercomputers of few years ago.

Network Trends. The next generation Internet (NGI) will connect sites at OC-3 (155 MBits/sec), speeds and higher. This is over 100 times faster than the connectivity provided by current networks. With this type of connectivity, it becomes possible to correlate distributed datasets using current algorithms and techniques. In addition, new protocol, algorithms, and languages are being developed to facilitate distributed data mining using current and next generation networks.

Scientific Computing Trends. As mentioned above, scientists and engineers today view simulation as a third mode of science. Data mining and knowledge discovery serve an important role linking the three modes of science: theory, experiment, and simulation, especially for those cases in which the experiment or simulation results in large datasets.

Business Trends. Today businesses must be more profitable, react quicker, and offer higher quality services than ever before, and do it all using fewer people and at lower cost. With these types of expectations and constraints, data mining becomes a fundamental technology, enabling business to more accurately predict opportunities and risks generated by their customers and their customer's transactions.

2.1.5 Research Challenges

In this section, we describe some of the major research challenges identified by the three workshops. The research challenges are divided into five broad areas: A) improving the scalability of data mining algorithms, B) mining nonvector data, C) mining distributed data, D) improving the ease of use of the data mining systems and environments, and E) privacy and security issues for data mining.

A. *Scaling data mining algorithms.* Most data mining algorithms today assume that the data fits into memory. Although success on large datasets is often claimed, usually this is the result of sampling large datasets until they fit into memory. A fundamental challenge is to scale data mining algorithms as
   the number of records or observations increases;
   the number of attributes per observation increases,

the number of predictive models or rule sets used to analyze a collection of observation increases; and,

as the demand for interactivity and real-time response increases.

Not only must distributed, parallel, and out-of-memory versions of current data mining algorithms be developed, but genuinely new algorithms are also required. For example, association algorithms today can analyze out-of-memory data with one or two passes, while requiring only some auxiliary data be kept in memory.

B. *Extending data mining algorithms to new types.* Today, most data mining algorithms work with vector-valued data. It is an important challenge to extend data mining algorithms to work with other data types, including 1) time series and process data, 2) unstructured data, such as text, 3) semistructured data, such as HTML and XML documents, 4) multimedia and collaborative data, 5) hierarchical and multiscale data, and 6) and collection-valued data.

C. *Developing distributed data mining algorithms.* Today most data mining algorithms require bringing all together data to be mined in a single, centralized data warehouse. A fundamental challenge is to develop distributed versions of data mining algorithms, so that data mining can be done while leaving some of the data in place. In addition, appropriate protocols, languages, and network services are required for mining distributed data to handle the metadata and mappings required for mining distributed data. As wireless and pervasive computing environments become more common, algorithms and systems for mining the data produced by these types of systems must also be developed.

D. *Ease of use.* Data mining today is at best a semiautomated process and perhaps destined to always remain so. On the other hand, a fundamental challenge is to develop data mining systems, which are easier to use, even by casual users. Relevant techniques include improving use interface, supporting casual browsing and visualization of massive and distributed datasets, developing techniques and systems to manage the metadata required for data mining, and developing appropriate languages and protocols for providing causal access to data. In addition, the development of data mining and knowledge discovery environments that address the process of collecting, processing, mining, and visualizing data, as well as the collaborative and reporting aspects necessary when working with data and information derived from it, is another important fundamental challenge.

E. *Privacy and Security.* Data mining can be a powerful means of extracting useful information from data. As more and more digital data becomes available, the potential for misuse of data mining grows. A fundamental challenge is to develop privacy and security models and protocols appropriate for data mining and to ensure that next generation data mining systems and designed from the ground up to employ these models and protocols.

2.1.6 Test Beds and Infrastructure

Experimental studies will play a critical role in advancing the field of data mining. Developed test beds for high performance and distributed data mining is essential for progress in the field.

The requirements for data mining test beds are different than those for general purpose high-performance computing test beds. For example, the computing resources for data mining test beds are as much disk oriented as processor oriented; the network resources must be able to move datasets and data elements between geographically distributed sites with guaranteed quality of service, and a variety of general purpose and specialized data mining software must be available.

Perhaps the two most difficult challenges in creating data mining test beds and national resources in data mining are assembling a) the appropriate datasets and b) the required interdisciplinary and multidisciplinary teams.

2.1.7 Findings and Recommendations

In this section, we list some of the major findings and recommendations.For all Interested Parties

Data mining and knowledge discovery is a new emerging discipline with both a scientific and an engineering component that is of strategic importance for the U.S.A. and of critical importance to future information access technologies. All interested parties are encouraged to work toward the maturation of data mining and knowledge discovery, its establishment as a scientific and engineering discipline in its own right, and the evolution of a community that includes the relevant traditions and disciplines and put them together in the proper context.

For the Federal Government

Create programs that encourage the emergence of data mining and knowledge discovery as an independent discipline. Support interdisciplinary and multidisciplinary research projects. Many advances in data mining require teams of mathematical and statisticians, computer scientists, and application domain scientists working together to create the appropriate datasets and the required algorithms and software to analyze them.

Support basic research in computer and information sciences that underlies mining, including machine learning, knowledge systems, databases' high-performance computing, high-performance networking, and digital libraries.

Support basic research in mathematics and statistics that underlies data mining, including statistics, probability, applied mathematics, logic, discrete mathematics, analysis and dynamical systems, linear algebra, and computational geometry and algebra.

Support data mining test beds:

The hardware, software, data and consulting requirements for data mining often outstrip the resources of individual scientists and small research groups. Supporting national resources and test beds for data mining is important in order to provide the proper experimental infrastructure required for next generation data mining experiments.

For Companies

Support applied research in data mining:

Work to develop, implement, and support appropriate privacy and security models for data mining systems.

Create sanitized versions of real datasets for use by data mining researchers.

Support joint research projects between industry and universities. Support collaborative test beds and demonstration projects.

For Scientists and Engineers:

As new data is collected and archived, support emerging protocols, languages, and standards to facilitate the future analysis and mining of the data, especially by scientists and engineers from other disciplines.

As new data is collected and new systems are built to manage it, ensure that the best available privacy and security models are used to protect inadvertent disclosures of private information.

Provide long-term maintenance and access to datasets created by scientists and engineers, as well as to the knowledge and information extracted from them.

Data mining and knowledge discovery are emerging as a new discipline with important applications to science, engineering, health care, education, and business. Data mining rests firmly on 1) research advances obtained during the past two decades in a variety of areas and 2) more recent technological advances in computing, networking, and sensors. Data mining is driven by the explosion of digital data and the scarcity of scientists, engineers, and domain experts available to analyze it.

Data mining is beginning to contribute research advances of its own, by providing scalable extension and advances to work in associations, ensemble learning, graphical models, techniques for on-line discovery, and algorithms for the exploration of massive and distributed datasets.

Advances in data mining require supporting a) single investigators working in data mining and the underlying research domains supporting data mining; b) interdisciplinary and multidisciplinary research groups working on important basic and applied data mining problems; and c) the appropriate test beds for mining large, massive, and distributed datasets. Appropriate privacy and security models for data mining must be developed and implemented.

2.2 Evolving Data Mining into Solutions for Insights

The capacity of digital data storage worldwide has doubled every nine months for at least a decade, at twice the rate predicted by Moore's law for the growth of computing power during the same period. This less familiar but noteworthy phenomenon, which we call storage law, is among the reasons for the increasing importance and rapid growth of the field of data mining.

The aggressive rate growth of disk storage and the gap between Moore's law and storage law growth trends represents a very interesting pattern in the state of technology evolution. The ability to capture and store data has far outpaced our ability to process and utilize it. This growing challenge has produced a phenomenon we call *data tombs*, or *data stores* that are effectively write-only; data is deposited to nearly rest in peace, since in all likelihood it will never be accessed again.

Data tombs also represent missed opportunities where the data might support exploration in a scientific activity or commercial exploitation by a business organization, the data is potentially valuable information. Without next generation data mining tools, most will stay unused; hence most of the opportunity to discover, profit, improve service, or optimize operations will be lost. Data mining – one of the most general approaches in reducing data in order to explore analyze and understood it – is the focus of this special section.

Data mining is defined as the identification of interesting structure in data. Structure designates patterns, statistical or predictive modes of the data, and relationships among part of the data. Each of these terms – patterns, models, and relationships – has a concrete definition in the context of data mining. A pattern is a parsimonious summary of a subset of the data (such as people who own minivans have children). A model of the data can be a model of the entire dataset and can be predictive; it can be used to, say, anticipate future customer behavior (such as the likelihood a customer is or is not happy, based on historical data of interaction with a particular company). It can also be a general model (such as a joint probability distribution or set of variables in the data). However, the concept of interesting is much more difficult to define.

What structure within a particular dataset is likely to be interesting to a user or task? An algorithm could easily enumerate lots of patterns from a finite database. Identifying interesting structure and useful patterns among the plethora of possibilities is what a data mining algorithm must do, and it must do it quickly over very large databases.

For example, frequent item sets (variable values occurring together frequently in a database of transactions) could be used to answer, say, which items are most frequently bought together in the same supermarket. Such an algorithm could also discover a pattern in a demographics database with exceptionally high confidence that, say, all husbands are males. While true, however, this particular association is unlikely to be interesting. This same method did uncover in the set of transactions representing physicians billing the Australian Government's medical insurance agency a correlation deemed extremely interesting by the agency auditors. Two billing codes were highly correlated: they were representative of the same medical procedure and hence had created the potential for double billing fraud. This nugget of information represented millions of dollars overpayment.

The quest for patterns in data has been studied for a long time in many fields, including statistics, pattern recognitions, and exploratory data analysis. Data mining is primarily concerned with making it easy, convenient, and practical to explore very large databases for organizations and users with lots of data but without years of training as data analysts. The goals uniquely addressed by data mining fall into certain categories:

Scaling analysis to large databases. What can be done to large datasets that cannot be loaded and manipulated in main memory? Can abstract data access primitives embedded in database systems provide mining algorithms with the information to drive a search for patterns? How might we avoid having to scan an entire very large database while reliably searching for patterns?

Scaling to high-dimensional data and models. Classical statistical data analysis relies on humans to formulate a model, and then use the data to access the models fit to the data. But humans are ineffective in formulating hypothesis when datasets have a large number of variables (possibly thousands in cases involving demographics and hundreds of thousands in cases involving retail transactions, Web browsing, and text document analysis). The model derived from this automated discovery and search process can be used to find lower dimensional subspaces where people find it easier to understand the aspects of the problem that are interesting.

Automating search. Instead of relying solely on human analysts to enumerate and create hypothesis, the algorithms perform much of this tedious and data-intensive work automatically.

Finding patterns and models understandable and interesting to users. Classical methodologies for scoring models focus on notions of accuracy (how well the model predicts the data) and utility (how to measure the benefit of the derived pattern, such as money saved). While these methods are well understood in decision analysis, the data mining community is also concerned with new measures, such as the understanding of a model or the novelty of a pattern and how to simplify a mode for interpretability. It is particularly important that the algorithms help end users gain insight from data by focusing on the extraction of patterns that are easily understood or can be turned into meaningful reports and summaries by trading off complexity for understandability.

2.2.1 Trends and Challenges

Among the most important trends in data mining is the rise of "verticalized," or highly specialized, solutions, rather than the earlier emphasis on building new data mining tools. Web analysis, customer behavior analysis, and customer relationship management all reflect the new trend; solutions to business problems increasingly embed data mining technology, often in a hidden fashion, into the application. Hence, data mining applications are increasingly targeted and designed specifically for end users. This is an important and positive departure from most of the field's earlier work, which is used to focus on building mining tools for data mining experts.

Transparency and data fusion represent two important challenges for the growth of the data mining market and technology development. Transparency concerns the need for an end user-friendly interface, whereby the data mining is transparent as far as the user is concerned. Embedding vertical applications is a positive step toward addressing this problem, since it is easier to generate explanations from models built in a specific context. Data fusion concerns a more pervasive infrastructure problem: where is the data that has to be mined? Unfortunately, most efforts at building the decision-support infrastructure, including data warehouses, have proved to be big, complicated, and expensive. Industry analysts report failure of a majority of enterprise data warehousing efforts. Hence, even though the data accumulates in stores, it is not being organized in a format that is easy to access for mining or even for general decision support.

Much of the problem involves data fusion. How can a data miner consistently reconcile a variety of data sources? Often labeled as data integration, warehousing, or IT initiatives, the problem is also the unsolved prerequisite to data mining. The problem of building and maintaining useful data warehouses remains one of the great obstacles to successful data mining. The sad reality today is that before users get around to applying a mining algorithm, they must spend months or years bringing together the data sources. Fortunately, new disciplined approaches to data warehousing and mining are emerging as part of the vertical solutions approach.

2.3 Knowledge Extraction Through Data Mining

Data mining is the process of sifting through and analyzing rich sets of domain specific data and then extracting the information and knowledge in the form of new relationships, patterns, or clusters for decision making purposes. Thus data mining is a form of knowledge discovery essential for solving problems in a specific domain.

The term *KDD* denotes the overall process of extracting the high-level knowledge from low-level data. The multitude of terms used for KDD includes data or information harvesting, data archeology, functional dependency analysis, knowledge extraction, and data pattern analysis.

Conventionally *data mining* refers to the act of extracting patterns or models from data (be it automated or human assisted). However many steps precede the data mining step: retrieving the data from large warehouse (or some other source), selecting the appropriate subset to work with, deciding on the appropriate sampling strategy, cleaning the data and dealing with missing fields, and applying the appropriate transformations, dimensionality reduction, and projections. The data mining step then fits models to or extracts patterns from, the preprocessed data. However, to decide whether this extracted information does represent knowledge, one needs to evaluate this information, perhaps visualize it, and finally consolidate it with existing (and possibly contradictory) knowledge. Obviously these steps are all on the critical path from data to knowledge. Furthermore any one step can result in changes in preceding or succeeding steps often requiring starting from scratch with new choices and settings. Hence in the definition, data mining is just a step in the overall KDD process.

The concept was born about ten years ago. The interest in the data mining field and its exploitation in different domains (marketing, finance, banking, engineering, health care, power systems, meteorology, etc.) has been increasingly recently due to a combination of factors. They include:

- the emergence of very large amount of data (terabytes – bytes – of data) due to computer-automated data measurement and/or collection, digital recording, centralized data archives, and software and hardware simulations.
- the dramatic cost decrease of mass storage devices
- the emergence and growth of fielded database management systems
- the advances in computer technology such as faster computers and parallel architectures
- the continuous developments in automatic learning techniques
- the possible presence of uncertainty in data (noise, outliers, missing information).

The general purpose of data mining is to process the information from the enormous stock of data we have or that we may generate, so as to develop better ways to handle data and support future decision making. Sometimes, the pattern to be searched for and the models to be extracted from data are subtle, and require complex calculus and/or significant specific domain knowledge. Or even worse, there are situations where one would like to search for patterns that humans are not well suited to find, even if they are good experts in the field. For example, in many power system-related problems one is faced with high-dimensional datasets that cannot be easily modeled and controlled on the whole, and therefore automatic methods capable of synthesizing structures from such data become a necessity.

This section presents the concept of data mining and aims at providing an understanding of the overall process and tools involved: how the process turns out, what can be done with it, what are the main techniques behind it, and which are the operational aspects. We aim also at describing a few examples of data mining applications, so as to motivate the power system field as a very opportune data mining application.

2.3.1 Data Mining Process

Data mining process consists of three major steps. Of course, it all starts with a big pile of data. The first processing step is data preparation often referred to as "scrubbing the data." Data is selected, cleaned, and preprocessed under the guidance and knowledge of a domain expert. The most time-consuming part of the data mining process is preparing data for data mining. This step can be streamlined in part if the data is already in a database, data warehouse, or digital library, although mining data across different databases, for example, is still a challenge. Second, a data mining algorithm is used to process the prepared data, compressing and transforming it to make it easy to identify any latent valuable nuggets of information. In the second step in data mining, once the data is collected and preprocessed, the data mining algorithms perform the actual sifting process. Many techniques have been used to perform the common data mining activities of associations, clustering, classification, modeling, sequential patterns, and time series forecasting. These techniques range from statistics to rough sets to neural networks.

The third phase is the data analysis phase, where the data mining output is evaluated to see if additional domain knowledge was discovered and to determine the relative importance of the facts generated by the mining algorithms.

The final step is the analysis of the data mining results or output. In some cases the output is in a form that makes it very easy to discern the valuable nuggets of information from the trivial or uninteresting facts. The relationships are represented in if-then rules form. With rules recast into textual form, the valuable information is much easier to identify. In other cases, however, the results will have to be analyzed either visually or through another level of tools to classify the nuggets according to the predicted value.

Whatever be the data mining algorithm used, the results will have to be presented to the user. A successful data mining application involves the transformation of raw data into a form that is more compact and more understandable, and where relationships are explicitly defined.

For most data mining applications, the relatively clean data that resides in the corporate data warehouse must usually be refined and processed before it undergoes the data mining process. This preprocessing might involve joining information from multiple tables, selecting specific rows or records of data, and it most certainly include selecting the columns or fields of data that need to be looked at in the data mining step. Often two or more fields are combined to represent ratios or derived values. This data selection and manipulation process is usually performed by someone with a good deal of knowledge about the problem domain and the data related to the problem under study. Depending on the data mining algorithm involved, that data might need to be formatted in specific ways (such as scaling of numeric data) before it is processed. Hence data preparation is crucial to successful data mining application.

The data mining process consists of three major steps.

(1) *Data Preparation:* Data is selected, cleaned, and preprocessed under the guidance and knowledge of domain experts who capture and integrate both the internal and external data into a comprehensive view that encompasses the whole organization.
(2) *Data mining algorithm:* Data mining algorithm is used to mine the integrated data to enable easy identification of any valuable information.
(3) *Data Analysis Phase:* Data mining output is evaluated to see if the domain knowledge discovered is in the form of rules extracted out of the network. The overall data mining process is shown in Fig 2.1.

In general, once the data is prepared, a neural network is used to build a model based on the data. When the model is ready then it opens a way to complete automation of the process as shown in Fig 2.2.

Data mining consists of five major elements:

Extract, transform, and load transaction data onto the data warehouse system.
Store and manage the data in a multidimensional database system.
Provide data access to business analysts and information technology professionals.
Analyze the data by application software.
Present the data in a useful format, such as a graph or table.
The typical KDD process is illustrated in Fig. 2.3.

By definition, data mining is the nontrivial process of extracting valid, previously unknown, comprehensible, and useful information from large databases and using it. It is an exploratory data analysis, trying to discover useful patterns in data that are not obvious to the data user.

What is a database (DB)? It is a collection of objects (called tuples in the DB jargon, examples in machine learning or transactions in some application fields), each one of which is described by a certain number of attributes, which provide detailed information about each object. Certain attributes are selected as input attributes for a problem, certain ones as outputs (i.e., the desired objective: a class, a continuous value, etc.). Table 2.1 shows some examples

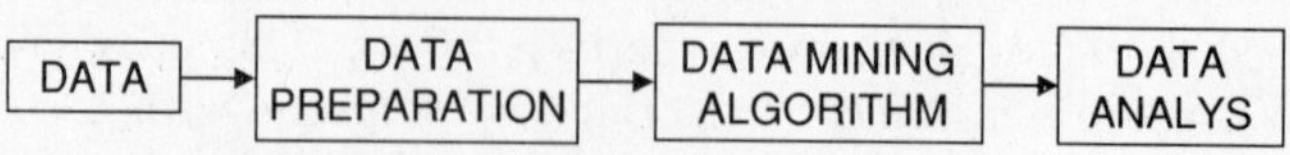

Fig. 2.1. Overall data mining process

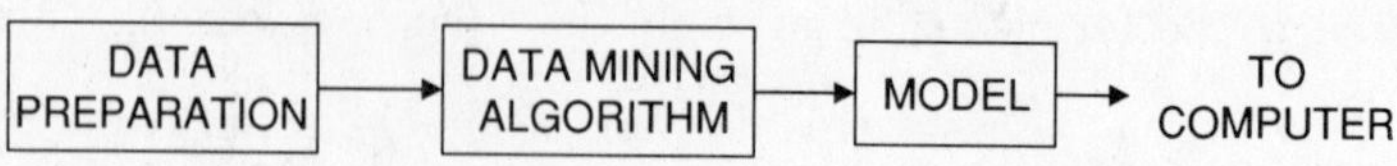

Fig. 2.2. Automation of data mining process

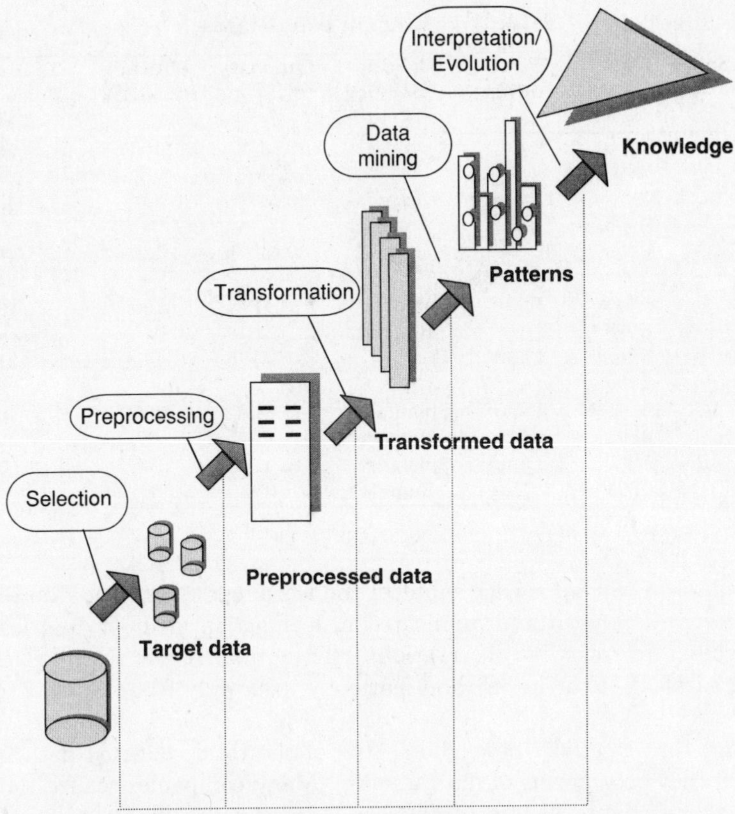

Fig. 2.3. A typical knowledge discovery process

of hourly energy transactions recorded in a data for a power market analysis application (each row of the table corresponds to an object and each column indicates one attribute, e.g., buyer, quantity, price). In such an application, the power system is considered as a price-based market with bilateral contracts (i.e., direct contracts between the power producers and users or broker outside of a centralized power pool), where the two parties, the buyer and the seller, could be utility distribution companies, utility and nonutility retailers (i.e., energy service providers), independent generators (i.e., independent power producers), generation companies, or end customers (as single customers or as parts of aggregated loads). The example will be used further in order to exemplify the techniques implied in the data mining.

Usually, one of the first tasks of a data mining process consists of summarizing the information stored in the database, in order to understand well its content. This is done by means of statistical analysis or query-and-reporting techniques. Then more complex operations are involved, such as identifying models, which may be used to predict information about future objects. The term *supervised learning* (known as "learning with a teacher") is implied in

Table 2.1. Example of a database

Buyer	Seller	Date	Hour ending	Product/ Service	Quantity	Unitary Price (%)	Transaction Price (Price Units)
A	B	23 Feb. 1998	9 a.m.	Energy	20 MWh	100	2000
A	C	23 Feb. 1998	11 a.m.	Energy	50 MWh	80	4000
D	A	5 Apr. 1998	9 a.m.	Energy	30 MWh	150	4500
A	B	9 Apr. 1998	2 p.m	Spinning Reserve	10 MW	100	1000
E	B	15 May 1998	4 a.m	Energy	30 MWh	70	2100
E	C	15 May 1998	5 a.m	Spinning Reserve	20 MW	200	4000
E	B	31 July 1998	8 a.m	Spinning Reserve	10 MW	100	1000

mining data in which for each input of the learning objects, the desired output objective is known and implicated in leaning. In unsupervised learning approaches ("learning by observation") the output is not provided or not considered at all, and the method learns by itself only from input attribute values.

Notice that generally only about 10% of the total collected data is ever analyzed (not only means of data mining). Many companies realize the poor quality of their data collection only when a data mining analysis is started on it. The databases are usually very expensive to create and maintain, and for a small additional investment in mining them, highly profitable information may be discovered hidden in the data. Thus, the classical scenario is as follows: a company realizing that there might be "nuggets" of information in the data they process starts building a long-term repository (a data warehouse) to store as much data as possible (e.g., by recording systematically all purchases by individual customers of a supermarket); then they would launch a pilot DM study in order to identify actual opportunities; finally some of the applications identified as interesting would be selected for actual implementation.

However, apart from the "cheaply" collected or already available data, there are some applications of data mining where the data is produced by computer simulations or expensive real experiments. For example, in the case where future yet unknown situations have to be forecast, or in fields where security aspects are analyzed for a system (computer system, power system, or banking system) when the history does not provide fortunately negative examples, one may use Monte Carlo simulations in order to generate a DB automatically and this is itself a nontrivial task.

The usual film when a company or a holder of a big amount of data decides that the information he has collected is worthy of being analyzed unfolds like this: he comes with the data to the data miner (e.g., a consultant), the data miner first gets familiar with the field of application and with the application specifics, then depending on the data mining software he has, he will select a portion of the available data and apply those techniques he expects to give him more knowledge in terms of some established objectives. In case the results of this combination of tools do not give to the interested one any improvement in the existing knowledge about the subject, either the miner gives up (it is indeed possible that this process yields only uninteresting results), or he tries to go further by implementing new methods for mining the specific data (e.g., for a temporal problem of early anomalies detection, a temporal decision tree may offer more valuable results than a decision tree).

Who is a data miner? – some person, usually with background in computer science or in statistics and in the domain of interest, or a couple of specialists, one in data mining, one in the domain of interest, who is able to perform the steps of the data mining process. The miner is able to decide how much iterative the whole process is and to interpret the visual information he gets at every substep.

In general the data mining process iterates through five basic steps:

- *Data selection.* This step consists of choosing the goal and the tools of the data mining process, identifying the data to be mined, then choosing appropriate input attributes and output information to represent the task.
- *Data transformation.* Transformation operations include organizing data in desired ways, converting one type of data to another (e.g., from symbolic to numerical), defining new attributes, reducing the dimensionality of the data, removing noise, "outliers," normalizing, if appropriate, deciding strategies for handling missing data.
- *Data mining step per se.* The transformed data is subsequently mined, using one or more techniques to extract patterns of interest. The user can significantly aid the data mining method by correctly performing the proceeding steps.
- *Result interpretation and validation.* For understanding the meaning of the synthesized knowledge and its range of validity, the data mining application tests its robustness, using established estimation methods and unseen data from the database. The extracted information is also assessed (more subjectively) by comparing it with prior expertise in the application domain.
- *Incorporation of the discovered knowledge.* This consists of presenting the results to the decision maker who may check/resolve potential conflicts with previously believed or extracted knowledge and apply the new discovered patterns.

Figure 2.3 presents schematically the whole process, by showing what happens with the data: it is preprocessed, mined, and postprocessed, the result being a refinement in the knowledge about the application. The data mining process is iterative, interactive, and very much a trial-and-error activity.

Visualization plays an important role. Because we find it difficult to emulate human intuition and decision making on a machine, the idea is to transform the derived knowledge into a format that is easy for humans to digest, such as images or graphs. Then, we rely on the speed and capability of the human user visual system to spot what is interesting at every step of the data mining process: preliminary representation of data, domain specific visualization, or result presentation.

From the point of view of software structure, there are two types of possible implementation:

- The one called data mining "in place": the learning system accesses the data through a database management system (DBMS) and the user is able to interact with both the database (by means of queries) and the data mining tools. The advantage is that the approach may handle very large database and may exploit the DBMS (e.g., the handling of distributed data).
- The one called data mining "offline": the objects are first loaded in the data mining software, with a translation into a particular form, outside the database, and the user interacts mainly with the data mining software. They allow to use the existing machine learning systems with only minor modification in implementation, and it may be faster but are generally limited to handle medium-sized datasets, which can be represented in main memory (up to several hundred Mbytes)

What can be done in the Data Mining step? - Depending mainly on the application domain and the interest of the miner, one can identify several types of data mining tasks for which data mining offers possible answers. We present them so that they are usually implied in the process. Possible results for each one of these tasks are provided by considering the example in table 1 as the database to be mined:

Summarization: Summarization is the abstraction or generalization of the data. A set of task relevant data is summarized and abstracted. This results in a smaller set, which gives a general overview of the data, usually with aggregate information. It aims at producing compact and characteristic description for a given set of data. It can take multiple forms: numerical (simple descriptive statistical measures like means, standard deviations, and so on), graphical (histograms, scatter plots, to name a few), or the form of "if-then" rules. It may provide descriptions about objects in the whole database or in selected subsets. *Example of summarization: "the minimum unitary price for all the transactions with energy is 70 price units" (see table.1).*

Clustering. A clustering problem is an unsupervised learning problem, which aims at finding in the data clusters of similar objects sharing a number of interesting properties. It may be used in data mining to evaluate similarities among data, build a set of representative prototypes, analyze correlations between attributes, or automatically represent a dataset by a small number of regions, preserving the topological properties of the original input space.

Example of a clustering result: "from the seller B point of view, buyers A and E are similar customers in terms of total price of the transactions done in 1998".

Clustering approaches address segmentation problems. These approaches assign records with a large number of attributes into a relatively small set of groups or "segments." This assignment process is performed automatically by clustering algorithms that identify the distinguishing characteristics of the dataset and then partition the n-dimensional space defined by the dataset attributes along natural leaving boundaries. There is no need to identify the groupings desired or the attributes that should be used to segment the dataset.

Clustering is often one of the first steps in the data mining analysis. It identifies groups of related records that can be used as a starting point for exploring further relationships. This technique supports the development of population segmentation models, such as demographic-based customer segmentation. Additional analyses using standard analytical and other data mining techniques can determine the characteristics of these segments with respect to some desired outcome. For example, the buying habits of multiple population segments might be compared to determine those segments that need to be targeted for a new sales campaign.

Classification. A classification problem is a supervised learning problem where the output information is a discrete classification, i.e., given an object and its input attributes the classification output is one of the possible mutually exclusive classes of the problem. The aim of the classification task is to discover some kind of relationship between the input attributes and the output class, so that the discovered knowledge can be used to predict the class of a new unknown object. *Example of a derived knowledge rule, which classifies sales made early in the day (a sale is said to be early if it was made between 6 a.m. and 12 a.m.): "if the product is energy then the sale is likely to be early (confidence 0.75)".*

This technique is used to classify database records into a number of predefined classes based on certain criteria. For example, a restaurant chain could mine customer purchase data to determine when customers visit and what they typically order. Classification involves associating an observation with one of several labels called classes. Classification provides a mapping from attributes to specified groupings. Once the data is classified the traits of these specific groups can be summarized. In this chapter also, pattern classification is done to identify the specific class of each test pattern.

Classification, perhaps the most commonly applied data mining technique, employs a set of preclassified examples to develop a model that can classify the population of records at large. Fraud detection and credit-risk applications are particularly well suited to this type of analysis. This approach frequently employs decision tree or neural network-based classification algorithms. The use of classification algorithms begins with a training set of preclassified example transactions. For a fraud detection application, this would include complete records of both fraudulent and valid activities, determined on a record-by-record basis. The classifier training algorithm uses these preclassified examples to determine the set of parameters required for proper discrimination. The algorithm then encodes these parameters into a model called *classifier*.

The approach affects the explanation capability of the system. Once an effective classifier is developed, it is used in a predictive mode to classify new records into these same predefined classes. For example, a classifier capable of identifying risky loans could be used to aid in the decision of whether to grant a loan to an individual.

Regression. A regression problem is a supervised learning problem of building a more or less transparent model, where the output information is a continuous numerical value or a vector of such values rather than a discrete class. Then given an object, it is possible to predict one of its attributes by means of the attributes, by using the built model. The prediction of numerical values may be done by classical or more advanced statistical methods and by "symbolic" methods often used in the classification task. *Example of a model derived in a regression problem: "when buyer A buys energy, there exists a linear dependence between the established unitary price and the quantity he buys."*

Dependency modeling. A dependence modeling problem consists in discovering a model, which describes significant dependencies among attributes. These dependencies are usually expressed as "if-then" rules in the form "if antecedent is true then consequent is true," where both the antecedent and the consequent of the rule may be any combination of attributes, rather than having the same output in the consequent like in the case of the classification rules. *Example: such a rule might be "if product is energy then transaction price is larger than 2000 price units."*

Deviation detection. This is the task focusing in discovering the most significant changes or deviations in the data between the actual content of the data and its expected content (previously measured) or normative values. It includes searching for temporal deviations (important changes in data with time) and group deviations (unexpected differences between two subsets of data). In our example, deviation detections could be used in order to find main differences between sales patterns in different periods of the year.

Temporal problems. In certain applications it is useful to produce rules, which take into account explicitly the role of time. There are databases containing temporal information, which may be exploited by searching for similar temporal patterns in data or learn to anticipate some abnormal situations in data. *Example*: *"a customer buying energy will buy spinning reserve later on (confidence 0.66)," or "if total quantity of daily transactions is less than 100 piece units during at least 1 month for a client, the client is likely to be lost."*

Causation modeling. This is a problem of discovering relationships of cause and effect among attributes. A casual rule of type "if-then" indicates not only that there is a correlation between the antecedent and the consequent of the rule, but also that the antecedent causes the consequent. *Example: "decreasing energy price will result in more sold energy daily."*

What techniques are behind all these tasks? - The enumerated types of data mining tasks are based on a set of important techniques originating in artificial intelligence paradigms, statistics, information theory, machine learning, reasoning with uncertainty (fuzzy sets), pattern recognition, or visualization. Thus, a data mining software package is supported to varying degrees by a set of technologies, which nearly always includes:

- *Tree and rule induction.* Machine learning (ML) is the center of the data mining concept, due to its capabilities to gain physical insight into a problem, and participate directly in data selection and model search steps. To address problems like classifications (crisp and fuzzy decision trees), regression (regression tree), time-dependent prediction (temporal trees), ML field is basically concerned with the automatic design of "if-then" rules similar to those used by human experts. Decision tree induction, the best known ML framework, was found to be able to handle large scale problems due to its computational efficiency, provide interpretable results and in particular, able to identify the most representative attributes for a given task.
- *Association rules.* These techniques identify affinities among the collection as reflected in the examined records. These affinities are often expressed as rules. Given a set of transactions, where each transaction is a set of literal, an association rule is a set of literals, the association rule is an expression of the form X $\Rightarrow$ Y, where X and Y are set of items. The intuitive meaning of such a rule is that a transaction of the database that contains X tends to contain Y. Association rule generators are a powerful data mining technique used to search through an entire dataset, for rules revealing the nature and frequency of relationships or associations between data entities. The resulting associations can be used to filter the information for human analysis and possibly to define a prediction model based on observed behavior.
- *Clustering methods.* They are used often in the data selection preprocessing step, due to the property of learning unsupervised similarities between objects and reducing the search space to a set of most important attributes for the application, or to a finite set of objects "alike." The most frequently used clustering method is the *k-means* method, which identifies a certain number of groups of similar objects; it may be used in combination with the *nearest-neighbor rule*, which classifies any new object in the group most similar (most near) to it. This method may also be used in order to identify outliers in a database. For example, by using this technique it might be possible in our example to identify groups of similar sales (large quantity and cheap unitary price versus small quantity and expensive unitary price) and to find out that some of the sales are outliers (e.g., small quantity and cheap). Then a supervised learning technique might be used in order to

find a rule to characterize these abnormal sales, in terms of attributes (seller, buyer, product, date, etc.).

- *Artificial neural networks.* Nonlinear predictive models learn through training and resemble biological neural networks in structure. Neural network has probably been of greater interest through the formative stages of data mining technology. True neural networks are biological systems that detect patterns, make predictions, and learn. Artificial neural networks are computer programs implementing sophisticated pattern detection. Although artificial neural networks cannot completely mimic the human brain and have some limitations, they have the advantage of being a highly accurate predictive model, which can be applied across a large number of problems.

They are recognized in the automatic learning framework as "universal approximators," with massively parallel computing character and good generalization capabilities, and also as black boxes due to the difficulty in obtaining insight into the relationship learned. They are used within the data mining step: to generate a regression model that can predict future behavior, on the basis of a database with input–output pairs of continuous numerical historical information (the neural network acts like a mapping, associating numerical outputs to any new object of known attributes values), and to automatically represent a dataset by a small number of representative prototypes, preserving the topological properties of the original attribute space (unsupervised learning).

- *Statistical techniques such as linear regression*, discriminant analysis, or statistical summarization. Classical statistical approaches include Bayesian network, regression analysis, correlation analysis, and cluster analysis. Modern statistical approach is the nearest neighbor method. An optimal model, based on a defined statistical measure, is searched among the patterns and regularities are then drawn from the model. K-nearest neighbor technique classifies each record in a dataset based on a combination of the classes of the k record(s) most similar to it in a historical dataset (where $k > 1$). It is sometimes called the *k-nearest neighbor* method.
- *Visualization techniques*: In this method data is transformed into visual objects such as dots, lines, and areas. The data is then displayed in a two- or three-dimensional space. Users can interactively explore the interesting spots by visual examination. In the visual interpretation of complex relationships in multidimensional data, graphic tools are used to illustrate data relationships.

Visualization of Massive Datasets: Massive datasets often generated by complex simulation programs, required graphical visualization methods for best comprehension. Recent advances in multiscale visualization have been allowing the rendering to be done far more quickly and in parallel, making these visualization tasks practical.

Histograms (estimate the probability distribution for a certain numerical attribute given a set of objects), scatter plots (provide information on the relation between two numerical attributes and a discrete one), three-dimensional maps, dendrograms (a correlation analysis between attributes or objects) help in visualization.

In addition, some DM packages include *genetic algorithms* (optimization techniques based on the concepts of genetic combination, mutation, and natural selection), *sequential patterns* discovery (group objects with the same succession of given attribute values over a time period), *time-series similarity* (detect similar time series over a period of time), *Bayesian belief networks* (graphical models that encode probabilistic relationships among variables of interest, systems able to learn causal relationships), *neurofuzzy* systems (fuzzy inference systems that incorporate the learning and generalization abilities of neural works).

Genetic Algorithms. Optimization techniques use processes such as genetic combination, mutation, and neural selection in a design based on the concepts of natural evolution. Although genetic algorithms can be classified generally as an emerging science, it has a wide variety of applications in real-life situations.

The three main areas are:

(i) Optimization: Genetic algorithms can be used to automatically determine the optimal values for the variables that optimize the function.

(ii) Prediction: Genetic algorithms have been used at metalevel operations that are used to help optimize other data mining algorithms. For example, in optimizing weights in a neural network.

(iii) Simulation: Sometimes a specific problem is not well defined in terms of what the result is or whether one solution is better than the other. In such cases simulation can be done through genetic algorithms. GAs are systems that can solve a complex problem by arriving at pretty good solutions without a complete understanding of the problem.

Trend analysis. Time series data are records accumulated over time. For example, a company's sales, a customer's credit card transactions, and stock prices are all time series data. Such data can be viewed as objects with an attribute time. The objects are the snapshots of entities with values that change over time.

Sequencing. This technique helps identify patterns over time, thus allowing different analysis methods. The input data is a set of sequence called *data sequence.* Each data sequence is a list of transactions where each transaction is a set of items. A sequential pattern consists of a list of items. The problem is to find all sequential patterns with a user-specified minimum support, where the support of a sequential pattern is the percentages of data sequence that contain patterns.

Even if we wish to consider data mining tools like toolboxes of multiple techniques able to perform a complete data analysis, the reality is not yet so, the market offering presently only partially equipped products.

DM techniques are different one from another in terms of problem representation, parameters to optimize, accuracy, complexity, run time, transparency, and interpretability. Making a compromise between accuracy and complexity (by means of pruning techniques), enhancing the comprehension of derived patterns, and fitting to avoid the over fitting (a problem that appears when the model to be extracted is too complex with respect to the information provided in the learning set) are common features for all techniques.

2.3.2 Operational Aspects

The success of mining some data is induced by a list of factors:

The right tools. A distinctive feature of data mining software is the quality of its algorithms, the effectiveness of the techniques, and sometimes their speed. In addition, the efficiency of using the hardware, the operating system, the database resource, and the parallel computing influence the process. Moreover, it turns out that the particular set of tools useful in a given applications are highly dependent on the practical problem. Thus at the prototyping step, it is useful to have available a broad enough set of techniques so as to identify interesting applications. However, in the final product used for actual field implementation it is often possible to use only a small subset of the latter tools. Customizing data mining techniques to the application domain and using methods that are reliable means to the proposed goal may enhance the process of extracting useful information.

The right data. The data to be mined should contain information worth mining: consistent, cleaned, and representative for the application. Of course, it is useless to apply data mining to an invalid database with high measurement or estimation data errors, or to try to precisely estimate numerical outputs, which present a priori high noise. A data mining tool ideally explains as much information as is stored in the data that is mined (a derived model is strongly dependent on the learning set used), and sometimes it is not what is in the data that matters for an application (wrong attributes, wrong selected sample).

An important part of data mining result errors are due to uncertainties in modeling and generation of objects in certain databases discordant with the real probabilities of phenomena appearances in the system. That is why the data mining errors often do not have a meaning by themselves; rather they just provide a practical means of comparing efficiencies of different criteria applied to the same database.

The right people. Regardless of what many producers of data mining tools claim, data mining is not (yet) an "automatic" operation with little or no human intervention. On the contrary, the human analyst plays an important role, mostly in the areas of data selection and data/knowledge interpretation. The data miner should have an understanding of the data under analysis and the domain or industry to which it pertains. It is more important for the mining process to embrace the problems of the application meant to solve, than to incorporate the hottest technologies in the data mining software.

The right application. Almost always a problem well posed is already a partially solved problem. It is important to clearly define the goals and choose the appropriate objectives so as to yield a significant impact on the underlying decision making process.

The right questions. An important issue is how the data mine structures a data analysis problem so that the right question can be asked, knowing how easy and useless it is to give the right answer to the wrong question.

The right sense of uncertainty. Data miners are more interested in understanding than accuracy or predictability per se. Often even the best methods of search will leave the data miner with a range of uncertainties about the correct model or the correct prediction.

2.3.3 The Need and Opportunity for Data Mining

Many of the techniques used by today's data mining tools have been around for many years, having originated in the artificial intelligence research of the 1980s and early 1990s. Yet these tools are only now being applied in the artificial to large-scale database systems. The confluence of several key trends is responsible for this new usage.

Widespread Deployment of High-Volume Transactional Systems. Over the past 15 to 20 years, computers have been used to capture detailed transaction information in a variety of corporate enterprises. Retail sales, telecommunications, banking, and credit card operations are examples of transaction-intensive industries.

These transactional systems are designed to capture detailed information about every aspect of business. Only five years ago, database vendors were struggling to provide systems that could deliver several hundred transactions per minute.

Information as a Key Corporate Asset. The need for information has resulted in the proliferation of data warehouses that integrate information from multiple, disparate operational systems to support decision making. In addition, they often include data from external sources, such as customer demographics and household information.

Widespread Availability of Scalable Information Technology. Recently, there has been widespread adoption of scalable, open systems-based information technology. This includes database management systems, analytical tools, and, most recently, information exchange and publishing through Intranet services.

These factors put tremendous pressure on the information “value chain.” On the source side, the amount of raw data stored in corporate data warehouses is growing rapidly. The “decision space” is too complex; there is too much data and complexity that might be relevant to a specific problem. And on the sink side, the knowledge required by decision makers to chart the course of a business places tremendous stress on traditional decision-support systems. Data mining promises to bridge the analytical gap by giving knowledge workers the tools to navigate this complex analytical space.

2.3.4 Data Mining Tools and Techniques

Data mining applications can be described in terms of three-level application architecture. These layers include applications, approaches, and algorithms and models. These three layers sit on top of the data repository. We discuss these three levels in the following sections the characteristics of the data repository are addressed in the implementation section that follows.

Applications

We can classify data mining applications into sets of problems that have similar characteristics across different application domains. The parameterization of the application is distinct from industry to industry and application to application. The same approaches and underlying models used to develop a fraud-detection capability for a bank can be used to develop medical insurance fraud detection applications. The difference is how the models are parameterized – for example, which of the domain specific attributes in the data repository are used in the analysis and how they are used.

Approaches

Each data mining application class is supported by a set of algorithmic approaches used to extract the relevant relationship in the data: association, sequence-based analysis, clustering, classification, and estimation. These approaches differ in the classes of problems they are able to solve.

Association. Association approaches address a class of problems typified by a market-basket analysis. Classic market-basket analysis treats the purchase of a number of items (for example, the contents of a shopping basket) as a single transaction. The goal is to find trends across large numbers of transactions that can be used to understand and exploit natural buying patterns. This information can be used to adjust inventories, modify floor or shelf layouts, or introduce targeted promotional activities to increase overall sales, or move specific products. While these approaches had their origins in the retail industry, they can be applied equally well to services that develop targeted marketing campaigns or determine common (or uncommon) practices. In the financial sector, association approaches can be used to analyze customers' account portfolios and identify sets of financial services that people often purchase together. They may be used, for example, to create a service "bundle" as part of a promotional sales campaign.

Association approaches often express the resultant item affinities in terms of confidence-rated rules, such as, "80% of all transactions in which beer was purchased also included potato chips." Confidence thresholds can typically be set to eliminate all but the most common trends. The results of the association analysis (for example, the attributes involved in the rules themselves) may trigger additional analysis.

Sequence-based analysis. Traditional market-based analysis deals with a collection of items as part of a point-in-time transaction. A variant of this problem occurs when there is additional information to tie together a sequence of purchases (for example, an account number, a credit card, or a frequent buyer/flyer number) in a time series. In this situation, not only may the coexistence of items within a transaction be important, but also the order in which those items appear across ordered transactions and the amount of time between transactions.

Rules that capture these relationships can be used, for example, to identify a typical set of precursor purchases that might predict the subsequent purchase of a specific item. In health care, such methods can be used to identify both routine and exceptional courses of treatment, such as multiple procedures over time.

Estimation. A variation on the classification problem involves the generation of scores along various dimensions in the data. Rather than employing a binary classifier to determine whether a loan applicant is a good or bad risk, this approach generates a credit-worthiness "score" based on a prescored training set.

Other techniques. Additional approaches used in conjunction with these and other analytical techniques include case-based reasoning, fuzzy logic, genetic algorithms, and fractal-based transforms. Fractal-based transforms (relatively new as data analysis tools) are interesting in that they are also used as aggressive, lossless data compression algorithms. Hence, there is the possibility that pattern-matching approaches based on these techniques could exploit substantially reduced dataset sizes to increase performance. Each of these has its own strengths and weaknesses in terms of the problem characteristics best addressed, discrimination capabilities, performance, and training requirements. The algorithms are often tunable using a variety of parameters aimed at providing the right balance of fidelity and performance.

Algorithms and Models

The promise of data mining is attractive for executives and IT professionals looking to make sense out of large volumes of complex business data. The promise that programs can analyze an entire data warehouse and identify the key relationships relevant to the business is being pushed as a panacea for all data analysis woes. Yet this image is far from reality.

Today's data mining tools have typically evolved out of the pattern recognition and artificial intelligence research efforts of both small and large software companies. These tools have a heavy algorithmic component and are often rather "bare" with respect to users for interfaces, execution control, and model parameterization. They typically ingest and generate Unix flat files (both control and data files) and are implemented using a single-threaded computational model.

This state of affairs presents challenges to users that can be summed up in a sort of "tools gap." The gap caused by a number of factors require significant pre- and postprocessing of data to get the most out of a data mining application. Preprocessing activities include the selection of appropriate data subsets for performance and consistency reasons, as well as complex data transformations to bridge the representation gap. Postprocessing often involves subselection of voluminous results and the application of visualization techniques to provide added understanding. These activities are critical to effectively address key implementation issues such as:

Susceptibility to "dirty" data: Data mining tools have no higher-level model of the data on which they operate. They have no application-oriented (semantic) structure and as such, they simply take everything that are given as factual and draw the resulting conclusions. Users must take the necessary precautions to ensure the data being fed to the discovery tools. However, if the company has a good data cleansing process that cleans up data going into a data warehouse, then data mining tools benefit from this cleansing effort.

Inability to "explain" results in human terms: Many of the tools employed in the data mining analysis use complex mathematical algorithms that are not easily mapped into human terms; for example, they do not always generate "if-then" rules that use the original data attributes by name, so the ability of these systems to "explain" their results is minimal. Even with approaches such as decision trees and rule induction that are capable of generating information about the underlying attributes, the volume and format of the information may be unusable without additional postprocessing and/or visualization.

The data representation gap: Most of the source data for today's data mining applications resides in large, parallel relational database systems. The information is typically somewhat normalized and the attributes being used in a data mining application may span multiple tables. The data mining engines typically operate over a set of attribute "vectors" presented through a Unix flat file. Conditioning code must be used to provide the denormalized representation the tools need. Large central fact tables in data warehouses designed using star schema often combine denormalized data into one flat table.

Many of the tools are constructed in terms of the types of data elements with which that can work. Users may have to categorize continuous variables or remap categorical variables. Time-series information may need to be remapped as well. For example, we might need to derive counts of the number of times a particular criterion was met in a historical database.

Although the 2-GB file is becoming less important with the advance of the 64-bit operating systems, many Unix implementations still have 2-GB file limitations. For flat file-based data mining tools, this limits the size of the datasets they can analyze, making sampling a necessity.

Parallel relational database systems store data that is spread across many disks and accessed by many CPUs. Current database architectures are such that result sets generated by the database engine are eventually routed through a single query coordinator process. This can cause a significant bottleneck in using parallel database resources efficiently. Because data mining applications are typically single-threaded implementations operating off Unix flat files, the process requires potentially large result sets to be extracted from the database.

Even if we are able to extract large datasets, processing then can be computation intensive. Although most data mining tools are intended to operate against data coming from a parallel database system, most have not been parallelized themselves.

This performance issue is mitigated by "sampling" the input dataset, which pose issues of its own. Users must be careful to ensure that they capture a "representative" set of records, lest they bias the discovery algorithms. Because the algorithms themselves determine those attributes that are important in the pattern matching, this presents a chicken-and-egg scenario that may require an iterative solution.

For algorithms that require training sets (classification problems), the training sets must adequately cover the population at large. Again this may lead to iterative approaches, as users strive to find reasonably sized training sets that ensure adequate population coverage.

Present-day tools are algorithmically strong but require significant expertise to implement effectively. Nevertheless, these tools can produce results that are an invaluable addition to a business's corporate information assets. As these tools mature, advances in server side connectivity, the development of business-based models, and user interface improvements will bring data mining into the mainstream of decision support efforts.

2.3.5 Common Applications of Data Mining

Data mining approach has a major advantage from the point of view of its applicability: almost all the domains of human activity may benefit from it, both the ones where a lot of data is already available and the ones where the data has to be simulated in order to extract some more profitable knowledge concerning the field. We mention further some particular broad domains of interest in the present data mining applications.

Market basket analysis refers to the process of examining point-of-scale data to identify affinities between products and services purchased by a customer. Data mining must deal in these applications with large volumes of transactional and spread data and must be performed in a time interval that will allow an organization to respond to market opportunity before competition does. Data mining techniques like association rules and sequential patterns discovery are involved in the automatic identification of important buying patterns, types of consumers exhibiting such patterns, customer characteristics that may be correlated to the consumer's choices.

Customer segmentation is the process of analyzing data about customers or general consumers to identify characteristics and behaviors that can be exploited in the market place. Clustering, statistical analysis, deviation detection, and modeling are implicated in reducing the customer attrition phenomenon, i.e., the loss of customers (searching for customers who exhibit characteristics typical of someone who is likely to leave for a competing company), or in target marketing (attracting other customers, identification of the risk associated with insurance, etc.).

Fraud detection. Data mining applications have demonstrated their benefits in the areas where many actions (transactions) are undertaken, making the respective system vulnerable to fraud: credit card services, telecommunications, computer systems, etc.

Detection of patterns in text, image, on the World Wide Web are broadly extensive areas of DM applications due to the impressive of information available: finding association amongst the keywords labeling items in a collection of textual documents, recognizing actions in video image sequences, helping users locate desired information in the Web, and so on

Medical diagnosis through means of data mining are intended to be helpful tools that can improve the physicians' performance and make the diagnosis process more objective and more reliable. From the descriptions of the patients treated in the past for which the final diagnosis were verified, diagnosis rules may be automatically derived by means of clustering machine learning, association rules, although the technology is not widely accepted in medical practice, encountering a resistance of the physicians to new diagnostic technology.

2.3.6 What about Data Mining in Power Systems?

Why would data mining tools be useful in the power system field? Like many other application areas, the power system field is presently facing an explosive growth of data. In power systems, irrespective of the particular application, there are three main sources of data: (i) field data, collected by various devices distributed throughout the system, such as digital records; (ii) centralized data archives, such as those maintained by control center SCADA systems, (iii) data from simulations, carried out in planning or operation environments.

In a power system there are a few DM-related aspects: large-scale character of power systems (thousands of state variables), temporal (from milliseconds to minutes, hours, weeks, years), and statistical nature of data, existence of a discrete (e.g., events such as topology changes or protection arming) and continuous (analog state variables) information mixture, necessity of communication with experts through means of visualization, online operation time restrictions for fact decision making existence of uncertainty (noise, outliers, missing information).

Engineers trying to solve power system-related problems should look at the whole toolbox of data mining methods and not hesitate to combine different techniques to yield a full, practical solution. Data selection step may be performed with a decision tree, a clustering approach, or a correlation analysis, and later on, the result may serve as input for other supervised techniques, possibly with the problem decomposed into simpler subproblems.

There are three dimensions along which data mining may complement classical system theory-oriented methods for power systems.

Computational efficiency. By using synthetic information extracted by DM, instead of numerical methods, much higher speed may be reached for real-time decision making. Further, in terms of data requirements, DM may require only significant and/or available input parameters database, instead of a full description of the system model.

Anticipative physical insight. The present-day practice generally handles new problems, though some undesirable consequences have already been observed on the system. Carrying out DM studies will allow the engineers to have a more anticipative view on potential problems.

Management of uncertainties. The behavior of a power system will always have some unexpected experiences (e.g., a relay that misoperated, an operator that did not behave as expected, a set point that was different from prescriptions, a load that was modeled inappropriately). DM copes with this problem by making use of more simulations carried out by relaxing assumption on the dynamic models used.

2.4 Data Warehousing and OLAP

2.4.1 Data Warehousing for Actuaries

Competition in the insurance industry has never been stronger. The emergence of bancassurance, the recent frenzy of insurance company demutualizations and mergers, and the potential of the Internet all add to this. To stay ahead in this market, companies must understand their business, in particular: Knowing who their customers are, understanding the profile of their business, identifying the most profitable customers and products, and understanding what makes producers successful. Insurance business is complex, characterized by large volumes of data, dating back many years, with complex data relationships. This data is often fragmented throughout the organization, stored on multiple systems, with the resulting information retained in small analytical niches. To address these data problems many companies have turned to data warehouses. Promising to be a single version of reality, a data warehouse has the obvious appeal of collecting all the data and storing it in a single place – somewhere it can be queried and manipulated, without affecting the day-to-day running of the business. But it is more than that, the data is structured in a way to optimize analytical applications, definitions are added to data elements to ensure a common understanding, and front-end tools have been developed that can efficiently drill into, manipulate, and distribute results. Data warehouses are now central to the systems strategy for companies in many industries. However, the life insurance industry has been slow to realize the full potential of this technology. The reasons for this are numerous, but at the forefront is the complexity of the business and the data relationships. Some companies have explored marketing data marts and reporting marts, but more complex analysis such as earnings by source and experience analysis is still rare.

2.4.2 Data Warehouse Components

The data warehouse will typically consist of several components. Data originates from various source administration systems, extracted as a series of snapshots at regular time intervals; typically monthly or quarterly for financial analysis, or more frequently for marketing analysis.

The ETL (Extraction, Transformation, and Loading) tool cleans and transforms the data, so that it meets the requirements of the warehouse; information about this process is stored as metadata. This metadata, together with business metadata (business rules and definitions), is available to end users of the system.

Data stored on the source systems is unlikely to be in the format required. Transformation can be as simple as converting "0" to "male" and "1" "female," or very complex involving multiple data fields to be evaluated simultaneously and logic applied to determine the desired value. This stage is critical as it provides users with confidence in the data they are using. It also provides system management benefits such as:

- An audit trail and documentation of the extract routine
- A formal loading process so that reloads are possible
- Better automation than ad hoc loading of data.

Data may also be fed to and from third-party systems, such as valuation systems, to make use of their specialist functions. OLAP or data cubes provide a simple and effective way to view data.

Data warehousing for actuaries cube can be thought of as the dimensions of the data – how the data is stored and viewed, such as by product line, time period, currency. The content of the cube is the item being measured, such as premium, reserve, or sum assured. It is easy to visualize how data is accessed, by identifying the content at the intersection of the selected dimensions. But here the analogy breaks down; data views are not limited to only three dimensions as physical cubes are. In addition, intelligence can be built in the dimensions to help the user by identifying natural drill down paths, such as product lines and time periods. Data is usually presented to users in the form of pivot tables, and may also consist of preformatted reports, Web portal, or graphical interface. Data warehousing tools bring a number of natural strengths to the sort of analysis that is required for insurance. Consistency is clearly enhanced with the introduction of a data warehouse. If the administration data is on multiple systems it is an opportunity to bring it together on one system, with consistent definitions for status, gender, and other demographic information that would be common across systems, but not necessarily stored consistently. Also, having assets and liabilities together on a single system allows consistent and controlled analysis of both sides of the balance sheet. Multiple hierarchies may be used to meet the requirements of various reports and reporting bodies. These will be consistent at the lowest level of detail; reconciliation is then simple with the help of metadata and drill down. Similarly demographic groupings can be predefined. With the ability to drill down into the demographic groupings reports should be clearer and more consistent.

2.4.3 Management Information

Most insurance companies have multiple sources of management information. The data warehouse provides an opportunity to combine these and at the same time ensure consistency between subject areas. These management information reports may include:

- Business profiles such as new business and in-force policy reports.
- Movement analysis.
- Experience studies.
- Financial reporting.

Even though financial reporting is basically the same every year or every quarter, it is often at a time when long hours are worked with last-minute surprises in the results. A data warehouse can strengthen any process and remove redundant processes, such as multiple data extracts. Ultimately, financial reporting consists of a few results presented in different ways and tested for trends and reasonableness. With a data warehouse these can prepared in a seamless fashion and can eliminate a lot of time in reconciliation between the various reports that have essentially the same numbers aggregated in different ways. This also reduces the chance of errors occurring in transferring data and helps simplify audit and process documentation. Previously, actuaries would have spent a significant amount of their time collecting and collating data.

With the data warehouse, actuaries can better spend their time understanding the impact of the results.

2.4.4 Profit Analysis

Profit analysis is more than the ability to present details of the published statement. Profit analysis should highlight the elements that make up the profit. Profit should be understood in terms of the source factors that contribute to it; the lapse contribution, for example, is not just the number of lapses for in the year, but also the impact on the reserves, mortality, and expenses.

The actuarial methods used for analyzing profit, when deployed in a data warehouse, allow profit to be viewed for any policy attribute such as product or product line, geographical sales area, premium range, etc. by simply limiting on the appropriate dimensions. Furthermore, with this implementation approach it is possible to drill in and constrain on any of the product attributes to give rich and informative reports on the business. This flexibility of presentation and detail can be extremely useful when communicating a difficult subject.

2.4.5 Asset Liability Management

Regular ALM reporting can be made quickly available with the use of the data warehouse. Just as with profit analysis the additional flexibility provided by the use of dimensional drill-down can really help communication. Insurance companies tend to have a significant amount of unstructured duplication of effort and data. By using a data warehouse and structuring data in way that suits analysis, organizations will quickly discover that reports and analysis become interrelated. Time and money are saved, as additional reconciliation is not required. More importantly, from marketing analysis, such as customer behavior to financial analysis, the depth of understanding is increased with the ability to quickly manipulate and share information.

Scorecards. Balanced scorecards and financial scorecards are becoming more common in organizations. A scorecard is basically a set of performance dials for measuring strategic health and performance. This approach can be used to translate vision, strategy, and tactics into concrete financial performance targets and measures. And in turn, for using these targets and measures for communication of what is important to the business.

Data Mining. Data mining in insurance companies has been used primarily for marketing purposes and also to assist with underwriting in avoiding antiselection. But for many insurance companies data mining is not practical; the quality of the data on the source system is likely to swamp any useful discovery. However, the data warehouse provides a natural platform on which to base any data mining activities. Any patterns that are discovered by the data mining tool are likely to require some justification. Often these are counterintuitive and time will have to be spent examining this before management will be persuaded to take action. However it could be argued that this is precisely the purpose of data mining.

2.5 Data Mining and OLAP

In the section that follows we give perspective of data mining such as: Why do we need data mining? How are other DM techniques are used? As an example of using data mining and OLAP, we have introduced in this section the definition of OLAP, which is short for Online Analytical Processing, a category of software tools that provide analysis of data stored in a database. OLAP tools enable users to analyze different dimensions of multidimensional data. For example, OLAP provides time series and trend analysis views. OLAP covers, including multiple hierarchies, rules analytical operations and the difference between OLAP and data mining.

2.5.1 Research

1. History Perspective – The Relational Model

Transaction processing systems. This is where clients store messages in a DBMS storing the data by using a number of files such as sequential files, indexed sequential access methods, variable sequential access methods, hierarchical databases and network databases. OLTP are good for putting data into databases.

The relational databases management system stores all sorts of business data by using the sequential files or indexed sequential access method or variable sequential access method, or hierarchical databases and network databases. Also it increases the electronic devices that enter all the data into database storage, such as a barcode scanner, EPOS, etc. It has become a useful accessory, which is cheaper and more powerful.

2. DBMS Problems

DBMS has given the access to store the data but there is no analysis to analyze the data. Because there is so much data, it would take a long time to find a specific data we require. With analysis there are all sorts of data that could be used such as decision support. Without the analysis, a new user would know nothing about the commercial business application and paralleled principles.

This is where data mining has become increasingly important to the organization, giving them more of an advantage. Data mining is the use of software that provides tools and applications (data analysis, decision support, and automation etc.) to gather reports and analyze the information, such as

- Survey results analysis
- Inappropriate practices
- Fraud detection
- Manufacturing process analysis
- Risk analysis and management
- Market and sales analysis
- Scientific data analysis
- Text document analysis

It is also used to collect all the necessary information and store it in electronic format for the future. Data mining involves capturing, storing, gathering, and analyzing the company information. There is a risk of losing such amounts of important data; hence the company will have to consider being careful about giving access or changing the system.

3. Data Mining

The definition of data mining is:

> "*Data mining is the search for relationships and global patterns that exist in large database but are 'hidden' among the vast amount of data, such as a relationship between patient data and their medical diagnosis. These relationships represent valuable knowledge about the database and the objects in the database and, if the database is a faithful mirror, of the real world registered by the database*".
>
> Marcel Holeshemier & Arno Siebes (1994)

Who needs Data Mining?

Every business needs data mining because companies are looking for new ways to access and to allow end users access to the data they need for making decisions, serving customers, and gaining the competitive edge.

We need data mining to collect data and allow us to analyze it. If it is not analyzed at that time the collected data could be important in the near future. As a database grows, the ability to support decisions by processing using traditional query languages is not feasible.

Data Mining Process

A data mining project consists of a life cycle that has six phases. It is not strict that these phases should be in order. It can either move backward or forward in between the phases, as they are required, in each phase it becomes the outcome of the task and is performed in the next phase. This indicates the most important and frequent dependencies between phases. The data mining process continues after a solution has been deployed. This has become new and often focuses more on business questions that subsequent data mining process will benefit from the experience of previous ones.

The steps below help to explain each stage of the data mining process:

- Business understanding

This focuses on understanding the project objectives and requirements from a business perspective and finds a data mining problem of definition. Then it is so designed to achieve the objectives.

- Data understanding

This subject deals with understanding the data before previous data is collected and proceeds with activities; so it will be familiar with the data to identify data quality problems, discover the data, and then start proceeding.

- Data preparation

The data preparation phase covers all activities to find the data that will be fed into modeling tools, from the initial raw data. In this phase the tasks are likely to be performed multiple times, and it is not required to do anything in sequence order. Tasks include table and record, while attribute selection includes transformation and cleaning of data for modeling tools.

- Modeling

This phase has various modeling techniques that are selected and applied, so their parameters are adjusted to optimal values. Sometimes, there are several techniques for the same data mining problem type. There are some techniques that have specific requirements in the form of data. So they step back to the data preparation phase, which is often needed.

- Evaluation

By this stage we would have built a model(s) for the project so that it appears to have high quality, hence we get data from analysis perspective. Before going to further phases of the model, it is important to thoroughly evaluate the model and review the steps to the end to construct the model – this can help in properly achieving the business objectives. This is a key objective to determine if there are some important business issues that need to be considered. At the end of this phase, this decision will find the use of the data mining results.

- Deployment

Creating the model is not the end of the project. Even if the purpose of the model is to increase knowledge of the data, the knowledge gained will need to be organized and presented in a way that the consumer can use it. On the requirements, this phase can be as simple as generating a report or as complex as implementing a repeatable data mining process. In many cases, it will be the customer, and not the data analyst, who will carry out the deployment steps. Therefore if the analyst does not carry out this deployment, it is important to make an effort for the customers to understand the action that is to be carried out in order to actually make use of the created models.

Main points of Data mining for the organization

- They will have large volumes of data
- Employees need to understand data
- There is a need to exist in a competitive market
- Data mining components (discovery and exploitation)
- Data mining discovers relationships between data

Data Mining Components (Discovery and Exploitation): Data mining makes discovery of facts easier as they are represented as information data. Then follows the exploitation or use of those facts for problem solving.

Data Exploration: This involves preparing the data to get a better pattern discovery and also to validate the result of the data preparation. But sometimes this involves examining the statistics (minimum, maximum, average, etc.) and the frequency distribution of individual data fields. Finally this involves the field versus field graphs to understand the dependency between fields.

Pattern Discovery: This is the stage of applying the pattern discovery algorithm to generate patterns. This allows business users to interact with and to impart their business knowledge to the discovery process. This stage also involves analyzing for understanding discovered patterns to predict the propensity of the business event, and for verification against an independent dataset.

Data Mining discovers relationships between data: All new data can be added easily to existing data so that if there is a relationship between any of them a solution may be worked out. For example if a woman asks her husband to buy apples in the supermarket, it is likely that he will want to buy soft drink for himself at the same time. So the management may suggest that the soft drink section be placed quite near to the baby goods shelves to encourage sales. This plan of action is used to promote most products in large shops.

Making choices – A decision has to be made as to which is a suitable choice from various alternatives.

Making predictions – A prediction is a choice about what may happen in the near future.

Improving processes – To improve the system, the current system must be updated and improved.

Mining For Golden Nuggets of Data.

Issues in Data mining

There are some problems. Data mining systems rely on databases to supply the raw data for input. This can cause problems in that the database tends to be dynamic, incomplete, noisy, and large. Then there are other problems that have arisen as a report of the adequacy and relevance of the information stored.

Limited Information

Sometimes the database is designed differently for the data mining; therefore the problem has to be solved before access to the information can be given. Undecided data causes problems; this can be because some attributes have useful knowledge about the application. This is not present in the data that may be impossible to discover the knowledge about a given domain. Databases are usually muddled up by errors, so it cannot be the correct way the data is contained from the originally inputted data.

Missing data can be treated by discovery systems in a number of ways such as:

- Simply disregarding missing values
- Omitting the corresponding records
- Inferring missing values from known values
- Treating missing data as a special value to be included additionally in the attribute domain.
- Averaging over the missing values using basic techniques.

Uncertainty refers to the strictness of the error and the noise in the data. Databases tend to be large and dynamic in that their contents change as information is added, modified, or removed.

Data mining Applications

Data mining in Government

- Develop advanced pattern recognition
- Identify problem in technologies
- Improve the project of their techniques

Data mining in Energy

- Identify loyal customer
- Fraud detection in energy industry
- Predict industry specific

4. OLAP

There are two technologies that are related to data mining: data warehouse and OLAP.

Data Warehouses

Here the data is collected and stored in the data warehouse. Data warehouse is the relational database management system, but it is specifically designed for a transaction processing system. These warehouses contain millions of pieces of information about customer's needs and distribution decisions. Data warehouse uses the data to analyze business needs and to make the decisions. They make amounts of data that span over many years. A data warehouse is not a transactional database.

Transactional databases use data to operate business needs. For instance, if an organization wants to look up data about 60–90 days old, they can. They are built for speed and efficiency to keep the company's day-to-day operations moving fast, unlike data warehouses, which runs very slowly.

This is where OLAP (On Line Analytical Processing) introduces the tools to analyze data warehouse information. We need data mining to collect data and to analyze it, or if not analyzed, collected. This could be important in the near future. As databases grow, the ability to support decision by processing traditional query languages is no longer feasible.

What is OLAP?

OLAP provides the users with multidimensional database to generate on-line description, or it compares the "views" of data and other analytic queries. OLAP gives the answers to multidimensional business questions quickly and easily. OLAP technology provides facts and efficient access to summarized data. Also OLAP is able to give control over global views of the business.

OLAP technology can be applied to:

- Sales and marketing analysis
- Financial reporting
- Quality tracking
- Profitability analysis
- Manpower and pricing applications
- Our unique data discovery needs

The users have found easy-to-understand graphical information that can be presented simply and acted upon quickly. Like for instance, when the clients have problems with their data mining process, they need to respond quickly for their requirement.

Relational OLAP approach builds such a system. Relational databases are used to build and query these systems. To analyze the query using in SQL

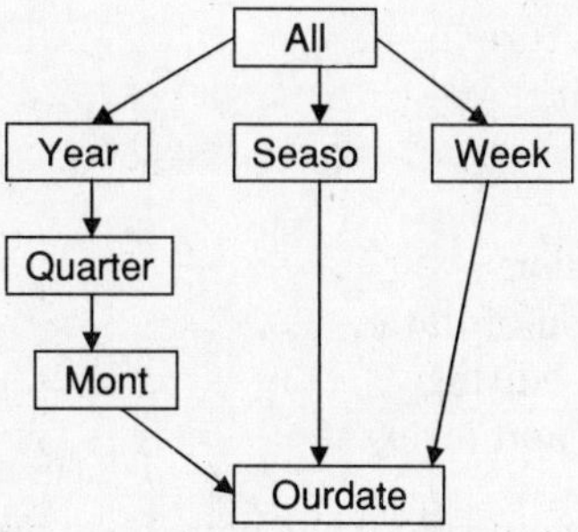

Fig. 2.4. Multiple dimensional hierarchies

might not be efficient to execute; therefore another away of using multidimensional database techniques has been applied to decision support applications. Data is stored in multidimensional structure, which is a more easy way of expressing the enterprise data and is more suited for analysis.

Multiple hierarchies

A dimension can consist of multiple hierarchies as shown in Fig. 2.4. This enables a consistent representation of the information in, e.g., data warehouse, yet provides flexibility to see the information using different perspectives.

There are three reasons why it is beneficial to create multiple hierarchies.

- Sharing a single key in the fact table. It can create a single key to represent both hierarchies in the fact table, thus reducing the fact table size and complexity.
- Usability. Multiple hierarchies in a single dimension provide better usability for the end user. If the front-end application supports multiple hierarchies logically grouped together.
- Sharing of aggregates. Suppose we have three hierarchies. One hierarchy is year, month, day, date; the second is season and date; and third is week and date. The three hierarchies share the aggregates that are built on the year level, thereby reducing the processing time necessary to create the same aggregate second and third times.

OLAP Rules

Originally, 12 OLAP rules were defined to characterize this technology. OLAP products tend to conform to these. The 12 rules are as follows:

- Multidimensional model
- Transparency of the server
- Accessibility
- Stable access performance

- Client server architecture
- Generic dimensionality
- Management of data
- Multiuser
- Operation on dimension
- Intuitive manipulation of data
- Flexible posting and editing
- Multiple dimensions and levels

Today these are the 12 main rules that have been expanded to 18 major rules and up to 300 in total.

Example of BI (Business Intelligence) Technology

By using ad hoc query application the user can have access to information on demand; this means that what they ask for is what they will get. For example a user creates and executes an ad hoc query by asking the question, "How much revenue was generated by each customer during this year?" The results from the query would ask the customer name and give the revenue for the year selected.

Then the customer asks a series of questions, "How much revenue was generated this year?" "Which customer generated the most revenue for the company?" "And which customer generated the least revenue for the company?"

BI technology could not identify the unusual patterns or reveal unusual relationships. This is the next stage where the OLAP comes in. OLAP analyses the summary and details the information. They could answer all the questions above. The user creates and performs an OLAP analysis that provides answers to the questions "What was the revenue for each quarter of this year by geographic region and customer?" The result from this analysis would contain geographic region, customer name, revenue, and quarter selected.

But the user needs to understand how to navigate the data, which must direct these processes. OLAP can only highlight the patterns within the data that was requested. So the BI technology will not identify unusual or reveal hidden relationships. The next stage is where the data mining comes in.

In BI technology data mining can extract the hidden and useful information from the data. Data mining discovers the hidden trends and patterns in large amounts of data. There are various techniques that can be deployed, each serving a specific purpose and varying amounts of user involvement.

2.5.2 Data Mining

Neural networks –detect potential fraudulent credit card transactions.

Induction – understands the relationship that exists, e.g., when people buy nappies, they also buy beer 50% of the time.

Statistics – requires highly skilled mathematicians to build and interpret the results.

Visualization – displays the data in a graphical map for the user to identify trends, patterns, and relationships. Images produced provide another perspective of data relationship. Visualization is often incorporated in data mining application. Ad hoc query applications stated the values that exist within a database, while OLAP provides users with understanding and gives more detailed information.

Data mining digs deeper and provides users with knowledge through the discovery of hidden trends and relationships. By using ad hoc query, OLAP application and data mining becomes powerful, it provides users with knowledge about the data that is analyzed and enables action on the knowledge that has been found.

Analytical Operations

The end users must get the data for analytical purposes by applying analytical operations such as ratios, cumulative totals, trends, and allocations across dimensions and hierarchical levels. OLAP functionally is described by dynamic multidimensional analysis of:

Consolidated data supporting the end user analytical navigational activities including

- Calculations and modeling applied across dimensions, through hierarchies, and across members
- Trend analysis over sequential time periods
- Slicing subsets for on-screen viewing
- Drill down to the deeper levels of consolidation
- Each through to underlying detail data
- Rotation to new dimensional comparisons in the viewing area

OLAP and Data Mining

OLAP is user driven, the analyst generates a hypothesis and uses OLAP to verify, e.g., "people with high debt are bad credit risks."

Data mining tool generates a hypothesis – Tool performs exploration, e.g., find risk factors for granting credit.

It also discovers new patterns that analysts did not think of, e.g., debt-to-income ratio. OLAP and DM need each other; we have mentioned this in the example about (BI Technology). OLAP and data mining should not remain separate components of decision support; therefore they should be fully merged and they really need each other. When dealing with OLAP and data mining it shows that decision support applications must consider data mining within multiple dimensions and that OLAP system needs to focus on discovery as much as on access. OLAP and data mining must work together within the framework of any theory to avoid getting incorrect results.

OLAP and OLTP

A major issue in information processing is how to process increasingly larger databases, containing increasingly complex data, without sacrificing response time. The client/server architecture gives organizations the opportunity to deploy specialized servers, which are optimized for handling specific data management problems. Until recently, organizations have been trying to target relational database management systems (RDBMSs) for the complete spectrum of database applications. It is however apparent that there are major categories of database applications that are not suitably serviced by relational database systems.

OLAP Example

An example of OLAP database may comprise sales data that has been aggregated by region, product type, and sales channel. A typical OLAP query might access a multigigabyte/multiyear sales database in order to find all product sales in each region for each product type. After reviewing the results, an analyst might further refine the query to find sales volume for each sales channel within region/product classifications. As a last step the analyst might want to perform year-to-year or quarter-to-quarter comparisons for each sales channel. This whole process must be carried out on-line with rapid response time so that the analysis process is undisturbed. OLAP queries can be characterized as on-line transactions, which:

- Access very large amounts of data, e.g., several years of sales data.
- Analyze the relationships between many types of business elements, e.g., sales, products, regions, channels
- Involve aggregated data, e.g., sales volumes, budgeted dollars, and dollars spent.
- Compare aggregated data over hierarchical time periods, e.g., monthly, quarterly, or yearly
- Present data in different perspectives, e.g., sales by region vs. sales by channels by product within each region.
- Involve complex calculations between data elements, e.g., expected profit as calculated as a function of sales revenue for each type of sales channel in a particular region.
- Are able to respond quickly to user requests so that users can pursue an analytical thought process without being stymied by the system.

Comparison of OLAP and OLTP

OLAP applications are quite different from on-line transaction processing (OLTP) applications, which consist of a large number of relatively simple transactions. The transactions usually retrieve and update a small number of records that are contained in several distinct tables. The relationships between the tables are generally simple.

A typical customer order entry OLTP transaction might retrieve all of the data relating to a specific customer and then insert a new order for the customer. Information is selected from the customer, customer order, and detail line tables. Each row in each table contains a customer identification number, which is used to relate the rows from the different tables. The relationships between the records are simple and only a few records are actually retrieved or updated by a single transaction.

The difference between OLAP and OLTP has been summarized as while OLTP servers handle mission-critical production data accessed through simple queries, OLAP servers handle management-critical data accessed through an iterative analytical investigation. Both OLAP and OLTP have specialized requirements and therefore require special optimized servers for the two types of processing.

OLAP database servers use multidimensional structures to store data and relationships between data. Multidimensional structure contains aggregated data relating elements along each of the dimensions. For example, a single cell may contain the total sales for a given product in a region for a specific sales channel in a single month. Multidimensional databases are a compact and easy-to-understand vehicle for visualizing and manipulating data elements that have many inter-relationships.

OLAP database servers support common analytical operations, including consolidation, drill-down, and slicing and dicing.

- Consolidation - involves the aggregation of data such as simple roll-ups or complex expressions involving inter-related data. For example, sales offices can be rolled up to districts and districts rolled up to regions.
- Drill-down - OLAP data servers can also go in the reverse direction and automatically display detail data, which comprises consolidated data. This is called drill-downs. Consolidation and drill-down are an inherent property of OLAP servers.
- Slicing and Dicing - Slicing and dicing refer to the ability to look at the database from different viewpoints. One slice of the sales database might show all sales of product type within regions. Another slice might show all sales-by-sales channel within each product type. Slicing and dicing is often performed along a time axis in order to analyze trends and find patterns.

OLAP servers have the means for storing multidimensional data in a compressed form. This is accomplished by dynamically selecting physical storage arrangements and compression techniques that maximize space utilization. Dense data (i.e., data exists for a high percentage of dimension cells) is stored separately from sparse data (i.e., a significant percentage of cells are empty). For example, a given sales channel may only sell a few products, so the cells that relate sales channels to products will be mostly empty and therefore sparse. By optimizing space utilization, OLAP servers can minimize physical storage requirements, thus making it possible to analyze exceptionally large amounts of data. It is also possible to load more data into computer memory, which helps to significantly improve performance by minimizing physical disk I/O.

In summary, OLAP servers logically organize data in multiple dimensions, which allows users to quickly and easily analyze complex data relationships. The database itself is physically organized in such a way that related data can be rapidly retrieved across multiple dimensions. OLAP servers are very efficient when storing and processing multidimensional data. RDBMSs have been developed and optimized to handle OLTP applications. Relational database designs concentrate on reliability and transaction processing speed, instead of decision support need. The different types of server can therefore benefit a broad range of data management applications.

2.6 Summary

Data mining is beginning to contribute research advances of its own, by providing scalable extension and advances to work in associations, ensemble learning, graphical models, techniques for on-line discovery, and algorithms for the exploration of massive and distributed datasets.

Advances in data mining require supporting a) single investigators working in data mining and the underlying research domains supporting data mining; b) interdisciplinary and multidisciplinary research groups working on important basic and applied data mining problems; and c) the appropriate test beds for mining large, massive, and distributed datasets. Appropriate privacy and security models for data mining must be developed and implemented.

There are three dimensions along which data mining may complement classical system theory-oriented methods for power systems. They are computational efficiency, anticipative physical insight, and management of uncertainties that are discussed earlier in this section.

Overall in this section more information is added to data mining and OLAP. There are many differences between data mining and OLAP and they have been used for the organization business (e.g., BT technology). Data mining and OLAP should work together; therefore the data has given faster information by using their techniques tools. Data mining has a lot of potential; it is increasing every year and is becoming very useful for organizations. It is fast and efficient along with OLAP. It has diversity in the field of application, which becomes more efficient in the database system. Now the estimated market for the data mining in the United States in $500 million.

2.7 Review Questions

1. What are the recent research achievements in data mining?
2. State some of data mining application areas.
3. Write about the various trends in the current scenario that affect data mining.

4. Give some of the research challenges in data mining area.
5. Define the KDD process.
6. Explain in detail about the process involved in data mining.
7. What are the major essential elements in data mining?
8. Who is a data miner and write on the approach of data miners in mining.
9. What factors influence the success of data mining?
10. Explain data mining in power systems with suitable example.
11. What are the components of data warehouse and data mining?
12. Define OLAP.
13. Give details on multidimensional hierarchies used in mining.
14. State the OLAP rules.
15. Explain about business information technology using OLAP.
16. Write short notes on OLAP and OLTP.
17. Compare and contrast OLAP and OLTP.

3

Data Marts and Data Warehouse: Information Architecture for the Millennium

Objectives:

- The new dynamics that characterize today's global business environments are unprecedented in their scope and impact, requiring organizations to adapt their strategies and business practices far more rapidly and intelligently than ever before.
- Data warehouses and marts give organizations the ability to consolidate information from multiple, heterogeneous production systems into a common, integrated database environment.
- In this section we explain the structure and value of today's enterprise data warehouse and data mart systems, positioning them within what we call the "corporate information factory."
- We describe how these systems are used by corporate decision makers to gain and maintain a strategic edge over their competition, and suggest criteria for selecting and evaluating different data warehouse and data mart technologies for different purposes. We highlight the benefits Informix® delivers with its Decision Frontier™.
- The data mart/enterprise data warehouse environment increasingly becomes important as the data warehouse environment evolves to the corporate information factory.
- To provide the modern communications company with an environment that supports the business, an architecture has been developed to address the technological and business issues it faces.
- To reduce the complexity of the decision support systems and the development of the data warehouse, it is necessary to understand the business issues, technical infrastructure issues, and data issues.
- The telecommunications lifecycle is a holistic approach that addresses the problems at the root, long before an invoice appears.
- To review the key features required for OLAP software as referenced in the OLAP Council benchmark specification.

S. Sumathi and S.N. Sivanandam: *Data Marts and Data Warehouse: Information Architecture for the Millennium*, Studies in Computational Intelligence (SCI) **29**, 75–150 (2006)
www.springerlink.com © Springer-Verlag Berlin Heidelberg 2006

- Data warehousing is a management tool that enables executives to access the information they need to make informed business decisions.
- A data warehouse coupled with a focused set of decision support applications provides seamless integration of communication services globally, better understanding of profitability, and a focused marketing effort.
- The future success or failure of telecommunication companies could well depend on a company's knowledge of the market and its customers, which will, in turn, depend on the successful implementation of application tools as part of a data warehouse solution.
- The telecommunications lifecycle is a holistic approach that addresses the problems at the root, long before an invoice appears.
- The telecommunications lifecycle approach is the best practices way to address the problems before they occur and avoid overpaying.
- This section looks into the different techniques used over the years to achieve the perfect balance between performance and security and then propose a security model that is mostly based on XML and related technologies.
- A single technology for implementing security meeting the required performance target is yet to be devised, but the existing technologies can be used in combination to produce satisfactory results.
- This section explores the business and technology issues related to both buying and building data warehousing solutions.
- We discuss requirements for effective insurance decision support systems and illustrate why we feel that buying a solution is ultimately more cost effective for insurance companies.
- We present all the main data mining techniques, giving explanations on how they work, and what can they be used for.
- The expression *data mining* includes a wide range of tools and techniques to extract information from a large collection of data.
- Another important issues is to design algorithms that can handle the huge dimension of modern data warehouses.

Abstract. In this section we explain the structure and value of today's enterprise data warehouse and data mart systems, positioning them within what we call the "corporate information factory." We describe how corporate decision makers use these systems to gain and maintain a strategic edge over their competitors, and suggest criteria for selecting and evaluating different data warehouse and data mart technologies for different purposes. We highlight the benefits Informix® delivers with its Decision Frontier™. The integrated products and services that make up decision frontier provide a complete, high-performing and highly scalable data warehousing environment—including query, reporting, and data analysis capabilities, along with industry-specific templates and consulting services – which enables the rapid deployment of business-critical analytic applications.

Data warehousing is a management tool that enables executives to access the information they need to make informed business decisions. Information is the greatest weapon in the competitive arsenal of healthcare organizations. Information can tell organizations how well providers are performing, what treatments are working, and if clients are satisfied with the services they receive. That may sound simplistic, but once the data is manipulated and interpreted, it is like cracking the enemy's code and being privy to information that reveals the lay of the battlefield and the conditions of the war. In the competitive healthcare arena, this extra information can mean the difference between survival and demise.

The raw material (data) needed for the development of an enterprise-wide data warehouse is already abundant in the deregulated telecommunications industry. Customer data is both the most plentiful and most underused resource currently available. However, data alone will not generate meaningful information and knowledge unless the company's business objectives and information technology (IT) are aligned. For technology to become a key enabler in the implementation of new strategies it must first become integrated into the company's business functions. The ability to acquire, manage, and retain customers will be driven by the speed and efficiency with which the carriers are able to utilize their architectural infrastructure. The future success or failure of telecommunication companies could well depend on a company's knowledge of the market and its customers, which will, in turn, depend on the successful implementation of application tools as part of a data warehouse solution. Effective data warehousing often results in very large databases, reaching multiple terabytes of storage. Currently, a major issue within the telecommunications companies is that the typical IT infrastructure is unable to deliver the level and quality of data to support the business unit new processes.

Data warehousing has increasingly become an important and rather essential phenomenon in the world of enterprise data storage. The idea is to allow a single infrastructure to be used in the functionality of data warehouse and to facilitate the data distribution and manipulation by the users/clients of the data in a simplified way. Different techniques are used to achieve this target with the main emphasis resting with the performance of the data warehouse. It is not easy to handle an enterprise-sized data and yet met the target performance as well. Performance may be the aim but the fact remains that with the data of an entire enterprise resting in a collection, the security has to be the critical issue. This section looks into the different techniques used over the years to achieve the perfect balance between performance and security and then propose a security model that is mostly based on XML and related technologies. In addition, the data warehousing concept for insurers is also discussed.

3.1 Data Marts, Data Warehouse, and OLAP

3.1.1 Business Process Re-engineering

The new dynamics that characterize today's global business environments are unprecedented in their scope and impact, requiring organizations to adapt their strategies and business practices far more rapidly and intelligently than ever before. Powerful market forces like mass customization in the manufacturing and retail sectors, deregulation in banking, utilities, communications, and insurance, and electronic commerce across virtually all industries are driving change within the organization at an extraordinary rate and down to its very foundations. What is more, this trend toward the complete reinvention of core business structures and processes will only accelerate over the coming years.

In the face of such profound change, organizations are concluding that high-quality information—about markets, competitors, economic conditions, resources, and their own business—has gone beyond being a success factor and has become a survival issue. And even as decision makers struggle to make sense out of the data relevant to their business the quantity, complexity, and scope of that data is growing at an exponential rate.

3.1.2 Real-World Usage

At the same time, however, many companies—including some of Informix's leading customers—are already developing innovative ways of doing business that are creating for them unprecedented opportunities. For example:

- A bank can consolidate information from checking, savings, mortgage, and investment management applications to understand, often for the first time, who are its most valuable customers. Once these customers have been identified, the institution can proactively market to them in new, innovative ways.
- A telecommunications company can analyze call records and identify customers who are likely to churn—that is, defect to a competing vendor. With this information the company can determine how much of an investment is required to keep those customers.
- A manufacturing firm can combine customer support and shipping data in a data warehouse to identify vendors that are supplying goods that cause downstream product failures. This information enables the manufacturer to work with its suppliers to help those companies improve their processes and procedures, helping each company produce a higher quality product, reduce waste, and improve customer satisfaction. So the challenge for today's organizations has become one of getting ahead of the rapidly moving business intelligence curve through the smart acquisition and use of data and information technologies.

3.1.3 Business Intelligence

Business intelligence is a term that has emerged to define the data-analysis and decision-making environment within corporations. It is a vehicle that can enable these organizations to define, establish, and maintain competitive advantage even as the volume and complexity of critical business data continues to explode. Transforming data into business intelligence has become a trend that is increasing at a very sharp rate, as demonstrated by greater than 50% annual growth in budgets allocated by the Global 2000 corporations for data warehouses, data marts, and business intelligence software solutions. Data warehouses and marts give organizations the ability to consolidate information

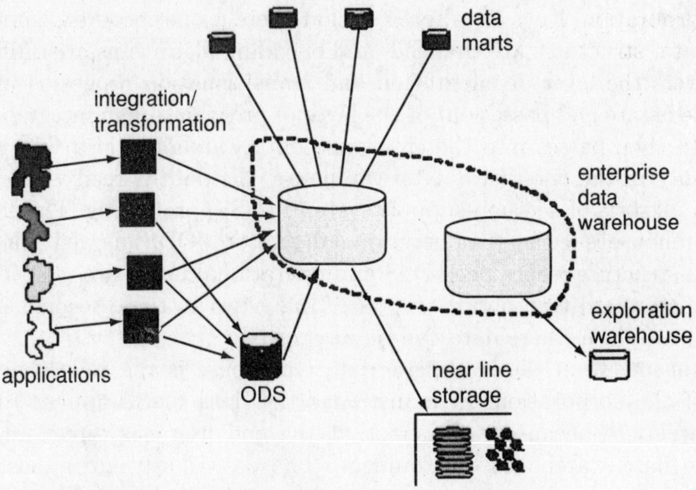

Fig. 3.1. The Corporate Information Factory

from multiple, heterogeneous production systems into a common, integrated database environment. Information in this environment can then be organized, summarized, regularly refreshed, and used to great advantage by a wide variety of powerful analytical applications. The latter give decision makers critically important tools for making strategic decisions quickly and confidently.

From the humble origins of a system that was something other than transaction processing capabilities, data warehousing has grown into a full-blown architecture. Today, data warehousing has different components that serve different purposes and communities. The architecture that has grown from the origins of data warehousing is today something that can be called the *corporate information factory.* Figure 3.1 depicts today's mature corporate information factory.

The corporate information factory provides the starting point for data warehousing—the legacy application systems. In the legacy applications, transactions are executed in which detailed transaction data is collected. The detailed data that is collected is more of a by-product of transaction processing than anything else. But this raw data forms the very detailed foundation needed for the data warehouse. The transaction data that is collected passes through a layer of programs whose purpose is to integrate and transform the data to make it fit to enter the data warehouse. The layer of programs can be called *integration and transformation* programs or *ETL*—extract/transform/load programs. These programs can be written manually or in an automated manner by program generators specifically designed to suit the needs of this interface.

Once the legacy data passes through the integration and transformation programs, the data is fundamentally changed to meet the informational needs of the corporation. Keys are altered so that there is one corporate understanding of data, structures are changed, and encoding algorithms are unified. The data enters the layer of integration and transformation programs in an unintegrated state and passes out of the layer of programs in an integrated state. The data then passes into the enterprise data warehouse in a very granular form. Once in the enterprise data warehouse, the data is ready to serve as a basis for all sorts of decision-support systems (DSS) processing. The enterprise data warehouse is fed by the operational data store (ODS) as well. The ODS is a hybrid structure where both DSS and operational needs are met. The ODS provides standard transaction response time—two to three seconds—besides serving as a place where data can be integrated.

Emanating from the enterprise data warehouse is the collection of data marts of the corporation. In many ways the data marts appear to be the data warehouse because it is here that the end user has direct interaction with the data warehouse environment. The data marts are almost always oriented to the needs and requirements of the department. The data marts are requirements-driven in that they are shaped by the unique and specific needs of the department. The data marts are fed granular data from the enterprise data warehouse, which is then reshaped to meet the requirements of the department.

In some organizations there is more than one enterprise data warehouse. In these cases, data warehousing takes on a nontraditional form. The appearance of multiple enterprise data warehouses is a double-edged sword. On one hand, it means data warehousing has been successful. But on the other hand, multiple enterprise data warehouses present special challenges to the designer and user of the data warehouse. For example, it often happens that in a multiple enterprise data warehouse environment there is overlap of data between two or more enterprise data warehouses. Done properly, this overlap is very beneficial. Done improperly, the overlap can cause architecturally severe problems. Another feature of the corporate information factory is that of the exploration warehouse. This is a separate, specialized facility designed solely for the needs of the corporate explorer. With the exploration warehouse, the very ad hoc queries created by the explorer can be run with no damage to the normal warehouse processing.

And, finally, there is the near-line storage facility. This facility extends the capacity of the warehouse effectively to infinity. No longer is the designer constrained by the technological and economic limitations of disk storage. With near-line storage, the designer can take data down to the lowest level of detail desired—something that realistically is not a possibility with disk storage. This section focuses on the enterprise data warehouse and the data marts. In recent vintage, no other part of the warehouse environment has created as much confusion and as much controversy as these parts of the corporate information factory. The general architectural positioning of the data marts and the data warehouse is shown in Fig. 3.2, which shows how data marts are fed directly from the data warehouse.

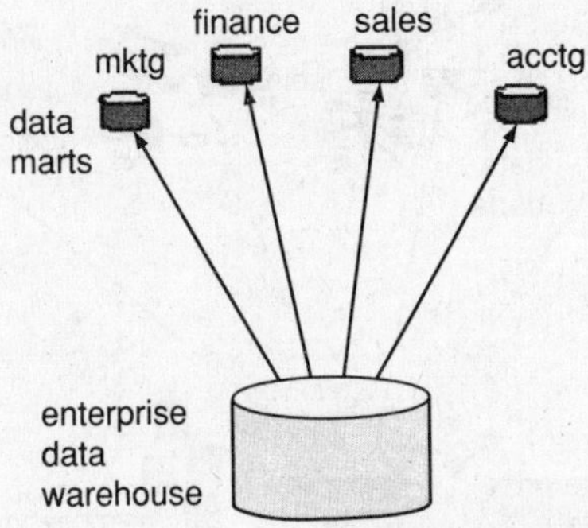

Fig. 3.2. The General structure of Data Marts and Data warehouse

The data warehouse exists to serve corporate functions, while the data marts exist to serve departmental functions. The departments that the data marts serve are typically finance, sales, marketing, and accounting. In other organizations, data marts are found in engineering, human resources, actuarial, and other places. Strictly speaking, the data marts shown in Figure 3.2 are called *dependent data marts* because they have data loaded into them ONLY from the enterprise data warehouse. Had the data marts been built independent of the enterprise data warehouse (and it is possible to build such structures) they would have been called *independent data marts*. But independent data marts are not part of the corporate information factory because there are some major architectural difficulties with them. The deficiencies of independent data marts will be discussed later in this section. Because of their deficiencies, independent data marts are not considered to be a part of the corporate information factory.

In general, enterprise data warehouses are larger than data marts. How much larger? Usually, a lot larger. There are several reasons why enterprise data warehouses are larger:

- enterprise data warehouse structures contain very detailed data; data mart structures contain aggregated and summarized data;
- enterprise data warehouse databases contain a generous amount of history; data marts contain a limited amount of history;
- enterprise data warehouse structures are very simple; data mart structures are much more complex; and,
- enterprise data warehouse structures are designed for unknown future usage; data mart structures are designed for specifically known requirements.

Figure 3.3 shows the differences between structures in the data warehouse and the data marts.

The data found in the enterprise data warehouse is for the entire corporation. The data found in the data marts is specific to the needs and the requirements of the department that owns the data mart. Because of

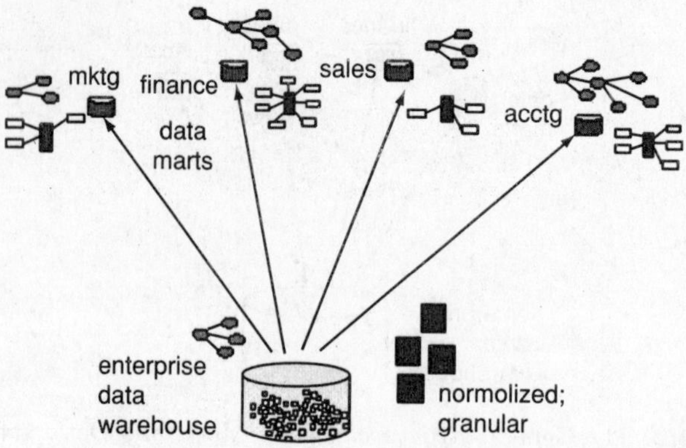

Fig. 3.3. The structure of data marts are requirements and can be called *star joins*

this basic difference, the data warehouse is "data driven" while the data marts are "requirements driven." The different types of structures found in the different environments reflect this very basic difference.

3.1.4 Different Data Structures

Data in the enterprise data warehouse is granular and simple. It needs to be this way because the enterprise data warehouse must support multiple and varied needs across the corporation. The data in the enterprise data warehouse can best be described as being normalized. For the most part, normalization fits the needs of the enterprise data warehouse quite well. However, the data found in the enterprise data warehouse is not perfectly normalized because if it were, it would not be efficient to access. The normalized data found in the enterprise data warehouse goes through a design process that can be described as *light denormalization.* Light denormalization of the enterprise data warehouse data restructures the normalized data into something less than a perfectly normalized structure. The kind of design considerations that make up light denormalization includes:

- creating a single place where multiple occurrences of data are stored when it is anticipated that the occurrences of data will normally and regularly be used together. For example, the data for the months January, February, March, ..., December may be physically aggregated into a single location for ease of access;
- creating redundant data where it is known that the redundant data will be used regularly in conjunction with other data. For example, the description field for a part might be included with data from the bill of materials, materials requirements processing, the parts inventory file, and so forth;
- creating summarized data when the data is calculated once and used many times. For example, an employee's annual ending pay may be calculated and stored as its own field; and,

- further separating data where the record of data includes data fields where there is a great disparity of probability of access of the data. For example, a bank account record might contain date of account opened, account balance, and domicile of the account. Separating date of account opened and domicile of the account from account balance accomplishes the separation of elements of data with a wide is parity in the probability of usage.

After the data in the enterprise data warehouse goes through the process of light denormalization, it still retains its strongly normalized character. The light denormalization does not reflect the requirements of any one application or department. Instead, the light denormalization generally improves the speed and ease of access of ALL users and ALL departments.

The structure of the data mart is fundamentally different from the structure of data found in the enterprise data warehouse. Data marts contain structures of data that can best be described as snowflake structures or star join structures.

Multidimensional technology best supports these types of data structures. Star joins and snowflake structures are the result of detailed requirements' analysis for a department or a like-minded set of end users. Different departments collect their own unique requirements and the result is an aggregation of data that resembles a star or a snowflake (if we use our imagination a little!).

Different departments will have different designs for their stars or snowflakes because the requirements are different for each department. The data that is aggregated together for the star joins comes from the enterprise data warehouse. The granular data that is found at the enterprise data warehouse is gathered and reshaped to make the star and the snowflakes.

The enterprise data warehouse then provides the detailed foundation needed to create the many different flavors of stars and snowflakes. There are many ways by which the granular data found in the enterprise data warehouse is reshaped in order to create the stars and snowflakes. For example:

- Granular data is summarized
- Granular data is resequenced
- Granular data is aggregated based on data found at the granular level
- New key/foreign key relationships are found
- New fields are created as the granular data is brought together

Another aspect of the data mart/enterprise data warehouse environment is that of the data models that reflects the contents and structure inside each component. The data model that is made for the enterprise data warehouse is small and simple. The simplicity reflects the granular, normalized demeanor of the enterprise data warehouse. The data models for the data marts are much more elaborate. These data models are different for each data mart and are a reflection of the requirements that have been gathered for each data mart. There is a unique mapping from each data model to each other data model.

3.1.5 Different Users

But there is another perspective on the data mart/enterprise data warehouse environment that is relevant and has to do with the different users found in each of the environments. Figure 3.4 shows two types of users who are of interest: "farmers" and "explorers."

Farmers are those users who are the predictable, regular users of the data warehouse. Farmers know what they want before they ever submit a query. Farmers submit small queries, but they submit them frequently throughout the day. Farmers usually find what they are looking for. Farmers find small flakes of gold frequently, but seldom find huge nuggets. Farmers know what their requirements are before they start out on a pursuit of information. There usually are lots of farmers.

Explorers are the corporations' "out of the box" thinkers. Explorers are very unpredictable. Explorers may go six months submitting no queries, then may submit six queries in the same day. The queries that the explorers submit are very large. The explorers look for patterns of activity that can be exploited by the corporation. The explorers look for relationships between types of data that have never before been examined.

Explorers often find nothing for their efforts. But occasionally an explorer makes a huge and significant discovery that has been previously unexpected and has been overlooked by everyone else in the corporation. Explorers find huge nuggets of information when they find anything at all.

There are reasons why the queries the explorers submit are so large. The first reason for large queries is that explorers need details. If the explorers are going to look in-depth at data that no one else has examined, then the explorers must have detail. The second reason why the explorers' queries are lengthy is because the explorers look at large amounts of history. The most current information a corporation has may or may not prove to be of use to the explorers. In order to do a thorough analysis and to detect subtle patterns, the explorers must look at information over a long period of time.

Fig. 3.4. The users in the DSS community

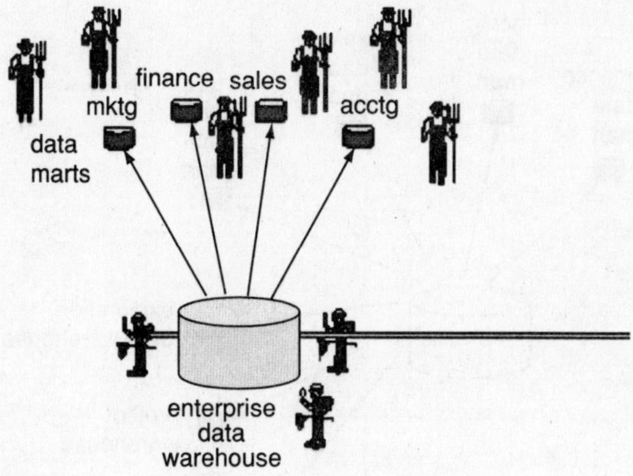

Fig. 3.5. Farmers are attracted to the data mart environment while explorers are attracted to the enterprise data warehouse

The third reason why the explorers submit lengthy queries is because the explorers look at data differently from other people. This means the data needs to be restructured. Restructuring data for the purpose of the satisfaction of a query implies that joins of data must be done—a lot of joins. So it is no accident that explorers submit what is termed the *72-hour query* or the *query from hell*. When we stop to consider the elements of the queries submitted by the explorers, we find that: *detail x history x fourteen way joins* adds up to large queries.

There is a peculiar affinity of farmers for data marts and explorers for the enterprise data warehouse, as seen in Fig. 3.5.

That farmers should find data marts attractive is no surprise. Data marts are shaped around predetermined requirements. Indeed, without requirements we cannot cogently design and build a data mart. And who are the users who know what their requirements are before they see a data warehouse database? Clearly, those are the farmers. On the other hand, which kinds of users have no idea going into an exploration of what their requirements are? These are the explorers of the world. An enterprise data warehouse is something that is versatile, if nothing else. The versatility of the enterprise data warehouse is just what the explorers are looking for. There is, therefore, a strong affinity of farmers for the data mart and of explorers for the data warehouse.

Farmers operate almost exclusively on data mart data while explorers operate on enterprise data warehouse data. Given that data marts are very requirements-driven, the attraction of the farmer community to them is very understandable. And it is likewise easy to understand the attraction of the explorer community to the enterprise data warehouse. But multiple data marts are not the only type of structure serviced by the enterprise data warehouse. Figure 3.6 shows that other types of architectural components found in the

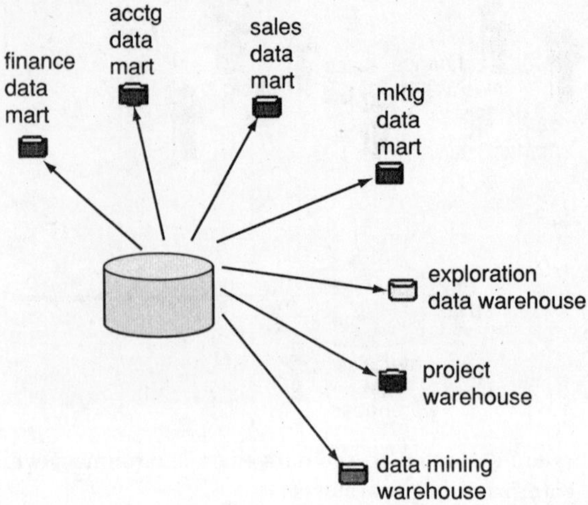

Fig. 3.6. The enterprise data warehouse needs to have a granular, corporate perspective because it supports many different kinds of DSS efforts

corporate information factory are serviced by the enterprise data warehouse as well.

The enterprise data warehouse serves as a basis for exploration warehouses, project warehouses, and data mining warehouses. Each of these types of data warehouses finds the granular data that resides in the enterprise data warehouse to be particularly attractive. One very important aspect to the notion of the enterprise data warehouse serving different communities is that there is a "single version of the truth." When the architecture looks like the one shown in Fig. 3.6, there is always a basis for reconcilability if any two analytical communities have a disagreement. In fact, the analysis created by different analytical communities will differ over time. However, when the enterprise data warehouse exists, there is the opportunity to reconcile the figures that are in disagreement. However, if the different analytical communities do not have a single basis of truth from which to operate, then there is almost no chance that there will be a reconciliation should a difference of opinion arise.

3.1.6 Technological Foundation

Not surprisingly, the world of data marts is served by different technology than the world of enterprise data warehouses. Figure 3.7 shows some of the technological considerations of each of these worlds.

The world of data marts is an on-line analytical processing (OLAP) multidimensional world. The volumes of data found in the data marts allow data to be accessed in a very flexible manner. In addition, relational technology is sometimes used for data marts when the data mart is large and/or of a general purpose nature. The essence of the data mart is flexibility and elegance of

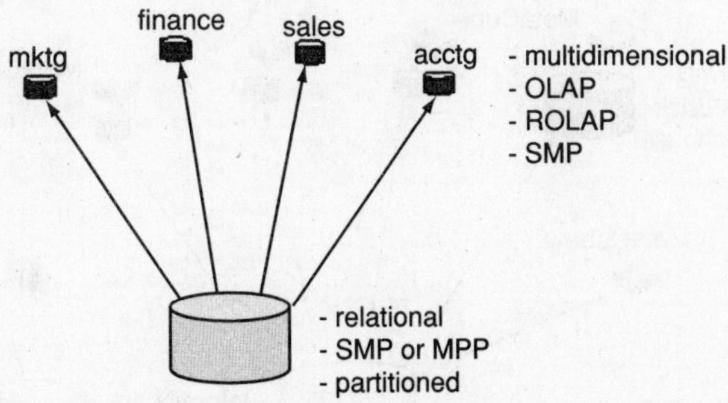

Fig. 3.7. The technologies found in the data mart are different from the technologies found in the enterprise data warehouse

presentation. To many end users, the data mart IS the data warehouse. Symmetric multiprocessor (SMP) architecture is often useful for the data mart environment because of its extreme flexibility and because of the ability to economically handle the volumes of data found in the data mart.

3.1.7 Data Warehouse

The enterprise data warehouse environment is substantively different from the data mart environment when it comes to technology. The enterprise data warehouse environment operates against very large amounts of data. Therefore, the number one priority of the database management system (DBMS) and hardware platform found at the enterprise data warehouse level is the ability to handle industrial strength amounts of data. The ability must be both economical and technological. In the data mart environment, the emphasis was on elegance of presentation of data and flexibility. It is a mistake to choose the enterprise data warehouse platforms based on the same criteria. The enterprise data warehouse platforms are typically relational with an emphasis on a massively parallel processing (MPP) hardware architecture. This combination allows the technology to be scalable up to very large volumes of data.

3.1.8 Informix Architecture

Informix offers an interesting and powerful architectural solution to the technology required for the data mart/enterprise data warehouse environment. Figure 3.8 shows this architecture.

The Informix relational DBMS provides a very solid foundation for the enterprise data warehouse. It offers the ability to manage very large amounts of data while providing the performance and scalability required by a truly enterprise-class data warehouse. Users should choose Informix as the foundation of their enterprise data warehouse principally because of the soundness of

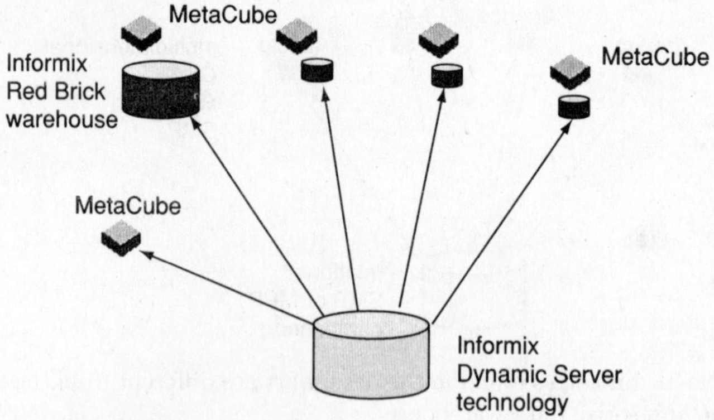

Fig. 3.8. The Informix approach to the data mart/data warehouse infrastructure

its parallel, shared-nothing relational architecture—required for highly scalable database configurations.

Informix Red Brick® Warehouse provides an ideal solution for the industrial strength data mart by virtue of its underlying star join architecture. Informix Red Brick Warehouse allows requirements to be defined then optimized in terms of design. But not only does Informix Red Brick Warehouse offer the ability to accommodate star joins and snowflake structures, but it also allows those structures to be managed on a grand scale. In this regard, Informix Red Brick Warehouse surpasses the capacity capabilities of any of the multidimensional vendors. And, finally, Informix MetaCube® rounds out the Informix offering by providing a robust solution for smaller-scale data marts that do not require all the capabilities provided by Informix Red Brick Warehouse.

The Informix solution has technology that is optimal for each component of the data mart/enterprise data warehouse environment. This is in contrast to its competitors who try to stretch a single technology into a "one-size-fits-all" solution. In fact, different organizations will need different combinations of the capabilities offered by the Informix DBMS, Informix Red Brick Warehouse, and MetaCube solutions. By having an architecturally diverse offering, Informix can provide true mix-and-match capabilities to its users. Supplying this kind of flexibility and complementary technology enables Informix to deliver an optimal solution for any organization's needs.

3.1.9 Building the Data Warehouse/Data Mart Environment

The data mart architecture/enterprise data warehouse architecture that has been described can be built in many ways. The classical way to build the architecture is to build the enterprise data warehouse first, then to build the data marts next. Figure 3.9 depicts this classical approach.

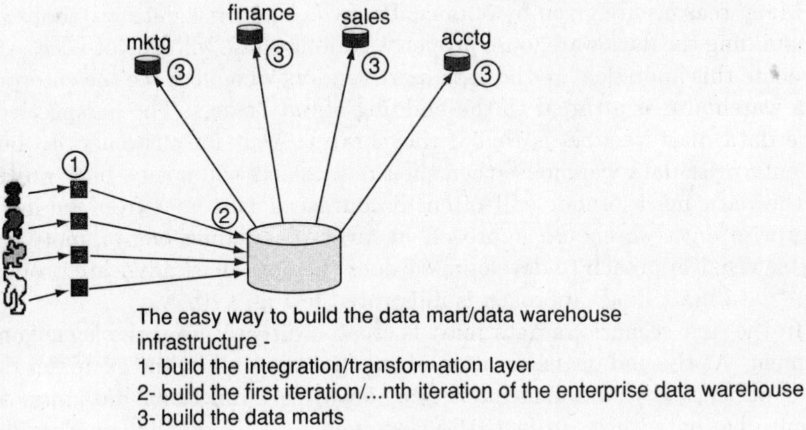

Fig. 3.9. The easy way to build the data mart/data warehouse infrastructure

In the classical approach, the integration and transformation programs are first built. Once built data is loaded into the enterprise data warehouse from the legacy applications, then the data marts are built. There is an interesting aspect to the pattern of development shown in Fig. 3.9, and that is the enterprise data warehouse is built iteratively. First, one part of the enterprise data warehouse is built, then another part of the enterprise data warehouse is populated. The enterprise data warehouse is not built in an all-at-once manner. This means the data marts can be built very quickly after the enterprise data warehouse is first populated. And there is nothing to say that farmers cannot use the enterprise data warehouse directly after it is built and before the data marts are built. As long as the data is in the enterprise data warehouse, and as long as there are not too many users, there is nothing to say the farmers cannot start their analysis on data found in the enterprise data warehouse. It is only after the enterprise data warehouse has grown and the enterprise data warehouse contains a lot of data that the farmers find the data mart attractive.

The development pattern shown in Fig. 3.9 is very straightforward. It is well documented in various methodologies for data warehouse development. But there are those who would argue that building an enterprise data warehouse and data marts, as shown in Fig. 3.9, is not a good approach. The arguments used by opponents of the classical approach are:

- the enterprise data warehouse is big and requires a lot of resources;
- the enterprise data warehouse will take two to five years to build;
- management wants immediate feedback from the data warehouse effort and does not have time to "do it right"; and,
- an enterprise data warehouse requires integration, and integration is not easy to do, and so forth.

Many reasons are given by opponents of the enterprise data warehouse for not building the data warehouse properly. Among those who do not necessarily advocate this approach are the data mart vendors who often see the enterprise data warehouse as a threat to the building of data marts. The perspective of some data mart vendors is that if they have to wait for customers to build the enterprise data warehouse then their data marts will not be built quickly. So the data mart vendor will often discourage a customer from taking an enterprise data warehouse approach in favor of building one or more data marts. What approach to development does the data mart advocate propose? The "data mart first" approach is illustrated in Fig. 3.10.

In the first scenario, a data mart is created directly from the legacy environment. At the end of the creation of the first data mart, the customer does not experience any pain. Instead, the customer has a functional data mart and is quite happy with it. In fact, the customer is so happy that another data mart is quickly built. Now the customer notices that each data mart needs the same detailed data as each other data mart. The customer remembers that redundancy of data was not a good thing. But the second data mart brings on a lot of happiness, so no one is particularly bothered by the redundancy of data between two data marts. Soon another data mart is on its way. When the third data mart arrives, it is noticed that the redundancy of detailed data between the data marts continues to increase. This has a very negative effect on the amount of money spent on each data mart because each data mart is quite large. But there is another ill side effect and that is that there is no reconcilability of data among data marts. When management asks how much revenue the company made, each department uses its data mart to produce a completely different value. The company now has stove pipe DSS applications with no integration among them. But the progress of data marts continues. A new data mart is built. Now the organization has four data marts. There is massive redundancy of detailed data among each data mart. This means that the cost of each data mart is high since each data mart is quite large. And there is no reconcilability of data among the data marts. When management asks the new data mart how much revenue was made, the answer is yet another un-integrated, irreconcilable opinion. And now someone notices that the interface between the legacy applications and the different data marts is starting to grow into its own nightmare.

The number of programs that must be built and maintained, the competition for resources in accessing the legacy data, and the inability to coordinate data among different data marts becomes unbearable. Someone stops and shouts, "This is not what data warehousing is all about!" And, indeed, the architecture described is not what data warehousing is all about. What must be done to solve the problems shown in Fig. 3.10 is to build an enterprise data warehouse, and it must become the source of data for the different data marts. So the data mart approach ends up at precisely where the data mart vendors had said a corporation did not want to be. The "data mart first" approach

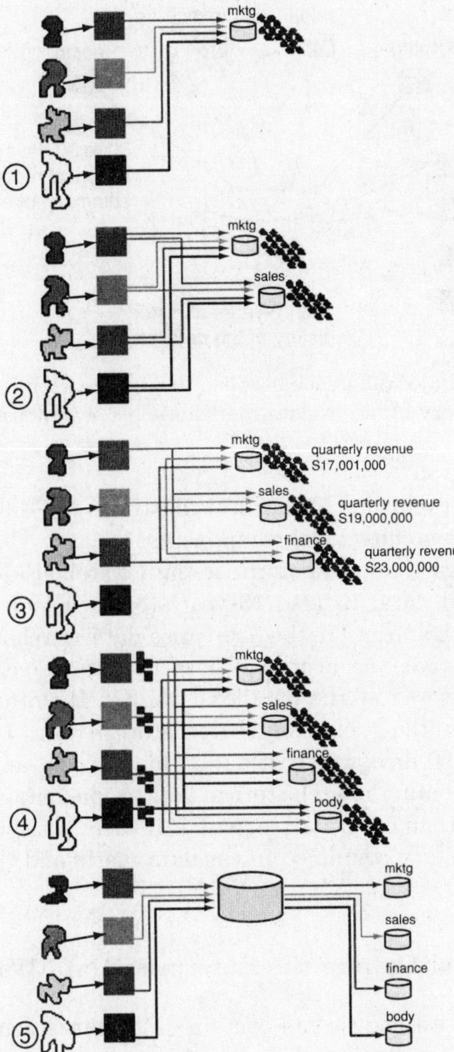

Fig. 3.10. The "long way around" to the building of the data mart/data warehouse infrastructure

caused the customer to take the long way around. There are two very different ways to arrive at the same place.

3.1.10 History

One of the major distinguishing characteristics of the data mart and the enterprise data warehouse environment is that of the amount of history stored

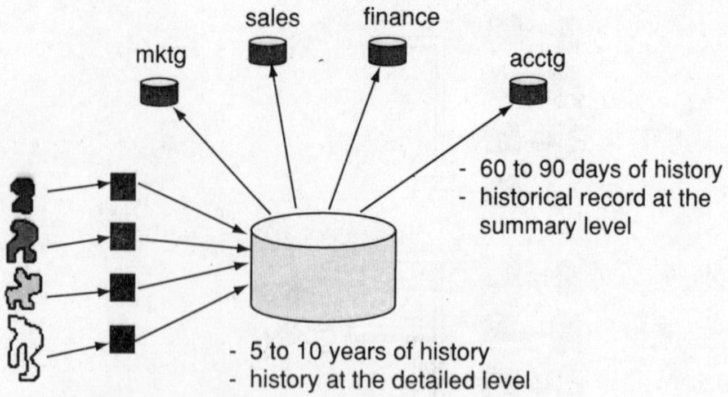

Fig. 3.11. Another major difference between the data mart and the data warehouse is in the amount history found in data mart and data warehouse

in the different environments. Figure 3.11 illustrates the differences in history as it is stored in the architectural components.

Figure 3.11 shows that the enterprise data warehouse contains a robust amount of historical data. Five to 10 years (or even longer), depending on the application, is the norm for the enterprise data warehouse. Furthermore, in order to be effective, the historical data that is stored in the enterprise data warehouse is stored at the detailed level. On the other hand, the data mart environment contains relatively little historical data. The data mart may contain from 60 to 90 days worth of data. On occasion, at a summary level, the data mart will contain more historical data. Some data marts will contain much more history than other data marts. But there simply is no comparison to the amount of history contained in the data marts and the enterprise data warehouse.

3.1.11 Nondetailed Data in the Enterprise Data Warehouse

The enterprise data warehouse can contain data that is nondetailed on occasion. As a rule, the nondetailed data found in the enterprise data warehouse makes up a very small percentage of the data found there.

Figure 3.12 shows that the enterprise data warehouse can contain corporately used, nondetailed data. In this case, the enterprise data warehouse serves as a convenient central location for the corporate data. The corporate data can take on any flavor and form. Another related type of data sometimes found in the enterprise data warehouse is public summary data or data created by a department and made available to the general public. Again, the enterprise data warehouse serves in the capacity as a convenient central location for such data. And, finally, a corporation may have benchmark data that is used widely across the corporation. The corporation will have to keep the benchmark data somewhere, and the enterprise data warehouse is a good place.

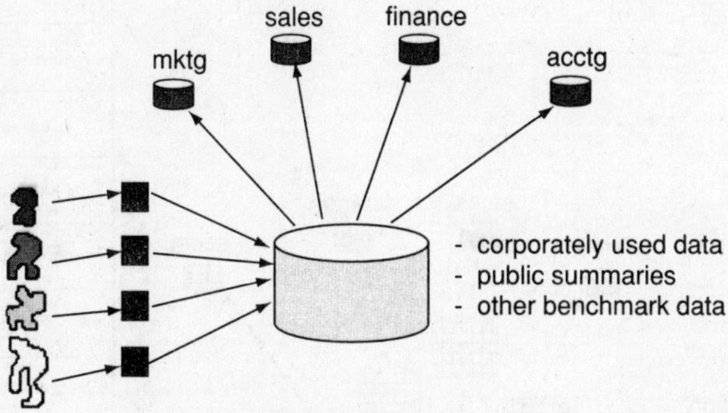

Fig. 3.12. Data other than detailed data is found in the enterprise data warehouse

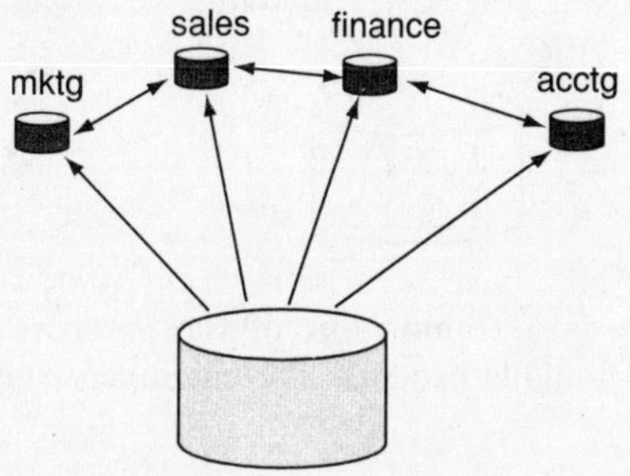

Fig. 3.13. The direct sharing of data among the departmental data marts is not a good idea

3.1.12 Sharing Data Among Data Marts

In the same vein, the question arises as to whether departments should directly share data among themselves. Figure 3.13 indicates that departments have the opportunity to share data.

The scenario shown in Fig. 3.13 is not a good idea. If there is to be sharing of data among departments, the sharing needs to be done by placing the data to be shared in the enterprise data warehouse, then allowing access from there. The scenario shown in Fig. 3.13 is the start of what has popularly become known as the "spider web" scenario. In the spider web scenario, data is passed in an almost random manner among different analysts. Soon data loses its validity and its meaning.

3.1.13 The Manufacturing Process

As data passes from the enterprise data warehouse to the data marts, the data undergoes what can be termed a *manufacturing* process. Figure 3.14 shows this process.

- the selection of data from the enterprise data warehouse;
- the aggregation of disparate data into a single structure, usually a star join or a snowflake structure;

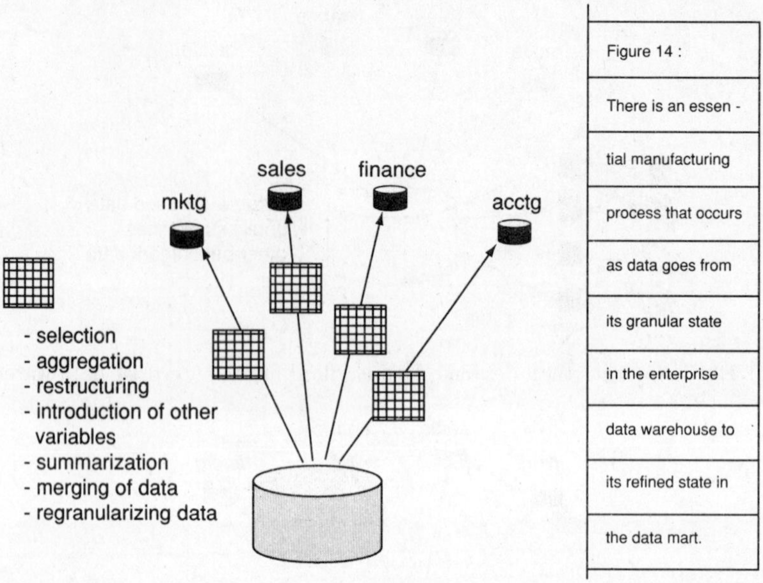

Fig. 3.14. shows that data coming out of the enterprise data warehouse goes through a complex and tedious process. The manufacturing process includes such things as:

- the restructuring of data to meet the needs of the data mart users (such as the restructuring of data into a multidimensional format);
- the creation and introduction of other variables. For example, as data is being read into the data mart, there may be two variables: Product A revenue and Product B revenue. The database designer creates a new variable—Combined Revenue—by adding Product A and Product B revenue together. There may be many reasons for the creation of such a variable. Ease of analysis once the data mart is populated is probably the primary reason for the creation of new variables at the moment of manufacture;
- summarization of detailed data;
- merging of data as it passes into the data mart;
- granularizing data to meet the needs of the data mart user, and so forth.

In short, there is a lot of work that takes place as data moves from the enterprise data warehouse to the data marts. Furthermore, the manufacturing done for one data mart is almost always different from the manufacturing done for another data mart. It is absolutely normal, however, for there to be some amount of overlap from one data mart to the next. The same data element, the same calculation, the same definition often appear across the different data marts that a corporation creates.

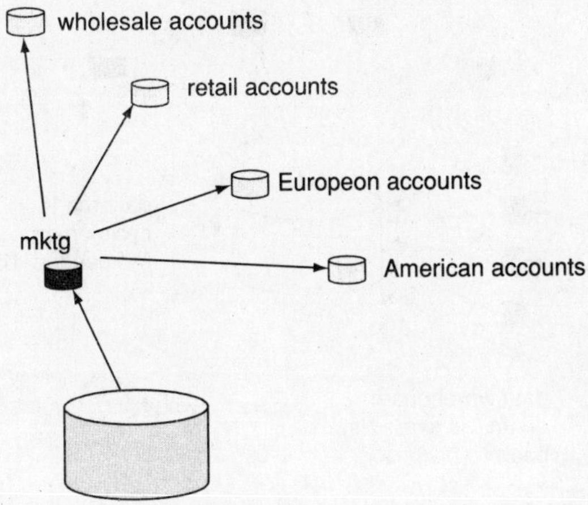

Fig. 3.15. For a large data mart, there are often subdata marts

3.1.14 Subdata Marts

On occasion a corporation will have a very large, general-purpose data mart. On these occasions there may start to appear what can be termed *subdata marts.* A subdata mart is one that emanates from a data mart.

Figure 3.15 shows how a marketing data mart has subdivided itself into several "subdata marts," one for wholesale accounts, one for retail accounts, one for European accounts, and one for American accounts. The subdata marts take on a life of their own, much like the data marts did as they sprang from the enterprise data warehouse. The subdata marts may or may not:

- share the same technology as the data mart;
- coordinate their data with other sub data marts;
- participate in a common infrastructure, such as a common metadata infrastructure, and so forth.

Subdata marts are neither good nor bad, but are simply a reality that must be dealt with. Subdata marts have the same problems as data marts, including integrity of data, cost, and performance. But, in a way, subdata marts are a sign of success since they indicate that users have adopted the concept of data marts to the point where they want to create their own "mini" version.

3.1.15 Refreshment Cycles

One of the marked differences between data marts and the enterprise data warehouse is in the refreshment cycles in which each participates. Figure 3.16 shows the differences between the two environments.

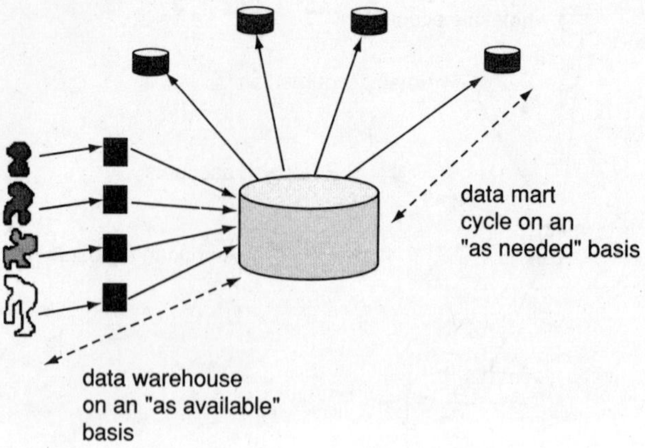

Fig. 3.16. Two very different refreshment cycles found in the data mart/data warehouse infrastructure

Figure 3.16 shows that the refreshment cycle for the data warehouse is relaxed and is "availability" driven. Availability driven refers to the movement of data into the enterprise data warehouse once the data is available. In other words, when data is transacted in the legacy environment, it is moved to the enterprise data warehouse. The enterprise data warehouse needs fresh data because it serves the needs of many types of DSS communities. However, the data refreshment in the data mart environment is "need" driven. Need driven refers to the fact that refreshment of the data mart is not made until the data is needed in the data mart.

In general, this means that the data mart schedule of refreshment is not done nearly as often as the refreshment of the enterprise data warehouse. One of the most important aspects of the data mart/data warehouse environment is that of refreshment. The amount of data needed to be refreshed, the integration and transformation that occurs during the refreshment process, the profound effect that the cycle of refreshment has on the usability of data and ultimately on the analysis that can be done on the data found in the data mart/data warehouse environment all lead to the conclusion that refreshment needs to be taken seriously as a cornerstone of the corporate information factory.

3.1.16 External Data

External data is data whose origin is outside of the corporation's systems. External data can be some of the most useful data that a corporation has. For example, it is often useful for a company to include general industry sales to be compared to internal corporate sales. The executive can sit back and compare the trends and the realities of corporate sales to the entire industry.

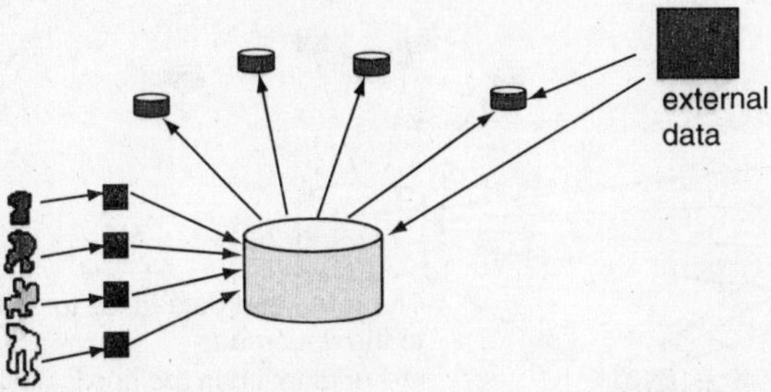

Fig. 3.17. External data is data that can be blended with the data mart or with the enterprise data warehouse

Such a benchmark is a good indication of how good or bad a corporation is doing.

As another example of the inclusion of external data with internal data, demographic growth is often useful to compare to sales or other corporate measures. Demographic changes often help to precisely address the issue of market penetration. Comparing internal results to demographic changes can lead to insights otherwise not obvious. And external data can be entered into either the data mart environment or the enterprise data warehouse environment. Figure 3.17 shows the possibilities for external data.

When external data is applicable and useful to only a single department, it can be entered into a data mart. But when external data has applicability or usefulness over more than one department, then the external data needs to be entered into the enterprise data warehouse. External data has some unique characteristics. The first is that the external data must be accepted on its face value. External data, once acquired by the corporation, is not easily altered. In the same vein, external data must have its key structure made compatible with the key structures found in the enterprise data warehouse and/or the data marts. If the external data is allowed to have different key structures, then no meaningful comparison can be made between the external data and the data found in the DSS environment.

3.1.17 Operational Data Stores (ODS) and Data Marts

Not all data goes from the legacy environment to the enterprise data warehouse, then on to the data marts. Some data makes a detour through the ODS. In particular, data needed for integrated online, real-time processing passes from the legacy environment to the ODS. Then, after it has served its purpose, it passes into the data warehouse.

Figure 3.18 shows that some data is routed through the ODS to the enterprise data warehouse. This data can ultimately arrive in the data mart environment. The data mart does not make a distinction between data that does or does not pass through the ODS. However, data never bypasses the enterprise data warehouse and goes directly to the data mart from the ODS.

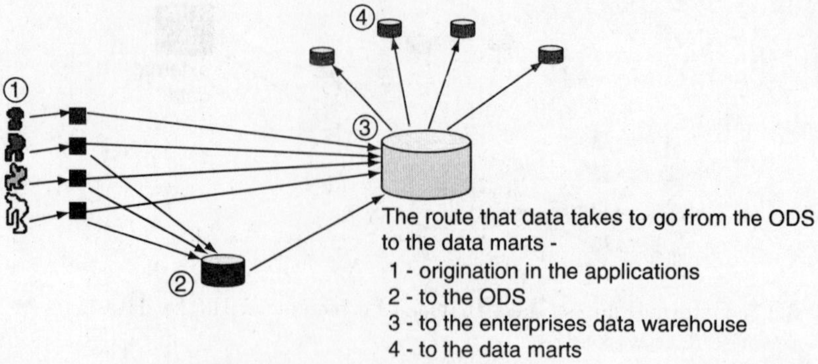

Fig. 3.18. The route that data takes to go from the application environment through ODS to the data marts

To omit placing the ODS data in the enterprise data warehouse would mean robbing the architecture of its integrity. Such a movement is the first step toward spider web systems.

3.1.18 Distributed Metadata

As we have already seen, the data mart environment and the enterprise data warehouse environment are two very different architectural structures that are often supported by different technologies. If there is to be cohesion—true integration of data and systems across the environments—how exactly is this cohesion achieved? The best way to achieve an enterprise-wide perspective is to employ metadata. Metadata is the facility by which different architectural components can exchange information.

What is metadata? The old definition of metadata is data about data. But that definition is so broad as to be almost meaningless. Instead, a more practical definition of metadata is that it is the information about the contents of systems, rather than the actual contents of the systems themselves. As a simple example, what does the number 458 represent in an inventory system? The answer is that the actual content does not mean much until you know that you are talking about 458 canisters. In this case, 458 is the content and "canisters" is the metadata that gives meaning to the content. Metadata in years past has always had a centralized orientation for the following reasons:

- metadata began in the days of the mainframe, where everything was centralized; and,
- the notion that in order for data to be shared, there needed to be a single, centralized definition and description of the terms of metadata.

But metadata in the world of data marts and the enterprise data warehouse is quite different. In this world, everything is distributed, not centralized.

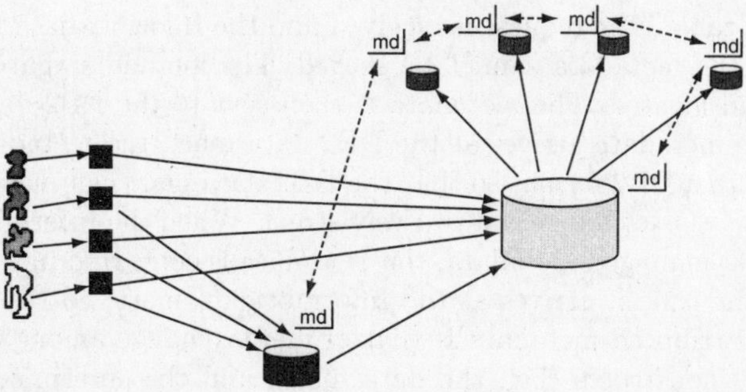

Fig. 3.19. Distributed metadata provides cohesion for the data mart/enterprise data warehouse infrastructure

Therefore, a centralized approach to metadata is not appropriate to the world of the data warehouse.

Figure 3.19 depicts a simplified view of distributed metadata and shows that metadata resides at every data mart and at the enterprise data warehouse. The first condition of distributed metadata is that there be local control and management of the metadata. What does local control and management of metadata mean? It means that each data mart and the enterprise data warehouse can:

- create, delete, and alter its own metadata;
- determine who can and cannot have access to the metadata;
- specify that metadata is private to only the data mart or enterprise data warehouse at hand;
- expect the metadata that is managed by the data mart or the enterprise data warehouse to reside in the technology that is used. For example, if one data mart were in Informix, then the metadata for that data mart would exist as an Informix table. And if another data mart had Teradata as its DBMS, then the metadata for that data mart would reside in the data mart as a Teradata table, and so forth.

The first condition then for distributed metadata is that it resides in the local architectural construct (i.e., the data mart or the enterprise data warehouse). But merely distributing metadata across the enterprise is not sufficient. All this distribution accomplishes is the solidification of the stove pipes that separate data marts and the enterprise data warehouse. The next condition of distributed metadata is the condition that metaobjects be able to be distributed across data marts and the enterprise data warehouse. This means that one data mart can easily share metadata with another data mart. Or the enterprise data warehouse can share metadata with another data mart.

As an example of the ability to share distributed metadata, consider two data marts, one in Informix and one in DB2. The Informix data mart receives a request for metadata that it manages and which resides in the data mart as an Informix table. The request is received and the Informix data mart sees no reason why the metadata cannot be shared. The metadata that is requested is sought and located. The metadata is then sent to the DB2 data mart. By the time the metadata arrives at the DB2 data mart, the metadata has been converted into a DB2 format so that the DB2 data mart can make sense of it.

Of course, the exchange is a two-way street. When the Informix data mart requests DB2-managed metadata, the metadata is converted into an Informix format by the time it arrives at the Informix data mart. So the second condition for distributed metadata is that of metaexchange among the different architectural constructs (i.e., the data marts and the enterprise data warehouse). But there is a third important condition: namely that there be integrity of metadata across the enterprise. This principle can be expressed as the "system of record for metadata," and can be stated as follows:

- every unit of metadata can have one and only one owner across the enterprise.

In other words, if an Informix data mart owns a unit of metadata, then no other data mart or enterprise data warehouse (or anything else, for that matter) can own the unit of metadata. This means that as metadata is added to the system and before the addition can be confirmed, the distributed metadata manager must check to see if the metadata is owned elsewhere. If the metadata is owned elsewhere, ownership cannot be assigned again. Only if the metadata is not previously owned can the metadata be assigned to the new owner. Of course, the ownership of metadata can be reassigned. But a reassignment entails organizational and political decisions, where a consensus of opinion needs to be reached.

Owning metadata implies that the organization that owns it has exclusive rights to the creation, deletion, and alteration of the metadata, as well as the right to determine who can and cannot access the metadata. Once the system of record across the enterprise is introduced, sharability of metadata with integrity becomes a reality. As a unit of metadata is passed around the enterprise, the record of the ownership is carried with the metadata to clearly mark the metadata as being shared.

3.1.19 Managing the Warehouse Environment

Another interesting aspect of the data mart/enterprise data warehouse environment is that of the ongoing management and administration of the environment. Figure 3.20 illustrates a monitor that is used in the management of the environment.

The activity that passes into the data warehouse and the data marts is of keen interest to the data warehouse administrator. With an activity monitor, the data warehouse administrator can tell such important things as:

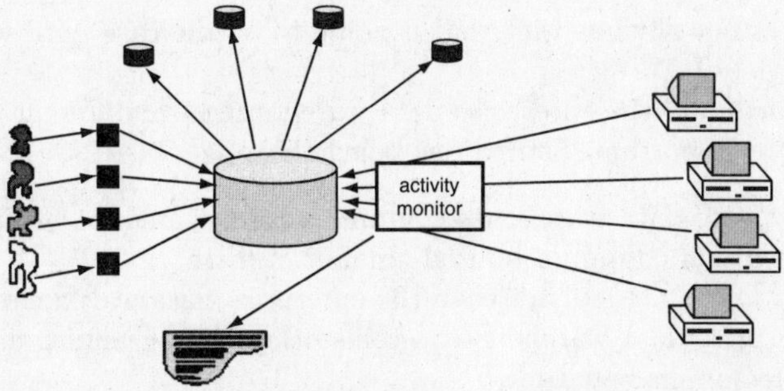

Fig. 3.20. Monitoring the enterprise data warehouse is an important part of the infrastructure

- who is using the warehouse;
- what queries are the most common;
- what resources are being used the most; and,
- when the warehouse is being used.

There are other important aspects of monitoring as well, such as the determination of what data is and is not being used. This determination is most important when considering what data is safe to be purged from the warehouse. If an organization does not take seriously the purging of data from the warehouse, then several very bad (and totally unnecessary!) things can happen:

- the cost of the warehouse goes sky high;
- the performance in the warehouse grows worse, as data that is used hides behind masses of data that is not being used; and,
- the designer must increasingly aggregate and summarize data in order to accommodate the small amount of data available for new data in the data mart or enterprise data warehouse.

Multidimensional, OLAP technology is quite suitable for the data mart environment, while industrial strength, scalable technology does well for the enterprise data warehouse environment. There are two ways to build the enterprise data warehouse/data mart environment. Build the enterprise data warehouse iteratively, then extend the data marts from the enterprise data warehouse, or build the data marts first, then go back and build the enterprise data warehouse and extend the data marts from the enterprise data warehouse. In the long run, building the enterprise data warehouse is the least expensive and the cleanest approach.

However, some organizations take what they believe to be a more expedient approach and build data marts first. In the long run, they use more development resources than if they had built the data warehouse first, although it is not obvious that that is going to be the case until well into the DSS exercise.

Data marts and the enterprise data warehouse have different characteristics, the most important distinctions being that:

- the enterprise data warehouse contains a robust amount of history while the data marts contain a limited amount of history;
- the data marts are smaller than the enterprise data warehouse; and,
- the enterprise data warehouse contains other than granular data, such as corporate summary data.

There is an essential "manufacturing process" that occurs as data moves from the enterprise data warehouse to the data marts. The manufacturing process reshapes the corporate, granular data found in the enterprise data warehouse into a requirements-driven departmental structure. In addition, "subdata marts" are often developed off of large, actively used data marts. The subdata marts experience all the same problems and challenges that the data marts do. There is a very different refreshment cycle for the enterprise data warehouse and data marts. Data marts are refreshed on the basis of need by the departments that own the data mart. Data warehouses are generally refreshed on the basis of availability of new data ready to enter the data warehouse. External data can be loaded into either the data marts or the enterprise data warehouse, depending on the usage of the external data. The comparison of external data to internal data can lead to very important insights at the executive level.

Distributed metadata is required for the management of the distributed data mart/enterprise data warehouse environment. Informix has an architectural edge over the competition in that they have specific and unique products to fit the architecture. The Informix DBMS is well suited for the needs of the data warehouse. Informix Red Brick Warehouse is well suited for large, industrial strength data marts. And MetaCube is positioned for start up and casual data marts. No other vendor has technology that is unique and optimal across the architecture of the corporate information factory.

3.1.20 OLAP

The purpose of the section that follows is to define On-Line Analytical Processing (OLAP), who uses it and why, and to review the key features required for OLAP software as referenced in the OLAP Council benchmark specification.

What Is OLAP?

During the last ten years, a significant percentage of corporate data has migrated to relational databases. Relational databases have been used heavily in the areas of operations and control, with a particular emphasis on transaction processing (for example, manufacturing process control, brokerage trading). To be successful in this arena, relational database vendors place a premium on the highly efficient execution of a large number of small transactions and near fault tolerant availability of data.

More recently, relational database vendors have also sold their databases as tools for building data warehouses. A data warehouse stores tactical information that answers "who?" and "what?" questions about past events. A typical query submitted to a data warehouse is: "What was the total revenue for the eastern region in the third quarter?"

It is important to distinguish the capabilities of a data warehouse from those of an OLAP (On-Line Analytical Processing) system. In contrast to a data warehouse, which is usually based on relational technology, OLAP uses a multidimensional view of aggregate data to provide quick access to strategic information for further analysis.

OLAP enables analysts, managers, and executives to gain insight into data through fast, consistent, interactive access to a wide variety of possible views of information. OLAP transforms raw data so that it reflects the real dimensionality of the enterprise as understood by the user.

While OLAP systems have the ability to answer "who?" and "what?" questions, it is their ability to answer "what if?" and "why?" that sets them apart from data warehouses. OLAP enables decision making about future actions. A typical OLAP calculation is more complex than simply summing data, for example: "What would be the effect on soft drink costs to distributors if syrup prices went up by $.10/gallon and transportation costs went down by $.05/mile?" OLAP and data warehouses are complementary. A data warehouse stores and manages data. OLAP transforms data warehouse data into strategic information. OLAP ranges from basic navigation and browsing (often known as "slice and dice"), to calculations, to more serious analyses such as time series and complex modeling. As decision makers exercise more advanced OLAP capabilities, they move from data access to information to knowledge.

Who Uses OLAP and Why?

OLAP applications span a variety of organizational functions. Finance departments use OLAP for applications such as budgeting, activity-based costing (allocations), financial performance analysis, and financial modeling. Sales analysis and forecasting are two of the OLAP applications found in sales departments. Among other applications, marketing departments use OLAP for market research analysis, sales forecasting, promotions analysis, customer analysis, and market/customer segmentation. Typical manufacturing OLAP applications include production planning and defect analysis.

Important to all of the above applications is the ability to provide managers with the information they need to make effective decisions about an organization's strategic directions. The key indicator of a successful OLAP application is its ability to provide information as needed, i.e., its ability to provide "just-in-time" information for effective decision making. This requires more than a base level of detailed data.

Just-in-time information is computed data that usually reflects complex relationships and is often calculated on the fly. Analyzing and modeling complex relationships are practical only if response times are consistently short. In addition, because the nature of data relationships may not be known in advance, the data model must be flexible. A truly flexible data model ensures that OLAP systems can respond to changing business requirements as needed for effective decision making.

Although OLAP applications are found in widely divergent functional areas, they all require the following key features:

- Multidimensional views of data
- Calculation-intensive capabilities
- Time intelligence

Multidimensional Views

Multidimensional views are inherently representative of an actual business model. Rarely is a business model limited to fewer than three dimensions. Managers typically look at financial data by scenario (for example, actual vs. budget), organization, line items, and time; and at sales data by product, geography, channel, and time.

A multidimensional view of data provides more than the ability to "slice and dice"; it provides the foundation for analytical processing through flexible access to information. Database design should not prejudice those operations that can be performed on a dimension or how rapidly those operations are performed. Managers must be able to analyze data across any dimension, at any level of aggregation, with equal functionality and ease. OLAP software should support these views of data in a natural and responsive fashion, insulating users of the information from complex query syntax. After all, managers do not have to understand complex table layouts, elaborate table joins, and summary tables.

Whether a request is for the weekly sales of a product across all geographical areas or the year-to-date sales in a city across all products, an OLAP system must have consistent response times. Managers should not be penalized for the complexity of their queries in either the effort required to form a query or the amount of time required to receive an answer.

The OLAP Council APB-1 performance benchmark tests a server's ability to provide a multidimensional view of data by requiring queries of varying complexity and scope. Basic aggregation is performed on some of the dimensions (product, customer, and channel). More complex calculations are performed on other dimensions. The measure dimension computes ratios and averages. Variances are computed along the scenario dimension. A complex model based on historical performance is used to compute the forecast scenario. Consistently quick response times to these kinds of queries are key to establishing a server's ability to provide multidimensional views of information.

Complex Calculations

The real test of an OLAP database is its ability to perform complex calculations. OLAP databases must be able to do more than simple aggregation. While aggregation along a hierarchy is important, there is more to analysis than simple data roll-ups. Examples of more complex calculations include share calculations (percentage of total) and allocations (which use hierarchies from a top-down perspective).

Key performance indicators often require involved algebraic equations. Sales forecasting uses trend algorithms such as moving averages and percentage growth. Analyzing the sales and promotions of a given company and its competitors requires modeling complex relationships among the players. The real world is complicated – the ability to model complex relationships is key in analytical processing applications.

OLAP software must provide a rich tool kit of powerful yet succinct computational methods. To make developers more efficient and business users more self-sufficient, the vehicle for implementing computational methods should be clear and nonprocedural. If the method for creating the desired calculations is not clear, development time and/or usage will suffer. If the calculation method is procedural, changes to the system cannot be done in a timely manner, effectively eliminating access to just-in-time information.

Whereas transaction processing systems are judged on their ability to collect and manage data, analytical processing systems are judged on their ability to create information from data. The OLAP Council APB-1 performance benchmark contains a representative selection of calculations, both simple and complex. An example of a simple calculation contained in the performance benchmark is the calculation of margin (sales minus costs). The computation of the forecast is the most complex calculation contained in the current version of the performance benchmark. Historical data is used to project the future and aggregate data is used to estimate input data. Other more complex calculations, such as allocations and trend analysis, are also often found in OLAP systems.

Time Intelligence

Time is an integral component of almost any analytical application. Time is a unique dimension because it is sequential in character (January always comes before February). True OLAP systems understand the sequential nature of time. Business performance is almost always judged over time, for example, this month vs. last month, this month vs. the same month last year.

The time hierarchy is not always used in the same manner as other hierarchies. For example, a manager might ask to see the sales for May or the sales for the first five months of 1995. The same manager might also ask to see the sales for blue shirts but would never ask to see the sales for the first five shirts. Concepts such as year-to-date and period-over-period comparisons must be easily defined in an OLAP system.

In addition, OLAP systems must understand the concept of balances over time. For example, if a company sold 10 shirts in January, five shirts in February, and 10 shirts in March, then the total balance sold for the quarter would be 25 shirts. If, on the other hand, a company had a head count of 10 employees in January, only five employees in February, and 10 employees again in March, what was the company's employee head count for the quarter? Most companies would use an average balance. In the case of cash, most companies use an ending balance.

The OLAP Council APB-1 performance benchmark contains examples of how time is used in OLAP applications. Smoothed sales are computed as a three-month moving average. Inventory is aggregated as an ending balance. The forecast calculation uses this year's vs. last year's knowledge, year-to-date knowledge, and annualization factors.

OLAP Benefits

Successful OLAP applications increase the productivity of business managers, developers, and whole organizations. The inherent flexibility of OLAP systems means business users of OLAP applications can become more self-sufficient. Managers are no longer dependent on IT to make schema changes, create joins, or worse. Perhaps more importantly, OLAP enables managers to model problems that would be impossible using less flexible systems with lengthy and inconsistent response times. More control and timely access to strategic information equal more effective decision making.

IT developers also benefit from using the right OLAP software. Although it is possible to build an OLAP system using software designed for transaction processing or data collection, it is certainly not a very efficient use of developer time. By using software specifically designed for OLAP, developers can deliver applications to business users faster, providing better service. Faster delivery of applications also reduces the applications backlog.

OLAP reduces the applications backlog still further by making business users self-sufficient enough to build their own models. However, unlike standalone departmental applications running on PC networks, OLAP applications are dependent on data warehouses and transaction processing systems to refresh their source level data. As a result, IT gains more self-sufficient users without relinquishing control over the integrity of the data.

IT also realizes more efficient operations through OLAP. By using software designed for OLAP, IT reduces the query drag and network traffic on transaction systems or the data warehouse.

Lastly, by providing the ability to model real business problems and a more efficient use of people resources, OLAP enables the organization as a whole to respond more quickly to market demands. Market responsiveness, in turn, often yields improved revenue and profitability.

3.2 Data Warehousing for Healthcare: The Greatest Weapon in your Competitive Arsenal

3.2.1 A Data Warehousing Perspective for Healthcare

The dramatic changes underway today in the healthcare industry are the result of ambitious industry and government efforts to curb sky-rocketing healthcare costs. Two key issues, cost containment and efficiency, will be the cornerstones of the healthcare industry's new infrastructure. To contain costs, payers (i.e., insurance companies, employers, and government agencies) are turning to health maintenance organizations (HMOs), preferred provider organizations (PPOs) and other managed care entities that have demonstrated the ability to cut costs. Through efficiencies that rely on economies of scale, tough bargaining methods, and the elimination of unnecessary services, these organizations have become the care providers of choice for government and private industry. As early as 1992, roughly half of all workers in the United States were enrolled in managed care plans, up from 27% five years before, according to a study by Foster & Higgins. It is no wonder in the age of cost control, capitated payments and the need for outcomes research (monitoring of clinical procedures to help determine the most clinically effective and cost effective treatments), senior executives are now looking toward information technology (IT) to give them the strategic/competitive advantage within the marketplace.

Data warehousing is a management tool that enables executives to access the information they need to make informed business decisions. A data warehousing system pulls together information from disparate sources to construct an integrated view of business activities as shown in Fig. 3.21. These systems then transform the data into a consistent, easily accessible format and distribute the data to where it is needed for decision making. Although the actual data warehouse itself may be the responsibility of the IT department, the project should be driven by business management because a data warehouse is only as good as its ability to guide management in making strategic and informed business decisions. Business users can then efficiently analyze vast quantities of data, uncover hidden opportunities, and enhance their organizations' competitive advantage.

3.2.2 Adding Value to your Current Data

Today's healthcare organizations gain competitive advantage by making smart decisions based on complete information. Developing the best method to man-

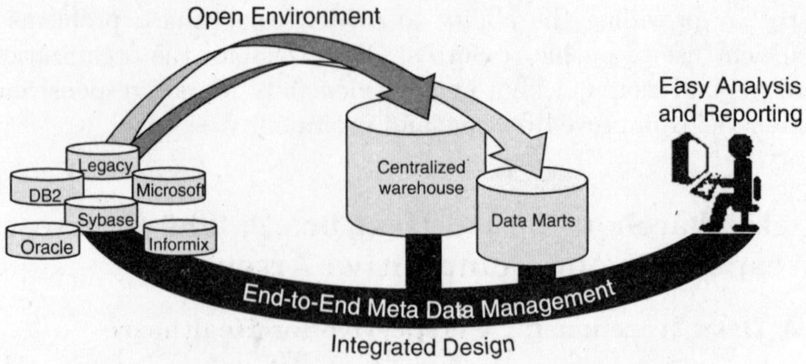

Fig. 3.21. A data warehousing system

age customer relationships or contain costs can mean sifting through massive amounts of data from multiple geographic locations, company departments and providers. The key to unlocking many healthcare issues, such as how to manage provider issues, eliminate fraud, control costs, and maintain efficiency, can be found in the data that already exists within the organization. The trick is being able to extract the relevant information from the data to build conclusions about customers and guide marketing efforts. Although data warehousing has been around for many years, healthcare organizations are just beginning to understand and realize the benefits this technology has to offer the industry. As a point of reference, it is believed that the data warehousing market will grow from a $4 billion industry in 1998 to $6.5 billion in the year 2000. Data warehousing is an effective tool for a competitive industry like healthcare. But why? What can data warehousing do to help a healthcare organization compete and survive?

- Enhance customer relationship management
- Improve provider management
- Prepare for HEDIS reporting
- Perform disease management
- Reduce fraud

3.2.3 Enhance Customer Relationship Management

"Cradle-to-grave" care is today's industry buzzword. Healthcare organizations understand that in order to keep patients for life, they need to satisfy customer groups and the individual patient. Customer relationship management is, therefore, a focus on member retention as well as customer satisfaction. Data warehousing can be used for a multitude of programs to effectively and efficiently enhance customer relationship management. For example, to keep customer groups informed of the high level of service they are receiving, provider outcomes and cost analysis reports are made easily accessible. On a much more personal level, a data warehouse can trigger a flag so that a birthday card can be sent to individual patients. Much can be learned about a patient population from analyzing basic member information.

Such information as service utilization patterns and demographics of member groups can help organizations make targeted business decisions that help them with more strategic marketing campaigns and better customer understanding. One of the largest healthcare providers in the country is planning to enhance its data warehousing system to enable managed-care providers and members to check on claim status and member eligibility and to receive wellness information via voice response. This superior customer service feature demonstrates the organization's commitment to its members as well as its providers.

3.2.4 Improve Provider Management

The cornerstone of managed care is provider management. It is imperative that an organization accurately gauge how each provider is doing in terms of cost of care, outcomes, as well as their contract and credentialing status. Information that "slips through the cracks" in this area could mean the loss of hundreds of thousands of dollars over time. A data warehouse provides a single source where users can analyze claims-based utilization in detail. Excessive spending can be pinpointed and providers can be alerted to both negative as well as positive results, then an appropriate course of action can be developed. Currently, a national healthcare management company is using data warehousing to support provider network management and referral queries, which assists with NCQA accreditation. Other healthcare organizations have utilized data warehousing to monitor ancillary department activities, such as pharmacy cost trends and mental healthcare analysis. As data warehousing becomes prevalent throughout the enterprise, costs will be better managed, quality of care will increase and greater provider satisfaction will be realized.

3.2.5 Reduce Fraud

Fraudulent claims cost the healthcare industry billions of dollars every year. By using a data warehouse, a healthcare organization can detect potential fraud and take steps to resolve the situation in a timely manner. One healthcare organization has combined artificial intelligence with their data warehouse. Claims are compared to the organization's business rules and discrepancies are reported. The organization can then look at each discrepancy on a case-by case basis. This system has reported a savings of $1 million every month! This technology saves money for the organization and helps to manage resources more effectively.

3.2.6 Prepare for HEDIS Reporting

Not only can organizations use a data warehouse for HEDIS reporting, they can use this information to proactively evaluate how they are doing. For example, an organization can run a report six months prior to needing to file official reports. This can identify areas that need improvement. The organization can then use the six months to implement improvements in order to score even higher on the actual HEDIS report.

3.2.7 Disease Management

A data warehouse can transform a patient's medical record into an analytical tool for optimizing the care of each patient individually. This enables clinicians to continuously monitor and analyze the condition of each patient to facilitate timely and effective clinical decisions. Workflow efficiencies are enhanced when the "big picture" can be seen by integrating clinical, administrative, and financial data. For example, in one healthcare organization's kidney dialysis ward, a provider was having greater success with his patients than the other provider. His care plans and costs were analyzed and it was discovered that he had consistently found an iron deficiency when he cleansed his patient's blood, so he prescribed iron supplements on a regular basis. The care plan in the ward was adjusted and it became a clinical practice that all providers prescribe iron supplements to their patients in this ward. The quality of care throughout the ward has increased, and the costs have gone down.

3.2.8 What to Expect When Beginning a Data Warehouse Implementation

In today's world of distributed, disparate data sources, changing user needs, and demanding turn-around schedules, building and owning a data warehouse is often a complex challenge. The development, implementation, and use of a data warehouse can be eased by considering the following factors:

Integration– Technology should be chosen that works in tandem to provide a single, simple data warehouse solution. Data must be able to be extracted from a variety of sources and transformed into information that makes sense to the business user. A data warehouse project is complex enough without having to fight with proprietary systems that cannot easily retrieve data. Open technology is the key.

Fast Implementation– An integrated data warehousing solution will enable rapid deployment and lower total cost of ownership. Rapid deployment equates to a faster return on investment (ROI). If planned correctly, and implemented appropriately, ROI for a data warehouse should be very quick. Most organizations can expect to realize their return within a couple of months.

Performance– Query speed is important, so look for data stores that are optimized for decision support. Performance should be maintained regardless of what is asked and how many people are asking it. Regardless of format, data should be able to be consolidated into a single repository.

Enterprise-wide metadata management– Look for integrated tools for capturing, synchronizing, managing, and using metadata throughout the data warehouse. Quick access to information across the entire enterprise allows users to make informed business and clinical decisions and do their jobs more efficiently.

Ease of Use– Once the warehouse is populated, an intuitive query interface should open the warehouse to a wide audience. (The warehouse will lose its value if it is not easy to use.) Business users need to be able to enter queries in simple English—not complex coding. Additionally, users need to be able to easily create professional reports and use third-party reporting and analysis tools. This increases productivity because people get the information they need quicker and resources are better managed because the IT department does not have to be called.

Replication– Whatever an organization's replication requirements—efficient regular updates, data that is within seconds of real time or sophisticated change-based warehouse updates—make sure the vendor's solution can access and synchronize diverse data sources through the nonintrusive capture of the operational transactions generated.

Expertise and Experience– Buying the right technology is only the first step toward implementing a data warehousing solution. You need to confirm that the vendor you have selected has the expertise and the experience to successfully implement the data warehousing solution. Make sure the vendor has a master plan for building, using, and managing an enterprise data warehouse. Demand a road map for meeting short-term tactical goals and for extending the data warehouse to meet your organization's long-term strategic objectives.

3.2.9 Definitions

As with any technology revolution, the industry terms and acronyms run rampant. To create a level playing field, here are some definitions:

Data Warehouse– A large-scale, central database with read-only electronic files that is loaded with information from multiple operational databases for the purpose of end-user access and decision support. Data warehouses accommodate random, ad hoc queries and let users drill down to minute levels of detail. Data warehouses differ from an operational system, where data is static and updated in a scheduled manner through massive data loading procedures. Data in the warehouse generally covers a longer time period and integrates information from more sources than do operational systems. A warehouse is designed to integrate data from the operating systems while retaining a basic relational structure. Simply put, a data warehouse puts raw data into context for product and marketing strategies.

Data Mart– Data marts are "segments" of a data warehouse presenting a specific data set to individual business units, such as vertical marketing departments or finance. Data marts are a flexible approach to data warehousing that allow individual business units to have strong control over the type of information they have access to and how they can manipulate it for analysis, while still maintaining a single centralized data warehouse.

Operational Data– Data filtered from operational systems, such as billing, lead management tracking systems and customer support systems, are entered into a data warehouse for analysis.

Data Mining– Querying a data warehouse's operational data to "dig up" information about customer preferences.

MetaData– Information about data, such as a piece of data's origin. For example, did this information come from marketing or finance, or did it come from some external source, such as a purchased mailing list?

3.3 Data Warehousing in the Telecommunications Industry

There's turmoil in the telecommunications industry—worldwide turmoil. In the U.S., the Telecommunications Act of 1996 has ended government rules that maintained barriers between local and long-distance calling, cable TV, broadcasting and wireless services. In Europe, deregulation has opened markets in England, Sweden, and Finland. In 1998, 15 European countries are due to liberalize theirs. Developing nations such as Chile, Malaysia, and Peru have dismantled state-owned phone monopolies. This restructuring will create new market opportunities and will end the exclusive access to customers that has been enjoyed by the incumbent monopolies.

As competition intensifies and the network becomes a commodity, service providers are faced with the need to reduce costs, add value to their services, and ensure differentiation. Once a customer can choose among different carriers, customer care and marketing intelligence become strategic functions in retaining and gaining market share. All this change demands that a carrier's supporting infrastructure be responsive and flexible, in one word—agile. Agile in its approach to the market, the way it responds to its customers, and its delivery of new services. Making decision-support information instantly accessible to the knowledge worker creates the agile enterprise: quick, competitive, and in control of business. When done right, a data warehouse coupled with a focused set of decision-support applications provides for seamless integration of communication services globally, better understanding of profitability, and a focused marketing effort.

The introduction of data warehousing and decision-support technology addresses the following issues in the organization:

Market Expansion. As was seen in the interexchange and cellular markets, competition will result in lower prices, new services, and, ultimately, a larger available market. The introduction of new technologies, as seen in the past with cellular, faxes, paging, and the Internet, has resulted in the growth of total network usage.

Integration of New Lines of Business. The opportunity to market more services and offerings to existing and new customers is moving telecom companies into new businesses such as long-distance service, cable TV, video-on-demand, and Internet access. The delivery of a worldwide support network with the capability to connect everyone in the world for the movement of voice, video, and data on a massive scale will provide immense market opportunity.

Mergers and Acquisitions. As the deregulated markets stabilize, many of the new entrants, particularly the smaller carriers, will be acquired by either larger communications companies or new investors. Some of these new competitors will have very different business models, e.g., resellers of local and/or interexchange services, than the acquiring carrier; these new business models must be quickly and efficiently integrated into the existing structure.

Market Environment. The telecommunications industry is one of the leading users of technology in the world. Historically, this technology has provided service and competitive advantage to early adopters (e.g., British Telecom providing specific discount structures through logging calls over a rolling 15-month period). Utilizing technology can enhance a company's market share and contribute to profitable growth. This is likely to be achieved by:

Focusing on the Customer. Products will be increasingly targeted at the individual in this new customer centric era. The better understanding of customers and marketing based on that knowledge then becomes essential. Communications companies are using tools such as marketing databases, customer information systems, enhanced customer service capabilities, predicative behavior models, and integrated marketing systems to improve and solidify their relationship with the customer. This will lead to mass customization through almost infinite rating structures providing service on a customer-by-customer basis, thereby keeping the customer base while maintaining revenues and margins.

Another aspect of focusing on the customer is the need to simplify things for the consumer. To meet this need, communications companies are moving toward offering bundled services or "one-stop shopping," providing customers with all their telecom needs: local telephone, long distance, wireless, Internet, and cable TV services—all in one bill. MCI is already providing bundled services through its MCI One program.

Reducing Time-to-Market. As the market becomes more competitive, the speed with which a carrier deploys new offerings will become more critical. The time to market of products becomes crucial as any technology leadership becomes transient. This will require activities such as prelaunch analysis, market trials, trend analysis, service roll-out structures, and lifecycle predictions. Irrespective of market type, additional capabilities at the beginning of the planning cycle will enhance the likelihood of service success, maximize return on investment, and increase the ability to respond quickly to market changes.

Differentiating Business Processes. The business processes and associated IT support systems must add value and support revenue generation while minimizing operational costs. The data stored in the IT systems must be transformed into information to introduce and enhance market opportunities. The challenge is to determine the best means to coordinate, measure, and control the different functions that deliver services to the customer.

Developing an Agile Company. The winners in the telecom industry need a number of basic elements: critical size, adequate capital, a responsive culture, a passion for customer satisfaction, market knowledge, a customer-driven service portfolio, and agility. Strategic planning is required although the company must have the ability to react to market changes or opportune situations. The agile company has the support infrastructure necessary to react quickly and flexibly within the constraints of current business practices.

Agile Decision Making. Decision making based on the analysis of customer data is a requirement across all areas of the organization, from strategic planning to customer service. If the telecom of the future is to succeed, it must use technology to address its business challenges, thus adding value directly to the business. There are a number of key activities an agile company must address:

Customer Care/Customer Retention. Carriers are facing complex challenges. They must offer multiple new services and convergent billing while presenting a single, seamless presence to the customer. Customer service is often the only point of contact the carrier has with the customer and is often the responsible element in both minimizing churn and maximizing revenue through cross-selling. To perform this task successfully, a single view of the customer is needed with rapid access to call history, both billed and nonbilled calls. This makes it possible to turn any customer query or complaint into a sales opportunity.

Churn Analysis. The objective of churn analysis is to reduce customer turnover by using advanced modeling tools that analyze customers' patterns in connects, disconnects, and changes of service. The information in a data warehouse can be used with churn prediction applications to identify high-risk targets and to provide, through outbound marketing, appropriate counter measures to retain the customer. One U.S. wireless carrier reduced churn from 30% to 25% in 6 months by using this application solution. With the advent of local exchange competition and local number portability, churn is about to become a major concern for wireline carriers as well.

Sales and Marketing Analysis (Campaign Management). The objective of sales analysis is to maximize the revenue potential of the network through the optimum mix of products and services, revenue, and profit. Accuracy and speed in tracking services and sales is the key to informed and agile decision making. Sales analysis requires a representative volume of call detail records (CDRs) to allow information based on actual data in the enterprise to be annualized. This allows business analysts to determine the direction of services and their profitability.

Database Marketing. Service providers see the need to provide customized services, targeted at each demographically dispersed customer. Data warehousing is the key component to enable this capability. Database marketing closely tracks trends so that the customer base can be profiled, segmented, and targeted. This type of marketing can maximize the short-term revenue of new and existing products and services by leveraging knowledge of existing customers. Accurate information is also key to rapidly trial new products. For example, customers who match the target profile for a new tariff can be selected. These customers are then monitored and the impact of the new tariff examined on a near real-time basis. In addition, billing can be carried out directly to prevent complex amendments to the billing system until the tariff has proven successful.

Network Capacity Planning and Utilization. Network planners require accurate information about the network utilization and the ability to link network information to the associated revenue and profit contribution of network elements. For example, a number of network elements may require upgrading. It is advantageous to prioritize the upgrades based on their potential impact on profit. The ability to analyze call detail records based on network element criteria makes this possible.

Fixed Asset Management. Many companies are not minimizing their cost base due to poor management systems, making them unable to accurately detail and track their assets. To address this problem, one major U.S. communications company uses an asset data warehouse to effectively manage their asset base. It ties financial and asset systems together via asset serial numbers which streamlines asset tracking. Managing fixed assets has a significant impact on cost structure and returns notable annual savings.

Executive Information Systems. Managers require a consistent, enterprise-wide view of business information. This requires a single data source that ensures a single view of the customer across all service areas. A top-level view is required, although the ability to "drill down" into specific information is also necessary. Trend analysis and investigation of anomalies can show marketing and service performance.

Why are Existing IT Systems Inadequate?. In the past, companies invested in proprietary systems and made significant investments to develop applications to support the business. However, these systems often failed to deliver quality information to the business because in most cases data warehousing efforts utilized rehashed on-line transaction processing (OLTP) techniques resulting in data that was outdated, of poor quality, and not entirely relevant. With the advent of robust, scalable open systems; new applications and software tools were introduced. Today, advanced transactional analysis, workshop, and dimensional modeling techniques are available. Transaction-intensive systems are designed for the capture and processing of data—not for the provision of information.

Attempts to use them for information provision fail to address the following issues:

Query Performance without Impacting Transaction Processing. Transaction processing systems are optimized for small discrete transactions. DSS queries that are run against an OLTP system involve complex table joins and full table scans. A single query can consume all the available computing resources and prevent the critical transaction processing load from running. Restricting the queries to run at off-peak hours or on a duplicate copy of the transaction processing system, while eliminating the impact on the transaction processing load, is not an acceptable solution as the computing resource required is still too great.

Single View of the Business. Typically companies have multiple transaction processing systems to support the multiple services that they offer. Even if it is technically possible to run a distributed query against the multiple systems it is not practical to link a customer's data from one system with that of another system. A single, integrated database is required.

Data Legibility for Business Users. Transaction processing database schemas are typically very complex and consist of many hundreds of tables. Since simple business entities are stored across multiple tables, to get meaningful information, a detailed understanding of the schema is required. The average business user has neither the time nor inclination to acquire this level of knowledge. Moreover, naive use of such a system frequently leads to incorrect results being generated. It is necessary to provide a data model that reflects the business.

Integration with Appropriate Tools. To enhance competitiveness it is not only necessary to have the right information at the right time, it must also be presented in a useful way. Large paper-based reports are typically ignored with the exception of a small amount of key data that is manually entered into a spreadsheet. Data must be delivered in a format that is appropriate for its use. For managers this would typically be integrated with their standard PC-based spreadsheet. Analysts require advanced modeling utilities, while customer service representatives require integration with existing call handling systems.

Existing System Structure. The diagram shows a view of a typical telecommunications company that has several data repositories utilizing various applications. Often, the applications do not have standard interfaces and have not been integrated. This leads to a number of different views of the customer with no ability to integrate the information. Often the data is not complete and cannot be combined into one data model. From a business standpoint, it is impossible to build an accurate, comprehensive profile of the customer, which often leads to incorrect actions being taken.

The Agile Telecom. To provide the modern communications company with an environment that supports the business, an architecture has been developed to address the technological and business issues it faces. This architecture takes all the disparate systems that have been developed and overlays an infrastructure, which centralizes the market data, customer data, call data, and internal resources data. It integrates and consolidates IT systems and implements focused solutions, which address each of the businesses' issues through system and application deployment—seamlessly. This architecture consolidates and fully integrates the three key functions in the organization:

One Call: Many communications companies today have developed internal systems to support newly introduced services. The existing system structure is shown in Fig. 3.22. However, often these systems are not integrated causing problems for customer service representatives handling a customer's request. Modern applications allow the various systems to be integrated and allow the same representative one view of the customer whether involved with inbound or outbound calls. Another benefit of this system is that each agent can handle more calls, thereby improving productivity and reducing costs. Also important is the reduction in the cost of training new agents. A well-designed system can prioritize a caller. Customers can be segmented into different groups and targeted for various services and level of "VIP" treatment. Very good customers can call special numbers or the system could flag a caller as a very important customer by matching their phone number to a list.

One Bill: Technology has created the opportunity for companies to launch a variety of related communications services. Not only do customers want to obtain all their communications services from one company but they also want to receive a single all-inclusive bill. Customers' lives and businesses have become increasingly complicated, a consolidated bill is likely to ease the problem. Additionally, consolidation of this function provides the communications company with a tactical weapon for cross selling and bundled discounting. The bill in itself becomes a product differentiator.

One View: Few companies today have a single view of the customer. The development of various databases, often conflicting, poses significant problems for sales, marketing, and finance departments. Today, telecommunications business solutions address issues such as churn, network profitability, and least-cost routing. This can be achieved through the consolidation of

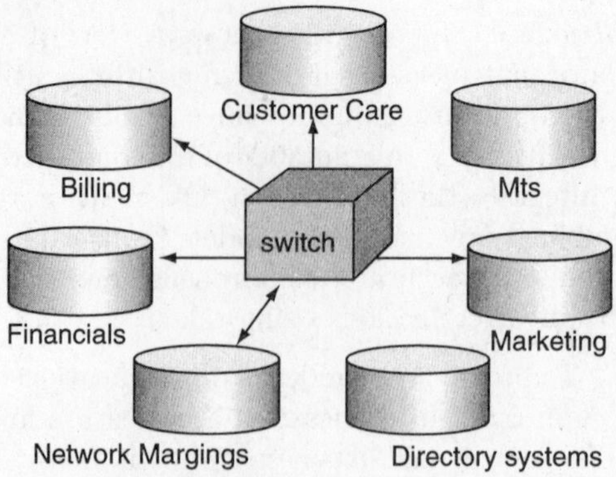

Fig. 3.22. Existing System Structure

the database, installing a data warehouse, and where appropriate, connecting satellite data marts that feed off the central data warehouse. The result of this architecture is the *agile telecom.* This architecture effectively addresses each of the following key business issues:

Minimize Customer Churn/Increase Customer Retention. This is a major issue for telecommunication companies. The agile structure is shown in Fig. 3.23 Providers offer significant incentives to get customers to move to their services, however, it costs many times more to win customers than to keep them. Between 20% and 40% of all communications customers churn annually.

Increase Services Sold To Existing Customers. Communications companies have considerable information about their customers' usage and purchase behaviors that can be leveraged to cross-sell existing services (e.g., call forwarding, single phone number) as well as new services (e.g., video-on-demand).

Gain New Customers. Telecoms have the ability to maintain an almost infinite number of rating structures, thereby allowing them to customize service almost on a customer-by-customer basis. This allows them to customize and target appealing services at new potential customers.

Increase Call Volume. Maximizing call volume allows companies to break even quicker (70% of all initial investment for new telecoms is infrastructure installation) and to utilize the fixed cost of interconnects (excluding call volume). Communications companies have considerable unused network capacity during certain periods of the day, week, and month since they must build enough capacity to handle peak loads. Understanding customer usage behaviors and unused network capacity allows communications companies to target service offerings to absorb unused network capacity.

3.3.1 Implementing One View

A key component of the agile telecom is the data warehouse. Many communications companies have been building data warehouses for years. What they

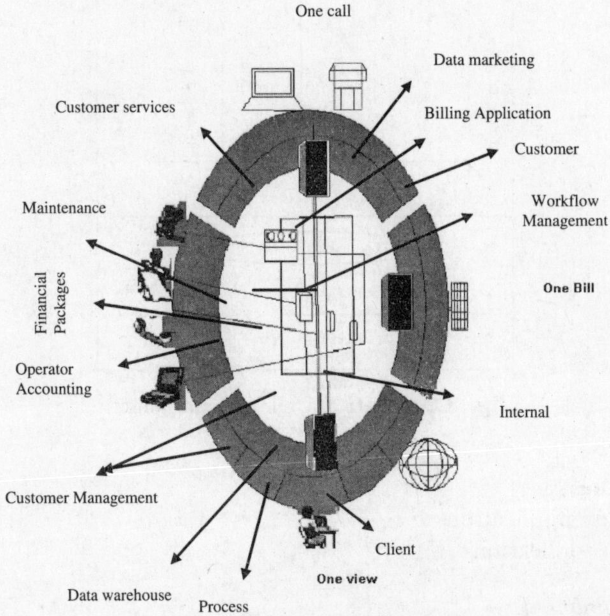

Fig. 3.23. Agile Structure

have not been doing is making them flexible, responsive, business focused—*agile*. To support the agile telecom, its data warehousing architecture must:

- have *flexibility* to cope with storing today's service transactions and the ability to accommodate tomorrow's
- give *one view* of the customer from all parts of the enterprise
- be responsive to today's business needs, not yesterday's
- use the latest tools to add intelligence and a business focus to particular issues for decision making
- be fast to implement for quick payback.

With the correct model, data schema, and integration expertise; implementation of the initial model can be achieved in as little as 3 to 6 months.

What are the components?

A business solution utilizing a data warehouse has two distinct components to it:

Physical Components

- Scalable open system platform
- Raw data
- Data model/schema

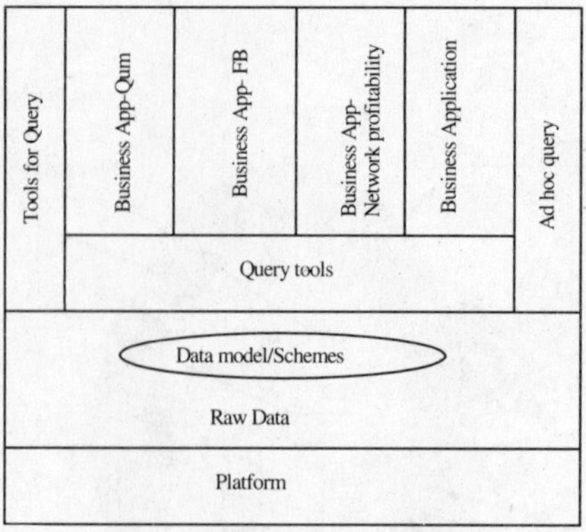

Fig. 3.24. Data warehouse components

- Query tools
- Warehouse applications
- Business applications

Human Component

- Knowledge of the data warehouse environment
- Necessary core skills to build the solution
- Sufficient telecom knowledge to apply the data and schema (providing a fast start to the project and early returns)
- Project management skills with a proven methodology

The development of a data warehouse has been depicted in Fig. 3.24. Experience has shown that a phased approach must be taken to ensure a successful implementation as illustrated in Fig. 3.25. An incremental methodology should be used, each phase addressing an area of business. This ensures that each business application is fully functional before the next phase is undertaken. This approach must also be integrated into an agile architecture, which provides support to each area of the business. Early implementation of the first phases helps generate an early return on investment.

The initial phase should involve a series of workshops or forums where the decision-support requirements of the business are discussed. These forums will help finalize and consolidate views, prioritize directions, and gain consensus on the subject at hand.

3.3.2 Business Benefit

Sequent provides a complete portfolio of Decision Advantage^{TM} Services to assist in the process of exploration, implementation, and deployment of DSS. A successful DSS strategy requires an informed analysis of a firm's current state, a strong business case (including projected financial impacts of better information), and a realistic implementation plan linked to the firm's business

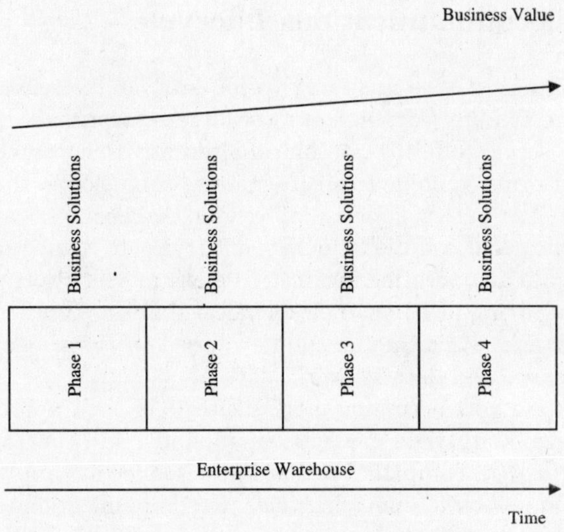

Fig. 3.25. A Phased Implementation

goals. Sequent's decision advantage business benefit analysis is an executive workshop, which meets the above goals and ultimately quantifies the benefits associated with DSS systems. Business benefit analysis combines traditional interviewing techniques and the workshop approach to answer key questions about the major opportunities and ideas, and the potential benefits and information requirements in implementing these. Each idea is plotted onto a matrix that notes technical feasibility and financial impact, suggesting high- and low-priority project phasing. The workshop concludes with a "consensus pledge" from the participants as to next steps.

3.3.3 A Holistic Approach

To reduce the complexity of the decision-support systems and the development of the data warehouse, it is necessary to understand the business issues, technical infrastructure issues, and data issues. An implementation methodology should encompass a phased approach and deliver benefits by:

- defining and *prioritizing* a high value business need that can be quickly solved
- building a cost *effective and timely* decision-support system via a phased approach that incrementally builds an organization's enterprise decision-support system
- *delivering real business value* with each phase
- providing *valuable experience* to the customer in developing decision-support systems
- providing the users with the *right information* for making critical decisions.

3.4 The Telecommunications Lifecycle

Telecommunications costs—including voice, data, and wireless services—are the second largest indirect expense at most large enterprises. The multiple service providers, service offerings, technical alternatives, contract complexities, and billing platforms combine to create massive confusion in the marketplace. This confusion has led to substantial vendor overcharges, rampant service abuse and misuse, and countless hours spent trying to manage it, creating an unnecessary drain on operating income. To make matters worse, your telecom and IT staff members, hired for their technical skills, are burdened with these administrative tasks when they should be focused on ensuring availability and quality of the voice and data network.

Enterprises are now becoming more strategic in their approach to resolving this problem. With tight budgets, small staffs, and constant pressure to cut costs even further, companies realize there is a need to apply a systematic, best practices approach to managing their telecommunications administrative functions. Therefore they are looking to outsource these noncore functions to telecom administrative outsourcing (TAO) services companies that can apply domain experience and intelligent technology while saving the enterprise money and reducing administrative hassles. They are also very carefully selecting their outsourcing partners, to ensure long-term success.

The telecommunications lifecycle is a holistic approach that addresses the problems at the root, long before an invoice appears. The terms of a telecom contract set the stage for errors, which grow exponentially through the order process. And without an up-to-date inventory of services and assets, the errors become increasingly harder to identify and correct. The telecommunications lifecycle approach addresses this by combining contract management, order management, inventory management, bill management, usage management, and information management all in one integrated solution.

It is clear that this problem is not going to go away and in fact will likely grow over the next several years. To ensure maximum success, it is imperative that the profit leaks are captured at their source and real-time tools are available to manage the complete lifecycle of telecommunications services on an ongoing basis. Therefore, enterprises need to implement a holistic, lifecycle solution to manage telecommunications administrative functions that will reduce hassles and improve the bottom line. It is imperative that the profit leaks are captured at their source and real-time tools are available to manage the complete lifecycle of telecommunications services on an ongoing basis.

3.4.1 Current Enterprise Environment

The telecommunications industry has seen dramatic change in the last ten years. The result has been a proliferation of service providers, new communications services, new communications devices, and more recently a decimation of customer service, as margins in the industry have vanished. During the same time period, telecom costs to the enterprise have been on the rise. According to Gartner Research, U.S. businesses will spend nearly $350 billion in 2003. The Aberdeen Group estimates the total cost per employee to be between $3,000 and $5,000 per year, which is double the amount it was just five years ago.

Yet most companies still have no systems or processes in place to tightly manage these costs. They still consider these indirect costs nonstrategic, therefore the purchasing and ongoing management of telecommunications services are a widely distributed (typically to someone on the IT staff who would much rather be working on more mission critical, technical tasks), poorly controlled, and heavily paper-based activity. As a result, this unwanted administrative burden does not get the level of attention commensurate with its importance to the company's bottom line. Traditionally, companies have attempted to attack the problem by hiring consultants who would identify savings opportunities and develop optimization strategies. More often than not, these consultants were paid a share of the savings achieved, so there was no motivation to fix the root of the problem, just to find immediate, one-time savings.

In more recent years, technologies and outsourcing companies have emerged to help enterprises automate and streamline the procurement process and the invoice management process. Though better than doing nothing, these initiatives have typically been shortsighted and fragmented, focusing on a narrow aspect of telecommunications costs. They have also focused more on the symptoms of the problem than the problem itself, resulting in one-time savings that are difficult to sustain over time. But the rapid decline of the global economy in 2001 was a painful reminder to the enterprise that cost control drives profitability in difficult times and supports growth in the good times. As companies have scrutinized their costs during this recession, it is becoming more and more apparent that the costs for voice, data, and wireless communications services offer one of the greatest opportunities for savings and process streamlining.

3.4.2 Getting to the Root of the Problem

No best practices approach to controlling voice, data, and wireless communications costs ever started with an invoice, yet that is what most companies focus on when they need a quick budget fix. They look to overcharges on their invoices that will likely give them a quick, one-time budget boost, but will not address the root of the problem since it begins the moment a vendor contract is signed.

Telecommunications contracts are extremely complex and heavily weighted in favor of the carriers that offer them. The sales pitch is that a custom contract means heavily negotiated, custom rates. But the reality is that buying telecom services is like buying a car, so unless you know exactly what price to ask for and know exactly how to structure your telecom agreement, you will lose at the negotiating table and you will pay more than necessary. After the contract is signed, the unnecessary costs go up exponentially when services begin to get ordered. Since the procurement process is typically very decentralized, and handled by people who understand a T-1 in detail but do not understand or care about what a purchase order is, order compliance is virtually nonexistent and the cost to complete an order through manual processes is enormous.

This results in services being ordered from multiple vendors, services being ordered that are not in conformance with company policy, services being ordered and not properly configured so they bill inaccurately from the first invoice, and wasted time as a large number of staff members are engaged in these burdensome administrative tasks. Once services are ordered, they need to be tracked through an inventory or asset management system, or costs will continue to escalate. The ever-changing enterprise, whether that change is growth or contraction, is constantly adding and deleting services to fit the current business need. Without knowing what you have and where it resides, it is impossible to keep it optimized or even terminate it when you no longer need it. Yet the typical enterprise has no inventory whatsoever of their telecommunications services. These core problems, along with the archaic vendor billing systems that are designed to bill you at the highest possible rate, cause billing errors that are virtually always overcharges. In fact, Gartner claims these billing errors an amount to an excess of 10% if there is no ongoing cost management process in place.

Therefore, there needs to be a process in place for validating billing as it is received to ensure it is your bill and it is coming from your chosen vendor. The system must also audit billing in real time to ensure the rates being charged are in line with the contracts negotiated. It must also handle charge backs to appropriate cost centers to ensure maximum visibility of the costs by the people who generate them. The old adage, "out of sight, out of mind," plays heavily here. If they do not get a bill, they do not know the costs and cannot do anything to help control them. But even this is not enough to address the complete problem, because it is also critical to enforce proper service utilization by end users. This requires that you have monitoring systems in place that capture the invoice usage detail and analyzes it for optimization opportunities that can be reported back to the abuser or misusers. Finally, you should be using all of the information generated from the contracts, orders, invoices and usage to better manage the business. The analytics available from these processes is invaluable if you are capturing it real time, at the point of entry.

3.4.3 The Telecommunications Lifecycle

To get complete control over voice, data, and wireless costs in the enterprise, you need to take a holistic approach that addresses the problem at its root, before it ever becomes a problem. You need a lifecycle approach that manages every aspect of each of the administrative functions in the telecom department so you can focus on keeping your voice and data networks up and running. An effective telecommunications lifecycle management system should include the following six elements: contract management, order management, inventory management, invoice management, usage management, and information management.

Contract Management – You need to negotiate world-class vendor agreements through an RFP process, renegotiate existing contracts on a regular basis, and ensure vendors comply with the negotiated contract terms on an ongoing basis. To accomplish this you need:

- Vendor Benchmarking – A knowledge base of supplier performance and compliance to benchmark vendors against one another for rates, contract terms, SLAs, technical suitability, financial stability, etc. thereby insuring a more fact-based selection process.
- RFP Management – Tightly managed RFPs, using a proven methodology, for voice, data, local, long distance, and wireless services to ensure best in class contract rates, terms, and conditions.
- Rate Reviews – Tightly managed contract renegotiations, using a proven methodology, to improve contract rates and terms at every opportunity throughout the contract term.
- Contract Compliance – To track existing contracts to ensure compliance with rates, terms, and commitment levels that helps identify overcharges and optimization opportunities.
- SLA Management – The ability to track and manage SLAs on a monthly basis and file claims for recoveries when they are not met.

Order Management – You need to manage the telecommunications services procurement process to enforce compliance and provide for fully automated approvals. You need to be able to track order status on a real-time basis. To accomplish this you need:

- Order Processing System – An automated interface for placing, tracking, and managing new orders, changes, and disconnects, eliminating expensive paper-driven processes and increasing enterprise-wide compliance with selected vendors. This should include workflow and approvals to expedite all orders.
- Order Integration with Vendor – Automated bonding with your vendors, including a catalog of services or devices (cell phones, pagers), where an

order can be placed, approved, then routed automatically to the appropriate vendor.

Inventory Management – You need an up-to-date inventory of your lines and device assets. To accomplish this you need:

- Line Inventory – An inventory of all lines, maintained current on a daily basis, to which you can apply specific charge back and other coding and perform ongoing maintenance. This enables accurate charge backs and greater accountability.
- Telecom Asset Management – An inventory of all cell phones, pagers, handhelds, desktop phones, etc. maintained on a daily basis, to allow visibility, accountability, and chargeback.

Invoice Management – A system is required to automatically process and audit your invoices on a real-time basis. To accomplish this you need: _ Invoice Processing – Automated systems to enter, code, validate, charge back, pay, and update your general ledger system. _ Invoice Auditing – Automated systems that audit your invoices real time, based on contract terms. The system should contain intelligence on vendor and service idiosyncrasies that routinely lead to overcharges. The components of the invoice management system are shown in Fig. 3.26.

Usage Management – You need visibility into usage detail. To accomplish this you need:

- Invoice Usage Visibility and Access – To have the ability to review usage from both your paper and electronic invoices on a monthly basis, which automatically reveals fraud, abuse, misuse, and optimization opportunities.
- Call Accounting – Real-time usage tracking to the desktop phone, which also automatically reveals fraud, abuse, misuse, and optimization opportunities.

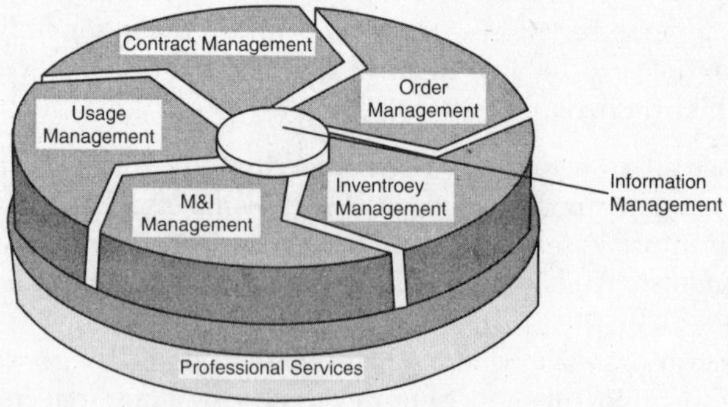

Fig. 3.26. Invoice management system

Information Management – You need to manage the extensive amount of information that comes from a telecommunications lifecycle system. To accomplish this you need:

- Reporting Functionality – A reporting tool that will create logical, easy to understand reports on each of the processes performed in the telecommunications lifecycle.
- Benchmarking – The ability to benchmark your total spend, spend by employee, rates, etc. against other firms in your industry.
- Analytics – The ability to measure this data against other business information to better manage the business.

3.4.4 Telecom Administrative Outsourcing

META Group claims that 70% of IT organizations currently outsource 10% to 50% of their IT operations and they predict the IT outsourcing market to grow by 20% in 2003. But the current trend in industry is to outsource noncore, administrative business processes, thus enabling the enterprise to focus on more strategic, mission critical tasks. But business process outsourcing ("BPO") is a long-term proposition and requires the forming of a partnership between the enterprise and the BPO provider. This warrants more detailed scrutiny of the BPO provider and an alignment of interests that should be enforced through contractual commitments. Telecommunications administrative functions are an ideal business process to consider for outsourcing. While the voice and data network are core to most enterprises, the administrative tasks required to support the networks are not. And the IT staff does not cover the tasks associated with this business process. In fact, the low man on the totem pole typically ends up with the responsibility, with hopes that they will quickly progress to a more challenging and exciting role within the department. Enterprises have attempted to develop internal systems and departments to gain control of telecommunications costs for decades, only to find that it is virtually impossible to keep pace with the rate of change. In the rare cases where companies have achieved a measure of success in this area, it has typically been as the result of one standout individual within the IT department with no great processes or systems, but a great eye for detail and a willingness to do the dirty work. However the success typically comes to a quick end when that all-star individual moves on, forcing the company to start from scratch.

3.4.5 Choose your Outsourcing Partner Wisely

The ideal partners for this business process are telecommunications administrative outsourcing companies. But with all the new entrants in this sector, you need to choose wisely. You should look for a firm with several years of experience in managing the telecommunications lifecycle. If you have been trying to solve this problem for decades, how is a company with one or two years' of experience in the industry going to do a better job? You should look for a firm that uses leading-edge technology that addresses the complete telecommunications lifecycle. Limited solutions lead to short-term, limited results.

Additionally, you should look for a firm that shares your company's values. Since outsourcing is a long-term commitment that is expensive to change on a regular basis, it is best if you work with a firm with people you like. You should ensure you are retaining a firm that is financially secure. A solid balance sheet that shows the ability to be there for the long haul is critical in this difficult economy. You should also look for a firm that charges you an easy-to-understand, fixed fee and allows you to keep 100% of your savings. Some pricing models are unpredictable, which creates risk and could lead to unaligned motivations. But most important of all, you should expect these companies to provide a contractual commitment to reduce telecommunications costs. Eliminating administrative hassles is a clear benefit of outsourcing this business process, but in today's difficult economic climate, it must come with a guaranteed ROI.

The current problem is clear. There is enormous confusion brought on by dramatic changes in the telecommunications industry. The outlook for improvement is bleak. With telecommunications companies still struggling, margins at all-time low levels, and vendor priorities still on selling, there is no expectation that this problem will be resolved on its own or with the help of the telecom providers. A solution needs to be found. Enterprise companies that have not addressed the problem are wasting hundreds of thousands, if not millions, of dollars a year. The solution must be a holistic approach that addresses the problem at the root, not the symptoms after it is too late. The telecommunications lifecycle approach is the best practices way to address the problems before they occur and avoid overpaying. Telecommunications administrative functions are not core to the enterprise's business. Therefore, outsourcing to a firm with proven expertise will offload a burdensome job and maximize long-term savings. Your finance department figured out about twenty years ago that there are certain noncore, highly administrative jobs that just do not make sense to perform internally. So they started outsourcing the biggest source of pain—payroll—and gave birth to and industry now led by ADP. Telecommunications administrative outsourcing is the next big opportunity in this area for the enterprise. Therefore, there is no reason not to begin increasing savings and reducing administrative hassles now.

3.4.6 Security in Web-Enabled Data Warehouse

Data warehouses represent implementation to simplify the storage of data of an entire enterprise. It will be not only easy to manage the stored data but also for the users and data manipulators to use this data more effectively and without having to tackle the underlying complexity of data storage format and the different extraction procedures specialized to handle only one type of data storage and management system. Unfortunately this convenience in data manipulation does not come without a price. The use of data after building a successful data warehouse may be much more easy and efficient, but to achieve the objectives laid down in the scope of the data warehouse, a proper design and infrastructure have to built and implemented. In the design and implementation strategy the aspect be, which is not given its due importance, is the security.

Security not only of the stored data but also of flowing data and the data transferred to the users is very important. With the introduction of Internet in accessing the data warehouse, the security becomes even more critical issue. Security roughly covers two aspects: firstly the data should be safe from the malicious party. Malicious party does not only mean the people looking to break into the data warehouse but also points to the clients or parties either accidentally or willfully get access to data that they are not meant to access. The second aspect of security deals with the stoppage of accidental spillage of data to ensure consistency and quality in data and safe conveyance of data to rightful party. This section gives a brief account of strengths and weaknesses of different techniques and practices being used and proposes a security model using XML and its related technologies to ensure a better and secure data warehousing experience.

3.5 Security Issues in Data Warehouse

In any cases of data warehouse implementation, the problem of security remains a critical issue. Without a proper infrastructure for secure distribution of information, it is like presenting all the important information required by a competitor or exploiter on a plate. It would be so easy to just put a lock on the entire data warehouse and sleep easy at night. Unfortunately the real world is much more complex then that and this complexity means inherited concerns about the security. The most common security concerns in data warehouse are the following:

Security of Static Data. The most obvious security concern lies at the core storage of data. The data should be safe from unauthorized users and access should be restricted to the users or group of users, which are "supposed" to get access to the data.

Security of Network. Another security weak point arises when the data is being transmitted on a request from the user. The static data is much more easier to secure but once the data leave the safe storage facilities its open for attack from sniffers, etc. and hence require much greater effort on the part of the data warehouse security implementer.

User Classification and Privileges. The user classification and assigning the respective privileges also represent a marked issue in data warehouse functionality and has a direct bearing on the security of the entire data warehouse. Users are arranged in groups and assigned privileges accordingly. Although it is much easier to assign a single privilege level to each user, but that is not really a very good idea considering the security of the entire data is at stack and not every user should be allowed to access the entire data warehouse just because he wants to use say only 1% of the data stored.

Human Factor. Perhaps the most overlooked security threat is that of the human factor. If an "insider" or employee leaves the team to join some other rival then how to stop him from exploiting the security infrastructure that perhaps he himself helped erecting in the first place.

Throwing in a New Problem: Web-Enabled Data Warehouse. Things get more complicated with the introduction of the Internet. With the increasing use of Internet in every aspect of life, the task of ensuring security for the Web-enabled data warehouse does not at all get easier. Although the Internet represents a very good ready made medium to access the enterprise data from anyway in the world but like with any other easy way out, this also presents the biggest threat to the security. Implementing security in Intranet is not as complicated an exercise than tackling Internet security threat where every kind of malicious people are looking for a way to hack into the data warehouse. With the advancements in hacking techniques and ever-increasing number of industrial spies, the Internet security threat is by no means to be taken lightly.

Metadata Security Threat. A big problem with a Web-enabled data warehouse is the proper management of metadata. Metadata is, in simple words, data about data. For a Web-enabled data warehouse, it is to a great advantage to arrange metadata in a way to facilitate the searching of the required data efficiently and unfortunately this also is the biggest drawback as far as security is concerned. Metadata can be used as a marker to locate the proper information required and this can also play into the hands of a person trying to hack into the data warehouse.

3.5.1 Performance vs Security

In any data warehouse the main emphasis is laid on the two most important aspects *Security* and *Performance.* Unfortunately these two mostly are on the opposite ends of the stick. While implementing a data warehouse, security is very important and should be given due consideration in the infrastructure, but the fact remains that the main concept of building a data warehouse is to provide easy and efficient access to its user; hence its not advisable to provide security at a level that compromises the performance too much. On the other hand, there is no point in providing greater performance at the cost of exposing the data warehouse to unnecessary security risks. So in the implementation of a data warehouse a delicate balance has to be found between the performance required and the security necessary. This balance is not so easy to achieve, performance will be compromised with increasing security but depending upon the requirements the maximum possible security should be in place.

3.5.2 An Ideal Security Model

Theoretically speaking, a perfect security model is a very simple one. It should not allow any unauthorized user to access data; data should be secure and any attempts to breach the security of the data warehouse should fail without exception. Authorized users should interact with the data with the confidence that there is no way anyone can break into the warehouse and that the data they are receiving is free of errors and by no means inaccurate. Unfortunately this is only possible in a perfect world and we mortals do not live in one. Security threats will always be there. Whenever an important data about a rival company is within grasp, people will not hesitate to get their hands dirty and as long as there are such threats, malicious users will always be looking for a way to break into the data warehouse and in a way help in exposing security vulnerabilities in the so called fool proof security.

3.5.3 Real-World Implementation

In the real world although it is not possible to achieve total perfection, that never held anyone back before. There are different techniques available to come close, if not achieve, the dream of a perfectly secure data warehouse. Depending upon the security issues stated above, different techniques and technologies are used to tackle the respective problems:

Static Data Security

As pointed out the data storage should be secure enough to stop any unauthorized intrusion exploiting the data. There are different techniques to achieve that:

Traditional DBMS security

A tempting thought, it will be much easier to manage security if only the database management system security mechanism can also be used to provide security at data warehouse level and DBMS do offer some form of data security. Unfortunately the data warehouse environment is rather different from the traditional database system. Also the fact that in almost every case DBMS security can be entirely bypassed by accessing the data managed under the DBMS in a lower level native access mode. In other words, a simple disk dump that accesses and off loads data at the physical block level suffices to bypass the strongest of DBMS security. Even though the data that has been dumped is in a "raw" state, it nevertheless is easily and totally available, making it very insecure.

Application-Based Security

Another method to enforce security is to use application-based security systems. The data is stored in a normal way without much emphasis on security but the accessing mechanism for the data, e.g., UI (user interface), provides the required security. The application acts as a filter to allow only the authorized requests to be processed and carried out. Data can enter and leave the data warehouse only through the application enforcing security. But this approach limits the flexibility necessary in data warehouse design and imposes restrictions that negatively impact the overall performance and the extra overhead of developing and implementing such an application to tailor it to the requirements of a particular data warehouse is also not desired.

Encryption of Data

Keeping the inside-out paradigm in mind, a reasonable approach is to provide the security from within the data warehouse storage by encrypting the data. This involves encryption of data before it is stored in the scrambled form into the data warehouse and when user requests data, it is again decrypted first and then is delivered to the required destination. Although this is a simple and yet a very powerful approach it does involve performance compromise.

Encryption of data increases the size of the data warehouse; also encryption requires machine cycles and hence can have a negative bearing on the performance. But in a data warehouse the main performance matrix is the I/O cycles and encryption of data does not require excessive I/O cycles; and hence encryption represents a logical approach to data warehouse security.

Selective Encryption

A more reasonable approach to security is to produce selective encryption. A data warehouse features a collection of data that can be in terabytes, and not all the users are supposed to have access to the entire data. To facilitate this, the users are divided into different groups and given privileges and permissions depending upon their requirements. But to increase the security further, the data should be fragmented and each slice of data should be encrypted using a separate encryption key. Then these encryption keys are distributed to the different users depending upon their privilege level. Symmetric cryptography is an obvious candidate for this type of system to work. A single secret key can do the trick, but it is more reasonable to use asymmetric keys with private keys issued to the user used for decryption of data, while public keys used to encrypt the data. This way if a malicious user does get his hands on the data she will require a private key to decrypt the scrambled data. In either of the two cases of cryptographical techniques used, it is understood that authentication of users will be carried out first before any of the processes are allowed.

Network Security: Intranet and Internet Security

The importance of securing the static data in the data warehouse cannot be overlooked but the biggest security threat remains with the inevitable introduction of networks and their inherited venerability. The famous saying *"never trust your network"* is still as true as ever and in fact due to the advancements in the hacking field this spells the biggest nightmare of every organization working to secure their networks from intruders. Different techniques are used in making the network highways safe.

Secure Socket Layer

SSL (Secure Socket Layer) security represents the security implementation at the presentation layer. SSL is a protocol developed by Netscape for transmitting private documents via the Internet. SSL works by using a public key to encrypt data that's transferred over the SSL connection. SSL works by using authentication of the user from a trusted third party. The SSL provided by the Internet Engineering Task Force (IETF), is the industry standard, and most commonly available security mechanism on the Internet. The IETF has renamed this mechanism to be "Transport Layer Security," and it is now usually referred to as SSL/TLS. SSL/TLS also provides data security for applications that communicate across networks. This protocol sits between various application protocols and TCP/IP. The SSL/TLS protocol provides connection security that has three basic properties:

Authentication. The connection can be authenticated using asymmetric, or public key, cryptography.

Privacy. The connection is private. During the initial handshake, public key encryption is used to define a symmetric secret session key. Symmetric cryptography is used for data encryption (for example, DES, Triple DES, RC4, and so on). Different encryption strengths, including 40-bit, 56-bit, 128-bit, and 112-bit encryption, are all supported (triple DES key efficiency is 112 bits rather than 168 bits in order to protect against a theoretical "man-in-the-middle" attack).

Integrity. The connection is reliable. Message transport includes a message integrity check, using a keyed Hashed Message Authentication Code (HMAC). Secure hash functions (for example, SHA-1 and MD5) are used for HMAC computations. The SSL protocol data encryption strength depends on the length of the symmetric cryptographic algorithm keys, which range from 40 to 128 bits.

S-HTTP

Another protocol for transmitting data securely over the World Wide Web is Secure HTTP (S-HTTP). Whereas SSL creates a secure connection between a client and a server, over which any amount of data can be sent securely, S-HTTP is designed to transmit individual messages securely. SSL and S-HTTP, therefore, can be seen as complementary rather than competing technologies.

Firewalls

A firewall is a device used to implement a security policy between networks. A firewall can have multiple network interfaces and is typically used to create a secure boundary between untrusted external networks and trusted internal networks. The security policy defines what type of access is allowed between the connected networks. There are three methods commonly used in firewalls to enforce the local security policy. The three methods are commonly used together, each complementing the others' weaknesses.

Packet Filters. Packet filters operate on the IP level, scanning the headers of each IP packet crossing the firewall and comparing its characteristics to a fixed set of rules. These rules determine whether the packet is allowed to pass unhindered or not. Characteristics recognized by packet filters are the source and destination IP addresses, the source and destination port numbers, various status bits in the header, and the direction the packet is traveling across the firewall. Packet filters do not know anything about protocols above the TCP/IP layer – they are fast and simple, but not very flexible.

Application Layer Proxies. Application layer proxies are applications running on the firewall, which users on one or both sides of the firewall can communicate with. The proxies forward the users' requests to the actual servers that can give a response, possibly imposing rules on what sort of traffic is acceptable. From the viewpoint of the "actual servers," it appears as if the firewall is making the requests – not the client. Application layer proxies are in general the most flexible type of fire walling software, but they frequently require added configuration or skills from the user. A common example of this is an HTTP proxy, which allows users to request Web pages from anywhere but may refuse some requests or rewrite pages based on rules defined by the administrator.

Network layer proxies. Network layer proxies are a cross between application layer proxies and packet filters. Like packet filters, they scan the headers of the IP packets crossing the firewall – but are able to respond in more ways than packet filters, which generally only "accept" or "reject." To make these "dynamic tunnels" as narrow as possible, a list of active communication channels (TCP, UDP, etc.) is maintained, and only packets (from the "outside"), which exactly match an active connection, are "proxied" back into the hidden network. Only internal, trusted machines can open such a tunnel through the firewall. In addition to this basic "translation" of network addresses and ports, some implementations contain built-in support for common protocols (such as File Transfer Protocol, FTP), which are dependent on external, untrusted hosts being able to initiate connections to the client. In general though, such protocols will not work through most firewalls based on these techniques.

Another type of network layer proxy is based on cooperation between specialized application proxies and the operating system's network code. In addition to rewriting the IP addresses on the network packets, some packets may be passed for further processing to an application proxy on the firewall machine. This proxy can monitor and filter the flow of data between the client and the external server in a much more complicated (and therefore error-prone) manner than is usually allowed within an operating system's low-level network code. These proxies are generally called *transparent proxies*, because they operate in a manner completely invisible to the user.

XML Firewalls. XML firewalls are designed to protect enterprises against the unique dangers posed by Web services. These firewalls examine SOAP (Simple Object Access Protocol) headers and XML tags, and based on what they find, block any dangerous or unauthorized content or services from getting inside a corporation. Traditional firewalls cannot do this, because they can only filter on the packet level, not on the content level. XML firewalls, on the other hand, examine the XML content of the incoming traffic, understand the content, and based on that understanding, take the required action.

XML Security

The increasing use of XML has prompted the developer to look into the possibilities of using XML in order to increase security. The following XML security specifications are being used to implement security of data:

XML Signature. Defines an XML schema for cryptographically authenticating data. The authenticated data may consist of a complete XML document, individual elements in an XML document, or an external data object referenced by an XML document.

XML Encryption. Defines an XML schema for encrypting data. The encrypted data may consist of a complete XML document, individual elements in an XML document, or an external data object referenced by an XML document.

XML Key Management Specification. The XMK Key Management Specification (XKMS) specifies the protocols for distributing and registering public keys:

- XML Key Information Service Specification (X-KISS): the protocol by which an application delegates, to a service, the processing of Key Information associated with an XML signature, XML encryption, or some other public key.
- XML Key Registration Service Specification (X-KRSS) The protocol for registration of a key pair by a key pair holder, with the intent that the key pair subsequently be usable in conjunction with the XML Key Information Service Specification.

3.5.4 Proposed Security Model

It is evident that none of the above options alone is capable of providing a solution to all the security problems of a data warehouse. So to build a secure data warehouse these techniques have to be used in combination. Figure 3.27 shows a proposed model to implement security in data warehouse. XML features prominently in the proposed model.

Implementing Security

The different steps of implementing security are given below:

i) The security is implemented right from base of the data warehouse, i.e., by using encryption to scramble the data. Using asymmetric keys for encryption and decryption. This means that the size of data warehouse will increase and also an increase of data warehouse costs to cater for the increased number of hardware cycles required, but nonetheless there will not be a big depreciation in the performance. The data is stored in scrambled format using triple DES and generating the encryption – decryption keys accordingly.

ii) Firewall is put around the data warehouse to provide a further safety barrier to the secured area from the intruders. The traditional firewalls are not good enough for that matter because they can only filter at the packet level, and cannot examine the contents of messages. Traditional firewalls protect a network's perimeter by blocking incoming Internet traffic using several different means. Some block all TCP ports except for port 80 (HTTP traffic), port 443 (HTTPS traffic), and port 25 (email traffic). Some ban traffic from specific IP addresses, or ban traffic based on the traffic's usage characteristics. A better option is to use XML services in the network and allow usage of XML firewalls. XML firewalls typically work by examining SOAP message headers. The header may have detailed information put there specifically for the firewall to examine, and if so, the firewall can take an action based on that information.

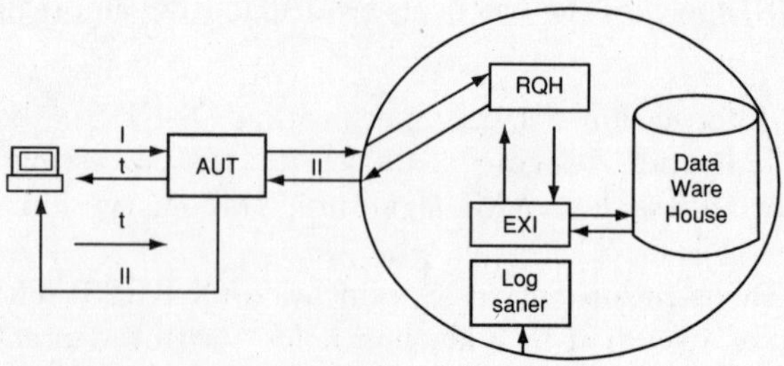

Fig. 3.27. Working steps of the Proposed Security Model

Even if the header does not have this information, however, XML firewalls can still take actions based on what is in the header. The header, for example, might have information about the recipients of the message, security of the overall message, or the intermediaries through which the message has passed. In addition, XML firewalls can look into the body of the message itself and examine it down to the tag level. It can tell if a message is an authorized one, or coming from an authorized recipient and then take action based on that – for example, blocking traffic, sending it to a secure environment where it can be further examined, or allowing it to pass through. XML firewalls have other methods of protection as well. They can understand metadata about the Web service's service requestor as well as metadata about the operation itself. They can gather information about the requesting user; such as understanding what role the user plays in the current request, for example. XML firewalls can also provide authentication, decryption, and real-time monitoring and reporting. So when a request comes from a user the XML firewall can look into the contents as well.

iii) The messages are passed using SOAP protocol over HTTP in communication between users and the data warehouse. SOAP messages are fundamentally one-way transmissions from a sender to a receiver, but SOAP messages are often combined to implement patterns such as request/response. Also the SOAP header is used to communicate information regarding the services requested, so that these headers can also be checked by the implemented XML firewall as a further security precaution.

iv) To secure the network, XML signatures are also used. Each user is identified by its signatures. SOAP Security Extensions proposes a standard way to use XML-Signature to sign SOAP 1.1 messages by defining the <SOAP-SEC:Signature> SOAP header entry. In addition, SOAP Security Extensions reuses two existing SOAP header items. These are the "actor" header item, which indicates the recipient of a header element, and the "mustUnderstand" header item, which indicates whether an application must attempt the validation of the enclosed XML Digital Signature.

Components of Security Model

The components in the security model are as follows:

Authentication Server

The authentication server is used to authenticate the users. This is the first part of security model with which the user first comes in contact with the request for manipulation of data.

XML Firewall

The firewall represents the protected boundary around the data warehouse and its core components. Authentication server is not placed inside the firewall, because a malicious user, if allowed to interact with the authentication server inside the firewall, will compromise the integrity of the firewall itself. The XML firewall works by looking into the payload of the SOAP packet, hence allow another way to stop malicious user.

Request Handler

This is a vital part of the data warehouse security model. When the authentication server authenticates the user, the authentication server hands over the request to the request handler. The request handler checks to see if the user has the privileges to access the data it is requesting. If so, it sends the request to the extractor otherwise sends back the request to the authentication server with the error message defining the user's privileges.

Extractor

Extractor represents the mechanism to extract the data depending upon the valid request handed to it by the request handler. The data extracted is no doubt in encrypted form and handed over to request handler.

Data Warehouse

This represents the collection of data for all types of users. The data is divided into different partitions and encrypted using different keys for each partition. The data in encrypted form is passed to the user and depending upon its privilege level the users are assigned private keys for the partitions for which it has access. This encrypted data is placed under a tag in XML format to be sent to the user.

User

The users are the customers, partners, clients using the data warehouse. Each user is issued the following:

- Username and password: to identify itself to authentication server in initial interaction.
- XML Signatures: to authenticate it to the authentication server.
- Private keys for each partition it has access: these private keys are used to encrypt the data when it wants to store the data back into the data warehouse (depending upon whether the user has this privilege or not) and also for decrypting the data extracted from the data warehouse.

Log Server

The log server stores all the information about the processes taking place.

Working Steps of the Security Model

The working steps of the proposed security model are given below:

1. The user interacts with the authentication server by presenting its issued user name and password as an initial step.
2. If the password is correct, as a second authentication method, the authentication server sends a document to be digitally signed by the user with its XML signatures. A part of the document is digitally signed by the authentication server itself to confirm its own identity and at the same time writes a log file with the date and time of interaction along with the user name. Otherwise the server generates an error message and logs that into the log file. The document to be sent to the user is also generated randomly every time any user interacts.
3. The user signs the document after verifying the digital XML-Signature of the Authentication Server itself. This verification of both Authentication Server and user takes place using XKMS and its two protocols X-KISS and XKRSS. After verifications the user sends the document back to the authentication server along with the request.
4. Authentication server verifies the XML signatures of the user and checks the time stamp, and if found valid, sends the request to the request handler inside the firewall along with the log information placed in the SOAP message.
5. The firewall checks the information in the header to verify that the message is from authentication server itself and allows only the request part to be transmitted to the request handler.
6. The request handler checks whether the request is within the privilege level of the user, if so it sends the request to the extractor; otherwise it generates an error and returns the request to the firewall, which in turn attaches the error information and sends it back to the authentication server.
7. The extractor extracts the required data from the data warehouse, which is in encrypted form
8. The extractor sends the data back to the request handler.
9. The request handler then creates a SOAP message placing the required data in scrambled form in it. And sends it to the firewall.
10. The firewall sends the data back to the authentication server along with a log of the requested information and the log from authentication server being sent to the log server. To achieve this the firewall must read the messages and be allowed to extract the information required by it.
11. The authentication server sends the required information to the user.
12. The user then extracts the required data, no doubt in scrambled form, and then uses his private keys to decrypt the data.

The security of a data warehouse is the most critical issue. The extra added security may increase the costs and decrease the performance by a fraction, but the fact remains that it is better to be safe than sorry. The data of the entire enterprise is at stack in a data warehouse and if a security breach does take place the effects can be devastating.

3.6 Data Warehousing: To Buy or To Build a Fundamental Choice for Insurers

3.6.1 Executive Overview

It is now well known that data warehousing offers an approach that solves the information access challenge faced both by users and information systems management. It can greatly increase the level of productivity and improve or hasten the decision-making process of knowledge workers within an organization. Business intelligence is quickly becoming a standard business practice, and the insurance industry is no exception. Given that data warehousing is "the price to play" in insurance, the focus then shifts from determining necessity to determining the best implementation method. There are two main methods in acquiring business intelligence in any firm: building a solution or buying one. Our position is that building a data warehouse from the ground up is often a risky, expensive, and resource-consuming effort. Whenever possible, it is better to buy a solution from an expert and customize it to a particular environment.

SQLiaison Inc. has been in the business of building decision-support systems for the insurance community for many years. This level of expertise is reflected in FellowDSS™, a decision-support environment for the insurance industry based on data warehousing technology. FellowDSS™ will dramatically reduce the time, cost, resource requirements, and risk associated with implementing and deploying a data warehouse for different types of insurance companies. This section explores the business and technology issues related to both buying and building data warehousing solutions, discusses requirements for effective insurance decision-support systems, and illustrates why we feel that buying a solution is ultimately more cost-effective for insurance companies.

3.6.2 The Fundamental Choice

Assuming that an insurance company wants to go ahead with data warehousing, they are now faced with one fundamental choice in methodology. A company can either build their data warehouse from scratch, or they can buy a packaged solution from a vendor. Any enterprise-wide (or localized) decision-support solution should have certain requirements and properties. It must:

- Have a facilitated deployment system. This allows for a fast and simple implementation that avoids problems;
- Produce a high return on investment with a low total cost of ownership;
- Support extensibility in the architecture that allows it to grow with the business. It should perform certain critical business analyses (depending on the type of firm) such as:
- Sales analysis;
- Policy, risk, coverage, and billing analysis;
- Line-of-business segmentation;
- Marketing campaign analysis;
- Business performance analysis/underwriter profitability or results analysis by different business metrics;
- Expense control;
- Pricing (by demographic or postal code for example)
- Customer profile and loyalty analysis;
- Impact of cross- and up-selling;
- Accident year data analysis;
- Claims analysis;
- Loss analysis;
- Annuities analysis.

It should allow you to *automatically* capture key data gathered in your operational and administrative systems and move them into the data warehouse, without constant maintenance or scripting. The most important thing is to make sure that your data warehousing solution is modeled after YOUR business.

3.6.3 Analyzing the Strategic Value of Data Warehousing

To begin, let us examine the different concepts in business intelligence. Every day, a company gathers *data* from various sources: the claims system gathers data on the names, addresses, and claims of claimants. These basic elements are *data*. For example: "Claimant Bill Reed lives at 38 Green Terrace and has insurance Plan B." This is data. The fact that ABC Insurance Company collected \$10 million in premiums and paid \$10 million in claims is also data. What companies are looking for is a way to leverage the enormous amounts of data to gather useful *information*. Information gives you a more global idea of what is happening in some area of the business that you want to examine. "36.7% of policyholders in Iowa have Plan B, and we collected \$3.67 million in premiums from them but paid out \$7.34 million in claims. Plan B is our most popular." Once the information is gathered, it is now completely up to the company to turn it into *wisdom*. Wisdom allows a company to make key decisions based on the information they have at their disposal. "Our loss ratio in Iowa as a whole is 100%, but on Plan B, it's 200 percent! That's atrocious! Let's take a close look at Plan B to see why it isn't profitable." These concepts make up business intelligence.

The best way to access the information required is through the packaged solution. There are four points to consider:

1) The wisdom to be gained from the Plan B scenario will not come from the data warehouse. In fact, it will come from the methodology developed by ABC's management staff to act on certain situations. The role of the data warehousing staff, internal or external, ends when the information is delivered by the decision-support system; it is up to the knowledge workers to make the decisions.
2) The less resources involved in building an infrastructure, the more available to work out the details of the actual decision-making methodology. There is little point in waiting longer than is necessary on the construction of a data warehouse; it is simply a way to get that required information.
3) If increased market share in insurance companies is compared to the jackpot in poker, then the data warehousing infrastructure has become the "ante." The packaged data warehouse alternative simplifies and accelerates the ability to pay that ante. With the state of the insurance marketplace, data warehousing is necessary to get to that information faster.
4) Besides the money involved in building an infrastructure, the *time* required to do so can be staggering. It seems better to avoid that altogether. For a firm that wants competitive advantage through the timely delivery of information to employees, it makes sense that the implementation of the required decision-support system be done in the shortest time at the least cost, and with a measurable return on investment. In other words, it would appear that the "blood, sweat, and tears" method involved in designing, developing, and deploying such infrastructure is the less desirable option. The task at this point is to determine if it really is faster, less expensive, and ultimately more effective to buy the infrastructure rather than build it. This is discussed in the next subsection.

3.6.4 Addressing your Concerns

Conservative industries such as insurance are very apprehensive acquiring new infrastructures and embracing new business concepts. Some of these concerns lie in the reluctance to accept business intelligence as a must-have practice. Others are more basic. We attempt to uncover what some of the common issues with buying are, expose the underlying realities, and overcome them.

If you have Already Started Building

Changing direction in the middle of a project is one of the most common deterrents to accepting a packaged solution. Many insurance companies start building their solutions because they just do not know that packaged solutions exist. Usually, the central question here is: "How do we justify the money we've already spent on a custom solution if we change to a packaged one?" Companies tend to look inside to find out what technology and human resources they have at their immediate disposal, and this is an excellent cost-saving measure. It does not preclude using an outside packaged solution by any means. Certainly there may be some unrecoverable costs incurred in the design, development, and deployment (D3) of a custom environment. These include human resources (time) spent in D3 or the opportunity cost of not spending those resources elsewhere.

When a company starts to build a decision-support system, they almost certainly acquire a number of skills and technologies that can be reused in future endeavors, such as a bought solution. For example, the knowledge gained by a company in D3 is usually reusable. Acquired hardware or software may also be reused, depending on the solution. This minimizes overlap of technology and provides a more accurate way to describe how much a packaged solution will cost in the end. Often, finding the reusable can mean the difference between a lengthy and expensive D3 process and a short, inexpensive one.

When the costs to bring a company to a certain point in their custom solution are known, the next task is to find out how much it will cost to complete the custom project, then compare that total cost to the cost of buying a new solution. Using the tallied amount of reusable assets, a firm can get a clear indication of what path to take. It is clear that when the cost of buying a solution with equivalent or better functionality and a faster time-to-market is less than it is to complete the custom project, it makes sense financially to switch to the bought solution. The things to consider are:

- What is the projected total per day D3 cost for the rest of the project?

While it is impossible to say for certain for all companies what that might be, it is fairly easy to calculate it for a particular company. If a company has been working internally on a custom project for an amount of time, t, then a full-service data-warehousing firm should be able to do the following:

1) Assess the current business and technological environment at time t,
2) Determine how much it is to complete the project from time $t$to the project's conclusion using different possible scenarios,
3) Add that cost to the cost of bringing the project to time $t$and finally,
4) Compare that total to the projection for the bought solution.

- How much longer will it take?

It is not difficult to estimate how long a packaged solution will take to deliver after a professional assessment of the environment. From beginning to end, a packaged solution takes a fraction of the time to deploy compared to a custom one. Even when a custom project is partly or mostly complete, a bought solution may still be faster to deploy. Furthermore, the amount of D3 is little to zero in the packaged solution.

- How much will extra hardware/software for the custom project cost and will that cost more than what is already offered in the packaged solution?

Again, maximizing the amount of reusables tends to equalize, or at least minimize the difference between the cost to complete a custom solution and the cost to purchase a packaged one. This is another metric that can be measured using an assessment.

- How much work will I need to do with this custom solution when the business expands? Will it be lesser/cheaper with a packaged alternative?

Unless the custom solution is being built with the extensibility of a packaged solution, it will not be cheaper to do extra work extending the custom solution when the business or data sets expand or change. Very often, companies find themselves having to rework their solution to accommodate expansion into other states, mergers, acquisitions, or corporate structure changes. This is almost the equivalent of building another warehouse from scratch. Furthermore, the kind of robust extensibility found in some packaged solutions took quite a while to develop. Doing it internally would delay the delivery of value-added decisional applications to knowledge workers. A needless delay, since that functionality is already available with a provider. Packaged solutions, by their very nature, are designed to be flexible enough to cover most existing and future business situations. Also, this flexibility is strengthened by any professional services used to customize the solution to a particular firm. Because of this, packaged solution providers have an easier time planning for and delivering such changes since they have the experience in managing many evolving business scenarios.

Does Building Save Money?

Despite all the technological advancements in the last 10–15 years, the insurance sector has historically been slow to embrace many of them. Years ago, policies, claims, and other administration systems were built by internal or outsourced development teams. A team was hired to develop it over a period of 2 years or so, a little at a time, and then implementation also took another 1–2 years. After all was said and done, final deliverables only happened after 3–4 years and millions spent.

Undertaking new technology in a business setting can be risky, but it is sometimes necessary. Insurers realized the importance of these operational systems to the running of their business, but when the opportunity to cut that development and implementation time came they were still reluctant to invest in the then-new technology. They were afraid of a lot of new infrastructure in a short amount of time. What if it did not deliver? Insurers were faced with a choice:

They could:

1) Continue to develop their own systems over a very long period of time without sufficient returns on their investments while other companies were leaping ahead in terms of vision, service, and results;

2) Forget about investing in these operational systems and continue the old way of doing things, hoping that the market will respond favorably to a company that saved money (in the short term, at least), or;
3) Buy a solution, customize it as necessary, and spend most of their efforts doing more business in a shorter amount of time using fewer resources.

In the end, the last option turned out to be the best one. Bought solutions on the operational side are standard in most of the insurance community today. On the decisional side, things are similar. The META Group reports that many data warehousing projects follow the 2:2:50 pattern. That is the average DW project takes two years to deploy, costs an average of $2 million, and has an expected failure rate of 50%. Other estimates cite overall failure rates of more than 60%.

The Data Warehouse Institute also reports rates higher than 70% across all industries. This is partly due to people not having all the technical expertise required for such a project. Companies often do not have the resources experienced in data warehousing. There is confusion in what the information systems staff can do and what they cannot. There is even confusion in what data warehousing means. In a decision-support context, the risks in building are very similar, if not identical, to the risks in building operational systems. Decision support and data warehousing are in the same position as claims and policy management systems were 10 years ago. The same arguments apply now as they did then. It no longer makes sense to develop internally. The commitment involved is not so great when you consider the alternative.

Do Internal Staff Members have Inside Knowledge not Available to Vendors?

This is partly true. While some vendors know the insurance industry very well, there may be some things in a particular company that are not standard. There may even be things that a bought solution may not address in its raw form. But there are ways around this. Using a vendor can allow more information systems projects to be completed in a shorter amount of time. An assessment where the vendor works with your company's internal business and IT staff ensures that the solution fits your business correctly. After that, your staff is free to do all the things that they have expertise in. For example, sometimes there are internal systems that need to be maintained, built, or rebuilt. Often, these projects can take precedence over data warehousing. Leveraging the vendor's expertise allows both initiatives to take place at the same time; therefore you leverage external expertise in the decision-support area.

It is always wise to focus on core competencies. Chances are the right vendor does nothing but data warehousing. It is their job to make your data useful to your knowledge workers. In that respect, this is often a good reason to allow such a vendor to support the business in decision support rather than internal resources. Your internal resources would not become obsolete in such an environment; in fact they would be better used doing what they do best.

Is it More Beneficial to Work with Familiar Technology?

It is not necessarily the best idea to work with familiar technology, especially from a business point of view. Consider a company that uses technologies A, B, and C. The actuarial department realizes that in order to be more efficient in their rate making, they need a certain functionality that cannot be found in these technologies. It is found, however, that technology D will allow that functionality. While it may make sense initially for the IT department to favor the familiar technology, it should be noted that there may be a significant opportunity cost associated with not providing the analytical capabilities required by the actuaries. This may be measured in terms of lost revenue or time not optimized by knowledge workers. It is the business need that drives technology, not the other way around. In cases where there is an absolute requirement for certain in-house technologies to be in place, it may be possible for components in the packaged solution to be changed, depending on the flexibility of the solution chosen. Using certain technologies mainly for comfort to the exclusion of what may be better technology for the business can ultimately defeat certain business goals. The following Table 3.1 summarizes these arguments.

3.6.5 Introducing FellowDSS™

As stated, it is no longer necessary for an insurance company to use internal resources to deliver an effective administrative system to run the business (or line of business). Now the same insurers no longer have to spend the time, money, and other resources to deliver a robust decision-support system to *drive* the business.

FellowDSS™ is a decision-support system based on data warehousing technology. It has been developed with the broad needs of insurance companies in mind. It will drastically reduce the time, cost, resource requirements, and risk associated with implementing and deploying a data warehouse for insurance companies. In particular, this decision-support system has all the functionality, right out of the box, of a robust decision-support system. The next subsection demonstrates how this is done.

How FellowDSS™ Stacks Up

We mentioned that there are a number of requirements in a decision-support system; Table 3.2 summarizes how FellowDSS™ addresses these requirements and more.

Table 3.1. Summary

Problem	Origin	Reality
We have already started building a solution	Concern of not being able to cost justify the solution	The total size of all recoverable costs may be quite high compared to the size of the nonrecoverable costs. The cost to complete may be more than the cost to switch to a packaged solution with equivalent or superior functionality, when "time-to-market" is factored in. If the ability of the custom project to adapt to new business solutions is in question, then the packaged alternative is worthy of consideration, especially in this marketplace.
We will save money by doing it internally, and we do not have to allocate all of the funds inmediately	Reluctance to commit	May be true in the short run, but not necessarily. Can start development and maintenance would be required. A change in the business could mean a complete reworking of the model, infrastructure and so on. Stretching out an implementation and missing out on business opportunities or not being able to effectively analyze practices can cost far more than any packaged data warehousing application.
Our developers have inside knowledge of our business that vendors could not possibly have.	Myth of internal staff obsolescence	They can work with your internal technical and business resources to customize the solution to reflect your business with little disruption to everyday operations. Then your resources can be better used in other projects.
Business intelligence is not that important anyway. Why get a whole new infrastructure anyway.	Resistance to change	Business intelligence is the price to pay in insurance today. See "the challenge of business intelligence across in the insurance industry."
We are very familiar with the existing technical environment	Resistance to change	The right vendor can teach you the technical environment, work with yours (where applicable), or a combination of both.

Table 3.2. How FellowDSS^{TM} handles data warehousing requirements, by module

Requirements	**Fellow DSS^{TM} Module (s)**
Customer profile, loyalty, and care analysis	Market Analysis
Marketing campaign analysis	Sales Management
Sales analysis	Sales Management
Policy, risk, coverage, and billing analysis	Market Analysis
Line of Business segmentation	Market Analysis
Business performance analysis/underwriter profitability or results analysis by business metric	Sales Management/Claims Analysis
Expense control	Agent Remuneration Analysis
Pricing (by demographic, postal code, impact of cross, and up-selling	Pricing Scenario Analysis/ Market Analysis
Accident year data Analysis	Claims Analysis
Claims Analysis	Claims Analysis
Channel Effectiveness	Distribution channel Effectiveness
Insurance Products Analysis	Insurance Products Movement
Financial Products Analysis	Annuities and Investment Products Movement
Loss Analysis	Claims Analysis

To learn more about FellowDSS^{TM} business models, metadata definitions, database structures, transformation/loading scripts, advanced reporting and other algorithms, see the FellowDSS^{TM} product brief.

3.7 Summary

The data mart/enterprise data warehouse environment increasingly becomes important as the data warehouse environment evolves to the corporate information factory. Data marts serve departments while the enterprise data warehouse serves the entire corporation. The enterprise data warehouse is shaped by very granular, simple data. Typically the enterprise data warehouse structure is normalized or "lightly denormalized." The data mart structures are shaped by the requirements of the departments that own them. The predominant structure for the world of data marts is the star join or the snowflake structure.

The community of users for the data mart/enterprise data warehouse environment includes farmers and explorers. Farmers and explorers have very different characteristics. Farmers have a strong affinity for data marts while explorers have a strong affinity for the enterprise data warehouse. The enterprise data warehouse serves many other architectural structures, other than the data marts. The enterprise data warehouse is served by very different technology than the data marts.

In order to compete in today's competitive healthcare marketplace, organizations must listen to everything that is going on around them—they just cannot focus on their own corner of the world. Data warehouses, are in essence, a sophisticated hearing aid that allows healthcare organizations to hear what their customers and their providers are saying, thereby offering enhanced service and cultivating long-term customer relationships. By using data warehouses to make data-driven decisions, organizations will no longer have to rely on hit-and-miss tactics to guess at what makes effective outcomes and what customers need. In this marketplace, data warehouses are not merely a luxury, or even just a competitive advantage. They are a matter of survival.

Also in this section the data warehousing in telecommunication industry is discussed in detail. A single technology for implementing security meeting the required performance target is yet to be devised but the existing technologies can be used in combination to produce satisfactory results. In this section an attempt is made at trying to devise such a security model. XML with the proper extensions to its different technologies looks a likely candidate to be used in solving the security problems in data warehouse but still it is a long way from perfection; nonetheless the future does looks promising.

This section has explored two main methods of establishing a business intelligence environment in an insurance company: building a solution from the ground up and buying an existing solution. We conclude that buying the solution and then customizing it to a particular environment is the most effective method strategically, technologically, and financially. FellowDSS™ is the method of choice for accessing that critical information. For more information, read "The Challenge of Business Intelligence Access for the Insurance Industry" as well as the FellowDSS™ product brief, both available from SQLiaison representative.

3.8 Review Questions

1. Explain in detail about business process re-engineering.
2. With a star cube, explain data mart technology.
3. State some of the different structures and uses of data mart.
4. Explain about data warehouse in the enterprise technology.
5. Draw the Informix architecture and state its advantages.
6. How to build the data mart/data warehouse infrastructure?
7. Explain about the amount of history stored in data warehouse.
8. How to share data among data marts and what is the efficiency of sharing?

9. What are subdata marts and state its uses?
10. What is refreshment cycle as used in warehouse technology?
11. Give details on external data and operational data stores based on their usage.
12. What is distributed metadata?
13. How to manage warehouse environment based on applications?
14. Explain in detail on data warehousing for health care.
15. What to expect when beginning a data warehouse implementation?
16. What are the issues in industry that data warehouse addresses?
17. Write about the issues and challenges faced when data mining is applied to telecom industries.
18. Explain in detail about the various management involved in telecom life-cycle.
19. What are the security issues in data warehouse technology?
20. Compare performance versus security.
21. What are techniques and technologies used to tackle the security issues in data warehouse?
22. Explain the steps involved in implementing security with suitable model structure.
23. What are the components and the working steps involved in security model?
24. State the fundamental choices for insurers in data warehousing.
25. How to analyze the strategic value of data warehousing as applicable in insurance?
26. Explain in detail on Fellow DSS™ stack up with its advantages.

4

Evolution and Scaling of Data Mining Algorithms

Objectives:

- Data mining is an application-driven field where research questions tend to be motivated by real-world data sets.
- While many data mining applications focus on commercial applications, such as credit scoring, fraud detection, and Web personalization, data mining as a tool for scientific discovery also motivates research interest.
- The data mining innovations being implemented worldwide often involve collaborations among domain experts, computer scientists, and statisticians.
- These application-driven developments will continue to proliferate as data owners seek new and better ways to gain insight into their data.
- The research effort on scaling data mining algorithms to large databases has now given analysts the ability to model and discover valid, interesting patterns over these large data sets.
- General scaling principles include use of summary statistics, data compression, pruning the search space, and incremental computation.
- We described these principles in the practical context of mining algorithms, expecting they would be useful in other areas of computer science as well.

Abstract. Traditional modeling methods from statistics and machine learning, including linear regression, logistic regression, discriminate analysis, and naïve Bayes models, are often the first tools used to model multivariate data. Newer predictive models, including additive regression, decision trees, neural networks, support vector machines, and Bayesian networks, have attracted attention in data mining research and applications, as modern computing power has allowed data miners to explore more complex models. These predictive models often sacrifice interpretability for increased flexibility in the functional forms they accommodate. The trade-off between flexibility and the interpretability often drives the choice of method applied to particular multivariate data set.

Recent research has shown that combining different models can be effective in reducing the instability that results from predictions using a single model fit to a

S. Sumathi and S.N. Sivanandam: *Evolution and Scaling of Data Mining Algorithms*, Studies in Computational Intelligence (SCI) **29**, 151–164 (2006)
www.springerlink.com © Springer-Verlag Berlin Heidelberg 2006

single set of data. A variety of model-combining techniques (with exotic names like bagging, boosting, and stacking) combine massive computational search methods with variance-reduction ideas from statistics; the result is relatively powerful automated schemes for building multivariate predictive models. As the data miner's multivariate toolbox expands, a significant part of the art of data mining is the practical intuition of the tools themselves. Also in this section scaling the mining algorithms to large databases is discussed.

4.1 Data-Driven Evolution of Data Mining Algorithms

Data mining is an application-driven field where research questions tend to be motivated by real-world data sets. In this context, a broad spectrum of formalisms and techniques has been proposed by researchers in a large number of applications. Organizing them is inherently rather difficult; that is why we highlight the central role played by the various types of data motivating the current research.

We begin with what is perhaps the best-known data type in traditional data analysis, namely, d-dimensional vectors x of measurements on N objects or individual, or N objects where for each of which we have d measurements or attributes. Such data is often referred to as multivariate data and can be thought of as an N $\times$ d data matrix. Classical problems in data analysis involving multivariate data include classification (learning a functional mapping from a vector x to y where y is a categorical, or scalar, target variable of interest), regression (same as classification, except y, which takes real values), clustering (learning a function that maps x into a set of categories, where the categories are unknown a priori), and density estimation (estimating the probability density function, or PDF, for $x, p\ (x)$).

The dimensionality d of the vectors x plays a significant role in multivariate modeling. In problems like text classification and clustering of gene expression data, d can be as large as 10^3 or 10^4 dimensions. Density estimation theory shows that the amount of data needed to reliably to estimate a density function scales exponentially in d (the so-called "curse of dimensionality"). Fortunately, many predictive problems including classification and regression, do not need a full d dimensional estimate of the PDF $p(x)$, relying instead on the simpler problem of determining of a conditional probability density function $p(y|x)$, where y is the variable whose value the data miner wants to predict.

Recent research has shown that combining different models can be effective in reducing the instability that results from predictions using a single model fit to a single set of data. A variety of model-combining techniques (with exotic names like bagging, boosting, and stacking) combine massive computational search methods with variance-reduction ideas from statistics; the result is relatively powerful automated schemes for building multivariate predictive models. As the data miner's multivariate toolbox expands, a significant part of the art of data mining is the practical intuition of the tools themselves.

4.1.1 Transaction Data

A common form of data in data mining in many business contexts is record of individuals conducting "transactions," examples include consumers purchasing groceries in a store (each record describes a market basket) and individuals surfing a Web site (each record describes the pages requested during a particular session). Employing the multivariate viewpoint, we can conceptually view this data as a very sparse $N \times d$ matrix of counts, where each of the N rows corresponds to an individual basket or session, each of the d columns corresponds to a particular item, and entry (i, j) is 1 if item j was purchased or requested as part of session i and is 0 otherwise.

Both N and d can be very large in practice. For example a large retail chain or e-commerce Web site might record on the order of $\mathrm{N} = 10^6$ baskets per week and have $\mathrm{d} = 10^5$ different items in its stores available for purchase or downloading. These numbers pose significant challenges from both the point of view of being computationally tractable and being amenable to traditional statistical modeling. For example, in a store with 10^5 different items and 10^6 baskets per week simply computing a pairwise correlation matrix requires $\mathrm{O}(\mathrm{N}d^2)$ time and $\mathrm{O}\ (d^2)$ memory, resulting in numbers of 10^{16} for time and 10^{10} for memory.

However, data miners routinely take advantage of the fact that transaction data is typically sparse; for example, since the average grocery basket might contain only 10 items, having only a few items in a basket means that only 10/50,000 or 0.02%, of the entries in the $\mathrm{N} \times d$ transaction matrix are nonzero. A substantial body of work in data mining research focuses on the idea of using subsets of items represented in each market basket, the so-called item sets I as "information nuggets" in large high-dimensional transaction data sets; an example of an item set is the combination of products *bread, wine, and cheese* in baskets in a grocery store. Several variants of efficient algorithms are available to find all frequent item sets from a sparse set of transaction data. Frequent items are item sets I such that $f_1 > T$, where the frequency f_1 is the number of rows in which all the items in I were purchased and T is preselected count threshold, such that $\mathrm{T} = 0.001 \times \mathrm{N}$.

Another strand of research takes a more statistical view of market basket data as a density estimation problem rather than a search problem. A methodology for finding statistically significant item sets, i.e., item sets I whose empirical frequency varies significantly from the frequency expected by a base line model (see illustrative visualizations at www.ics.uci.edu/~smyth/cacm02/). Determining statistical significance in this contest is a subtle problem. A Bayesian approach can uncover complex multi-item associations ignored by more traditional hypothesis testing techniques. It has been used by the U.S. Food and Drug Administration to search large postmarket surveillance databases for significant but relatively rare adverse reactions, – a good example of the marriage of computationally oriented data mining ideas with more traditional inferential theories from statistics. Increasingly, much of the research work in data mining occurs at this interface of computational and inferential approaches.

Frequent item sets can also be viewed as constraining on the set of all possible high-order probability models for the data. The technique of maximum entropy estimation provides theoretical framework for estimating joint and conditional probability distribution from the frequent item sets that can then be used for forecasting and answering queries. Unfortunately the maximum entropy approach scales exponentially the number of variables as to model (in both time and memory), limiting the technique in practice to relatively short queries or low-dimensional models.

Viewing transaction data as a sparse $N \times d$ matrix is a gross oversimplification of the true structure of the data in most applications. Typically, real transaction data has significant additional structure at various levels of detail; for example retail items are usually arranged in product hierarchies, and Web pages can be related to each other (through hyper links) or can be instances of a more general database scheme. Thus the columns of a data matrix such as products and Web pages can themselves have attributes (such as price and content) as well as implicit inter item relationships. Similarly the rows in a transaction data set can also have significant structure manifested by hourly, weakly, and seasonal temporal patterns. While some of these techniques explicitly exploit this structure, many open research challenges remain. Clear, however, is that techniques exploiting special structure in the data are likely to produce much more valuable insights and predictions than techniques that choose to ignore this structure.

4.1.2 Data Streams

The term *data stream* pertains to data arriving over time in a nearly continuous fashion. In such applications, the data is often available for mining only once, as it flows by. Some transaction data can be viewed this way, such as Web logs that continue to grow as browsing activities occur over time. In many of these applications, the data miner's interest often centers on the evaluation of user activity; instead of focusing on the relationships of items (columns), the data miner focuses on modeling individuals or objectives (rows).

Data streams have prompted several challenging research problems, including how to compute aggregate counts and summary statistics from such data. A related problem is that of incremental learning, whereby a global model assumed for the data stream, and the model is estimated incrementally as data arrives. A good example of the approach for online adaptation of classification tree models uses analytical probabilistic bounds to guide the degree to which the model needs to be updated over time.

Another aspect of data stream research involves scaling traditional ideas in statistical data analysis to massive, heterogeneous, nonstationary environments. Using large streams of call-record data in, for example, the telecommunications industry, statistical models (called signatures) can be built for individual telephone customers. Note that the collection of customers' signatures resulting from this methodology can be viewed in a database context as a statistical view of the underlying transaction data. Thus the derived data can help provide approximate (statistical sense) answers to queries. Numerous applications of these techniques tackle problems in forecasting, fraud detection, personalization, and change detection.

4.1.3 Graph and Text-Based data

The possibility of discovering patterns in large graphs also motivates data mining interest. We can think of representing relationships among objects. Such "data graphs" appear in multiple settings; for example, the Web can be viewed as a graph where nodes are pages and hypertext links are edges. Similarly, user browsing can be viewed as a bipartite graph where nodes are either users or Web pages, and the edges or pages users have visited. An inevitable question arising from a graphical view of the Web is: What kind of structure can be automatically discovered from its topology? Research suggests, for example, the graph structure underlying the Web is distinctly nonrandom and possesses many interesting proprieties.

Graphs can be represented by an adjacency matrix conveying the nodes as row/column labels and edges as cell entries. Such matrices are indeed large, and fortunately, sparse. That is all nodes in a real graph are not created equal; some have an extremely high degree, outgoing or incoming edges, while the vast majority barely have degree 1. If the nodes are sorted according their degree, the result is often "laws" of the form.

Degree α 1/rank^a, where a is often termed as the "degree" exponent.

The matrix representation of a graph suggests that many classical methods in linear algebra are likely to be extremely useful for analyzing the properties of graphs. Indeed, the singular value decomposition is the engine behind many powerful tools, including latent semantic indexing, the hubs and authorities algorithm, and Google's PageRank algorithm. Reflecting what can be discovered from connectivity information alone, page rank uses a recursive system of equations defining the importance of each in terms of the importance of the pages pointing to it. The importance (or page rank) of each page can then be determined by solving this set of linear equations. Once again, sparseness is important from a computational point of view. Since the number of out links per page is on average extremely low relative to the total number of pages on the Web, this system of linear equations is sparse, and an iterative algorithm typically converges on a solution rather quickly.

Hyperlink connectivity represents only one type of Web data. The navigation patterns of Web surfers, obtained from Web logs, also represent opportunities for prediction, clustering, personalization, and related techniques, often referred as to as "Web mining."

Web content, including text documents, is another vast and ready available data source for data mining. Considerable progress in text classification and clustering has been made by representing text as *term vectors* (a vector where component j is 1 if the document contains term j and 0 otherwise). Nevertheless, modeling documents at a richer semantic level is clearly worthwhile for, say, trying to identify the relations among sets of objects such as documents.

4.1.4 Scientific Data

While many data mining applications focus on commercial applications, such as credit scoring, fraud detection, and Web personalization, data mining as a tool for scientific discovery also motivates research interest. For example, data in the form of DNA and protein sequences, microarray-based gene expression measurements, and biological images have revolutionized the fields of biology and medicine. Biologists often spend more time looking at data than through a microscope. Since much biological research is data-rich and relatively theory-poor, data mining research promises significant opportunities for assisting biologists pursuing new scientific discoveries. Rather than viewing the field of computational biology as just applications, data miners find themselves confronted with interesting and fundamental research challenges from a number of perspectives, including modeling, interference, and algorithmic. For example, the discovery of "motifs" in DNA sequences is an example of a biologically motivated data mining problem. Motif discovery can involve prior knowledge as to the number of motifs (such as one per sequence) and their exact or expected lengths. However, little knowledge is typically available as to where the motifs occur in each sequence or what symbols they contain. Related research is driven by development of both score functions for patterns (to be interesting, pattern must differ from the back ground in a systematic way) and efficient search techniques to locate the likely candidates from the combinationally large space of possible patterns in a set of sequences. Ideas from a systematic search, heuristic search, and stochastic search have all proved useful in this context. Several publicly available algorithms are used in computational biology for motif discovery, each combining basic statistical models with massive search capabilities.

Scientists in other disciplines also have an increased awareness of the importance of data mining; for example, in astronomy, the Sloan Digital Sky Survey generates 5TB of data annually, leading to significant data engineering challenges (see www.sdss.org). An important research topic concerning such data is how to develop efficient algorithms to perform common data analysis tasks, including clustering and density estimations, on massive data sets. Multiresolution *kd*-trees, or a flexible data structure for indexing data in multiple dimensions, can provide orders of magnitude speed-ups in the density estimations of astronomical data using mixture models.

One research area conspicuous by its absence in data mining research though tremendously important in practically any scientific context, is human operator interaction for discovery; for example how can the algorithm designer and the scientist represent prior knowledge so the data mining algorithm does not just rediscover what is already known? And how can scientists "get inside" and "steer" the direction of a data mining algorithm? While some research on this topic has been pursued in a number of areas, including artificial intelligence and statistics, it has had relatively little effect on data mining in general.

4.2 Scaling Mining Algorithms to Large DataBases

Data mining is increasingly recognized as a key to analyzing, digesting, and understanding the flood of digital data collected by business, government, and scientific applications. Achieving this goal requires the scaling of mining algorithms to very large databases. Many classic mining algorithms require multiple database scans and/or random access to database records. Research today focuses on overcoming limitations imposed when it is costly or impossible to scan large databases multiple times or access records at random, while developing innovative algorithms and data structures to speed computation.

4.2.1 Prediction Methods

Predictive modeling is often a high-level goal of data mining in practice. After outlining the predictive modeling problem, we focus on two classes of algorithm: decision tree methods and support vector machines. Input into predictive modeling algorithms is a data set of training records. The goal is building a model that predicts a designated attribute value from the values of the other attributes (see Fig. 4.1). Many predictive models have been proposed in the literature, including neural networks and Bayesian methods.

Decision Tree Construction. Decision trees are especially attractive in data mining environments since human analysts readily comprehend the resulting models. Their construction does not require an analyst to provide input parameters: prior knowledge about the data is also not needed. A record can be associated with a unique leaf node by starting at the root and repeatedly choosing a child node based on the splitting criterion, which evaluates a condition on the input records at the current node.

Decision tree construction algorithms consist of two stages: tree building and pruning. In the former, most decision tree construction algorithms grow the tree top down in the following greedy way. Starting with the root node, the database is examined by "split selection method" for selecting the split condition at each node. The database is then partitioned and the procedure applied recursively. In the pruning stage, the tree constructed in the tree-building phase is pruned to control its size, and sophisticated pruning methods select the tree in a way that minimizes prediction errors.

The training database is accessed extensively while the tree is constructed; if the training database does not fit in memory, an efficient data-access method

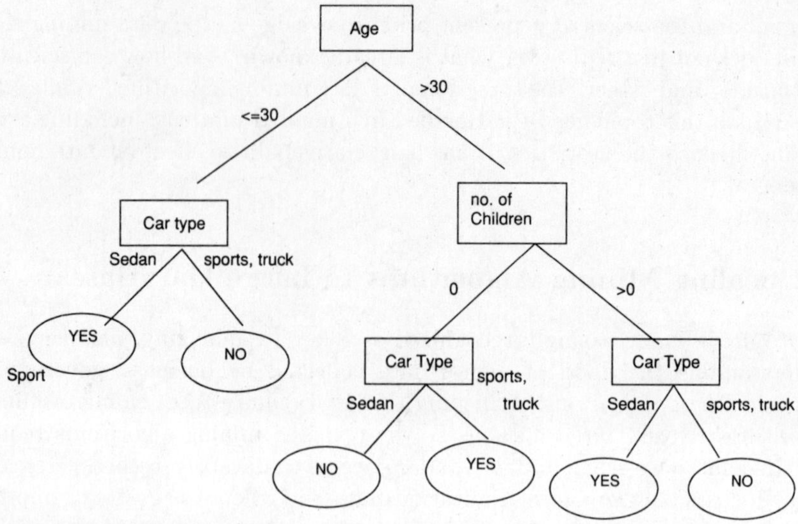

Fig. 4.1. Magazine subscription example classification tree

is needed to achieve scalability. Many scalable algorithms incorporate the observation that only a small set of sufficient statistics (such as aggregate measures, like counts) is necessary for applying popular split selection methods. The aggregated data is much smaller than the actual data. The statistics can be constructed in memory at each node in a single scan over the corresponding database partition that is, it satisfies the splitting criteria, leading to the node.

Although the sufficient statistics are often quite small, there are situations where the sufficient statistics are about as large as the complete data set. One way to deal with the size of the sufficient statistics is to observe that a large class of split selection methods searches over all possible split points and all attributes. The sufficient statistics at each step of the search are small – small enough to fit in memory. One way to utilize this observation is to create index structures over the training data set, thus permitting fast incremental computation of the sufficient statistics between adjacent steps of the search. For example, one class of algorithm vertically partitions the data set, sorts each partition by attribute value, then separately searches splitting criteria for each attribute by scanning the corresponding vertical partition.

Another way to deal with the size of the sufficient statistics is to split the problem into two phases. In the first, the algorithm scans the data set, constructing sufficient statistics in memory at a coarse granularity. Using the in-memory information, the algorithm prunes large parts of the search space of possible splitting criteria due to smoothness properties (such as bounds on the derivative) of the splitting criteria. In the second, the data set is scanned a second time, constructing exact sufficient statistics for only the parts of the search space that could not be eliminated in the first phase. Variations of

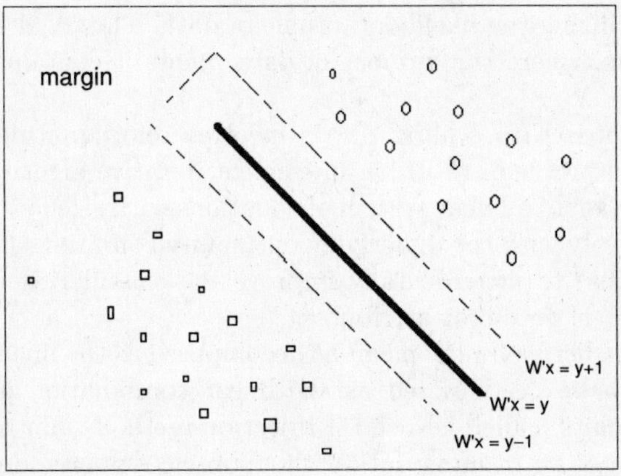

Fig. 4.2. SVM classification data points with dependent attribute = 0 are labeled with o's, data points with dependent attribute = 1 are labeled with s □

this idea eliminate only part of search space with high probability in the first phase; the algorithm then checks decisions in the second phase. Algorithms based on this two-phase approach appear to be the fastest known methods for classification tree construction.

Support Vector Machines. Support vector machines (SVMs) are powerful and popular approaches to predictive modeling with success in a number of applications, including hand written digit recognition, charmed quark detection, face detection, and text categorization.

SVMs fit in the context of classification where the attribute whose value is to be predicted (the dependent attribute) has two possible values: 0 or 1. SVM classification is performed by a surface in the space of predictor attributes separating points with dependent attribute = 0 from those with dependent attribute = 1. An optimal separating surface is computed by minimizing the margin of separation (see Fig. 4.2). The margin of separation is the distance between the boundary of the points with dependent attribute = 0 and the boundary of those with dependent attribute = 1. The margin is a measure of "safety" in separating the two sets of points - the larger the better. In the standard SVM formulation, computing the optimal separating surface requires solving a quadratic optimization problem.

The burden of solving the SVM optimization problem grows dramatically with the number of training records. To reduce this burden, a method called "chunking" iteratively updates the separator parameters over chunks of training cases; the size of a chunk is chosen, so it fits into main memory. To obtain optimal classification chunks often need to be revisited, implying, multiple passes over the data.

A data compression approach called *squashing* is also applicable to SVM training, where the training records are summarized by a smaller data set emulating the distribution of the original training records. The training records are clustered utilizing the likelihood profile of data. The SVM is then trained over the clusters, where the number of data points it contains weights each cluster.

Another approach to scaling SVMs involves reformulating the underlying optimization problem, resulting in efficient iterative algorithms. Methods require the solution of a linear system of equations with size $m+1$at each iteration or deal directly with the underlying optimality conditions (Karush-Kuhn-Tucker conditions) to incrementally improve the classifier at each iteration; m is the number of predictor attributes.

The SVM predictive function can be decomposed as the linear combination of functions can be decomposed as the linear combination of functions of training data points (called kernels). Projection methods aim to approximate the combination of all training data with a subset of points. Some projection methods use m randomly selected points on which to base the separating surface. Sparse, greedy matrix approximations try to determine the best m points to use.

4.2.2 Clustering

Clustering aims to partition a set records into several groups such that "similar" records are in the same group according to some similarity function, identifying similar subpopulations in the data. For example a cluster could be a group of customers with similar purchase histories, interactions, and other factors.

One scalability technique for clustering algorithms is to incrementally summarize dense regions of the data while scanning a data set. Since a cluster corresponds to a dense region, the records within this region can be summarized collectively through a summarized representation called a *cluster feature* (CF), such as the triple consisting of the number of points in the cluster, the cluster centroid, and the cluster radius. More sophisticated cluster features are also possible.

CFs are efficient for two reasons: they occupy less space than maintaining all objects in a cluster; and if designed properly, they are sufficient for calculating all intercluster and intracluster measurements for making clustering decisions. Moreover, these calculations can be performed faster than all objects in clusters. Distances between clusters, radii of clusters and CFs – and hence other properties of merged clusters can all be computed quickly from the CFs of individual clusters.

CFs have also been used to scale iterative clustering algorithms, such as the k-means algorithm and expectation-maximization algorithm. When scaling iterative clustering algorithms, the algorithm identifies sets of discardable points, sets of compressible points, and a set of main-memory points. A point is discardable if its membership in a cluster can already be ascertained with high confidence; only the CF of all discardable points in a cluster is retained while the actual points are discarded. A point is compressible if it is not discardable but belongs to a tight subcluster consisting of a set of points that always move between clusters simultaneously. The remaining records are designated as main-memory records, as they are neither discardable nor compressible. The iterative clustering algorithm then moves only the main-memory points and the CFs of compressible points between clusters until a criterion function is optimized.

Other research on scalable clustering focuses on training databases with large attribute sets. The search methods involve discovering the appropriate subspace of attributes in which the clusters are most likely to exist. These methods help analysis trying to understand the results, as they focus only on the attributes associated with a given cluster.

4.2.3 Association Rules

Association rules capture the set of significant correlation's present in a given data set. Given a set of transactions, where each transaction is a set of items, an association rule is an implication of the form X $\Rightarrow$ Y, where X and Y are sets of items. This rule has support s if $s\%$ of transactions include all the items in both X and Y, and confidence c if $c\%$ of transactions containing X also contain Y. for example, the rule "[carbonated beverages] and [crackers]$\Rightarrow$ [milk]" might hold in a supermarket database with 5% support and 70% confidence. The goal is to discover all association rules with support and confidence greater than the user-specified minimum support and minimum confidence, respectively. This formulation has been extended in many directions, including the incorporation of taxonomies, quantitative associations, and sequential patterns.

Algorithms for mining association rules usually have two distinct phases. First, they find all sets if items with minimum support (in other words, the frequent item sets). Since the data may consist of millions of transactions, and the algorithm may have to count millions of potentially frequent (candidate) item sets to identify the frequent ones, this phase can be computationally expensive. Next, rules can be generated directly from the frequent item sets, without having to go back to the data. The first step usually consumes most of the time: hence, research on scalability focuses here. Scalability techniques can be partitioned into two groups: those that reduce the number of candidates that need to be counted: and those that make the counting of candidates more efficient.

In the first group, identification of the antimonotonicity property that all subsets of a frequent item set must also be frequent proved to be a powerful pruning technique that dramatically reduces the number of item sets that need to be counted. Subsequent research has focused on variations of original problem. For instance, for data sets and support levels where the frequent item sets are very long, finding all frequent item sets is intractable, since a frequent item set with 'n' items has 2^n subsets. However, the set of maximal frequent item sets can still be found efficiently by looking again: once an item set is identified as frequent none of its subsets need to be counted. The key is to maximize the probability that item sets counted by looking ahead are actually frequent. A good heuristic is to bias candidate generation, so the most frequent items appear in the most candidate groups. The intuition behind this heuristic is that items with high frequency are more likely to be part of long frequent item sets.

In the second group of techniques, nested hash tables can be used to efficiently check those candidate item sets that are contained in a transaction. This is very effective when counting shorter candidate item sets less so for longer candidates. Techniques for longer item sets include database projection, where the set of candidate item sets is partitioned into groups such that the candidates in each groups share a set of common items. Then before counting each candidate group the algorithm first discards the transactions that do not include all the common items for the remaining transactions, discards the common items (since it knows they are present), as well as items not present in any of the candidates. This reduction in the number and size of the remaining transactions can yield substantial improvements in the speed of counting.

4.2.4 From Incremental Model Maintenance to Streaming Data

Real-life data is not static, evolving constantly through additions and deletions of records; in some applications, such as network monitoring, data arrives in such high-speed data streams it is practically impossible to store it for offline analysis. A framework we call "block evolution" illustrates these models of evolving data. The input data set to the data mining process is not static, as it is updated with a new block of tuples at regular time intervals, say, everyday at midnight (a "block" is a set of tuples added simultaneously to the database). For large blocks, this model captures the common practice in many data warehouse installations, whereby updates from operational databases are batched together and performed in the block update. For small blocks (in the extreme, a single record), the model captures streaming data.

For evolving data, two classes of problem are of particular interest: data mining model maintenance and change detection. Model maintenance aims to maintain a data mining model undergoing insertion and deletion of blocks of data. Change detection aims to qualify the difference in terms of data characteristics between two blocks of data.

Recent research has focused on mining evolving data. Incremental model maintenance has also received attention, since it is desirable to go from incremental updates of the data warehouse to only incremental updates of existing data mining models, especially in the light of very large size of data warehouses. Incremental model maintenance algorithms concentrate on computing exactly the same model as if the original model construction algorithm was run on the combined collection of old and new data. One scalability technique widespread in model maintenance algorithms is the localization of changes for inserting new records. For example, for density-based clustering algorithms, inserting a new record affects only clusters in the neighborhood of the record; thus efficient algorithms "localize" the change to the model without having to recompute the complete model. Another example involves decision tree construction, whereby the split criteria at the tree might change only within certain confidence intervals under insertion of records, assuming the underlying distribution of training records is static.

When working with high-speed data streams, algorithms have to be able to construct data mining models while looking at the relevant data items only once and in a fixed order (determined by the stream arrival pattern) with a limited amount of main memory. Data stream computation has given rise to several recent (theoretical and practical) studies of online and one-pass algorithms with limited memory requirements for data mining and related problems; examples include computating quantiles and order statistics; estimating frequency moments and join sizes; clustering and decision tree construction; estimating correlated aggregates and computing one dimensional, or single-attribute, histograms; and Haar wavelet decompositions. Scalability techniques include: sampling, summary statistics, sketches (small random projections with probable performance guarantees), and online compression of sufficient statistics.

Scalability in data mining is an active area of research, though many challenging questions remain, including the following:

- Can we mine patterns from huge data sets while preserving the privacy of individual records and the anonymity of the individuals who provided the data?
- What are suitable data mining models for high-speed data streams, and how can we construct them? and
- In light of the plethora of huge and growing sets of linked data available today, including in the Internet, newsgroups, and news stories, what type of knowledge can we mine from these resources, and can we design scalable algorithms for them?

4.3 Summary

The data mining innovations being implemented world wide often involve collaborations among domain experts, computer scientists, and statisticians. We expect these application-driven developments will continue to proliferate as data owners seek new end better ways to gain insight into their data. We can hope that more synergistic view of data mining, combining ideas from computer science and statistics will gradually emerge to provide a unifying theoretical framework for many of these efforts.

Large and growing databases are commonplace in business organizations, governments, and scientific applications. Prior to the invention of scaling techniques, sampling was the primary method for running conventional machine learning, statistical, and other analysis algorithms on these databases. However, this approach also involved having to determine sufficient sample size, as well as the validity of discovered patterns or models (the patterns may be an artifact of the sample). We view sampling as orthogonal and complementary to scaling techniques, since scaling techniques allow the use of much larger data sets.

The research effort on scaling data mining algorithms to large databases has now given analysts the ability to model and discover valid, interesting patterns over these large data sets. General scaling principles include use of summary statistics, data compression, pruning the search space, and incremental computation. We have described these principles in the practical context of mining algorithms, expecting they would be useful in other areas of computer science as well.

4.4 Review Questions

1. Write short note on transaction data used in the mining algorithms.
2. Briefly describe on the data streams applied in mining process.
3. Explain about graphical, text-level, and scientific data used in mining.
4. Define predictive modeling and explain various techniques involved in this modeling.
5. Discuss clustering and about the association rules used in scaling.

5

Emerging Trends and Applications of Data Mining

Objectives:

- Discuss the technology and enterprise – adoption trends associated with business analysis.
- Recent innovations and trends is business analytics – spanning organizations and technical processes, new technologies, user interface design, and system integration – are all driven by business value.
- We discuss the technology and enterprise – adoption trends associated with business analysis.
- In order to achieve the greatest possible business value, analytic solutions have to produce results that are actionable, along with ways to measure the effects of key changes.
- Data mining applications have proved highly effective in addressing many important business problems.
- We outline the related challenges in several emerging domains.
- Biology is in the midst of a revolution with an unprecedented flood of data forcing biologists to rethink their approach to scientific discovery.
- Data mining flourishes in telecommunications due to the availability of vast quantities of high quality data.
- The scope, coverage and volume of digital geographic data seers have grown rapidly in recent years due to the progress in data collection and data processing technologies.
- Earth science data mining consists of two main components: the modeling of ecological data and the design of efficient algorithms for finding spatiotemporal patterns.
- It is needed to develop highly automated, scalable, integrated, reliable data mining systems, and tools.
- It is important to promote information exchange among users, data analysts, system developers, and data mining researchers to facilitate the advances available from data mining research, application development, and technology transfer.

S. Sumathi and S.N. Sivanandam: *Emerging Trends and Applications of Data Mining*, Studies in Computational Intelligence (SCI) **29**, 165–183 (2006)
www.springerlink.com © Springer-Verlag Berlin Heidelberg 2006

Abstract. The field of business analytics has improved significantly over the past few years, giving business users insights, particularly from operational data stored in transactional system. An example is e-commerce data analysis, which has recently come to be viewed as a killer appropriate for the field of data mining. The data sets created by integrating click stream records generated by Web site activity with demographic and other behavioral data dwarf, in size and complexity, the largest data warehouses of a few years ago. The result is massive database requiring a mix of automated analysis techniques and human effort to give business users strategic insight about the activity on their sites, as well as about the characteristics of the sites visitors and customers. With many millions of click stream records generated every day, aggregated to customer focused records with hundreds of attributers, there is a clear need for automated techniques for finding patterns in the data. Here, we discuss the technology and enterprise – adoption trends associated with business analysis.

Also in this section, emerging scientific applications in data mining like biomedical engineering, Geospatial data, and Telecommunications are discussed.

5.1 Emerging Trends in Business Analytics

The key consumer is the business user, whose job possibly in merchandising, marketing, or sales, is not directly related to analytical tools to improve the results of some business process along one or more dimensions (such as profit and time to maker). Fortunately, data mining analytic applications, and business intelligence systems are now better integrated with transactional systems than they were once, creating a closed loop between operations and analysis that allows data to be analyzed and the results reflected quickly in business actions. The mined information today is deployed to a broader business audience taking advantage of business analytics in its everyday activities. Analytics are now routinely used in sales, marketing, supply chain optimization, and fraud detection.

5.1.1 Business Users

Even with these advances, business users, while expert in their particular areas, are still unlikely to be expert in data analysis and statistics. To make decisions based on the data collected by and about their organizations, they must either rely on data analysis to extract information from the data or employ analytic applications that blend data analysis technologies with task – specific knowledge. In the former, business users impart domain knowledge to the analyst then wait for the analyst to organize and analyze it and communicate back the results. These results typically raise further questions; hence several iterations are necessary before business users can actually act on the analysis. In the latter, analytic applications incorporate not only a variety of data-mining techniques but provide recommendations to business users as to how to best analyze the data and present the extracted information. Business users are expected to use it to improve performance along multiple metrics. Unfortunately, the gap between relevant analytics and users strategic business needs is significant. The gap is characterized by several challenges.

Cycle time. The time needed for the overall cycle of collecting analyzing, and acting on enterprise data must be reduced. While business constraints may impose limits on reducing the overall cycle time, business users want to be empowered and rely less on other people to help with these tasks.

Analytic time and expertise. Within the overall cycle, the time and analytic expertise necessary to analyze data must be reduced.

Business goals and metrics. Unrealistic expectations about data mining "magic" often lead to misguided efforts lacking clear goals and metrics.

Goals for data collection and transformations. Once metrics are identified, organizations must collect and transform the appropriate data. Data analysis is often an afterthought, limiting the possible value of any analysis.

Describing analysis results. Most analysis tools are designed for quantitative analysts, not the broader base of business users who need the output translated into language and visualizations appropriate for business needs.

Integrating data from multiple sources. The extract transform load (ETL) process is typically complex and its cost and difficulty are often underestimated.

5.1.2 The Driving Force

The emerging trends and innovations in business analytics embody approaches to these business challenges. Indeed, it is a very healthy sign for the field that regardless of the solution – process, technology, system integration, or user interface, business problems remain the driving force.

"Verticalization"

In order to reduce discovery cycle time, facilitate the definition and achievement of business goals, and deploy analysis results to a wider audience, developers of analytics solutions started vertical zing, their software, or customizing applications within specific industries. The first step toward verticalization is to incorporate task – specific knowledge; examples include: knowledge about how to analyze customer data to determine the effectiveness of a marketing campaign; knowledge of how to analyze click stream data generated by a web site to reduce shopping cart abandonment and improve ad effectiveness; knowledge about how an investment bank consolidates its general ledger and produces various types of forecasts; and how an insurance company analyzes data in order to provide an optimally priced policy to an exiting customer.

In the process of incorporating industry-specific knowledge, companies are also able to optimize the performance of their applications for specific industries. For example, a company that developed an analytic application for budgeting and forecasting targeted at the financial services industry determined that its online analytical processing, or OLAP, engine's execution speed could be optimized by limiting to nine the number of dimensions it had to handle, a number deemed sufficient for the particular application in that industry.

The use of industry – specific knowledge is not limited to the data mining components of analytics applications but also effects how the extracted information is accessed and presented. For example, organizations in the financial services, retail, manufacturing, utilities, and telecommunications industries increasingly want their field personnel to have access to business analytics information through wireless devices. Analytics application vendors are now developing technologies to automatically detect wireless devices and their form factors, automatically tailoring analysis results to fit the capabilities of a particular device. For example, if the information is to be displayed on a phone supporting the Wireless Access Protocol (implying small screen size), it may be necessary to automatically summarize text, abbreviate words, and limit the use of graphics by automatically selecting only the most relevant figures.

Comprehensible Models and Transformations

In light of the need to let the business users analyze the data and quickly gain insight, and aiming for the goal of reducing reliance on data mining experts, comprehensible models are more but popular than the opaque models. For example, in KDD-cup 2000 a data mining competition in which insight was important, the use of decision trees, generally accepted as relatively easy to understand, outnumbered the other methods more than two to one. Business users do not want to deal with advanced statistical concepts; they want straightforward visualizations and task relevant outputs.

Part of the Larger System

The needs of data analysis are being designed into systems, instead of being an afterthought, typically addressing the following area:

Data Collection: We cannot analyze what we do not collect, so collection of rich data is critical. For example, e-commerce systems can collect attributes ranging from the users local time, screen resolution (useful for determining the quality of images to send), and network bandwidth.

Generation (and Storage) of Unique Identifiers: In order to help merge information from several records and remove duplicate records, systems must generate unique keys to join data and store them. For example, all click stream records in the same session should store the session IDs so they can be joined later to session records stored in other tables.

Integration with Multiple Data Sources: Analysis is more effective when data is available from multiple sources. For example, in customer analytics, data should be merged from multiple touch points, including the web, call centers, physical store, wireless access, and ad campaigns (both direct and online). Behavioral data can be more powerful when overlaid with demographic and socioeconomic data from other sources.

Hardware Sizing: Analysis requires hardware capable of dealing with large amounts of data. Some organizations have traditionally underestimated the need for sophisticated IT infrastructure and the hardware needed to make timely analysis feasible.

In New Areas

During the past few years, recognition of the strategic value of business analytics has led to significant development in business application that analyzes the customer data. They have been used to reduce customer attrition, improve customer profitability, increase the value of e-commerce purchases, and increase the response of direct mail and email marketing campaigns.

This success has paved the way for the new applications; three are particularly promising; supply chain visibility, price optimization, and work force analysis. Organizations have automated portions of their supply chains; enabling collection of significant data about inventory, supply performance and logistics of materials, and finished goods. Newer applications analyze this data to provide insights about the performance of the suppliers and partners, material expenditures, accuracy of sales forecasts for controlling materials inventory, accuracy of production plans, and accuracy of plans for order delivery.

The wide adoption of customer relationship management or CRM, and supply chain management software has allowed enterprises to fully interface and integrate their demand and supply chains. Based on this integration they are better able to capture up to the minute data about demand for a particular product as well as data of similar granularity about the supply of corresponding data. Analyzing these two data streams, organizations optimize the price of a particular product along several dimensions so demand meets available supply; for example, the price of a product may be different through one channel (such as the web) than through another (such as retail store). Price optimization allows any type of organization to maximize the profit margins for each item sold while reducing inventory.

Once organizations are able to analyze data about their customers and their suppliers, they begin analyzing data about their employees too. A new generation of analytic applications allows enterprises to identify work force trends (such as attrition rates) and perform HR management tasks (such as compensation and benefits analyses). Companies whose cost or revenue model is dependent on hourly models (such as contact centers and systems integrators) use it to optimize staffing levels and skill requirements while minimizing the number of employees who are not able to bill.

Integration with action and measurement. With increased understanding of and experience in analytics, business users become more demanding and discerning, particularly when it comes to action based on insight and return on investment (ROI). Increasingly, analytics users ask two key questions: How do we turn discovered information into action? And how can we determine the effect of each action on the organization's business performance? Tales of data mining applications used to end with some novel analytical result; today, however, it is increasingly necessary that solutions use analytic results as a starting point toward the critical next steps of action and measurement. It is no longer enough for, say, cluster discovery algorithms to uncover interesting groups of customers. The successful analytic solution must make it easier for the user to grasp the significance of these clusters in the context of a business action plan; for example, these people have a propensity for purchasing new fashions. Achieving these results requires nontrivial transformations from the base statistical model. Traditionally, achieving these results necessitated the participation of expert human analysts.

Integration analytics with existing systems is a key to both action and measurement. For example, if the analytic application identifies customers likely to respond to promotion, but it takes a cadre of IT specialists to incorporate the relevant data into the advertising system to run the promotion, the results are unlikely to be used, as IT specialists are likely to be in short supply. Similarly, if promotion-targeting solutions enable distribution of catalogs with optimized promotions, but the order submission system is not closely tied back into the customer analytics, the resulting lag in ROI reports inhibits timely adjustment in the next catalog mailing. Efforts to integrate operations and analytic systems have seen major initiatives over the past five years including entire product lines whose value proposition is the optimization of the collect–analyze–act–measurement cycle.

Broadening the effects of analytics in the business process, solutions go beyond customer-centric applications to support sales, marketing, supply chain visibility, price optimization, and work force analysis. Finally, in order to achieve the greatest possible business value, analytic solutions have to produce results that are actionable, along with ways to measure the effects of key changes.

5.2 Business Applications of Data Mining

The traditional approach to data analysis for decision support has been to couple domain expertise with statistical modeling techniques to develop handcrafted solutions for specific problems. More recently, several trends have emerged to challenge this approach. One is the increasing availability of large volumes of high-dimensional data occupying database tables with millions of rows and thousands of columns. Another is the competitive demand for the rapid construction and deployment of data-driven analytics. Third is the need to give end users analysis results in a form they readily understand and assimilate, helping them gain the insights they need to make critical business decisions. Moreover, knowledge discovery in databases (KDD) techniques emphasizing scalable, reliable, fully automated, explanatory structures have shown that in data analysis, such structures supplement, and sometimes supplant, existing human-expert-intensive analytic techniques for improving decision quality.

Measurable Benefits

KDD applications deliver measurable benefits, including reduced cost of doing business, improved profitability, and enhanced quality of service. Industries, in which such benefits have been demonstrated, include insurance, direct-mail marketing, telecommunications, retail, and health care.

Risk management and targeted marketing. Insurance and direct mail are two industries that rely on data analysis to make profitable business decisions. For example, insurers must be able to accurately assess the risks posed by their policyholders to set insurance premiums at competitive levels. For example, overcharging low-risk policyholders would motivate them to seek premiums elsewhere; undercharging high-risk policyholders would attract more of them due to lower premiums. In either case, costs would increase and profits inevitably decrease. Effective data analysis leading to the creation of accurate predictive models is essential for addressing these issues.

In direct-mail targeted marketing, retailers must be able to identify subsets of the population likely to respond to promotions in order to offset mailing and printing costs. Profits are maximized by mailing only to those potential customers most likely to generate net income to a retailer in excess of the retailer's mailing and printing costs.

Businesses relying on data-driven analysis for decision making typically construct data warehouses to capture as much information as possible about their customers. Examples of such information include details of past customer transactions, as well as additional information obtained from third-party data providers, including credit scores and demographics, for targeted marketing purposes and motor vehicle records for insurance purposes.

To aid decision-making, analysts construct predictive models using warehouse data to predict the outcomes of variety of decision alternatives. For example, in order to set policy premiums, insurers need to predict the cost of claims filed by policyholders annually, what is known about each policyholder. In order to select customers for a targeted marketing campaign, retailers need to predict revenues or gross profits that would be generated the customers receiving the mailings.

A popular approach to predictive modeling used by many data analysts and applied statisticians involve partitioning the data records for a population of customers (or other entities) into segments, then developing separate predictive models for each segment. Typically, data is partitioned through a combination of domain knowledge, simple heuristics, and clustering algorithms. Predictive models are constructed once segments are identified. The drawback is that this sequential approach ignores the strong influence segmentation exerts on the predictive accuracies of the models within each segment. Good segmentations tend to be obtained only through trial and error by varying the segmentation criteria.

A better approach is to simultaneously perform segmentation and predictive modeling within each segment, optimizing the segmentation so as to maximize the overall predictive accuracy of the resulting model. This approach is built into the IBM Probabilistic Estimation (ProbE) data mining server, making it possible to automatically construct high-quality segmentation-based predictive models from very large high-dimensional data sets. A top down tree-based algorithm is used to construct the segmentations. A collection of other algorithms is incorporated for constructing segment models, including stepwise linear regression and stepwise native Bays algorithms for general-purpose modeling and a joint Poisson/log-normal algorithm for insurance risk modeling. A key feature of the ProbE server is it is readily extended to incorporate different types of predictive modeling algorithms for the segments, as well as different types of segmentation algorithms.

Two different client applications have been developed by IBM's Data Abstraction Research Group that utilizes the ProbE data-mining server. One is called IBM Advanced Targeted Marketing for Single Events (ATM-SE), built jointly with the Business Intelligence group at Fingerhut, Inc., a large US catalog and Internet retailer based in Minnetonka, MN, for constructing customer-profitability and response likelihood models for targeted marketing in the retail industry. The other is the IBM Underwriting Profitability Analysis (UPA) application, co-developed with Farmers Insurance Group, a large automobile and home insurance company based in Los Angles, for discovering homogeneous insurance risk groups.

Fingerhut's 2000 evaluation of the ATM-SE application for direct-mail response modeling demonstrated the application produced segmentation-based response models that either equaled or slightly outperformed Fingerhut's own proprietary models. This evaluation was significant because numerous vendors and consultants had previously failed to beat Fingerhut's in-house modeling capability. If these results ultimately hold across all of Fingerhut's models, the ATM-SE models would yield an estimated increase in annual profits directly to Fingerhut of more than $1 million. Moreover, the ProbE server achieved its result in a fully automated mode of operation, with no manual intervention.

The UPA application configures the ProbE server so as to use a point Poisson/log-normal statistical model within each segment to simultaneously model both frequency with which insurance claims are filed by policyholders and the amounts, or severities, of these claims for each segment. Using this class of segment model, the identified segments correspond to distinct risk groups whose loss characteristics, such as claim frequency and severity, are estimated in accordance with standard actuarial practices.

The Farmers Group's 1997 evaluation of the application's ability to analyze insurance policy and claims data for all policyholders in one state involved mining runs for 18 unique combinations of customers with specific insurance products and coverage, including explanatory variables. Each run generated about 40 rules, from which 43 combinations were identified as "nuggets," or previously unknown rules with significant potential value. Six nuggets were selected for a detailed benefits assessment which indicated that implementing just these six is a single state could potentially yield a net profit gained several million dollars in the first year alone.

Although insurers know that drivers of high-performance sports cars are more likely to have accidents than drivers of other types of cars, the UPA fours that if a sports car was not the only vehicle in the household, the accident rate is not much greater than that of a regular car. One estimate determined that "just letting Corvettes and Porsches into [the insurer's] 'preferred premium' plan could bring in an additional $4.5 million in premium revenue over the next two years without a significant rise in claims. Another publicly disclosed nugget related to experienced drivers, who tend to have relatively low case frequencies. However, the UPA also turned up the particular segment of experienced drivers who are unusually accident prone.

ProbE's segmentation-based predictive modeling capability permits construction of mining applications optimized for specific problems. Indications are that the ProbE server can consistently produce high quality on a fully automated basis without requiring costly manual adjustments of the models of the mining parameters by data mining experts. These characteristics will make data mining increasingly attractive to mid-size business as well as to be much lager counterparts.

Customer profiles and feature construction. An important ingredient for obtaining highly predictive models is to use highly predictive features, or attributes and variables, as model input. Although a database might contain sufficient information to construct highly predictive models, it is not always stored in a form that permits the data to be uses directly as input to a model. In such cases, the data must be transformed to obtain accurate models.

Transaction data is notorious for requiring transformation before it can be used for data mining applications. Such data consists of records of pairs of individuals and events. An example is a set of overall items purchased by a customer and grouped into a "market basket." Another is a set of Web pages requested from a Web site by a particular suffers and grouped by session. The ability of companies' worldwide to collect vast amounts of such transaction data has far outpaced their ability to analyze it. Transaction data is especially challenging from data mining respective due to several factors:

Massive numbers of records. Large retail chains generate millions of transaction per day.

Sparseness. A typical basket contains only a small fraction of the total possible number of items; individual customers may have few baskets, perhaps only one.

Homogeneity. Purchasing behavior varies consider holy depending on individual tastes and means, along with individual purchasing patterns over time.

These factors combine to make transaction data highly nontrivial when using traditional data analysis techniques. The related challenges, along with the transaction data itself, motivated much of the early work in data mining, including development of association-rule algorithms for efficiently searching for correlations among items in retail transaction data. While the association-rule approach can be useful for exploratory analysis of transaction data, such as discovering combinations of products purchased together, it is less well suited for predicting individual customer behavior.

A recently developed framework called predictive profiting handles transaction detain predictive modeling. A predictive profile is a model that predicts future purchasing behavior of an individual customer, given historical transaction data for both the individual and for the larger population of all of a particular company's customers. The predictive profiling approach is based on a flexible probabilistic model that works in the following way: Let y be a randomly chosen market, where y is a d-dimensional vector describing how many of each of the d items were purchased in the basket. The high-dimensional joint distribution on baskets $p(y)$, is approximated by a linear combination of K simpler models. Each of the K simpler models in effect captures" prototype combinations" of products in baskets.

In the first phase of modeling, the K prototype combinations are learned from the data through a well-known expectation-maximization procedure for statistical estimation. In the second phase, each customer is "mapped" onto the product space represented by the K prototypes, where the mapping is based on individual past purchasing patterns. The mapping effectively transforms the transaction data for each customer into a set of feature values that are then used to make prediction about future purchasing behavior. The transformation is not defined prior to data mining but is inferred by the mining algorithm.

This model is not designed to capture all aspects of individual customer behavior but to extract useful first-order characteristics in terms of how customers shop. Figure 5.1 outlines how the prototypes are used to support exploratory visualizations of the data, providing an interpretable description of the heterogeneity of customer behavior as reflected by different basket prototypes. The predictive profiling method was tested by the university of California, Irvine, research team on two large real-world transaction data sets collected over several years.

The data sets involved several million baskets and about 500,000 customers. Models were trained using historical data from the early years of each data set, then tested on data from later years, typically using from K = 20

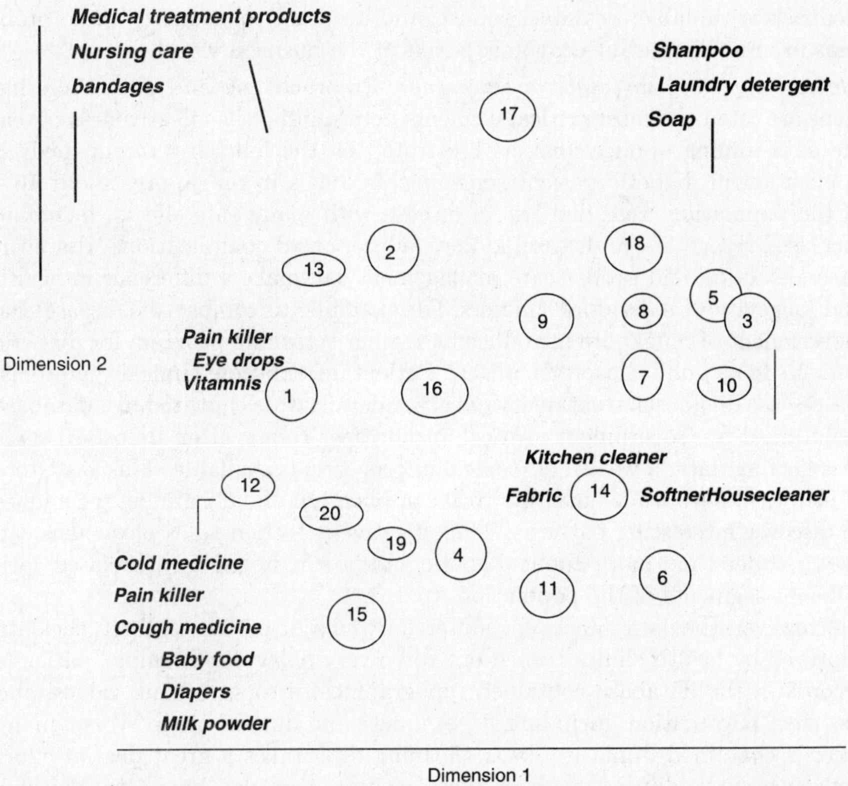

Fig. 5.1. A set of K = 20 prototypes represented here in a 2D space by using multidimensional scaling. The prototype baskets were learned from a set of about six million baskets from a chain of drugstores in Japan. The numbers in each circle refers to different prototypes; the area of each circle represents how likely a randomly chosen basket belongs to that prototype. The names of three items with the greatest lift (defined as p{item\prototype})/p(item)) are also displayed for some of the prototypes. Prototypes close together are also statistically close in the data.

to K = 100 prototypes. The models demonstrated systematic improvement in out of sample predictive performance compared to more standard alternatives. Empirically, the time taken to fit the models was found to scale linearly with both number of baskets and number of fitted prototypes K. The wall clock time to learn all the prototypes and customers profiles took only a few hours on a standard PC.

Such methods for handling transaction data are likely to prove useful across a variety of business applications, including customer segmentation, personalization, forecasting, and change detection, especially in e-commerce environments, where real time modeling of an individual customer and personalized feedback is valuable. Scalable, robust, and accurate solutions to these problems promise significant economic payoff in the business world.

Medical applications (diabetic screening). Preprocessing and postprocessing steps are often the most critical elements determining the effectiveness of real life data mining applications, as illustrated by the following recent medical application in diabetic patient screening. In 1990s in Singapore, about 10% of the population were diabetic, a disease with many side effects, including increased rescue of eye disease, kidney failure, other complications. However, early detection and proper care management can make a difference in health and longevity of individual sufferers. For example, to combat the disease, the Government of Singapore introduced a regular screening program for diabetic patients in its public hospitals in 1992. Patient information, clinical symptoms, eye disease diagnosis, treatment, and other details were captured in a database maintained by Government medical authorities. Today, after almost 10 years of collecting data, a wealth of medical information is available. This vast store of historical data leads naturally to the application of data mining techniques to discover interesting patterns. The objective is to find rules physicians can use to understand more about diabetic and how it might be associated with different segments of the population.

However, the data miners encountered two major problems. First, the data captured by health clinics turned out to be very noisy; for example, patient's records in the database contained typographical errors, missing values, and incorrect information, including street names and date of birth. Worse, many records contained duplicate data. Cleaning data takes a great deal of effort and time. In addition, many of these records were not in a form suitable for data mining; they had to be transformed to more meaningful attributes before mining could proceed. The second problem was that some state of the art association rule algorithms generate too many rules from the data, no matter how clean it is. Because physicians are busy seeing patients, they cannot take the time to shift through large numbers of rules. It was therefore important to present the discovered rules in some easy to understand form.

To overcome the problem noisy data, a data mining team at National University of Singapore developed a semiautomatic data cleaning system to reconcile database format differences by allowing physicians to specify the mapping between attributes in different format styles and/or different encoding schemes. Reconciling the format differences addressed the problem of identifying and removing duplicate records.

To resolve the problem of too many rules generated by the mining algorithms, the same team developed a user-oriented approach providing step-by-step exploration of both the data and the discovered patterns. Data visualization is used as an integral part of the process to give users a general view of findings. During rule mining, the mining algorithm employs a pruning method to remove insignificant rules. The final rules were also organized into general rules and exceptions to facilitate browsing and analysis. This rule mining approach to organizing mining results is useful to Singapore's medical authorities because it allows them to view the general patterns that are discovered as well as the detailed patterns. Because it is also a common strategy people employ in everyday learning, the mining results are easy to interpret.

The physicians confirmed that many of the rules and causal relationships the data mining algorithms discovered conformed to the trends they observed in their practices. However, they were surprised by many of the exceptions they did not know before. As a result of data mining they gained a much better understanding of how diabetes progresses over time and how various treatments affect its progress.

Data mining applications have proved highly effective in addressing many important business problems. We expect to see the continued construction and deployment of KDD applications for crucial business decision support systems. Exemplary applications employing data mining analytical techniques will require the KDD technical community to keep improving the underlying techniques for model building and model understanding. The emphasis in model building will be on developing mining techniques that are automated, scalable, and reliable. For domain understanding, the challenge is to keep developing sophisticated techniques that assist users in analyzing discovered knowledge easily and quickly.

5.3 Emerging Scientific Applications in Data Mining

Recent progress in scientific and engineering applications has accumulated huge volumes of high dimensional data, stream data, unstructured and semi-structured data, and spatial and temporal data. Highly scalable and sophisticated data mining tools for such applications represent one of the most active research frontiers in data mining. Here we outline the related challenges in several emerging domains.

5.3.1 Biomedical Engineering

Biology is in the midst of a revolution, with an unprecedented flood of data forcing biologists to rethink their approach to scientific discovery. First, large-scale data-collection techniques have emerged for a number of data sources limited by throughput, of the amount of available data. Examples of the data glut include: systematic genome DNA sequencing of organisms; high throughput determination of small molecule structures, as well as large macromolecular structures (such as proteins, RNA, and DNA); large scale measurements of molecular interactions; and simultaneous measurement expression level of all genes (thousands to tens of thousands) in a population of cells. Second, the availability of this data requires the biologists to create system for organizing, storing, and disseminating it, thus creating a need for standard terminologies and development of standards of interchange and annotations. Third, because of the apparent opportunities for automated learning from the data sets, a market for robust machine learning and data mining algorithms has emerged to take advantage of previous knowledge without being overly biased in the search for new knowledge. As a result, biology has changed from a field dominated by an attitude of "formulate hypothesis, conduct experiment, evaluate results" to more of a big science attitude of "collect and store data, mine for new hypothesis, confirm with data or supplemental experiment." the long-term significance of new data of molecular biology is that it can be combined with clinical medical data to achieve a higher resolution understanding of the causes for and treatment of disease. A major challenge for data mining in biomedicine is therefore the organization of the molecular data, cellular data, and clinical data in ways allowing them to integrate for the sake of knowledge extraction.

A major additional source of information is the published medical literature, increasingly available online in full text form or as useful (but unstructured) summaries of the main data and biomedical hypothesis.

5.3.2 Telecommunications

Data mining flourishes in telecommunications due to the availability of vast quantities of high quality data. A significant stream of it consists of call records collected at network switches used primarily for billing; it enables data mining applications in toll-fraud detection and consumer marketing.

Perhaps the best-known marketing application of data mining, albeit via unconfirmed anecdote, concerns MCI's "Friends & Family" promotion launched in the domestic US market in 1991. As the anecdote goes, market researchers observed relatively small subgraphs in this long-distance phone company's large call-graph of network activity, revealing the promising strategy of adding entire calling circles to the company's subscriber base, rather than the traditional and costly approach of seeking individual customers one at a rime. Indeed, MCI increased its domestic US market share in the succeeding years by exploiting the "viral" capabilities of calling circles; one infected member causes others to become infected. Interestingly, the plan was abandoned some years later (not available since 1997), possibly because the virus had run its course but more likely due to other competitive forces.

In toll-fraud detection, data mining has been instrumental in completely changing the landscape for how anomalous behaviors are detected. Nearly all fraud detection systems in the telecommunications industry 10 years ago were based on global threshold models; they can be expressed as rule sets of the form "If a customer makes more than X calls per hour to country Y; then apply treatment Z." The placeholders X, Y, and Z are parameters of these rule sets applied to all customers.

Given the range of telecommunication customers, blanket application of these rules produces many false positives. Data mining methods for customized monitoring of land and mobile phone lines were subsequently developed by leading service providers, including AT&T, MCI, and Verizon, whereby each customer's historic calling patterns are used as a baseline against which all new calls are compared. So, for customers routinely calling country Y more than X times a day, such alerts would be suppressed, but if they ventured to call a different country Y', an alert might be generated.

Methods of this type were presumably in place for the credit card industry a few years before emerging in telecom. But the size of the transaction streams is far greater in telecom, necessitating new approaches to the problem.

It is expected that algorithms based on call-graph analysis and customized monitoring will become more prevalent in both toll-fraud detection and marketing of telecommunications services. The emphasis on so-called "relational data" is an emerging area for data mining research, and telecom provides relational data of unprecedented size and scope.

These applications are enabled by data from the billing stream. As the industry transforms itself from a circuit-switched to a packet-switched paradigm, the data mining community could well experience a dearth of data, since billing is likely to be increasingly insensitive to usage. Moreover, the number of records that could potentially be recorded in a packet-switched network (such as packet headers) is orders of magnitude greater than today's circuit-switched networks are. Thus, unless a compelling business need is identified, the cost of collecting, transmitting, parsing, and storing this data will be too great for the industry to willingly accept. A dearth of data could well spell the end to future significant data mining innovations in telecommunications.

However, this view might yet be altered by the following scenarios:

New network architectures. New-generation network infrastructure will have to adapt to changes in demand yet be more reliable and secure; for example, capacity in mobile networks will have to be assigned dynamically, necessitating development of new data mining techniques for understanding and predicting network load. Similarly, network-intrusion detection will continue to be important to data mining, helping ensure that artificially induced traffic cannot cripple a network.

Mobility and microbilling. In Europe and Japan, merchants use mobile handsets for selling (and billing) a variety of consumer goods, including vending machine purchases and parking tees. Since these consumer activities correspond to "billable events," data will certainly be collected and maintained for such services.

Mobile services. Ease of use is crucial for enticing customers to adopt new mobile services. Data mining will probably play a major role in the design of adaptive solutions enabling users to obtain useful information with relatively few keystrokes.

Homeland security. Federal regulations require US telecommunications companies to maintain call records for two years. With the recent emphasis on homeland security, along with the role telecom data can play identifying and tracking terrorist cells and activities, data will continue to be collected and maintained, even though the records may not be used for billing.

5.3.3 Geospatial Data

The scope, coverage and volume of digital geographic data seers have grown rapidly in recent years due to the progress in data collection and data processing technologies. These data sets include digital data of all sorts, created, processed, and disseminated by government- and private-sector agencies on land use and socioeconomic infrastructure; vast amounts of georeferenced digital imagery and video data acquired through high-resolution remote sensing systems and other monitoring devices; geographic and spatiotemporal data collected by global positioning systems, as well as other position-aware devices, including cellular phones, in-vehicle navigation systems, and wireless Internet clients; and digital geographic data repositories on the Web. Moreover, information infrastructure initiatives, including the US National Spatial Data Infrastructure, facilitate data sharing and interoperability, making enormous amounts of space-related data sharable and analyzable worldwide.

The increasing volume and diversity of digital geographic data easily overwhelm traditional spatial analysis techniques that handle only limited and homogeneous data sets with high-computational burden. To discover new and unexpected patterns, trends, and relationships embedded within large and diverse geographic data sets, several recent studies of geospatial data mining have developed a number of sophisticated and scalable spatial clustering algorithms, outlier analysis techniques, spatial classification and association analysis methods, and spatial data-cleaning and integration tools.

Nevertheless, considering the challenges posed by the already enormous and increasing amount of spatial data, geospatial data mining is in its infancy. Lots of research needs to be done, especially concerning the following pressing issues.

Developing and supporting geographic data warehouses. Although data warehouses are central to the knowledge discovery process, no true geospatial data warehouse exists today. Creating one requires solutions to problems in geographic and temporal data compatibility, including reconciling semantics, referencing systems, geometry, accuracy, and precision. Creating a warehouse might also need to solve the problems of efficient computation of sophisticated spatial aggregations, as well as how to handle spatial-related data streams. However, spatial data warehouses are likely to eventually play an essential role in geospatial information exchanges and data mining, so it is critical that we develop and support such an infrastructure today.

Exploring and mining richer geographic data types. Geographic data sets are moving beyond the well-structured vector and raster formats to include semi-structured and unstructured data, especially georeferenced stream data and multimedia data. Techniques have to be developed to handle spatiotemporal data, robust geographic concept hierarchies and granularities, and sophisticated geographic relationships, including non-Euclidean distances, direction, connectivity, attributed geographic space (such as terrain), and constrained interaction structures (such as networks).

Reaching a broader user community. Geospatial data mining needs to go beyond researchers to also deliver its benefits to general users. This requires high-level user interfaces and visualization tools that aid diverse users in geospatial data mining. Moreover, these interfaces and tools have to be integrated with existing geographical information systems and database systems to guide users searching for geographic knowledge, interpreting and visualizing discovered knowledge, and using the discovered geographic knowledge in their decision-making.

5.3.4 Climate Data and the Earth's Ecosystems

The large amount of climate data acquired through NASA's Earth observation satellites, terrestrial observations, and ecosystem models offers an unprecedented opportunity for predicting and preventing future ecological problems by managing the ecology and health of the planet. Such data consists of a sequence of global snapshots of the Earth, typically available at monthly intervals, including various atmospheric, land, and ocean variables (such as sea surface temperature, precipitation, and net primary production, or the net photosynthetic accumulation of carbon by plants). Due to the nature and scale of this data, data mining techniques can play a major role in the automatic extraction and analysis of interesting patterns, thus complementing existing statistical techniques.

Earth science data mining consists of two main components: the modeling of ecological data and the design of efficient algorithms for finding spatiotemporal patterns. An important goal is the discovery of teleconnection patterns, or recurring and persistent climate patterns spanning vast geographical regions. They manifest themselves as spatiotemporal relationships among ecological variables observed at various locations on the Earth and are critical for understanding how the ecosystem's various elements interact with one another. Clustering techniques, which divide data into meaningful or useful groups, help automate the discovery of teleconnections. Specifically, clustering identifies regions of the Earth whose constituent points have similar short- and long-term climate characteristics. By analyzing correlations among climate variables across these regions, it is possible to rediscover existing patterns (such as the El Nino periodic ocean–atmosphere disruption in the tropical Pacific Ocean), as well as new, previously unrecognized teleconnections. An alternative approach is to convert the time series into sequences of events, then apply existing association-rule techniques to discover interesting patterns in the sequences.

The difficulty of mining Earth science data is illustrated by the following examples of issues arising during the various stages of data mining analysis:

Preprocessing. It is often beneficial to aggregate data into a smaller number of points, easing computational requirements and (typically) reducing the amount of noise. However, it can be difficult for researchers to choose the proper level of aggregation, since too much limits the patterns that can be directed, while too little results in noisy data in which only the strongest patterns can be discovered. Event definition is another necessary but ill-defined task. In the spatial domain, the problem is too many events, and in the temporal domain, events are rare; for example, El Nino events occur only every four to seven years. Yet another concern is integrating data from heterogeneous sources (such as data covering different time periods). Earlier data may come from manual, earth-based observations, while later data may originate from satellites.

Similarity of time series. The "proper" measure of similarity between time series is fraught with challenges. Linear correlation works well with standard clustering algorithms and lends itself to statistical tests. Nevertheless, alternate measures of time series similarity would be beneficial if they allowed the detection of patterns that could not be detected via linear correlation, and might, for example, be based on either dynamic time warping or cepstral coefficients representing the frequency spectrum of a time series. An "ideal" measure of similarity would account for time lag and the fact that only extreme events are usually correlated.

Identifying interesting patterns. Once patterns are discovered, it is difficult to distinguish the spurious ones from the significant ones. For example, given 40,000 time series recording the sea surface temperature at various points on the ocean's surface and 60,000 time series representing precipitation on land, some of these series might, just by chance, have strong correlations. While a number of statistical approaches estimate significance levels, it is not possible to apply such approaches directly due to spatial and temporal autocorrelation. When genuine patterns are identified, domain-specific knowledge is inevitably still needed to identify patterns of interest to Earth scientists.

5.4 Summary

Recent innovations and trends in business analytics – spanning organizations and technical processes, new technologies, user interface design, and system integration – are all driven by business value. Business value is measured in terms of progress toward bridging the gap between the needs of the business user and the accessibility and usability of analytic tools. In order to make analytics more relevant and tangible for business users, solutions increasingly focus on specific vertical applications tailoring results and interfaces for these users, yielding human-level insight. For ease of use, simpler and more effective deployment, and optimal value, analytics are also increasingly embedded in larger systems. Consequently, data collection, storage, processing, and other issues specific to analytics are incorporated into overall system design. Thus the emerging applications involve great data management challenges.

These emerging applications involve great data-management challenges that also represent new opportunities for data mining research, Methods for mining biomedical, telecommunication, geospatial, and climate data are under active development. However, in light of the tremendous amount of fast-growing and sophisticated types of data and comprehensive data analysis tasks, data mining technology may be only in its infancy, as the technology is still far from adequate for handling the large-scale and complex emerging application problems. Research is needed to develop highly automated, scalable, integrated, reliable data mining systems, and tools, Moreover, it is important to promote information exchange among users, data analysts, system developers, and data mining researchers to facilitate the advances available from data mining research, application development, and technology transfer.

5.5 Review Questions

1. Define cycle time
2. Explain in detail on the driving forces used in the business strategies
3. Justify – KDD applications deliver measurable benefits
4. State in detail on the scientific applications in data mining

6

Data Mining Trends and Knowledge Discovery

Objectives:

- The rapidly emerging field of knowledge discovery in databases (KDD) has grown significantly in the past few years.
- Modern database technology enables economical storage of these large streams of data.
- We do not yet have the technology to help us analyze, understand, or even visualize this stored data.
- These challenges are central to KDD and need urgent attention.
- KDD is an interdisciplinary field that brings together researchers and parishioners from a wide variety of fields.
- The major related fields include statistics, machine learning, artificial intelligence, and reasoning with uncertainty, databases, knowledge acquisition, pattern recognition, information retrieval, visualization, intelligent agents for distributed and multimedia environments, digital libraries, and management information systems.

Abstract. The rapidly emerging field of knowledge discovery in databases (KDD) has grown significantly in the past few years. A mix of daunting practical needs and strong research interest drives this growth. The technology for computing and storage has enabled people to collect and store information from a wide range of sources at rates that were considered unimaginable only a few years ago. Although modern database technology enables economical storage of these large streams of data, we do not yet have the technology to help us analyze, understand, or even visualize this stored data.

Examples of this phenomenon around in a wide spectrum of fields, finance, banking, retail sales, manufacturing, monitoring and diagnosis (be it of humans or machines), health care, marketing, and science data acquisition, among others. In science, modern instruments can easily measure and collect terabytes (10^{12} bytes) of data. For example, NASA's Earth Observing System is expected to return data at rates of several gigabytes per hour by the end of this century. Quite appropriately, the problem of how to put the torrent of data to use in analysis is often called "drinking

S. Sumathi and S.N. Sivanandam: *Data Mining Trends and Knowledge Discovery*, Studies in Computational Intelligence (SCI) **29**, 185–194 (2006)
www.springerlink.com © Springer-Verlag Berlin Heidelberg 2006

from the fire hose." What we mean by analysis is not well defined because it is highly context and goal dependent. However, it typically transcends by far anything achievable via simple queries, simple string matching, or mechanisms for displaying the data. The data mining trends and knowledge discovery are the subjects of this section.

Prolific sources of data are not restricted to esoteric endeavors involving spacecraft or sophisticated scientific instruments. Imagine a database receiving transactions from common daily activities such as supermarket or department store checkout-register sales, or credit card charges. On the other hand, think of the information reaching one's home television set as a stream of signals that, to be properly managed, need to be cataloged and indexed, and perhaps searched for interesting content at a higher level-channels, programs, genre, or mood, for example. The explosion in the number of resources available on the global computer network – the World Wide Web – is another challenge for indexing and searching through a continually changing and growing "database."

6.1 Getting a Handle on the Problem

Why are today's database and automated match and retrieval technologies not adequate for addressing the analysis needs? The answer lies in the fact that the patterns to be searched for and the models to be extracted are typically subtle and require significant specific domain knowledge. For example, consider a credit card company wishing to analyze its recent transactions to detect fraudulent use or to use the individual history of customers to decide online whether an incoming new charge is likely to be from an unauthorized user. This is clearly not an easy classification problem to solve.

One can imagine constructing a set of selection filters that trigger a set of queries to check if a particular customer has made similar purchases in the past, or if the amount or the purchase location is unusual, for example. However, such a mechanism must account for changing tastes, shifting trends, and perhaps travel or change of residence. Such a problem is inherently probabilistic and would require reasoning with uncertainty scheme to properly handle the trade-off between disallowing a charge and risking a false alarm, which might result in the loss of a sale (or even a customer).

In the past, we could rely on human analyst to perform the necessary analysis. Essentially, this meant transforming the problem into one of the simply retrieving data, displaying it to an analyst, and relaying an expert knowledge to reach a decision. However, with large databases, a simple query can easily return hundreds or thousands (or even more) matches. Presenting the data, letting the analyst digests it, and enabling a quick decision becomes infeasible. Data visualization techniques can significantly assist this process, but ultimately the reliance on the human in the loop becomes a major bottleneck. (Visualization works only for small sets and a small number of variables. Hence, the problem becomes one of finding the appropriate transformations and reductions – typically just as difficult as the original problem.)

Finally, there are situations where one would like to search for patterns that humans are not well suited to find. Typically, this involves statistical modeling followed by "outlier" detection, pattern recognition over large data sets, classification, or clustering. (Outliers are data points that do not fit within a hypothesis's probabilistic mode and hence are likely the result of interference from another process.) Most database management systems (DBMSs) do not allow the type of access and data manipulation that these tasks require; there are also serious computational and theoretical problems attached to performing data modeling in high dimensional spaces with large amount of data.

These challenges are central to KDD and need urgent attention. Without heavily emphasizing KDD development and research, we run the risk of forfeiting the value of most of the data that we collect and store. We would eventually drown in an ocean of massive (but valuable) data sets that are rendered useless because we cannot distil the essence from the bulk. To draw the data mining analogy: the precious nuggets of knowledge need to be extracted and the massive raw material needs to be managed appropriately (and preferably recycled effectively). Before proceeding further, let us define what we mean by KDD and data mining.

6.2 KDD and Data Mining: Background

We use the term KDD to denote the overall process of extracting high-level knowledge from low-level data. Others might use the term's *data mining and KDD* interchangeably. The multitude of names used for KDD includes data or information harvesting, data archeology, functional dependency analysis, knowledge extraction, and data pattern analysis. Historically, in statistics especially, the term *data mining or fishing* refers to sloppy exploratory data analysis with no a priori hypotheses to verify.

A simple definition. A simple high-level definition of KDD is as follows:

Knowledge discovery in databases in the nontrivial process of identifying valid, novel, potentially useful, and ultimately understandable patterns in data. Given the scope of this short section, we will not go into the definitions of each term in this high-level statement. Note, however, that the term knowledge is (and has a long history of being) difficult to define in the abstract. We adopt the view that knowledge is in the eye of the beholder, so one person's knowledge could easily be another's junk. We define knowledge in domain-dependent terms relating strongly to measures of utility, validity, novelty, and understandability. The term *patterns* in this definition loosely denote either models or patterns. In general, it designates some abstract representation of a *subset of the data.* A significant term in the definition is *process*, which indicates that knowledge discovery often involves experimentation, iteration, user interaction, and many design decision and customizations. Extracting knowledge from data can easily turn into a complicated and sometimes arduous process. But the payoffs for success can be dramatic and rewarding, sometimes enabling people and organizations to achieve tasks that would not otherwise be possible.

We adopt the convention that data mining refers to the act of extracting patterns or models from data (be it automated or human-assisted). However, many steps precede the data-mining step: retrieving the data from a large warehouse (or some other source); selecting the appropriate subset to work with; deciding on the appropriate sampling strategy; cleaning the data and dealing with missing fields; and applying the appropriate transformation, dimensionality reduction, and projections. The data-mining step then fits models to, or extracts patterns from, the preprocessed data. However, to decide whether this extracted information does represent knowledge, one needs to evaluate this information, perhaps visualize it, and finally consolidate it with existing (and possibly contradictory) knowledge. Obviously, these steps are all on the critical path from data to knowledge. Furthermore, any one-step can result in change in the preceding or succeeding steps, often requiring starting from scratch with new choices and settings. Hence, in the definition we adopt, data mining is just a step in the overall KDD process.

Figure 6.1 outlines the KDD process; it is perhaps deceptive because it gives the impression that the steps are well defined. In fact, the interactions between the choices of techniques used in the various steps, the parameters used for those techniques, and the choice of problem representation are extremely complex. Small changes in one part can dramatically affect the rest, and consequently can make the difference between success and failure of a KDD enterprise.

The KDD Process: Table 6.1 expands on the steps outlined in Fig. 2.1. A few items in Table 2.1 warrant further comment. Step 4 is critical and can be quite involved. Indeed, in many cases, some sophisticated searching and cataloging problem must be solved before the actual subsequent analysis is performed. This transformation could require solving a significant problem in its own

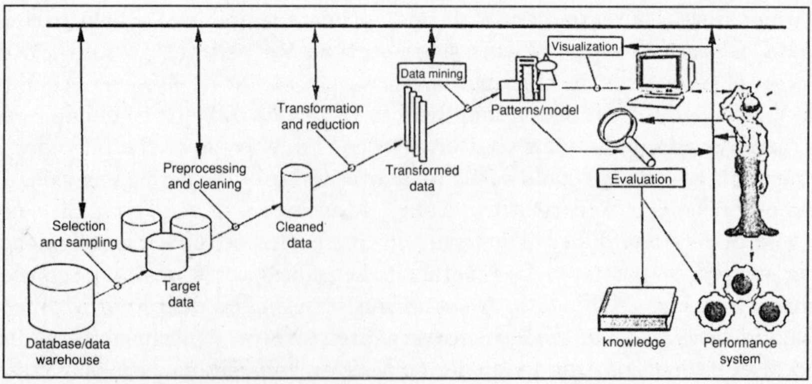

I .An overview of the KDD process.2 (For simplicity, the illustration omils arrows indicuting the multilude of potential loops and iterations.)

Fig. 6.1. An overview of the KDD process. (For simplicity, the illustration omits arrows indicating the multitude of potential loops and iterations.)

Table 6.1. Steps involved in the KDD Process

Step	Explanation
1. Developing an understanding of the application domain, the relevant prior knowledge, and the goals of the end user	With today's technology, this step requires a fair bit of reliance on the user/analyst. Factors to consider include: • What are the bottlenecks in the domain? What is worth automating and what is best left for processing by humans. • What are the goals? What performance criteria are important? • Will the final product of the process be used for classification, visualization, exploration, summarization, or something else? • Is understandability an issue? What is the trade-off between simplicity and accuracy of the extracted knowledge? Is a black box model appropriate for the performance element of the system?
2. Creating a target data set, selecting a data set, or focusing on a subset of variables or data samples, on which discovery is to be performed.	This involves considerations of homogeneity of data, any dynamics and change over time, sampling strategy (such as uniform random versus stratified), sufficiency of sample, degrees of freedom, and so forth.
3. Data cleaning and preprocessing.	Involved here are basic operations such as the removal of noise or "outliers," if appropriate; collecting the necessary information to model or accounting for noise; deciding on strategies for handling missing data fields; accounting for time sequence information, known changes, and appropriate normalization; and so forth.
4. Data reduction and transformation	This involves finding useful features to represent the data, depending on the goal of the task' using dimensionally reduction or transformation methods to reduce the effective number of variables under consideration or to find invariant representations for the data; and projecting the data onto spaces in which a solution is likely to be easier to find.
5. Choosing the data mining task	This involves deciding whether the goal of the KDD process is classification, regression, clustering, summarization, dependency modeling, or change and deviation detection.
6. Choosing data mining algorithms	Here we select the methods to be used fro searching for patterns in or fitting models to the data. The choice of which models and parameters mat be appropriate is often critical. In addition, the data mining method must be compatible with the goals; the end user may be more interested in understanding the model than its predictive capabilities.

7. Data mining	This involves searching for patterns of interest in a particular form or a set of such representations: classification rules or trees, regression, clustering, and so on. The user can significantly aid the data mining method by correctly performing the preceding steps.
8. Evaluating output of step 7.	Here we decide what is to be deemed knowledge, which can be fairly a difficult task. Achieving acceptable results may involve using options (possibly in combination): • Defining an automated scheme using measures of "interestingness" and others to filter knowledge from other outputs. Such measures might be statistical measures, goodness of fir, or simplicity, among others. • Relying on visualization techniques to help the analyst decide the utility of extracted knowledge or reach conclusions about the underlying data/phenomena. • Relying entirely on the user to sift through derived patterns in hope of coming across items of interest. The outcome of this step might result in changes to any of the preceding steps and a restart of the entire process.
9. Constructing discovered knowledge: incorporating this knowledge into the performance system, or simply documenting it and reporting it to users.	This also includes checking for and resolving potential conflicts with previously believed (or extracted) knowledge.

right. In classical pattern-recognition work, this is called *feature extraction* problem. In general, its solution requires a good bit of domain knowledge and strong intuition about the problem. It typically makes the difference between success and failure of the data mining (step 7).

However not to discourage the reader, feature definition and extraction in many applications is not terribly difficult (especially for a motivated domain expert who is involved in the process). Humans find it easier to define features than to solve the data-mining problem. For instant, an expert can observe a set of low-level variables and reach initiative decision. For example, the low-level may consist of a stream of readings of voltages, currents, capacitances, loads, and so forth from a power plant; a set of pixels in multispectral images from a remote sensing instrument; or a set of transactions for a given group of bank accounts. However, the expert might not at all be capable of elucidating the reasons for reaching some decision about the state of the system being observed. This is typically a reasonable setting to use classification (supervised learning) techniques to derive classifiers from examples (the data) directly. Hence, the expert presents the system with training data consisting of classified examples.

For a nonexpert (especially a machine), using the raw observed data to classify events is likely to result in failure: knowledge of time sequence, of properties, of instruments, of noise, of what is an important quantity, and so forth is simply a prerequisite. Experts can be asked to define features from the lower-level data. In effect, feature definition by experts lets them decompose the problem into small parts and encode significant prior knowledge implicitly in their choice of representation. This can easily result in a large number of features. Typically, the expert would not know how to use these features to solve the classification (discrimination) or modeling problem. Data mining techniques provide a way to get the solution in this feature space.

In step 8, the reliance on visualization is simply a work-around of the fact that we find it difficult to emulate human intuition and decision-making on a machine. The idea is to transform the derived knowledge into a format that is easy for humans to digest (such as images or graphs) and then rely on the speed and capability of the highly evolved human visual system to spot what is interesting. Of course, this only works in low dimensional spaces, so the choice of what to show the user to facilitate the discovery is still critical and typically not an easy problem to circumvent.

6.3 Related Fields

By definition, KDD is an interdisciplinary field that brings together researchers and parishioners from a wide variety of fields. The major related fields include statistics, machine learning, artificial intelligence and reasoning with uncertainty, databases, knowledge acquisition, pattern recognition, information retrieval, visualization, intelligent agents for distributed and multimedia environments, digital libraries, and management information systems. The remainder of this section briefly outlines how some of these relate to the various parts of the KDD process. We focus on the main fields and hope to clarify to the reader, the role of each of the fields and how they fit together naturally when unified under the goals and applications of the overall KDD process. A detailed or comprehensive coverage of how they relate to the KDD process would be too lengthy and not very useful because ultimately one can find relations to every step form each of the fields.

Statistics. Statistics plays an important role primarily in data selection and sampling, data mining, and evaluation of extracted knowledge steps. Historically, most statistics work has focused on evaluation of model fit to data land on hypothesis testing. These are clearly relevant to evaluation the results of data mining to filter the good from the bad, as well as within the data mining step itself in searching for, parametrizing, and fitting models to data. On the front end, sampling schemes play an important role in selecting which data to feed to the data-mining step. For the data-cleaning step, statistics offers techniques for detecting "outliers," smoothing data when necessary, and estimating noise parameters. To a lesser degree, estimation techniques for dealing with missing data are also available. Finally, for exploratory data analysis, some techniques in clustering and design of experiments come into play. However, the focus of research has dealt primarily with small data sets and addressing small sample problems.

On the limitations front, work in statistics has focused mostly on theoretical aspects of techniques and models. Thus, most work focuses on linear models, additive Gaussian noise models, parameter estimation, and parametric methods for a restricted class of models. Search has received little emphasis, with emphasis on closed-form analytical solutions whenever possible. While the latter is very desirable both computationally and theoretically, in many practical situations a user might not have the necessary background statistics knowledge (which can often be substantial) to appropriately use and apply the methods. Furthermore, the typical require an a priori model and significant domain knowledge of the data as well as of the underlying mathematics for proper use and interpretation. In addition, issues having to do with interfaces to databases, dealing with massive data sets, and techniques for efficient data management have only recently begun to receive attention in statistics. John Elder and Darryl Pregibon in 1996 provide an excellent exposition of statistical perspective on KDD.

Pattern Recognition, Machine Learning, and Artificial Intelligence. In pattern recognition, work has historically focused on practical techniques with an appropriate mix of rigor and formalism. The major applicable techniques fall under the category of classification learning and clustering. There are several texts on the topic; Pattern Classification and Scene Analysis provide a good start. Hence, most pattern-recognition work contributes to the data-mining step in the process. Significant work in dimensionality reduction, transformations, and projections has relevance to the corresponding step in the KDD process.

Within the data-mining step, pattern-recognition contributions are distinguished from statistics by their emphasis on computational algorithms, more sophisticated data structures, and more search, both parametric and nonparametric. Given its strong ties to image analysis and problems in 2D signal processing, work in pattern recognition did not emphasize algorithms for dealing with symbolic and categorical data. Classification techniques applied to categorical data typically take the approach of mapping the data to a metric space (such as nearest-neighbor norms). Such a mapping is often not easy to formulate meaningfully: Is the distance between the values "square" and "circle" for the variable *shape* greater than the distance between "male" and "female" for the variable *sex*?

Techniques originating in AI have focused almost exclusively on dealing with data at the symbolic (categorical) level, with little attention paid to continuous variables. In machine learning and case-based reasoning, algorithms for classification and clustering have focused heavily on heuristic search and nonparametric models. Emphasis on mathematical rigor and analysis of results has not been as strong as in statistics or pattern recognition, with the exception of computational learning theory, which has focused on formal general worst-case bounds for a wide class of representations (a good starting point here is Computational Learning Theory). Machine learning work contributes mainly to the data – mining step of the process, with some contributions in the area of representation and selection of variables through significant search. In addition, the machine discovery community has focused on techniques for discovering structure in data as well as empirical laws to describe observations as in scientific discovery of laws.

AI techniques for reasoning, especially techniques form the Uncertainty in AI community and graphical models for Bayesian modeling and reasoning provide a powerful alternative to classical density estimation in statistics. These techniques have the advantage of allowing prior knowledge about the domain and data to be included in a relatively easy and natural framework. Other areas of AI, including knowledge-acquisition techniques, knowledge representation, and search, are relevant to the various steps in the process, including data mining, data transformation, data selection, and preprocessing.

Databases and Data Warehouses. The relevance of the field of databases to KDD is obvious from the name. Databases provide the necessary infrastructure to store, access, and manipulate the raw data. With parallel and distributed database management systems, they provide the essential layers to insulate the analysis for the extensive details of how the data is stored and retrieved. We focus here only on the aspects of database research relevant to the data-mining step. A strongly related term is on-line analytical processing, which mainly concerns providing new ways of manipulating and analyzing data using multidimensional methods. This has been primarily driven by the need to overcome limitations posed by SQL and relational DBMS schemes for storing and accessing data. The efficiencies achieved via relational structure and normalization can pose significant challenges to algorithms that require special access to the data; in data mining, one would need to collect statistics and counts based on various partitioning of the data, which would require excessive joins and new tables to be generated. Supporting operations form the data-mining perspective is an emerging research area in the database community. In the data-mining step itself, new approaches for functional dependency analysis and efficient methods for finding association rules directly from databases have emerged and are starting to appear as products. In addition, classical database techniques for query optimization and new object-oriented databases make the task of searching for patterns in databases much more tenable.

6.4 Summary

An emerging area in databases is data warehousing, which is concerned with schemes and methods of integrating legacy databases, on-line transaction databases, and various no homogeneous RDBMSs so that they can be accessed in a uniform and easily managed framework. Data warehousing primarily involves storage, data selection, data cleaning, and infrastructure for updating databases once new knowledge or representations are developed.

6.5 Review Questions

1. Define the term Knowledge Discovery Data mining Process (KDD)
2. Describe KDD process and the steps involved in the process with a neat sketches
3. Draw the block diagram of the overall KDD Process

7

Data Mining Tasks, Techniques, and Applications

Objectives:

- To compete effectively in today's marketplace, business managers must take timely advantage of high return opportunities.
- This chapter explores two applications of current data mining techniques: market basket analysis and customer segmentation.

Data mining takes two forms. *Verification-driven data mining* extracts information in the process of validating a hypothesis postulated by a user.

Discovery driven data mining uses tools such as symbolic and neural clustering, association discovery, and supervised induction to automatically extract information.

- Once a data warehouse has been developed, the data mining process falls into four basic steps: data selection, data transformation, data mining, and result interpretation.
- Seven operations are associated with data mining: three with verification-driven data mining and four with discovery-driven data mining.
- **Verification-driven data mining operations.** These include query and reporting, multidimensional analysis, and statistical analysis.
- **Discovery-driven data mining operations.** These include predictive modeling, link analysis, database segmentation, and deviation detection.
- Data mining adopted its techniques from many research areas, including statistics, machine learning, database systems, neural networks, rough sets, and visualization.

Abstract. In this chapter a reality check for data mining is proposed. *Data mining* – the process of extracting the valid, previously unknown, comprehensible, and actionable information from large databases and using it to make crucial business decisions – currently performs this task for a growing range of businesses. After presenting an overview of current data mining techniques, this chapter explores two particularly noteworthy applications of those techniques: *market basket analysis* and *customer segmentation*.

S. Sumathi and S.N. Sivanandam: *Data Mining Tasks, Techniques, and Applications*, Studies in Computational Intelligence (SCI) **29**, 195–216 (2006)
www.springerlink.com © Springer-Verlag Berlin Heidelberg 2006

Mountainous amounts of data records are now available in science, business, industry, and many other areas. Such data can provide a rich resource of knowledge discovery and decision support. For example, when we shop at a supermarket, the cashier scans the bar codes of items and stores your shopping transaction into a database. The supermarket can find valuable information for product selection by analyzing the sales data in its transaction database. (The store can also sell this valuable information.). To understand, analyze, and eventually use this data, a multidisciplinary approach called data mining has been proposed. Data mining is the process of identifying interesting patterns from large databases.

7.1 Reality Check for Data Mining

To compete effectively in today's marketplace, business managers must take timely advantage of high return opportunities. Doing so requires that they be able to exploit the mountains of data their organizations generate and collect during daily operations. Yet the difficulty of discerning the value in that information – of separating the wheat from the chaff – prevents many companies from fully capitalizing on the wealth of data at their disposal.

For example, a bank account manager might want to identify a group of married, two-income, affluent customers and send them information about the bank's growth mutual funds, before a competing discount broker can lure them away. The information surely resides in the bank's computer system and has probably been there in some form for years. The trick, of course, is to find an efficient way to extract and apply it.

7.1.1 Data Mining Basics

Data mining takes two forms. *Verification-driven data mining* extracts information in the process of validating a hypothesis postulated by a user. It involves techniques such as statistical and multidimensional analysis. *Discovery-driven data mining* uses tools such as symbolic and neural clustering, association discovery, and supervised induction to automatically extract information. The extracted information from both approaches takes one of several forms:

Facts – for example, sales increased by 12%,
Trends – sales of the northeastern stores are decreasing,
Regression or classification models,
Relations between database records – such as when consumers buy lettuce and tomatoes, they also buy bacon 80% of the time, and
Deviations from norms, among others.

To be effective, a data mining application must do three things. First, it must have access to organization-wide views of data instead of department-specific ones. Frequently, the organization's data is supplemented with open-source or purchased data. The resulting database is called the *data warehouse*.

During data integration, the application often cleans the data – by removing duplicates, deriving missing value (when possible) – and establishing new derived attributes, for example. Second, the data mining application must mine the information in the warehouse. Finally, it must organize and present the mined information in a way that enables decision making.

Systems that can satisfy one or more of these requirements range from commercial decision-support systems such as Lotus 1-2-3 and Pilot Software's Lightship, to customized decision-support systems and executive information systems such as SAS/EIS. As one example, Lotus 1-2-3 can access data from a data warehouse, allow its users to perform variety of statistical operations, and present the results using simple business graphics.

The overall objective of each decision-making operation determines the type of information to be mined and the ways for organizing the mined information. For example, by establishing the objective of identifying good prospective customers for mutual funds, the bank account manager mentioned earlier implicitly indicates that she wants to segment the database of bank customers into groups of related customers – such as urban, married, two income, mid-thirties, low-risk, high-net-worth individuals – and establishes the vulnerability of each group to various types of promotional campaigns.

7.1.2 The Data Mining Process

Once a data warehouse has been developed, the data mining process falls into four basic steps: data selection, data transformation, data mining, and result interpretation (see Fig. 7.1).

Data selection. A data warehouse contains a variety of data, not at all of which is needed to achieve each data mining goal. The first step in the data

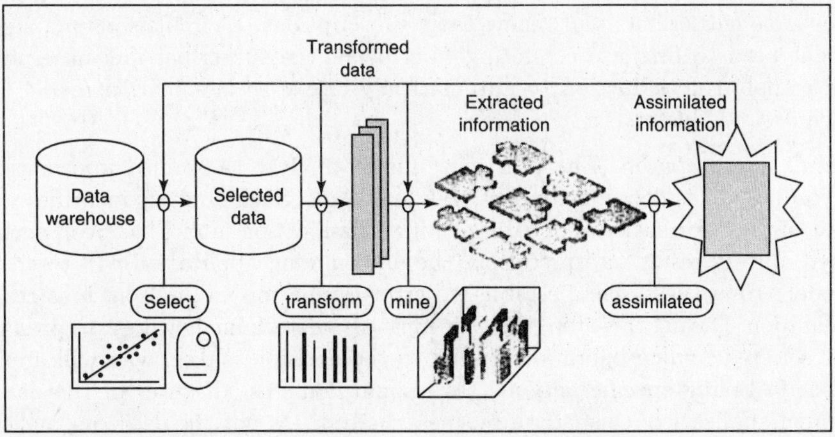

The data-mining process.

Fig. 7.1. The Data Mining Process

mining process is to select the target data. For example, marketing databases contain data describing customer purchases, demographics, and lifestyle preferences. To identify which items and quantities to purchase for a particular store, as well as how to organize the items on the store's shelves, a marketing executive might need only to combine customer purchase data with demographic data. The selected data types may be organized along multiple tables: during data selection, the user might need to perform table joints. Furthermore, even after selecting the desired database tables, mining the contents of the entire table is not always necessary for identifying useful information. Under certain conditions and for certain types of data mining operations (such as when creating a classification or regression. model), it is usually a less expensive operation to sample the appropriate table, which might have been created by joining other tables, and then mine only the sample.

Data transformation. After selecting the desired database tables and identifying the data to be mined, the user typically needs to perform certain transformations on the data. Three considerations dictate which transformation to use: the task (mailing-list creation, for example), the data mining operations (such as predictive modeling), and the data mining technique (such as neural networks) involved. Transformation methods include organizing data in desired ways (organization of individual consumer data by household), and converting one type of data to another (changing nominal values into numeric ones so that they can be processed by a neural network). Another transformation type, the definition of new attributes (derived attributes), involves applying mathematical or logical operators on the values of one or more data base attributes – for example, by defining the ratio of two attributes.

Data mining. The user subsequently mines the transformed data using one or more techniques to extract the desired type of information. For example, to develop an accurate, symbolic classification model that predicts whether magazine subscribers will renew their subscriptions, a circulations manager might need to first use clustering to segment the subscriber database, and then apply rule induction to automatically create a classification model for each desired cluster.

Result interpretation. The user must finally analyze the mined information according to his decision-support task and goals. Such analysis identifies the best of the information. For example, if a classification model has been developed, during result interpretation, the data mining application will test the model's robustness, using established error-estimation methods such as cross validation. During this step, the user must also determine how best to present the selected mining-operation results to the decision maker, who will apply them in taking specific actions. (In certain domains, the user of the data-mining application – usually a business analyst – is not the decision- maker. The latter may make business decisions by capitalizing on the data mining results through a simple query and reporting tool.) For example, the user might decide that the best way to present the classification model is logically in the form of if-then rules.

Three observations emerge from this four-step process:

Mining is only one step in the overall process. The quality of the mined information is a function of both the effectiveness of the data-mining technique used and the quality, and often sizes, of the data being mined. If users select the wrong data, choose inappropriate attributes, or transform the selected data inappropriately, the results will likely suffer.

The process is not linear but involves a variety of feedback loops. After selecting a particular data-mining technique, a user might determine that the selected data must be preprocessed in particular ways or that the applied technique, did not produce the results of the expected quality. The user then must repeat earlier steps, which might mean restarting the entire process from the beginning.

Visualization plays an important role in the various steps. In particular, during the selection and transformation steps, a user could use statistical visualizations – such as scatter plots and histograms – to display the results of exploratory data analysis. Such exploratory analyses often provide preliminary understanding of the data, which helps the user select certain data, subsets. During the mining step, the user employs domain-specific visualizations. Finally, visualizations – either special landscapes or business graphics – can present the results of a mining operation.

7.1.3 Data Mining Operations

Seven operations are associated with data mining: three with verification-driven data mining and four with discovery-driven data mining.

Verification-driven data-mining operations: These include query and reporting, multidimensional analysis, and statistical analysis.

Query and reporting. This operation constitutes the most basic form of decision support and data mining. Its goal is to validate a hypothesis expressed by the user, such as "sales of four-wheel-drive vehicles increase during the winter." Validating a hypothesis through a query-and-reporting operation entails creating a query, or set of queries, that best expresses the stated hypothesis, posing the query to the database, and analyzing the returned data to establish whether it supports or refutes the hypothesis. Each data interpretation or analysis step might lead to additional queries, either new ones or refinements of the initial one. Reports subsequently compiled for distribution throughout an organization contain selected analysis results, presented in graphical, tabular, and textual form, and include a subset of the queries. Because these reports include the queries, analyses can be automatically repeated at predefined times, such as once a month.

Multidimensional analysis. While traditional query and reporting suffices for several types of verification-driven data mining, effective data mining in certain domains requires the creation of very complex queries. These often contain an embedded temporal dimension and may also express change between two stated events. For example, the regional manager of the department store chain might say "Show me weekly sales during the first quarter of 1994 and 1995, for Midwestern stores, broken down by department." Multidimensional databases, often implemented as multidimensional arrays, organize data along predefined dimensions (time or department, for example), have facilities for taking advantage of sparsely populated portions of the multidimensional structure, and provide specialized languages that facilitate querying long dimensions while expediting query-processing performance. These databases also allow hierarchical organizations of the data along each dimension, with summaries on the higher levels of the hierarchy and the actual data at the lower levels. Quarterly sales might take one level of summarization and monthly sales a second level, with the actual daily sales taking the lowest level of the hierarchy.

Statistical analysis. Simple statistical analysis operations usually execute during both query and reporting, as well as during multidimensional analysis. Verifying more complex hypothesis, however, requires statistical operations (such as principal component analysis regression modeling) coupled with data visualization tools. Several statistical analysis tools (SAS, SPSS, S+) incorporate components that can be used for discovery-driven modeling (such as CHAID in SPSS and S+). To be effective statistical analysis must rest on a methodology, such as exploratory data analysis. A methodology might need to be business or domain dependent, so statistics tools such as SAS and SPS are open ended, providing function libraries that can be organized into larger analysis software systems.

Discovery-driven data mining operations: These include predictive modeling, link analysis, database segmentation, and deviation detection.

Predictive modeling. This is the most commonly used of the discovery-driven data mining operations, primarily because of the proliferation of automatic model development techniques. (These include symbolic-induction techniques such as CART or C4.5, and neural network techniques such as back propagation.) This operation uses the database records that reflect historical data information about past behavior – to automatically generate a model that can predict the future behavior. For example, an insurance underwriter might want to predict the likelihood that a customer will let her policy lapse. A marketing executive might want to predict whether a particular customer would switch brands for a specific product. The value of discovery-driven modeling techniques, especially symbolic induction techniques, is that the models are expressed as sets of if-then rules, and are therefore comprehensible and explainable.

Database segmentation. It is often necessary to automatically partition a database into collection of related records. Such collections enable users to summarize each database or select a smaller portion of the database on which to apply a different data mining operation such as model creation or link analysis. For example, by segmenting a department stores point off scale data, a user can automatically create segments containing transactions from specific periods, such as back to school days or after Christmas sales.

Link analysis. While the modeling and segmentation operations aim to create the generalized description that characterizes database's contents, link analysis seeks to establish relations between the records in the database. For example, a merchandise buyer must determine what items sell together men's shirts sell together with ties and men's fragrances – so that he can decide what items to buy for a store (shirts, ties, and fragrances), as well as where to place these items in relation to each other (ties and fragrances must be displayed close to the store's men's shirts section). Certain algorithms that fall in this category can also trace connections between records overtime.

Deviation detection. This operation attempts to identify points that cannot be fitted into the segment and then explain whether each such point is noise or should be examined in more detail. This operation usually operates in conjunction with database segmentation and, because "outliers" express deviation from expected norms, often leads to true discovery.

7.1.4 Discovery-Driven Data Mining Techniques:

While there are only four basic discovery-driven mining operations, numerous data mining techniques support these operations. For example, supervised induction techniques support predictive model creation; association discovery and sequence discovery techniques support link analysis; and statistical techniques support deviation detection.

Supervised induction. This process involves automatically creating a classification model from a set of records (examples) called *the training set.* The training set might either be a sample of the database or a warehouse being mined, the entire database, or a data warehouse. The records in the training set must belong to a small set of classes that the analyst has predefined. The induced model consists of patterns – essentially generalizations over the records—that are useful for distinguishing the classes. Once induced, a model can help automatically predict the class of other unclassified records. Supervised induction methods can be either neural or symbolic. Neural methods such as back propagation represent the model as architecture of nodes and weighted links. Symbolic methods create models that are represented either as decision trees or as if-then rules. A supervised induction technique is particularly suitable for data mining if it has three characteristics.

It can produce high-quality models even when the data in the training set is noisy and incomplete.

The resulting models are comprehensible and explainable, so that the user can understand how the system makes the decision.

It can accept domain knowledge, which can expedite the induction task while simultaneously improving the quality of the induced model.

Association discovery. Given a collection of items and a set of records, each of which contains some number of items from the given collection, an association discovery function is an operation against this set of records that returns affinities existing among the collection of items. These affinities can be expressed by rules such as, "72% of all the records that contain items A, B and C also contains items D and E." The specific percentage of occurrences (in this case 72) is *the confidence factor* of the association. Also in this association, A, B, and C are said to be on opposite side of the association to D and E. Association discovery can involve any number of items on either side of the association.

Sequence discovery. In the transaction log discussed earlier, the identification of the customer who made the purchase generally remains unknown. If this information exists, analysis can be made on the collection of records of the same structure as above (that is consisting of the number of items drawn for a given collection of items). The records are related by the identity of the customer who made the repeated purchases.

Clustering. This is used to segment the database into subsets, or clusters with the members of each cluster sharing a number of interesting properties. The results of a clustering operation fulfill one of two functions.

Summarizing the contents of the target database by considering the characteristics of each created cluster, rather than those of each record in database.

Serving is an input to other methods, such as supervised induction. A cluster is smaller and more manageable data set to the supervised inductive learning component.

Clusters use neural and symbolic unsupervised induction methods. The various neural and symbolic methods are distinguished by the type of attributes values they allow the records in the target database to take (numeric, nominal, structured objects), the way they represent each cluster, and the way they organize the set of clusters. After clustering the database, the analyst can use visualization component to examine the created clusters to locate ones that are useful or interesting.

Neural clustering methods, such as feature maps, represent the cluster as a prototype with which they associate a subset of the instances in the data set being clustered. Symbolic clustering methods operate primarily on records with nominal-valued attributes. They consider all the attributes that characterize each instances and use AI-based search methods to establish the subset of these attributes that will desire each created cluster.

Characteristics of data mining: Let us begin our discussion about data mining applications by examining the available, albeit preliminary, statistics. 80% to 90% of the operational data mining applications today employ only a single operation. There are two reasons for this. First, because tool developers usually understand only one or two determining operations, the tools on which these applications are based usually contain only one technique. Second when initially applying data mining end-user organizations select narrowly focused problems that require only one operation to solve.

In 70% of the applications, users perform data mining using verification-driven operations. Data analysts and business analysts alike thoroughly understand query and reporting, multidimensional analysis, and statistical analysis. Because many of them are graduates of business schools, they usually receive training such operations, so they feel comfortable trying to solve their decision support problems with them. Neural and symbolic induction methods have only recently appeared in business school curricula.

To date, two factors have inhibited the broad deployment of applications that incorporate discovery-driven data mining techniques: the significant effort necessary to develop each data mining application and the inappropriate state of the data that application must mine.

Application development. Most deployed data mining applications are not developed by business analysts but through the collaboration of data mining tool vendors, data analysts, and end users. Because the tool vendors and data analysts usually first must develop an understanding of the end user's problem, such collaborations are time consuming. Furthermore, the current generations of data mining tools are aimed at the data analyst, not the business analyst.

Data. Data is often incorrect and incomplete, distributed across many departmental databases, organized using incompatible data models and formats, and using often-incomprehensible naming conventions. Because of such problems, and the lack of appropriate data integration and cleaning tools, the Gartner group has estimated that only about 10% of the collected data is ever analyzed. Many organizations realize the poor quality of the data they have been collecting only after they start developing a data mining application. The development of data warehouses should alleviate this problem. A data warehouse ensures that disparate data is integrated consistently under a single data model and is cleaned in the process, and that metadata is encoded explicitly. Over 40% of the organizations surveyed by the metagroup across 10 industries have started developing data warehouses during the past six months, with the intention of using them in conjunction with multidimensional databases and discovery-driven data mining tools. These warehouses will vary from departmental data marts, size ranging from 20 to 60 Gbytes, to corporate warehouses, which often hold several terabytes of data.

7.2 Data Mining: Tasks, Techniques, and Applications

Data mining is the core part of the knowledge discovery in database (KDD) process. The KDD process may consist of the following steps: data selection, data cleaning, data transformation, pattern searching (data mining), and finding presentation, finding interpretation, and finding evaluation. Data mining and KDD are often used interchangeably because data mining is the key to the KDD process. These steps are illustrated in Fig. 7.2.

7.2.1 Data Mining Tasks

The tasks of data mining are very diverse and distinct because many patterns exist in a large database. Different methods and techniques are needed to find different kinds of patterns. Based on the patterns we are looking for, tasks in data mining can be classified into summarization, classification, clustering, association, and trend analysis.

Summarization. Summarization is the abstraction or generalization of data. A set of task-relevant data is summarized and abstracted. This results in a smaller set, which gives a general overview of the data, usually with aggregate information.

For example, the long-distance calls of a customer can be summarized into total minutes, total spending, total calls, and so forth. Such high-level, summary information, instead of the individual elements of each call, is presented to the sales managers for customer analysis.

The summarization can go to different abstraction levels and can be viewed from different angles. For example, calling minutes and spending can

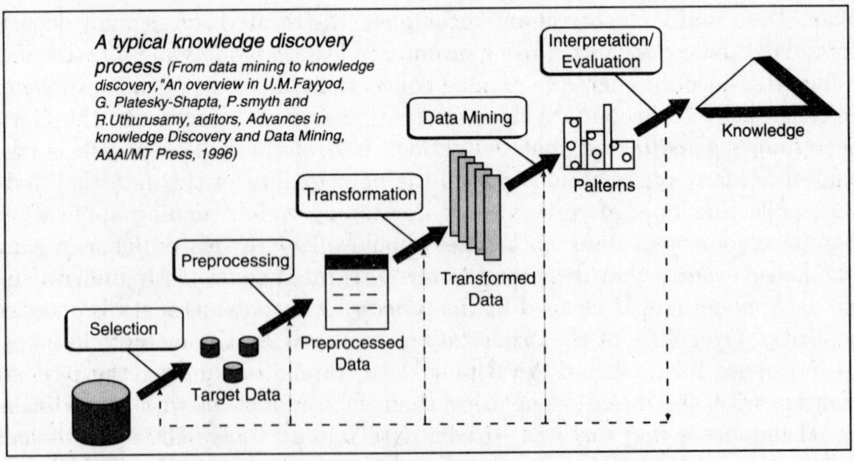

Fig. 7.2. A Typical Knowledge Discovery Process

be totaled along various calling periods: weeks, months, quarters, or years. Similarly, the calls can be summarized into in-state calls, state-to-state calls, calls to Asia, calls to Europe, etc. These groupings can be further summarized into domestic calls and international calls. Different combinations of abstraction levels and dimensions reveal various kinds of patterns and regularities.

Classification. Classification derives a function or model, which determines the class of an object, based on its attributes. A set of objects is given as the training set. In it, every object is represented by a vector of attributes along with its class. A classification function or model is constructed by analyzing the relationship between the attributes and the classes of the objects in the training set. This function or model can then classify future objects. This helps us develop a better understanding of the classes of the objects in the database.

For example, from a set of diagnosed patients who serve as the training set, a classification model can be built. This model concludes a patient's disease from his/her diagnostic data. The classification model can diagnose a new patient's disease based on data such as age, sex, weight, temperature, blood pressure, etc.

Association. Association is the discovery of togetherness or the connection objects. Such togetherness or connection is termed an *association rule*. An association rule reveals the associative relationships among objects; i.e., the appearance of a set of objects in a database is strongly related to the appearance of another set of objects.

For example, in a telecommunication database, we may find a rule that "call waiting" is associated with "call display," denoted as "call waiting → call display." It states that if a customer subscribes to the "call waiting" service, he or she very likely also has "call display."

Association rules can be useful for marketing, commodity management, advertising, and so forth. For example, a retail store may discover that people tend to buy soft drinks and potato chips together. Store personnel then place the potato chips near the soft drinks to promote the sale of both. They may even discount one to entice buying the other, since these shoppers now will be "saving money."

Clustering. Clustering identifies classes – also called *clusters* or *groups* – for a set of objects whose classes are unknown. The objects are so clustered that the interclass similarities are maximized and the interclass similarities are minimized. This is done based on some criteria defined on the attributes of the objects. Once the clusters are decided the objects are labeled with their corresponding clusters. The common features for object in a cluster are summarized to form the class description.

For example, a bank may cluster its customers into several groups based on the similarities of their age, income, and residence. The characteristics, the customers in a group share can be used to describe that group of customers.

These clusters help the bank understand its customers better and provide more suitable products and customized services.

Trend analysis. Time series data are records accumulated over time. For example, a company's sales, a customer's credit card transactions, and stock prices are all time series data. Such data can be viewed as objects with an attribute time. The objects are snapshots of entries with values that change over time. Finding the patterns and regularities in the data evolution along the dimension of time can be fascinating.

One topic in trend analysis is identifying patterns in an object's evolution, such as its ups and downs, or peaks and valleys. A model or a function is constructed to simulate the behavior of the object to predict future behavior. For example, we can estimate this year's profit for a company based on its profits last year and the estimated annual increasing rate.

Another topic in trend analysis is matching objects, changing trends, such as increasing streaks, decreasing streaks, etc. By comparing two or more objects historical changing curves or tracks, similar and dissimilar trends can be discovered. These can help us to understand the behavior of the objects. For example, a company's sales and profit figures can be analyzed to find the disagreeing trends. These trends can be further researched to discover the reasons behind such disagreements.

7.2.2 Data Mining Techniques

Data mining adopted its techniques from many research areas, including statistics, machine learning, database systems, neural networks, rough sets, and visualization.

Statistical approaches. Many statistical tools have been used for data mining, including Bayesian network, regression analysis and cluster analysis, and correlation analysis. Usually statistical models are built from a set of training data. An optimal model, based on a defined statistical measure, is searched among the hypothesis space. Rules, patterns, and regularities are then drawn from the models.

A Bayesian network is directed graph computed using the Bayesian probability theorem. It represents the causal relation ships among the variables. Regression is the function derivation, which maps a set of attributes of objects to an output variable. Correlation analysis studies the correspondence of variables to each other, such as the x^2. Cluster analysis finds groups from a set of objects based on distance measures.

A simple Bayesian network for a medical problem is given in Fig. 7.3. Nodes in a Bayesian network represent variables or states; edges represent the dependencies between nodes directed from the cause to the effect. The figure shows that a patient's age, occupation, and diet affect the disease, which causes the symptoms.

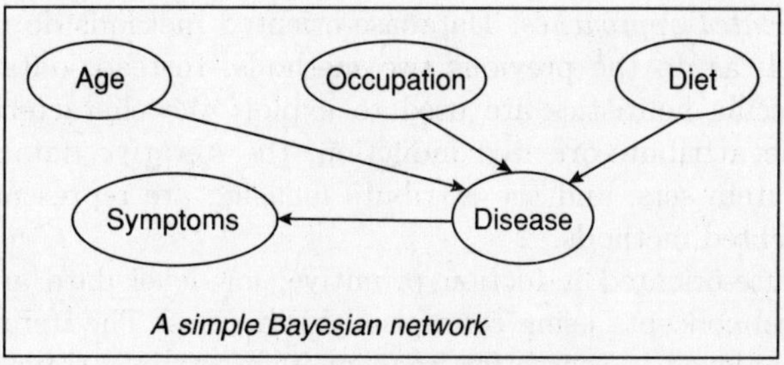

Fig. 7.3. A Simple Bayesian network

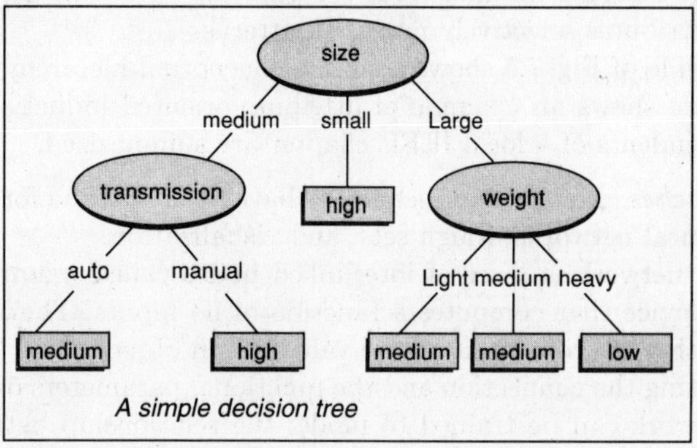

Fig. 7.4. A Simple Decision Tree

Machine learning approaches. Like statistical methods, machine-learning methods search for the best model that matches the testing data. Unlike statistical methods, the searching space is a cognitive space of n attributes instead of a vector space of n dimensions. Besides that, most machine learning methods use heuristics in the search process.

The most common machine learning methods used for data mining include decision tree, inductive concept learning, and conceptual clustering. A decision tree is a classification tree, which determines an object's class by following the path from the root to a leaf node. It chooses the branches according to the attribute values of the object. Decision trees are induced from the training set. Classification rules can be extracted from the decision trees.

Inductive concept learning derives a concise, logical description of a concept from a set of examples. Conceptual clustering finds groups or clusters in a set of objects based on conceptual closeness among objects.

A simple decision tree is given in Fig. 7.4. It determines a car's mileage from its size, transmission type, and weight. The leaf nodes are in square boxes that represent the three mileage classes. From the decision tree, we can conclude, for example, a medium size, automatic car will have medium mileage.

Database-oriented approaches. Database-oriented methods do not search for a best model, as do the previous two methods. Instead, data modeling or database specific heuristics are used to exploit the characteristics of data in hand. The attribute-oriented induction, the iterative database scanning for frequent item sets, and the attribute focusing are representatives of the database-oriented methods.

In attribute-oriented induction primitive, low-level data are generalized into high-level concepts using conceptual hierarchies. The iterative database method is employed to search for frequent item sets in a transactional database. The association rules are then derived from these frequent item sets. The attribute-focusing method looks for patterns with unusual probabilities by adding attributes selectively into the patterns.

The left side of Fig. 7.5 shows a simple conceptual hierarchy for students. The right side shows an example of attribute-oriented induction. In the example, the students of a local IEEE chapter are summarized.

Other approaches. Many other techniques have been adopted for data mining, including neural networks, rough sets, and visualization.

A neural network is a set of interlinked nodes called *neurons.* A neuron is a simple device that computes a function of its inputs. The inputs can be outputs of other neurons or attribute values of an object.

By adjusting the connection and the functional parameters of the neurons, a neural network can be trained to model the relationship between a set of input attributes and an output attribute. A neural network can be used, for example, in classification when the output attribute is the object class.

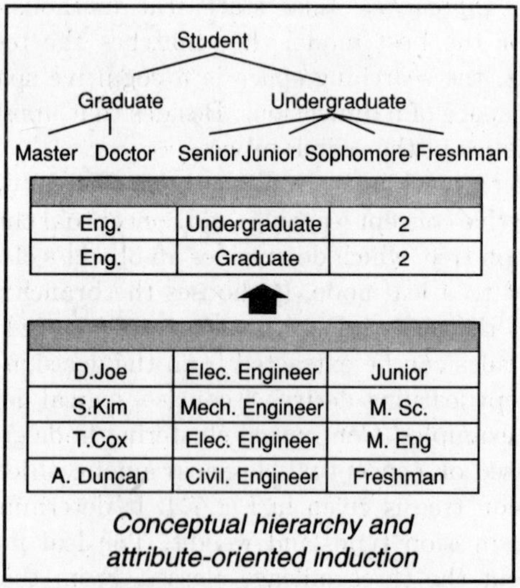

Fig. 7.5. Conceptual hierarchy and attribute-oriented induction

A rough set is a set whose membership is fuzzy. A set of objects can be arranged to form a group of rough sets for use, in say, classification and clustering.

Visual exploration is another interesting data mining technique. Data are transformed into visual objects such as dots, lines, and areas. The data is then displayed in a two- or three-dimensional space. Users can interactively explore the interesting spots by visual examination.

These methods can be integrated or combined to deal with complicated probabilities, or provide solutions. For example, data is visually summarized in charts, graphs, and such. This is done to help us understand the results and allow further examination. Indeed most data mining systems employ multiple methods to deal; with different kinds of data, different data mining tasks, and different application areas.

7.2.3 Applications

Data mining techniques have been applied successfully in many areas, from business, science, to sports.

Business applications. Many organizations now employ data mining as a secret weapon to keep or gain a competitive edge. Data mining has been used in data base marketing, retail data analysis, stock selection, credit approval, etc. For example, Mellon Bank, a Pittsburgh-based $40 billion financial services company, is applying IBM's intelligent Miner on its customer database in an effort to retain profitable customers.

- Database marketing is a very successful and popular business application of data mining. By mining historical customer databases, patterns and trends are extracted. Customer profiles are built from these results to produce more effective marketing.
- Retail databases contain customer-shopping transactions. Data mining can find customer-shopping patterns that can be used, for example, in a sales campaign.
- Using data mining techniques, investors can build models to predict the performance of stocks. By searching trends and patterns in stock data, data mining can also help investors to find stocks with good performances.
- Applications for credit or loan are decided based on the applicants' information. A decision support model for credit or loan approval may be constructed from historical data using data mining tools.

Science applications. Data mining techniques have been used in astronomy, molecular biology, medicine, geology, and many more. For example, the Jet Propulsion Lab at the California Institute of Technology has developed a data mining system, which can classify the sky objects, such as stars, in satellite images.

Other applications. Data mining techniques have also been used in health care management, tax fraud detection, money laundering monitoring, and even sports. For example, the Advanced Scout system developed by IBM has been used by coaches of more than a dozen teams in the National Basketball Association (NBA) to improve their games.

7.2.4 Data Mining Applications – Survey

Numerous verification data driven and some discovery-driven data mining applications and application prototypes are already available for domains that range from credit risk analysis, fraud detection, to astronomy. Several of these applications, especially those using verification-driven data mining techniques, are currently in daily use and have been described in detail elsewhere. Of the already-deployed applications that use discover-driven data-mining techniques, three warrant particular mention:

- The Falcon credit-risk assessment system from HNC Inc., used by a large percentage of retail banks to detect suspicious credit-card transactions.
- JPL's Skicat system, which astronomers use to identify and classify new types of celestial objects, and
- The FAIS system from the Financial Crimes Enforcement Network (FinCEN), which identifies financial transactions that might indicate money laundering.

These three applications use the model creation as their primary data-mining operation, often supplemented with clustering and visualization techniques. They balance high predictive accuracy with the ability to use mined knowledge to explain the actions they recommended to their users. They also took significant time to implement and deploy. The Falcon system builds upon HNC's neural network shell. The fact that only a few companies capture credit card transaction data facilitated Falcon's deployment. Even though each of the companies uses its own data format, every bank issuing credit cards uses one of these few formats. Therefore, an application that works effectively with even one format can be easily adopted by a large number of banks.

The Skicat system, developed from scratch for one particular group of astronomers at JPL, uses data collected from astronomical instruments. It is also finding a larger user community, because astronomers also use data with common formats.

The FIAS application uses data from common government forms and consists of a combination of off-the-shelf and custom-built components. Its use should expand to a variety of government agencies concerned with detecting suspicious financial transactions that might indicate money laundering. Of the three systems, this last one faces the hardest data quality problem because it mines data that is often entered by hand, frequently hand written.

Financial analysis applications that employ predictive modeling techniques – whether they are statistical regression, neural networks, or symbolic modeling techniques – run into a special problem. Such applications perform portfolio creation and optimization, trading model creation and so forth, and have been in use for several years. To maintain a competitive advantage, the users and developers of such applications do not publicize their use and effectiveness. In these applications, predictive accuracy is much more important than the ability to use the extracted knowledge to explain a recommended action. Thus, the modeling methods used must first ensure that they do not overfit the data. The term *data mining* has always been associated with model overfit and has consequently been rejected by the financial analysis community.

Developers have recently incorporated discovery-driven data mining techniques in two classes of decision support applications: Market basket analysis and customer segmentation. Off the shelf versions of these two applications currently assist strategic marketing personnel in gaining significant insight about consumer behavior.

Market basket analysis: MBA refers to the process of examining point-of-scale data to identify affinities between products and services purchased by the customer. For example, after analyzing POS data from a particular set of supermarkets, one might identify that when the sales of lettuce increase, so do the sales of tomatoes and salad dressing. MBA is usually one step in larger-scale analyses aimed at such tasks as planning the replenishment of a store, assessing the effectiveness of an organization's promotions, and fine-tuning manufacturing and distribution operations. Because retail organizations have traditionally used MBA, typical users of these applications include marketing analysis, merchandise buyers, and store managers, catalog managers, and brand managers.

Early implementations of MBA applications used query-and-reporting data mining techniques and applied them to summaries aggregations and specialized views of the captured POS data, as well as to data about the purchases of closely monitored customer groups. For example, A.C Nielsen monitors the purchase of 40,000 consumers and sells the data to corporations interested in performing MBA. By using query-and-reporting techniques on this type of data, organizations can extract item movement information such as "when the sales of brand X coffee increase so do the sales of brand Y sugar."

Extracting information from summarized data was, and often still is, necessary because of the volumes of the captured data. For example, because a supermarket **or** a department store routinely stocks over 100,000 type of merchandise, the resulting POS databases of retail chains routinely consume several gigabytes per week. Until parallel database technology recently emerged, the query processing times needed to execute each query deterred any one from posing queries against such databases.

More recently, tools that can perform multidimensional analysis, such as DSPlus from Kelly Information Systems and DSS Agent from Micro strategy, have emerged for further improving the quality of MBA results produced by query and reporting tools. Using multidimensional analysis tools, an analyst can easily ask queries such as "show me the three highest selling products of the produce department for the first quarter of 1994–1995, for the northeastern United States, organized store." The dimensions used in this query are department, time, geographic region, and organization (store). These tools continue to work with summarized data, and like any other verification-driven data mining tools rely on the analyst to postulate hypothesis and convert the returned data into information. The size of the multidimensional databases being deployed with MBA application varies from 10 to 30 Gbytes, to more than 100Gbytes. Corporations such as Mervyn's and Target Stores have recently reported successful deployment of multidimensional databases for MBA.

Two major retailing trends today are the emergence of megastore chains (home improvement chains and discount clothing chains being two examples), and the lack of differentiation in the customer's eyes between the offering of any two retail chains with in a particular segment, such as home improvement supermarkets. To improve their profits under such conditions, retailers attempt to identify micromarkets and exploit them before competition does. The recent mergers, Chase Bank with Chemical, for example, of large banks and insurance companies also create large financial institutions with little differentiation among themselves. Such organizations will start relying heavily on micromarket identification for improving their financial performance and growth. While the remainder of this section discusses how retailers use MBA, large financial institutions are also adapting MBA Applications. Summarized data that is often used today in conjunction with verification-driven data mining is not effective for micromarket identification; summarizing data removes individual consumer characteristics which once detected, could have helped identity micromarkets.

Micromarket identification requires:

- The use of the item's actual POS transactions, so that one can find out exactly what each customer is buying at the transaction level rather than at some aggregated level, whether that customer is buying an item on sale, at what time he or she is making a particular trip to the store, and so forth.
- Overlaying POS data with demographic and life style consumer data to identify other customer characteristics that may be correlated to the consumer's choices and
- The use of discovery-driven techniques (primarily tools that support database segmentation and link analysis operations) that can automatically identify important buying patterns and the types of consumers exhibiting such patterns.

For example, a marketing executive could use database segmentation tools on a set of supermarket POS transactions to identify product drivers ("lettuce and whole grain breads drive a large percentage of the sales in suburban supermarkets in Area X") in particular subset of the transactions. She could then use association rule discovery techniques on each of the identified clusters to automatically extract rules, such as "Customers who buy lettuce and tomatoes also buy bacon 80% of the time."

Recently, IBM consultants worked with a supermarket chain's business analysts to perform MBA using neural segmentation and association rule discovery techniques. Together, they discovered that shoppers use certain of the chain's supermarket as convenience stores during Friday and Saturday nights. Based on this discovery, a super market chain decided to start offering video rentals in these stores. This decision provided a chain with a good way to differentiate itself from competing supermarket chains and to identify a new source of revenue. The number of consumers going to the chain's supermarkets increased, there by increasing the general sales volume, and the sales of items such as popcorn, potato chips, and soft drinks also jumped significantly. Through this action the chain realized significant incremental sales and profit.

The application of discovery-driven data mining techniques for MBA analysis imposes two requirements on database management systems. First, they must deal with large volumes of transactional and overlaid data. In particular, while it has been a common practice to perform discovery-driven data mining on sampled data rather than the entire database, one might miss a particular micro market by doing so. Second, the systems must perform the mining operation within a time interval that will allow an organization to respond to a market opportunity before the competition does. For example, catalog companies send out a new issue of each catalog about every two months. During this eight-week period, they must collect the data about a set of consumers (collecting sales data from the previous issue of the catalog and combining it with demographic data), mine the collected data, determine the consumers to whom a particular catalog will be sent (identify the customers segments or micromarkets), and prepare and send the catalog.

Customer segmentation: Customer segmentation is the process of analyzing the data about the actual customers or general consumers to identify characteristics and behaviors that can be exploited in the market place. For example, an insurance company might want to identify the characteristics of customers with automotive policies who have let their policies lapse, so that it can establish a set of strategies for convincing the highest valued of these customers to renew their policies. Ultimately, customer segmentation allows an organization to view each of its customers as a segment of one, thereby forming a highly personalized relationship with each customer. Organizations apply customer segmentation to both actual customers and general consumers; consumer data analysis identifies characteristics that will allow the creation of strategies for tuning targeted consumers into actual customers.

Organizations use customer segmentation to address two basic marketing problems: understanding the causes of customer action so that it can be reduced and identifying new customers.

Customer attrition. Policy lapsing is a form of attrition. Others include switching credit cards, switching mortgage institutions, and buying merchandise from different direct-mail company. In order to identify the causes of an attrition problem, a customer segmentation application must also automatically establish which customers the organization should try to keep and which should be left to defect – most organizations derive their profits from only a percentage of their customer base.

Target marketing. Organizations are invested in two types of target market. First, they want to know how to offset particular product or service to a specific set of consumers to turn them into customers. Every direct-mail company addresses this problem each time it creates a mailing list of the households that receives the next issue of its catalog. Second, companies want to determine which customers of one of their other products they should offer a product or service. Also known as cross selling, this action aims to convince customers to also the offered product. For example, a company might try to convince certain of its credit card customers to apply for a second mortgage. The two questions that will be answered in cross- selling scenario are "To which customers should a particular product be offered?" and "Which products should be offered to a particular customer?"

Typical users of this application include marketing analysts, brand managers, risk managers, and other types of business analysts. As in MBA, early implementations of customer segmentation applications rely on verification-driven data mining techniques. In particular, such applications used statistical analysis techniques – for example, binary and logistic regression – to create sector models. Analysts then applied these models to larger databases to assess an individual risk or propensity to lapse for example, they selected the data used to create model either by sampling a database with customer or customer data, or by selecting a set of customers through queries.

The use of sampling led to models could only cover broad customer characteristics. The use of queries to select a customer could lead to more targeted models, but the analyst must determine the features to include in the query, and thus cannot detect and capitalize market shifts and micromarket characteristics. Models developed using linear and logistic regression techniques can only account for certain characteristics in the data; essentially these techniques cannot account for "every ounce of information" in the data. The data that was used was also not very rich itself. Limitations in database technology and in the means of obtaining customer and customer data prevented users from performing fine-grained analysis even with the available techniques. Finally because statistical analysis tools are primarily targeted to data analysis, early implementations of customer segmentation applications were used by organizations staffed primarily by statisticians. Such organizations reside in large corporations and specialized service bureaus.

The customer segmentation applications being developed and deployed today arise in response to several trends. First, the variety and quality of the captured a data have increased. In fact, organizations currently capture more data than they use, because the improvements in database technology have simplified data maintenance. Second, having solved the data capture and maintenance problem, business units increasingly want to take ownership of their data, rather than relying on specialized departments and service bureaus. Also early experimentation with discovery driven data mining techniques – primarily clustering, deviation detection, and modeling (from the symbolic and neural induction regimes) has demonstrated their advantages over traditional techniques.

The challenge that has to be addressed was making the applications embedding such techniques more appropriate to business users, not just statisticians. Symbolic learning techniques (modeling or clustering) have contributed significantly to this need, because they express the information they extract in forms that are comprehensible by business users. However, the new generation of customer-segmentation application goes a step further. By embedding this logic and knowledge into the application itself, they have begun eliminating the need for users to deeply understand data analysis and modeling processes.

Customer segmentation applications with these characteristics are now at work in the research and strategy departments of major banks, the customer-relations management departments of banks, insurance companies, retail organizations, as well as the marketing departments of major airlines. IBM recently reported that through the use of such an application, Legal and General, a major insurance company, conducted a successful target marketing campaign whose results improved significantly on similar campaigns run using more traditional customer segmentation applications.

The effectiveness of discovery-driven data mining techniques in extracting valuable and previously unknown information from large, complex databases is speeding acceptance of the two marketing applications now being deployed aggressively by various organizations. As the value these early adopters derive becomes apparent, applications using discovery-driven data mining techniques will proliferate, drawing data mining into the mainstream.

To summarize, data mining is the process of extracting interesting patterns from large databases. Data mining can be the solution to the data analysis problems faced by many organizations. More work is needed, although a great deal of progress has been made in research and development.

7.3 Summary

Data mining is becoming an integral part of the operations in organizations of varying sizes. Many organizations that only recently have begun analyzing their data have started to successfully use applications employing verifications-driven data mining techniques. Applications using discovery-driven techniques are also finding increased use. While many of the deployed applications primarily employ predictive modeling techniques, application developers and end users alike are beginning to recognize the need to use additional techniques from the discovery-driven data mining repertory. Applications with broad market appeal such as MBA and customer segmentation have successfully demonstrated the advantage of using such techniques.

Today, the development of data mining applications remains a tedious process. Applications recognized as successful invariably require the cooperation of business analysts and developers of generic data mining tools. At these early phases, this is not necessarily a drawback; most successful applications in any discipline require the collaboration of end users and developers. However, a phase must quickly ensue in which business analyst can quickly develop and specialize data mining applications, much as is happening with database technology.

7.4 Review Questions

1. Explain the data mining process and its operational techniques.
2. What are the characteristics of data mining?
3. Give some of the data mining tasks and techniques involved in processing of the data.
4. Give details on performance survey on data mining applications.

8

Data Mining: an Introduction – Case Study

Objectives:

- To study the importance of data mining in health care.
- We discusses how data mining can be packaged in such a way that professional can take part directly in data mining and thus assess the implications of using data mining in safely-critical or service-critical applications.
- Data mining offers a new approach to data analysis using techniques based on machine learning alongside the conventional methods.
- To discuss how data mining can be packaged in such a way that professionals can take part directly in data mining.

Abstract. This section outlines a case study describing the importance of data mining in health care taken from Clementine, SPSS Inc. USA. Health care and pharmaceutical companies face an explosion of data arising from clinical, administrative, commercial and scientific activities. There are many traditional techniques for analyzing data including statistics, management reporting and data display. Data mining offers a new approach to data analysis using techniques based on machine learning (algorithms derived from research in artificial intelligence), alongside the conventional methods.

These techniques work by "learning" patterns in data. They find patterns and make predictions, which elude all but the most expert users of conventional methods. In addition, they generate decision or prediction models, based on the actual historical data. These models are synthesized- not programmed explicitly by programmer or physician. Thus, they represent true evidence-based decision support.

Health care and pharmaceutical professionals have a special duty of care, as their decisions may be a matter of life and death for their clients. The esoteric nature of data mining can distance these professionals from the models. We discusses how data mining can be packaged in such a way that professional can take part directly in data mining and thus assess the implications of using data mining in safely-critical or service-critical applications.

S. Sumathi and S.N. Sivanandam: *Data Mining: an Introduction – Case Study*, Studies in Computational Intelligence (SCI) **29**, 217–229 (2006)
www.springerlink.com © Springer-Verlag Berlin Heidelberg 2006

8.1 The Data Flood

The proceedings of the 1995 conference in Knowledge Discovery in Databases (KDD) opens with the following quotations:

> *"It is estimated that the amount of information in the world doubles every 20 months. What are we supposed to do with this flood of raw data? Clearly little of it will ever be seen by human eyes."*
>
> *"Computers promised fountains of wisdom but delivered floods of data."*

We are familiar with this phenomenon of data explosion. Computerized systems collect data about a myriad of every day transactions: at supermarket checkouts, bank cash machines, airline tickets, phone calls, buying gasoline, etc. (the list is endless).

Health care providers, insurers and suppliers, and the pharmaceutical industries contribute their share

- Administrative systems log patient admissions and discharges, resource utilization in hospitals and practices, delivery and use of supplies, staff shift patterns and hours worked, cost of procedures and times taken for procedures.
- Every clinical act and its outcome are recorded. Patient records are being transferred to electronic form.
- Pharmaceutical knowledge increases daily, with new compounds, new dosage patterns, etc. Clinical trails generate huge amounts of data which must be analyzed and which, because of the controlled environment in which they are collected, should be an invaluable resource for future study.
- Marketing and sales by pharmaceutical manufacturers and pharmacies provide the return on the huge investments required bringing new drugs to the market. Every such sale can be captured as data and analyzed to help make more sales and predict future demand.
- Scientists in companies and hospitals gather mountains of experimental and laboratory data.

Information Systems managers are only too aware of this data explosion. They continually have to upgrade computers with more disk storage. The IT industry bombards them with offers of the latest databases, data warehouses, data marts and a host of data translation, transformation and reporting tools all aiming to tame the data explosion.

8.2 Data Holds Knowledge

Data holds the record of the organization's performance in all of its business areas. A hospital, which has been admitting patients for many years, has data it can use to estimate accurately the likely cost of treatment and length of stay of patient. A health care insurer has data about all of its subscribers, including those who defaulted on payments and those who generated above average claims. Today these organizations use statistical techniques to estimate the overall average figures to help plan the future. If only they could understand the data for each individual patient (or group of similar patients) *in detail* – they could use patient history as recorded in the data, to help make management or clinical decisions. As these decisions would be based on actual outcomes, this would truly be evidence-based care planning.

8.2.1 Decisions From the Data

Look at aspects of business knowledge we can discover, or the *"decisions from data"* which can be made:

- *Marketing.* One can tell from results of past sales campaigns who is likely to buy what and when
- *Forecasts.* Can past demands for a resource be used to forecast the pattern of demand for upcoming weeks or months?
- *Customer loyalty.* Some people transferred to another health insurer last month; can one detect the "signature" of those who defected? Can it be determined which are profitable clients whom one wishes to recapture? Can one identify others like them who may be at risk of detecting and thus plan pre-emptive offerings to keep the profitable ones?
- *Fraud.* When did it occur and what were the key signs? Do other cases show the same signs?

Provided the relevant data has been collected, is available and reasonably accurate, the answers to these questions, as well as host of others, may be hidden in the organization's data. Yet it is the volume of data collected that makes it impossible for humans to understand.

8.3 Data Mining: A New Approach to Information Overload

Many vendors, consultants and analysts make data mining appear complex, difficult, mysterious and expensive. It may sometimes be complex (involving many parts), but it need not be mysterious or difficult.

Data Mining simply means:

8.3.1 Finding Patterns in Data, which we can use to Better, Conduct the Business

Some people, especially in the United States, use the term *knowledge discovery* instead of data mining. In this section, the terms knowledge discovery and data mining are used interchangeably. Both describe the process of discovering a non-obvious pattern in data that can be used to for making better business decisions.

It turns out that vast majority of applications boil down to finding a relatively small number of types of data patterns. Here are some examples:

- *Classification.* To which set of predefined *categories* does this case belong? In marketing, when planning a mail shot, the categories may simply be the people who will buy and who will not buy. In health care, they may be high-risk and low-risk patients.
- *Association.* Which things occur together? For example, looking at shopping baskets, one may find that people who buy beer tend also to buy nuts at the same time.
- *Sequence.* Is essentially a time-ordered association, although the associated events may be spread far apart in time. For example, one may find that after marriage, people buy insurance.
- *Clustering or Segmentation.* Is like classification except that the categories are not normally known beforehand. We might look at a collection of shopping baskets and discover that there are clusters corresponding to health food buyers, convenience food buyers, luxury food buyers and so on.

Data mining is not mysterious; it is simply applied common or business sense.

8.3.2 Data Mining can be Breakthrough Technology

For the first 25 years of computing, the emphasis was on automating existing business processes. Order entry systems became online transaction processing systems, and existing manufacturing, accounting and other systems were implemented in the computer. During the last 10 years much of the emphasis has been on managing these automated processes better. Executive Information Systems (EISs), management reporting tools and more recently Online Analytical Processing (OLAP) tools have allowed management to observe the performance of the business as recorded by these computerized processes.

Data mining is *evolutionary,* in the sense that is an extension of the EIS. It can be used simply to uncover more subtle patterns in the data that help optimize the existing processes.

However, data mining can also be *revolutionary;* it can enable new, proactive business processes. The main reason for this is that in the era of the EIS, and with many OLAP tools, only summary information is available. These reports might indicate that fraud is a problem, but unless the detailed transactions are retained, there is no possibility to identify individual frauds. Another example is that until recently, banks charged one standard rate of interest on loans, depending on the circumstances of the borrower. This means that there are higher and lower risk borrowers in each interest rate band, the low risk borrowers are paying more than they should, effectively subsidizing the higher risk borrowers. With modern data ware houses and data mining tools, it is possible to segment customers according to the risks involved ultimately to charge an interest rate based on individual cases, rather than average risk. This model of lending is fundamentally different process from that based on average risk.

8.3.3 Data Mining Process in an Information System

Many enterprises now have a multilevel-information hierarchy. Business transactions are captured in an online database. This is optimized for high throughput and high security of transactions. This usage is incompatible with the requirements for analysis and management reporting. Every CIO dreads the killer SQL statement that could lock up the entire transaction database for minutes and end annoying and possibly losing the customer and staff.

For this reason, many have implemented a *data warehouse*. A data warehouse can be a complete copy of all the transaction information together with customer and product details. The warehouse is, however, free of the requirement to complete thousands of online transactions a second. It is organized to support queries, report writing and deeper analysis of information. Typically a warehouse is topped up with the most recent transactions from the online database (say) hourly, or overnight.

Such a warehouse rapidly becomes very large, and different departments tend to want to look only at different subsets of the data. For this reason, some companies are implementing data marts, effective departmental subsets of the enterprise data warehouse.

The only real prerequisite for data mining is a business problem plus relevant data. So data mining can be carried out on any data source. However, pattern finding is very demanding of computer power so it is unusual to mine the transaction database directly. More typically, mining is carried out on a data warehouse or data mart.

It is also common for data mining to require, or benefit from, additional data. This is often brought – in-geo-demographic or customer lifestyle data, which is combined with the organization's data about their own customer's behavior.

Successful data mining requires both business knowledge and some analytical ability. In the past experience, the former has proven most crucial, as business knowledge and common sense can go a long way toward steering the user into reasonable use of data mining tools.

In theory, the process of data mining is straightforward. Once a business problem has been identified, data, which is likely to have a bearing on that problem (or all data), is assembled. In practice, this is the most time-consuming part of data mining; data is rarely complete, and there may be tedious, but not usually too difficult work to be done in combining separate data sets.

Ideally, there will be cases that we think may contain a pattern related to the goal. Say we are trying to reduce the frequency of cases where there is an unplanned re-admission of a patient. We study a reasonable and representative selection of recent cases where such re-admissions occurred and also select at least an equivalent number of broadly similar cases where there was no unplanned re-admission. The approach is to divide this data into at least two subsets. One or more will be used for *learning* – "training" the models to distinguish between the re-admitted cases and the others. Different subsets of the data will be used for testing the ability of the model to correctly predict whether or not there is a re-admission. The model is used only if it achieves a consistent, reasonably accurate, prediction on these test cases.

Decisions from Data

The result of data mining is one or more decision models. These must be tested rigorously. Once tested, these decision models can be deployed in many different places throughout the enterprise. For example, a model may be trained to recognize cases of bad credit risk; this would be deployed at the sales transaction (or possibly the sales quotation) process. Another model might alert management to falling sales (or rising sales); this might be run immediately after the data warehouse is updated with the most recent transaction information. Both these models are *intelligent agents*, whose intelligence lies in applying the business rules and patterns learned during the modeling process.

In the health care domain, one might use a model when a patient is to be admitted for a particular treatment. Such a model, trained with information obtained from similar past cases, might estimate the total cost of treatment, or the likely length of stay. If there is enough detail in the past histories, this model may be able to provide forecasts for that admission, which are more accurate than assuming the average for the standard procedure in each case. Whether this is technically possible depends on the richness, accuracy, and completeness of the available data. Whether or not it is desirable might be a management or an ethical decision.

Also, as indicated above, models may suggest new processes. In a company, which sells insurance policies, for example, customer and product segmentation may indicate success in development of a new product and the customer group to which it could be profitably promoted.

8.3.4 Characteristics of Data Mining

As indicated above, data mining is normally about detail data (not summary data). If we want to identify individuals who will respond to a particular medication we have to look at details of the individual, not some average across many patients. Of course, if the business question is about summary data – for example, predicting the total annual costs of a department – then it may be appropriate to learn from such total costs. Data mining considers data at the appropriate level of detail for the business problem.

Data mining, despite the impression given by some is not blundering blindly about in data hoping to find something interesting. Many years ago gold miners worked by, washing away mountains. Today, effective miners know what they are looking for. They take samples before mining in earnest, and deploy business knowledge (geology, mineralogy and so on) before starting work.

The best data mining is like this intelligent prospecting. It works best if we have known patterns to learn from. These might to known clinical successes and failures, known re-admissions, bad debts, already detected fraud cases or known of operational failure. In each case, the goal is to learn the *signature* in order to recognize these cases again and take action to ensure good outcomes, prevent bad outcomes or mitigate their effect.

Although this technique of learning in *supervised* mode (i.e. with known cases whose signature we are trying to learn) is usually the simplest approach; it is possible to learn unsupervised. For example, we can apply data mining techniques to learn normal behavior (without known cases of normal and abnormal behavior) and so create a model, which will warn of deviations from the norm, whenever they occur. This is the basis for detecting frauds in cash transactions, for example.

Understand the Business or Clinical Problem First

As with most projects, the first requirement is to have a clear idea of what problem we are trying to solve; the more precise this is better. Avoid projects, which say, "Let's explore the data and see what we find." Projects which say "let's try and segment our customers into different types" are possible, though they are often much more tractable if the preferred groupings are first identified on business grounds (e.g. find high profit and low profit groupings). The easiest project has a precise focus. "Can the profile of cases which led to unplanned re-admissions be identified?" Why are they different from other similar cases? Can better predictions be made for unplanned re-admission?"

This overall approach is summarized as, first define a business goal, then deploy the data mining technology on the pertinent data and models, then deploy the models to reap the benefit.

8.3.5 Data Mining Technology

The idea of trying to turn data into useful information is not new. Traditional techniques include statistical analysis and data visualization (i.e. plotting graphs and charts), and these are the heart of the most Executive Information Systems (EIS) or Business Intelligence (BI) products. OLAP (Online Analytical Processing) allows users to interactively select subsets of the total data and produce on-demand management reports.

With these techniques, however sophisticated, the computer is acting in a *passive* role. The user says what set of attributes is to be displayed, and the computer "merely" computes the relevant graphs and reports. However visually appealing the graphical displays may become, and no matter how fluently the user to the can drill into the data to discover more detail, it remains the *responsibility of the user to identify the interesting trends*, or request the fruitful cross-correlation's. When faced with mountains of data, even the most diligent user may miss profitable opportunities.

These passive techniques worked well when only 10 or 20 pieces of information were available for each case. Then it was possible to examine all the cross-correlation's or plot all the combinations of four, five or even six factors that can be displayed on a computer screen. However, today, hundreds of pieces of information about each case are available, thousands if we are including transaction or case histories. Trying to find the appropriate combination of field values, which best predict a particular set of outcomes is like searching for a needle in a haystack.

While these passive techniques are important parts of the data-mining toolkit, the excitement arises from new technologies that *actively* support the discovery of new patterns. Using *machine learning techniques* from artificial intelligence (for example, neural networks and rule induction), the computer takes an active role. The user says "devise a method for predicting these business goals (it could be store profitability, likelihood of a prospect purchasing a product, the best treatment for a medical condition) taking given factors as input." Typical factors might be customer demographics, geographical location of sites, medical history – anything about the situation for which we have historical data .The algorithms "discover" which factors contribute to each goal and can consider all factors if necessary. The result is a decision procedure which, given the values for all the relevant input factors provides forecast (s) for the business goal(s).

8.3.6 Technology Limitations

It is not algorithms that discover knowledge. The function of an algorithm is to find a pattern, perhaps a *coincidence*, in the data. Only human expertise in the form of knowledge of the meaning and context of the data, can decide both, how the data might reasonably be analyzed and how to interpret and evaluate any results produced. It uniquely remains the reasonability of the professional to rigorously test any model or production and crucially to decide whether to act on its "recommendations." This is especially true in the health care and pharmaceutical area where decisions are literally matter of life or death.

But the people with the business or medical or expertise – often the "data owners" – are seldom skilled in analysis technologies and even less in machine learning. Until recently, they required an additional "technology" expert usually an external consultant, to operate the analysis technology on their behalf. As the sophistication, and hence the complexity, of analysis techniques increases and more different techniques are used together, the gap between the business professional and the technical analyst widens.

This situation is not ideal in a business sense because:

- The cost of such expertise is high.
- When modeling is complete, the data owner is left dependent on what may be an external, and still expensive, resource.

Many banks and credit card companies in the United States are under dilemma because they rely on models built at some considerable expense by external modeling companies for credit scoring direct marketing. As the pace of business increases and competitors repeatedly introduce new products to address gaps and niches in the market, customer behavior changes. Unless models are constantly updated they become obsolete. But the cost and time scales of external modeling (and also the costs of deployment) inhibit the banks from updating their models frequently enough.

The gap between business knowledge and analytical knowledge is even more important during modeling.

- The data owner is removed from direct participation in the analyses, limiting the use of his business knowledge to a time-consuming feedback process.
- Yet data mining is really more of an art than science. It is very hard work to follow train of thought hunches, if the business professional has to keep telling the technology expert what to do, and then wait minutes or often hours or days for the result.

8.3.7 BBC Case Study: The Importance of Business Knowledge

SPSS Inc. was retained by the BBC to develop a model that would predict the audience share that a proposed new TV program would achieve given it was transmitted at a particular time. This appears to be an ideal data mining application; the BBC has years of historical data showing what audience share watched each program. However, the context of a TV program is quite complex. The data includes:

- The proposed TV program (time of showing, genre, target audience type, star presenter, etc.)
- Preceding and following programs on the same channel
- Programs at the same time or overlapping with proposed program, on competitor channels
- The weather
- Major public or sporting events
- Time of year (e.g. winter, summer, etc.)
- Whether or not the program is syndicated across many regions

The project was carried out with one year's viewing data for prime time, between 6:00 and 10:00 in the evening. A neural network model was able to predict audience share within plus or minus four percent. It was not assumed to be particularly accurate, but the BBC was pleased. The prediction had similar accuracy to their best program planner's estimates. It took two years for these planners to become experts; with data mining the same accuracy is achieved in a few seconds. And as viewing habits can be changed we could retrain in a matter of hours to provide an updated model.

But there were some problem programs where prediction was much more than four percent wrong. This was disappointing and without explanation. It turned out BBC was very excited about many of results, as the neural network prediction had matched their own human expert prediction. Armed with this "confirmation" of their own estimate, they could go back to management with renewed confidence and look for other reasons why the programs had under-performed.

While the neural network can make predictions, it is very much a black box and provides little insight into why the prediction is being made. By contrast, another machine learning technique, rule induction, does provide an "explanation" in the form of explicit rules. For the BBC, the induced rules gave insight into "reasons" for the predictions and how much each factor contributes to the predicted audience share.

But it is the time for a health warning about these so-called explanations. Rules induced by data mining tools are merely descriptions of patterns found in the data. They may be explanations "which causal dependency"; equally they may be coincidences, and they may suggest causality where none is present. In general, knowledge of the subject is required to decide.

In the BBC project, the rule induction tools generated the following rule:

> *"Any program (X) which follows the UK soap opera will achieve six percent less share that if X is put on at any other time."*

We might be tempted to draw the conclusion that UK soap operas are so bad that they cause the audience to turn off (or switch channels), leaving them so irritated that they wont turn on (or switch back) for the next program, no matter how good it is.

However, the expertise of BBC planners immediately provided a much better explanation. UK soap opera figures were dominated by one program, East Enders. This achieved the highest market share of any BBC TV program. So the commercial channels specifically targeted this large audience by putting on one of their best programs (a crime serial called The Bill) immediately after East Enders. The data mining had found a compact, accurate description of a pattern in the data, but it was not an explanation.

This case illustrates the importance of having business knowledge and experience intimately associated with the data mining and interpretation of the results. It was the desire to bridge the gap between professional and business data owners and the technical analysts and modelers, that led SPSS to develop Clementine – an advanced, interactive, data mining system for end-users.

Project Clementine

In 1992, Integral Solutions Ltd. initiated "Project Clementine", aiming to build a comprehensive data mining system accessible to business and professional end users such as doctors. To make the advanced analysis techniques available to data owners requires:

- a variety of techniques including visualization , statistics and machine learning
- packaging with technology details hidden from the user
- delivery in an intuitive, easy – to-learn tool

The main technique to achieve this is "visual programming". From palettes, the user selects icons representing data sources, manipulation, graphs, reports, learning algorithms, etc. The icons are connected to define data flows, and edited through pop-up dialogues to define the details of the processing involved. The stream of icons provides a natural representation of the screen of the data mining process.

Users are encouraged to interact with data. Data features and patterns – a cluster, for example – can be identified using the mouse; users then generate icons that select cases corresponding to the defined regions. These facilities allow rapid data exploration, and formulation and testing of hypotheses based on observed features.

The configuration of the machine learning engines is automatic. Users express only high-level preferences – such as "favor generality over accuracy"– and Clementine configures the tools by considering both user input and by examining the structure of the data. The user is protected from consideration of the details of the technologies involved. (Tools can also be used in "expert" mode.)

Feedback from users suggests that this approach does indeed make sophisticated analyses accessible to professional and business users. Clementine is now in use in such areas as: banking, insurance, telecom, electricity, real-time financial trading, dentistry, childcare, pharmaceuticals and toxicology; with users ranging from IT specialists and dedicated data analysts to financial managers and biochemists.

In health care, the duty of care which physicians and management owe to patients requires that responsible professionals are fully involved in, and fully comprehend the implications of the decision models, which influence the choice of patient care. A user interface such as that of Clementine makes this possible.

8.3.8 Some Medical and Pharmaceutical Applications of Data Mining

Described briefly are two applications of data mining in science and medicine:

- *Drug discovery.* Very large databases now exist describing the composition and properties of countless chemical compounds. Data mining is now being used by pharmaceutical companies to research for promising substances that may from the basis of new drugs.
- *Skin corrosively.* Univeler's Colworth laboratory in the UK is responsible for assessing the effect on the human skin of possible new cosmetic, skin care and household products. At present, there is great reliance on animal testing. Data mining is being used to learn the known effect on the skin of known chemicals and hence, predict the effect of new preparations. The effect is to allow much more the screening to be done using computers and reducing the need for animal testing.

The range of applications is limited only by the availability of systematic data.

8.3.9 Why Does Data Mining Work?

Data mining can be remarkably effective and can find patterns and relationships that elude other techniques. Some of the reasons for this are:

- Since it works simply by learning from data it can (with right tools) be easy-to-use. This means business experts can contribute directly. This contribution often provides the magic ingredient.
- Results of data mining are very easy to test. Since the goal is to "predict" the past, the results are known; accuracy of the models can easily be determined.
- Models are rooted directly in the business experience. We do not have to depend on the skills or intuition of a programmer to synthesize a model. They are created automatically from the data and represent an unbiased distillation of the business experience.
- Decision models are generated automatically. This is a fast, cost-effective procedure and there should be no (programming) errors in the model.
- Models are easily updated by re-learning.
- Many of the techniques can handle many input factors (Clementine has used 7000/record) while the same techniques can ignore input factors, which don't contribute to a particular decision.
- Some of the techniques, notably neural networks, can discover complex non-linear models. It takes considerable expertise in classical statistics to deal with non - linear relationships.
- Good data mining tools (Such as Clementine) allow users to mix and match many techniques to solve problems. User's confidence in the results is increased if multiple techniques are deployed, and they all provide similar predictions.

8.4 Summary

Data mining is a new approach to finding patterns in data. In addition to using traditional techniques of data visualization and statistics the modern data mining toolkit includes machine-learning algorithms. These algorithms synthesize the models directly from data. The key to successful data mining is to define the business or clinical problem to be solved. New knowledge is not discovered by the algorithms, but by the user. The user's business or clinical knowledge is essential to determine what factors to consider for particular applications, and how the various input factors may need to be combined. Clinical, business or scientific expertise is always required to determine whether models can safely or usefully be employed. Some data mining tools like Clementine package the techniques so they can be used directly by and the results can be understood by business or clinical professionals.

8.5 Review Questions

1. What is data flood?
2. What are the data mining technology limitations?
3. Explain about the mining in BBC – Case study

9

Data Mining & KDD

Objectives:

- In this section, the ideas of Data Mining and KDD process are presented.
- The objective of data mining is to convert data and information into profitable knowledge for actionable strategic decisions.
- DMI: Data mining is traditional data analysis methodology updated with the most advanced analysis techniques applied to discovering previously unknown patterns.
- DMII: Data Mining is the activity of extracting hidden information (pattern and relationships) from large databases automatically: that is, without benefit of human intervention or initiative in the knowledge discovery process.
- DMIII: Data Mining is the step in the process of knowledge discovery in databases, that inputs predominantly cleaned, transformed data, searches the data using algorithms, and outputs patterns and relationships to the interpretation/evaluation step of the KDD process.
- KDDI: Knowledge discovery in databases in a process that requires hypothesis or model formulation, hypothesis or model testing, and derivatively all the data, techniques, and sub-processes necessary to bring hypothesis or model testing to a successful conclusion.
- KDDII: True knowledge discovery in databases is the process of automated data mining applied without benefit of human intervention or initiative.
- KDDIII: *"Knowledge discovery in databases* is the nontrivial process of identifying valid, novel, potentially useful, and ultimately understandable patterns in data."
- The historical roots of data mining come primarily from two different directions: from statistics and from artificial intelligence.
- The statistical culture in data mining emphasizes the role of predictive modeling (PM). The artificial intelligence culture emphasizes the role of knowledge discovery (KD).

S. Sumathi and S.N. Sivanandam: *Data Mining & KDD*, Studies in Computational Intelligence (SCI) **29**, 231–241 (2006)
www.springerlink.com © Springer-Verlag Berlin Heidelberg 2006

- We discuss these two data mining cultures and some key issues in data mining from this perspective.
- The PM tradition favors accuracy over understandability; the KD culture favors understandability over complexity.

Abstract. In this section, the ideas of Data Mining and KDD process and cultures of data mining taken from National Center for Data Mining, University of Illinois at Chicago are presented.

The objective of data mining is to convert data and information into profitable knowledge for actionable strategic decisions. The process consists of three elements: data, mining and business issue. The data involved are large databases of geo-demographic, demographic, transactional and relationship type, obtainable from inside as well as outside the organization. They are both current/instantaneous and historical in nature. The mining process can go back to query systems and OLAP (on line analytical processing). However, we shall limit ourselves to the *knowledge discovery* part of data mining. Knowledge discovery is driven by advances in technologies, in particular, in Information Technologies, computing power, software performance, machine learning, and optimization. It allows us to dig deep into masses of data to detect hidden patterns, interesting behaviors useful trends or niche profiles that we had not suspected to exist. In order to detect these useful elements for particular decision, the business issues have to be addressed. The strategy has to be clearly articulated with well-defined business objectives. In a practical approach to data mining, other factors such as data mining tools, infrastructure, data management system, human expertise, solution architecture, and the use of data also contribute to the success of data mining endeavor. In this section, we also discuss these two data mining cultures and some key issues in data mining from this perspective.

9.1 Data Mining and KDD – Overview

Data mining is an idea based on a simple analogy. The growth of data warehousing has created mountains of data. The mountains represent a valuable resource to the enterprise. But to extract value from these data mountains, we must "mine" for high-grade "nuggets" of precious metal – the gold in data warehouses and data marts. The analogy to mining has proven seductive for business. Everywhere there are data warehouses, data mines are also being enthusiastically constructed, but not with the benefit of consensus about what data mining is, or what process it entails, or what exactly its outcomes (the "nuggets") are, or what tools one needs to do it right.

Data mining as a field is not yet through with the process of definition and conceptualization of the scope of the field. There are at least three distinct concepts of data mining that are being used by the practitioners and vendors.

DMI: Data mining is traditional data analysis methodology updated with the most advanced analysis techniques applied to discovering previously unknown patterns. A specific instance of this concept is stated more explicitly and with a more commercial orientation is provided by the SAS Institute [http://www.sas.com/feature/4qdm/whatisdm.html].

SAS defines data mining *as the process of selecting, exploring, and modeling large amounts of data to uncover previously unknown patterns for a business advantage.*

In specifying its notion of data mining further, SAS describes it as involving a five-step process: Sample, Explore, Modify, Model, and Assess, or the acronym SEMMA. The difference between SEMMA and traditional methodology used in statistical analysis is hard to see with the naked eye, though we emphasize that methodology and tools or techniques are different things, and we are certainly not saying that because SAS's SEMMA methodological approach is traditional, but it would not or could not incorporate the most advanced data mining tools.

DMII: Data mining is the activity of automatically extracting the hidden information (pattern and relationships) from large databases: that is, without benefit of human intervention or initiative in the knowledge discovery process. In this view, data mining is knowledge discovery in databases or at least it is automated knowledge discovery in databases.

DMII is the data-mining concept implicit in the advertising collateral of many specialized data mining companies. Customers are promised an automatic process of model development that requires little or no human interaction from sophisticated data analysts. The data mining package supplies the necessary high quality analysis, and business users are promised that they can achieve knowledge discovery and predictive success on their own with little investment of time or effort compared to what is necessary with nondata mining (often labeled as traditional statistical) techniques. We do not see this concept as much outside of the vendor literature, but it is either present or closely approached in many articles on data mining. It advocates drawing a sharp distinction between data-driven tools using automated discovery-based approaches and user- or verification-driven tools using hypothesis-testing approaches. The hypothesis-testing tools are seen as the ones limited by the skill and experience of humans, while the data mining tools are seen as free of human initiative or assumption, and empowered by pattern-matching algorithms. Most importantly, the hypothesis-testing tools are seen as "verifiers", while the data mining tools are seen as "discoverers".

DMIII: Data mining is the step in the process of knowledge discovery in databases, that inputs predominantly cleaned, transformed data, searches the data using algorithms, and outputs patterns and relationships to the interpretation/evaluation step of the KDD process. DMIII is the statement of the view of data mining emerging from the 1994 AAAI workshop on KDD, the KD Mine [http://www.kdnuggets.com], and S*I*FTWARE [http://kdnuggets.com/siftware.html] web sites and the recent Advances in Knowledge Discovery and Data Mining volume (U. Fayyad, G. Piatetsky-Shapiro, P, Smyth, and R. Uthurusamy (eds.), 1995). The definition clearly implies that what the data mining (in this view) discovers is the hypotheses about patterns and relationships. Those patterns and relationships are then subject to interpretation and evaluation before they can be called knowledge.

The commitment to search algorithms in DMIII does not imply a completely automated data mining process. Data analyses must be algorithms with some degree of search autonomy to qualify as instances of data mining, but they will also use human initiative in the areas of background knowledge (a specification of which is required for applying some machine learning and case-based reasoning techniques), model selection and specification, input and output variable selection and specification, in constraining model parameters and in other ways. In short, the data mining process described by those adhering to DMIII is one in which automated search algorithms play a vital role in complex iterative "interaction, protracted over time, between a *human* and a database."

9.1.1 The Idea of Knowledge Discovery in Databases (KDD)

The three definitions of data mining are also closely associated with three apparently different concepts of KDD. DMI is associated with no explicit concept of KDD. But for the purposes of discussion, we will assume that supporters of DMI believe that KDD refers to a process that uses computer-based data analysis as a primary means of investigation, and that produces scientifically validated knowledge. Here are the three KDD concepts.

KDDI: Knowledge discovery in databases in a process that requires hypothesis or model formulation, hypothesis or model testing, and derivatively all the data, techniques, and sub-processes necessary to bring hypothesis or model testing to a successful conclusion. In this view, data analysis includes both exploratory and confirmatory data analysis, and the latter is necessary for hypothesis or model testing. That outcome of hypothesis or model testing is knowledge discovery, even if the knowledge discovered is a negative finding that some hypothesis or model is not knowledge.

KDDII: True knowledge discovery in databases is the process of automated data mining applied without benefit of human intervention or initiative. According to this view there is no distinction between data mining and KDD. The data mining does not just generate hypotheses. It produces valid knowledge that business can apply without fear of bad results.

KDDIII: "*Knowledge discovery in databases* is the nontrivial process of identifying valid, novel, potentially useful, and ultimately understandable patterns in data." Further, this process includes five steps: data selection, data preprocessing, data transformation, data mining and interpreting, and evaluating mined patterns and relationships. This process is interactive and interaction with KDD users is heavily involved at every step. Many loops may occur between steps. There is no deterministic progression assumed from one step to another. Also, the interpretative and evaluative step can involve returns to any of the previous steps, any number of times.

9.1.2 How Data Mining Relates to KDD

The relationships between data mining and KDD are different for the three approaches generated by the DM/KDD pairs, and are implicit in the definitions already presented. We will consider these implications in the same order the contrasting DM and KDD concepts were presented.

Traditional Data Mining

DMI and KDDI, equate data mining with KDD and since they do not distinguish it from previous investigative processes, they essentially equate both with previous investigative processes and methodologies of analytical and statistical modeling. SAS's SEMMA data mining process could be used equally well to describe traditional processes of analysis followed for years by SAS users. Some of the tools, such as Neural Networks and Tree-based models may be different, but the patterns of investigation, and more important, of validation, are essentially the same. It is hard to escape the conclusion that for this approach, data mining is traditional modeling and analysis updated with the addition of some new technique and incorporated into the commercially relevant data warehouse framework.

In drawing this conclusion, we do not mean to be pejorative or to express criticism for traditional approaches. If the DMI/KDDI explication of data mining makes the most sense for further development of the field, so be it. But it is important to recognize the approach for what it is, and to refrain from claiming methodological novelty, when we are really talking about progress in software and hardware tools for data analysis.

In the DMI/KDDI approach, also, data mining is not restricted to the step of hypothesis formation. The SEMMA model assessment step is a validation step. That is why data mining and KDD can be so easily equated. But though data mining and KDD are equated, the data mining/KDD process is not viewed as fully automated. The traditional approach recognizes the vital role of human-initiated hypothesis and model formation, and computer-based, partly automated, exploratory and confirmatory data analysis in data mining.

Automated Data Mining

The DMII and KDDII definitions of data mining and knowledge discovery in databases also equate the two. But in contrast to DMI/KDDI, they imply that ***automated data mining,*** as expressed by DMII, ***includes both hypothesis formulation and scientific validation in the algorithmic process.*** That is why, it is logically consistent for proponents of DMII to claim that data mining and knowledge discovery are one and the same, and it is also the reason why they view the knowledge discovery process as one that excludes human intervention or initiative. But are there any commercial data mining products, or even any research efforts that fit this definition? The answer is: it depends on what we mean by scientific validation of KDD, a matter on which there is far from universal agreement. All data mining products produce patterns and relationships from appropriate data input. So, all can produce hypotheses. But products may vary in the specific validation criteria they use empirical fit to sample data mining process. All data mining products seem to use empirical fit to sample data, as a general validity criterion, but the specific algorithms evaluating empirical fit may vary from product-to-product, and the criteria for measuring empirical fit may also vary. Also empirical fit of patterns to data is not the whole story, sometimes patterns can be over fit to data, and different products employ different tests of over fit.

Apart from empirical fit of patterns to data, there are many other validation criteria that can be applied to interpretation and evaluation of the results of data mining. To fully understand this, it is necessary to step back from the perspective of one or two products of a particular type, and to recognize that there are many data mining models out there, and they offer alternative theories of data. Which one is the right one for our data mine? If we compare the results of a number of alternatives, what criteria do we use to compare them? What if all fit the data well empirically, but the criteria of empirical fit are either not readily compared because of the use of different fit statistics, or a comparison is not meaningful because we cannot tell which model involves over fitting? What if the same data-mining product has generated alternative patterns based on slightly different input assumptions with number material difference in empirical fit? How do we choose then?

Recent KDD research has specified a number of criteria apart from empirical fit to sample data as relevant including: **predictive validity** on new data, **novelty** of the patterns discovered by the data-mining tool, **pragmatic utility** of the patterns as measured by some utility function, **ultimate understandability** of the patterns, and a composite of these called **"interestingness."** However, even though these criteria can be listed, research on applying them is not far advanced, and promises to be difficult to implement. Nor are these criteria in any sense exhaustive. Almost anyone in the KDD field today could specify additional criteria or alternatives for at least some of the criteria of validation and provide an equally plausible defense of these as reasonable validation criteria.

Validity criteria in KDD are a developing area of research, and there is no consensus yet on standards, procedures, or algorithms for measuring validity. Without such a consensus the DMII/KDDII concept of automated data mining is premature. The results of data-mining activities cannot now be validated by an algorithm or algorithms incorporating generally agreed upon validity standards. And there is no prospect that such validation will be available in the near future.

So, whatever the preferences of advocates of the DMII/KDDI position, the outcome of current automatic data mining investigations in the DMII/KDDII sense, must be viewed as highly hypothetical, exploratory in nature, and subject to a careful validation analysis before they are relied on for practical applications considering the exploratory nature of results using the automated data mining perspective, we believe that vendors and consultants who are selling data mining on the basis of the DMII/KDDII position, are overselling data mining.

Data Mining as Part of KDD

The DMIII/KDDIII position is probably the one with the most current momentum. It attempts to distinguish data mining from traditional analyses by emphasizing the automated character of data mining in generating patterns and relationships, but it also clearly distinguishes the data mining from knowledge discovery, by emphasizing the much broader character of KDD as an overarching process, including an interpretation and evaluation step distinct from data mining and relying more heavily on human interaction. In a very real sense DMIII/KDDII is a middle way between the other two positions.

But if DMIII/KDDIII is a middle way, that does not necessarily, mean it is the right way. Sometimes compromises are just unstable platforms for methodological development. DMMII/KDDIII seems to postulate no difference from the traditional data mining approach in the area of validation or confirmatory analysis. The difference is in the area of exploratory data analysis where practitioners holding this position emphasize the use of automated methods to generate patterns, while practitioners of DMI/KDDI do not talk quite so much about automation, but talk more about using a variety of techniques including human initiative to guide exploratory analysis. But is this difference a real methodological difference between the two camps, or just a way of maintaining a distinct identify of placing old wine in new bottles?

Current studies by participants in the KDD group make overwhelmingly clear the exhaustive interaction between human and machine that is part of the data mining process in a real KDD project. The iterative process to prepare for data mining and to implement it follows the careful investigative pattern of traditional analysis. The algorithmic techniques are more powerful than they were ten years ago, but there is no methodological requirement that pattern generation be guided solely by automated data mining techniques. Instead, the requirement is a focus on techniques with a certain degree of search autonomy – a small difference from the viewpoint of traditional data mining at best.

9.1.3 The Data Mining Future

The data mining foreseeable future will involve an appreciable human component, whether we are taking the viewpoint of either DMI or DMIIII. The problems inherent in model and variable selection, in measurement and dynamic model construction, and in pattern validation methodology all guarantee that.

But, it is also true that we will continue to make progress in the area of adaptive intelligence that underlies data mining. DMII/KDDIII may be an invalid construct now, but research on computational models of theory evaluation will eventually bring us much closer to having measurement models of theory evaluation will eventually brings us much closer to having measurement models of validity and to having agreement on both the models and the criteria they incorporate. Also, the new analysis techniques (Neural Networks, Genetic Algorithms, Machine Learning, Bayesian Belief Networks, Fuzzy Engineering, Chaotic Dynamics, etc.) that have come to prominence in the last 10 to 15 years, and that are now becoming fully commercialized, will continue to advance in power and sophistication and to become more fully integrated in analysis methodologies that we can partially automate. For now, the practical task at hand is to bring to bear the most powerful analytical techniques at our disposal to the problem of making private and public enterprises more adaptive. Practically speaking, this means analysis of enterprise performance in all its aspects through the use of the data in data warehouses and data marts. Exploratory analysis of this data is called data mining (DMI and DMII). Sometimes confirmatory data analysis is also included in data mining (DMI). The important thing is that, for the foreseeable future, good data mining cannot be done without significant human interaction between a human data miner and the computer-based software extension. That is because data mining is not automatic. And the dream of making it so is, at best, an ideal motivating long-term development.

9.2 Data Mining: The Two Cultures

Data mining is about finding patterns in data. The importance of data mining has grown dramatically as the amount of archived and warehoused digital data has grown. The historical roots of data mining come primarily from two different directions: from statistics and from artificial intelligence. The statistical culture in data mining emphasizes the role of predictive modeling (PM). The artificial intelligence culture emphasizes the role of knowledge discovery (KD). In this section, we discuss these two data mining cultures and some key issues in data mining from this perspective.

9.2.1 The Central Issue

The central issue is simple. To illustrate it, consider using data mining for fraud detection. In the PM tradition, given a credit card transaction, telephone call, or insurance claim x, the goal is to predict whether x is fraudulent as accurately as possible. This is usually considered to be a classification problem (0 means no fraud, 1 means fraud). A classifier examines the attributes of x (such as the number of transactions during the past hour) and returns as 0 or 1, indicating whether the transaction is fraudulent or not.

Generally, more accurate classifiers are more complex. For example, a good fraud classifier for a large data set using a tree-based classifier might contain thousands of nodes. At best this is difficult to interpret. This is a basic trade-off. In the PM tradition, increased accuracy is traded for ease of interpretation.

On the other is the KD tradition, the goal is to extract useful from large data sets. To be useful, these facts must be easily interpretable and easily actionable. For example, an algorithm for extracting shallow trees might reveal that low dollar transactions at each machines outside of certain retail stores is highly correlated with fraud. The action here might be to put in place a rule, which defers subsequent transactions for certain types of high priced retail goods. This illustrates another basic trade-off. In the KD tradition, ease of interpretation and implementation is trade for accuracy.

9.2.2 What are Data Mining and the Data Mining Process?

Data mining is one step in the data mining process. The definition of data mining and of the data mining process differs somewhat between the two cultures.

A standard definition of data mining from the KD perspective is given by Fayyad, Piatesky-Shapiro, and Symth (1996): "Knowledge discovery in databases in the nontrivial process of identifying valid, novel, potentially useful and ultimately understandable patterns in data."

Here is the PM perspective: Data mining is the automatic discovery of associations, clusters, changes, patterns, anomalies, and other significant structures in large data sets and the exploitation of these discoveries to improve predictive modeling.

Recall that the data mining process consists of a sequence of steps, which are usually repeated, in an iterative fashion. The process typically includes, 1) data preparation and cleaning, 2) data warehousing, 3) identifying relevant predictive attributes, 4) computing derived attributes, 5) data reduction and attribute projection, 6) extracting patterns relevant to the predictive attributes using one or more mining algorithms, 7-pm) predictive modeling, 7-kd) knowledge extraction 8-pm) scoring of operational and warehoused data 8-kd) interactive data analysis and discovery 9) validation, report preparation, and related activities, and 10) repeating the process as necessary.

Here Steps 7 and 8 are slightly different in the two traditions and is indicated by using the suffixes pm or kd. Data mining projects usually incorporate aspects of both the PM and KD cultures. A common strategy is for groups of analysts and modelers in an organization to focus on the KD aspects of data mining and for IT staff with operational responsibility to focus on the PM aspects of data mining.

9.2.3 Machine Learning

The essence of data mining is machine learning and this is what occurs in Step 6 above. Both the PM and DM traditions, start with a space of learning sets L. Each element of the space is a particular learning set, that is, some data set which is to be automatically analyzed.

The PM tradition requires a space of models M. In the fraud example above, M is the space of binary classifiers (0 for normal and 1 for fraud)

In the PM tradition, data mining can be thought of as a map from L to a space of models M:

L→ M (PM perspective)

The input is a data set and the outputs are (one or more) models: the goal is to produce as accurate a classifier as possible.

Two measures are relevant here: the detection rate of the model and the rate of false positive. In practice, increasing the detection rate of model usually is accompanied by increasing the false positive rate.

The KD tradition replaces the space of models M with a space of assertions of predicates P. To be more concrete, a simple type of assertion is a conditional: if X the Y. For example, if a credit transaction is for less than $2 and the transaction occurs at a gas station, then the transaction is fraudulent. Here X is the conjunction that the transaction is for less than $2 and occurs at a gas station and Y is claim that the transaction is fraudulent.

From the KD viewpoint, learning can be thought of as a map from L to a space of assertions or predicates P:

L→P (KD perspective)

The input is a data set and the outputs are (one or more) assertions. The goal is for the predicate or predicates discovered to be as relevant and useful as possible.

Two measures are relevant here. An assertion has confidence c% if c% of the transactions that contain X also contain Y. An assertion has support s% if s% of the transactions contain X or Y.

When someone talks about automatically extracting patterns or automatically discovering information, what they really mean is that there is an algorithm which takes a learning set and produces a model (in the PM tradition) or which takes a learning set and produce one ore more predicates (in the KD tradition). When there was less digital data and more analysts, automating this step was not important. Today, with so much more digital data than can ever be analyzed the automation of this step (Step 6 in the data mining process) is a key enabling technology for a variety of scientific, engineering, and business problems.

A deeper understanding of machine learning requires that the two maps above be understood in a probabilistic framework in the sense that one tries to understand the probability that the model or assertions produced is accurate.

9.2.4 Impact of Implementation

Here, we discuss some of the practical and organizational issues in data mining projects. An executive involved in a data mining project is responsible for making sure that the results of the project can be effectively exploited by the organization. For projects with a KD focus this means that the modeling or analyst group understands the role of data mining for assisting them and are not threatened by a new technology and that reports summarizing discover is reach the relevant decision makers. For projects with a PM focus this means that the operational managers are included in the early discussions so that the predictive models produced can be easily exploited by the organization.

When designing and implementing data mining systems, the Data-Mining Administrator (DMA) must understand whether the primary goal is 1) to improve predictive modeling of an important business process (PM) or 2) to give analysts and modelers new knowledge and insights (KD). Those offering professional services involving data mining must also be aware of the same distinction.

9.3 Summary

This section described an overview of the data mining and KDD. Also it discussed the two data mining cultures. Both the predictive modeling (PM) culture and the knowledge discovery culture (KD) are essential to data mining. In some sense, data mining is about the interaction of these two cultures and the scaling up of traditional algorithms and systems from small data sets to the large datasets, which are common today. Machine learning is a step in the data mining process. In the PM culture, this step takes a learning set and produces one or more predictive models. In the KD culture, this step takes a learning set and produces one or more assertions (which are interpreted as discovered knowledge). The essence of data mining is that data mining automates this. With the amount of data growing so quickly it is simply number longer practical to develop all predictive models or assertions by hand.

The PM tradition favors accuracy over understandability; the KD culture favors understandability over complexity. Developing a good practical solution to a data mining problem understanding both the PM and KD perspective and implementing a solution by incorporating technique from both cultures. Some problems benefit from viewpoint emphasizing the PM perspective, others from a viewpoint emphasizing the KD perspective. It is important to understand both cultures and the expectations and objectives of the project team if an appropriate data mining solution is to be successfully developed.

9.4 Review Questions

1. Describe the five step process involved in SAS
2. Differentiate DMI, DMII and DMIII
3. Differentiate KDI, KDII and KDIII
4. What is traditional and automated data mining process?
5. What is the importance of machine learning in the discovery process?

10

Statistical Themes and Lessons for Data Mining

Objectives:

- Data mining is a growing discipline, which originated outside statistics in the database management community, mainly for commercial concerns.
- We examine here data mining definitions, tools, and how data mining could be used in official statistics.
- Data mining is made on the interface of computer science and statistics, utilizing advances in both disciplines to make progress in extracting information from large database.
- This chapter highlights some statistical themes and lessons that are directly relevant to data mining.
- Attempts to identify opportunities where close cooperation between the statistical and computational communities might reasonably provide synergy for further progress in data analysis.
- This section briefly describes some of the central statistical ideas we think relevant to data mining.

Abstract. Data mining is a new field at the frontiers of statistics and information technologies (database management, artificial intelligence, machine learning, etc.), which aims at discovering structures and patterns in large datasets. We examine here its definitions, tools, and how data mining could be used in official statistics.

Data mining is on the interface of computer science and statistics, utilizing advances in both disciplines to make progress in extracting information from large database. It is an emerging field that has attracted much attention in a very short period of time. This section highlights some statistical themes and lessons that are directly relevant to data mining and attempts to identify opportunities where close cooperation between the statistical and computational communities might reasonably provide synergy for further progress in data analysis.

S. Sumathi and S.N. Sivanandam: *Statistical Themes and Lessons for Data Mining*, Studies in Computational Intelligence (SCI) **29**, 243–263 (2006)
www.springerlink.com © Springer-Verlag Berlin Heidelberg 2006

10.1 Data Mining and Official Statistics

Data mining is often presented as a revolution in information processing. Here are two definitions taken from the literature:

U.M. Fayyad: "Data mining is the nontrivial process identifying valid, novel, potentially useful, and ultimately understandable patterns in data."

D.J. Hand: "Data mining (consists in) the discovery of interesting, unexpected, or valuable structures in large data sets."

The development of data mining is related with the availability of very large databases and the need of exploiting these bases in a new way. As J. Friedman in 1997 pointed out, the database management community has become recently interested in using DBMS for decision support. They realized that data, which have been collected for management organization such as recording transactions, might contain useful information for, e.g., improving the knowledge of, and the service to, customers.

The metaphor of data mining means that there are treasures (or nuggets) hidden under mountains of data, which may be discovered by specific tools. Of course this is not a new idea and there has been many developments in statistical methodology, which were oriented toward the discovery of patterns or models through techniques like EDA (exploratory data analysis) and multivariate exploratory analysis (dimension reduction methods: component, correspondence and cluster analysis among many others).

10.1.1 What is New in Data Mining is:

The availability of very large databases, which makes obsolete classical definitions of large samples: billions of records, terabytes of data are not unusual. More and more data are collected automatically, the massive use of new techniques coming from the computer science community like neural networks, decision trees, induction rules; commercial interests in valorizing existing information in order to propose individual solutions to targeted customers; new software packages, more user friendly, with attractive interfaces, directed as much toward decision makers as professionals analysts, but much more expensive!

It is interesting to remark that statisticians (those who are not reluctant to this new trend?) prefer the expression *data mining*, while computer scientists use knowledge discovery, but it is more a matter of expressing a belonging to a community than a real difference.

10.1.2 Goals and Tools of Data Mining

The purpose of data mining is to find structures in data. Following D. Hand in 2000, there are two kinds of structures that are sought in data mining activities: models and patterns. Building models is a major activity of many statisticians and econometricians, especially in NSI [National Statistical Institutes], and it will not be necessary to elaborate too long on this. Let us say that a model is a global summary of relationships between variables, which both helps to understand phenomenons and allows predictions. Linear models, simultaneous equations are widely used. But a model is generally chosen on an a priori basis, based upon a simplifying theory. Exploration of alternative models, possibly nonlinear or in a nonclosed mathematical form, is made feasible by DM algorithms. DM appears then as a collection of tools presented usually in one package, in such a way that several techniques may be compared on the same dataset.

DM does not reduce to decision trees, neural networks, or genetic algorithms, as some software vendors claim. DM algorithms offers extensive possibilities of finding models relating variables together: we have mentioned neural networks, decision trees, which finds nonlinear models, graphical models (or Bayesian belief networks), etc. give a valuable representation of relations between variables.

In contrast to the global description given by a model, a pattern is often defined as a characteristic structure exhibited by a few number of points: for instance a small subgroup of customers with a high commercial value, or conversely highly risked. Tools involved here are often cluster analysis, which is a well-known statistical technique, and also Kohonen self-organizing maps, based on neural networks.

Association rule discovery, or market basket analysis, is one of the favorite and perhaps new tools of data mining: its origin is in analyzing purchases in supermarket: one is interested in the percentage of customers whose purchase simultaneously two goods, more precisely in identifying couples of goods. A, B with high conditional probabilities of purchasing A, given that B is purchased. Of course the result is interesting only if P (A/not B) is low. The problem in pattern discovery is to be sure that patterns are real and useful: the probability of finding any given pattern increases with the size of the database. It is thus necessary to develop validation rules, and in this respect processing the whole database is not the best thing to do: sampling (or splitting the base into several sets) will be better than exhaustive processing. One should test if a model or a pattern remains valid in another part of the base than the one that has been explored. Usefulness should also be tested: associations are only correlations, and not causation and it is sure that promoting B will mechanically increase the purchase of A.

10.1.3 New Mines: Texts, Web, Symbolic Data?

Besides classical database where data is usually presented in the form of a rectangular array, new kinds of data are now present: Symbolic data, such as fuzzy data or intervals: the data is not known precisely but belongs to an interval, with or without a probability distribution.

Text mining: Most of the information that circulates is now digitized as word processor documents, and powerful techniques are currently available for a wide variety of applications. For example, one can categorize information from news agencies; analyze patent portfolios, customer complaints; classify incoming e-mail according to predefined topics, group-related documents based on their content, without requiring predefined classes; or assign documents to one or more user-defined categories.

Web mining is a new concept for statistical analysis of Website information (text mining) or frequentation, as well as of the behavior of the Web surfers.

10.1.4 Applications in Official Statistics

Of course in National Statistical Institutes there has always been some use of exploratory Data analysis, or of model choice algorithms, but as far as we know it seems that there are few, if not none, known applications of data mining techniques in the meaning of trying to discover new models or patterns in their databases by using the new tools described before.

It is not surprising because the main task of NSI is data production, and analysis is often done by different institutes. Furthermore it seems that the idea of exploring a database with the objective of finding unexpected patterns or models is not familiar to official statisticians who have to answer precise questions and make forecasts. One may consider that official statisticians are experts and do not need DM, which is an automatic process for end users who do not know what is hidden under their data. Statistical analyses are done generally if they can be repeated in a production framework. But as far as NSI manage large databases on population, trade, agriculture, companies, there is certainly great potentialities in exploiting their mines of data.

Let us point out a few fields where data mining tools could be useful:

Business statistics with special mention for innovation policy, financial health household equipment and savings health statistics, more precisely mortality and morbidity, in order to detect unexpected risk factors analysis of metadata information by means of text mining.

10.2 Statistical Themes and Lessons for Data Mining

Sta-tis-tics (noun). The mathematics of the collection, organization, and interpretation of numerical data, especially the analysis of population characteristics by inference from sampling (*American Heritage Dictionary*).

Statistics is enjoying a renaissance period. Modern computing hardware and software have freed the statistician from narrowly specified models and spawned a fresh approach to the subject, especially as it relates to data analysis. Today's statistical toolkit draws on a rich body of theoretical and methodological research (Table 10.1).

Table 10.1. Statisticians have developed a large infrastructure (theory) to support their methods and a language (probability calculus) to describe their approach to quantifying the uncertainty associated with drawing inferences from data. These methods enable one to describe relationships between variables for prediction, quantifying effects, or suggesting causal paths.

Area of Statistics	Description of Activities
Experimental design and sampling	How to select cases if one has the liberty to choose.
Exploratory data analysis	Hypothesis generation rather than hypothesis testing
Statistical graphics	Data visualization
Statistical modeling	Regression and classification techniques.
Statistical inference	Estimation and prediction techniques.

The field of data mining, like statistics, concerns itself with "learning from data" or "turning data into information." The context encompasses statistics, but with a somewhat different emphasis. In particular, data mining involves retrospective analyses of data: thus, topics such as experimental design are outside the scope of data mining and fall within statistics proper. Data miners are often more interested in understandability than accuracy or predictability per se. Thus, there is a focus on relatively simple interpretable models involving rules, trees, graphs, and so forth. Applications involving very large numbers of variables and vast numbers of measurements are also common in data mining. Thus, computational efficiency and scalability are critically important, and issues of statistical consistency may be a secondary consideration. Furthermore, the current practice of data mining is often pattern focused rather than model focused, i.e., rather than building a coherent global model that includes all variables of interest, data mining algorithms (such as any of the many rule induction systems on the market) will produce sets of statements about local dependencies among variables (in rule form).

In this overall context, current data mining practice is very much driven by practical computational concerns. However, in focusing almost exclusively on computational issues, it is easy to forget that statistics is in fact a *core* component. The term *data mining* has long had negative connotations in the statistics literature. Data mining without proper consideration of the fundamental statistical nature of the inference problem is indeed to be avoided. However, a goal of this section is to convince the reader that modern statistics can offer significant constructive advice to the data miner, although many problems remain unsolved. Throughout the section we highlight some major themes of statistics research, focusing in particular on the practical lessons pertinent of data mining.

10.2.1 An Overview of Statistical Science

This section briefly describes some of the central statistical ideas we think relevant to data mining.

Probability distributions. The statistical literature contains mathematical characterizations of a wealth of probability distributions, as well as properties of random variables – functions defined on the "events" to which a probability measure assigns values. Important relations among probability distributions include marginalization (summing over a subset of values) and conditionalization (forming a conditional probability measure from a measure on a sample space and some event of positive measure). Essential relations among random variables include independence, conditional independence, and various measures of dependence, of which the most famous is the correlation coefficient. The statistical literature also characterizes families of distributions by properties that are useful in identifying any particular member of the family from data, or by closure properties useful in model construction or inference, for example conjugate families, closed under conditionalization, and the multinormal family, closed under liner combination. Knowledge of the properties of distribution families can be invaluable in analyzing data and making appropriate inference.

Estimation, consistency, uncertainty, assumptions, robustness, and model averaging. An estimator is a function from sample data to some estimated, such as the value of a parameter. When the data comprise a sample from a larger actual or a potential collection governed by some probability distribution, the family of estimators corresponding to all possible samples from that collection also has a probability distribution. Classical statistics investigates such distributions of estimators in order to establish basic properties such as reliability and uncertainty. A variety of resampling and simulation techniques also exist for assessing estimator uncertainty.

Estimation almost always requires some set of assumptions. Such assumptions are typically false, but often useful. If a model (which we can think of as a set of assumptions) is incorrect, estimates based on it can be expected to be incorrect as well. One of the aims of statistical research is to find ways to weaken the assumptions necessary for good estimation. "Robust statistics" looks for estimators that work satisfactorily for larger families of distributions and have small errors when assumptions are violated.

Bayesian estimation emphasizes that alternative models and their competing assumptions are often plausible. Rather than making an estimate based on a single model several models can be considered and an estimate obtained as the weighted average of the estimates given by the individual models. In fact, such Bayesian model averaging is bound to improve predictive performance, on average. Since the models obtained in data mining are usually the results of some automated search procedure, accounting for the potential errors associated with the search itself is crucial. In practice, this often requires a Monte

Carlo analysis. The impression is that the error rates of search procedures proposed and used in the data mining and in the statistical literature are far too rarely estimated in this way.

Hypothesis testing. Since statistical tests are widely used, some of their important limitations should be noted. Viewed as a one-sided estimation method, hypothesis testing is inconsistent unless the alpha level of testing rule is decreased appropriately as the sample size increases. Generally, an α level test of one hypothesis and an α level test of another hypothesis do not jointly provide an α level test of the conjunction of the two hypotheses. In special cases, rules (sometimes called contrasts) exist for simultaneously testing several hypotheses. An important corollary for data mining is that the α level of a test has nothing directly to do with the probability of error in a search procedure that involves testing a series of hypothesis. If, for example, for each pair of a set of variables, hypotheses of independence are tested at $\alpha = 0.05$, then 0.05 is not the probability of erroneously finding some dependent set of variables when in fact all pairs are independent. Thus, in data mining procedures that use a sequence of hypothesis tests, the alpha level of the tests cannot generally be taken as an estimate of any error probability related to the outcome of the search.

Data miners should note that while error probabilities of tests have something to do with the truth of hypotheses, the connection is somewhat tenuous. Hypotheses that are excellent approximations may be rejected in large samples; tests of linear models, for example, typically reject them in very large samples no matter how closely they seem to fit the data.

Model scoring. The evidence provided by data should lead us to prefer some models or hypotheses to others, and to be indifferent between still other models. A score is any rule that maps models and data to numbers whose numerical ordering corresponds to a preference ordering over the space of models, given the data. For the reasons just considered, scoring rules are often an attractive alternative to tests. Typical rules assign models a value determined by the likelihood function associated with the model, the number of parameters, or dimension, of the model, and the data. Popular rules include the Akaike information criterion, Bayes information criterion, and minimum description length. Given a prior probability distribution over models, the posterior probability on the data is itself a scoring function, arguably a privileged one. The Bayes information criterion approximates posterior probabilities in large samples.

There is a notion of consistency appropriate to scoring rules; in the large sample limit, almost surely the true model should be among those receiving maximal scores. AIC scores are not, in general, consistent. There are also uncertainties associated with scores, since two different samples of the same size from the same distribution may yield not only different numerical values for the same model, but even different orderings of models.

For obvious combinatorial reasons, it is often impossible when searching a large model space to calculate scores for all models; it is, however, often feasible to describe and calculate scores for a few equivalence classes of models receiving the highest scores.

In some contexts, inferences made using Bayesian scores can differ a great deal from inferences made with hypothesis tests. Raftery in 1995 gives examples of models that account for almost all of the variance of an outcome of interest, and have very high Bayesian scores, but are overwhelmingly rejected by statistical tests.

Markov Chain Monte Carlo. Historically, insurmountable computational difficulties forced data analysts to eschew exact analysis of elaborate hierarchical Bayesian models and complex likelihood calculations. Recent dramatic advances in Monte Carlo methods have, however, liberated analysts from some of these constraints. One particular class of simulation methods, dubbed Markov chain Monte Carlo, originally developed in statistical mechanics, has revolutionized the practice of Bayesian statistics. Smith and Roberts in 1993 provided an accessible overview from the Bayesian perspective; Gilks et al. in 1996 provided a practical introduction addressing both Bayesian and non-Bayesian perspectives. Simulation methods may become unacceptably slow when faced with massive datasets. In such cases, recent advances in analytic approximations prove useful.

Generalized model classes. A major achievement of statistical methodological research has been the development of very general and flexible model classes. Generalized linear models, for instance, embrace many classical linear models, and unify estimation and testing theory for such models. Generalized additive models show similar potential. Graphical models represent probabilistic and statistical models with planer graphs, where the vertices represent (possibly latent) random variables and the edges represent stochastic dependencies. This provides a powerful language for describing models and the graphs themselves make modeling assumptions explicit. Graphical models provide important bridges between the vast statistical literature on multivariate analysis and such fields as artificial intelligence, casual analysis, and data mining.

Rational decision and planning. The theory of rational choice assumes the decision maker has available a definite set of alternative actions, knowledge of a definite set of possible alternative states of the world, knowledge of the payoffs or utilities of the outcomes of each possible action in each possible state of the world, and knowledge of the probabilities of various possible states of the world. Given all of this information, a decision rule specifies which of the alternative actions ought to be taken. A large literature in statistics and economics addresses alternative decision rules – maximizing expected utility, minimizing maximum loss, etc. Typically, rational decision making and planning are the goals of data mining, and rather than providing techniques or methods for data mining, the theory of rational choice poses norms for the use of information obtained from a database.

The very framework of rational decision making requires probabilities and knowledge of the effects alternative actions will have. To know the outcomes of actions is to know something of cause-and-effect relations, and extracting such causal information is often one of the principle goals of data mining and statistical inference more generally.

Inference to causes. Understanding causation is the hidden force behind the historical development of statistics. From the beginning of the subject, in the work of Bernoulli and Laplace, the absence of causal connection between two variables has been taken to imply their probabilistic independence, and the same idea is fundamental in the theory of experimental design. Early in this century, Wright in 1921 introduced directed graphs to represent causal hypotheses (with vertices as random variables and edges representing direct influences), and they have become common representations of causal hypotheses in the social sciences, biology, computer science and engineering.

Kiiveri and Speed in 1982 combined directed graphs with a generalized connection between independence and absence of casual connection in what they called the *Markov condition*: provided Y is not an effect of X, X and Y are conditionally independent given the direct causes of X. They showed that much of the linear modeling literature tacitly assumed the Markov condition; the same is true for casual models of categorical data, and virtually all causal models of systems without feedback. Under additional assumptions, conditional independence therefore provides information about causal dependence. The most common, and most thoroughly investigated, additional assumption is that all conditional independencies are due to the Markov condition applied to the directed graph describing the actual causal processes generating the data, a requirement that has been given many names, including "faithfulness." Directed graphs with associated probability distributions satisfying the Markov condition are called by different names in different literatures: Bayes nets, belief nets, structural equation models, path models, etc. Nonetheless, causal inferences from uncontrolled convenience samples are liable to many sources of error, and data miners should proceed with extreme caution.

Sources of error peculiar to casual inference from a database include the existence of unrecorded causes of associations (confounders) between recorded variables, associations among recorded variables produced by the influence such variables themselves have on whether a unit is included in the database or the sample drawn from it (sample selection bias), missing data (which can be produced either by confounding or by sample selection bias), and samples composed of units with different causal structures. Assuming that all independencies result from the Markov property of directed graphs representing causal hypotheses, for multinormal and multinomial distributions there are procedures based on hypothesis tests of constraints that, in the large sample limit, give correct causal information under most of these conditions. Scoring search algorithms have been developed for the case where there are no confounders, mixtures, and sample selection bias, but the best performance in search seems to be obtained using hypothesis-test initial searches followed by a Bayes score postsearch. Work is under way developing feasible scoring searches for latent variable models.

Prediction. Sometimes one is interested in using a sample, or a database, to predict properties of a new sample, where it is assumed that the two samples are obtained from the same probability distribution. As with estimation, in prediction we are interested both in reliability and uncertainty, often measured by the variance of the predictor.

Prediction methods for this sort of problem always assume some structure in the probability distribution. In data mining contexts, structure is typically either supplied by human experts, or inferred from the database automatically. Regression, for example, assumes a particular functional form relating variables. Structure can be also be specified in terms of constraints, such as independence, conditional independence, higher order conditions on correlations, etc. On average, a prediction method that guarantees satisfaction of the constraints realized in the probability distribution – and no others – will be more accurate and have smaller variance than one that does not. Finding the appropriate constraints to satisfy is the most difficult issue in this sort of prediction. As with estimation, prediction can be improved by model averaging, provided the prior probabilities of the alternative assumptions imposed by the model are available.

Often the results of data mining are used to attempt to predict the effects of a policy change or intervention of some kind. Such predictions are essentially causal inferences, and are not in general correctly given by estimating a probability distribution and predicting by conditioning on the variable to be manipulated.

10.2.2 Is Data Mining "Statistical Deja Vu" (All Over Again)?

In the mid 1960s, the statistics community referred to unfettered exploration of data as *fishing* or *data dredging.* The community, enamored by elegant (analytical) mathematical solutions to inferential problems, argued that since their theories were invalidated by "looking at the data," it was wrong to do so. The major proponent of the exploratory data analysis (EDA) school, J.W. Tukey, countered this argument with the obvious retort that statisticians were putting the cart before the horse. He argued that statistical theory should adapt to the scientific method rather than the other way around. Thirty years hence, the statistical community has largely adopted Tukey's perspective, and has made considerable progress in serving both masters, namely acknowledging that model search is a critical and unavoidable step in the modeling process, and devising formal methods to account for search in their inferential procedures.

Three themes of modern statistics that are of fundamental importance to data miners are: clarity about *goals*, appropriate *reliability* assessment, and adequate accounting for sources of *uncertainty.*

Clarity about goals. Sometimes data analysis aims at finding a convenient, easily computable representation of how the data are distributed in a particular database. In other cases, data analysis aims at predicting features of new cases, or new samples, drawn from outside the database used to develop a predictive model (this is particularly challenging in dynamic situations). In yet other cases, data analysis aims at providing a basis for policy. That is, the analysis is intended to yield insight into casual mechanisms that are used to form predictions about new samples that might be produced by interventions or actions that did not apply in the original database from which the model (or models) were developed. Each of these goals presents distinct inference problems, with distinct hazards. Confusing or equivocating over the aim invites the use of inappropriate methods and may result in unfortunate predictions and inferences.

As an example, consider the observational study reported by Chasnoff et al. in 1989 comparing babies born to cocaine-using mothers with babies born to noncocaine-using mothers. The authors concluded: "For women who become pregnant and are users of cocaine, intervention in early pregnancy with cessation of cocaine use will result in improved obstetric outcome." Fortunately, there exists independent evidence to support this causal claim. However, much of Chasnoff et al. focus on a statistical analysis (analysis of variance) that has little, if anything, to do with the causal question of interest.

Hand in 1994 provides a series of examples illustrating how easy it is to give the right answers to the wrong question. For example, he discusses the problem of analyzing clinical trial data where patients drop out due to adverse side effects of a particular treatment. In this case, the important issue is which population is one interested in modeling. The population at large versus the population who remain within the trail. This problem arises in more general settings than in clinical trails, e.g., nonrespondents (refusers) in survey data. In such situations it is important to be explicit about the questions one is trying to answer.

In this general context an important issue is that of formulating statistical strategy, i.e., how does one structure a data analysis problem so that the right question can be asked? Hand's conclusion is that this is largely an "art" because it is less well formalized than the mathematical and computational details of applying a particular technique. This "art" is gained through experience (at present at least) rather than taught. The implication for data mining is that human judgment is essential for many nontrivial inference problems. Thus, automation can at best only partially guide the data analysis process. Properly defining the goals of an analysis remains a human centered, and often difficult, process.

Use of methods that are reliable means to the goal, under assumptions the user (and consumer) understands and finds plausible in the context . Statistical theory applies several meanings to the word *reliability*, many of which also apply to model search. For example, under what conditions does a search procedure provide correct information, of the kind sought, with probability one as the sample size increases without bound. Answers to such questions are often elusive and can require sophisticated mathematical analysis. Where answers are available, the data analyst should pay careful attention to the reasonableness of underlying assumptions. Another key data mining question is this: what are the probabilities of various kinds of errors that result from using a method in finite samples? The answers to this question will typically vary with the kinds of errors considered, the sample size, and the frequency of occurrence of the various kinds of targets or signals whose description is the goal of inference. These questions are often best addressed by Monte Carlo methods, although in some cases analytic results may be available.

A sense of the uncertainties of models and predictions. Quite often background knowledge and even the best methods of search and statistical assessment should leave the investigator with a range of uncertainties about the correct model, or the correct prediction. The data analyst must quantify these uncertainties so that subsequent decisions can be appropriately hedged.

Another example involves a current debate in the atmospheric sciences. The question is whether or not specific recurrent pressure patterns can be clearly identified from daily geopotential height records, which have been compiled in the Northern Hemisphere since 1948. The existence of well-defined recurrent patterns (or "regimes") has significant implications for models of upper atmosphere low-frequency variability beyond the timescale of daily weather disturbances (and, thus, models of the earth's climate over large timescales). Several studies have used a variety of clustering algorithms to detect in homogeneities ("bumps") in low-dimensional projections of the gridded data. While this work has attempted to validate the cluster models via resampling techniques, it is difficult to infer from the multiple studies whether regimes truly exist, and, if they do, where precisely they are located. It seems likely that 48 winters' worth of data is not enough to identify regimes to any degree of certainly and that there is a fundamental uncertainty (given the current data) about the underlying mechanisms at work. All is not lost, however, since it is also clear that one could quantify model uncertainty in this context, and theorize accordingly. In what follows we elaborate on these points and offer a perspective one some of the hazards of data mining.

10.2.3 Characterizing Uncertainty

The statistical approach contends that reporting a single number for a parameter estimate or a prediction is almost always inadequate. Quantification of the *uncertainty* associated with a single number, while often challenging, is critical for subsequent decision making. As an example, Draper in 1995, considered the case of the 1980 Energy Modeling Forum (EMF) at Stanford University where a 43-person working group of economists and energy experts convened to forecast world oil prices from 1981 to 2020. The group generated predictions based on a number of econometric models and scenarios, embodying a variety of assumptions about supply, demand, and growth rates of relevant quantities. A plausible reference scenario and model was selected as representative, but the summary report cautioned against interpreting point predictions based on the reference scenario as "[the working group's] 'forecast' of the oil future, as there are too many unknowns to accept any projection as a forecast." The summary report did conclude, however, that most of the uncertainty about future oil prices "concerns not whether these prices will rise.... but how rapidly they will rise."

In 1980, the average spot price of crude oil was around $32 per barrel. Despite the warning about the potential uncertainty associated with the point estimates, governments and private companies around the world focused on the last sentence in the quotation above, and proceeded to invest an estimated $500 billion dollars, on the basis that the price would probably be close to $40 dollars per barrel in the mid-eighties. In fact, the actual 1986 world average spot price of oil was about $13 per barrel.

Using only the information available to the EMF in 1980, along with thoughtful but elementary statistical methods, Draper in 1995 shows that a 90% predictive interval for the 1986 price would have ranged from about $20 to over $90. Note that this interval does not actually contain the actual 1986 price-insightful statistical analysis does not provide clairvoyance. However, decision makers would (and should) have proceeded more cautiously in 1980, had they understood the full extent of their uncertainty.

Correctly accounting for the different sources of uncertainty presents significant challenges. Until recently, the statistical literature focused primarily on quantifying parametric and predictive uncertainty in the context of a particular model. Two distinct approaches are in common use. "Frequentist" statisticians focus on the randomness in sampled data and summarize the induced randomness in parameters and predictions by the so-called *sampling distributions*. "Bayesian" statisticians instead treat the data as fixed, and use *Bayes theorem* to turn prior opinion about quantities of interest (always expressed by a probability distribution), into a so-called *posterior distribution* that embraces all the available information. The fierce conflicts between previous generations of frequentists and Bayesians, have largely given way in recent years to a more pragmatic approach; most statisticians will base their choice of tool on scientific appropriateness and convenience.

In any event, recent research has led to increased awareness that *within-model* uncertainty (as discussed in the previous paragraph) may often, in practice, be dominated by *between-model* uncertainty. It is common practice nowadays for statisticians and data miners to use computationally intensive model selection algorithms to seek out a single optimal model from an enormous class of potential models. The problem is that *several* different models may be close to optimal, yet lead to different inferences. Intuitively, ambiguity over the model should dilute information about effect parameters and predictions, since "part of the evidence is spent to specify the model." Promising techniques for properly accounting for this source of uncertainty include

Bayesian model averaging and resampling methods. The main point here is that data miners need to think carefully about model assessment and look beyond commonly used goodness-of-fit measures such as mean square error.

10.2.4 What Can Go Wrong, Will Go Wrong

Data mining poses difficult and fundamental challenges to the theory and practice of statistics. While statistics does not have all the answers for the data miner, it does provide a useful and practical framework for which to search for solutions. In this section, we describe some lessons that statisticians have learned when theory meets data.

Data Can Lie

Data mining applications typically rely on observational (as opposed to experimental) data. Interpreting observed associations in such data is challenging; sensible inferences require careful analysis, and detailed consideration of the underlying factors. Here we offer a detailed example to support this position.

Wen et al. in 1995 (WHN) analyzed administrative records of all Ontario general hospital separations (discharges, transfers, or in-hospital deaths) from 1981 to 1990, focusing specifically on patients who had received a primary open cholecystectomy. Some of these patients had in addition received an incidental (i.e., discretionary) appendectomy during the cholecystectomy procedure. Table 10.2 displays the data on one outcome, namely in-hospital deaths. A chi-square test comparing this outcome for the two groups of patients shows a "statistically significant" difference. This "finding" is surprising since long-term prevention of appendicitis is the sole rationale for the incidental appendectomy procedure – no short-term improvement in outcomes is expected. This "finding" might lead a naïve hospital policy maker to conclude that all cholecystectomy patients should have an incidental appendectomy to improve their chances of a good outcome! Clearly something is amiss – how could incidental appendectomy improve outcomes?

WHN did separately consider a subgroup of low-risk patients. For these patients (using ten different definitions of "low-risk"), incidental appendectomy indeed resulted in poorer outcomes. Paradoxically, it could even be the case that appendectomy adversely affects outcomes for both high-risk patients and low-risk patients, but appears to positively affect outcomes when the low-risk

Table 10.2. In-hospital survival of patients undergoing primary open cholecystectomy with and without incidental appendectomy

	With appendectomy	Without appendectomy
In-hospital deaths, No. (%)	21 (0.27%)	1,394 (0.73%)
In-hospital survivors, No. (%)	7,825 (99.73%)	190,205 (99.27%)

Table 10.3. Fictitious data consistent with the Wen et al. (1995) data

	With appendectomy		Without appendectomy	
	Low risk	High risk	Low risk	High risk
Death	7	14	100	1294
Survival	7700	125	164009	26196

Table 10.4. Proportion of in-hospital deaths cross-classified by incidental appendectomy and patients risk grouping for the fictitious data of Table 10.3.

	With appendectomy	Without appendectomy
Low risk	0.0009	0.0006
High risk	0.10	0.05
Combined	0.003	0.007

and high-risk patients are combined. WHN did not provide enough data to check whether this so-called "Simpson's paradox" [Simpson, 1951] occurred in this example. However, Table 10.3 presents data that are plausible and consistent with WHN's data.

Table 10.4 displays the corresponding proportions of in-hospital death for these fictitious data. Clearly the risk and death categories are directly correlated. In addition, appendectomies are more likely to be carried out on low-risk patients than on high-risk ones. Thus, if we did not know the risk category (age) of a patient, knowing that they had an appendectomy allows us to infer that they are more likely to be lower risk (younger). However, this does not in any way imply that having an appendectomy will lower one's risk. Nonetheless, when risk is omitted from the table, exactly such a fallacious conclusion appears justified from the data.

Returning to the original data, WHN provide a more sophisticated regression analysis, adjusting for many possible confounding variables (e.g., age, sex, admission status). They conclude that "there is absolutely no basis for any short-term improvement in outcomes" due to incidental appendectomy. This careful analysis agrees with common sense in this case. In general, analyses of observational data demand such care, and come with no guarantees. Other characteristics of available data that connive to spoil casual inferences include:

- Associations in the database may be due in whole or part to unrecorded common causes (latent variables).
- The population under study may be a mixture of distinct causal systems, resulting in statistical associations that are due to the mixing rather than to any direct influence of variables on one another or any substantive common cause.
- Missing values of variables for some units may result in misleading associations among the recorded values.

- Membership in the database may be influenced by two or more factors under study, which will create a "spurious" statistical association between those variables.
- Many models with quite distinct causal implications may "fit" the data equally or almost equally well.
- The frequency distributions in samples may not be well approximated by the most familiar families of probability distributions.
- The recorded values of variables may be the result of "feedback" mechanisms, which are not well represented by simple "nonrecursive" statistical models.

There is research that addresses aspects of these problems, but there are few statistical procedures yet available that can be used "off the shelf" – the way randomization is used in experimental design – to reduce these risks. Standard techniques such as multiple regression, and logistic regression may work in many cases, such as in the appendectomy example, but they are not always adequate guards against these hazards. Indeed, controlling for possibly confounding variables with multiple regression can in some cases produce inferior estimates of effect sizes. Procedures recently developed in the artificial intelligence and statistics literature address some of the problems associated with latent variables and mixing, but so far only for two families of probability distributions, the normal and multinomial.

Sometimes It Is Not What Is In The Data That Matters

Classical statistical methods start with a random sample, yet in practice, data or the institutions that give rise to data, can be uncooperative. In such cases, inferences that ignore how the data were "selected" can lead to distorted conclusions.

Consider, for example, the Challenger Space Shuttle accident. The Rogers Commission concluded that an O-ring failure in the solid rocket booster led to the structural breakup and loss of the Challenger. In reconstructing the events leading up to the decision to launch, the Commission noted a mistake in the analysis of thermal-distress data whereby flights with no (i.e., zero) incidents of O-ring damage were excluded from critical plots of O-ring damage and ambient launch temperature since it was felt that they did not contribute any information about the temperature effect. This truncation of the data led to the conclusion that no relationship between O-ring damage and temperature existed, and ultimately, the decision to launch. Dalal et al. in 1989 throw statistical light on the matter by demonstrating the strong correlation between O-ring damage and temperature, and quantifying the risk (of catastrophic failure) at 31°F. Had the original analysis used all of the data, it would have indicated that the decision to launch was at best a risky proposition.

In the above case, the selection bias problem was one of "human error" and could easily have been avoided. In most problems, selection bias is an inherent characteristic of the available data and methods of analysis need to deal with it. It is our experience that every dataset has the potential for selection bias to invalidate standard inferences. The lessons to be learned here are

- that any technique used to analyze truncated data as if it was a random sample, can be fooled, regardless of how the truncation was induced;
- the data themselves are seldom capable to alert the analyst that a selection mechanism is operating – information external to the data at hand is critical in understanding the nature and extent of potential biases.

The Perversity of the Pervasive P-Value

P-values and associated significance (or hypothesis) tests play a central role in classical (frequentist) statistics. It seems natural, therefore, that data miners should make widespread use of P-values. However, indiscriminate use of P-values can lead data miners astray in most applications.

The standard significance test proceeds as follows. Consider two competing hypotheses about the world: the *Null Hypothesis*, commonly denoted by H_o, and the *Alternative Hypothesis*, commonly denoted by H_A. Typically H_o is "nested" within H_A; for example, H_o might state that a certain combination of parameters is equal to zero, while H_A might place no restriction on the combination. A *test statistic*, T is selected and calculated from the data at hand. The idea is that T (Data) should measure the evidence in the data against H_o. The analyst rejects H_o in favor of H_A if T (Data) is more extreme than would be expected if H_o were true. Specifically, the analyst computes the P-value, that is, the probability of T being greater than or equal to T (Data), given that H_o is true. The analyst rejects H_o if the P-value is less than a preset *significance level*, α.

There are three primary difficulties associated with this approach:

1. The standard advice that statistics educators provide, and scientific journals rigidly adhere to, is to choose α to be 0.05 or 0.01, *regardless of sample size*. These particular α-levels arose in Sir Ronald Fisher's study of relatively small agricultural experiments (on the order of 30–200 plots). Textbook advice has emphasized the need to take account of the power of the test against H_A when setting α, and somehow reduce α when the sample size is large. This crucial but vague advice has largely fallen on deaf ears.
2. Raftery in 1995 points out that the whole hypothesis testing framework rests on the basic assumption that only two hypotheses are ever entertained. In practice, data miners will consider very large numbers of possible models. As a consequence, indiscriminate use of P-values with "standard" fixed α-levels can lead to undesirable outcomes such as selecting a model with parameters that are highly significantly different from zero, even when the training data are pure noise. This point is of fundamental importance for data miners.

3. The P-value is the probability associated with the event that the test statistic was as extreme as the value observed, or more so. However, the event that actually happened was that a specific value of the test statistic was observed. Consequently, the relationship between the P-value and the veracity of H_o is subtle at best. Jeffreys in 1980 puts it this way:

   > I have always considered the arguments for the use of P absurd. They amount to saying that a hypothesis that may or may not be true is rejected because a greater departure from the trail value was improbable; that is, that it has not predicted something that has not happened.

Bayes Factors are the Bayesian analog of the frequentist P-values and admit to a more direct interpretation – the Bayesian analyst computes the posterior probability that a hypothesis is correct. With fixed α-levels, the frequentist and the Bayesian will arrive at very different conclusions. For example, Berger and Sellke in 1987 show that data that yield a P-value of 0.05 when testing a normal mean, result in a posterior probability for H_o that is at least 0.30 for any "objective" prior distribution. One way to reconcile the two positions is to view Bayes factors as a method for selecting appropriate α-levels.

Intervention and Prediction

A specific class of prediction problems involves interventions that alter the probability distribution of the problem, as in predicting the values (or probabilities) of variables under a change in manufacturing procedures, or changes in economic or medical treatment policies. Accurate predictions of this kind require some knowledge of the relevant causal structure, and are in general quite different from prediction without intervention, although the usual caveats about uncertainty and model averaging apply. For graphical representations of causal hypotheses according to the Markov condition, general algorithms for predicting the outcomes of interventions from complete or incomplete causal models were developed.

Consider the following example, Herbert Needleman's famous studies of the correlation of lead deposits in children's teeth with their IQs resulted, eventually, in removal of tertraethyl lead from gasoline in the United States. One dataset Needleman examined included more than 200 subjects and measured a large number of covariates. Needleman et al. in 1985 reanalyzed the data using backwards stepwise regression of verbal IQ on these variables and obtained six significant regressors, including lead. Klepper in 1988 reanalyzed the data assuming that all of the variables were measured with error. Their model assumes that each measured number is a linear combination of the true value and an error, and that the parameters of interest are not the regression coefficients but rather the coefficients relating the unmeasured "true value" variables to the unmeasured true value of verbal IQ. These coefficients are in fact indeterminate – in econometric terminology, "unidentifiable." An interval estimate of the coefficients that is strictly positive or negative for each coefficient can be made, however, if the amount of measurement error can be bounded with prior knowledge by an amount that varies from case to case. Klepper found that the bound required to ensure the existence of a strictly negative interval estimate for the lead-IQ coefficient was much too strict to be credible; thus he concluded that the case against lead was not nearly as strong as Needleman's analysis suggested.

Allowing the possibility of latent variables, Scheines in 1996 reanalyzed the correlations (using TETRAD methodology) and concluded that three of the six regressors could have no influence on IQ. The regression included the three extra variables only because the partial regression coefficient is estimated by conditioning on all other regressors, which is just the right thing to do for linear prediction, but the wrong thing to do for causal inference using the Markov condition. Using the Klepper model, but without the three irrelevant variables, and assigning to all of the parameters a normal prior probability with mean zero and a substantial variance, Scheines that used Markov chain Monte Carlo to compute a posterior probability distribution for the lead-IQ parameter. The probability is very high that lead exposure reduces verbal IQ.

10.2.5 Symbiosis in Statistics

Easy access to data in digital form and the availability of software tools for statistical analyses have made it possible for the man in street to set up shop and "do statistics." Nowhere is this more true today than in data mining. Based on the arguments in this section, let us assume that statistics is a necessary but not sufficient component in the practice of data mining. How well will the statistics profession serve the data mining community? Hoerl et al. in 1993, for example, assert that:

We are our own best customers. Much of the work of the statistical profession is intended for other members of the statistical profession.

Despite this rather negative view of the relevance of statistical research, real-world applications do in fact drive much of what goes on it statistics, although often in a very indirect manner.

As an example, consider the field of signal processing and communications, an area where a specialized set of relatively sophisticated statistical methods and models have been honed for practical use. The field was driven by fundamental advances from Claude Shannon and others in the 1940s. Like most of the other contributors to the field, Shannon was not a statistician, but possessed a deep understanding of probability theory and its applications. Through the 1950s to the present, due to rapid advances in both theory and hardware, the field has exploded and relevant statistical methods such as estimation and detection have found their way into everyday use in radio and network communications systems. Modern statistical communications reflects the symbiosis of statistical theory and engineering practice. Engineering researchers in the field are in effect "adjunct" statisticians: educated in probability theory and basic statistics they have the tools to apply statistical methods to their problems of interest. Meanwhile statisticians continue to develop more general models and estimation techniques of potential applicability to new problems in communications.

This type of symbiosis can also be seen in other areas such as financial modeling, speech recognition (where for example hidden Markov models provide the state of the art in the field), and most notably, epidemiology. Indeed, if statistics can claim to have revolutionized any field, it is in the biological and health sciences where the statistical approach to data analysis gave birth to the field of biostatistics.

The relevance of this symbiosis for data mining is that data-miners need to understand statistical principles, and statisticians need to understand the nature of the important problems that the data mining community is attacking or being asked to attack. This has been a successful model in the past for fields where statistics has had considerable impact and has the potential to see ongoing success.

10.3 Summary

Data mining is a growing discipline, which originated outside statistics in the database management community, mainly for commercial concerns. The gap is now filled and data mining could be considered as the branch of exploratory statistics where one tries to find unexpected and useful models and patterns, through an extensive use of classic and new algorithms. The expression *unexpected* should not be misleading: one has greater chances to discover something interesting when one is previously acquainted with the data.

Caution is of course necessary to avoid drawing wrong conclusions and statisticians, because they are familiar with uncertainty and risk, and are the right people to derive validation tests. A completely automatic process of knowledge extraction is also a misleading idea: even with very efficient software, human expertise and interventions are necessary. Official statistics should be a field for data mining, giving a new life and profitability to its huge databases, but it may implies a redefinition of the missions of NSI.

The statistical literature has a wealth of technical procedures and results to offer data mining, but it also has a few simple methodological morals: prove that estimation and search procedures used in data mining are consistent under conditions reasonably thought to apply in applications; use and reveal uncertainty, do not hide it; calibrate the errors of search, both for honesty and to take advantages of model averaging; do not confuse conditioning with intervening; and finally, do not take the error probabilities of hypothesis tests to be the error probabilities of search procedures.

10.4 Review Questions

1. What are the goals and tools of data mining?
2. Give an overview of statistical science for application to data mining.
3. Justify – data mining is all over gain.
4. Define intervention, prediction, and P-value.

11

Theoretical Frameworks for Data Mining

Objectives:

- Research in data mining and knowledge discovery in databases has mostly concentrated on developing good algorithms for various data mining tasks.
- In this section we present some possible theoretical approaches to data mining.
- The area is at its infancy, and there probably are more questions than answers in this section.
- We start by discussing reductionist approaches, i.e., ways of looking at data mining as a part of some existing area, such as statistics or machine learning.
- We discuss the probabilistic approach, which is of course closely linked to statistics: it views data mining as the activity aimed at understanding the underlying joint distribution of the data
- A theoretical framework should also be able to model typical data mining tasks and discuss the probabilistic nature of the discovered patterns and models.
- Able to talk about data and inductive generalizations of the data, and accept the presence of different forms of data.

Abstract. Research in data mining and knowledge discovery in databases has mostly concentrated on developing good algorithms for various data mining tasks. Some parts of the research effort have gone to investigating data mining process, user interface issues, database topics, or visualization. Relatively little has been published about the theoretical foundations of data mining. In this section we present some possible theoretical approaches to data mining. The area is at its infancy, and there probably are more questions than answers in this section.

First of all a question such as, "Why look for a theory of data mining? Data mining is an applied area, why should we care about having a theory for it?" Probably the simplest answer is to recall the development of the area of relational databases. Databases existed already in the 1960s, but the field was considered to be a murky

S. Sumathi and S.N. Sivanandam: *Theoretical Frameworks for Data Mining*, Studies in Computational Intelligence (SCI) **29**, 265–270 (2006)
www.springerlink.com © Springer-Verlag Berlin Heidelberg 2006

backwater of different applications without any clear structure and any interesting theoretical issues. Codd's relational model was a nice and simple framework for specifying the structure of data and the operations to be performed on it. The mathematical elegance of the relational model made it possible to develop advanced methods of query optimization and transactions, and these in turn made efficient general purpose database management systems possible. The relational model is a clear example of how theory in computer science has transformed an area from a hodgepodge of unconnected methods to an interesting and understandable whole, and at the same time enabled an area of industry.

Given that theory is useful, what would be the properties that a theoretical framework should satisfy in order that it could be called a theory for data mining? The example of relational model can serve us also here. First of all, the theoretical framework should be simple and easy to apply; it should (at least some day) give us useful results that we could apply to the development of data mining algorithms and methods.

A theoretical framework should also be able to model typical data mining tasks (clustering, rule discovery, classification), discuss the probabilistic nature of the discovered patterns and models, talk about data and inductive generalizations of the data, and accept the presence of different forms of data (relational data, sequences, text, Web). Also, the framework should recognize that data mining is an interactive and iterative process, where comprehensibility of the discovered knowledge is important and where the user has to be in the loop, and that there is not a single criterion for what an interesting discovery is. (We could also ask. "What actually is a theory?" For that we have the simple answer: we recognize a theory when we see it).

We start by discussing reductionist approaches, i.e., ways of looking at data mining as a part of some existing area, such as statistics or machine learning; in this case, of course, there is little need for new theoretical frameworks. Then we discuss the probabilistic approach, which is of course closely linked to statistics: it views data mining as the activity aimed at understanding the underlying joint distribution of the data. After that, we review the data compression approach to the theory of data mining. The very interesting microeconomic viewpoint on data mining is considered after that, and finally we look at the concept of inductive database and show how it can perhaps be used to understand and develop data mining.

11.1 Two Simple Approaches

A simple approach to the theory of data mining is to declare that data mining is statistics (perhaps no larger data sets that previously), and thus the search for a theoretical framework for data mining can stop immediately: we just have to look at the appropriate statistics literature. The theory of data mining is statistics (as a science).

Data mining obviously is very close to statistics, and data mining researchers with computer science backgrounds typically have too little education differences between the areas. However, one can argue that there are important differences between the areas. The volume of the data is probably not a very important difference: the number of variables or attributes often has a much more profound impact on the applicable analysis methods. For example, data mining has tackled with problems such as what to do in situations where the number of variables is so large that looking at all pairs of variables is computationally infeasible. Overall, the issue of computational feasibility is has a much clearer role in data mining than in statistics. Another difference pointed out to me by David Hand is that data mining is typically secondary data analysis: the data has been collected for some other purpose that for answering a specific data analytical questions.

Several other differences between the areas could be pointed out. For example, the emphasis on database integration, simplicity of use, and the understandability of results are typical of data mining approaches. For the purpose of this section it is sufficient to point out that at least currently the theoretical framework of statistics seems to be relatively distant from the actual development of data mining methods. Also, statistical theory does not seem to pay a lot of attention to the process character of data mining. However, in the next section we describe a closely related approach.

A similar (but weaker) case of reducing data mining to an existing area has been made from the viewpoint of machine learning. One could say that data mining is applied machine learning, and thus the theory of data mining is equal to the theory of machine learning. Again, this approach fails for two reasons. First, there are important differences between machine learning and statistics, and second, the theoretical machine learning approaches (such as the PAC model) do not really address the special requirements that we made for the theory of data mining.

11.1.1 Probabilistic Approach

A possible theoretical approach to data mining is to view data mining as the task of finding the underlying joint distribution of the variables in the data. Typically one aims at finding a short and understandable representation of the joint distribution, e.g., Bayesian network or a hierarchical Bayesian model.

This approach is obviously closely related to the reductionist approach of viewing data mining as statistics. The advantages of the approach are that the background is very solid, and it is easy to pose formal questions. Tasks such as clustering or classification fit easily into this approach. What seems to be lacking, as in most of the approaches, are ways for taking the iterative and interactive nature of the data mining process into account.

Hierarchical Bayesian models seem a very promising statistically sound approach to data mining. Such a model describes the structural part of the distribution independent of the actual functional form of the distribution. For example, if we have information about the supermarket-buying behavior of a group of people, we could describe the model using the diagram in Fig. 11.1. That is, for each customer there is a set of visits to the shop, and for each visit of a customer there is a set of products bought during the visit. To give a probabilistic model for this phenomenon we have defined how the number

Fig. 11.1. Hierarchical Bayesian models

of visits is distributed for each customer, and how the selection of products is distributed for each visit. What is interesting to notice is that the figure could be interpreted either as part of the probabilistic model or just simply as an ER schema for the data set. When statisticians and database people talk about data modeling, they are talking about different things, but there still is lots of commonality in the meanings. For hierarchical models such as the one shown below there exist good tools for approximating the posterior distribution of a model. It seems to me that exploring the usefulness of such models for data mining is a promising area of research: issues such as scalability etc. need to be addressed.

11.1.2 Data Compression Approach

The data compression approach to data mining is simply to state: the goal of data mining is to compress the data set by finding some structure for it. That is, data mining looks for knowledge, where knowledge is interpreted as a representation that makes it possible to code the data using few bits. If desired, the minimum description length (MDL) principle can be used to select among different encodings [J. Rissanen et al. 1999]. To yield structure that is comprehensible to the user, we have to specify compression methods that are based on concepts that are easy to understand.

Several simple data mining techniques can be viewed as instances of this approach. For example, association rules can be viewed as ways of providing compression of parts of the data [R. Agrawal et al.]. In the same way, an accurate decision tree can be considered a compression method for the target attribute. A clustering of the data can also be viewed as a way of compressing the data set.

This approach has a nice formal foundation. It is connected to Bayesian approaches for modeling the joint distribution: any compression scheme can be viewed as providing a distribution on the set of possible instances of the data. As in the probabilistic approach, the process view of data mining is not so easy to capture in this framework. An interesting opening in this direction is given in S. Chakrabani et al., 1998, where it is shown how to mine for novel nuggets using the MDL principle.

11.2 Microeconomic View of Data Mining

The microeconomic view of data mining introduced by J. Kleinberg et al., in 1998, is a very interesting approach. The starting point is that data mining is about finding actionable patterns: the only interest is in patterns that can somehow be used to increase utility. Kleinberg et al. give a decision theoretic formulation of this principle: the goal of the organization is to find the decision x that leads to the maximum utility $f(x)$. The form of the utility $f(x)$ is typically a sum of utilities $f_i(x)$ for each customer i. The function $f_i(x)$ is actually a complex function of the decision x and the data y_i on customer i, and can often be represented using a single function, i.e., as $f_i(x) = g(x, y_i)$. Thus the task is to find the decision x maximizing the sum of the terms $g(x, y_i)$ over the customers i. The basic observation of Kleinberg et al. is that data mining is useful if and only if the function g is nonlinear. They are able to describe pattern discovery, clustering, etc. as instantiations of the framework, and demonstrate also interesting connections to sensitivity analysis. Furthermore, they also show how the framework gives useful suggestions for research problems. We cannot do justice to this delightful page in this section, but the approach is clearly very promising.

11.3 Inductive Databases

As mentioned above, relational database theory has been remarkably successful. One of the basic concepts in that theory is the powerful notion of a query. Instead of thinking of accessing the database as a special entity, relational database theory just considered queries as functions mapping databases to databases. This made it possible to speak about composing queries, etc. and in many ways was instrumental to the development of relational databases.

The basic idea of inductive databases is that the query concept should be applied also to data mining and knowledge discovery tasks. In the slogan form from T. Lmielinski et al.: *there is no such thing as discovery, it is all in the power of the query language.* That is, one can benefit from viewing the typical data mining tasks not as dynamic operations constructing new nuggets of information, but as operations unveiling hitherto unseen but pre-existing pieces of knowledge. The term *inductive database* refers to a normal database plus the set of all sentences from a specified class of sentences that are true of the data [H. Mannila et al.]. In model-theoretic terms, the inductive database contains the data and the theory of the data.

The approach can be compared to the idea of deductive databases, which contain a normal database plus a set of rules for deriving new facts from the facts already existing in the database. The user of a deductive database can act as if all the facts derivable from the database would be actually stored there. Of course, this set might be infinite, or finite by very large, so in practice it cannot be represented. But the idea of treating stored and derived facts in the same way is crucial for deductive databases.

In the same way, an inductive database does not contain all the rules that are true about the data stored in it; the user is just able to assume that all these rules are there. In practice, the rules are constructed on demand.

The schema of an inductive database consists of a normal relational database schema plus a schema for the generalizations. It is relatively easy to design a query language that works on such schemas. The result of a query on an inductive database is again an inductive database, so we have the closure property that has been so useful for relational databases.

The process view on data mining is directly built in to the concept of inductive databases. It also suggests architecture for data mining systems. Association rules and other simple pattern formalism fit quite easily into the framework, and there are some good partial solutions that can be viewed as partial implementations of inductive databases. However, e.g., clustering is harder to describe in a useful way. The probabilistic nature of data mining can be incorporated by having the underlying concept class support probabilistic concepts.

11.4 Summary

We required that a good theory for data mining should consider the process of data mining, have a probabilistic nature, be able to describe different data mining tasks, be able to allow for the presence of background knowledge, etc. None of the above candidates satisfies all the requirements, unfortunately. The current favorite approach would be to combine the microeconomic view with inductive databases (or some other database oriented approach): these two aspects would seem to satisfy most of the requirements, and both directions suggest a wealth of interesting research issues.

11.5 Review Questions

1. Explain the probabilistic approach and data compression approach in theoretical framework.
2. What are inductive databases employed in mining?

12

Major and Privacy Issues in Data Mining and Knowledge Discovery

Objectives:

- In this section, the major data mining issues, and OECD personal privacy guidelines, some privacy issues in knowledge discovery and data mining are enumerated.
- The scope of this section addresses major issues in data mining regarding mining methodology, user interaction, performance, and diverse data types.
- We discuss new privacy threats posed by knowledge discovery and data mining (KDDM).
- KDDM poses the following new challenges to privacy: stereotypes, guarding personal data from KDDM researchers, protecting privacy of individuals from training sets, and combination of patterns. We discuss the possible solutions and their impact on the quality of discovered patterns.
- KDDM technology includes massive data collection, data warehouses, statistical analysis and deductive learning techniques, and uses vast amounts of data to generate hypotheses and discover general patterns.
- KDDM is now moving to other domains where privacy issues are very delicate.
- The Organization for Economic Cooperation and Development (OECD) has provided probably the best known set of guidelines. This section focuses on the OECD guidelines since many nations have adopted then as statutory law.
- This section provides some insight into the problems concerning personal privacy and data faced by those who wish to employ knowledge discovery.

Abstract. In this section, the major data mining issues and OECD personal privacy guidelines, some privacy issues in knowledge discovery and data mining are enumerated. The major issues concerned with that of the data mining are individual privacy, which is a social one, issue, related to data integrity and technical issue whether to set up a relational database structure or a multidimensional one.

S. Sumathi and S.N. Sivanandam: *Major and Privacy Issues in Data Mining and Knowledge Discovery*, Studies in Computational Intelligence (SCI) **29**, 271–291 (2006)
www.springerlink.com © Springer-Verlag Berlin Heidelberg 2006

The scope of this section addresses major issues in data mining regarding mining methodology, user interaction, performance, and diverse data types.

Recent developments in information technology have enabled collection and processing of vast amounts of personal data, such as criminal records, shopping habits, credit and medical history, and driving records. The information is undoubtedly very useful in many areas, including medical research, law enforcement, and national security. However, there is an increasing public concern about individuals' privacy. Privacy is commonly seen as the right of individuals to control information about them. The appearance of technology for knowledge discovery and data mining (KDDM) has revitalized concerns about the following general privacy issues: secondary use of the personal information, handling misinformation, and granulated access to personal information. These issues demonstrate that existing privacy laws and policies are well behind the developments in technology, and no longer offer adequate protection. We discuss new privacy threats posed by KDDM. KDDM technology includes massive data collection, data warehouses, statistical analysis and deductive learning techniques, and uses vast amounts of data to generate hypotheses and discover general patterns. KDDM poses the following new challenges to privacy: stereotypes, guarding personal data from KDDM researchers, protecting privacy of individuals from training sets, and combination of patterns. We discuss the possible solutions and their impact on the quality of discovered patterns.

The Organization for Economic Cooperation and Development (OECD) has provided probably the best known set of guidelines. A number of countries have adopted these guidelines as statutory law, in whole or in part.

12.1 Major Issues in Data Mining

One of the key issues raised by data mining technology is not a business or technological one, but a social one. It is the issue of individual privacy. Data mining makes it possible to analyze routine business transactions and glean a significant amount of information about individuals' buying habits and preferences.

Another issue is that of data integrity. Clearly, data analysis can only be as good as the data that is being analyzed. A key implementation challenge is integrating conflicting or redundant data from different sources. For example, a bank may contain credit cards accounts on several different databases. The addresses (or even the names) of a single cardholder may be different in each. Software must translate data from one system to another and select the address most recently entered.

A hotly debated technical issue is whether it is better to set up a relational database structure or a multidimensional one. In a relational structure, data is stored in tables, permitting ad hoc queries. In a multidimensional structure, on the other hand, sets of cubes are arranged in arrays, with subsets created according to category. While multidimensional structures facilitate multidimensional data mining, relational structures thus far have performed better in client/server environments. And, with the explosion of the Internet, the world is becoming one big client/server environment.

Finally, there is the issue of cost. While system software costs have dropped dramatically within the past five years, data mining and data warehousing tend to be self-reinforcing. The more powerful the data mining queries, the greater the utility of the information being gleaned from the data, and the greater the pressure to increase the amount of data being collected and maintained, which increases the pressure for faster, more powerful data mining queries. This increases pressure for larger, faster systems, which are more expensive.

The scope of this section addresses major issues in data mining regarding mining methodology, user interaction, performance, and diverse data types. These issues are introduced below:

Mining methodology and user interaction issues: These reflect the kinds of knowledge mined, the ability to mine knowledge at multiple granularities, the use of domain knowledge, ad hoc mining, and knowledge visualization.

- *Mining different kinds of knowledge in databases:* Since different users can be interested in different kinds of knowledge, data mining should cover a wide spectrum of data analysis and knowledge discovery tasks, including data characterization, discrimination, association, classification, clustering, trend and deviation analysis, and similarity analysis. These tasks may use the same database in different ways and require the development of numerous data mining techniques.
- *Interactive mining of knowledge at multiple levels of abstraction:* Since it is difficult to know exactly what can be discovered within a database, the data mining process should be *interactive*. For databases containing a huge amount of data, appropriate sampling techniques can first be applied to facilitate interactive data exploration. Interactive mining allows users to focus the search for patterns, providing and refining data mining requests based on returned results. Specifically, knowledge should be mined by drilling down, rolling up, and pivoting through the data space and knowledge space interactively, similar to what OLAP can do on data cubes. In this way, the user can interact with the data mining system to view data and discovered patterns at multiple granularities and from different angles.
- *Incorporation of background knowledge:* Background knowledge, or information regarding the domain under study, may be used to guide the discovery process and allow discovered patterns to be expressed in concise terms and at different levels of abstraction. Domain knowledge related to databases, such as integrity constraints and deduction rules, can help focus and speed up a data mining process, or judge the interestingness of discovered patterns.
- *Data mining query languages and ad hoc data mining:* Relational query languages (such as SQL) allow users to pose ad hoc queries for data retrieval. In a similar vein, high-level data mining query languages need to be developed to allow users to describe ad hoc data mining tasks by facilitating the specification of the relevant sets of data for analysis, the domain knowledge, the kinds of knowledge to be mined, and the conditions and constraints to be enforced on the discovered patterns. Such a language

should be integrated with a database or a data warehouse query language, and optimized for efficient and flexible data mining.

- *Presentation and visualization of data mining results:* Discovered knowledge should be expressed in high-level languages, visual representations, or other expressive forms so that the knowledge can be easily understood and directly usable by humans. This is especially crucial if the data mining system is to be interactive. This requires the system to adopt expressive knowledge representations techniques, such as trees, tables, rules, graphs, charts, crosstabs, matrices, or curves.
- *Handling noisy or incomplete data:* The data stored in a database may reflect noise, exceptional cases, or incomplete data objects. When mining data regularities, these objects may confuse the process, causing the knowledge model constructed to overfit the data. As a result, the accuracy of the discovered patterns can be poor. Data cleaning methods and data analysis methods that can handle noise are required, as well as outlier mining methods for the discovery and analysis of exceptional cases.
- *Pattern evaluation – the interestingness of problem:* A data mining system can uncover thousands of patterns. Many of the patterns discovered may be uninteresting to the given user, representing common knowledge or lacking novelty. Several challenges remain regarding the development of techniques to assess the interestingness of discovered patterns, particularly with regard to subjective measures that estimate the value of patterns with respect to a given user class, based on user beliefs or expectations. The use of interestingness measures to guide the discovery process and reduce the search space is another active area of research.

Performance issues: These include efficiency, scalability, and parallelization of data mining algorithms.

- *Efficiency and scalability of data mining algorithms:* To effectively extract information from a huge amount of data in databases, data mining algorithms must be efficient and scalable. In other words, the running time of a data mining algorithms must be predictable and acceptable in large databases. From a database perspective on knowledge discovery, efficiency and scalability are key issues in the implementation of data mining systems. Many of the issues discussed above under *mining methodology and user interaction* must also consider efficiency and scalability.
- *Parallel, distributed, and incremental mining algorithms:* The huge size of many databases, the wide distribution of data, and the computational complexity of some data mining methods are factors motivating the development of parallel and distributed data mining algorithms. Such algorithms divide the data into partitions, which are processed in parallel. The results from the partitions are then merged. Moreover, the high cost of some data mining processes promotes the need for incremental data mining algorithms that incorporate database updates without having to mine the entire data again "from scratch." Such algorithms perform knowledge

modification incrementally to amend and strengthen what was previously discovered.

Issues relating to the diversity of database types:

- *Handling of relational and complex types of data:* Since relational databases and data warehouses are widely used, the development of efficient and effective data mining systems for such data is important. However, other databases may contain complex data objects, hypertext and multimedia data, spatial data, temporal data, or transaction data. It is unrealistic to expect one system to mine all kinds of data, given the diversity of data types and different goals of data mining. Specific data mining systems should be constructed for mining specific kinds of data. Therefore, one may expect to have different data mining systems for different kinds of data.
- *Mining information from heterogeneous databases and global information systems:* Local- and wide-area computer networks (such as the Internet) connect many sources of data, forming huge, distributed, and heterogeneous databases. The discovery of knowledge from different sources of structured, semistructured, or unstructured data with diverse data semantics poses great challenges to data mining. Data mining may help disclose high-level data regularities in multiple heterogeneous databases that are unlikely to be discovered by simple query systems and may improve information exchange and interoperability in heterogeneous databases. Web mining, which uncovers interesting knowledge about Web contents, Web usage, and Web dynamics, becomes a very challenging and highly dynamic field in data mining.

The above issues are considered major requirements and challenges for the further evolution of data mining technology. Some of the challenges have been addressed in recent data mining research and development, *to a certain extent*, and are now considered *requirements*, while others are still at the research stage. The issues, however, continue to stimulate further investigation and improvement.

12.2 Privacy Issues in Knowledge Discovery and Data Mining

Not surprisingly, data is treated today as one of the most important corporate assets of companies, governments, and research institutions supporting fact-based decision-making. It is possible to have fast access, correlate information stored in independent and distant databases, analyze and visualize data on-line, and use data mining tools for automatic and semiautomatic exploration and pattern discovery. Knowledge discovery and data mining (KDDM) is an umbrella term describing several activities and techniques for extracting information from data and suggesting patterns in very large databases.

Marketing applications have adopted and expanded KDDM techniques. KDDM is now moving to other domains where privacy issues are very delicate. Recent developments in information technology have enabled collection and processing of vast amounts of personal data, such as criminal records, shopping habits, credit and medical history, and driving records. This information is undoubtedly very useful in many areas, including medical research, law enforcement and national security. For the analysis of crime data, KDDM techniques have been applied by the FBI in the US as a part of the investigation of the Oklahoma city bombing, the Unabomber case, and many lower-profile crimes. Another example is an application of KDDM to analyzing medical data.

Despite its benefits to social goals, KDDM applications inspire reservations. Individuals easily imagine the potential misuses from unauthorized tapping into financial transactions or medical records. Survey in the US reveal growing concern about the use of personal information. The newest Equifax/Harris Consumer Privacy Survey shows that over 70% of respondents are against unrestricted usage of their medical data for research purposes. As much as 78% believe that computer technology represents a threat to personal privacy and that the use of computers must be restricted sharply in the future, if privacy is to be preserved. Up to 76% believe they have lost control over their personal information and that their privacy is threatened. Time/CNN reveals that at least 93% of respondents believe companies selling personal data should be required to gain permission from individuals. In another study 96% of respondents believe that private information should never be used for another purposes without permission, and over 20% had personally experienced a privacy invasion. By contrast, in 1970 the same Equifax/Harris Survey found only 33% considered computer technology a threat to their privacy.

Massive amounts or personal data are available. In 1990, Lotus Development Corporation announced the release of a CD-ROM with the data on 100 million households in the US. The data was so detailed that it generated string public opposition and Lotus abandoned the project. However, this mostly affected small business, as large companies already had access and continued to use Lotus datasets. At least 400 million credit records, 700 million annual drug records, 100 million medical records and 600 million personal records are sold yearly in the US by 200 superbureaus. Among the records sold are bank balances, rental histories, retail purchases, criminal records, unlisted phone numbers, and recent phone calls. Combined, the information helps to develop *data images* of individuals that resold to direct marketers, private individuals, investigators, and government organizations. These data images are now the subject of analysis by automatic and semiautomatic knowledge discovery and mining tools. In this section we address revitalized general privacy issues and new threats to privacy by the application of KDDM. We distinguish them from threats to privacy or security resulting from the expansion of computer networks and on-line distributed information systems.

12.2.1 Revitalized Privacy Threats

Secondary Use of the Personal Information

Recent surveys have shown a great concern about the use of personal data for purposes other than the one for which data has been collected. An extreme case occurred in 1989. Despite collecting over $16 million USD by selling the driver-license data from 19.5 million Californian residents, the Department of Motor Vehicles in California revised its data selling policy after Robert Brado used their services to obtain the address of actress Rebecca Schaeffer and later killed her in her apartment. While it is very unlikely that KDDM tools will reveal precise confidential data directly, the exploratory KDDM tools may correlate or disclose confidential, sensitive facts about individuals resulting in a significant reduction of possibility. In fact, this is how they were applied in the investigation of the Unabomber case and other criminal investigations. They facilitated filtering large volumes of reports form informants so resources could be concentrated on much fewer promising leads and suspects. Thus, we would not expect that detailed personal addresses would be disclosed by a KDDM analysis; however, with enough data about the patterns of behavior of young actresses in Los Angeles, the potential addresses may be reduced to a few possibilities making it feasible to visit them all. A simple application of link analysis can correlate phone and banking records to determine, with a certain degree of accuracy, if bank customers have a fax machine at home and how this impacts the likelihood of accepting offers on equity loans. Most individuals consider the use of information for secondary analysis a direct invasion of privacy, and perhaps even more if this reveals aspects like what does a person have inside its home. Individuals understand that phone companies need to monitor length of phone calls for billing purposes, and that their bank must keep track of transactions in their accounts, but consider unauthorized secondary use of their data to be a violation of privacy.

Handling Misinformation

Misinformation can cause serious and long-term damage, so individuals should be able to challenge the correctness of data about themselves. For example, District Cablevision in Washington fired James Russell Wiggings on the basis of information obtained from Equifax, Atlanta, about Wiggings' conviction for cocaine possession; the information was actually about James Ray Wiggings, and the case ended up in court. This illustrates a serious issue in defining property of the data containing personal records. While individuals and legislators supporting the right of privacy favor the view that a person's data is the person's property, data collectors favor the view that the data collector owns the data. These ethical issues have been illustrated by some cases of celebrities and other public figures who have been able to obtain rights on reproduction of their photographed image. However, for the average citizen the horizon is not so promising. The existing market of personal data postulates that the gathering institution owns the data. However, the attitude of data collectors and marketers toward privacy is more moderate than 20 years ago when marketers believed that there was "too much privacy already."

The reason for this change, apart from the fact that privacy is under much bigger threat now, is probably the fear of massive public opposition and losing the trust of customers. Many data owners acknowledge that there is a "Big Brother" aspect in the exploitation of personal datasets, and data collectors now take some measures to preserve the customers trust. Others imply that the "sinister purpose" of data mining is the "product of junk science and journalistic excess," but nevertheless believe that "marketers should take a proactive stance and work to diffuse the issue before it becomes a major problem,"

Granulated Access to Personal Information

The access to personal data should be on a need-to-know basis and limited to relevant information only. For example, employers are obliged to perform a background check when hiring a worker but it is widely accepted that information about diet and exercise habits should not affect hiring decisions. There seem to be two directions in this issue. Some advocate the removal of variables and even prohibiting their collection. Others support the release for very detailed data so research and analysis can advance. The new privacy laws in Germany illustrate the first approach. These regulations have dramatically reduced the number of variables in the census and the microcensus. The second approach is illustrated by personal data placed on large on-line networked databases, like the Physician Computer Network in the US, with the intention to build and expand knowledge. Another example of this approach are more precise geo-referenced datasets in geographical information systems and their databases that include cadastral data, aerial photography, and detailed features of private properties.

Scholars form diverse backgrounds in history, sociology, business, and political science have concluded that the existing privacy laws are far behind the developments in information technology and do not protect privacy well even for the issues mentioned above. The OECD [Organization for Economic Cooperation and Development] Principles on Data Collection addresses some of the concerns raised by the issues in this section. However, only 24 countries have adopted them, thus far with varying degrees. Twelve nations have adopted all OECD's principles in statutory law. Australia, Canada, New Zealand, and the US do not offer protection to personal data handled by private corporations. In Australia, the Privacy Act 1988 predates the growing on-line purchasing and other massive networked data-collection mechanisms. Australia's Privacy Commissioner, Moira Scollay, has taken steps to simplify privacy regulations and provide a single national framework for data-matching systems such as FLY-BUYS cards. However, so far, she has only proposed for discussion a draft of principles for fair handling of personal information.

12.2.2 New Privacy Threats

Stereotypes

General patterns may be used for guessing confidential properties. Also, they may lead to stereotypes and prejudices. If the patterns are based on properties such as race, gender, or nationality, this issue can be very sensitive and controversial. Examples are debates over studies about intelligence across different races. The issue raises debate because KDDM tools may allow the application of different commercial standards based on race or ethnic group. Banks may use KDDM tools to find a different pattern of behavior between two racial or ethnic groups and then deny credit or apply a different policy based on the attribute. An example of different commercialization based on ethnic group occurs in marketing of petrol in Canada because it is well know that French-Canadians are more likely to buy premium petrol than Anglo-Canadians. While these examples seem harmless, in the current situation where more and more business consider data analysis as necessary for ensuring efficiency and competitiveness, wide application of KDDM patterns can easily spread other kinds of stereotypes. For example, data analysis has made it possible for car insurance companies to adjust their premiums based on the age of the driver. It has allowed health insurance companies also adjust their premiums based on some patterns of behavior of the policyholder (for example, the practicing of sports like scuba driving or hand gliding). Stereotypes may have serious implications into the type of data mining technology applied to some task. For example, an inductive learner may be used to create a classifier out of previous credit card applications. If the learner is of the type that can explain why it is denying or granting some credit level, the corporation using it may detect that it is using controversial attributes (i.e., gender, nationality) to perform the credit ranking. However, if the tools represents knowledge in implicit forms that are hard to interpret (a la artificial neural networks), it may be possible that credit assignment is been mainly based on controversial attributes without the awareness of the credit-card company and the application. Many legal issues may be raised if the applicant is to discover this.

Another problem is that KDDM tools use parametric and nonparametric methods from the fields of statistics and machine learning. Their limitations are difficult to explain to a lay person. However, it may be easy to claim that the pattern was derived by a sophisticated computer program, and thus, accepted as truth about individuals. In particular, the fact that they are obtaining rules of thumb and generalization that work in most cases (and are about groups) but are false of each individual is one of the hardest concepts to grasp. This is also true for statistics. One toss of a fair coin will end up in heads or in tail and not in between, but such in between is the "expected value" of such a single event. Similarly, statistics allows us to predict accurately values about populations, like how many smokers will develop cancer this year, but it will not guarantee than a Mr. X will develop cancer, and despite him being a smoker.

Guarding Personal Data From KDDM Researchers

The protection of privacy cannot simply be achieved by restricting data collection or restricting the use of computer and networking technology. How could planning decisions be taken, if census data was not collected? How could epidemics be understood if medical records were not analyzed? Individuals benefit from data collection efforts in the process of building knowledge that guide's society. Researchers feel that privacy regulations may enforce so many restrictions on data, that it would make the data useless. On the other hand, researchers who apply KDDM tools to datasets containing personal information should not be given access to individual data. But such a restricted access can make KDDM tasks very difficult or even impossible. Thus an appropriate balance between a need for privacy and a need for knowledge discovery should be found.

Individuals From Training Sets

The classification task in KDDM takes as input a set of cases and their classes (training set); the output is a classifier that is, an operator that assigns classes to new, unclassified cases. For example, the cases may correspond to paints and classes to diagnoses. There are two security problems regarding the training set. The first is how to provide the analyst with a training set. If such a set is provided from real data, then each record of a training set is a discloser of the information of the individual corresponding to the record. If the training set contains artificial or perturbed data, then the quality of a classifier may be seriously affected. The second problem is how to protect privacy if somebody has a classifier and a record that is known to belong to the training set that built the classifier. The KDDM classifiers are typically very accurate when applied to case from the training set. Thus a classifier and knowledge that case A is in the training set reveals the class of case A. In this section, we argue that a classifier should be modified in such a way so as to have similar accuracy when applied to the case from the training set, as when applied to the new cases.

Combination of Patterns

Combining two or more general patterns may lead to a disclosure of individual information, either with certainty, or with a high probability. Also, knowledge of totals and other similar facts about the training data may be correlated to facilitate compromising individual values. For example consider a dataset where there are 10 people: 2 females and 8 males; there are 8 cases of disease A, and none of the females has disease A. if it is known that Mr. $X's$ information is part of the data, it is possible to infer that Mr. X has disease A. The problem here is how to protect individual values while preserving values of parameters that are calculated by statistical inference. These parameters usually described a probabilistic model and their accurate inference hardly reveals anything about individual records. Thus, perturbation to the data that preserve the estimators of statistical inference technique could potentially be sufficient for the problem.

How to hide what can be inferred by all possible methods? This implies in a sense that we discover first everything that is to be discovered about a dataset in order to be sure that we conceal it. Or in other words, to understand how to protect values in individual records, we first must understand very well the capabilities of KDDM tools and technology. We have clearly illustrated that tools for knowledge discovery pose new threat to privacy. However, researchers in KDDM seem to have contradictory views on privacy issues. Some believe that KDDM is not a threat to privacy, since the derived knowledge is only about and from groups. Clearly, this is only true if the identified group is large, but otherwise, information about an individual is partially disclosed. Also, performing inferences about groups does not prevent the creation of stereotypes. Others clearly support our views saying that KDDM deals mainly with huge amounts of microdata. Some fear there would be different academic standards: "statutory limitations that vary from country to country ... suggests that the practice ... varies from country to country." Europe has adopted the OECD directives and investigators across all fields of scholarly research now require "the subject's written consent or data may not be processed." Others think that subject approval may not be sufficient for data miners to refer or disclose incidentally discovered patterns or relationships. That is, how can one ask approval from an individual about something yet unknown?

12.2.3 Possible Solutions

We discuss the possible solutions to the privacy problems imposed by KDDM and their impact on the quality of discovered patterns. The issue of how data may create or reinforce stereotypes is more an issue of the application of the technology than a problem of the technology. It is similar to considering the impact of and biased reporting or political propaganda on large audiences as a disadvantage of TV. This is not a disadvantage of the technology, but a very complex issue of the social structures that allow the production and broadcasting of materials, later distributed through the powerful technology. Thus, the issue of created or reinforced stereotypes with KDDM tools falls in the sociological, anthropological, and legal domain. The interpretation of patterns from KDDM is as difficult as the interpretation of statistics. For example, the times/distance/etc. achieved by female athletes vs. male athletes in Olympic competitions constitute data that statistically distinguishes two groups. When does finding two groups justify creating different policies is not an issue of the technology that discovered the two clusters. In what follows, we concentrate on technological solutions that could alleviate some of the risk to privacy associated with the other issues listed in previous section. Those threats to privacy listed in previous section are not totally new and there is some understanding of them from pervious attempts to ensure privacy in the facts if the need to perform data analysis. In particular, the field of statistical database has done much progress in this direction and we believe it can provide much insight into new issues like those listed previously.

The field of statistical database has developed methods to guard against the disclosure of individual data while satisfying request or aggregate statistical information. In particular, the experience form this field indicates that removing identifiers such as names, addresses, telephone numbers, and social security numbers is a minimum requirement for privacy but it is enough. Re-identification based on remaining fields may still be possible and removing identifiers is considered the weakest approach, and should never be used on its own. In fact, the information models used in KDDM and statistical database is quite similar. The main objectives of a security control mechanism in a statistical database are to provide statistical users with a sufficient amount of high-quality statistics (statistics quality is measured by the consistency, bias, and precision) and at the same time, to prevent disclosure of confidential individual information. Similar to the context of statistical databases, the objective of a security control mechanism for KDDM is to prevent the disclosure of confidential individual values. However, unlike in statistical databases, another objective is not to minimize the distortion of all statistics, but rather to preserve general patterns; in order words, the objective is to preserve those statistics based on attributes that have a high impact on the patterns and rules to be discovered.

Various techniques have been proposed for security protection in statistical database, but none of them is both effective and efficient. All methods trade-off privacy of individual values for statistics and/or pattern distortion. The methods can be classified into two main categories, "query restriction" and "noise addition." *Query restriction* techniques provide exact answers to some queries and reject others that may lead to a database compromise. Hence, the statistical quality of released information is high, but these techniques are typically overly restrictive. They are inadequate against skilled users, who have previous knowledge about information stored in the database, and/or require high initial implementation effort. When applied to KDDM, query restriction techniques may deny some particularly important information and obscure general patterns. The idea here is to supply a subset of the data so restricted that is not useful for the data miner. This has been much criticized. First, why would a miner acquire or investigate data guaranteed not to have anything useful? Second, the only way to guarantee that the dataset contains no patterns is to find them all (which require infinite computational time) or to provide a very small set. Third, for this scheme to work, it is assumed that each miner will not co-operate with other miners (and in particular, that nobody gains access to more data). Since the objective of query restriction is not to maximize the number of answerable queries, but rather to answer all, or most of, the important queries partitioning is a potential approach. This requires the development of methods for grouping record on the basis of their values of important attributes. These attributes may be identified using rough set theory. *Noise addition* techniques prevent compromise by introducing an error either to data or to result of queries. These techniques are robust (resistant to users' supplementary knowledge) and provide answers to all queries, but at the expense of allowing partial disclosure and/or providing information with less statistical quality. O'Leary has suggested that noise addition by itself could provide protection from all those KDDM techniques that are very sensitive to noise. In particular, probability distribution data perturbation techniques referred to as *data swapping*, seems suitable for privacy protection in KDDM. Data swapping refers to interchanging the values in the records of the database in such a way that low-order statistics are preserved. Statistics that employ exactly k attributes are called k-order statistics. A database D is said to be d-transformable if there exists a database D', but has no records in common with D, but has the same $k-$order counts as D, for $k \in \{0, \ldots., d\}$. However, finding a general data swap is thought to be an intractable problem.

Recent investigations have shown that it is possible to transform the data by data swapping, so the rules produced by decision trees (a ubiquitous KDDM technique) on the new set corresponds to finding a-transformation of a given database, but relaxing the condition that D and D' have no records in common. The exchange of values of the confidential attribute consists of randomly shuffling within heterogeneous leaves of the decision tree. Thus all the statistics, which do not involve this attribute, are preserved. Similarly, the statistics that involve the confidential attribute and whose query sets are defined by nodes of a decision tree will also be preserved. All the perturbed statistics are those whose query sets could be obtained by the repeating process of splitting heterogeneous leaves, until homogeneous leaves are reached. Since the heterogeneous leaves have a vast majority of records belonging to a single class and no straightforward way for further splitting, we can argue that the most seriously disturbed statistics will be those that involve a small number of records. Furthermore, we can balance the statistical precision against the security level by choosing to perform the swapping in the internal nodes, rather than in the leaves of the decision tree: the closer to the root, the higher the security but lower the precision.

12.3 Some Privacy Issues in Knowledge Discovery: The OECD Personal Privacy Guidelines

Several countries have generated principles to protect individuals from the potential invasion of privacy that data collection and retrieval poses. The Organization for Economic Cooperation and Development (OECD) has provided probably the best known set of guidelines. A number of countries have adopted these guidelines as statutory law, in whole or in part. The OECD has specific guidelines pertaining to data privacy that directly affect those performing knowledge discovery generally, and those who use the so-called "personal data" in particular. In this section we address such questions as

- What are the implications of the existing privacy guidelines, especially those of the OECD, for knowledge discovery?
- What are the limitations of these guidelines?
- How do the restrictions on knowledge discovery about individuals affect knowledge discovery on groups?
- How do legal systems influence knowledge discovery?

The answers we pose to these and other related issues will be helpful in generating a larger dialog on this important subject. There has been relatively little investigation into the privacy and security issues relevant to knowledge discovery, in particular and intelligent systems in general. Developers have proposed and used intrusion-detection systems as the basis of security systems designed to protect privacy. Typically, intrusion-detection systems determine if a user is an intruder to a legitimate user, generally by way of various internal systems profiles. Earlier studies of security issues in intelligent systems included issues of privacy and the security of systems knowledge. There has been some concern about knowledge discovery as a different kind of threat to data base security as well.

12.3.1 Risks Privacy and the Principles of Data Protection

The classic definition of the invasion of privacy refers to the "abuse or disclosure of intimate personal data." Recently, there has been an effort to expand this definition to include other issues, such as the protection of general privacy and protection from the unauthorized use of one's "personal" data taken from computer databases.

Increasingly, companies and organizations are using computer-based systems to capture personal data. Although this method typically increases both efficiency and productivity, there are a number of risks to individual privacy. In particular, those risks include the following:

- the data can be used for some purpose other than that for which it was collected;
- the data can be inaccurate, incomplete, or irrelevant;
- there are risks of unauthorized access to personal information;
- individual databases can be linked, increasing the range of information about individuals;
- The increased ability to construct individual profiles from multiple sources may affect "decisions concerning the individual's qualifications, credit eligibility, health, insurance consumption patterns social security, employment and so on."

As both the amount of information and number of users on the Internet grows, these risks become increasingly likely to manifest themselves. This is particularly true for joining previously disparate databases. Hence, many feel that additional guidelines and statutory-based controls are necessary to prevent the invasion of personal privacy. These concerns have led organizations to generate guidelines to mitigate these privacy risks, including the OECD and the council of Europe. This section focuses on the OECD guidelines since many nations have adopted then as statutory law.

OECD principles of data collection. The following are the OECD principles of data protection:

1. *Collection limitation:* Data should be obtained lawfully and fairly, while some very sensitive data should not be held at all.
2. *Data quality:* Data should be relevant to the stated purposes, accurate, complete, and up-to-data; proper precautions should be taken to ensure this accuracy.
3. *Purpose specification:* The purpose for which data will be used should be identified, and the data should be destroyed if it no longer serves the given purpose.
4. *Use limitation:* Use of data for purposes other than specified is forbidden, except with the consent of the data subject or by authority of the law.
5. *Security safeguards:* Agencies should establish procedures to guard against loss, corruption, destruction, or misuse of data.
6. *Openness:* It must be possible to acquire information about the collection, storage, and use of personal data.
7. *Individual Participation:* The data subject has a right to access and challenge the data related to him or her.
8. *Accountability:* A data controller should be accountable for complying with measures giving effect to all these principles.

The OECD principles arose to help nations cope with the shipment of data outside the country of origin. They attempt to ensure that when data is transported across country borders the data subjects enjoy the same level of privacy as in the original country. Thus far 24 countries have adopted the OECD guidelines to varying degrees, including Australia, Austria, Belgium, Canada, Denmark, Finland, France, Germany, Greece, Iceland, Ireland, Italy, Japan, Luxembourg, the Netherlands, New Zealand, Norway, Portugal, Spain, Sweden, Switzerland, Turkey, the UK, and the US. Not all countries employ the OECD guidelines as statutory law, and not all countries have adopted all eight guidelines. Instead, the level of participation – the number of guidelines adopted – varies from country to country.

Twelve nations have adopted all eight of the principles in statutory law; Japan adopted seven of the principles (excluding #7), and UK has adopted six of the principles (excluding #7 and #8), as statutory law. Alternatively, Australia, Canada, New Zealand, and the US do not offer protection to personal data handled by private corporations. However, those for four countries have similar statutory constraints on personal data held in the public sector.

Scope of application: personal data. The OECD developed the primary protective guidelines for personal data. Consequently, the KDD community must determine what kinds of data fall under the heading of personal. According to Neisingh and de Houwer, personal data is data gathered by corporations and government, including financial, educational, economic, social, political, medical, criminal, welfare, business, and insurance data. As a result, it is easy to see that these principles affect many different kinds of datasets.

12.3.2 The OECD Guidelines and Knowledge Discovery

Now, we discuss the impact, implications, and limitations of the guidelines for knowledge discovery.

Collection limitation. This principle states that, "some very sensitive data should not be held at all," thereby limiting the scope of knowledge discovery from data. If the data is "very sensitive," knowledge discovery researchers should probably not have access to the data, as it could lead to repercussions. Such sensitive data is likely to include information about religious beliefs, race, national origin, and other issues.

However, it is not clear what it means for data to be sensitive. What may be deemed sensitive depends on the context and country in which the authorities develop the legislation. What is sensitive in one country may not be sensitive in another, suggesting that knowledge discovery could differ from country to country. Accordingly, such cultural differences could from the international differences in computer science practices.

Data quality. Knowledge discovery may influence the data quality principle. For example, knowledge discovery may lead to questions about additional categories of information, such as derived data. The data quality principle suggests that researchers differentiate derived data from original data and not include it in the database, since its accuracy could not be assured. Over time the data on which derived data is based may change, thereby changing the derived data as well. As a result, researchers should not store this data, as it could be outdated. If the derived data is kept, researchers should treat it with the same concerns as the original data.

Also, this principle's recommendation that proper precaution be taken suggests that there be quality standards in knowledge discovery. However, since the discipline is still evolving, it may be premature to talk about generating standards.

Purpose specification limitations. This principle indicates that databases be used only for the declared purposes. Goals for the use of data should be generated, and the data should be used to accomplish those goals exclusively. Any other uses would require the consent of the data subject. Consequently, it is critical that if a database is planned for knowledge discovery, then the use of knowledge discovery is specified.

In addition, if knowledge discovery is only done on databases for which knowledge discovery has been declared, then only those databases generated since the gathering of purpose information began are available for this activity. Accordingly, legacy and existing databases are probably outside the scope of knowledge discovery. Users may have to declare the specific knowledge discovery tasks when gathering the data, instead of declaring anticipated knowledge discovery for some general purpose.

The purpose principle is critical for knowledge discovery using multiple databases. If the data was gathered for use in a single database, analysis across multiple databases generally would violate the purpose principle. This could limit knowledge discovery using individual personal data to particular databases.

This principle threatens knowledge discovery's potential to expand on its on discoveries as well. Feedback can play a very important role in knowledge discovery tasks. As the system generates more knowledge, that knowledge can form the basis of the search for additional knowledge. Therefore, if the principle limits the knowledge discovery task to the first-level findings specified in the original purpose, it limits the power of knowledge discovery significantly.

Another possible limitation is the required level of detail in the statement of purpose. It is possible, in an extreme case, that authorities would require researchers to elicit each specific knowledge discovery activity, not just the fact that knowledge discovery would be done.

Use limitation. This principle is closely related to the purpose specification principle, as it specifies that if data is to be used for some purpose other than the originally specified purpose, the data subject must provide consent. By extension, the data subject will need to provide consent when his or her personal data is to be used for knowledge discovery. The purpose specification principle requires users to identify the original use of the information, and the use limitation principle constraints data use to the original purpose. Both principles require data subject consent if changes in the use of the data occur.

The use limitation principle has a direct impact on performing knowledge discovery from related databases. Generally, expanding the analysis of knowledge discovery from one database to multiple databases would require data subject consent, since the interaction of multiple, previously unconnected databases would suggest alternative uses beyond the original scope.

Acquiring data subject consent may be very difficult, in part because most data subject would have difficulty understanding the technology of knowledge discovery. Further, it is unclear what level of detail data subjects would need. For example, would the awareness that knowledge discovery was being done be sufficient or would users need to explain the individual task level?

Security safeguards. This principle calls for establishing safeguards against the misuse of data. In some cases, knowledge discovery may qualify as a misuse of data, especially if unauthorized users perform knowledge discovery, or if knowledge discovery occurs without gathering consent. As a result, authorization procedures for knowledge discovery must be established. The limitations associated with the statement of purpose also influence the security safeguard principle. A particular concern is how to secure a database from knowledge discovery without eliminating access to virtually all users.

Openness. Taken to one extreme, the openness principle suggests that data subjects should be able to acquire information about the uses of knowledge discovery and the specific knowledge discovered about them. Requiring analysis to inform individuals about particular derived data could limit the general use of knowledge discovery and thereby inhibit its use. If knowledge discovery does not lead to inferences about individual data subjects, there would not necessarily be an openness issue.

However, since it is virtually impossible to deter users of a database from performing knowledge discovery, it will be equally difficult to knowledge discovery, it will be equally difficult to know for certain whether knowledge discovery is being done using information about a particular data subject. Thus, the individual participation and accountability principles play a critical role in controlling inappropriate knowledge discovery.

Individual participation. This principle suggests that data subjects should be able to challenge knowledge discoveries related to them. These discoveries might pertain to the individual only or to the individual's relationship to specific groups. The knowledge discovered may directly influence how the users perceive and treat the data subject, possibly adversely affecting that person's available options.

In light of the right to challenge knowledge discoveries, it is critical to document the development of conclusions. Substantiating the quality of different knowledge discovery approaches and algorithms will become increasingly important. The development and use of standards will help mitigate the challenges to knowledge discovery findings.

Accountability. This principle calls for a data controller who is accountable for user compliance with the OECD measures. Thus, a knowledgeable data controller should authorize and be responsible for the adherence of knowledge discovery activity to the OECD measures. In addition, the data controller should inform data subjects of the use and findings from knowledge discovery.

However, due to the decentralization of databases and the difficulty of controlling knowledge and the difficulty of controlling knowledge discovery activity by those who have access to databases, data controllers will have great difficulty monitoring knowledge discovery effectively. Accordingly, it will be important for the data controller to inform database users and maintenance personnel about the policies regarding knowledge discovery activities, including the consequences of inappropriate use.

12.3.3 Knowledge Discovery about Groups

This section has thus far focused primarily on privacy issues associated with individual personal data. The OECD guidelines do not refer explicitly to knowledge discovery about groups. As a result, unless the knowledge discovered directly affects the individual personal data, there would be no general application of the guidelines. Instead alternative legislation or guidelines could be used to guide knowledge discovery about groups. For example, in the US it is illegal to discriminate against certain groups based on sex, race, religion, or national origin. Knowledge discovery about groups, then, could comply with these laws in its use of knowledge pertaining to these categories.

Further, the OECD guidelines suggest that individuals have the right to control the use of data about themselves, even in apparently innocuous knowledge discovery about groups. As a result, individuals could request that they not be included in the generation of knowledge about groups of which they may be a member.

One drawback of these individual privacy constraints is that they could interfere with important knowledge discoveries. For example, certain diseases seem to strike some groups and not others. As a result, information relating to groups could be the key to the discovery of cures, or other important kinds of knowledge.

12.3.4 Legal Systems and other Guidelines

The OECD guidelines form one basis of analysis. This could also be extended to investigate alternative sets of guidelines and statutory laws. The Council of Europe issued a similar set of guidelines for the European community that included the eight OECD principles and some additional constraints relating to the so-called transborder data flows. As alternative legal structures develop, researchers could analyze them for their impact on knowledge discovery.

Legal systems offer bases for the interpretation of different terms and situations in knowledge discovery as well. Many states, for the purposes of protecting litigants from undue invasions of privacy by adverse parties, have statutes defining personal or consumer information. For example, the California Code of Civil Procedure, section 1985.3, provides detailed definitions about personal records; they are the "original or any copy of books, documents, or other writings pertaining to a consumer and which are maintained by any 'witness' which is a physician, chiropractor..."

In the specific case of litigation, there are laws regarding the disclosure of information. For example, the California Code of Civil Procedure, section 1985.3, deals with "Sub-poena for production of personal records," while section 1985.4 summarizes the law regarding "production of consumer records maintained by state or local agency."

Further, in many cases different levels of government regulate certain industries, to a certain extent. Such industries include insurance, law, accounting, and medicine. As a result, these industries are likely to have regulations on limitations of disclosure of information. In other cases, the industries are self-regulated.

The limitations to the use of knowledge discovery that we have discussed are not limited to the new methods of knowledge discovery developed by the artificial intelligence community. Rather, they apply to all methods used to generate knowledge, including more traditional statistical and database approaches. The OECD guidelines limit the knowledge that can be obtained using any process, including direct examination, classic database updates, queries, or statistical methods.

This section provides some insight into the problems concerning personal privacy and data faced by those who wish to employ knowledge discovery. When it comes to personal data, there often are statutory limitations that vary from country to country. As a result, it suggests that the practice of computer science and artificial intelligence varies from country to country, based on different cultural and legal differences. However, it is clear that there are some general principles of data collection and maintenance that a number of countries adhere to. Those principles influence what data can be used in knowledge discovery and how users process and maintain discovered data.

Many of the above limitations are a result of the OECD personal privacy legislation predating knowledge discovery's widespread use in the artificial intelligence community. The OECD guidelines could not anticipate or address many important issues regarding knowledge discovery, and thus several principles are too general or unenforceable.

12.4 Summary

The mentioned above issues are considered major requirements and challenges for the further evolution of data mining technology. Some of the challenges have been addressed in recent data mining research and development, *to a certain extent*, and are now considered *requirements*, while others are still at the research stage. The issues, however, continue to stimulate further investigation and improvement.

Knowledge discovery and data mining revitalizes some issues and posses new threats to privacy. Some of these can be directly attributed to the fact that this powerful technique may enable the correlation of separate datasets in other to significantly reduce the possible values of private information. Others can be attributed more to the interpretation application and actions taken from the inferences obtain with the tools. While this raises concerns, there is a body of knowledge in the field of statistical databases that could potentially be extended and adapted to develop new techniques to balance the rights to privacy and the needs for knowledge and analysis of large volumes of information. Some of these new privacy protection methods are emerging as the application of KDD tools moves to more controversial datasets.

The OECD guidelines could not anticipate or address many important issues regarding knowledge discovery, and thus several principles are too general or unenforceable.

12.5 Review Questions

1. What is mining methodology and state some of the user interaction issues?
2. Discuss in detail on performance issues and issues relating to the diversity of database types.
3. Explain about privacy issues on data mining.
4. Give details on the Organization for Economic Cooperation and Development(OECD) personal privacy guidelines.
5. What are the OECD principles of data collection in the mining technology?

13

Active Data Mining

Objectives:

- Introduce an active data mining paradigm that combines recent work in data mining with the rich literature on active database system.
- In this paradigm, data is continuously mined at a desired frequency.
- To be able to specify shape queries, we describe the constructs for defining shapes.
- Discuss how the shape predicates are used in a query construct to retrieve rules whose histories exhibit the desired trends.
- Describe how this query capability is integrated into a trigger system to realize an active mining system.
- The query language described provides the capability to discover interesting information by analyzing rules and their histories in novel ways.
- As the data mining technology is applied in the production mode, the need for active mining arises.

Abstract. We introduce an active data mining paradigm that combines the recent work in data mining with the rich literature on active database system. In this paradigm, data is continuously mined at a desired frequency. As rules are discovered, they are added to a rule base, and if they already exist, the history of the statistical parameter associated with the rules is updated. When the history starts exhibiting certain trends, specified as shape queries in the user-specified triggers, the triggers are fired and appropriate actions are initiated.

To be able to specify shape queries, we describe the constructs for defining shapes and discuss how the shape predicates are used in a query construct to retrieve rules whose histories exhibit the desired trends. We describe how this query capability is integrated into a trigger system to realize an active mining system. This case study is taken from R. Agrawal and G. Psaila, IBM Almaden Research Center, CA.

Data mining (also called *knowledge discovery*) in databases is the efficient discovery of previously unknown patterns in large databases and is emerging as a major application area for databases. In three classes of data mining problems involving associations, sequences and classification were introduced and it was argued that

S. Sumathi and S.N. Sivanandam: *Active Data Mining*, Studies in Computational Intelligence (SCI) **29**, 293–302 (2006)
www.springerlink.com © Springer-Verlag Berlin Heidelberg 2006

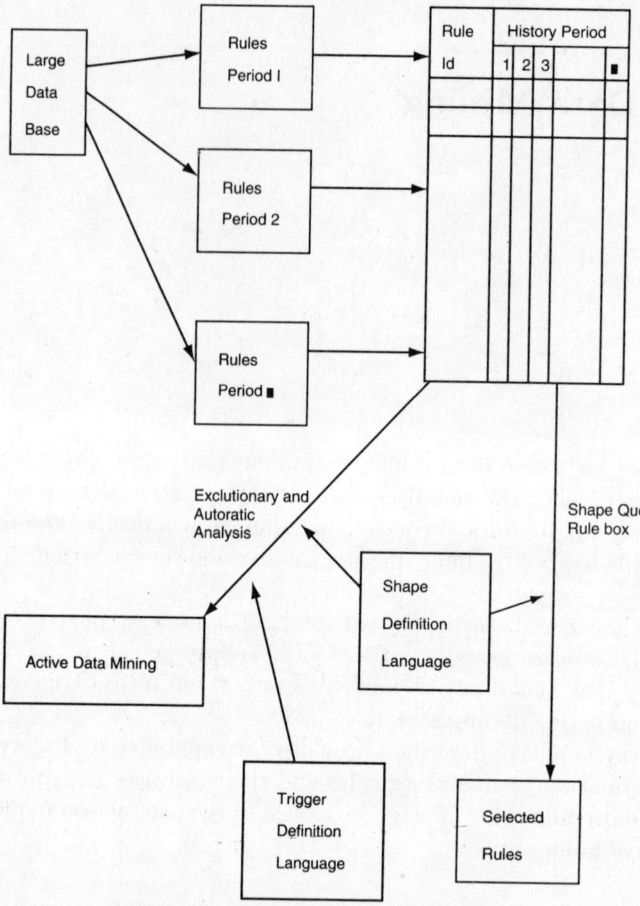

Fig. 13.1. Active Data Mining Process

these problems can be uniformly viewed as requiring discovery of rules embedded in massive data. Attached to every discovered rule are some statistical parameters, such as confidence or support of the rule.

As the data mining technology is applied in the production mode, the need for *active mining* arises. Figure 13.1 shows a schematic of the active data mining process. The basic idea is as follows. Rather than applying a mining algorithm to the whole data, the data is first partitioned according to time periods. The granularity of the time period is application dependent. The amount of data available is large (generally in gigabytes and more), so that this partitioning does not lose significance of the rules discovered. The mining algorithm is now applied to each of the partitioned data set and rules are obtained for each time period. These rules are collected into a rule base. In this rule base, each statistical parameter of a rule will have a sequence of values, called the *history of the parameter* for that rule. We can now query the rule base using predicates that select rules based on the shape of the history of some or all parameters.

The user can specify triggers over the rule base in which the triggering condition is a query on the shape of the history. As the fresh data comes in for the current time period, the mining algorithm is run over this data, and the rule base is updated with the generated rules. This update causes the histories of the rules to be extended. (A history of a new rule is initialized with zero values for the past time periods.) This, in turn, may cause the triggering condition to be satisfied for some rules and the corresponding actions to be executed.

Such active systems can be used, for instance, to build early warning systems for spotting trends in the retail industry. For example, if we were mining association rules, we will have histories for the support and confidence of each rule. [An association rule is an expression of the form A $\Rightarrow$ C, where A and C are sets of literals. In a database of transactions, where each transaction is a set of literals, the rule A $\Rightarrow$ C signifies that very often when A appears in a transaction, so does C. How often this happens is captured by the "confidence" parameter and is indicative of the strength of the rule. The "support" parameter gives the fraction of transactions in the database in which the given rule is present, and is indicative of the prevalence of the rule]. Following the promotion for an item X, the user may specify a notification trigger on the rule X $\Rightarrow$ Y; the triggering condition being that the support history remains stable, but the confidence history takes the shape of a downward ramp. Firing of this trigger will signify that if the goal of promoting X was to drag the sale of Y, it was not fulfilled. The loyalists continued to buy X and Y together, but the new buyers cherry picked X.

13.1 Shape Definitions

A shape found in a history can be described by considering the transition of values assumed by the shape at the beginning and end of each unit time period. We let the user define classes of transitions and assign symbols to them. These symbols are called *elementary shapes*. We do not have a pre-canned set of elementary shapes; the user can add or delete shapes or change the definition of any of them. However, to keep the discussion concrete, assume a user-defined set of elementary shapes, consisting of *up, Up, down, Down, appears, disappears, stable and zero.* The shape *up* could be a slightly increasing transitions with a minimum and maximum variation of 0.05 and 0.19 respectively; *up* could be a highly increasing transition with a minimum and maximum variation of 0.20 and 1.00, respectively; *appears* could be a transition from a zero value to a nonzero value; *stable* could be a transition in which the absolute difference between the initial and final value is not more than 0.04; etc.

Complex shapes can be derived by recursively combining elementary shapes and previously defined derived shapes, using the *shape operators.* These operators are summarized in Table 13.1. We give syntax for each operator and how the corresponding derived shape is matched in a history H.

Using these operators, one can describe a wide variety of shapes found in a history, including "blurry," shapes where the user cares about the overall shape but does not care about specific details. The syntax for defining a shape is:

(Shape name (parameters) descriptor).

For example, here is a definition of a double peak:

```
(Shape spike (upcnt dncnt)
        (Concat (at least upcnt (any up Up))
        (atleast dncnt
        (any down Down))))
(Shape doublespeak (width ht1 ht2)
        (in width (in order spike (ht1 ht1)
        spike (ht2 ht2))))
```

We first define *spike* to be a shape that has at least upon number of either *up or Up* transitions followed by at least *dncnt* number of either *down or Down* transitions. Then *doublespeak* is a shape width wide that has two nonoverlapping spikes. Note that width may be wider than the sum of the width of the two spikes and there may be noise on either side of them. As another example, the shape bullish:

```
(Shape bullish (width upcnt dncnt)
        (in width
            (and
            (noless upcnt (any up Up))
            (nomore dncnt
                    (any down Down)))))
```

is defined to have at least upcnt ups (either up or Up) and at most dncnt downs (either down or Down) in width time periods. Finally, the shape drift:

```
(shape drift (width)
    (in width (precisely 0
            (any up Up down Down))))
```

has no ups or downs in width time periods.

Earlier languages based on regular expressions for finding patterns in sequences were not targeted at defining shapes [Seshadri et al. *Proc. IEEE*

Table 13.1. Shape Operators

Multiple Choice	(any $P_1, P_2, \ldots P_n$)
Match all subsequences of H that match at least one of the P_1 shapes.	
Concatenation	(concat $P_1, P_2, \ldots P_n$)
First, match the shape P_1 if a matching subsequence s is found, match P_2 in the subsequence of H immediately following the last element of s. Accept the match if it is strictly contiguous to s, etc.	
Multiple Occurrences	(exact n P) (atleast n P) (atmost n P)
Match all subsequences of H that contain exactly (at least/at most) no contiguous occurrences of the shape P. In addition, the resulting subsequences must neither be preceded nor followed by a subsequence that matches P.	
Bounded Occurrences	(in length Shape occurrences)
Do "blurry" matching. Here length specifies the length of the shape in number of transitions. The shape-occurrences has two forms given below:	
Shape-Occurrences: Logical combination Using and and or.	(precisely n P) (noless n Q) (nomore n R)
Match all length long subsequences of H that contain precisely (no less than/no more than) n occurrences of the shape P (Q/R). The n occurrences of P (Q/R) need not be contiguous in the matched subsequence; there may be overlap or arbitrary gap between any two.	
Shape-occurrences: Ordered shapes.	(in order $P_1 P_2 \ldots P_n$)
Match all length long subsequences of H containing the shapes P_1 through P_n in that order. P_1 and P_{i+1} may not overlap, but may have an arbitrary gap.	

Int. Conf. on Data Engg. 1995]. The difference in design focus influences which expressions are easy to write, understand, optimize, and evaluate.

13.2 Queries

With the machinery for defining shapes in hand, we are ready to specify how we can retrieve rules whose one or more histories contain the desired shapes. The syntax for defining a query is:

(query (shape history-spec))

Here, shape is the descriptor for the shape to be matched. The history-spec is of the form:

history-name start-time end-time

Here history-name specifies the name of the history in which the shape should be matched. The portion in which the matching occurs is constrained by the interval specified by start time and end time. Matching over the complete history can be specified by using the keywords start and end for start time and end time, respectively.

The result of the execution of a query is the set of all rules that contain the desired shape in the specified history. In addition, the result also contains the list of subsequences of the history that matched the shape. If no subsequence matches the specified shape, the result is an empty set.

Here is an example of a query:

```
(shape ramp( ) (concat Up Up))
        (query ((ramp) (confidence start end)))
```

We have defined a simple shape *ramp,* consisting of two consecutive Ups, and we want to retrieve all the rules whose confidence history contains a ramp. Instead of the shape name, we could have alternatively written its definition in the above query. We also could have limited the range of confidence history in which the shape should be matched. Here is a modified query:

```
(query ((Concat Up Up)
  (confidence start 10)))
```

The user can also retrieve combinations of several shapes in different histories by using the logical operators and and or. Here is an example of a query that is looking for different shapes in the two histories of a rule – an *upramp* in support but a *downramp* in confidence:

```
(shape upramp (len cnt)
  (in len (noless cnt (any up Up))))
(shape dnramp (len cnt)
(in len
  (noless cnt (any down Down))))
(query
  (and
  (upramp 5 3) (support start 10))
  (dnramp 5 3) (confidence start 10))
```

13.3 Triggers

The query language we just described provides the capability of discovering interesting information by analyzing rules and their histories in novel ways. Consider a user who is periodically collecting rules in the rule base and wants to discover rules that are assuming critical (or interesting in some other way) behavior. For instance, the user may be interested in rules that have started exhibiting increasing trend. Rather than running queries every time the data for a new period comes and rules are added to the rule base, it will be preferable to post these queries as triggers and let the system initiate appropriate actions (e.g., notification) when the trigger conditions are satisfied.

We use the ECA (Event Condition Action) model by Chakravarthy et al. 1989 as the basis for the trigger system. The interesting aspects of our trigger system are what can be specified as trigger conditions, the semantics of the trigger execution, and how it is used in the active mining process.

The syntax for specifying a trigger is:

```
(trigger trigger-name
        (events events-spec)
        (condition (shape history-spec))
        (actions actions-spec)
```

A trigger definition has three sections: *events, condition, and actions.* Let us examine each of them. The trigger system reacts to predefined and user-defined events. The predefined events describe an external update of the rule base. These events are: *createrule and updatehistory.* They occur when a new rule is added to the rule base and the history of rule is updated, respectively. A user-defined event is introduced to the system as:
(event event-name)
where event-name is the name of the event.

The *events-spec* in the events section specifies the events to which the trigger being defined reacts. Predefined and user-defined events and their logical combinations using the logical operators or and can appear in the *events-spec.* A trigger is considered fired if the event specification is true for at least one rule in the rule base. That is, the specified event combination has occurred for some rule.

The condition section is syntactically and semantically similar to the query construct discussed in Section Queries. The difference is that the condition is evaluated only on rules present in the *affected set* produced by the events section, instead of the whole rule base. A condition selects the affected set of rules from the rule base and performs the specified shape query on the relevant histories of those rules. The condition is true if the output set resulting from the query is not empty. In that case, the action section is executed on the query output.

The *action-specs* in the actions section is a list of actions that are executed for all (and only those) rules that belong to the output set produced by the condition evaluation. An action can be an execution of a function, such as *notify or show,* which can be defined by the user or system supplied. An action can also be a user-defined event name, in which case an occurrence of the specified event is generated. An action does not change the state of the rule base; the goal of model is only to notify that the properties expressed by the condition of a trigger holds for some rules and is accomplished by generating predefined and user-defined events that alert any possibly interested trigger.

13.3.1 Wave Execution Semantics

Several semantics have been proposed for trigger systems in active databases. The execution semantics for trigger system follows what we call the *wave execution model.* This semantics is close to what is known as the deterministic semantics for Datalog like rules in S. Ceri et al. 1990. The attractiveness of this semantics is its simplicity and a good match for our application.

A *wave* is a set of event occurrences that come together to the active system. The trigger execution process starts when a new wave is ready. First, the event specification of every trigger is checked to determine if it will fire. Triggers for which this evaluation is true are selected for firing and their affected set (rules affected by events-spec) is produced. When the event specification has been checked for all the triggers, the current wave has been used up and any event generated as a consequence of the triggers fired will belong to a new wave.

The selected triggers are fired now. For each fired trigger, its condition is evaluated only on rules in the affected set. If the condition is true, the output set is passed to the actions section. The actions section is immediately executed for each rule in the output set. If any event is generated as a consequence, it is added to the new wave.

After the conclusion of the condition evaluation and the eventual actions execution for all the fired triggers, the process is repeated considering the events generated as belonging to the new wave. The process terminates if the evaluation of all the event specifications determines that no trigger needs to be fired.

Example: We now give a simple example to illustrate the trigger facility. Suppose that a user wants to be notified if the support for a rule is increasing but its confidence is decreasing at the same time. The following definitions show how the user can accomplish this goal:

```
(shape uptrend (width upcnt)
        (in width
                (noless upcnt (any up Up))))
        (shape dntrend (width dncnt)
                (in width (noless
                        dncnt (any down Down))))
        (event upward)
        (trigger detect_up
    (events updatehistory)
    (condition
    (uptrend (5 4)
        (support (- end 5) end)))
    (actions upward)
            )
            (trigger detech_dn
    (events upward)
    (condition
        (dntrend (5 4)
                (confidence (- end 5) end)))
        (actions notify)
            )
```

We first specify what is meant by support is increasing and confidence is decreasing by defining two shapes: uptrend and dntrend. We introduce an event named upward to the system using the event construct. We then define the trigger detect_up. This trigger can be fired by the predefined event update history. If this trigger is fired, the condition section of this event checks if the rules that were updated (affected set of the event update history) contain up trend in the last five periods of their support history. If this condition is evaluated as true for some rules, the user-defined event upward is generated for each of these rules.

The second trigger detect_dn reacts to the generation of the occurrences of the upward event and it checks for dntrend in the last five periods of the confidence history of only those rules for which the trigger has been fired (affected set of the upward event). Thus, the user is notified of only those rules that simultaneously had an uptrend in support and dntrend in confidence in the last five time periods.

13.4 Summary

In this chapter an active data mining paradigm that combines the recent work in data mining with the rich literature on active database system is described. Here data is continuously mined at a desired frequency. As rules are discovered, they are added to a rule base, and if they already exist, the history of the statistical parameter associated with the rules is updated. Thus this section gives details on active data mining.

13.5 Review Questions

1. Explain active data mining process.
2. Write note on shape definitions with its operators in the active mining.
3. Write short note on queries, triggers, and wave execution semantics employed in mining.

Decomposition in Data Mining - A Case Study

Objectives:

- Decomposition is a tool for managing complexity in data mining and enhancing the quality of knowledge extracted from the large databases.
- A typology of decomposition approaches applicable to data mining is presented.
- One of the decomposition approaches, the structured rule-feature matrix, is used as the backbone of a system for informed decision making.
- As a new discipline, data mining draws from other areas such as statistics, machine learning, database retrieval, pattern recognition, and high-performance computing.
- In this section, numerous decomposition approaches are defined and applied for effective knowledge discovery and decision making.
- Decomposition has been discussed in the data mining literature, however, largely in the context of distributed learning.
- There are two basic approaches to data mining
- Direct mining of data sets
- Mining of transformed data sets
- The following two forms of decomposition in space are considered in this section
- Feature set decomposition
- Object set decomposition
- One of the most meaningful applications of decomposition in data mining is hybrid modeling.

Abstract. Decomposition is a tool for managing complexity in data mining and enhancing the quality of knowledge extracted from the large databases. A typology of decomposition approaches applicable to data mining is presented. One of the decomposition approaches, the structured rule-feature matrix, is used as the backbone of a system for informed decision making. Such a system can be implemented as a

S. Sumathi and S.N. Sivanandam: *Decomposition in Data Mining - A Case Study*, Studies in Computational Intelligence (SCI) **29**, 303–313 (2006)
www.springerlink.com
© Springer-Verlag Berlin Heidelberg 2006

decision table, a decision map, or a decision atlas. This case study is taken from A. Kusiak, Intelligent Systems Laboratory University of Iowa, 2001.

Data mining is concerned with discovery of patterns, associations, rules, and other forms of knowledge in data sets. This knowledge is automatically extracted from data rather being formulated by a user as it is done in traditional modeling approaches, e.g., statistical or optimization modeling. As a new discipline, data mining draws from other areas such as statistics, machine learning, database retrieval, pattern recognition, and high-performance computing.

In many applications, data is automatically generated and therefore the number of objects to be mined can be large. The time needed to extract knowledge from such large data sets is an issue, as it many easily run in days, weeks, and beyond. One way to reduce computational complexity of knowledge discovery with data mining algorithms and decision making based on the acquired knowledge is to reduce the volume of data to be processed at a time, which can be accomplished by decomposition. In this section, numerous decomposition approaches are defined and applied for effective knowledge discovery and decision-making. Besides simplifying competition, decomposition facilities dynamic extraction of knowledge that can be used for real-time decision making.

14.1 Decomposition in the Literature

Decomposition has been discussed in the data mining literature, however, largely in the context of distributed learning. This research emphasizes the use of decomposition to enhance decision making rather than learning. Grossman *et al.* 1999 outlined fundamental challenges for mining large-sale databases, with one of them being the need to develop distributed data mining algorithms. Guo and Sutiwaraphun 1988 described a meta-learning concept called *knowledge probing* to distributed data mining. In knowledge probing, supervised learning is organized in two stages. At the first stage, a set of base classifiers is learned in parallel from a distributed data set. At the second stage, the relationship between an attribute vector and the class from all of the base classifiers is determined. Zaki *et al.* 1999 discussed a project called SPIDER that uses shared-memory multiprocessors systems (SMPs) to accomplish parallel data mining on distributed data sets. Cluster analysis provides the basic theory and algorithms for decomposition. Some of the most efficient clustering algorithms are presented in Kusiak *et al.* 2000.

14.1.1 Machine Learning

Bazan 1998 categorized the existing learning algorithms as follows:

- Decision tree

Decision rule
Inductive logic programming and
Rough set algorithms

14.2 Typology of Decomposition in Data Mining

There are two basic approaches to data mining

Direct mining of data sets
Mining of transformed data sets

The first approach is most often applied for mining data sets that can be processed in an acceptable time by the existing data mining algorithms. Transforming data sets before mining is intended either for large data sets or data sets with special properties, e.g., hybrid data discussed next. One of the most useful forms of data transformation is decomposition, which may take place in space and time. The area of decomposition in time is extensive and has received rather broad coverage in the literature though it is beyond the scope of this section.

The following two forms of decomposition in space are considered in this section

Feature set decomposition
Object set decomposition

The feature set decomposition is further classified into:

- Content-based decomposition: The feature set is decomposed into mutually exclusive or partially overlapping subsets with the same decision D used for each subset. The feature origin, availability, and any other criterion could drive the content of each feature set.
- Intermediate-decision decomposition: In some applications feature values are generated over time. In addition the downstream features may dependent on the upstream features.
- Feature-type decomposition: some of the existing rule extraction algorithms are intended for specific types of feature, e.g., discrete value features.
- Feature relevance decomposition: Features may show various degree of relevance to the outcome, measured with statistical metrics (e.g., correlation) and context relationship, which is more tacit and difficult to measure (e.g., the impact of outside temperature on computer energy consumption).

The object set decomposition is further classified as:

- Object content decomposition: Objects are grouped according to time interval, origin, applicability, and so on.
- Decision value decomposition: the set of objects is split into subsets according to the decision value.
- Feature value decomposition: the objects (and possibly features) are partitioned into subsets based on the value of selected features.

14.3 Hybrid Models

One of the most meaningful applications of decomposition in data mining is hybrid modeling. A hybrid model is a collection of models of different types, for example, models developed on the first principles and models constructed from the knowledge extracted by machine learning algorithms. Hybrid models are often built because of different degree of understanding of the modeled process, availability of data, and other application specific limitations.

The need for hybrid has been motivated by numerous engineering and medical applications. A typical process, e.g., a semiconductor manufacturing process or disease management process, involves stages that are well understood due to available models and stages that are only loosely known. This lack of process knowledge is likely behind unwanted events in industrial processes (e.g., products below expected quality level) and in medicine (e.g., premature patient's death).

This section meets the need for hybrid models involving both well-defined models and knowledge models derived from the data collected while observing an actual process. During the process operations, both desirable and adverse events occur, and therefore, the data collected is transformed into knowledge that can be used to minimize or even prevent adverse events from happening.

A domain of interest captured by the hybrid decision making model in Fig. 14.1 is decomposed into three stages. The model allows for bi-directional reasoning, which is important for in-depth understanding of the model process.

The rule-structuring algorithm in Kusiak 2000 organizes the knowledge extracted by different learning algorithms into decision tables that will be integrated with analytical models as well as models of other types are illustrated in Fig. 14.2. The structured knowledge is "packaged" in the form of decision tables that can be combined in other constructs such as decision maps, and those in turn in decision atlases. These constructs will increase transparency and effectiveness of the decision-making process.

A decision table provides decision basis (e.g., decision rules, rule support, rule coverage, etc.) and justification for the decision (e.g., historical cases supported by the decision rule) as symbolically illustrated in Fig. 14.2.

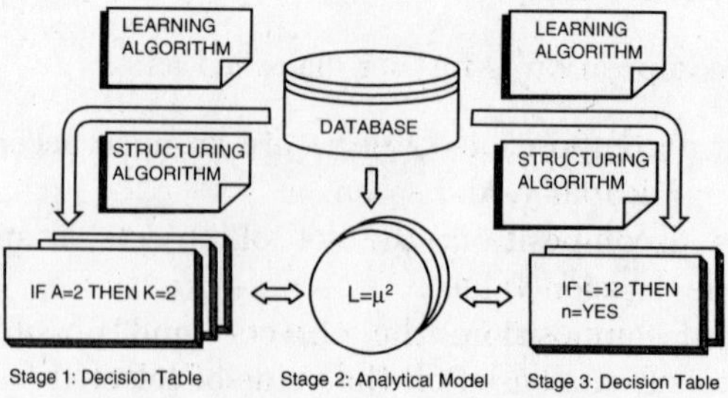

Fig. 14.1. Hybrid decision-making process

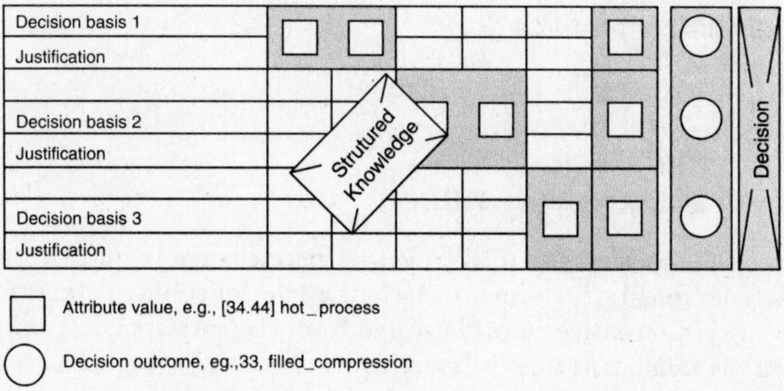

Fig. 14.2. Example of a decision table

The entries of the decision table in Fig. 14.2 are attribute values generated by a learning algorithm. Each entry may represent a singular numerical or symbolic value, a bounded range of values, an unbounded value range (inequality), and so on. Each decision basis contains feature values, while justification characterizes the decision basis, for example, it may contain the decision confidence and risks associated with a decision. The simplest way of making decisions with a decision table is to match the values of an object with unknown outcome to an appropriate row of the decision table. Other ways of making robust decisions with orthogonal algorithms are discussed in Kusiak *et al.* 2000.

An Example for a typical hybrid process:

Consider a five-stage process in which the models at stages 1, 3, and 5 are unknown, while stages 2 and 4 are modeled with known functions, F_1 and F_2 (see the functions below and Figure). During a three-month period, for process stages, 1, 3, and 5 three data sets containing numerous observations (objects) on selected features $f_i, I = 1, \ldots, 9$ and the decision D have been collected. A learning algorithm has extracted three sets 1, 2, and 3 of decision rules shown next.

Rule set 1
IF $f_1 = 2$ AND $f_2 =$ Low THEN $f_3 = 4$
IF f_2 in [2.1,4] THEN $f_3 = 5$
Function F1
$F_4 = 3.1 + (f_3 - 3.1)^3$
Rule set 2
IF $f_4 < 8.5$ AND $f_5 =$ High THEN $f_6 = 8.4$

IF f_5= Low THEN $f_6 = 12.4$
Function F2
$f_7 = \text{In}\ (f_6 + 2.9)^{1/2}$
Rule Set 3
IF $f_7 < 1.3$ THEN D = Good
IF $f_8 >= 3.3$ and $f_{9=}$ Positive THEN D = Bad

The content of each of the three rule sets may change frequently while the two functions remain the same. In fact, the two functions, F_1 and F_2 could interface with alternative rules generated from the corresponding data sets.

When the number of rules is large, their collective interactions are difficult to understand. For better usability of decision rules they will be organized by the rule-structuring algorithm in Kusiak 2000 and represented as decision tables, decision maps, or decision atlases.

The decomposition approach discussed in this section offers the following advantages.

- Ease of model construction and understanding: the data set is partitioned into independent subsets (data sets 1, 3, and 5 in Fig. 14.3) and therefore the rule induction is simplified.
- Support of the evolutionary computation concept: the basic the premise of the discussed approach is that the models are separable and they change with different frequency. Evolutionary computation algorithms can be involved in individual models as well as controlling the evolution of the overall model;
- Increased model structure stability: Only one sub-model (a rule set or a function in Fig. 14.3) at a time is usually modified.
- Ease of data acquisition and model maintenance: As the scope of data at each stage is limited it is easier to acquire the data and maintain the component models;
- Reuse of known models and dependencies: Rather than building a new model from scratch, the model is built around existing component models, e.g., functions and neural network models;
- Representation of alternative solutions with positive and negative rules: The role of negative rules in decision making is discussed in Tsumoto *et al.* 2000.

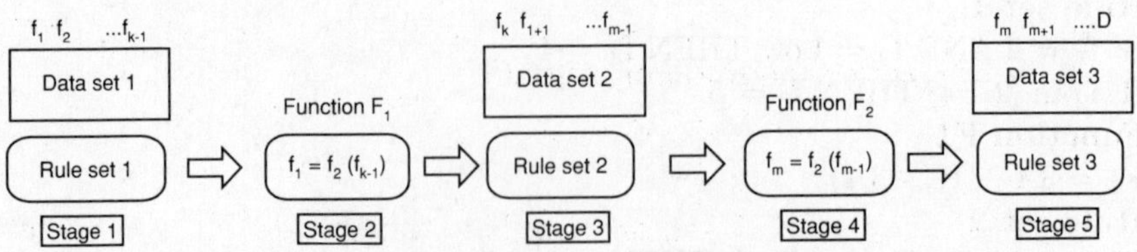

Fig. 14.3. Example process with three rule sets linked by two functions

14.4 Knowledge Structuring

One of the main reasons for extracting knowledge from data sets is to use it for decision making, which has not received sufficient attention in the literature. Most of decision-making algorithms are rather simplistic and usually based on partial or full-matching schemes. Many users have difficulty accepting decision rules that are nonintuitive and algorithms making decisions based on nontransparent matching. This section addresses this important gap in the presentation of knowledge for effective decision making.

The knowledge extracted by a learning algorithm is usually in the form of decision rules that make predictions at some level of accuracy, typically far from perfect. Decision rules might be numerous, the relationships discovered may be flawed, and their meaning might be difficult to understand, and so on. In other words, users have certain expectations for the knowledge discovered that are outside of the scope of learning algorithms. The rule-structuring models and algorithms to be developed in this research will meet these expectations. They will be used both for supporting the user view as well as autonomous decision making.

An example for the rule-structuring concept

Three different learning algorithms A1–A3 were used to extract eight decision rules R1–R8 from a data set. These rules R1–R8 are represented as the rule-feature matrix in fig.14.4. To simplify our considerations the information pertinent to each rule such as support, classification quality and so on has not been included. Though the rule set in the Fig. 14.4 is small, its analysis is not simple. Transforming the rule-feature matrix into the structured matrix significantly improves interpretation of the rule set. The rule-structuring algorithm Kusiak 2000 generated the matrix in fig. 14.4A from the one in fig. 14.4 by removing two rules R7 and R8 due to their dissimilarity with the rules R1 through R6 and changing the sequence of the remaining rows and columns.

The content of the matrix in fig. 14.4A is structured and it allows drawing numerous conclusions, for example:

- The decisions D = High, Medium, and Low are totally separated by features;
- The rules R3 and R6 generated by algorithms A2 and A3 are equivalent;
- The decision D = Low can be reached in two alternative ways, using the features values f1 = {B, C, D} and f2=a, or f1= {E, F} and f2= b.

The example illustrates only a few of numerous user's requirements that can be incorporated in the rule-structuring algorithm, such as:

- Matrix structure: different structures of the rule-feature matrix may be considered, e.g., block-diagonal (see fig. 14.4A), block-diagonal matrix with overlapping features, block-diagonal matrix with overlapping

f_1	f_2	f_3	f_4	f_5	D	Rule	Algorithm
{B, C, D}	a				Low	R1	A1
			>9		Medium	R2	A2
			<2	(2, 5]	Low	R7	A3
		(2, 6]		=<8	High	R3	A2
{E, F}	b				Low	R4	A3
{C, F}		<4			Medium	R8	A1
			>9		Medium	R5	A1
		>=2		[1, 3]	High	R6	A3

Figure 4. Rule-feature matrix.

f_3	f_5	f_2	f_1	f_4	D	Rule	Algorithm
(2, 6]	=<8				High	R3	A2
>=2	[1, 3]				High	R6	A3
		a	{B, C, D}		Low	R1	A1
		b	{C, F}		Low	R4	A3
				>9	Medium	R2	A2
				>9	Medium	R5	A1

Structured rule-feature matrix

Fig. 14.4. &A: Rule-Feature matrix and Structured rule-feature matrix

rules, triangular (for dependency analysis among rules), L-shape matrix. T-shape, etc.

- Differentiation of decision on features: Each decision value is associated with an independent subset of features;
- Differentiation of decision on features values: Any two decision values are discernable on unique subset of feature values;
- Inclusion of user-selected features: A user may have her/his preferences in terms of the features to be included in the selected rules, exclusion of some features, presence of the minimum set of features, and so on;
- Contrasting positive rules against negative ones is valuable in decision making in some applications.

14.5 Rule-Structuring Model

The rule-structuring problem can be represented as m-partite graph with each node representing a set of rules called *candidate rule set.* The rules contained in each candidate rule set share some common property, e.g., same decision value, common features set, which depends on the objective of the rule-structuring problem. The candidate rule sets could be defined for the rules extracted a

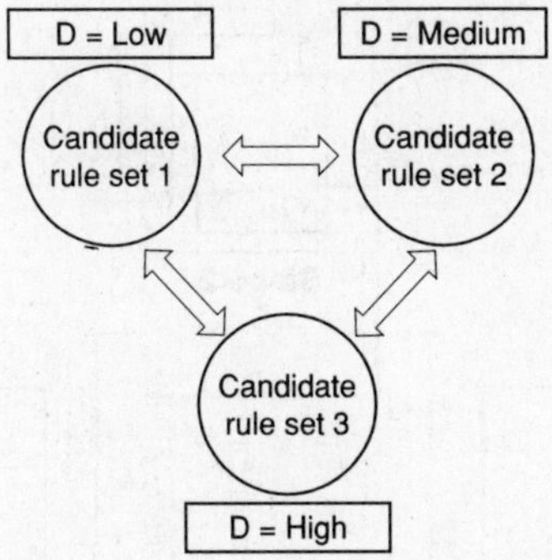

Fig. 14.5. Graph representation of the rule-structuring problem for the rules in Fig. 14.4.

specific data set, e.g., data set 1 in fig. 14.3 or across different data sets, e.g., data set 1 and 3 in fig. 14.3. Candidate rules sets may include rules extracted by different learning algorithms. An arc of the$m$-partite graph represents relationships between the corresponding nodes, e.g., a distance.

Transforming the unstructured matrix in fig. 14.4 into the structured matrix in fig. 14.4A calls for development of a rule-structuring model that could be solved with the standard or a specialized algorithm. Both modeling and solution alternatives are explored in this section. The rule-structuring model is illustrated in the next example.

Example: Represent the eight rules in fig. 14.5 with the 3-partite graph in fig. 14.5.

Each of the three candidates rule sets in fig. 14.5 contains rules with the same outcomes. For example, the candidate rule set 1 includes the rules R1, R4, and R7 from fig. 14.4. The arrows in fig. 14.4 symbolize relationships (e.g., distances) among the rules belonging to different candidate rules sets. In fact these relationships can be of different types thus leading to a hypergraph [Berge & Shi 1992].

Based on the m-partite graph (or hypergraph) representation of the rule-structuring problem different models and algorithms can be developed. The rule-structuring problem could be loosely formulated as follows:

Minimize the total distance between any two rules belonging to different nodes of the m-partite graph subject to constraints, for example, limiting the number of rules selected from one node, limiting the number of clusters.

14.6 Decision Tables, Maps, and Atlases

A decision table is a collection of knowledge needed to make decision in a particular domain. It generalizes the structured matrix introduced in fig. 14.4A by:

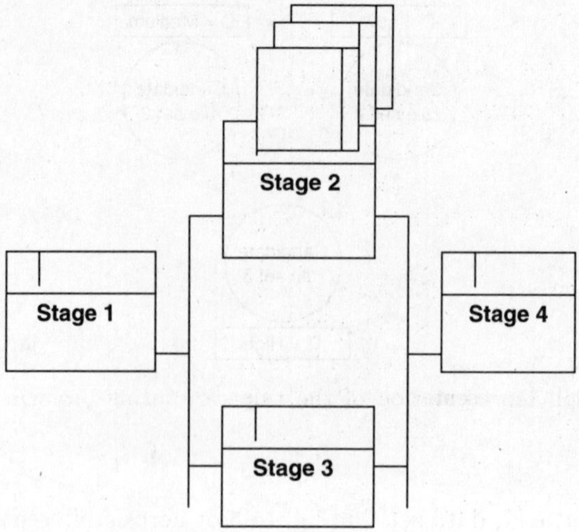

Fig. 14.6. Simple decision map

- Including decision rules generated by different learning algorithms;
- Transforming rules that may combine different decision rules, and so on;
- Content organization determined by the rule-structuring algorithm.

The actual decision can be made using one or more decision tables at any stage of the decision map (see fig. 14.6). Note that some decision tables may reduce to functions, neural network models, etc. The collection of decision tables distributed over all decision stages constitutes a decision map. Maps in turn can be combined in an atlas and multiple atlases make up a library, etc. There are two primary reasons for alternative decision tables. One is that a decision may follow one or more paths contained in one of those tables. The notion of independence has a profound impact on decision accuracy and user's confidence in the algorithmically generated result.

14.7 Summary

The decomposition has been discussed in the data mining literature, however, largely in the context of distributed learning. A typology of decomposition approaches applicable to data mining has been presented. One of the decomposition approaches, the structured rule-feature matrix, is used as the backbone of a system for informed decision making. Such a system can be implemented as a decision table, a decision map, or a decision atlas.

14.8 Review Questions

1. What is the typology of decomposition in data mining?
2. With an example explain a typical hybrid process and rule-structuring concept.
3. Form a simple decision map with its stages of operation.

15

Data Mining System Products and Research Prototypes

Objectives:

- Although data mining is a young field with many issues that still need to be researched in depth, there are already a great many off-the-shelf data mining system products and domain-specific data mining application software available.
- We describe the features to consider when selecting a data mining product and offer a quick introduction to a few typical data mining systems.
- To choose data mining system that is appropriate for our task, it is important to have a multiple dimensional view of data mining systems.
- Due to infancy and rapid evolution of the data mining market, we briefly outline a few typical data mining systems in order to help the reader to get an idea of what can be done with current data mining products.
- Many data mining systems specialize in one data mining function, such as classification, or just one approach of a data mining function, such as decision tree classification.
- Other systems provide a broad spectrum of data mining functions.
- Here we introduce a few systems that provide multiple data mining functions and explore multiple knowledge discovery techniques.

Abstract. Although data mining is a young field with many issues that still need to be researched in depth, there are already a great many off-the-shelf data mining system products and domain-specific data mining application software available. As a young discipline, data mining has a relatively short history and is constantly evolving – new data mining systems appear on the market every year; new functions, features, and visualization tools are added to existing systems on a constant basis; and efforts toward the standardization of data mining language have only just begun. We describe the features to consider when selecting a data mining product and offer a quick introduction to a few typical data mining systems.

To choose data mining system that is appropriate for our task, it is important to have a multiple dimensional view of data mining systems. In general, data mining systems should be assessed based on the discussed multiple dimensional features in

S. Sumathi and S.N. Sivanandam: *Data Mining System Products and Research Prototypes*, Studies in Computational Intelligence (SCI) **29**, 315–320 (2006)
www.springerlink.com © Springer-Verlag Berlin Heidelberg 2006

this section. The section also discusses on the examples of commercial data mining systems.

15.1 How to Choose a Data Mining System

With many data mining system products available on the market, we may ask, "what kind of system should we choose?" Some people may be under the impression that data mining systems, like many commercial relational database systems, share the same well-defined operations and a standard query language, and behave similarly on common functionalities. If such were the case, the choice would depend more on the system's hardware platform, compatibility, robustness, scalability, price, and service. Unfortunately, this is far from reality. Many commercial data mining systems have little in common with respect to data mining functionality or methodology and may even work with completely different kinds of data sets.

It is important to have a multiple dimensional view of data mining systems, so a data mining system that is appropriate for our task has to be chosen. In general, data mining systems should be assessed based on the following multiple dimensional features.

Data types: Most data mining systems that are available on the market handle formatted, record-based, relational-like data with numerical, categorical, and symbolic attributes. The data could be in the form of ASCII text, relational database data, or data warehouse data. It is important to check what exact format(s) each system we are considering can handle. Some kinds of data or applications may require specialized algorithms to search for patterns and so their requirements may not be handled by off-the-shelf, generic data mining systems. Instead specialized data mining systems may be used that mine either text documents, geospatial data, multimedia data, time series data, DNA sequences, Web log records or other web data, or are dedicated to specific applications (such as finance, reality industry, or telecommunications). Moreover many data mining companies offer customized data mining solutions that incorporate essential data mining functions or methodologies.

System issues: A given data mining system may run on only one operating system, or on several. The most popular operating systems that host data mining software are UNIX and Microsoft Windows (including 95, 98, 2000, and NT). There are also data mining systems that run on OS/2, Macintosh, and Linux. Large industry-oriented data mining systems should ideally adopt a client/server architecture, where the client could be a personal computer running on Microsoft Windows, and the server could be a set of powerful parallel computers running on UNIX. A recent trend has data mining systems providing Web-based interfaces and allowing XML data as input and /or output.

Data sources: This refers to the specific data formats on which the data mining system will operate. Some systems work only on ASCII text files, whereas many others work on relational data, accessing multiple relational data sources. It is important that a data mining system supports ODBC connections or OLEDB for ODBC connections. These ensure open database connections, i.e., the ability to access to any relational data (including those in DB2, Informix, Microsoft SQL Server, Microsoft Access, Microsoft Excel, Oracle, Sybase, etc.), as well as formatted ASCII text data. A data mining system that operates with a data warehouse should follow the OLE DB for OLAP standard, since this helps ensure that the system is able to access the warehouse data provided not only by Microsoft SQL server 7.0 but also by other data warehouse products supporting the standard.

Data mining functions and methodologies: Data mining functions from the core of a data mining system. Some data mining systems provide only one data mining function, such as classification. Others may support multiple data mining functions, such as description, discovery-driven OLAP analysis, association classification, prediction, clustering, outlier analysis, similarity search, sequential pattern analysis, and visual data mining. For a given data mining function (such as classification), some systems may support only one method, while others may support a wide variety of methods (such as decision tree analysis, Bayesian networks, neural networks, genetic algorithms, case based reasoning, etc.). Data mining systems that support multiple data mining functions and multiple methods per functions provide the user with greater flexibility and analysis power. Many problems may require users to try a few different mining functions or incorporate several together, and different methods can be shown to be more effective than others for different kinds of data. In order to take the advantage of the added flexibility, however users may require further training and experience. Thus such systems should also provide novice users with convenient access to the most popular function and method, or to default settings.

Coupling data mining with data mining and/or data warehouse systems: A data mining system should be coupled with a database and/or a data warehouse system, where the coupled components are seamlessly integrated into a uniform information-processing environment. In general there are four forms of such coupling: no coupling, loose coupling, semitight coupling, and tight coupling. Some data mining systems work only with ASCII data files and are not coupled with database or data warehouse systems at all. Such systems have difficulties handling large data sets and using the data stored in database systems. In data mining systems that are loosely coupled with database and data warehouse systems, the data are retrieved into a buffer, main memory by database, or warehouse operations, and then mining functions are applied to analyze the retrieved data. These tend to have poor scalability and may be inefficient when executing some data mining queries. The coupling of data mining system with a database or data warehouse system may be semitight, providing the efficient implementation of only a few essential data mining primitives. Ideally, a data mining system should be tightly coupled with a database system in the sense that data mining and data retrieval process are integrated by optimizing data mining queries deep into the iterative mining and retrieval process. Tight coupling of data mining with OLAP-based data warehouse systems is also desirable so that data mining and OLAP operations can be integrated to provide OLAP mining features.

Scalability: Data mining has two kinds of scalability issues: row scalability and column scalability. A data mining system is considered row scalable if the number of rows is enlarged ten times, it takes no more than ten times to execute the same data mining queries. The system is considered column scalable if the mining query execution time increases linearly with the number of columns due to the course of dimensionality, it is much more challenging to make a system column scalable than row scalable.

Visualization tools: "A picture is worth a thousand words"—this is very true in data mining. Visualization in data mining can be categorized into data visualization, mining result visualization, mining process visualization, and visual data mining. The variety, quality, and flexibility of visualization tools may strongly influence the usability, interpretability, and attractiveness of a data mining system.

Data mining query language and graphical user interface: Data mining is an exploratory process. An easy-to-use and high-quality graphical user interface is essential in order to promote user guided, highly interactive data mining. Most data mining systems provide user-friendly interfaces for mining. However unlike relational database systems where most graphical user interfaces are constructed on top of SQL that serves as a standard, well-designed database query language, most data mining systems do not share any underlying data mining query language. Lack of standard mining language makes it difficult to standardize data mining products and to ensure the interoperability of data mining systems

15.2 Examples of Commercial Data Mining Systems

As mentioned earlier, due to infancy and rapid evolution of the data mining market, we briefly outline a few typical data mining systems in order to help the reader to get an idea of what can be done with current data mining products.

Many data mining systems specialize in one data mining function, such as classification, or just one approach of a data mining function, such as decision tree classification. Other systems provide a broad spectrum of data mining functions. Here we introduce a few systems that provide multiple data mining functions and explore multiple knowledge discovery techniques.

- **Intelligent Miner** is an IBM data mining product that provides a wide range of data mining algorithms, including association, classification, regression, predictive modeling, deviation detection, sequential pattern analysis, and clustering. It also provides an application toolkit containing neural network algorithms, statistical methods, and data visualization tools. Distinctive features of Intelligent Miner include the scalability of its mining algorithms and its tight integration with IBM's DB2 relational database system.
- **Enterprise Miner** was developed by SAS Institute, Inc. It provides multiple data mining algorithms, including association and classification, and statistical analysis packages. A distinctive feature of Enterprise Miner is its variety of statistical analysis tools, which are built based on the long history of SAS in the market of statistical analysis.
- **MineSet** was developed by Silicon Graphics Inc. (SGI). It also provides multiple data mining algorithms, including association and classification, as well as advanced statistics and advanced visualization tools. A distinguishing feature of MineSet is its set of robust graphics tools (using powerful using powerful graphics features of SGI computers), including rule visualizer, tree visualizer, map visualizer, and (multidimensional data) scatter visualizer, for the visualization of data and data mining results.
- **Clementine** was developed by Integral Solutions Ltd. (ISL). It provides an integrated data mining development environment for end users and developers. Multiple data mining algorithms, including rule induction, neural nets, classification, and visualization tools, are incorporated in the system. A distinguishing feature of Clementine is its object-oriented, extended module interface, which allows users' algorithms and utilities to be added to Clementine's visual programming environment. Clementine has been acquired by SPSS Inc.
- **DBMiner** was developed by DBMiner Technology Inc. It provides multiple data mining algorithms, including discovery-driven OLAP analysis, association, classification, and clustering. A distinctive feature of DBMiner is its data-cube-based on-line analytical mining, which includes efficient frequent-pattern mining functions, and integrated visual classification methods.

There are many other commercial data mining products, systems, and research prototypes that are also fast evolving. Interested readers may wish to consult timely surveys on data warehousing and data mining products. More detailed introduction to the commercial data mining system is presented in Appendix IV.

15.3 Summary

This section discussed on the method of choosing a data mining system and also gave examples of the commercial data mining systems.

15.4 Review Questions

1. What are the choices required to choose a data mining system?
2. Give few examples of commercial data mining systems.

16

Data Mining in Customer Value and Customer Relationship Management

Objectives:

- To understand the concept of data mining by comparing traditional marketing research with relationship marketing.
- To create separate portraits of individual customers.
- To secure a database and to examine the capabilities and effectiveness of various data mining techniques to determine the usefulness as related to the customer relationship management.
- To help organizational decision makers identify their most profitable customers.
- To deliver "ideal" customer-defined value at each customer interaction using CVM.
- To use the data mining for incentive allocation.
- To give an overview of what the (future) worker in the field of ERP can expect and has to have heard of when dealing with CRM.
- Linking a call center to the enterprise's network is to enable the agent to access, create, and change customer data.
- To feed a CRM system in order to make it effective, such as materials billing and order history, is often stored all over the enterprise and in the best case embedded in an ERP system.
- To provide high-quality information for the CRM programs, we must discover first rate customer-related data and evaluate them with suitable data mining techniques.

Abstract. Since data mining is a young discipline with wide and diverse applications, there is still a nontrivial gap between general principles of data mining and domain specific, effective data mining tools for particular applications.

Current technology has made available enormous amounts of multimedia data such as audio, image, video, sequence, and hypertext data that need to be analyzed. The main objective of the analysis is to extract (or mine) interesting patterns, associations, rules, changes, anomalies, and general regularities from the data to improve the process of decision making.

S. Sumathi and S.N. Sivanandam: *Data Mining in Customer Value and Customer Relationship Management*, Studies in Computational Intelligence (SCI) **29**, 321–386 (2006)
www.springerlink.com © Springer-Verlag Berlin Heidelberg 2006

Data mining is a promising and relatively new area of current research and development, which can provide important advantages to the users. It can yield substantial knowledge from data primarily gathered for a wide range of quite different applications. Financial institutions have derived considerable benefits from its application and other industries and disciplines are now applying the methodology to increasing effect.

The background of data mining, data warehousing, knowledge discovery in databases, a concept of customer relationship marketing, customer acquisition, customer value management, customer lifetime value, customer relationship management concepts are discussed in this section.

This section describes the concept of data mining by comparing traditional marketing research with relationship marketing. The background of data mining is discussed with special emphasis paid to the various terms in data mining such as *data warehouses* and *data marts* as well as *knowledge discovery in databases (KDD)* and *customer relationship management (or) marketing (CRM)*. Steps necessary for companies to implement successful data mining projects are enumerated and directions for further research are suggested.

For most businesses, the primary means of growth involves the acquisition of new customers. This could involve finding customers who previously were not aware of the product, were not candidates for purchasing the product (for example, baby diapers for new parents), or customers who in the past have bought from their competitors. Some of these customers might have been the customers previously, which could be an advantage (more data might be available about them) or a disadvantage (they might have switched as a result of poor service). In any case, data mining can often help segment these prospective customers and increase the response rates that an acquisition marketing campaign can achieve.

Also the features of the customer relationship management and customer value management are discussed. Even the comparison and various information regarding both are given. The way in which companies interact with their customers has changed dramatically over the past few years. A customer's continuing business is no longer guaranteed. As a result, companies have found that they need to understand their customers better, and to quickly respond to their wants and needs. In addition, the time frame in which these responses need to be made has been shrinking. It is no longer possible to wait until the signs of customer dissatisfaction are obvious before action must be taken. To succeed, companies must be proactive and anticipate what a customer desires.

16.1 Data Mining: A Concept of Customer Relationship Marketing

16.1.1 Traditional Marketing Research

Today the majority of companies that consider themselves market driven are still organized around their products. These companies position their products to a carefully researched segment of customers whose wants are unfulfilled. To virtually guarantee success, these companies believe that they must give additional value to the chosen segment by differentiating their product in some unique way. Companies of this type emphasize the refining of internal processes and outputs to meet the needs of the mass market and customers are treated as a homogeneous and basically passive mass.

A number of companies attempted to change or redirect their efforts in the late 1980s and early 1990s. At that time "customer service" became a "hot" topic. Everyone from CEOs to brand managers to hourly employees was admonished to "Take Care of the Customer." Traditional surveys of what the customers want or the service they have received are what many companies rely on today. This traditional survey gives the company reliable information on what customers think they think or what they think they want, but it may not be what they really think or want. If we are only supplying what their customer wants or think they want today, we are not tapping into the unspoken needs and un-served markets that may be the key to the customer of today and the potential customers of tomorrow.

Companies that consider themselves market driven spend an inordinate amount of time differentiating their product through quality improvement. It is estimated that focusing on quality improvements is only about 10l of what we should be doing in the company. This overriding strategy of the past was to acquire customers and respond to their aggregate needs.

16.1.2 Relationship Marketing – the Modern View

Forward-looking companies of today believe that customers are what sustains any business and that they have "lifetime value" not just the value of a single sale. It is believed that customer groups, if managed and maintained, cannot be easily copied by the competition, i.e., they are one of the few "sustainable" competitive advantages open to the company.

Progressive companies of the future will know and understand the difference between knowledge of the customer and customer knowledge. For instance, knowledge of the customer is knowing how many hits a browser makes on their Web site, whereas customer knowledge is knowing what to do with the hits. To benefit from this "new" philosophy a company must change the entire business operation, so that research and development and marketing work seamlessly, and financial resources are allocated in the "right" places.

The producers and suppliers must be able to put together the right mix of service and information surrounding the differentiated or personalized products of the future. This mix will be customized, by creating very separate portraits of individual customers.

The technology to develop these portraits exists in today's data mining technology. Companies are able to take information from their own company's database and augment it with enhancement information provided by a data compiler and then apply a predictive model to the augmented data set using sophisticated data mining techniques. In this way we can understand some of the things the individuals in the year 2020 will want to achieve as customers. Namely:

(a) *To make better and easier decisions.* Data mining technology can help the supplier make use of more intimate knowledge to better target their offers. Goods or services, price, distribution channels, and communication tools can be adapted to give a near exact offer for a very targeted group of consumers. The marketer can tailor demographic data and response data to get close to one-on-one marketing.

(b) *To better manage pressure and anxieties.* It is expected by the year 2020 companies may respond to customer needs for stress reduction and time management by developing such innovative strategies as regular direct home delivery of necessities that customers will not spend time shopping for. This process may be so transparent that consumers will receive bathroom rolls, paper towels, detergents, and specific food items the very way we receive heating, oil, and electricity today.

(c) *To fulfill their "measured hopes".* Customers will be wanting to experience more things in their short amount of leisure time. Data mining techniques will enable service-oriented companies to provide engineered experiences for a consumer's leisure time. These engineered experiences will be similar to the ones provided today by Disney World and Disneyland.

(d) *To benefit from better and faster innovation and, above all, be treated as individuals.* By the year 2020 individual and personalized products will be so highly customized that they will adapt to changing needs and habits. The customer may not even be aware of the changing need, but the sophisticated data mining system will be able to detect them. It is predicted that in the next 30 to 50 years, customers will have such an enormous range of new products built on natural, biotech, and atomically manipulated materials that they (customers) will be in control of essentially everything in the marketplace.

16.1.3 Understanding the Background of Data Mining

Data Warehouse

With the reduction in cost of computing power, companies are collecting all kinds of data about their customers. The repository for this large amount of data has become known as a *data warehouse.* A data warehouse is designed for decision-makers strategic design support and is made up largely from parts of an operational database. This data warehouse can contain billions of records. Wal-Mart's warehouse maintained by NCR Corporation has 101 terabytes of information. With a terabyte containing 250 million pages of text, this data warehouse contains more than 25 billion pages of text. This data warehouse runs more than 30 business applications, supports more than 18,000 users, handles 120,000 complex queries a week, and receives 8.4 million update every minute during peak times. A complex data warehouse like the one described above can cost in excess of $10 million and take from one to three years to complete.

Data Mart

A specialized repository of data used by specific departments such as finance or sales and fed from an enterprise-wide data warehouse is called a *data mart.* Average cost to build a data mart is usually between $10,000 and $1 million and can be up and running in less than six months.

Knowledge Discovery in Databases (KDD)

In its broadest scope data mining is referred to as KDD. However, data mining is generally thought of as a particular activity of KDD that applies a specific algorithm to extract patterns that help convert data into knowledge. KDD has been performed in some form since the first business enterprise, but usually on an ad hoc, catch-as-catch-can role that supported decision makers. The difference in this past role and today is that the process is being continuously operated and is becoming central to the core of business operations.

Data Mining

Data mining has been defined as the process of sifting through large amounts of data to spot patterns and trends that can be used to improve business functions. Simply put, it is prospecting for profits in the depths of the company's database or "like looking for gold in our computer." It combines techniques from statistics, databases, machine learning, and pattern recognition to extract (mine) concepts, concept interrelations, and interesting patterns automatically from large business databases.

The difference between data mining and other analytical methods is the approach they use in exploring the data. Most analytical tools use the verification-based method – the user hypothesizes about specific relationships and tries to prove or refute the presumptions. Data mining uses what is called *discovery-based approaches* in which pattern matching and other algorithms are employed to determine the key relationships in the data. Actually it is nothing more than the analysis of existing data to extract new, previously unknown, or unrealized information. This analysis of existing data benefits both businesses and consumers as the growing capabilities of the new technique are realized.

Data mining is often referred to as having two scopes. The narrow scope is defined as the automated discovery of "interesting" non-obvious patterns hidden in a database that have a potential for contributing to the overall profit of the firm. This narrow definition encompasses computer-based or "machine learning: methods such as neural networks, genetic algorithms, and decision trees to extract patterns of information from data while requiring only limited human involvement."

The broad scope of data mining encompasses "confirmation" or testing of relationships revealed in the narrow scope. These relationships are confirmed to support the theories, models, and hypotheses formulated within the narrow scope definition of data mining. Examples of procedures used include exploratory data analysis, ordinary least squares, regression, logistical regression, and discriminate analysis. The broad scope involves managers and analysis identifying important variables and structuring the investigation.

Data Mining: From Data to Knowledge

Collecting business or scientific data is a basic step toward achieving competitive advantage. In order for data to become a strategic resource it must be possible to extract knowledge from them about the system under study. Data mining techniques, rooted in the fields of artificial intelligence and statistics, jointly with appropriate computational power, allow knowledge to be extracted and synthesized by means of automatic learning from raw data. Knowledge, expressed in terms of decision trees, association rules, dependency structures, probability, etc. allows questions such as the following to be answered: What goods should be promoted to the customer? What is the probability that a certain patient suffers a given disease? How should a medical image be interpreted? What is the avalanche risk in a certain area? Will this customer default on a loan? What is the good that a certain customer is likely to buy given that he bought another? The data mining approach is viable when a convenient database is available and is largely independent of the specific application domain, not requiring experts' knowledge.

16.1.4 Continuous Relationship Marketing

As we consider data mining in its narrow and broad scope, it is important to understand the relationship of database marketing and data mining with the personal touch. Companies such as Hertz, USAA, Wal-Mart, and Nordstrom are successful, not because they have a gigantic data warehouse, but because they have figured out practical ways to gather information and act on it quickly. These successful companies understand their customer their competitive position and they understand profitability.

With the lower price of information technology (IT) markets can offer real customer relationships of the past, before mass markets, combined with greater variety and lower prices. This type of combination is known as the *continuous relationship marketing* (CRM) strategy. Several key rules are important to consider for use in the implementation of this strategy.

a) *Use the information that is gathered to serve the customers better.* Marketers can arrive at a customer lifetime value (CLV) calculation that sums the profitability of individual purchases to arrive at current customer value and factors in time to reflect the importance of customer retention.
b) *Continuous relationship marketing (CRM) strategy* is most effective when it concentrates on building relationships with customers who offer attractive lifetime value.

c) *Build customer relationships, not just databases.* It is not enough to have a customer's name in the database; this information must be used to build a stronger relationship with the customer.
d) *Be willing to treat customers differently.* Some customers may have a customer's lifetime value (CLV) often or even a hundred times greater than other customers, yet the company may not treat the more valuable customer differently from any other customer. Width dedicated ticket lines, priority upgrades and "early" boarding, airlines have perfected this "class treatment" better than most industries.
e) *Compete with skills, not capital.* The successful CRM practitioner analyzes data to understand customer behavior and identifies ways to serve customers better.

In short, continuous relationship marketing (CRM) is an approach in which a company seeks to build close relationships with its potential and current customers, so that both segments will be encouraged to concentrate a disproportional high share of their value with the company.

16.1.5 Developing the Data Mining Project

Companies that are successful in data mining efforts need (a) careful planning, (b) careful selection of the right data, (c) to be certain that data is in the proper format to be analyzed and (d) to have a clearly defined business objective. There are several steps necessary in implementing a data mining project. Namely:

1). A company needs to establish a research objective for the project. The researchers may ask themselves: why are we doing this? what problems are we addressing? what do we hope to accomplish? Data mining, like any other research, is designed to provide information that can be used to improve the current situation. A firm cannot just decide to mine its data and expect solutions to present themselves. It is necessary to decide what issues need to be addressed and then determine if data mining techniques are an appropriate solution.
2). After establishing an objective, it is necessary to select an appropriate data set. Many data sets include transaction data, demographic data, and lifestyle data. Just because data mining packages can handle large data sets, it is not necessary or prudent to include variables that have absolutely no relationship to the objective.
3). The next step in the data mining process is to cleanse and transform the data set. This step is vital to ensure accuracy and effectiveness of the outcome. Often times in large data sets customer records are incomplete or the same customer appears multiple times. Cleansing the data set includes deleting fields where data is missing or deleting duplicate records. Transforming data can involve converting data from one type to another

such as numeric to character or currency. The point of this stage in the process is to remove or transform any data that could lead to "dirty" or inaccurate results.

4). The last stage in the data mining project is to actually mine the data. After the research objective is determined and the data is cleansed appropriately and transformed, the researcher must select the appropriate way to mine the data. This involves deciding what type of data mining operation to use, selecting the data mining technique to support this operation and ultimately mining the data. Once the data is mined and the designed information is extracted it can be analyzed and interpreted with respect to the original research objective.

16.1.6 Further Research:

Here, we discussed the concept of data mining by comparing traditional marketing research with relationship marketing, providing an understanding of the background of data mining and guidelines for developing the data mining projects.

The next step for the further research in this area is to actually secure a database and to examine the capabilities and effectiveness of various data mining techniques to determine the usefulness as related to the customer relationship management. Value to the researcher could be obtained by attempting to evaluate a realistic managerial issue that a marketer might face. Data mining techniques could be employed to determine their effectiveness in addressing the issue. For example, historical transactions data could be examined utilizing data mining techniques to develop custom clusters and predictive models that could then be used to help organizational decision makers identify their most profitable customers.

16.2 Introduction to Customer Acquisition

The traditional approach to customer acquisition involved a marketing manager developing a combination of mass marketing (magazine advertisements, billboards, etc.) and direct marketing (telemarketing, mail, etc.) campaigns based on their knowledge of the particular customer base that was being targeted. In the case of a marketing campaign trying to influence new parents to purchase a particular brand of diapers, the mass marketing advertisements might be focused in parenting magazines (naturally). The ads could also be placed in more mainstream publications whose readership demographics (age, martial, status, gender, etc.) were similar to those of new parents.

In the case of traditional direct marketing, customer acquisition is relatively similar to mass marketing. A marketing manager selects the demographics that they are interested in (which could very well be the same characteristics used for mass-market advertising), and then works with a data vendor (sometimes known as a service bureau) to obtain lists of customers who meet those characteristics. The service bureaus have large databases containing millions of prospective customers who can be segmented based on specific demographic criteria (age, gender, interest in particular subjects, etc.) to prepare for the "diapers" direct mail campaign, the marketing manager might request a list of prospectives from a service bureau. This list could contain people, aged 18 to 30, who have recently purchased a baby stroller or crib (this information might be collected from people who have return warranty cards for strollers or cribs). The service bureau will then provide the marketer with a computer file containing the names and addresses for these customers so that the diaper company can contact these customers with their marketing message. It should be noted that because of the number of possible customer characteristics, the concept of "similar demographics" has traditionally been an art rather than a science. There usually are no hard-and-fast rules about whether two groups of customer share the same characteristics. In the end, much of the segmentation that took place in traditional direct marketing involves hunches on the part of the marketing professional. In the case of 18-to-30 year old purchasing baby strollers, the hunch might be that people who purchase a stroller in this age group are probably making the purchase before the arrival of their first child (because strollers are saved and used for additional children). They also have not yet decided the band of diapers to be used. Seasoned veterans of the marketing game know their customer well and are often quite successful in making these kinds of decisions.

16.2.1 How Data Mining and Statistical Modeling Change Things

Although a marketer with a wealth of experience can often choose relevant demographic selection criteria, the process becomes more difficult as the amount of data increase. The complexities of the patterns increase both with the number of customers being considered and the increasing detail of each customer. The past few years have seen tremendous growth in consumer databases, so the job of segmenting prospective customer is becoming overwhelming.

Data mining can help this process, but it is by no means a solution to all of the problems associated with customer acquisition. The marketer will need to combine the potential customer list that data mining generates with offers that people are interested in. Deciding what is an interesting offer is where the art of marketing comes in.

16.2.2 Defining Some Key Acquisition Concepts

Before the process of customer acquisition begins, it is important to think about the goals of the marketing campaign. In most situations, the goal of acquisition marketing campaign is to turn a group of potential customers into actual customers of the product or service. This is where things can get a bit fuzzy. There are usually many kinds of customers, and it can often take significant amount of time before some one becomes a valuable customer. When the results of an acquisition campaign are evaluated there are often different kinds of responses that need to be considered.

The responses that come in as a result of a marketing campaign are called *response behaviors*. The use of the word *behavior* is important because the way in which different people respond to a particular marketing message can vary. How a customer behaves as a result of the campaign needs to take this variation into consideration. A response behavior defines a distinct kind of customer action and categorizes the different possibilities so that they can be further analyzed and reported on.

Binary response behaviors are the simplest kind of response. With a binary response behavior, the customer response is either a *yes* or *no*. If someone has sent a catalog, did they buy something from the catalog or not? At the highest level, this is often the kind of response that is talked about. Binary response behaviors do not convey any subtle distinctions between customer actions, and these distinctions are not always necessary for effective marketing campaigns.

Beyond binary response behaviors are categorical response behaviors. As one would expect, a categorical response behavior allows for multiple behaviors to be defined. The rules that define the behaviors are arbitrary and are based on the kind of business we are involved in. Going back to the example of sending out catalogs, one response behavior might be defined to match if the customer purchased women's clothing from the catalog, whereas a different behavior might match when the customer purchased men's clothing. These behaviors can be refined a far as deemed necessary (for example, "purchased men's red polo shirt.")

It should be noted that it is possible for different response behaviors to overlap. A behavior might be defined for customers that purchased over $100 from the catalog. This could overlap with the "purchased men's clothing" behavior if the clothing that was purchased cost more than $100. Overlap can also be triggered if the customer purchases more than one item (both men's and women's shirt's, for example) as a result of a single offer.

Although the use of overlapping behaviors can tend to complicate analysis and reporting, the use of overlapping categorical response behaviors tends to be richer and therefore will provide a better understanding of the customers in the future.

There are usually several different kinds of positive response behaviors that can be associated with an acquisition marketing campaign. (This assumes that the goal of the campaign is to increase customer purchases, as opposed to an informational marketing campaign in which customers are simply told of the company's existence.) Some of the general categories of response behaviors are the following:

- *Customer inquiry*. The customer asks for more information about the products or services. This is a good start. The customer is definitely interested in the products—it could signal the beginning of a long-term customer relationship. We might also want to track conversions, which are follow-ups to inquiries that result in the purchase of a product.

- *Purchase of the offered product or products.* This is the usual definition of success. We offered the products to someone, and they decided to buy one or more of them. Within this category of response behaviors, there can be many different kinds of responses. As mentioned earlier, both "purchased men's clothing" and "purchased women's clothing" fit within this category.
- *Purchase of a product different that the ones offered.* Despite the fact that the customer purchased one of the products, it was not the one we offered. We might have offered the deluxe product and they chose to purchase the standard model (or vice versa). In some sense, this is a very valuable response because we now have data on a customer/product combination that we would not otherwise have collected.

There are also typically two kinds of negative responses. The first is a nonresponse. This is not to be confused with a definite refusal of the offer. For example, if we contacted the customer via direct mail, there may be any number of reasons why there was no response (wrong address, offer misplaced, etc.). Other customer contact channels (outbound telemarketing, e-mail, etc.) can also result in ambiguous nonresponses. The fact there was no response does not necessarily mean that the offer was rejected. As a result, the way we interpret a nonresponse as part of additional data analysis will need to be thought out.

A rejection (also known simply as a "no") by the prospective customer is the other kind of negative response. Depending on the offer and the contact channel, we can often determine exactly whether or not the customer is interested in the offer (for example, an offer made via outbound telemarketing might result in a definitive "no, I'm not interested" response). Although it probably does not seem useful, the definitive "no" response is often as valuable as they positive response when it comes to further analysis of customer interests.

16.2.3 It all Begins with the Data

One of the differences between customer acquisition and most other marketing applications of data mining revolves around the data that is used to build predictive models. The amount of information that we have about people is that we do not get relationship with his customer, with limited information. In some cases, the data might be limited to their address and/or phone number. The key to this process is to have a relationship between the information that we do have and the behaviors that we want to model.

Most acquisition marketing campaigns begin with the prospect list. A prospect list is simply a list of customers who have been selected because they are likely to be interested in the products or services. There are numerous companies around the world that will sell lists of customers, often with a particular focus (for example, new parents, retired people, new car purchasers, etc.).

Some times it is necessary to add additional information to a prospect list by overlaying data from other sources. For example, consider a prospect list that containing only names and addresses. In terms of a potential data mining analysis, the information contained in the prospect list is very weak. There might be some patterns in the city, state, or zip code fields, but they would be limited in their predictive power. To augment the data, information about customers on the prospect list could be matched with external data. One simple overlay involves combining the customer's ZIP code with U.S. census data about average income, average age, and so on. This can be done manually or, as is often the case with overlays and the list provider can take care of this automatically.

More complicated overlays are also possible. Customers can be matched against purchase, response, and other detailed data that the data vendors collect and refine. This data comes from a variety of sources, including retailers, state and local governments, and the customers themselves. If we are mailing out a car accessories catalog, it might be useful to overlay information (make, model, year) about any known cars that people on the prospect list might have registered with their department of motor vehicles.

16.2.4 Test Campaigns

Once we have a list of prospect customers, there is still some work that needs to done before we can create predictive models for customer acquisition. Unless we have data available from previous acquisition campaigns, we will need to send out a test campaign in order to collect data for analysis. Besides the customers we have selected for the prospect list, it is important to include some other customers in the campaign, so that the data is as rich as possible for further analysis. For example, assume that the prospect list (that we purchased from a list broker) was composed of men over age 30 who recently purchased a new car. If we were to market to these prospective customers and then analyze the result, any patterns found by data mining would be limited to subsegments of the group of men over 30 who bought a new car. What about women or people age 30? By not including these people in our test campaign, it will be difficult to explain further campaign to include segments of the population that are not in our initial prospect list the solution is to include a random of selection of customers whose demographics differ from the initial prospect list. This random selection should constitute only a small percentage of overall marketing campaign, but it will provide valuable information for data mining. We will meet to work with our data vendor in order to add a random sample to the prospect list.

More sophisticated techniques than random selection, do exist such as those found in statistical experiment design and multivariable testing (MVT). Deciding when and how to implement these approaches is beyond the scope of this book, but there are numerous resources in the statistical literature that can provide more information.

Although this circular process (customer interaction? data collection? data mining? customer interaction) exists in almost in every application of data mining to marketing, there is more room for refinement in customer acquisition campaigns. Not only do the customers that are included in the campaigns change over time, but the data itself can also change. Additional overlay information can be included in the analysis when it becomes available. Also, the use random selection in the test campaigns allows for new segments of people to be added to our customer pool.

16.2.5 Evaluating Test Campaign Responses

Once we have started our test campaign, the job of collecting and categorizing the response behaviors begins. Immediately after the campaign offers go out, we need to track responses. The nature of the response process is such that responses tend to trickle in over time, which means that the campaign can go on forever. In most real-world situations, though, there is threshold after which we no longer look for responses. At that time, any customers on the prospect list that have not responded are deemed "nonresponses." Before the threshold, customers who have not responded are in state of limbo, somewhere between a response and a nonresponse.

16.2.6 Building Data Mining Models Using Response Behaviors

With the test campaign response data in hand the actual mining of customer response behaviors can begin. The first part of this process requires one to choose which behaviors one is interested in predicting, and at what level of granularity. The level at which the predictive models work should reflect the kinds of offers that one can make, not the kinds of responses that one can track. It might be useful (for reporting purposes) to track catalog clothing purchases down to the level of color and size. If all catalogs are the same, however, it really does not matter what the specifics of a customer purchase for the data mining analysis. In this case (all catalogs are the same), binary response prediction is the way to go. If separate men's and women's catalogs are available, analyzing response behaviors at the gender level would be appropriate. In either case, it is a straightforward process to turn the lower-level categorical behaviors into a set of responses at the desired level of granularity, If there are overlapping response behaviors, the duplicates should be removed prior to mining.

In some circumstances, predicting individual response behavior might be an appropriate course of action. With the movement toward one-to-one customer marketing, the idea of catalogs that are custom produced for each customer is moving closer to reality. Existing channels such as the Internet or outbound telemarketing also allow one to be more specific in the ways we target the exact wants and needs of their prospective customers. A significant drawback of the modeling of individual response behaviors is that the analytical processing power required can grow dramatically because the data mining process needs to be carried our multiple times, once for each response behavior that one is interested in. How one-handle negative responses also need to be thought out prior to the data analysis phase. As discussed previously, there are two kinds of negative responses: rejections and nonresponses. Rejections, by their nature, correspond to specific records in the database that indicate the negative customer response. On-responses, on the other hand, typically do not represent records in the database. Nonresponses usually correspond to the absence of a response behavior record in the database for customers who received the offer. There are two ways in which to handle non-responses. The most common way is to translate all nonresponses into rejection, either explicitly (by creating rejection records for the nonresponding customers) or implicitly (usually a function of the data mining software used). This approach will create a data set comprising all customers who have received offers, with each customer's response being positive (inquiry or purchase) or negative (rejections and nonresponses).

The second approach is to leave nonresponses put of the analysis data set. This approach is not typically used because it throws away so much data, but it might make sense if the number of actual rejections is large (relative to the number of nonresponses); experience has shown that nonresponses do not necessarily correspond to a rejection of one's product or services offering. Once the data has been prepared, the actual data mining can be performed. The target variable that the data mining software will predict is the response behavior type at the level we have chosen (binary or categorical). Because some data mining applications cannot predict nonbinary variables, some finessing of the data will be required if we are modeling categorical responses using noncategorical software. The inputs to the data mining system are the input variables and all of the demographic characteristics that we might have available, especially any overlay data that we combined with our prospect list. In the end, a model (or models, if we are predicting multiple categorical response behaviors) will be produced that will predict the response behavior that we are interested in. The models can then be used to score lists of prospect customers in order to select only those who are likely to response to our offer. Depending on how the data vendors we work with operate, we might be able to provide them with purchasing overlay data in order to aid in the selection of prospects, the output of the modeling process should be used to determine whether all of the overlay data is necessary. If a model does not use some of the overlay variables, we might want to save some money and leave out these unused variables the next time we purchase a prospect list.

16.3 Customer Relationship Management (CRM)

Knowledge of CRM and Customers: Considering the key issues that relate to how business interpret and make use of data about customers

Definition of CRM
How CRM data is used
Strategic data analysis for CRM
Data warehousing and data mining, how they can be used in CRM
Sharing customer data within the value chain

Strategy and Technology:

(CVM) Customer Value Management
The re-emergence of smart cards as a technology for customer management

16.3.1 Defining CRM

Traditionally, companies have developed databases to capture customer information, including such details as customer profiles, demographics, products purchased, and other items of interest. This data is used by management salespeople, service personnel, and others to determine market trends, customer preferences, service, and maintenance required by customers. However, there was a question whether companies had managed to transform this data into value.

16.3.2 Integrating Customer Data into CRM Strategy

Today, technologies such as data warehousing and data mining allow companies to collect, store, analyze, and manipulate enormous volumes of data. This can be important for marketers trying to provide better service and more satisfaction to the customer than competitors.

16.3.3 Strategic Data Analysis for CRM

Many companies have not determined how to deal with the rapidly increasing volumes of data about customers now being recorded in and about their business, through research, operations, or external data suppliers. Examples here are firstly, consumer goods retailers. Retailers can now obtain reports about the purchasing of individual products by specific customers, usually within a few hours of purchasing taking place and sometimes online. In its extreme form this can include the results of promotions mounted through mobile phone. Utilities, financial services, and B2B companies – managers in these industries now have comprehensive databases, which record not only purchase but also sales enquiries, responses to promotions, and a mass of detailed data about customers. All the examples illustrate two of the key dimensions that exist in marketing data analysis-product and customer.

Product Dimension

Companies such as retailers, which do not know the identity of their final customers, while applying customer management disciplines to their immediate direct customers, must use product management to get the best result. This is the so-called "product optimizers." Their own and market data on price, promotion, inventory levels and movements, and shipments is used to determine the optimum marketing and distribution policy. Their data is organized along the product dimension, and the key analysis task is to make sense of the possibly daily millions (e.g., P&G and Nestle) of transaction in which their products are involved. There are many companies that organize data by product and focus on marketing efforts that aim to increase demand for specific products – Gillette is one example, aiming to increase regular shaving, perhaps to twice a day.

Customer Dimension

Companies that can manage their customers as individual or small groups need to become expert in analyzing customer data, and be able to answer questions such as: Which customers do we want to market to market to, and we do not want to? How do we want to manage our customers? At what price would we like to sell, through which channels of distribution, and when? We will be very surprised that almost 901 of the companies in UK do not apply this basic principle to their companies.

The classification of management actions can be combined with a categorization of decision areas to produce a clear statement of requirement for the analyses to be undertaken. For example, for an FMCG supplier, the decision areas might be categorized to include: packaging, manufacturing and inventory volumes, distribution strategy and tactics, standard costing and pricing, promotional pricing, promotional offers, media advertising, individual brand/product definition and positioning.

Companies should normally need to consolidate and synthesize data from different sources (internal, market research, generally available market data, etc.) and functions (marketing, sales, finance, customer service, manufacturing, etc.). A successful approach to analyze the company is by defining the key analysis dimensions in some detail, and then to ensure focus on a few areas, which can be backed by management action.

The needs of different users vary, and different users will require different tools to support their different management actions, but this must not lead to data being dispersed to different users so that the capability for coordinated analysis is lost.

16.3.4 Data Warehousing and Data Mining

What is Data Mining?

A simple definition of data mining in marketing is: extraction of previously unknown, comprehensible and actionable information from large repositories of data, and using it to make crucial business decision and support their implementation, including formulating tactical and strategic marketing initiatives and measuring their success.

Why Using Data Mining in Marketing?

The aim of data mining is to obtain a sufficient understanding of a pattern of market behavior to allow quantifiable benefits to be derived from changes in behavior suggested by the analysis. This involves learning previously unknown facts about market behavior; answering specific questions, including forecasting questions.

In deciding whether to use data mining we must develop a clear view as to how we are going to use the output, in policy terms. Far too many data mining projects have been undertaken on the initiative of IT departments. These usually result in lots of interesting findings but number action. Companies with large customer bases, with a reasonable degree of stability in the marketing staff and in their marketing services suppliers, should have a wealth of knowledge about customers in the heads of their people and in reports on their past activities. Data mining will often confirm their beliefs, but add a more quantitative or practical edge – e.g., specifying the size of a long-suspected segment, identifying which customers belong to it. Some deductions are reasonable, such as buying behavior depending on socio-economic status, and these are the ones where one should collect data to investigate.

But data mining is more than simple data analysis. It is the understanding of a business environment, such that relevant questions can be answered by the use of the appropriate data analysis tool on the properly selected data. As such, data mining requires:

Understanding of the industry conditions
Appreciation of specific factors that apply to the company
Familiarity with a wide range of analytical tools
Ability to present extracted information in informative ways

16.3.5 Sharing Customer Data Within the Value Chain

How products and services (prices, availability, etc.) can more easily flow up and the supply chain using e-technology, helping customers toward the end of the chain better. However, it is clear that the more that suppliers further up the supply chain known about customers further down it, the better they can serve those customer needs. Put simply, customer data is valuable more than the supplier immediately facing the customer. An example of this is the willingness of suppliers in a market to share negative data about their customers, including those that do not pay in their activities. While there are many such examples of negative (or risk avoidance) data being shared, there are few occasions where positive data is shared within a market or value chain.

Data protection legislation

Data protection legislation has become a key management issue for all-new and existing processes and systems. A poor understanding of options can constrain a company's ability to use customer data competitively, while infringing legislation can have dramatic and expansive impact on company brand, create legal action. So the rules of importance for value chain management include such as: how the data was collected, for what purpose and with what expectation and approvals by the customer, data sharing within the company or group and with value chain partners.

Arguments for sharing customer knowledge

Improved targeting of marketing strategy.
Improved targeting of marketing communications.
Improved more relevant content of marketing communications.
Improved product planning.
Improved pricing.
Reduced costs of data acquisition.
Reduced costs of data processing.
Reduced media advertising.
Reduced direct mail expenditure.
Increased responsiveness to changing market conditions.
Gaining an advantage over competition at same level of the value chain.
Reducing market risk.
Learning/skills transfer.

Arguments against sharing customer knowledge

Increased complexity of the marketing process.
Increased problems with data management.

Conflict caused by mismatch in objectives/types/pace of marketing/sales process.
General conflict of interest.
Conflict of interest over customer ownership.
Conflict of interest over data ownership.
Systems incompatibilities.
Legal complexities (regulatory, data protection).
Data security.
Skills shortage-data analysis.

16.3.6 CVM – Customer Value Management

One of the biggest problems facing the company today is how to continue to attract customers in attaining growth, often in an environment where products and prices among competitors are moving closer and closer together. Traditional bases for differentiation, as product features or cost, are becoming less tangible and senior management are forced to look for new ways to be attractive to a target market. Many companies now should start thinking of the concept of customer value management (CVM) in order to identify the "value" that can be delivered, not only by products but also by processes and services, then engineering their business capabilities to deliver "ideal" customer-defined value at each customer interaction.

Due to the rapid introduction of new technologies and resultant rapid changes in customers' perceived "needs" and "values," in applying CVM companies can continuously monitor and maintain an alignment between the customer's vision of 'ideal value delivery' and the capabilities of the business to deliver that value.

The goal of CVM is to deliver optimal value to customers.

Measuring Up and Brands Have Relationships Too

The idea here is: lasting customer relationship. This results in long-term payback to a company. Customers want to be able to trust brands and rely upon them; they want a sense of commitment. In order to know how well they are doing in creating an emotional connection with their brands, managers must be able to measure such a concept by carrying out qualitative research, and one step further to benchmark data. They must know where these relationships are weak and where they are strong.

According to research by the Bristol Group, customers whose relationship with a company is 80 and higher on the 100-point scale will provide approximately 101 higher share of their business than those whose relationship is between 60 and 80. These customers are also 341 more likely to say that they are "very likely" to remain a customer and are more than twice as likely to recommend the company or firm to a friend. Such results point to the value to a firm's relationships with their customers, rather than merely making it easy for customers to deal with them.

GCRM and Implementation

This part investigates some of the problems large companies have in transferring customer management knowledge, expertise, and process between countries. Customer management consists of a range of business practices, ranging from full-fledged customer relationship management to its contributing in disciplines such as direct mail, telemarketing, database management, campaign management, and e-business. Many global companies have initiated new activities in the last few years, often in individual countries. Some of these companies are now looking to consolidate their gains and ensure that good practice in one country or division is transferred across the business.

16.3.7 Issues in Global Customer Management

CRM – A Graveyard for Marketing?

We have already expressed a strong skepticism about the blind acceptance of some of the ideas promoted under the guise of CRM. We stress that there are many other models of customer management, many of them classic – some of them working much more profitable than CRM – and the newer e-business-based models actually help customers get excellent value by cherry-picking, spot-buying, and resisting relationships.

Nevertheless, CRM became a global fashion in 1999. However, as the more cynical among us suspected, the popularity of this fashion (which in its most naïve form translated CRM into "'CRM system") is already starting to evoke yawns among senior management as they see the problems caused.

A study in October 1999 by the Meta Group has confirmed this. The study confirms that many CRM initiatives in the world's largest companies are at "serious risk of failure." The Meta Group in conjunction with its research affiliate, IMT strategies, interviewed 50 end users, including Eastman Kodak/Nortel Networks and Sprint. Although 80% of companies polled said they have at least one CRM application up and running, they failed to benefit fully from them.

For example, they are not using data warehouses for in-depth analysis. Few of them are using application that allows proper collaboration with customers. While many companies are employing customer-facing applications such as call centers and Web sites integrated with front, back, and mobile officers, Meta estimates the world's largest companies will have to spend as much as US$ 250 million over the next two to three years to achieve tangible returns on CRM investments.

Any model of customer management requires a balanced approach, covering strategies, people, systems, processes, data, measurement, and the like.

Improving customer management globally is an even greater task – yet some companies are working on it, and some are succeeding. One of these reasons why global capacity is important is that e-business has given immense impetus to the global capacity of customer's management.

Global Commerce, Global CRM, and the Death of Distance

One of the current dominant influences on management thinking is the idea that "competition is only a click away." Although this simplifies the nature of the competition, there is no doubt that what Francis Cairns crosses calls "the death of distance" is a reality of many managers. The telecommunications and IT revolution that we are still experiencing means: distance is now only a minor factor in the cost communicating, and becoming less and less important as more communication moves off from conventional voice into data.

This makes location less relevant to business success. Businesses need less and less to be located near the market they serve provided that the logistics network is in place to deliver all the products/services and information required to ensure smooth transition through the value chain. Size is also less relevant. Once, only larger companies could afford the complex IT, telecommunications, and logistics infrastructure required to service global market. Now these are all available, cheaply, to almost any company. These accelerated infrastructure developments mean that any kind of "customer can be served anywhere in the world, customizing the offerings of the world to their needs."

Global Knowledge and Information Management

A critical part of improving customer management performance across the globe seems to be having some kind of approach to transferring knowledge and skills. However, it is clear that there are two very different approaches. One is to formalize and codify knowledge (in case about how to manage customers) and transfer it through global systems and processes. The other is to recruit very skilled people and assigned them to the creation of improved customer management in a particular country are region. There is no golden rule here and much depends upon the nature of the product and service.

There is evidence, for example, that where the relationship with customers is very ideas based (e.g., in management consultancy), the latter approach is likely to be more profitable, but where the relationship with customers is highly automated and functional, the more formal approach is likely to be more profitable.

16.3.8 Changing Systems

Just as we are beginning to settle down on the idea that the Web as we know it is revolutionizing customer management in ways that we are learning to predict, along comes another revolution – the mobile phone as the location of customer management activity. The signs are already clear that it will have an effect that vary by country. However let us consider why it poses both a problem and an opportunity for global customer management programs.

Teenagers, among the most avid users of mobile phones, already ask why they need to be in a particular room to receive a pone call. However, many business customers already know that they do not need to be physically connected to anything in order to carry out any transaction, receive any information, or keep in touch with customers, suppliers, or colleagues. They share the vision of the mobile phone companies, in which each individual is in communication with whatever he or she wants to communicate with, with permanent or temporary connection according to that individual's wishes, but at relatively low cost. That means that the customer is often connected by a mobile phone to a computer, not a person (except in cases where the computer cannot do the job).

The mobile phone vision is supported by the widening gap between mobile access and Web access in most countries. This gap is widening very rapidly in those countries where the Web had the slowest uptake (e.g., in South East Asia). In the United States, on the other hand, the relatively poor development of mobile standards and networks has caused the country to fall behind Japan.

Of course it is the development of mobile phones as a means of sending and receiving information, rather than just mobile phones, which is bringing the reality of individual customer management to the fore. There will be even more of a need to ensure that the preferences and requirements of customers are managed and targeted in the right way as the more things become this is close to the individual, the more the potential for infringing on the customers privacy.

16.3.9 Changing Customer Management - A Strategic View

If this vision is to become a reality, what must change in customer management? We see the following as the key areas of change.

The Supply Chain

In many situations this is sticky. Transferring data along the supply chain and managing the eventual relationships with customers is difficult even using established technology. Companies will need to adopt a different vision of how they interact with customers; free up data within their supply chain to make it available to customers; and free up their supply chain to make it possible for customers to influence the chain itself.

The Customer Management Model

Companies will need to reconsider the nature of their customer management model. If it is based on privileged service to privileged customers, and if that privilege is based on access to and influence of the supply chain the position is vulnerable, as the other companies will be able to make what was privileged access available to all validated customers.

It may be seen that companies that have very frequent transactions with their customers may be the first to benefit from this technology. However frequency of transaction may not be the issue, if value of transaction is high enough. A used-car buyer can request mobile notification of when an appropriate vehicle comes into stock, or can review the stock availability of a nearby dealer. A new-car buyer can request mobile notification of delivery status. A customer waiting for annual maintenance can receive notification of a problem and request for authorization of additional work, or that the car is ready for collection or delivery.

Content Management

As customers are able to access or receive the latest information from companies, they will need to combine information from company and other sources. For example knowing the latest share price of transaction balance is one thing, doing something about it is another.

The UPS sponsored a research analysis that was carried out by IBM recently, and the study was carried out using the following.

An e-mail questionnaire was sent to 50 members of the customer management group, an IBM-sponsored group of large companies that are interested in improving how they manage their customers.

Formal discussions during the period of the research with around 20 individual companies and 7- conference presentation around the world.

Discussions with both leading corporate and individual customers and suppliers.

The outcomes are:

Awareness of the customer management issue
Lack of local awareness of benefits
Global programmers
Recognition of requirement of explicit transfer
Effect of global systems
Preferences for own systems, processes
Awareness that particular marketing strategies are not the determining factor
Skills transfer strategy
Understanding importance of measure across countries and divisions
Use of internal benchmarks
Cultural issues
Senior management skills, motivation
Conflict with ideas of delegation of authority/devolution
Reliance of internal networking
Importance of IT support to the transfer process

Implementing Customer Management

One of the most difficult questions facing all companies interested in improving their customer management is "We have decided what customer management strategy to put in place. We have decided which models of customer management to use. We have bought our systems. We have decided on our implementation program. Now where on earth do we get the people to manage it all?"

This is the wrong question. Leaving the people until last is not a good idea. Indeed, companies that succeed in implementing new approaches to customer management tend to appoint much of the team first and leave the team to develop plans. This follows the excellent principles of not expecting people to implement plans that they had no hand in shaping.

Of course, it is not possible to appoint a complete team before deciding what to do, as the decisions about what to do, and where and when to do it, affect decisions about the nature and size of the customer management team.

Customer management research suggests a very high correlation between having the right people and people programmers in place, with achieving results for their customers and company. It is perhaps not surprising that some successful managers extol the mantra "the customer comes second," meaning that where there is internal focus on employees they in turn will be able to understand customer needs and manage relationships effectively.

Key Skill Requirements in CRM

Strategic skill requirements are many. Few companies would claim that they had all the skills they required in the quantities they require them. Developing and keeping the right mix and level of skills is a constant battle. The phrase "two steps forwards, one step back" rings very true here. As soon as a company recruits and or trains to the right levels, strategies needs may change or key staff may leave to join other companies. This section discusses the ideal mix of skills and what a company can do to move toward the ideal, while recognizing that the ideal will never be reached.

Core Competencies

Setting out and understanding the core competencies we need for customer management is critical. Some examples of new competencies that organizations have to develop are:

Understanding customers – customer analysis and research
Customer strategy design
Contact strategy design and management
Understanding business impact profitability

Key account management
Managing, developing, and coaching people
Managing customer information
Innovating for and with customers
Researching the market
Customer program design and management
Customer proposition
Customer product rating
Web site contact design
Web community design
Customer process and service level design

Management, Personal and Technical Skills

In training we usually distinguish three fundamentally different types of skill as follows:

Management skills
Personal skills
Technical skills

Fundamental Marketing Skills

We might be surprised by the number of people working in customer management whose level of knowledge of marketing and certain customer management discipline is weak. If they are specialists who have been drawn from an agency they would have had little chance of developing broader marketing skills.

What kind of general marketing skills would we like in our customer management staff?

Customer orientation
Understanding of different basic approaches to marketing
Understanding of the basic marketing approach

Personal Skills

Personal disciplines are usually at a premium in customer management, two skills in particular. The first is communication – the amount of teamwork required to develop and launch customer management campaigns means that those involved need to be good communicators. This is not just for formal presentations, but also in the sense of keeping the team informed. The second is team working.

Data and Database Skills

Customer management is probably the most-needed quantity for marketing. Direct marketers need to understand how customer databases are built and maintained. They must know what sort of data is required as the basis for successful customer management, how to analyze the data to find out which approaches work, and so forth. They do not necessarily need to be able to carry out the analyses themselves, as these are often contracted out to third parties. But they must understand the basis of data analysis in order to make sense of the results.

Different Strategies for Different Companies

Although the broad management skill requirements are common across most industries, different types and size of companies need different mixes of customer management skills. The major differences are likely to be as follows

Smaller Companies

These are unlikely to be able to afford many or any dedicated customer management staff. Staff responsible for customer management activity will have as their main task the orchestration of external suppliers to achieve effective campaign at low cost. The skills of supplier management are likely to be at a premium as are the efficient personal skills. At the same time, the staff concerned will need to be closely involved with the development of overall marketing policy and probably be expected to contribute to it, rather than taking it as given. In such situations, the marketing all rounder, who is personally an effective worker, is likely to be premium.

Larger Companies

These can afford and need specialists. Their tasks are likely to be more precisely allocated as part of an overall marketing plan. As specialists of various kinds, they will be "pitting their wits" against their opposite numbers in competitive companies, to gain an advantage over them. This degree of precision in job definition implies that these staff will be working as members of a large in-house team. The team needs to be communicated with, listened to, and influenced, rather than told what to do. Even the external suppliers may not be appointed directly, but as part of a wider corporate even international policy. So in this case the skills mix needs to be richer in the areas of team working, as well as there being depth in the particular specialism concerned.

Consumer Marketers

Companies marketing mainly to consumers need to have skills relating to the more "mass-market" media-mail, inbound telemarketing, published and broadcast media, Web marketing – as these are likely to be used more intensively. They also need to understand the kinds of consumer data available from third parties and the kinds of analysis that can be carried out on such data to segment the market.

B2B Companies

These will need mass market skills when they are marketing to small business. However if their market is mainly to larger organizations they will need specific strengths in the areas of telemarketing, Web management and using customer management in support of sales, staff, or large agents. In that case they will need to be very strong in the skill of "working with." These include communicating, influencing, negotiation, functioning as part of a team. Sales forces are rightfully suspicious of new approaches to marketing, which involve addressing people they see as their customers.

Long-Term Relationship Marketers

Companies marketing to customers who maintain a long-term relationship with the company, e.g., if the purchase is frequent or if there are additional products and services, which can be sold after the main sale, require the skills of database marketing. This is because they will probably find the development of an in-house database cost effective.

Top Ten Reasons for CRM and the Five Key Elements of Success

We have identified ten reasons for CRM and five key elements required for success in CRM as follows.

Ten Reasons

Information flow improves sales force productivity
Improved sales force productivity means more deals
Better customer knowledge results in better win rates
Improved communication between staff gives the company a united voice
Accurate communications between customers makes more marketing more effective
Enhanced market and competitor intelligence brings better business decisions
Customer care brings competitive advantage
Back office integration produces business efficiency
Integrated customer touch points present an integrated customer experience
Better sales forecasting means more accurate business decision.

Five Critical Factors for Success

Customer management strategy - which customers are to be managed, for what products or service through which channels
Customer management models- how these customers are to be managed
Infrastructure- systems, data, operational customer management
People- who is to develop the new capability, who is to manage it, and how
Programmers- how new customer management capabilities are to be installed

16.4 Data Mining and Customer Value and Relationships

The way in which companies interact with their customers has changed dramatically over the past few years. A customer's continuing business is no longer guaranteed. As a result, companies have found that they need to understand their customers better, and to quickly respond to their wants and needs. In addition, the time frame in which these responses need to be made has been shrinking. It is no longer possible to wait until the signs of customer dissatisfaction are obvious before action must be taken. To succeed, companies must be proactive and anticipate what a customer desires.

It is now a cliché that in the days of the corner market, shopkeepers had no trouble understanding their customers and responding quickly to their needs. The shopkeepers would simply keep track of all of their customers in their heads, and would know what to do when a customer walked into the store. But today's shopkeepers face a much more complex situation. More customers, more products, more competitors, and less time to react means that understanding one's customers is now much harder to do. A number of forces are working together to increase the complexity of customer relationships:

- *Compressed marketing cycle times.* The attention span of a customer has decreased dramatically and loyalty is a thing of the past. A successful company needs to reinforce the value it provides to its customers on a continuous basis. In addition, the time between a new desire and when we must meet that desire is also shrinking. If we do not react quickly enough, the customer will find someone who will.
- *Increased marketing costs.* Everything costs more. Printing, postage, special offers (and if we do not provide the special offer, our competitors will).
- *Streams of new product offerings.* Customers want things that meet their exact needs, not things that sort of fit. This means that the number of products and the number of ways they are offered have risen significantly.
- *Niche competitors.* Our best customers will also look good to our competitors. They will focus on small, profitable segments of our market and try to keep the best for themselves.

Successful companies need to react to each and every one of these demands in a timely fashion. The market will not wait for our response, and customers that we have today could vanish tomorrow. Interacting with our customers is also not as simple as it has been in the past. Customers and prospective customers want to interact on their terms, meaning that we need to look at multiple criteria when evaluating how to proceed. We will need to automate:

- The Right Offer
- To the Right Person
- At the Right Time
- Through the Right Channel

The right offer means managing multiple interactions with customers, prioritizing what the offers will be while making sure that irrelevant offers are minimized. The right person means that not all customers are cut from the same cloth. Our interactions with them need to move toward highly segmented marketing campaigns that target individual wants and needs. The right time is a result of the fact that interactions with customers now happen on a continuous basis. This is significantly different from the past, when quarterly mailings were cutting-edge marketing. Finally, the right channel means that we can interact with our customers in a variety of ways (direct mail, email, telemarketing, etc.). We need to make sure that we are choosing the most effective medium for a particular interaction.

The purpose of this section is to provide with a thorough understanding of how a technology like data mining can help solve vexing issues in our interaction with our customers. We describe situations in which a better understanding of our customers can provide tangible benefits and a measurable return on investment. It is important to realize, though, that data mining is just a part of the overall process. Data mining needs to work with other technologies (for example, data warehousing and marketing automation), as well as with established business practices.

16.4.1 What is Data Mining?

Data mining, by its simplest definition, automates the detection of relevant patterns in a database. For example, a pattern might indicate that married males with children are twice as likely to drive a particular sports car than married males with no children. If we are a marketing manager for an auto manufacturer, this somewhat surprising pattern might be quite valuable. However, data mining is not magic. For many years, statisticians have manually "mined" databases, looking for statistically significant patterns.

Data mining uses well-established statistical and machine learning techniques to build models that predict customer behavior. Today, technology automates the mining process, integrates it with commercial data warehouses, and presents it in a relevant way for business users. The leading data mining products are now more than just modeling engines employing powerful algorithms. Instead, they address the broader business and technical issues, such as their integration into today's complex information technology environments.

In the past, the hyperbole surrounding data mining suggested that it would eliminate the need for statistical analysts to build predictive models. However, the value that an analyst provides cannot be automated out of existence. Analysts will still be needed to assess model results and validate the plausibility of the model predictions. Because data mining software lacks the human experience and intuition to recognize the difference between a relevant correlation and an irrelevant correlation, statistical analysts will remain in high demand.

An Example

Imagine that we are a marketing manager for a regional telephone company. We are responsible for managing the relationships with the company's cellular telephone customers. One of our current concerns is customer attention (sometimes known as "churn"), which has been eating severely into our margins. We understand that the cost of keeping customers around is significantly less than the cost of bringing them back after they leave, so we need to figure out a cost-effective way of doing this.

The traditional approach to solving this problem is to pick out our good customers (that is, the ones who spend a lot of money with our company) and try to persuade them to sign up for another year of service. This persuasion might involve some sort of gift (possibly a new phone) or maybe a discount calling plan. The value of the gift might be based on the amount that a customer spends, with big spenders receiving the best offers. This solution is probably very wasteful. There are undoubtedly many "good" customers who would be willing to stick around without receiving an expensive gift. The customers to concentrate on are the ones that will be leaving. Do not worry about the ones who will stay.

This solution to the churn problem has been turned around from the way in which it should be perceived. Instead of providing the customer with something that is proportional to their value to our company, we should instead be providing the customer with something proportional to one's value to them. Give our customers what they need. There are differences between the customers, and we need to understand those differences in order to optimize our relationships. One big-spending customer might value the relationship because of our high reliability, and thus would not need a gift in order to continue with it. On the other hand, a customer who takes advantage of all of the latest features and special services might require a new phone or other gift in order to stick around for another year. Or they might simply want a better rate for evening calls because their employer provides the phone and they have to pay for calls outside of business hours. The key is determining which type of customer we are dealing with.

It is also important to consider timing in this process. We cannot wait until a week before a customer's contract and then pitch them an offer in order to prevent them from churning. By then, they have likely decided what they are going to do and we are unlikely to affect their decision at such a late date. On the other hand, we do not start the process immediately upon signing a customer up. It might be months before they have an understanding of our company's value to them, so any efforts now would also be wasted. The key is finding the correct middle ground, which could very well come from one's understanding of their market and the customers in that market. Or, as we discuss later, we might be using data mining to automatically find the optimal point.

16.4.2 Relevance to a Business Process

For data mining to impact a business, it needs to have relevance to the underlying business process. Data mining is part of a much larger series of steps that takes place between a company and its customers. The way in which data mining impacts a business depends on the business process, not the data mining process. Take product marketing as an example. A marketing manager's job is to understand their market. With this understanding comes the ability to interact with customers in this market, using a number of channels. This involves a number of areas, including direct marketing, print advertising, telemarketing, and radio/television advertising, among others.

The issue that must be addressed is that the results of data mining are different from other data-driven business processes. In most standard interactions with customer data, nearly all of the results presented to the user are things that they knew existed in the database already. A report showing the breakdown of sales by product line and region is straightforward for the user to understand because they intuitively know that this kind of information already exists in the database. If the company sells different products in different regions of the country, there is no problem translating a display of this information into a relevant understanding of the business process.

Data mining, on the other hand, extracts information from a database that the user did not know existed. Relationships between variables and customer behaviors that are nonintuitive and are the jewels that data mining hopes to find. And because the user does not know beforehand what the data mining process has discovered, it is a much bigger leap to take the output of the system and translate it into a solution to a business problem.

This is where interaction and context comes in. Marketing users need to understand the results of data mining before they can put them into actions. Because data mining usually involves extracting "hidden" patterns of customer behavior, the understanding process can get a bit complicated. The key is to put the user in a context in which they feel comfortable, and then let them poke and prod until they understand what they did not see before.

How does someone actually use the output of data mining? The simplest way is to leave the output in the form of a black box. If they take the black box and score a database, they can get a list of customers to target (send them a catalog, increase their credit limit, etc.). There is not much for the user to do other than sit back and watch the envelopes go out. This can be a very effective approach. Mailing costs can often be reduced by an order of magnitude without significantly reducing the response rate.

Then there is the more difficult way of using the results of data mining: getting the user to actually understand what is going on so that they can take action directly. For example, if the user is responsible for ordering a print advertising campaign, understanding customer demographics is critical. A data mining analysis might determine that customers in New York City are now focused in the 30-to-35-year-old age range, whereas previous analyses showed that these customers were primarily aged 22 to 27. This change means that the print campaign might move from the *Village Voice* to the *New Yorker*; there is no automated way to do this. It is all in the marketing manager's head. Unless the output of the data mining system can be understood qualitatively, it will not be of any use.

Both of these cases are inextricably linked. The user needs to view the output of the data mining in a context they understand. If they can understand what has been discovered, they will trust it and put it into use. There are two parts to this problem: 1) presenting the output of the data mining process in a meaningful way, and 2) allowing the user to interact with the output so that simple questions can be answered. Creative solutions to the first part have recently been incorporated into a number of commercial data mining products. Response rates and (probably most importantly) financial indicators (for example, profit, cost, and return on investment) give the user a sense of context that can quickly ground the results in reality.

16.4.3 Data Mining and Customer Relationship Management

Customer relationship management (CRM) is a process that manages the interactions between a company and its customers. The primary users of CRM software applications are database marketers who are looking to automate the process of interacting with customers. To be successful, database marketers must first identify market segments containing customers or prospects with high-profit potential. They then build and execute campaigns that favorably impacts the behavior of these individuals.

The first task, identifying market segments, requires significant data about prospective customers and their buying behaviors. In theory, the more data the better. In practice, however, massive data stores often impede marketers, who struggle to sift through the minutiae to find the nuggets of valuable information. Recently, marketers have added a new class of software to their targeting arsenal. Data mining applications automate the process of searching the mountains of data to find patterns that are good predictors of purchasing behaviors. After mining the data, marketers must feed the results into *campaign management software*, which, as the name implies, manages the campaign directed at the defined market segments.

In the past, the link between data mining and campaign management software was mostly manual. In the worst cases, it involved "sneaker net," creating a physical file on tape or disk, which someone then carried to another computer and loaded into the marketing database.

This separation of the data mining and campaign management software introduces considerable inefficiency and opens the door for human errors. Tightly integrating the two disciplines presents an opportunity for companies to gain competitive advantage.

16.4.4 How Data Mining Helps Database Marketing

Data mining helps marketing users to target marketing campaigns more accurately and also to align campaigns more closely with the needs, wants, and attitudes of customers and prospects.

If the necessary information exists in a database, the data mining process can model virtually any customer activity. The key is to find patterns relevant to current business problems.

Typical questions that data mining addresses include the following:

Which customers are most likely to drop their cell phone service? · What is the probability that a customer will purchase at least $100 worth of merchandise from a particular mail-order catalog? · Which prospects are most likely to respond to a particular offer? Answers to these questions can help retain customers and increase campaign response rates, which, in turn, increase buying, cross selling, and return on investment (ROI).

Scoring

Data mining builds models by using inputs from a database to predict customer behavior. This behavior might be attrition at the end of a magazine subscription, cross-product purchasing, willingness to use an ATM card in place of a more expensive teller transaction, and so on. The prediction provided by a model is usually called a *score*. A score (typically a numerical value) is assigned to each record in the database and indicates the likelihood that the customer whose record has been scored will exhibit a particular behavior. For example, if a model predicts customer attrition, a high score indicates that a customer is likely to leave, whereas a low score indicates the opposite. After scoring a set of customers, these numerical values are used to select the most appropriate prospects for a targeted marketing campaign.

The Role of Campaign Management Software

Database marketing software enables companies to deliver timely, pertinent, and coordinated messages and value propositions (offers or gifts perceived as valuable) to customers and prospects. Today's campaign management software goes considerably further. It manages and monitors customer communications across multiple touch points, such as direct mail, telemarketing, customer service, point of sale, interactive Web, branch office, and so on.

Campaign management automates and integrates the planning, execution, assessment, and refinement of possibly tens to hundreds of highly segmented campaigns that run monthly, weekly, daily, or intermittently. The software can also run campaigns with multiple "communication points," triggered by time or customer behavior such as the opening of a new account.

Increasing Customer Lifetime Value

Consider, for example, customers of a bank who use the institution only for a checking account. An analysis reveals that after depositing large annual income bonuses, some customers wait for their funds to clear before moving the money quickly into their stock brokerage or mutual fund accounts outside the bank. This represents a loss of business for the bank.

To persuade these customers to keep their money in the bank, marketing managers can use campaign management software to immediately identify large deposits and trigger a response. The system might automatically schedule a direct mail or telemarketing promotion as soon as a customer's balance exceeds a predetermined amount. Based on the size of the deposit, the triggered promotion can then provide an appropriate incentive that encourages customers to invest their money in the bank's other products. Finally, by tracking responses and following rules for attributing customer behavior, the campaign management software can help measure the profitability and ROI of all ongoing campaigns.

Combining Data Mining and Campaign Management

The closer data mining and campaign management work together, the better the business results. Today, campaign management software uses the scores generated by the data mining model to sharpen the focus of targeted customers or prospects, thereby increasing response rates and campaign effectiveness. Ideally, marketers who build campaigns should be able to apply any model logged in the campaign management system to a defined target segment.

Consideration of Customer Value in the Data Mining Process

One of the most important issues for business-oriented use of data mining is the incorporation of value considerations into the analysis process. *Value* is a general term that may mean different things in different settings, such as: the average monthly revenue from the customer, number of lines the customer owns or other combination value we would like to consider certain point of time. In context of churn management, some of the tactics and ideas often employed are:

- Predata-mining segmentation of the customers by their "value" and separate analysis for the various segments.
- Postdata-mining analysis of the results by the value of the customers – e.g., considering the coverage of prediction rules only for the highest value customers.
- Customer lifetime value analysis that combine estimation of the customer "lifetime" with a revenue estimation during this period.

We propose an original approach in which value is integrated into the data mining algorithm, in a way that the process of data partitioning is considering the distribution of value at the same time as the size of populations.

Effective Incentive Allocation

In several applications data mining is used for analysis followed by countermeasure reaction. For example, in the churn management, the analysis of churning customers will normally result with incentive campaigns. This means that we will accord incentives to valuable customers who are predicted to churn. There are two main areas of interaction between the incentive component and the data mining component in such application: the attribution of incentives to population segments and the measurements of their effects in future analysis.

We propose to use the data mining results (in the implementation-induced rules) for incentive allocation. The generated data mining rules and their related customer segment can be useful symptom descriptors for matching effective incentives. For example, following a rule quoting that young customers in a certain area are massively disconnecting, the analyst may design a campaign that will propose an attractive price plan for customers with young customer usage profile through an aggressive targeting media in this area.

In the following, we address the question how to consider incentives' attribution in future data mining, on one hand, and how to a posteriori evaluate their effectiveness on the other.

Incorporation of External Events into the Data

A churn prediction model would usually be constructed from data extracted from the corporate data warehouse, such as: usage history and trends (number of calls, duration, services used, destinations, etc.) and social-demographic data (income, city, education, profession, etc.). At the same time, there are some implicit punctual events in history or at the present, such as a competitor coming out with a promotion campaign, or a major financial crisis in a certain area – which may have a major effect on the behavior of the customers. This may have effect on different segments of the customer population, depending on: geographical area, usage patterns, etc. Ignoring these events may lead to wrong prediction models. Therefore, any successful analysis cannot ignore their existence and must incorporate them into the model. In general, we see distinguish three relevant approaches to this issue.

- The ideal situation, where an expert can quantify the impact of these singular events for the different segments of the customer population. This enables the inclusion of "external effects" as an additional input field.
- When such quantification is not available, some discrete input fields indicating "competitor promotion," "financial crisis," etc. may be added. This is much less desirable and can be counterproductive when for example different promotions have completely different effects.
- To some extent, the effects of singular events may be diminished by "time randomization." In such solution the model is built from samples at different points in time. In this manner, the effects of external events are averaged but the resulting models are inferior to those obtained with incorporation of those events.

16.5 CRM: Technologies and Applications

It is supposed to give an introduction to and overview of CRM (customer relationship management) and its connections with ERP. This section sheds some light on applications, enabling technologies, users, and providers of CRM.

We put the emphasis on CRM technologies – ones currently in use as well as upcoming technologies – and the connection between ERP and CRM. By doing so we wish to give a useful overview of what the (future) worker in the field of ERP can expect and has to have heard of when dealing with CRM. Section 3 shows some implications of implementing CRM. In this light it might be surprising that we paid little attention to CRM as ***the*** enabler of e-commerce. Everyone still calls any business on the Internet e-commerce. Paired with the fact that CRM is a hype word itself, this made finding useful information like looking for a needle in a haystack. E-commerce would be an interesting topic for further study, though.

This section is not meant to be a profound, let alone a complete view on CRM. We chose to omit a discussion of possible disadvantages of using CRM as well as the pitfalls of implementing CRM. We do however take a look at difficulties that can be encountered in the integration of ERP and CRM. Furthermore, we intend to show the possibilities of CRM, not necessarily the average implementation. For these reasons, the discussion can appear to be somewhat euphoric. Our decision of scope is a result of the scarcity of articles critical of CRM.

16.5.1 What is CRM ?

Introduction and brief history of CRM

CRM stands for customer relationship management and is the term used to describe any methodologies, strategies, software, and Web-based capabilities that help an enterprise organize and manage customer relationships.

The idea is to have the same information available to all people and departments in the company so that every product or service need of the customer is met. CRM makes it possible that everyone in the enterprise is focused on the customer.

CRM is one of many ICT developments in the past decade to come forth from the growing awareness of information as a key strategic business asset. Today this is a nearly universally accepted fact.

CRM came up in the early 1990s, when global competition was becoming an increasingly important issue for enterprises. It was this competition that made it necessary to find new ways of looking at business. An important development was the shift from a product-centric view to a customer-centric view.

Enterprises with a product-centric view have a splintered view of their customers, because they have several customer contact points and several separate systems and databases that could not interact at all or very little, whereas enterprises with a customer-centric view have a complete picture of their customers and can focus on them better.

In order to realize an enterprise with a customer-centric view, it was necessary to develop new strategies and technologies. These are explained in the rest of this section.

16.5.2 What is CRM Used for?

The basic idea behind CRM is to consolidate all contact points with the customer as well as all customer information into a single system. This enables an enterprise to do the following as illustrated in Table 16.1:

CRM can also help acquire new customers to a certain degree, by sharing customer information with business partners or by simply knowing what kind of customer you should be on the look out for. However, the emphasis in CRM is clearly on maintaining a good relationship with existing customers. Research has shown that it costs an enterprise ten times as much to obtain a new customer than to retain an existing one.

All of these applications of CRM can directly or indirectly lead to an increase of profitability. Without CRM, many of the benefits explained above would be impossible. For example, a customer could register a complaint about his treatment by one department and shortly after be treated the same way by another department, as a result of inconsistent or incomplete customer information.

Table 16.1. Functions of an Enterprise

Use	Goal
Gather and access information about customers:	
- Purchase history	Increase marketing and selling opportunities by offering similar or related products and services as well as updates of the purchased product itself
- Wishes and requirements	Develop new products and services
- Complaints	Improve products and services, prevent mistakes in the future and give (highly profitable) customers special treatment
- Preferences	
Customize Web pages and other information services based on previous visits, preferences, and other information on the customer	Finetune services and cater to customers' special requests
Enhance and optimize help-desk functions and performance	Increase customer satisfaction and reduce service-related complaints
Respond more quickly to customer inquiries	
Gain a better understanding of their customers' wants and needs	Create new products and services in the best interest of the customer
Increase customer loyalty by creating exit barriers (making the customer feel so much "at home" that he does not want to switch to another company)	Secure a market position in the long term
Better anticipate what customers will want	Be ahead of the competition and reduce time to market
Increase efficiency through automation	Reduce production and/or operation costs and reduce production times
Identify the most profitable customers	Concentrate marketing efforts on them and drop unprofitable customers
Obtain suitable information and share it with business partners via Electronic Data Interchange (EDI)	Acquire new customers
Leverage every role, device, channel, and customer contact points within an enterprise	Enhance profitability

Who uses CRM?

CRM is being employed in industries where a one-to-one relationship with customers offers competitive advantage. Companies in consumer service industries as financial services, banking, travel, and telecommunications rely heavily on CRM solutions to stay competitive. A special group of enterprises in which the one-to-one relationship with customers is important are those doing e-business. The Internet being an IT itself can be connected directly to an enterprise's CRM applications, opening up even more possibilities.

CRM is used successfully by companies from extremely diverse branches of industry. SAP alone already implemented CRM functionality in enterprises from the automotive, metal, food, retail, process, ICT and media industries.

On the one hand, small- to medium-sized enterprises can use CRM to compete with or even gain an advantage above bigger enterprises. On the other hand, CRM yields great benefits to large enterprises as well, by solving the problems of customer data that is spread over several "island" systems and databases, or simply over large geographic distances. At the same time, with increasing size and complexity of an enterprise, the consequences of the implementation of CRM become more severe and difficult to handle.

16.5.3 Consequences of Implementation of CRM

CRM cannot be implemented without a customer strategy, because that is exactly what it is designed to support. That means that the first step in any CRM implementation effort is to develop such a strategy to optimally support the enterprise's mission. A customer strategy includes the choice of what kind of customer the enterprise wants, how to find them, and how to provide service to them, as well as the choice of a preferred communication medium (mail, phone, Internet, etc.), how often to contact the customer, about what to contact the customer, and so on.

Once a customer strategy has been established, the enterprise must ensure that it will be carried out, some might even say lived by. This is all but straightforward, because the whole view on doing business shifts from a product-centric view to a customer-centric view. Staff has to be trained and motivated to work within the chosen strategy. In practice, they have to be able to solve customers' problems and work with the new information technology. This training in itself (not including the CRM system) is already a major investment, which should not be underestimated.

Another important task is connecting existing systems to the new system or component. Depending on what type of system has to interact with the CRM package, this task can range from reconfiguration to full-blown development of a special interface.

On the one hand, if an enterprise system is already in place, the enterprise might only have to reconfigure it or add a CRM component from the same provider. Some enterprise systems do not offer CRM functionality, but do have an interface ready to use with a CRM package from a different provider.

On the other hand, existing systems that were not designed to do so, might have to interact with the new system. Then an interface must be tailor-made to fit the needs of the enterprise precisely. The complexity, costs, and risks of such a project might even give reason to consider doing away with the old system altogether and replacing it with a standard package.

16.5.4 Which Technologies are Used in CRM?

CRM Software

16.5.5 Business Rules

As stated in section 3, an enterprise needs to develop a strategy for handling its customers. The strategy can be formulated as business rules. Business rules define criteria for processing transactions with customers and are meant to enforce efficient, predefined business structures and to control and influence the type of transactions the enterprise makes. They align day-to-day operations with corporate goals. In an ideal situation, business rules are enforced in real time, i.e., the very moment an employee or the customer himself uses the system to do a transaction. Business rules are not technology in themselves. The implementation is realized with a wide variety of technologies, including for instance Corba, XML, Java, HTTP, etc.

16.5.6 Data Warehousing

The most important technology used in CRM is data warehousing. A very simple explanation of data warehousing would be that there is a single central database that contains all relevant customer information. The database model underlying a data warehouse system typically corresponds more with reality than the models underlying the systems it collects data from, such as order processing, product inventory, and marketing. That way, the data warehouse gives a total and more natural view of the customer. If the data warehouse also integrates all applications that deal with customer information, it is sometimes called an *integration framework*.

The fundamental concept of data warehousing is that data for business analysis is combined from more than one source application and is stored and analyzed independently from those applications and their operational data. The main advantage of this strategy is that data analysis and queries, which can be very computation intensive, run on the data warehousing system and not on other, possibly critical systems. Computational load can be further avoided by constantly keeping up standard reports of the customer data that is queried most often.

Another advantage is the possibility to cross-reference data from the different applications. This in turn makes it possible to do queries over a certain time, such as a query about a customer's purchases and the services provided to him in a certain month. Data warehousing is typically used in connection with a company's Intranet, which is often connected to the Internet. This combination enables worldwide access to consistent data and data analysis tools at a low cost, compared to the use of legacy systems.

16.5.7 Data Mining

Data mining is the discovery of new information in terms of patterns or rules from (very) large amounts of data that would not become evident in standard queries to a database or data warehouse.

There are different types of patterns to be found by data mining:

- Associations: e.g.: It seems that when men buy diapers, they often also buy beer. Or when someone buys a certain service, he also buys another, related service.
- Sequential patterns: e.g.: Someone buys a camera and every few months photo supplies and every once in a while photo accessories. Or when a customer buys a software package, he regularly buys updates and new plug-ins.
- Customer classification trees: e.g.: Customers can be classified by how often or what kind of products and services they buy, by how much they buy, or by their preferred method of paying, etc.

Data mining can be carried out in order to reach a number of goals:

- Prediction: To make prediction, data mining is combined with business knowledge to show how certain attributes of the data will behave in the future. For example, one can predict what a customer will buy when made a certain special offer, how much a certain store will sell in a given period of time, or whether discontinuing a product or service will increase profits.
- Identification: Data patterns can be used to identify the existence of an item or an entry in the database. An example application is authentication, which checks whether a user is really a specific user. This is important for privacy in applications, especially if the customer interacts directly with the CRM software.
- Classification: Customers can be partitioned into different classes or categories based on combinations of properties stored in a CRM system. An enterprise can then concentrate its (marketing) efforts on groups of customers that are especially profitable, loyal, or show an increasing consumption pattern.
- Optimization: The use of limited resources such as time, space, money, materials, and bandwidth can be optimized in order to maximize sales or profits.

Two of the most important applications of data mining within CRM are marketing and finance. Based on analysis of customer behavior, an enterprise can determine marketing strategies, including advertising, outlet location, design of catalogs, and campaigns. Financial applications include analysis of creditworthiness, segmentation of clients, and evaluation of financing options.

Data mining is a technology still in development. As a result, it not quite clear those techniques and technologies that are encompassed by the term *data mining* and which are not.

16.5.8 Real-Time Information Analysis

It is possible to reduce response time to customers' e-mails by having them automatically processed. They are parsed to discover important information such as customer number, order number, and key phrases. Based on the result of the analysis, the messages are scored with a priority level, from which the CRM software can either automatically generate a precomposed response to the inquiry where no customer review is necessary, as in the case of order status inquiries, or route the e-mail to the appropriate employee for response.

Another application is the real-time analysis of online customer behavior. Conclusions can be drawn as to what is of most interest to the customer and how products and services should be presented in order to satisfy him.

When a customer places an order, a CRM application can check whether the requested item is sufficiently on stock and estimate what the delivery time will be. If the item is out of stock or the delivery time exceeds the customer's order constraints, an alternative product or service can be offered.

Real-time information analysis can be combined with call centers (technology) by presenting call-center agents (near) real-time information on a customer's buying habits. The agent can then determine those products that the customer is likely to be interested in.

Call Center Technologies

A call center is a place where telephone calls are either placed or received in high volumes for the purpose of sales, marketing, customer service, telemarketing, technical support, or other specialized business activity. Customers' calls are most often on orders, questions, complaints, etc.

Call centers are almost always linked to the enterprise's network in a client-server architecture. The most important reason for linking a call center to the enterprise's network is to enable the agent to access, create, and change customer data.

CTI or Computer-Telephony Integration is an increasingly widespread technology that combines voice and data flows in the call center. When a call is received from a known customer, his data, including name, purchase history, etc., automatically appears on the screen of the call center agent. During the phone call, the agent can make notes and appointments and further process the customer's information. When an agent makes a call to a customer, he can do so by selecting a customer on screen instead of having to dial a number or a speed dial. In both cases, after the call is terminated, the CRM software automatically processes any changes in the customer's data and passes on orders, appointments, etc. to the responsible applications of the ERP system.

IVR or Interactive Voice Response is a term used for several different applications. IVR can take over the task of asking routine questions, so that qualified personnel can devote its time to real services. Some services can even be realized (around the clock) completely without a human counterpart, so the customer more or less helps himself with the possibility of speaking to an employee if he feels the need to do so. The customer can not only access services, but also his data by phone. IVR does this by transforming data into speech. Data can be entered or change by means of voice recognition.

Another application of IVR is the replacement of waiting tunes with information relevant to the waiting customer, who before has been identified by his phone number.

Other call center technologies are Private Branch eXchange (PBX), Voice over Internet Protocol (VoIP), and Automatic Call Distribution (ACD).

16.5.9 Reporting

CRM produces two kinds of reporting: customer reporting and internal reporting:

Customer reporting can notify the user in the case of an order that cannot be fulfilled or beforehand that a regular item will be unavailable in a certain period. Furthermore, automatic order tracking or regular account information reports can be generated and sent to the customer.

Internal reporting can supply managers with real-time analyses of customer data and statistics to support identification of trends and customers' value, support strategic decision making, and monitor overall performance of the enterprise. Many reporting tools automatically make a graphical representation of the analysis, which further aids the manager.

16.5.10 Web Self-Service

CRM offers customers the opportunity for Web self-service. They can access their back-office information or account from anywhere at any time without the assistance of an employee. At the appropriate times it allows access to front-office contact channels as well. An upcoming Web self-service is Electronic Bill Presentment and Payment or EBPP. Customers are presented their bill online, where they can also pay the bill. An example could be that when a telephone customer's bill contains more international calls than domestic calls, within short notice he receives a special offer for an international calling plan.

Application Service Providers (ASP)

An application service provider, or ASP, hosts applications that are accessed through the Internet or private networks and sold as services, typically with a monthly subscription fee. ASPs allow an enterprise to outsource CRM activities in order to concentrate on its core activities.

It is currently possible to outsource all CRM functionality to an ASP, but it is still an upcoming technology and service. Some analysts even say that CRM hosting is already over its peak, because it makes little sense to outsource important functionality while the main advantage of CRM and ERP lies in integration (Lee). ASPs often do not have as much experience with CRM than providers of CRM software. Some enterprises therefore choose to outsource only some CRM components, especially operational components such as sales automation, marketing automation, and customer service support.

The appeal of ASPs for outsourcing of analytical or even all CRM applications is limited for many organizations by issues pertaining to CRM application customization, control, and security. The use of an ASP inherently offers less possibilities for customization and control than the installation of an own CRM system or CRM component of an ERP system. The security of customer information is a crucial issue in CRM and the fact that an ASP stores this information remotely can make an enterprise be reserved.

There is a trend, however, that (experienced) providers of CRM software offer ASP services themselves or closely work together with existing ASPs. Companies that have already chosen this direction are for example Peoplesoft together with Agilera and Corio amongst others, and Siebel together with ManagedOps and Corio as well.

16.5.11 Market Overview

The following section gives an overview of companies that sell CRM software. The overview also contains some ERP system retailers who offer CRM functionality as well.

According to an IDC study, the CRM market is quite fragmented, and beyond the top five suppliers in the market, no company holds more than 2l market share.

Major suppliers are Siebel and BOPS: Baan, Oracle, Peoplesoft (Vantive), and SAP. These vendors all offer complete CRM packages and an advanced integration possibilities with ERP systems. They all aim at the broad market and not at vertical industries, so the actual functionality of the packages is similar.

Other vendors of CRM software worth mentioning are: Pivotal, E.piphany, eGain, Ivensys, Nortel, Clarify, SalesLogix, Onyx Software, Kana, Silknet Software.

Some of these CRM vendors specialize in a certain discipline:

Pivotal specializes in midsize enterprises and demand chain management and cooperates with Microsoft's.NET initiative. E.piphany specializes in customizable extractors and integrating data from disparate source systems, while eGain's specializes in integration of unstructured data sources.

Important integration cooperations in the CRM (and ERP) market are Peoplesoft with Vantive, SAP and J.D. Edwards with Siebel, Baan with Ivensys and Oracle with CSC (Computer Sciences Corporation).

There is also a high degree of cooperation between CRM and ERP vendors with consulting firms like KPMG consulting, Deloitte & Touche, PriceWaterhouseCoopers, and Cap Gemini.

16.5.12 Connection between ERP and CRM

Brief History of ERP and CRM

ERP systems in the widest sense have been around since the 1960s. First, for inventory handling in the 1970s for MRP (Material Requirement Planning) and in the 1980s for MRP-II (Manufacturing Resource Planning). Since the 1990s and especially with the upcoming of the Internet as a mass media, ERP is continually being enhanced with capabilities in the fields of engineering, project management, HRM, finance, and CRM.

The trend of ERP vendors integrating best of breed functionalities into their products is not so advanced in the field of CRM as it is in other fields of business. Today, some ERP packages still do not contain CRM capabilities. That is why there are still several best of breed vendors of CRM applications as well.

Integration of ERP and CRM

While both ERP and CRM arose in order to solve integration problems, neither was created to integrate with the other. ERP concentrates on back-office functions and data sources, while CRM concentrates on the front-office with collection and management of customer and sales information. Their coexistence and the differences between the goals and the functionality are a challenge in creating a single integrated system.

The data that is necessary to feed a CRM system in order to make it effective, such as materials billing and order history, is often stored all over the enterprise and in the best case embedded in an ERP system. Even then, the collection of that information is far from straightforward. In practice, enterprises often choose the ERP and CRM systems that fit them best and hire yet another party to hook up everything.

There is however a growing consciousness of the need for close cooperation between ERP and CRM. The result is that ERP vendors cooperate with CRM vendors, or simply buy them, like Peoplesoft bought Vantive. CRM vendors for their part develop more integration tools and APIs (Application Programming Interface) to support integration.

Other important challenges in the integration process are data volume and speed. The amount of data an ERP system has to handle increases significantly when extended with CRM. The system has to be able to process it all in a short time; otherwise the aimed benefits cannot be obtained. If a customer calls a company and speaks to a representative, and then calls back several minutes later and speaks to someone else, the customer expects that person to know about the previous call and its outcome.

To summarize the challenge of integration: ERP needs to acquire the immediacy of CRM, while CRM must get access to and use more of the back-office information stored in the ERP system. Even though sharing data is important, ERP and CRM systems should connect at several points, including the financial, order entry, purchasing, inventory, manufacturing management, human resources management, and logistics functions, as each of these components plays an integral role in the entire customer interaction and fulfillment process.

Ways of Integrating CRM into an ERP system

Depending on how CRM is integrated in the ERP package, the customer's data can be included in the ERP database or kept in a separate database. In the latter case, there should still be a close connection with the ERP database.

The following paragraphs explain different possibilities of integrating CRM functionality and data into an ERP system.

Full integration: Full integration is something like the holy grail of CRM. The ideal is that the ERP vendor delivers a product in which you configure specific CRM functionality just like current components. There is a single, central database, on which the system operates. The borders between ERP and CRM become more and more vague toward this ideal. Currently there is no vendor of a true front-to-back solution, according to Web/CRM analyst Dennis Pombriant, even though some vendors claim to have reached (near) full integration, such as Oracle and Peoplesoft.

Modules: Some ERP vendors offer a module for CRM next to their standard ERP packages. The module is developed to work solely with the corresponding ERP. As a result, the integration level is quite high. The CRM functionality operates on the database already in use. SAP for example offers mySAP CRM as an extension of its standard package.

Bolt-ons and interfaces for the 3rd party product: Best-of-breed CRM vendors are becoming more and more aware of the fact that their clients use ERP. For quite some time now, they design or tailor their products to connect with certain third-party ERP systems as bolt-ons. There are also co-operations between ERP and CRM vendors for developing an interface, which lead to a shorter implementation time and better integration. For example, Siebel's CRM integrates with Epicor ERP.

Existing ERP with CRM from ASP: There are some ASPs who can handle the integration of hosted CRM into an existing CRM system, but most are still oriented toward simply providing turnkey CRM applications. Analysts Caruso and Pombriant warn that trying to integrate CRM from an ASP adds challenges to connecting with the back-office and burdens end-user enterprise with the integration or costs for yet another firm to do the integration for them.

16.5.13 Benefits of CRM to the Enterprise

To avoid too much repetition, this paragraph only handles benefits that come forth from integration with an ERP system.

The central storage of CRM data within the ERP system enables every employee to access consistent and complete customer data at any time.

At the same time, it enables a client to access his data for a longer time as well. The customer order decoupling point (CODP) is brought closer to the point of delivery. The CODP is a virtual point in the supply chain where a client cannot control his order anymore. The closer the CODP is to delivery, the more customized a product or service can be delivered. When an enterprise can offer assembled-to-order or even engineered-to-order products and services, it gains an advantage against competitors who only offer off-the-shelf products and services.

The supply chain can also be supported by information exchange and cooperation between order management and resource scheduling. Information exchange and cooperation between customer service, bill payment, and financial management support administration and management.

16.5.14 Future of CRM

Trends

For a large part of this section, we leave forecasting of future trends to the experts.

Increasing integration: An important trend already taking place is the effort to increase integration of CRM in enterprises' businesses, so of course in their ERP systems as well. Another development that goes hand in hand with integration is "migration of functions that were previously performed in the back office out to the front office at a point closer to the customer – very often at the point of customer contact," according to Dick Lee, principal of High-Yield Marketing.

Denis Pombriant, Web/CRM analyst with Boston's Aberdeen Group, warns that, for the most part, no one has yet offered a full front-to-back CRM-ERP solution. "It's still very much a best-of-breed marketplace." This holds true at the moment, but vendors are working hard from several sides (ERP, CRM, consulting, etc.) to offer completely integrated packages that enterprises will gladly buy. ERP vendors cooperate with CRM vendors or simply buy them, CRM vendors develop more integration tools and APIs to support integration and (IT-) consulting firms' advice goes into the direction of more integration as well.

Vertical CRM: CRM software vendors will begin crafting packages targeted at distinct, vertical markets in order to fit different types of customers, business models, and sales strategies. The vendors will offer more and more special solutions for certain industries, instead of forcing the industries to customize the software products to fit their needs. Lee believes that the number of CRM vendors will shrink significantly due to the disappearance of many companies that are looking for fast profit from the hype and offer the same type of products.

Less CRM via ASP: Erin Kinikin, vice president of the research firm Giga Information Group of Cambridge, states why CRM hosting came up in the first place: "We look at outsourcing as filling two key functions: as a way for a beginning company to meet a competitive need to get going quickly, and as a method of cost-reduction at companies that've decided that customers are not very important. But it's not a strategic move for most businesses."

Section 4.2 mentions the trend that CRM vendors will offer ASP services themselves or closely work together with existing ASPs, but CRM hosting will become less and less important. Enterprises are discovering that good relationships with their customers as a strategic asset. Outsourcing customer relationship management would mean separating the customer from one's business.

Common practice CRM: The lines between CRM and other information-based management disciplines, particularly ERP, are blurring. "There is definitely a broadening of the definition," concedes Jon Wurfl, CRM evangelist at SAP. Dick Lee's vision on the future of CRM seems to summarize the current upbeat attitude toward CRM: "I don't think the term 'CRM' is going to be as predominant several years from now. It's going to simply be what we need to do."

Market growth prognosis

Businesses and research institutes alike are enthusiastic about CRM. There are a large number of reports on the growth of both ERP and CRM markets. Quite often, forecasts vary between them, depending on which segments of the markets were included (software, services, consulting, etc.).

ERP Market

The top five ERP vendors, SAP, Oracle, Peoplesoft, Baan, and J.D. Edwards, account for 61l of total ERP market revenue. Their predominance in the ERP market is likely to persist in the coming years.

AMR Research predicts that the enterprise resource planning (ERP) market will grow at a compound annual growth rate of 32l over the next three years. Total company revenue will reach $66.6 billion by 2003.

Another AMR's research predicts that ERP market growth will slow to 5l a year, increasing from $16.9 billion in 1999 to $21.4 billion in 2004. The reasons for the decrease of the ERP market are the Y2K problems and the introduction of the Euro, which induced enterprises to invest heavily in ERP systems, in order to tackle several issues at once. They are now refocusing strategic technology plans to extend ERP throughout global supply chains to gain competitive advantage.

CRM Market

The CRM market is booming as a whole, which is also a result of increased average budgets for single CRM initiatives. *Dataquest* says the average had reached $1 million by the end of 2000 and that it is expected to have doubled until the end of 2001.

A report from the Meta Group predicts a 50l annual growth rate for the global CRM market. Meta Group analysts expect demand for CRM software, consulting, and systems integration services to increase from $13 billion in 2000 to $67 billion in 2004.

IDC (International Data Corp) reports the CRM segment is expected to grow at a compound annual growth rate of 29l through 2004. In addition, the demand for CRM solutions among midmarket companies will result in a market uplift for implementation services in the second half of 2001, continuing through the year 2004.

IDC estimates that $40 billion was spent on CRM consulting, systems integration, and outsourcing in 1999, and predicted the market would grow to $90 billion by 2003.

AMR Research expects the CRM market to grow to more than $20 billion by 2004 making it as large as the ERP market.

North America is the largest CRM market and held more than 70l of the sector's revenues in 1999. Revenue growth in North America is expected to slow, however, and the region's market share will drop to 64l by 2004, the study found. Western Europe's CRM market will see the most rapid revenue growth, at about 36l, in the next five years, compared to 30l for the sector overall.

16.6 Data Management in Analytical Customer Relationship Management

Customer relationship management (CRM) is a strategy to acquire new customers, retain them, and recover them if they defected. The corresponding CRM goals can only be achieved if the right data sources are combined. This section discusses what external and internal data are available along the CRM process and how they support the achievement of the specific CRM goals. Starting with defining a CRM process model and the belonging goals within the CRM programs – acquisition, loyalty, and recovery – we explain the internal data situation. From here, we derive the need for external data and how one can merge and manage the information along the CRM process. The aim is to provide a rough guideline for the selection and combination of data sources among the CRM programs and to give hints how to overcome possible problems.

Uniform products, along with individualization of customers, have brought pressure for change in marketing practices. In the automotive industry CRM generates additional product benefits by means of communication and services that are designed and delivered to match the individual needs of customers. This is one of the main goals of customer relationship management (CRM).

Although CRM is an advanced concept, its implementation still requires the development of feasible approaches. From a practical point of view, a crucial issue concerns the systematical collection, storage, usage, and continuous improvement of customer data. The only companies able to construct lasting relationships with their customers are those that properly process and maintain an adequate volume of customer information.

The overall aim of this section is to discuss the different data situations along the CRM process and to give hints on how to overcome data problems. We try to explain how one has to consider the whole process when handling data within a certain stage. These suggestions are valid for many CRM-oriented companies. But, for the ease of comprehension and to cover the specific issues of automotive companies, we restrict ourselves to this industry. In addition, we focus solely on private customers.

In this section we aim at generating a homogeneous and complete understanding of the CRM process, since there is no generally accepted approach concerning this topic. Afterward, we describe what data sources can be used in the car industry and how their importance varies along the CRM process. Also we discuss options for handling these sources in order to fulfill the activities of operational CRM (oCRM). Here we talk about both single activities and long-term strategy. We conclude by summarizing the main points and outlining open issues for further research.

16.6.1 The CRM Process Model

According to one of the earliest definitions **R**elationship **M**arketing (RM) is a "... strategy to attract, retain and enhance customer relationships." The term CRM is a later version of RM, having similar meaning, but used differently in literature. Within this section we use these terms synonymously. Operational CRM includes all activities concerning the *direct customer contact*, such as campaigns, hotlines, or customer clubs. Every oCRM activity is generally implemented in one of the three enterprise processes: sales, marketing, or service, since these are the processes concerned with direct customer contact.

Analytical CRM (aCRM) provides all components to *analyze customer characteristics* (behaviors) in order to accomplish oCRM activities, with respect to the customers' needs and expectations. There, the idealistic goal is to provide all information necessary to create a tailored cross-channel dialog with each single customer on the basis of his or her actual reactions. To reach

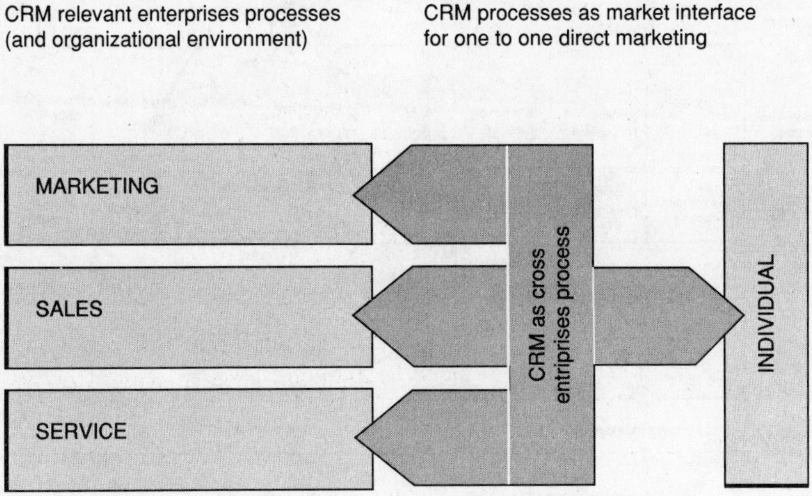

Fig. 16.1. CRM as cross-functional process

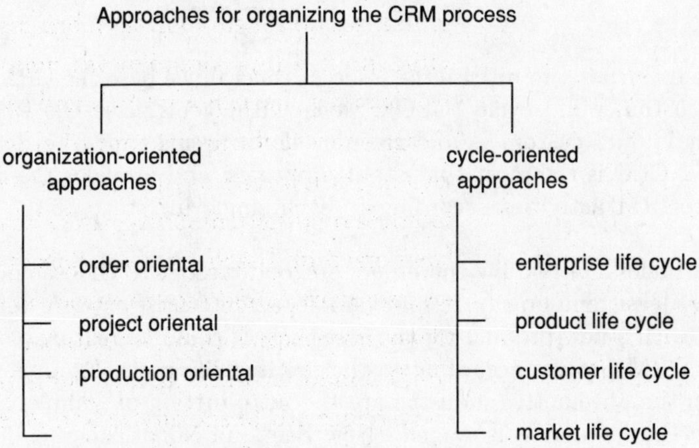

Fig. 16.2. General possibilities of the CRM process

this goal and aiming to show only one company image to the customer, it is necessary to look at CRM (oCRM and aCRM) as a cross-enterprise process. Marketing, sales, and service departments have to coordinate their responsibilities, activities, information systems, and data. Figure 16.1 illustrates this demand.

There are many theoretical ways to organize the cross-functional CRM process within an enterprise. Generally, we divide the approaches into two broad clusters: *organization-oriented* approaches and *cycle-oriented* approaches. Figure 16.2 gives an overview.

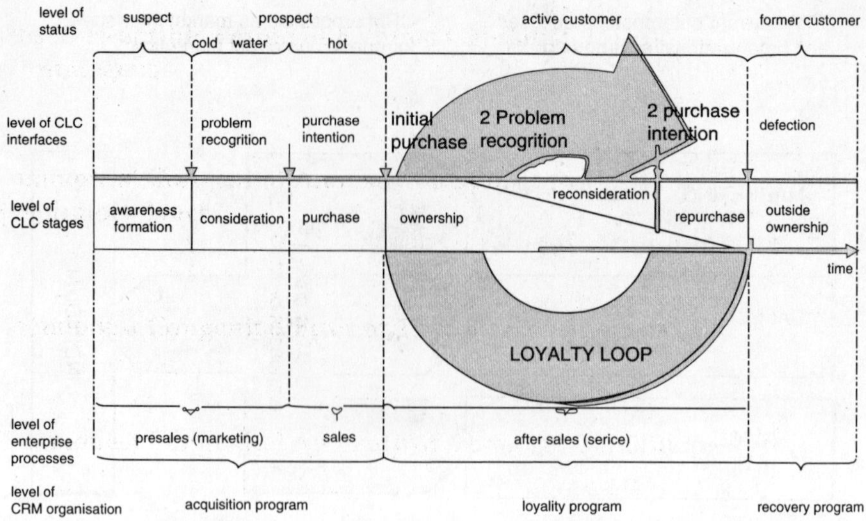

Fig. 16.3. CRM process based on customer life cycle

For our purpose, to discuss the issue in general, we base the CRM process on the customer life cycle (CLC). Since CRM centralizes the satisfaction of the individual customers and their needs, it seems logical to do so. Furthermore, CLC is the most time-stable approach and provides the ability to arrange oCRM activities according to each single life stage of the targeted subjects.

Nevertheless, in real life there may be good reasons to choose another approach, at least temporarily. If, for example, an enterprise enters a new market segment with a new product for the first time, it could be reasonable to organize the CRM process according to the product life cycle. Examples of such events in the automotive industry are the introduction of DaimlerChrysler's Smart in Europe and Volkswagen's New Beetle in North America.

Before enterprises can develop marketing or CRM strategies, they must understand how consumers make their purchase decisions. This decision process is called customer buying cycle (CBC). We assume that the chain of all CBCs a single customer runs through is his or her customer life cycle. The process ends with the final stop of consumption. Figure 16.3 illustrates the overall system and is described below.

We start with explaining the level of CLC stages. As mentioned before, the CLC refers to the various stages of the relationship between businesses and a (potential) customer. Now we take a closer look at these stages, assuming that the product in question is a car.

Each relationship starts theoretically long before the first "physical" contact. During the awareness formation stage a person is called *suspect* and gets in touch with lots of information on available brands in a determined market.

From there the suspect builds a set of product brands known to him and suitable to satisfy his needs in general. This set is called the *awareness set*.

The first stage ends when the suspect recognizes a problem, an unsatisfied need, or desire and starts to seriously consider several alternatives. Based on the individual criteria and their importance, he evaluates and ranks the brands. Therefore, he typically gets in touch with the respective companies and turns into a "prospect." The result of this stage is the prospects-evoked set.

By forming a concrete purchase intention, the person becomes a "hot prospect" (sometimes called hot lead) and enters the purchase phase. Here, he decides what kind of car or brand to buy, chooses where to buy the product, and negotiates the terms of sale (e.g., price, delivery, credit arrangement). This stage is closed by the initial purchase. Now the prospect has become an active customer and starts to use the car. He evaluates it in order to determine whether it is performing as expected. The outcome of this evaluation lies in a range between satisfaction and dissatisfaction, which influences subsequent CBCs.

After a certain time, the customer begins to consider replacement. If he is satisfied with his car experiences, he is most likely to remain an active customer. With the loss of the actual car (selling or stop of usage) the ownership stage of the first CBC ends and the person either stays with the company (loyalty loop) or ceases being a customer. For the latter, he might come back during any subsequent CBC.

As mentioned above, the CRM-relevant enterprise processes are cross-functional to the CLC-based CRM process. But they can be differentiated depending on their target groups and depending on the events "direct customer contact" and "purchase." This enables us to relate them to our process model, at least for one CBC. The first direct customer contact might be in the awareness, consideration, or purchase stages. To simplify the model, we assume that the first direct customer contact is identical to the purchase intention point. Therefore, we can construct the relationship shown on the level of enterprise processes of Fig. 16.3.

In order to organize the oCRM and aCRM activities along the CRM process, they are implemented as separate programs (as illustrated in Fig. 16.3 on the level of CRM organization) with clear interfaces, special goals, and corresponding direct marketing activities, like acquisition campaigns or road shows. Nevertheless, the programs have to be coordinated closely, as we discuss later. Figure 16.4 illustrates how the program-related CRM goals fit into the general goal pyramid of the enterprise.

Within the operational CRM goals we distinguish between goals related to single programs, e.g., increase of usage or cross selling for the loyalty program, and overlapping goals, such as efficient use of the budget or creating customer insights. Both require strong interactions among the CRM programs, since program-related goals partly need data gathered at some time in the process. For example, customers with a high probability to defect can be better

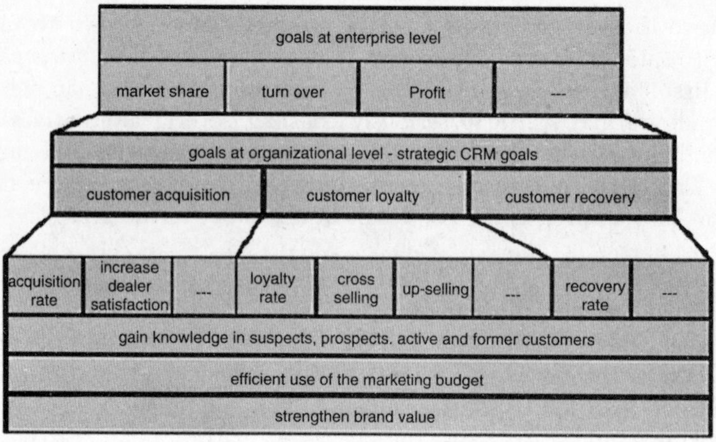

Fig. 16.4. Systematic CRM goals

detected in the loyalty program, if patterns of already defected customers and reasons for defection are known from the recovery program. Summing up, we may say that the synergetic effects of program interactions should not be ignored. We refer to this point in later part of this section again.

16.6.2 Data Sources for Analytical CRM

Analytical CRM supports the oCRM activities, as discussed before, through systematical collection, storage, and evaluation of data. Data management is, therefore, the starting point for effective CRM and will continue to grow in scale and importance. We have to consider three aspects: the quantity, the quality, and the actuality of data available. Based on the continuous cycle of data management, we learn about people's behavior and needs. We want to use this knowledge to optimize our oCRM activities through targeting the right people with the right information and offers at the right time.

In order to provide high-quality information for the CRM programs, we must discover first-rate customer-related data and evaluate them with suitable data mining techniques. Even if it seems that there is a wide variety of data, it is hard to find and merge the "right" data, even within the own company. Generally, we can find data sources within and outside the enterprise. Typically, internal data is considered more valuable data, because it reveals true insights belonging to our company and products. Additionally, we aim to generate a competitive internal data situation in the long run. Since competitors have the same access to external databases as we do, it seems to be the only way to generate real advantages.

In the first column of Fig. 16.5 we list the main information categories concerning customer characteristics. The most important examples for external sources, as used in automotive industry, are shown on the right-hand side of

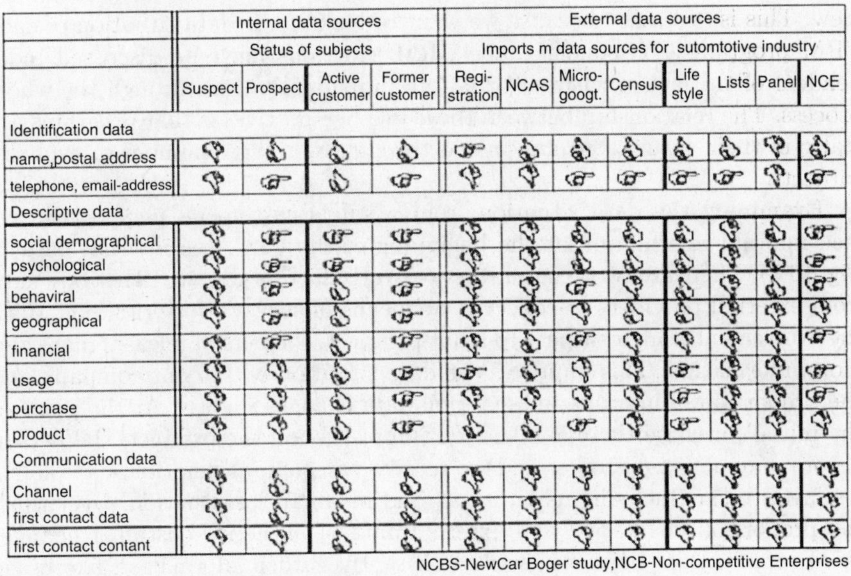

Fig. 16.5. Overview of general data sources

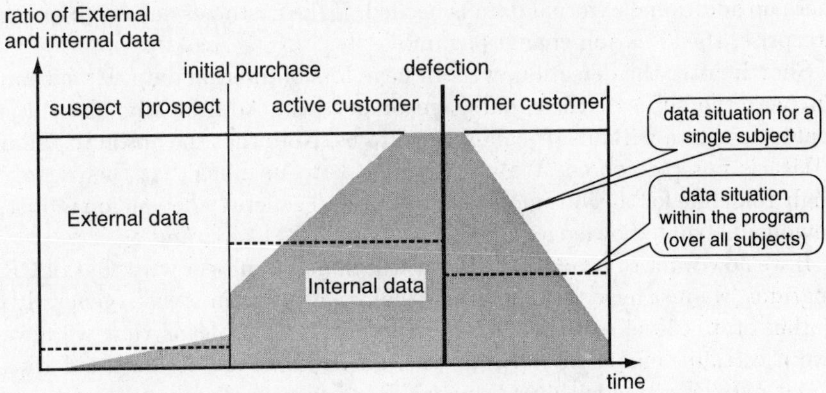

Fig. 16.6. The general data situation along the CRM process

the figure. For internal data we show only the availability during the CLC, as they can be stored at different databases in any company. The hands indicate the internal data, when the data categories are most likely to be available. For external data they give evidence as to which data sources these categories can be found in. Because it is very difficult to measure information quality, they do not indicate the excellence of information.

After the short overview, we now want to relate the potential information sources to the CRM process. In order to do so, we start by describing the data situation for the CLC of a *single person* (illustrated by the gray area in Fig. 16.6). After that, we change the perspective to the *enterprise*'s point of view. This is necessary because we want to look at the data situation of each CRM program. Consequently, the CRM programs must be discussed independent of the fact of whether or not an individual passes through the whole process. The relationship between these two perspectives is that programs are static in their nature, but are passed through by a large number of evolving subjects.

Examining the data situation from a single customer's perspective, we lack internal information at the beginning of the CLC (see internal data in Fig. 16.5). There are two general ways to overcome this gap: we either buy data from external providers or collect it by means of surveys, lotteries, etc. With the evolvement from a suspect to a hot prospect, the person releases more and more information. This improves the data situation within our company and the proportion of internal and external data changes slightly. At the point of the initial purchase there is a sudden jump. Buying the product, the person becomes an active customer and we receive personal information, e.g., about his favorite car (including equipment) and his financial situation. Over time, the proportion of internal and external data of an active customer changes more rapidly than before. Now we constantly gather information like usage behavior or service interest. If the customer stays active for several buying cycles we are theoretically able to develop the internal data up to a point where no additional external data is needed. If the customer defects to another enterprise, the situation changes again.

Shortly after the defection, we still have lots of internal data. If we cannot win back the subject, these data age or get lost. Additionally, the subject creates new data without revealing them to us. Note that the absolute amount of data is not considered. We also have to keep in mind that this is only a rough guideline for the automotive industry, and generally depend on industry, previous internal knowledge, CRM goals, and CRM programs.

If we now want to connect the data situation of a subject with that of CRM programs, we just have to understand that each program focuses generally on another status level of the CLC (see Fig. 16.3). This means that we always have a certain number of subjects with an identical status level and with a similar data situation within a specific CRM program. Consequently, the data situation of each program is almost static over time. If we take all subjects belonging to a certain CRM program for a certain spot in time and calculate the data average (in terms of quality, quantity, and actuality), the results are the dotted lines in Fig. 16.6. As it becomes clear, we have different information levels of internal data amongst the programs. Therefore, we need more or less external data and for that reason the ration of external and internal data varies also.

16.6.3 Data Integration in Analytical CRM

In this subsection, we explain how the data described before can properly support the CRM process. To do so, we discuss the single programs as well as overlapping issues. Section 4.1 clarifies basic requirements and assumptions. From Sections 4.2 to 4.4 we look at each CRM program. Before we outline special questions related to the corresponding program, we first restate the specific goals and then point out the internal and external data situation. Section ?? deals with the program overlapping goals and the resulting interactions between CRM programs.

Basic Remarks

The most likely situation is that the CRM programs have to be developed for companies that already exist. In this case, the company must first capture the status quo of the internal data. Depending on their quality in terms of accuracy, actuality, completeness, etc. external data have to be purchased. Two or even all three CRM programs could be started simultaneously. Then, benefits of interactions between the programs are possible from the very beginning. But in companies that already exist, we face difficulties with existing (nonoptimal) conditions, e.g., in terms of internal data quality or given information systems.

If a company starts with CRM for the first time, in theory it could start from scratch. This would mean that there are no existing customers and, therefore, no customer data. Naturally, the company in question would first develop an acquisition program, immediately buy external data, and form the relationship with the new customers along their CLC. Although this situation is possible (e.g., when smart AG entered the car market for the first time in the 1990s), it is rare, especially in established industries. For the purpose of this section we discuss both situations if there are any significant differences.

Acquisition

Acquisition mainly aims at establishing a dialog with the suspects belonging to predefined target groups and at gradually converting them from suspects into customers. When a suspect goes through the acquisition program, the major challenge is to become our brand part of the awareness and later the evoked set of the subject. If the subject does not leave the acquisition program unplanned (e.g., because of no interest or unrecognized change of address), the program ends with the initial purchase (see Fig. 16.3).

As shown in Fig. 16.6, external data sources play an important role, especially in the acquisition program. Therefore, we would want to take a closer look at their suitability. In order to rank data sources in terms of their usefulness we have to find ranking criteria. The researchers explained that the measurement of information quality is a complex task with various facets. Aiming to make a first rating of the general suitability of an external data source for acquisition, we try to restrict ourselves to four main aspects here. These are the number and the quality of descriptive variables, the market coverage, the costs per contact, and the expected response rate for suspects.

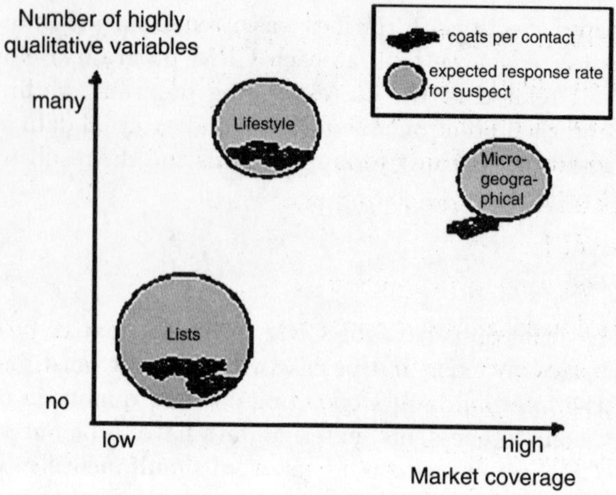

Fig. 16.7. Ranking portfolio for most important external data sources

In order to run *continuous acquisition programs*, we need many new addresses every time and hence, the *market coverage* is a fact to take into account. Because we *lack internal data* about suspects as well as prospects we have to buy this information. For that, the *number and quality of descriptive variables* is important in acquisition. Especially at the beginning of a CRM project we are often obliged to proof its *usefulness and efficiency* within the company. Thus, we must consider the *expected response rate* and of course the overall *costs per contact*.

Considering the market coverage for passenger cars, registration data (for legal aspects), NCBS data (New Car Buyer Survey), panel data, and data from noncompetitive enterprises (NCE) can only be used as supplements here. Census data is not only highly aggregated, but also mostly available within microgeographical data sources. If we focus on the remaining data sources (cf. Fig. 16.5), we receive the portfolio illustrated in Fig. 16.7.

Now we want to see how these data sources fall in place during acquisition. An ongoing aim of acquisition programs is the generation of new suspects. As we have learned in section 3 (Fig. 16.6), usually we have only internal information about them. As a result, in acquisition one aims to buy external data continuously.

If a company starts acquisition with direct marketing for the first time or is new in business, normally it lacks data about responders or customers, and thus it misses a proper target variable for the use of data mining techniques like classification or predictive modeling. Consequently, the company does not need any descriptive variables for prediction. Furthermore, because it is just starting acquisition campaigns and therefore did not use lots of publicly available addresses in the past, market coverage is not very relevant. Considering these two aspects as well as the highly expected response rates for properly selected lists, we recommend using them. This will help to achieve quick results in terms of generated prospects (responders) and to strengthen the position of CRM within the enterprise.

Nevertheless, there are two aspects to bear in mind about lists. First, addresses generated from lists are contacted frequently by many enterprises and therefore show a certain saturation. And second, suitable lists are limited and certain social classes are neglected. In the case of running acquisition programs the potential value of lists is less. Because of their small market coverage, appropriate lists will at times be nearly exhausted. We can still use lists as frequent sources for the acquisition program, because there will appear new needs for the subjects already on the lists, new subjects in existing lists or completely new lists, but we also have to look for alternatives.

After running the first campaigns, the internal data situation is improved. The reactions to our mailings and the identification data (name, address, etc.) of the contacted suspects from the lists are the core of our internal database. Additional data can be added, if we send questionnaires with the mailings. They inquire, e.g., about the next replacement need, the product, and the communication preferences of the prospect and give hints how to correspond with him properly. Since during acquisition we aim mainly at distinguishing between potential customers and noncustomers, we focus on data mining techniques like classification and prediction.

Because of the now-improved data situation we have positive responders and therefore a valid target variable. This can be used to predict the response behavior of subjects from other data sources. In order to do so, we have to find data that refer to both our responders and the targeted subjects. For this reason, it is advisable to refer to external data sources again. If we look at the dimensions of our portfolio from Fig. 16.7, we now face new requirements. For generating predictive models we clearly need descriptive variables, although the specific requirements vary for different data mining approaches and techniques.

Additionally, we call for high market coverage. This is essential for maintaining ongoing acquisition campaigns. Microgeographical and lifestyle data mostly fulfill the two requests. Because of the higher expected response rates, it makes sense to start using lifestyle data. If the data is exhausted, microgeographical data can take their place. The more knowledge we possess about the responders (and responders that become customers), the finer we can select new subjects to contact. Therefore, the usually lower response rates of microgeographical data compared to lists and lifestyle data can be partly compensated by better targeting (and lower address costs).

Loyalty

Loyalty clearly focuses on establishing a permanent dialog with active customers. The aim is to build a mutually profitable long-term relationship. Therefore, we pursue operational goals like increasing usage, upselling, and repurchase rates. A person becomes a customer and therefore a potential target for the loyalty program, if he signs a contract (initial purchase). He stays within the program as long as he remains an active customer. With defection to another company (voluntary or forced) he will be handed over to the recovery program (Fig. 16.3).

From a data point of view, loyalty is the key program of CRM. Neither during acquisition nor recovery are there so many and such close contacts to customers. As a result we can gather lots of internal data with supposedly high quality. Therefore, the ratio between internal and external data will change over time, in favor of the internal data (cf. section 3). If many customers stay with our company and pass through several CBCs, theoretically there is a point where no additional external data is needed.

Taking a closer look at the quantitative and qualitative growth of internal data, we discover three general ways for improvement. First, we can collect information (like sociological or financial data) freely given by the customer. This data has very high reliability, if the customer has to prove the accurateness (e.g., for leasing or credit purposes). The data is less reliable if it is based on volunteer statements (e.g., from questionnaires). The second way for obtaining data is observing customer behavior. Here we get hard facts like mileage, product characteristics, or use of services. As the third category there is data derived from the internal database. For deriving the data, we use basic statistical or data mining techniques like regression or neural networks (e.g., for building averages, or estimating and prediction of variables).

Because of the growing internal data, the need for external data decreases the longer the loyalty program runs and it is clearly lower than in the acquisition program. Nevertheless, there may always be a good reason to buy external data. As we explain here, there are two basic applications for external data in the loyalty program.

First, we can buy external data on a regular basis and thus extend the customer database with valuable information, *permanently.* For example, we could add socio-demographical data from microgeographical sources in order to generate selection criteria for direct marketing activities. For applications like these, we still need data that refers to most of the customers, and consequently, we need high market coverage as found by microgeographical or lifestyle data sources. Lists do not represent an additional value here, because they lack the descriptive variables and market coverage.

Second, we can use external data on a *nonpermanent* base, e.g., for special tasks in market research. For these tasks we work mostly with samples. Consequently, we do not need high market coverage. Data sources with special topic-related information (e.g., NCBS or panel data) have higher priority.

After talking about the general data situation in loyalty, we now want to look at the main tasks to fulfill. During his ownership, the customer is exposed to a variety of influences that can have an impact on his attitudes concerning our company, brands, or products. Such influences are called *triggering events.* There are two groups of triggering events we would like to point out, because

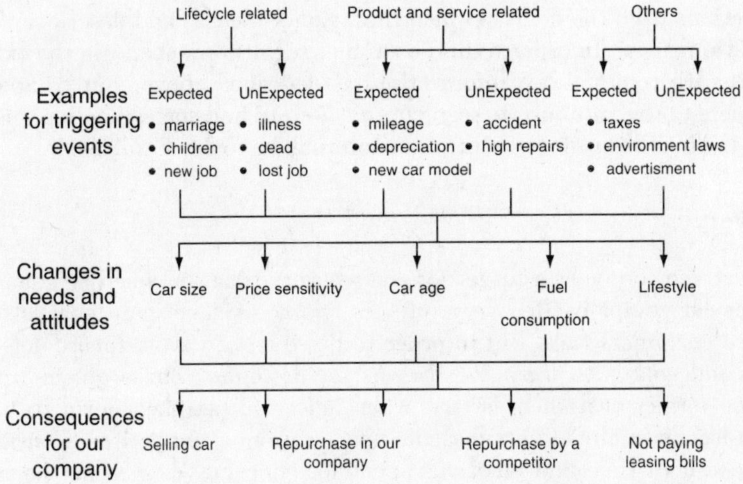

Fig. 16.8. Examples of triggering events and their consequences

of their high impact on the customer-to-business relationship. The first group contains events related to the *lifecycle* of the customer himself, such as marriage, death, or unemployment. The second group encloses *product (car)- or service*-related events (e.g., accidents, car maintenance, or mileage).

Product- and service-related events can sometimes be caused by lifecycle events. As shown in Fig. 16.8, all triggering events can be distinguished by expected and unexpected events as well. A single triggering event or a combination of events can result in fundamental changes in the customer's needs and attitudes and therefore, his expectations and actions regarding the relationship to us may change, with serious consequences (see Fig. 16.8).

One of the main tasks of aCRM for loyalty programs is the prediction (e.g., use of decision trees for predictive modeling) and recognition of triggering events (e.g., use of residual analysis for derivation recognition). Not only do we have to *prevent undesired consequences*, but additionally, triggering events often provide opportunities for *enhancing customer relationships.* The goal of prediction is hard to accomplish for expected events and barely possible for unexpected events (How to predict an accident?). For the latter we are mostly restricted to event recognition. Another approach could be the evaluation of likelihood classes as used, e.g., for car insurance policies. For the detection of triggering events we refer mostly to the use of internal data. This should be more detailed and have higher actuality than most external sources (Fig. 16.5). Examples for such triggering events are a sudden change of annual mileage that indicates different usage habits or a modified email address that refers to a new company. External data might be used additionally for explaining such indicators (e.g., microgeographical data can show new housing conditions).

Another important task for aCRM in loyalty is to provide customer insights for the tailored customer dialog. For that, we primarily use data mining applications from the fields of clustering (e.g., k-means) and description (e.g., association rules). In order to find customer segments we can use the external sources with topic-related information, as described above. But to apply all customers to the appropriate segments we need to find corresponding selection criteria, which is available within our internal customer database.

Recovery

The customer becomes a target for the recovery program when he changes to another car company. Recovery pursues two goals. The main goal is to *win defected customers back*. But in order to do so and to avoid future defection, the second goal is to *learn the reasons for defection*. The problem with recovering former customers is that when defection become known to us, the person has often already purchased a new car from a competitor. Considering the amount of time and emotions spent for choosing, as well as the money spent for acquiring the car, the customer will not sell the actual car in the near future. The average length of car ownership, e.g. in Germany, is approximately 3.5 years. That is why the goal of the dialog cannot consist in winning back the customer immediately, but in establishing and keeping contact with him until his next repurchase.

The final goal of the dialog is to gain the top position in his evoked set again. The reasons for defection must be questioned as one of the first steps of the recovery program. We need this information for planning the content and timelines of the recovering program and it is also helpful to prevent churn of actual customers in the loyalty program. Because the recovery program deals with former customers, we generally have detailed internal information about the targeted subjects, at least more information than in acquisition. But this is true only for subjects who defected recently. The more time that has elapsed since people defected from our company, the less valuable their data becomes (see Fig. 16.5). More concretely, new information about car characteristics, as well as car usage and financial data emerge, but not in our enterprise.

Unfortunately, such information is rarely available in public data sources. This implies that the lack of actual internal data cannot be compensated by external data. Additionally, the identification data becomes more and more obsolete and one day it is impossible to purchase external data because the customer's identification data is no longer valid. Regarding this debasing data situation, a recovery dialog must be established as soon as the defection becomes clear. Only when the customer has defected for a short time do we possess a lot of actual data about his contact address and his communication preferences. This is the precondition for a successful dialog. At this point we now know what do to when the defection is known already. In practice the customer rarely informs us of his defection.

For that, it is necessary to find proper identification variables. There are two fundamentally different types of churn. People can churn *voluntarily* if they do not want to be customers anymore (e.g., because they are unsatisfied with the product or got a good offer from a competitor) or *involuntarily* if they are forced to stop consumption (e.g., because of unemployment or illness). Both types of churn can occur for many different reasons and can be recognized by different outcomes.

Involuntary churn is recognized more easily (e.g., people stop to pay their bills or the car is wrecked in an accident). For the recovery program we want to recognize mainly voluntary defection, because voluntary churners are more likely to return and stay good customers. For that we mostly have to use indirect indicators. Examples of such indicators are the abrupt stop of service bills or if loyalty measures get unanswered suddenly. In order to find these hints within our internal data we use the techniques of deviation recognition or basic statistics. Before ending this section we want to point out that recognizing the reasons for defection is mainly part of the recovery program, but it is important for the loyalty program as well. The behavior patterns and indicators discovered in recovery are the basis for churn prevention in loyalty.

Interactions Between the CRM Programs

As mentioned already within the preceding subsections, there can be several interactions between the different programs. In this section, we want to examine them from different perspectives. Looking at the interactions of the CRM programs we have to consider the *objects* of interaction. Naturally, subjects evolve from suspects to customers and therefore pass through the different program levels. These subjects are represented through their data and consequently the objects of the exchange are data. The exchanged data can differ concerning its *reference level.* First, data can be exchanged on an *individual level.* This means that data about a specific customer gained in the acquisition program is used in loyalty as well. Second, the data exchange can take place on an *aggregated level.* This implies that insights gained about a certain group of people, for instance in loyalty, are used for marketing activities in the acquisition program.

Because external data is available for each program and can be bought any time, normally, we do not include it in the program interchange. But we must keep in mind that, when purchasing external data for one program, the external data needs in other programs can be taken into account. More concretely, it makes sense to consider data needs in loyalty and recovery programs when buying external data for the acquisition program and the other way around.

Individual internal data is exchanged mostly *forward oriented,* meaning that data gained in acquisition is used in loyalty and recovery. In fact the individual data is not transferred, but the customer file is enriched with all information gained about the subject during his CLC. Prospect data acquired during the acquisition program is retained and enriched with new customer data. If the customer churns, the inquired reasons for defection are stored in the same customer file as well. This type of data exchange is already assumed in the explanations within the programs and is not examined in detail here.

Aggregated data can also be exchanged forward oriented. But the insights gained through the evaluation of aggregated data are especially advantageous in *backward direction* as we explain.

For that, we want to consider the *reasons* for the data exchange. Of course, data is exchanged to gain further knowledge, but there can also be interaction to compare and adjust data. This means that individual data with similar content may be gathered in different programs. For example, we may inquire as to the preferred car characteristics of a prospect via a questionnaire in acquisition phase. When the prospect becomes a customer, we gain reliable knowledge about his chosen car characteristics. As discussed before, the latter are better than the volunteer nonproofed statements.

We can update information about preferred car characteristics with data about chosen car characteristics. This update, again, is forward oriented on an individual level. Backward oriented we can analyze the differences between announced and true product preferences in order to make these insights available in precedent programs. This point refers to aggregated data. Another example of backward-oriented exchange of aggregated and individual data is the use of defection data to identify customers likely to churn. Therefore, we need information about the typical profile of churners. The corresponding data emerges within the recovery program and is transferred to the loyalty program. There, the descriptive variables about customers are used as input variables in order to generate a predictive model. The output variable is the information regarding whether a customer has defected or not. The generated predictive model assigns a score to new customers that indicate their fit with the churner's profile.

It becomes clear that backward-oriented exchange of aggregated data often goes along with the use of data mining methods like classification, prediction, and clustering. Similar to the prediction methods just described, clustering permits distinguishing between different groups of responders, active customers, or churners. Deviation analysis must also be established on an aggregated level in order to incorporate a certain standard deviation in the model.

16.6.4 Further Research

This section has not focused on the question of how to distinguish between different data providers that offer the same or similar kind of external data. This problem arises after the needed data within the programs is determined. The selection of a wrong provider often can destroy the benefits of the precedent steps.

The thoughts were oriented on CRM programs in automotive industry. But the CRM process model and the data suggestions are applicable for other industries as well. Nevertheless, there are differences in goals and data between industries depending on their product (industrial, consumer, or utility goods) and their customers (businesses or end users). These differences will have an impact on the concrete design of the CRM process model, the CRM goals, the internal data situation, and the resulting need for external data. Further research should examine the differences and discuss the resulting consequences.

16.7 Summary

So far, the selection and integration of data within the CRM programs was rarely mentioned and hardly ever treated in marketing or data mining research. But for both domains, combining the right internal and external data is a guarantee for marketing success and for the use of data mining techniques within analytical CRM.

Most research has been restricted either to CRM strategy without discussing the practical issues of deploying the strategy or to characterize data independent on the business problem. Despite being really important in evaluating data sources, criteria like actuality, accuracy, and completeness resulted, and they do not give evidence as to which categories of data should be used depending on the marketing goal and on the internal data situation. In this chapter, we connected the CRM goals within acquisition, loyalty, and recovery to the internal data situation. Based on that, we derived the need for external data and the need for data exchange between the CRM programs.

We showed several levels at which the CRM process can be described. First, we pointed out the CLC level showing the possible evolvement from a suspect over a prospect to a (former) customer. The amount and quality of internally available data grows when the subject becomes an active customer. Second, we illustrated the organizational level with its programs. We showed the modest internal data situation in acquisition, which make additional external data a must. Loyalty and recovery also require external data – but to a lower extent and with different focal points.

Depending on the CRM goals we stated different requirements on the external data. While in acquisition market coverage is an important criterion, specific data about purchase history and behavior are much more important in loyalty and recovery. Furthermore, in order to reach the corresponding CRM goals the programs have to be coordinated closely especially for exchanging data and knowledge derived.

16.8 Review Questions

1. Define customer relationship management (CRM) and state its uses.
2. What are the key rules used in implementing CRM?
3. What are the general categories of separate behavior in CRM?
4. How are data mining models built using response behaviors?
5. Explain the application of data mining in marketing.

6. What are arguments for and against the sharing of customer knowledge?
7. Define customer value management (CVM).
8. Define global customer relationship management (GCRM). What are its issues?
9. Write a short note on content management and the core competencies in CRM.
10. Explain in detail on the fundamental skills in CRM.
11. State the reasons for CRM and the key elements required for success in CRM.
12. How does data mining help in database marketing?
13. What are the consequences of implementation of CRM?
14. State some of the technologies used in CRM.
15. Explain the connection between enterprise resource planning (ERP) and CRM.
16. What are the benefits of CRM to the enterprise?
17. Differentiate the market strategy of CRM and ERP.
18. Explain CRM as cross-functional process.
19. Explain CRM process based on customer life cycle.
20. How is data integrated in analytical CRM?

17

Data Mining in Business

Objectives:

- To help data engineers in a large corporation investigate the bad debts database and uncover useful patterns in selecting targets for debt recovery, thereby dramatically improving the corporation's debt recovery.
- To understand the difference between the results of the "average" practitioner and the "quality" practitioner
- To find the right balance between software, intellectual property, and so forth, is all part of the evolution of the industry.
- The business expert not only uses the results of data mining but also *evaluates* them, and this evaluation should be a continual source of *guidance* for the data mining process.
- The process must be thoroughly domain-oriented rather than technically oriented, and the tools must support an interactive, incremental, and iterative style of work.
- Data mining techniques can be implemented rapidly on existing software and hardware platforms across D&B to enhance the value of existing resources, and can be integrated with new products and systems as they are brought on-line.
- The commercial success of data mining lies in providing true value to the businessperson in a form that can be used and understood by the business community.
- Data mining tool aims to empower the business analyst to explore and understand the dataset in relation to his/her own knowledge, rather than aiming to replace the analyst with some automated data-discovery algorithm.

Introduction

In this section, data mining for business problems, business focus on data engineering, overview on business applications, and data mining in business case studies are presented.

S. Sumathi and S.N. Sivanandam: *Data Mining in Business*, Studies in Computational Intelligence (SCI) **29**, 387–409 (2006)
www.springerlink.com
© Springer-Verlag Berlin Heidelberg 2006

Data engineering is inherently a multidisciplinary field, because of the number of technologies involved: visualization, data analysis, knowledge engineering, perhaps databases, and courses the subject matter of the application. So there we have the technical background of the community, and some idea of the range of applications. What are the business implications here? A number of factors have emerged in the consulting works that are beginning to give a better understanding of the business nature of the discipline. First, the community has a number of different focuses. The business expert plays a critical role in data mining, both as an essential source of input (business knowledge) and as the consumer of the results of data mining. Data mining can reveal patterns in data, but only the business expert can judge their usefulness. It is important to remember that the data is not the business, but only a dim reflection of it. This gap, between the data and the business reality it represents, we call "the chasm of representation" to emphasize the effort needed to cross it. Also this section discusses a case study on "*An Overview of Data mining at Dun & Bradstreet (D&B)*".

17.1 Business Focus on Data Engineering

Data mining has emerged this decade as a key technology for areas such as business intelligence, marketing, and so forth. For the purposes of discussion, application, and business domains we consider here include telecommunications, medical devices, space science (vehicle health management and scientific instrumentation), targeted marketing, and mining.

From a technical view, we do not consider data mining to be a new field, but rather another discipline in the lengthy history of engineering sciences that use data is a core focus for developing knowledge. This family of disciplines we consider here comes under the term *data engineering.* Some traditional and nontraditional examples follow: Data engineers work with physicists in analyzing spectral data measured from a high-resolution imaging spectrometer develop sophisticated models of the spectrometer's complex error modalities (registration, response function, calibration, measurement glitches) so that a high-fidelity model of the spectrometer's measurements can be developed. Data engineers investigating the performance of an industrial strength place-and-route package uncover useful characteristics of the optimization process and thereby improve the performance of the algorithm. Data engineers work with astronomers in analyzing infrared data from an electronic star catalog. The analysis, in concert with the astronomer's interpretations reveals new, publishable classes of stars and also uncovers troublesome, never-before recognized artifacts with the original instrument. Data engineers in a large corporation investigate the bad debts database and uncover useful patterns in selecting targets for debt recovery, thereby dramatically improving the corporation's debt recovery.

At the time of the development, the individuals performing these tasks may have considered themselves applied machine learning researchers, decision analysts, statisticians, or neural network researchers; however they were all performing data engineering. We may have also head of the terms data mining and knowledge discovery, exploratory data analysis, intelligent data analysis, and so forth. These areas performing similar tasks however have a particular emphasis that distinguishes their origins, whether it be the applications they serve of the algorithms for data analysis that they use.

First, the community has a number of different focuses.

- First, there are more common turnkey applications. A well-known example is HNCs pushing into credit applications. We do not consider this data mining, but rather an application with a small component of data mining technology embedded.
- Second, there are general-purpose tools. Like the tools market everywhere, and as Kohavi (SGI – Silicon Graphics Institute) has said in his recent talks, this is a consolidating market.
- Then there are special purpose developments, which involve the technology we have referred to above as data engineering.

The experiences in this third focus period an interesting conundrum for the business manager. We find that in this third focus, there is a big difference between the results of the "average" practitioner and the "quality" practitioner. Every good software manager would know that a really good programmer can produce 100 times more code than an average programmer, partly due to the net result of subsequent maintenance, reduction in overhead and systems validation, and so forth. We find the same with data engineering. Except with data engineering, we find there are few key insights made in a project that make all the difference. Mundane use of the "usual tools" in the "usual manner" by the average practitioner gets one so far. But a big difference in performance is gained by the quality practitioner who makes a few key insights to change the project.

What are the business implications here? There are several.

- We often find businesses are reluctant to put outside specialists onto key technology development regarding perceived ownership of their intellectual property. If there core business involves software, they do not like to be seen as needing outside consultants developing part of their proprietary software. Network management companies are more reluctant to seek outside consultants, than say, medical devices companies. Because advances in these areas often need some key technology insights, the development in the network management company simply remains untouched unless they have an in-house, experienced, research and development group knowledgeable with the right methods (such as Microsoft, HP, etc.) We see this as damaging to the application domains themselves. Most importantly, we believe this effect favors the multinationals over the small businesses, usually considered the engine of innovation. This is a problem with the business handling of intellectual property.

- Second, we fear, but have not yet seen a backlash against data mining as had occurred for expert systems. Large corporations with ideal and potentially lucrative applications throw inexperienced staff together with the usual tools in poorly conceived projects. The broad experience in applications has shown that except in circumscribed vertical markets where complete systems can be developed by professionals the key insights required to make the difference can require the right professionals on the task. Thus many may suffer. We have anecdotal evidence here from data mining consultants. From a distance, business managers would blame the technology and not their implementation.
- Third, we are developing a software environment where we can address the kinds of developments we believe are necessary in those domains where specialist R & D is needed. MindSet from SGI, for instance, is successful because it provides an easy-to-use but powerful environment for visualizing data. Tools such as SAS rely on the large number of SAS-trained users in industry, for instance prior to the growth in data mining. The in-house focus for supporting consulting is to provide common platforms and tools, and plug compatible algorithms with modular components so that we can efficiently customize systems for the clients to focus on their needs.

Regardless of what happens to data mining as a community, we know that data engineering in one form or another will continue to remain a key enabling technology for many businesses, and thus finding the right balance between software, intellectual property, and so forth, is all part of the evolution of the industry.

17.2 Data Mining for Business Problems

As data mining and machine learning techniques are moving from research algorithms to business applications, it is becoming obvious that the acceptance of data mining systems into practical business problems relies heavily on the integration of the data mining system in the business process. Some key dimensions that data mining developers must address include understanding the business process from the end user perspective, understanding the environment in which the system will be applied, including end users throughout the lifecycle of the development process, and building user confidence and familiarity of the techniques.

One critical aspect of building a practical and useful system is showing that the techniques can tackle the business problem. Traditionally, machine learning and data mining research areas have used classification accuracy in some form to show that the techniques can predict better than chance. While this is necessary, it is not sufficient to sell data mining systems. The evaluation methods needs to more closely resemble how the system will work while in place.

One way to more closely evaluate data mining software in their intended setting is to incorporate time into the evaluation process. Although the time issue makes the prediction scenario more complicated, many data warehouses have data that is time dependent. For example, billing data stores billing, payment, and usage data for each customer indexed by time. As Kurt mentions in "Some thoughts on the current state of data mining software applications," few if any data mining systems deal with the time variable indirectly stored in the data.

Although the time dimension is left out during the prediction process, data mining systems should be evaluated with some aspects of time kept in mind. Researchers and developers can simulate time-dependent evaluation by evaluating models on historical data stored in the data warehouse. For example, suppose we are building a model to predict whether a customer will churn in a given month. Suppose we have the data for the independent variables at time t and we make a prediction for person x saying that they will churn. When will x churn? Will they churn immediately, in the next two weeks, in the next month? For what time period should we evaluate the model? An intuitive guess would say that the model is most accurate at predicting churn the closer it occurs in time to the independent data. When does the model predict the same as the background churn rate? While accuracy is a valid evaluation criterion, determining how long the model is valid is also important information. Instead of showing the accuracy of a model as a single number, the accuracy could be shown as a function of time. This information can also be used to compare different data mining techniques. The characteristics of the model should be tested while increasing time to better simulate the data mining software while used in a business process. The accuracy of the model in given time and other evaluation criteria should be provided to the end users, so they can determine those characteristics that are more important to their business task.

Given that models have some accuracy function implies that models should be relearned or refreshed after some amount of time. If the accuracy of a model comes close to random chance after some time, a new model should be learned. The older model and the newer model should be somewhat consistent and similar. For instance, if a model at time t is based on attributes A, B, and C, we would expect the refreshed model to use a similar set of attributes. If they change radically, the models may be overfitting the data or the models may reflect seasonal trends. The end users of the system expect the models to be somewhat consistent. They might lose confidence if the models change radically, because intuitively the radical change may not make sense.

As stated earlier, the evaluation should reflect the business process it will be applied in. For example, if the churn system is used to identify churners monthly for targeted campaigns, then an interesting question from an end user may be to ask what percentage of churners in month y would the data mining software predict to churn ahead of time. The results of running experiments on historical data to answer this question may give some indication of how often campaigns need to be run to capture a certain percentage of churners. In addition, by noting those customers that end up being on the predicted churn list for successive months we also find out more about the consistency of the models. These types of questions come about by interacting with end users and by looking at the task through their perspective.

Unique Perspective on Data Mining Problems

The combined experience in applying data mining technology to many domains over the years has taught several lessons that are not commonly discussed in the community, by either vendors, researchers, or business users. The important three problems are presented as follows:

1). Before business problems can be solved with data mining, they must be transformed to match existing tools.

Data mining tools perform a small set of basic tasks such as classification, regression, and time-series analysis. Rarely is a business problem exactly in one of these forms. Usually it must be transformed into (or rephrased as) one of these basic tasks before data mining tool can be applied. Often, in order to solve a problem it must be decomposed into a series of basic tasks. Indeed, much of the art of data mining involves the creative decomposition of a problem into a sequence of such subtasks that are solvable by existing tools.

For example, the work on cellular phone fraud detection transformed the problem of fraud detection into a sequence of knowledge discovery, regression, and classification tasks (mining for indicators of fraud, profiling customer behavior, combining evidence to classify behavior as fraudulent). No single type of task was adequate to solve the problem.

2). Evaluation of data mining results is more complex than either developers or users believe.

Most data mining tools, like the research prototypes from which they were derived, measure performance in terms of accuracy or classification error. A tacit assumption in the use of classification accuracy as an evaluation metric is that the class distribution among examples is constant and relatively balanced. In the real world this is rarely the case. Classifiers are often used to sift through a large population of normal or uninteresting entities in order to find a relatively small number of unusual ones; for example, looking for fraudulent transactions or checking an assembly line for defective parts. Because the unusual or interesting class is rare within the classification is rare within the general population, the class distribution is very skewed.

Evaluation by classification accuracy also assumes equal error costs. In the real world this is unrealistic because classifications lead to actions, which have consequences, sometimes grave. Rarely are mistakes evenly weighted in their cost. We have yet to encounter a domain in which they are. The class skew (as well as error costs) may change over time, after a data mining solution is deployed. Indeed, error costs and class distribution in the field may never be known exactly.

Unfortunately, the importance and difficulty of evaluation is often not appreciated by business users either. The business user usually knows the general problem to be solved, but may not be able to specify error costs or even advise in their calculation. Sometimes the business user does not know how well current procedures solve the problem, and has no mechanisms in place to evaluate their performance. We are sympathetic to this, since evaluating performance often tasks time and effort away from the task itself. However, it makes measuring the efficacy of a data mining solution difficult or impossible. These recurring difficulties with evaluation have directed the research at Science & Technology Center Bell Atlantic. They have developed a technique based on ROC analysis that greatly facilitates comparison and evaluation of data mining results. The technique is especially useful when error costs and class distribution are only known approximately, or may change.

3). Data preparation and data cleaning are more time consuming and knowledge intensive than is acknowledged.

In our experience, understanding the data, reducing noise, and converting the data to an appropriate representation is the most time-consuming part of the data mining process. Furthermore, the process is usually iterative and knowledge intensive: as the project progresses, we learn more about the process that generates the data and we have to go back and re-clean them based on the new knowledge. Although the provider usually has information about the data, we are often the first people ever to analyze the data carefully. We have uncovered errors, idiosyncrasies, and artifacts of the data gathering process that were unknown to the provider. These discoveries sometimes end up changing how we approached the data mining task. Data preparation and cleaning are often tedious, uninteresting tasks. However, over the life of a data mining project, these tasks account for far more time than the taken by applying the machine learning algorithms.

Benefits of a Standard Data Mining Process Model

Data mining success stories have triggered increased interest within the business community, particularly in large corporations with vast stores of data about their customers and business operations. Their interest appears to be following a path similar to that of early research in machine learning: the tendency to view data mining as the isolated application of a data mining algorithm to a pre-existing dataset, where the key determinant of success is selecting (or creating) the "best" model-building algorithm.

As businesses continue to use data mining technology, they are likely to discover, as experienced practitioners and researchers already have, that:

- There is usually little difference in accuracy between modeling algorithms.
- Availability of useful data, dataset preparation, and user skill are more important than which algorithm is chosen.
- Model development is more properly viewed as a multistep *process*, of which application of the modeling algorithm is only a small part (practitioner's informal estimates tend to range from about 10% to 30% of the total effort).

While business customers would eventually learn these lessons through experience, it is hazardous for the health of the industry to allow this tool-centric focus to continue. At this early stage, too many disaster stories would be fatal to the field. Tool vendors have found it effective to sell tools based on the premise that they are like golden geese – feed data at one end and golden nuggets of knowledge will magically come out the other. However, when someone's goose produces the wrong kind of nuggets, their first impulse is not to question whether they are properly skilled in raising golden geese or providing the right food. They will scream "it's a hoax – data mining doesn't work." If so, data mining will be discredited like other technologies that were overhyped and failed to meet the inflated expectations.

For the data mining industry to prosper on a wide scale, it is necessary to create the perception that "data mining works." It is fine if the perception is "data mining works – if done properly." Of course, then the obvious question is "how?" For acceptance to spread beyond the "early adopters," customers must feel confident that they will know how to manage a data mining project to ensure success. A major part of the comfort level is to understand what the stages of a project are, what issues will be need to be dealt with, what tasks will need to be performed, etc. Having a process model goes a long way toward creating this initial comfort level. Customers would not feel as if they are wandering into completely uncharged territory. They realize they will encounter many difficult situations that will need to be dealt with, but at least they known what they are likely to encounter. A service provider that can communicate a convincing process will have a big advantage over one that cannot.

The existence of a process model is a big improvement over the situation without one. However, a great deal of confusion will result if a customer is presented with several different process models. It might be that in reality they are very similar. However, they might sound quite different to a customer who is not a data mining expert, and is then presented with the dilemma of trying to decide if one is better then the other. Faced with such a situation, the conservative customer may well delay data mining project, preferring to wait until the picture becomes clearer.

Contrast this with the situation where nearly even prospective service provider describes the same, standard, data mining process model. This removes a major obstacle to the decision to do a project. The customer is still confronted with which service provider to hire. But he/she will no longer need to get tangled up in confusing arguments about which process models are really different and which is best. The key issue is that if a sufficient comfort level is achieved, the customer will be much more willing to proceed with the project.

Benefits of a Common Process Model

By adopting and promoting a common view of the data mining process, the data mining community would benefit in a number of ways:

- A standard would increase the comfort level of potential customers of data mining products and service.
- A greater proportion of efforts would be successful, because they followed established principles, increasing the perception of data mining as an effective technology and a high-return investment.
- Projects are more likely to proceed smoothly if all parties understand the normal course of data mining project.
- A standard process improves communication about data mining topics by providing a common reference point for customers and clients to compare products and services.

However, different groups are affected differently by standard model. The major communities, and their expected relation to a process model, are discussed below.

Tool Vendors

For tool vendors the key question is the market's verdict of the value of data mining. If the perception is positive, they will buy lots of tools. Of course, some vendors will succeed more than others. Naturally, the prospective market for tools is significantly larger if the belief of the market is that data mining can be successfully performed by less experienced people. However, if the verdict is that data mining is another overhyped hoax, the market will shrink, and no one will sell many tools. In this case, tool vendors will suffer the fate of any company whose habitat disappears: most will become extinct, others will somehow find a way to adapt to a new environment. The conflict for tool vendors is that there is a short-term benefit to downplaying the amount of effort and skill needed to achieve high-quality results.

Service Providers

Service providers have a different conflict. Certainly, they would benefit if the perception becomes "Data mining can produce great rewards, but only if performed by experts who know what they're doing," i.e., data mining service providers. In this situation, a service provider with a high-quality, proprietary process has a distinct competitive advantage. However, prospective customers confronted with arguments over conflicting process descriptions might decide to stay out of the market. Adoption of a common view of the data mining process should increase the total market for data mining services, though it might be harder to differentiate offering. In such a world, a propriety process model could turn from an asset into a liability; prospective customers will question why your process is different form the industry consensus. It will be up to the service provider to justify excluding a standard task or to explain the added benefit of additional tasks.

End Users

Less experienced users would probably be most eager to embrace a process model, for the guidance it would provide. At first glance, experienced modelers might view a standard process model as a threat, because it would provide greater benefit to less experienced practitioners. However, the greater the demand for data mining services overall, the greater will be the demand for experts. In addition, the existence of a widely known model should make it much easier for the analyst to communicate what they are doing to a client. In addition, the client is much less likely to question about the necessity of tasks that are described in a standard process.

Business Customers

Probably the most enthusiastic advocates of a standard process model. These are the poor folks who, despite limited technical understanding of data mining, must somehow sift through conflicting definitions and marketing claims and decide how their organization will use data mining. A common version of the data mining process provides them with a framework for structuring their projects and for evaluating tool and service offerings.

17.3 Data Mining and Business Intelligence

Data mining is about finding *useful* patterns in data. This word *useful* can be unpacked to expose many of the key properties of successful data mining.

The patterns discovered by data mining are useful because they extend existing business knowledge in useful ways. But new business knowledge is not created "in a vacuum"; it builds on existing business knowledge, and this existing knowledge is in the mind of the business expert. The business expert therefore plays a critical role in data mining, both as an essential source of input (business knowledge) and as the consumer of the results of data mining.

The business expert not only uses the results of data mining but also *evaluates* them, and this evaluation should be a continual source of *guidance* for the data mining process. Data mining can reveal patterns in data, but only the business expert can judge their usefulness. It is important to remember that the data is not the business, but only a dim reflection of it. This gap, between the data and the business reality it represents, is called *the chasm of representation* to emphasize the effort needed to cross it.

Patterns found in the data may fail to be useful for many different reasons. They may reflect properties of the data, which do not represent reality at all, for example when an artifact of data collection, such as the time a snapshot is taken, distorts its reflection of the business. Alternatively, the patterns found may be true reflections of the business, but they merely describe the problem that data mining was intended to solve – for example arriving at the conclusion that "purchasers of this product have high incomes" in a project to market the product to a broader range of income groups. Finally, patterns may be a true and pertinent reflection of the business, but nevertheless merely repeat "truisms" about the business, already well known to those within it.

It is all too easy for data mining, which is insufficiently informed by business knowledge to produce useless results for reasons like the above. To prevent this, the business expert must be at the very heart of the data mining process, spotting "false starts" before they consume significant effort. The expert must either literally "sit with" the data miner, or actually perform the data mining. In either case, the close involvement of the business expert has far-reaching consequences for the field of data mining.

Data Mining as a Nontechnical Process

Business experts are seldom also technical experts, and their deep involvement in data mining has a fundamental effect on its character. The process of data mining is one in which the business expert interprets the data, a simple extension of ordinary learning by experience. In such a framework, technology must as far as possible remain hidden, while revealing the patterns in the data.

The organizers rightly state that successful business data mining does not come down to "hot algorithms." Equally irrelevant to the core process of data mining are database support, application integration, business templates, and scalability; data mining tools may usefully have such attributes but they are essentially *technological* properties. The business user must be able to approach data mining as a window on the business and engage *with the data* without the distraction of technological detail.

Data Mining Tools

The requirement of data mining to be accessible to business experts also shapes the requirements for data mining tools. These end-user-oriented requirements can be described in many different ways, but here we focus on three key properties: data mining tools must be *interactive, incremental*, and *iterative*.

Interactive: Modern "desktop" applications are highly interactive as a matter of course, but here we focus on a deeper interpretation of "interactive": the user must be enabled to *interact with the data*, and not just with the technology. The user interface of data mining tools should be designed to highlight the properties of data and play down the details of technology, whether that technology be database links, efficient indexing, visualization display parameters, or machine learning algorithm.

Incremental: The data mining process is incremental, each successive investigation builds on the results of the previous one; thus the principle learning from experience applies not just to the data mining exercise as a whole but also to each step within it. Data mining tools must be designed to encourage this re-use of results as the data miner, in a step-by-step manner, *builds up a picture* of the patterns in the data. This means that data mining tools must be highly integrated; query must lead naturally to visualization, visualization to data transformation and modeling, and modeling to visualization or further queries. These transitions are merely examples; overall the process must appear seamless, with the effective methods of investigation at any point being also the most obvious, and without the intervention of technological barriers or distraction.

Iterative: Data mining is seldom a simple linear process; successive steps not only build on one another's results, but also refine the approach of earlier steps. For example the results of modeling may show that the data should be further refined and the modeling repeated, or may point to areas for closer examination in an earlier data exploration phase. Any result may point to earlier steps, refining not only the data but also the process itself; each step also has the potential to open up entirely new avenues of inquiry. It should be emphasized that the process is not organized into discrete steps concerned with different types of knowledge; rather the discovery of detailed properties of the data proceeds alongside a gradual refinement of the business concept involved, and the unfolding of key patterns to be utilized.

The iterative nature of data mining is apparent at a variety of levels. For example at the detailed level, a modeling process may be repeated many items (and gradually transformed), for example in the space of a day. Many models are built over this time, and each contributes a small "nugget" of knowledge to the overall process, we might call these "throw-away" models – they are formed to be read, digested, and then thrown away. At the overall project level, the data mining process is also iterative, and should for a project of significant duration contain "planned in" iterations for the production of improved models or other results.

Data mining tools must be designed to support this iterative property of the data mining process. The requirements here are similar but not identical to those relating to the "incremental" property. Data mining operations, once configured, must not be "set in stone" – they should be *designed to be refined* in the light of subsequent events.

17.4 Data Mining in Business - Case Studies

Case Study 1 – An Overview of Data mining at Dun & Bradstreet (D&B): This case study is taken from Data Intelligence Group (DIG), D&B, 1995.

Executive Summary

Data mining, the extraction of hidden predictive information from large databases, is a powerful new technology with great potential to help (D&B) "pre-emptively define the information market of tomorrow." Data mining tools predict future trends and behaviors, allowing businesses to make proactive, knowledge-driven decisions. The automated, prospective analyses offered by data mining move beyond the analyses of past events provided by retrospective tools. Data mining answers business questions that traditionally were time consuming to resolve. Data mining tools scour databases for hidden patterns, finding predictive information that experts may miss because it lies outside their expectations.

D&B companies already know how to collect and refine massive quantities of data to deliver relevant and actionable business information. In this sense, D&B has been "mining" data for years. Today, some D&B units are already using data mining technology to deliver new kinds of answers that rank high in the business value chain because they directly fuel return-on-investment decisions.

Data mining techniques can be implemented rapidly on existing software and hardware platforms across D&B to enhance the value of existing resources, and can be integrated with new products and systems as they are brought on-line. When implemented on high-performance client-server or parallel processing computers, data mining tools can analyze massive databases while a customer or analyst takes a coffee break, then deliver answers to questions such as, "Which clients are most likely to respond to the next promotional mailing, and why?"

In the D&B units DIG surveyed, we found strong interest and a wide range of activities and research in data mining. Groups are engaged in data mining to varying degrees, from experimentation by individual analysts to the deployment of completed projects. We also found a wealth of potential business opportunities that could open up through data mining technology.

The breadth of D&B's collected data places the company in a unique position to take advantage of the fact that data mining tools produce better results with larger, broader databases. By integrating data mining into its products and services, D&B can leverage its exiting resources to achieve new revenue.

A Strategic opportunity for D&B

D&B units are connected through the common goal of delivering integrated, global solutions to support business decisions. In accomplishing this goal across a broad spectrum of markets, D&B units face similar market pressures and opportunities. For example, customers urgently require tools to help them keep pace with accelerating growth in the size and complexity of business data. At the same time, customers demand ever more timely, sophisticated, and widely integrated data analyses.

D&B units work hard today to maintain their leadership against a growing competitive threat from other vendors. These vendors have often aggressively exploited new technology to capture market advantage. While D&B units have responded successfully to these competitive challenges in the past, the question remains: What new technology is becoming available today the D&B can leverage proactively?

Data mining is such a technology. D&B is in a unique position to take the lead in delivering the benefits of data mining technology to customers. The company has a wealth of data unrivaled in its breadth and depth, and the understanding of the relevant markets that is necessary to bring this technology to customers successfully. D&B units are engaged in markets where data mining can have significant impact. These markets use large databases and need the power of data mining to achieve a better understanding of their data.

The Scope of Data Mining

Data mining derives its name from the similarities between searching for valuable business information in a large database – for example, finding linked products in Nielsen's gigabytes of store scanner data – and mining a mountain for a vein of valuable ore. Both processes require either sifting through an immense amount of material, or intelligently probing it to find exactly where the value resides.

Given databases of sufficient size and quality, data mining technology can generate new business opportunities by providing these capabilities:

- ***Automatic prediction of trends and behaviors.*** Data mining automates the process of finding predictive information in large databases.

Questions that traditionally required extensive hands-on analysis can now be answered directly from the data – quickly. A typical example of a predictive problem is targeted marketing. Data mining uses data on past promotional mailings to identify the targets most likely to maximize return on investment in future mailings. Other predictive problems include forecasting bankruptcy and other forms of default, and identifying segments of a population likely to respond similarly to given events.

- ***Automatic discovery of previously unknown patterns.*** Data mining tools sweep through databases and identify previously hidden patterns in one step. An example of pattern discovery is the analysis of retail sales data to identify seemingly unrelated products that are often purchased together. Other pattern discovery problems include detecting fraudulent credit card transaction and identifying anomalous data that could represent data entry keying errors.

Data mining techniques can yield the benefits of automation when implemented on existing software and hardware platforms at D&B, and can be implemented on new systems, as existing platforms are upgraded and new products developed. When data mining tools are implemented on high-performance parallel processing systems, they can analyze massive databases in minutes. Faster processing means that users can automatically experiment with more models to understand complex data. High speed makes it practical for users to analyze huge quantities of data. Larger databases, in turn, yield improved predictions. Databases can be larger in two senses:

- ***Higher dimensionality.*** In hands-on analyses, analysts must often limit the number of variables they examine because of time constraints. Yet variables that are discarded because they seem unimportant may carry information about unknown patterns. High-performance data mining allows users to explore the full dimensionality of a database, without pre selecting a subset of variables.
- ***Larger samples.*** Larger samples yield lower estimation errors and variance, and allow users to make inferences about small segments of a population.

A recent Gartner Group Advanced Technology Research Note listed data mining and artificial intelligence at the top of the five key technology areas that "will clearly have a major impact across a wise range of industries within the next three to five years." Gartner also listed parallel architectures and data mining as two of the top ten new technologies in which companies will invest during the next five years. According to a recent Gartner HPC Research Note, "With the rapid advance in data capture, transmission and storage, large-systems users will increasingly need to implement new and innovative ways to mine the after-market value of their vast stores of detail data, employing MPP [massively parallel processing] systems to create new sources of business advantage (0.9 probability)".

At D&B, data mining technology provides a basis for new products and for enhancements to existing offerings. For example, at DBIS, data mining tools can be used to automate more elements of the process of building risk models for a variety of markets. Data mining can present a Nielsen customer with the top ten most significant new buying patterns each week, or present an IMS customer with patterns of sales calls and marketing promotions that have significant impact within certain market niches.

The most commonly used techniques in data mining are:

- ***Artificial neural networks:*** Nonlinear predictive models that learn through training and resemble biological neural networks in structure.
- ***Decision trees:*** Tree-shaped structures that represent sets of decisions. These decisions generate rules for the classification of a dataset. Specific decision tree methods include classification and regression trees [CART] and chi square automatic interaction detection (CHAID).
- ***Genetic algorithms:*** Optimization techniques that use processes such as genetic combination, mutation, and natural selection in a design based on the concepts of evolution.
- ***Nearest neighbor method:*** A technique that classifies each record in a dataset based on a combination of the classes of the k record(s) most similar to it in a historical dataset (where k > 1). Sometimes called the *k-nearest neighbor technique.*
- ***Rule induction:*** The extraction of useful if-then rules from data based on statistical significance.
- ***Data Visualization:*** The visual interpretation of complex relationships in multidimensional data.

Many of these technologies have been in use for more than a decade in specialized analysis tools that work with relatively small volumes of data. These capabilities are now evolving to integrate directly with industry-standard data warehouse and OLAP platforms.

The Foundations of Data Mining

Data mining techniques are the result of a long process of research and product development. This evolution began when business data was first stored on computers, continued with improvements in data access, and more recently, generated technologies that allow users to navigate through their data in real time. Data mining takes this evolutionary process beyond retrospective data access and navigation to prospective and proactive information delivery.

Data mining is ready for application in the business community because it is supported by three technologies that are now sufficiently mature:

Massive data collection and storage
Powerful multiprocessor computers
Data mining algorithms

Commercial databases are growing at unprecedented rates. The accompanying need for improved computational engines can now be met in a cost-effective manner with parallel multiprocessor computer technology. Data mining algorithms embody techniques that have existed for at least ten years, but have only recently been implemented as mature, reliable, understandable tools that consistently outperform older statistical methods.

In the evolution from business data to business information, each new step has built upon the previous ones. For example, dynamic data access is critical for drill-through in data navigation applications, and the ability to store large databases is critical to data mining.

This survey of a limited number of D&B units points to quantifiable business benefits achievable through the integration of data mining technology with D&B products and services. Data mining is a powerful technology with great potential for adding value to the offerings of D&B units. D&B is in a remarkable position to take advantage of this technology: it has the data, and it has the infrastructure to support units in collaborating to solve shared problems.

Case Study 2: Success of Data Mining in Business-Data Cruncher Design Aspects - This case study is taken from Data Mind Corporation, San Mateo, CA.
Awesome data mining tools, fantastic algorithms, rapidly converging neural networks, highly accurate classification methods, clustering methodologies, etc. are neat and useful tools for the knowledge discovery professional, but they are far from demonstrating significant value to the business person. The key to commercial success of data mining lies in providing true value to the businessperson in a form that can be used and understood by the business community. We present here several of the most important aspects of how the data cruncher was designed to accomplish the business users goals.

Recognize who the customer is

To be commercially successful the first thing to realize is that the customer is not the fellow scientist. The customer is not the statistical analyst and the customer is not the mathematician. Sure, we can sell tools to all of them, but in order to be a commercial success we have to sell to the businessperson. Then we need to realize that business people spend their money in tools that help to solve specific business problems. So the tools need to demonstrate that they are useful in business situations and that they have a visible impact in the business.

Speak Business

The tools have to make themselves comprehensible to the business users. The language used has to be simple and business oriented. Results should be explained in terms that are comprehensible to the business user. It is important to note that this is not just replacing a statistical concept by the English sentence that describes it. Rather it is realizing the communication with the user in the terms and concepts that are familiar to the user.

Task and Goal Oriented

The business user does not want to create a neural network, a decision tree, or an agent network. The user wants to solve a specific problem, to find a specific answer. The user wants to know into what segments it makes sense to divide the customers. The user wants to know what customers are most likely to churn. The user want to know how likely a specific customer will respond to a product being currently promoted.

The "data mining tools" user interface should reflect the problems that the user is trying to solve. The specific approach used in the data cruncher to attack this issue is the concept of assistants. Assistants somewhat resemble wizards in the sense that they guide the user through a set of steps, but they are more complete than wizards. They provide random access to the steps and they follow the user through the whole process, always being there, accessible, and well documented. They guide the user through the mining process. The assistants also allow the user to step outside the assistant and do things using the full flexibility of the tools when necessary, and then go back to the assistant. The data cruncher assistants are customizable to many different business situations through the use of a scripting language.

Bridge the gap between analysis and deployment

Data mining models are developed for a purpose. The data mining tools should help in allowing the user to apply the model to its purpose. For example, if the model is developed with the goal of identifying the best customers for a mailing campaign, then the model should be available where the mailing lit is built. Approaches to solve this issue include providing APIs that enable other applications to make use of the model or adding capabilities to build a mailing list to the data mining tool itself, or integrating the data mining pieces into a mailing list generation.

Another approach is to provide access to the data mining models through a service-oriented interface, where the models are published to a centralized server and then can be used by any application wanting to evaluate specific models against specific records. For example, several models maybe built to determine customer segmentation, likelihood to churn, customer value, cross-selling opportunities, etc. Then these models can be made available through the server and any number of applications can apply the models to different customers by sending the appropriate messages to the server. For example, a mailing list building application may consult a model to score the likelihood of the customers to respond to the mailer.

Combine explicit with implicit knowledge

An important aspect of bridging the deployment gap is to understand that data mining models alone cannot take decisions. The models represent the implicit or learned knowledge. Model results have to be filtered through business rules – which represents the explicit knowledge – before they are put to work. These business rules may contain overrides, additional targeting criteria, geographic or time restriction, etc. For example, a company that sells video cassettes may want to avoid offering a rated R film to a customer who is a minor, even if the cross selling model says that the customer's profile indicates that this is a good title to offer. Another example maybe targeting a churn avoidance campaign to the residents of California, in this case even if the data mining model may indicate that a customer is about to churn, the offer should not be made because the customer does not live in California.

An additional advantage of having business rules combined with the data mining models is that the same rules and models can be used at the many different points where a decision is made. For example, a marketing campaign targeted at attracting new customers may use several models, like customer segmentation, value, and likelihood to accept the offer, combined with some business logic maybe used to decide whether the offer should be made. If these models are deployed to a central server together with the business logic, then the same selection criteria can be used at the many points of contact between the company and the customer. For example, the company's call center, the mailing of the next bill, or the customer's visit to the company's Web pages.

Make it responsive and easy to change

Business situations change very rapidly. It is very important for the business user to react quickly to changing business conditions. In today's competitive world it is not acceptable to have the answer to a question be delivered three months after it was asked. For example, a marketing person developing a promotion maybe interested in modeling the customer's behavior to finetune the targeting. It is important for this model to be available very soon. In addition, the whole package that includes the several data mining models and business logic should be readily available and easy to modify to adapt to the necessary changes in the promotion.

Decision Delivery Systems

Decision delivery systems are designed as a vehicle for bringing decisions to different applications. These systems can typically combine the results of different data mining models with business logic to generate the decisions. As a centralized facility they provide a focused point for the deployment of models and knowledge, helping bridge the gap between the development of useful data mining models and putting them to work.

Business Applications- Overview

Many businesses are interested in data mining because of the falling cost of data storage, the increasing ease of collecting data over networks, and the immense computational power available at low prices. The development of robust and efficient data mining algorithms has caused most businesses to create huge databases containing as much information about their activities as possible. Already available on the market are generic multitask data mining tools to perform a variety of discovery operations. Examples include Clementine, IMACS, MLC++, MOBAL, and Recon.

Making data mining programs useful to businesses requires several elements. First, the problem needs to be started in the business users' terms, including viewing the data in a business model perspective. Second, the program needs to support specific key business analyses such as segmentation, which is very important in marketing application. Third, the results of the data mining need to be presented in a form geared to the business problem being solved. Finally, there has to be a support for protracted data mining on an increasing data set, since business databases are continually growing to store increasing numbers of business transactions.

Data mining applications have been developed for a variety of businesses, including marketing, finance, banking, manufacturing, and telecommunications. Data mining in marketing falls into the broad area called *database marketing.* It consists of analysis of consumer databases to select the best potential customers for a particular product. *Business Weekly* estimated that more than 50% of all U.S. retailers use or plan to use database marketing. American Express has had good results from database marketing, experiencing a 10% to 15% increase in credit card use.

The BBC of the U.K. hired Integral Solutions Ltd. developed a system for predicting the size of television audiences. Integral Solutions Ltd program used neural networks and rule induction to determine the factors playing the most important roles in relating the size of a program's audience to its scheduling slot. The final version performed as well as human experts but adapted more quickly to changes because it was constantly retrained with current data.

Early developments in data mining included Cover story and Spotlight, programs that analyzed supermarket sales data and generated reports on the most significant changes in volume and share broken down by region, product type, and other qualities. Causal factors like price changes and distribution channels were analyzed and related to changes in volume and share. Spotlight later grew into the opportunity explorer system, which includes support for sales representatives of consumer packaged-goods companies to examine their business with individual retailers. This is accomplished by creating presentations showing the advantages of stocking additional products or having special promotions. It even generates interactive reports using hyperlinks for easy navigation.

The Management Discovery Tool (MDT) developed by AT&T and NCR Corporation, incorporates a set of business rules so that users can easily set up monitors for detecting significant changes in important business indicators. MDT also allows automatic HTML report generation, making it easier for users to understand the causes of changes while allowing deeper analysis through point and click links. To appeal more to mainstream business users, MDT provides a limited set of analysis types, including summarization, trend analysis, change analysis, and measure and segment comparison.

The Fidelity Stock Selector fund uses a neural network data mining system to select investments. It performed quite well in the stock market overall. A human fund manager evaluates the output of the system before the action is taken however, so it is not ascertainable how to divide the credit between human and machine.

A data mining system developed by Carlberg & Associates using neural networks was used to predict the Standard & Poor's 500 index. It incorporated interest rates, earnings, dividends, the dollar index, and oil prices in its analysis. The system was amazingly successful, accounting for 96% of the variation in the Index from 1986 to 1995.

The Clone detector system developed by GTE Corporation uses customer profiles to detect cellular cloning fraud. If a particular user suddenly starts calling in an unusual way, the Clone detector automatically informs GTE security. A similar system was developed by AT&T to detect international calling fraud, but that system is much more interactive with and reliant upon human operators.

AcknoSoft developed a data mining program called CASSIOPEE for General Electric and SNECMA. CASSIOPEE is being used by three European airlines to diagnose and predict technical problems in Boeing 737 aircraft. Clustering methods are used to derive families of faults.

Not All In the Data

This case study is taken from QuadStone Ltd, Edinburgh, Scotland. Whatever be the various competing definitions of "data mining" used, it is clear that it has something to do with finding patterns in data. There would also probably be fairly broad agreement where desirable properties of patterns found would include

- Meaningful
- Comprehensible
- Expressible
- Actionable

It is important to distinguish between various different kinds of information that can be relevant to data mining. Perhaps the most important of these might be classed as:

1. Information contained in the dataset being mined in a form that is reasonably accessible to automatic pattern detection, whether with traditional statistical methods, machine learning methods, or other automated procedures (e.g., a strong, meaningful correlation between one or more "independent" variables that exist as fields in the data and a dependent variable being modeled that also exists as a field in the data);
2. Information that is expressed by automated data mining but is in a form that is not readily available for exploitation by automated data mining methods (e.g., a relationship between the ratio of two obscure customer aggregates that could be derived from a transaction stream and an outcome of interest in a customer table, such as a fraud tag);
3. Patterns that exist in the dataset being mined in a form that is accessible to automated mining, but are either incorrect, open to misinterpretation, or in some other way misleading (e.g., a strong but spurious correlation between a variable that is thought to be independent of an outcome of interest, but which is in fact causally dependent on the outcome);
4. Information that is not expressed in the data set being mined at all, but is either essential or highly relevant to producing and understanding meaningful patterns in the data (e.g., information about competitor activity, which is hard to capture in an analysis dataset, or knowledge that the basis of aggregation for a particular quantity in the dataset changed on some data during an observation window, giving rise to apparently changed behavior, when in fact number real behavioral change occurred).

This states that the consequences of the existence of these four kinds of information have direct and concrete implications for the construction of useful data mining tools, and that while most current tools concentrate strongly on information of the type that we listed above, most of the business value is likely to be lost if types 2–4 cannot be handled. This section states that

- Useful data mining tools must allow users to look at the raw data in powerful ways to allow the business analyst the opportunity to understand the data and spot deficiencies.
- Useful data mining tools must allow that business analyst to manipulate and preprocess the data in powerful ways to expose patterns that may otherwise be inaccessible to the mining methods available
- Useful data mining tools must express patterns found in ways that are comprehensible to the business analyst to reduce the likelihood of producing statistically sound patterns in the data that in reality are meaningless, false, or require careful interpretation.
- Useful data mining tools must allow the business analyst a variety of methods for controlling the mining process to allow information not present in the data either to be incorporated in the mining or to influence the mining in ways to compensate for the lack of information in the data.

- Useful data mining tools must express patterns in ways that maximize their action ability, if analysis is to be of practical business value, rather than just a source of "did you know...?" statements.

The above suggests that a useful data mining tool should aim, first and foremost, to empower the business analyst to explore and understand the dataset in relation to his/her own knowledge, rather than aiming to replace the analyst with some automated data discovery algorithm. Such a tool would necessarily provide an integrated set of functionality to facilitate the full range of activities necessary in a meaningful analysis, with an emphasis on interactivity, visualization, and flexibility. These characteristics are at least as important in determining the utility and power of data mining package as the selection of a suitable set of methods from the ever-growing plethora of automated classification, discrimination, and clustering algorithms.

Summary

We contend that data mining techniques should be evaluated according to the business task. This requires knowledge of the business process and interaction with the end users. Although most traditional evaluation has held time constant, the time variable cannot be forgotten when data mining software is put into the business process. The learned models can be evaluated and compared along the time dimension. By understanding the characteristics of the learned models, developers as well as end users can make more informed decisions.

The data mining process must be driven by those with expert knowledge of the domain. This has many implications for the process and for the tools, which support it: the process must be thoroughly domain oriented rather than technically oriented, and the tools must support an interactive, incremental, and iterative style of work. Data mining, because of its interactivity and domain orientation, has sometimes been described as a "hunch machine." The key to commercial success in data mining is *hunching, not crunching.*

Review Questions

1. Explain how data mining is used for business problems.
2. What are the benefits of a standard and common data mining process model?
3. Describe data mining and business knowledge strategy.
4. Explain the success of data mining in business with a case study.

18

Data Mining in Sales Marketing and Finance

Objectives:

- Data mining enables companies to identify trends within the data warehouse.
- The recent advances in technology have enabled companies to couple these technologies (data mining and campaign management) more tightly, with the following benefits: increased speed with which they can plan and execute marketing campaigns, increased accuracy and response rates of campaigns, and higher overall marketing return on investment.
- The key to making a successful data mining software product is to embrace the business problems that the technology is meant to solve, not to incorporate the hottest technology.
- Intelligent information agents scan the Internet for relevant information and aggregate it into an information service to automate the process of monitoring and aggregate relevant online sources and capturing significant content for decision makers.
- The concept of a sector-oriented agent-based online information system is formed, which focuses on the external information needs of enterprises in individual sectors.
- Data mining helps the company to understand the patterns behind past purchase transactions, thereby helping in the design and launch of new offerings, in an expeditious and cost-effective manner.
- "Market basket" analysis is formed that suits the application of affinity rules to analyzing consumer purchases.
- Data mining applied in finance supports financial asset management and risk management. Its use in computational finance will have a major impact in the modeling of currency markets, in tactical asset allocation, bond and stock valuation, and portfolio optimization.
- Financial data collected in the banking and financial industries is often relatively complete, reliable, and of high quality, which facilitates systematic data analysis and data mining.

S. Sumathi and S.N. Sivanandam: *Data Mining in Sales Marketing and Finance*, Studies in Computational Intelligence (SCI) **29**, 411–438 (2006)
www.springerlink.com
© Springer-Verlag Berlin Heidelberg 2006

Abstract. The data mining applications in various fields like sales database marketing, integrating customer value and campaign management software, and financial data analysis are described in this chapter.

The popular types of applications that leverage companies' investments in data warehousing are data mining and campaign management software. Data mining and campaign management have been successfully deployed by hundreds of Fortune 1000 companies around the world, with impressive results. But recent advances in technology have enabled companies to couple these technologies more tightly, with the following benefits: increased speed with which they can plan and execute marketing campaigns, increased accuracy and response rates of campaigns, and higher overall marketing return on investment.

The data mining process results in the creation of a model. A model embodies the discovered patterns and can be used to make predictions for records for which the true behavior is unknown. These predictions, usually called *scores*, are numerical values that are assigned to each record in the database and indicate the likelihood that the customer will exhibit a particular behavior. These numerical values are used to select the most appropriate prospects for a targeted marketing campaign. Campaign management software enables companies to deliver to customers and prospects timely, pertinent, and coordinated offers, and also manages and monitors customer communications across all channels.

The market for data mining will be billions of dollars by the turn of the century. Unfortunately, much of what is now considered data mining will be irrelevant, since it is disconnected from the business world. In general, marketing analysts predictions that the technology of data mining will be very relevant to businesses in the future are correct. The key to making a successful data mining software product is to embrace the business problems that the technology is meant to solve, not to incorporate the hottest technology. In Section. 18.2 we address some of the issues related to the development of data mining technology as it relates to business users.

Information agent technology is an approach, which is used to present to improve the worse situations occurring. Section 18.3 introduces the concept of a sector-oriented agent-based online information system that focuses on the external information needs of enterprises in individual sectors of the agrifood industry.

Data mining software can help find the "high-profit" gems buried in mountains of information. However, merely identifying the best prospects is not enough to improve customer value. We must somehow fit the data mining results into the execution of marketing campaigns that enhance the profitability of customer relationships. Unfortunately, data mining and campaign management technologies have followed separate paths – until now. The organization stands to gain a competitive edge by understanding and utilizing this new union. Section 18.4 describes how we can profit from the integration of data mining and campaign management technologies.

Section 18.5 discusses some of the problems of the consumer package goods industry, a case study of some of the challengers presented to data miners within this industry, and critiques current knowledge discovery research in these areas.

Data mining is being increasingly applied in finance, especially to support financial asset management and risk management. It is considered by many financial management institutions as an innovative technology to support conventional quantitative techniques. Its use in computational finance will have a major impact in the modeling of currency markets, in tactical asset allocation, bond and stock valuation, and portfolio optimization. In addition the application of data mining for scoring tasks delivers valuable support for the management of client credit risk and fraud detection. The last subsection of this chapter discusses how data mining is applied for financial data analysis

18.1 Data Mining can Bring Pinpoint Accuracy to Sales

Data warehousing – the practice of creating huge, central stores of customer data that can be used throughout the enterprise – is becoming more and more commonplace. But data warehouses are useless if companies do not have the proper applications for accessing and using the data.

Two popular types of applications that leverage companies' investments in data warehousing are data mining and campaign management software. Data mining enables companies to identify trends within the data warehouse (such as "families with teenagers are likely to have two phone lines," in the case of a telephone company's data). Campaign management software enables them to leverage these trends via highly targeted and automated direct marketing campaigns (such as a telemarketing campaign intended to sell second phone lines to families with teenagers).

Data mining automates the detection of patterns in a database and helps marketing professionals improve their understanding of customer behavior, and then predict behavior. For example, a pattern might indicate that married males with children are twice as likely to drive a particular sports car than married males with no children. A marketing manager for an auto manufacturer might find this somewhat surprising pattern quite valuable.

The data mining process can model virtually any customer activity. The key is to find patterns relevant to current business problems. Typical patterns that data mining uncovers include those customers who are most likely to drop a service, who are likely to purchase merchandise or services, and who are most likely to respond to a particular offer.

Unfortunately, for most companies today, the use of data mining models within campaign management is a manual, time-intensive process. When a marketer wants to run a campaign based on model scores, he or she has to call a modeler (usually a statistician) to have a model run against a database so that a score file can be created. The marketer then has to solicit the help of an IT staffer to merge the scores with the marketing database. This disjointed process is fraught with problems and errors and can take weeks. Often, by the time the models are integrated with the database, either the models are outdated or the campaign opportunity has passed.

The solution is the tight integration of data mining and campaign management technologies. Under this scenario, marketers can invoke statistical models from within the campaign management application, score customer segments on the fly, and quickly create campaigns targeted to customer segments offering the greatest potential. Here is how it works:

Step 1: Creating the Model

A modeler creates a predictive model using the data mining application. He or she then exports the model to a campaign management application, possibly by simply by dragging and dropping the data from one application to the other. This process of exporting a model tells the campaign management software that the model exists and is available for later use.

Step 2: Dynamically Scoring the Data

Once a model has been put into the campaign management system, marketers can then reference the model's score just as they would reference any other piece of data. Records can be selected based on the score, in conjunction with other characteristics in the data. When the campaign is run, the records in the database are scored dynamically using the model.

Dynamic scoring avoids manual integration of scores with the database and eliminates the need to score an entire database. Instead, dynamic scoring marks only relevant customer subsets and only when needed. This shrinks marketing cycle times and assures fresh, up-to-date results. Once a model is in the campaign management system, the user can start to build marketing campaigns based upon it simply by choosing it from a menu of options.

Any company that is creating or has created a data warehouse should be considering the use of integrated data mining and campaign management applications, which unlock the data and put it to use. By discovering customer behavior patterns and then acting upon them quickly, companies can stave off competition; and increase customer retention, cross selling and up-selling, all of which ultimately contribute to higher overall revenues.

18.2 From Data Mining to Database Marketing

18.2.1 Data Mining vs. Database Marketing

The current state-of-the-art analysis of databases is done by high-tech analysts (typically statisticians) using sophisticated tools, e.g., SAS or S-Plus. In essence these analysts are manual data miners. In contrast, data mining software technology promises to automate that analysis, allowing business users (who do not have a Ph.D. in statistics) to develop a more accurate and sophisticated understanding of their data.

Before we go any further, it is probably a good idea to discuss the terminology found in much of the data mining literature. There seems to be a multitude of terms related to the process of analyzing information contained in a database: data mining, database mining, and database marketing. Is there a difference between these terms?

Let us start with the technology. The technology is *data mining*. Data mining is, in some ways, an extension of statistics, with a few artificial intelligence and machine learning twists thrown in. Like statistics, data mining is not a business solution – it is just the underlying technology. Statistics does not by itself solve business problems. Unfortunately, data mining is being touted as a business solution when it is simply the base technology upon which business solutions might be built.

Database mining, which incorporates the ability to access directly data stored in a database, is one step beyond the core technology of data mining. The distinction (database rather than data) might seem to be a trivial improvement, but like most transitions from technology to solutions it requires a major leap for developers. For example, at a recent data mining conference, only one presenter discussed how their work interacted with a database. All the other presenters assumed that the data was available in flat files or that any interaction with a database was so irrelevant as to be not worth mentioning. However, anyone familiar with commercial information processing knows the critical impact of interacting with data stored in relational databases (RDBMS).

Database marketing, on the other hand, supports a variety of business processes. It involves transforming a database into business decisions. For example, consider a catalog retailer who needs to decide who to send a new catalog to. The information incorporated into the database marketing process is the historical database of previous mailings and the features associated with the (potential) customers, such as age, zip code, their response in the past, etc. The database marking software would use this information to build a model of customer behavior that would generate a mailing list of customers most likely to respond to the new catalog. In the end, any models of the database the data mining software might create are irrelevant – what matters is the list of potential customers who receive the catalog and the accuracy of the list.

18.2.2 What Exactly is Data Mining?

OK, now we know that data mining, the technology, is not the solution to our problems. But what is the technology? How does it differ from statistics and other time-proven techniques? And what is the end product from the technology? In a field filled with hype, the answers to these questions can often be vague or misleading. In this section we ground some expectations.

The phrase *discover interesting patterns* often comes up during discussions of data mining. A pretty vague statement since "interesting" usually depends on a specific vertical market and "pattern" is irrelevant without some specific of business problem. For most problems, a pattern is some set of measurable characteristics that can be correlated with some other characteristic. For example, a pattern that might be discovered by a data mining application could be something like this: if the age is between 16 and 20 and the zip code is 90210, then one probably drives a car costing greater than $50,000. What this pattern does not say is everyone matching this pattern must drive an expensive car. Usually a pattern is associated with an "accuracy," which specifies the percentage of pattern matches where the correlated characteristic is correct. As far as "interesting" is concerned that would depend on the business problem. If we are trying to market luxury products, this sort of pattern might very well be interesting. But if we are trying to predict medical insurance fraud, this pattern is unlikely to be useful, and therefore uninteresting.

Coverage is also an important concept. In the previous example, the discovered pattern only applies to some fraction of people living in one zip code. If the business is national, a pattern that includes only one zip code is not enough. In that case the database marketing system would need to discover many more patterns. Coverage relates the total number of possible pattern matches to the number of records that do match some pattern for a desired characteristic. If a collection of patterns matches all records with the desired characteristic, the coverage is 100%. The tradeoff is between accuracy and coverage. A pattern that matches everyone in the US would naturally match all people who buy luxury cars. The pattern would have 100% coverage but very low accuracy.

Another word that often shows up in data mining is "model." A model is simply a collection of patterns for some desired characteristic (models usually come in a form more complicated than a simple list of characteristics to match). For example, one common model is known as ARMA (autoregressive moving average). Recently neural network and other models based on biological concepts have come into vogue. There are lots of model types out there, but in the end they are irrelevant to the business problem. A model should never be confused with a solution.

Given that the model and the business solution are two different things, how can a model be turned into a business solution? To start, there are some things that apply to nearly all database marketing applications. For instance, actionable characteristics, those characteristics that the business has some control over, are usually more important than those that are nonactionable. An example of an actionable characteristic is whether or not someone is sent a catalog. A nonactionable characteristic might be the amount of their last order. A business can decide to send or not send a catalog but it cannot control the amount of a customer's last order. This is especially important when targeting new customers. A pattern that says, "If someone is sent a catalog with a 10% off coupon, they will order $100 worth of merchandise" is much more useful than the pattern "if someone ordered $100 before, they will order $100 again." In the first case the catalog retailer can take action to target potential customers while in the second they must simply wait for the order to come in.

18.2.3 Who is Developing the Technology?

Researchers, primarily in the fields of computer science and statistics, have been responsible for the development of most of the data mining technology currently available. From a business standpoint, this has been a problem since (academic) researchers are good at developing and evaluating data mining technologies, but they tend to get caught up in minute details of the technology. They are not interested (nor, should they be) in the fact that the core technology is only a small part of delivering a business solution, and that compromise must be made in order to deliver a usable piece of software. Another group of data mining researchers are who we call *downsized data miners.* These are people, primarily with research backgrounds, who worked on data mining research until cutbacks and company downsizing forced them into product development. When downsized data miners develop software, the end product is usually a complex tool (as opposed to a problem-solving application) or intermediate software product. Lately some downsized data miners have claimed that they will be deploying business solutions; however most software is currently in some form of pre-release (beta, alpha, even prealpha!). These complex data mining tools compete with other high-end analysis tools (e.g., SAS or S-Plus) that require users to have sophisticated skills. Ultimately very few of these researchers will directly impact the development of database marketing as a business solution.

On the other side of the coin the researchers are the developers who are trying to create database marketing software applications for business users. Unlike data mining tools, these applications do not require users to know how to set up statistical experiments or build data models. The developers of database marketing applications start with the business problems and try to determine if some piece of data mining technology might be useful in solving the problem. The technology associated with a data mining software application, just one small part of the overall product, will be built using techniques developed by researchers. Although current software products could be more sophisticated, the future for these software companies is the future of data mining.

18.2.4 Turning Business Problems into Business Solutions

The technology commonly referred to, as data mining, already exists in at least cursory form. Unfortunately, for business users, the data mining community is currently focusing on refining the technology, without attempting to validate it in business applications. From a practical standpoint, who cares if some algorithm is a 5% improvement over the best data mining technique if it only works from a command line interface on some supercomputer? If it is not easily usable, it is irrelevant to most users.

To deliver data mining technology into the hands of business users, several changes from the current state of the technology will be required. These changes can be broken down into three key areas:

- A built-in understanding of business problems
- Ease of use (a.k.a. executive level)
- Integration with relational database products

The first point is the most important. A database marketing software product will not succeed if it does not start with an understanding of real-world business problems. Ultimately the transition between model and business solution will require a thorough understanding of the marketplace to formulate the problem in a way that will affect a business. The ability of a database marketing application to make use of this information will determine if it is truly useful to a business. Therefore, industry-specific value-added solution providers will probably have an important place in the field of database marketing. They should be able to contribute vertical market-specific templates and metadata that will guide the database mining technology toward solutions to the business problems.

Once the business problem has been taken into consideration, the process and results need to be conveyed to the businessperson who needs to make a decision. It cannot be assumed that the person who makes the decision will understand how to work with a neural network model or how to interpret the results from such a model. User-friendly graphical user interfaces (GUIs) are a necessity. These GUIs must integrate smoothly into the business user's overall decision support (DSS) application environment. This environment is usually client/server, with a PC running Windows as the preferred client platform. Technologically related input parameters must be avoided at all costs. A decision tree database mining application should not require the user to specify search width, search depth, amount of training records, etc. The user will not understand what these terms mean, let alone know what to provide as input values. Instead the user should be asked for things related to his or her world. How much time can the process take? How much "effort" should be dedicated to the problem? The application will need to translate between the user-specified parameters and the parameters required by the technology. A feedback process by which the application provides the user information related to their input parameters would be very useful. For example, the system might tell the user that when the "effort" knob is set to 5, the process will take about three hours and will look at 40% of the database. By increasing the setting to 7, the time might increase to five hours but 75% of the database will be analyzed. Such tradeoffs are within the scope of knowledge of business users.

Finally, database marketing applications must be smoothly integrated with standard relational database products. Business users do not want to deal with dumping an RDBMS as a flat file or translating between different data formats. Database marketing applications need to work with ODBC (Open Database Connectivity) and leading relational database interfaces so that they can interact directly with the databases. When an application speaks to a database, it will probably be in SQL, the standard for the relational database industry. These things would be obvious to developers of business software, but not necessarily to those in the research-oriented field of data mining.

18.2.5 A Possible Scenario for the Future of Data Mining

What does the future have in store for data mining? In the end, much of what is called data mining will likely end up as standard tools built into database or data warehouse software products. As a indication for this statement, we would like to use the field of spell checking software as an example. Just look back ten years to the infancy of computer word processing. Many companies made spell-checking software. We would usually buy a spell checker as a separate piece of software for use with whatever word processor we might have. Sometimes the spell checker would not understand a particular word processor's file format. Some spell checkers might have even required to dump the document as an ASCII file before it would check the spelling (on the ASCII file). In that case, we would have had to manually make corrections in the original document. Eventually the spell checkers became more user friendly and understood every possible document format. Functionality also increased. The future of spell checking probably looked pretty rosy.

So, where are the spell-checking companies today? Where is the spell-checking software? If we look at the local computer store we will not find much there. Instead we will find that the new word processor comes with a built-in spell checker. As word processor software increased in sophistication and functionality, it was a natural progression to include spell checking into the standard system.

The future of data mining may very well parallel the history of spell checking. The functionality of database marketing products will increase to integrate with relational database products (no more dumping a RDBMS into a flat file!) and with key DSS – Decision Support Systems application environments, it will stress the business problem rather than the technology, and present the process to the user in a friendly manner. Database marketing will start losing some of the hype and begin to provide real value to users. This will make database marketing an important business in and of itself. The larger RDBMS and data warehouse companies have already expressed an interest in integrating data mining into their database products. In the end, this new market and its business opportunities will drive mainstream database companies to database marketing. Ten years from now there may be only a few independent data mining companies left in existence. The real survivors will likely be the ones with the foresight to develop a strong relationship with the mainstream database industry.

18.3 Data Mining for Marketing Decisions

The Internet and the information available on it provide a new source for competitive market monitoring of agrifood companies. Intelligent information agents that scan the Internet for relevant information and aggregate it into an information service, offer the possibility to automate the process of monitoring and relevant online sources and capture significant content for decision makers. The combination of information agents for sector-specific information search on the Internet, with filtering techniques for enterprise-focused information retrieval from the search results, provides an efficient information system alternative for agrifood sectors with small-and medium-sized enterprises. Ongoing research deals with the formulation of appropriate search directives for the agents and the differentiation of appropriate filter templates for information retrieval.

In today's fast-paced world, the availability of relevant information to serve critical success factors of business management is a critical success factor in itself. The magnitude of information sources on the Internet could, in principle, greatly improve the information situation. However, the size of the Internet and the limits in the search efficiency of standard search engines reduce the success of processes in information search and information retrieval. Information agent technology is an approach, which could improve the situation. The section introduces the concept of a sector-oriented agent-based online information system that focuses on the external information needs of enterprises in individual sectors of the agrifood industry. This approach supports sectors with small- and medium-sized enterprises, which lack the resources to establish enterprise-specific systems. The sector orientation requires system features, which allow enterprises to "personalize" information retrieval from the system as far as possible. The discussion is separated into three parts. After an introduction into the principles of information agents (part 1), the section discusses the design of agent-based information retrieval systems (part 2) and, specifically, systems for sectors in the agrifood industry (part 3).

The business information circle. The focus of any management information system is the information needs of management. They are linked to management's information system are the information needs of management. They are linked to management's critical success factors and define the basic information complexes, which could serve management's information needs through appropriate delivery systems. This is a well-established and tested information circle and could involve information from internal and external information sources.

The Information circle. With the advent of the Internet with its magnitude of information sources, the delivery of business external information receives increased attention. The focus of interest includes the

1. Search for appropriate information in anonymous information sources and
2. Adjustment of information complexes (documents) to individual needs through information personalization.

In principle, personalization is the link between the specific information needs of any individual member of management and the delivery of information for management's critical success factors from anonymous information sources.

The integration of information search and personalization is one of the major challenges in the utilization of Internet information sources for management information needs. An approach to reach this integration is the utilization of intelligent information agents, which use knowledge about management's critical success factors to search information sources on the Internet for appropriate information complexes.

Intelligent Information Agents. Computer programs, which are able to carry out a given task in an open environment using certain human characteristics such as autonomy, social cooperation, intelligence, and learning capacity, may be referred to as "intelligent software agents." This is the definition we build on in this section. However, it is not the only definition in literature. As a fast growing area of research, the characteristics of intelligent software agents, and, in consequence, their definition, reflect the actual level of advancements in research.

"Intelligent information agents" represent a specific category of "intelligent software agents," which access heterogeneous and potentially geographically distributed information sources as, e.g., on the Internet, to proactively acquire, mediate, and maintain information with specific relevance for an interest group of users with predetermined information needs. The search usually builds on lists of prespecified keywords. Intelligent information agents act autonomously through the so-called robots, which

(a) communicate with Web servers without human intervention,
(b) monitor Internet sites for relevant contents or changes in contents, and
(c) may utilize the Internet's hypertext structure to reach beyond specified Internet sites and to search for information along identified hyperlinks.

To improve search efficiency, the space of robots could be confined to predefined information sources or sources that are characterized by certain industries or certain domain-relevant content. Intelligent information agents are being delivered through a number of software solutions. A direct comparison is difficult, as most differences are more gradual than principal. Differences concern primarily

(a) the search directive for the robots,
(b) the information retrieval from the documents identified through the Internet search, which may reach beyond keyword search and include pattern recognition or context search, and
(c) the presentation of results.

18.3.1 Agent-Based Information Retrieval Systems

Principles

The knowledge of users' critical success factors and the availability of robots with their advanced ability to search information sources on the Internet for appropriate documents are not sufficient for the development of an information retrieval system. It requires, in addition, the availability of

(a) a taxonomy model, which maps users' information needs and provides the links between users' Critical success factors (CSF), the information topics, which could serve the CSFs, and the search results,
(b) a search directive for the robots which is usually provided through lists of search keywords and Internet sites,
(c) functions (tools) for the extraction of relevant information items from collected documents, and
(d) functions for the classification of information and its linkage to the taxonomy model for users' retrieval support.

The taxonomy model directed toward information needs of business management employs a multilevel hierarchical structure in which, e.g., critical success factors (first level) are linked to one or several levels of information topics with relevance for the CSF. The taxonomy model provides a structural basis for the efficient organization and access of the magnitude of information documents that might be provided by the robots. In principle, the extraction of information form documents involves the attachment of an appropriate content descriptor to a document (indexing) and the elimination of nonrelevant information retrieval systems. The indexing follows a three-step procedure,

(a) the (automatic) identification of phrases of interest (keywords) in textual data,
(b) the aggregation of relevant information phrases in content descriptors that represent a document's topic and
(c) the attachment of the content descriptor to the documents for retrieval support.

The classification of information links documents to the information topics in the hierarchical taxonomy model. The "quality" of information extraction, and, subsequently, the effectiveness of an information retrieval system, is determined by the users' impression of the fit between the information need in a certain information domain, the formulation of the content descriptors, and the content of the document.

Retrieval Support

The taxonomy model based on users' critical success factors and their linkage to information topics and search results provides the basic infrastructure for information retrieval. However, users could have advanced retrieval knowledge, which might allow them to bypass the taxonomy model's search structure and to identify the required document more directly. As an example, a user might know that certain information should be in one of a few documents retrieved from a certain information source or include, as one of a few documents; a certain keyword. In these cases, retrieval systems based on source or keyword identification would be superior. To capture such situations, it is suggested to complement the "taxonomy retrieval system" by additional retrieval alternatives, which required their own (hierarchical) retrieval infrastructure and document indexing.

Personalization

With appropriately focused robots, agent-based information retrieval systems could be designed to best meet the information needs of any individual user. However, the broad-based realization of such an approach is not only limited by available resources for information search at least in small- and medium-sized enterprises (SMEs), but it disregards the efficiency potential provided by similarities in information needs among enterprises within any specific sector of the economy. It adds to efficiency in information search, if robots focus on a sector's aggregated information needs and leave further individualization (personalization) to a subsequent selection of classified documents. This approach requires the development of

(a) a taxonomy model, which best maps the joint user needs of all or at least the majority of enterprises in the sector and
(b) filters, which allow the separation of, personalized information subsets for individuals.

The personalization could be based on multi-tier templates, which provide filters for a hierarchical individualization scheme, which stepwise narrows down information accessibility. As an example, while the robots may have collected information for aggregated information needs of a certain sector of the economy, the filters may be directed toward the information needs of

(a) subsectors within the sector (first level of filter templates),
(b) enterprises within any of the subsectors (second level of filter templates), and
(c) functional departments within any of the enterprises (third level of filter templates).

Retrieval Systems for Agribusiness Supply Chains

In the development of agent-based retrieval systems for agribusiness production chains, the taxonomy models would need to capture user needs of enterprises on the same and on different stages of the chain (horizontal and vertical information needs). User needs ask for as many as possible individualized taxonomy models, efficiency considerations for as few as possible aggregated ones.

Research suggests that agribusiness production chains could build on a common list of critical success factors and information topics. Differences between stages do not affect the list but priorities within the list. The topics include competitors, products, personnel, market development, research, food law, regional news, and sector-related politics. For agent-based information search, the main differences in information interest concern

(a) the keyword list in the search directive and
(b) the selection of information sources.

As both subjects are not part of the taxonomy model, one could build an information retrieval system for agribusiness supply chains on a common taxonomy model, and consider these differences in the indexing or documents and the implementation of appropriate filter templates. Ongoing research deals with the identification and evaluation of keyword lists, information sources, and filter templates for different groups of users (e.g., in enterprises, functional departments, etc.) on different stages of the supply chains.

In summary, since the availability of information has become a critical success factor for enterprise management, access to competitive intelligence is essential for enterprises. In general, small- and medium-sized enterprises cannot afford to establish an individual information service that monitors their competitive environment.

18.3.2 Applications of Data Mining in Marketing

Data mining has been applied in many areas of marketing such as direct marketing, customer acquisition, customer retention, cross selling, trend analysis, affinity grouping, and customer lifetime value analysis.

Many firms have successfully implemented data mining methodologies to gain sustainable competitive advantage. For instance, Capital One designed a highly successful direct marketing campaign by investing in data mining technology. Through the user of these methodologies, the company was able to accurately predict the response patterns of potential customers and the types of offers best suited for specific types of customers. Numerous other companies have adopted data mining methodologies for customer relationship management. These include Marriott, *Reader's Digest*, Bank of America, Reuters, etc. Wal-Mart is considered among the pioneers in using data mining technology to analyze and predict market trends.

Data mining can provide companies with accurate information on which customers are likely to purchase new products, how much are they willing to spend, and what kinds of products are likely to be purchased together. It enables a company to understand the patterns behind past purchase transactions, thereby helping in the design and launch of new offerings, in an expeditious and cost-effective manner. Indeed, with data mining technology, company data that was previously perceive as "junk" suddenly becomes valuable. The use of data mining methodologies often results in a strategic marketing advantage that cannot be easily duplicated by the competition (unless they adopt similar methodologies themselves).

18.4 Increasing Customer Value by Integrating Data Mining and Campaign Management Software

As a database marketer, we understand that some customers present much greater profit potential than others. But, how will we find those high-potential customers in a database that contains hundreds of data items for each of millions of customers?

To be successful, database marketers must, first, identify market segments containing customers or prospects with high profit potential and, second, build and execute campaigns that favorably impact the behavior of these individuals.

The first task, identifying market segments, requires significant data about prospective customers and their buying behaviors. In theory, the more data, the better. In practice, however, massive data stores often impede marketers, who struggle to sift through the minutiae to find the nuggets of valuable information.

Recently, marketers have added a new class of software to their targeting arsenal; data mining applications automate the process of searching the mountains of data to find patterns that are good predictors of purchasing behaviors. After mining the data, marketers must feed the results into campaign management software that, as the name implies, manages the campaign directed at the defined market segments.

In the past, the link between data mining and campaign management software was mostly manual. In the worst cases, it involved "sneaker net," creating a physical file on tape or disk, which someone then carried to another computer, where they loaded it into the marketing database.

This separation of the data mining and campaign management software introduces considerable inefficiency and opens the door for human errors. Tightly integrating the two disciplines presents an opportunity for companies to gain competitive advantage.

18.4.1 Some Definitions

A Data warehouse is a repository for relevant business data. While traditional databases primarily store current operational data, data warehouses consolidate data from multiple operational and external sources in order to attain an accurate, consolidated view of customers and the business.

Database Marketing uses information in computerized databases to target offerings to customers and prospects.

Data Mining uses technologies such as neural networks, decision trees or standard statistical techniques to search large volumes of data. In doing so, data mining builds models for patterns that accurately predict customer behavior.

Scoring uses a model to predict future behavior. The score assigned to each individual in a database indicates that person's likelihood of exhibiting a particular customer behavior.

Campaign Management uses information in a data warehouse or marketing database to plan, manage, and assess marketing campaigns designed to impact customer behavior.

A Customer segment is a group of prospects or customers who are selected from a database based on characteristics they possess or exhibit.

Scoring on the fly or *dynamic scoring* is the ability to score an already defined customer segment within a campaign management tool. Rather than scoring an entire database, dynamic scoring works with only the required customer subsets, and only when needed.

Attrition, sometimes known as *churn*, occurs when a customer terminates his or her relationship with a service provider. Marketing efforts usually focus on minimizing churn because the cost of bringing a customer back is usually much greater than the cost retaining the customer in the first place.

18.4.2 Data Mining Defined

Data mining, by its simplest definition, automates the detection of relevant patterns in a database. For example, a pattern might indicate that married males with children are twice as likely to drive a particular sports car than married males with no children. If we are a marketing manager for an auto manufacturer, this somewhat surprising pattern might be quite valuable.

However, data mining is not magic. For many years, statisticians have manually "mined" databases looking for statistically significant patterns. Today, data mining uses well-established statistical and machine learning techniques to build models that predict customer behavior. The technology enhances the procedure by automating the mining process, integrating it with commercial data warehouses, and presenting it in a relevant way for business users.

The leading data mining products, such as those from companies like SAS and IBM, are now more than just modeling engines employing powerful algorithms. Instead, they address the broader business and technical issues, such as their integration into today's complex information technology environments.

In the past, the hyperbole surrounding data mining suggested that it would eliminate the need for statistical analysts to build predictive models. However, the value that an analyst provides cannot be automated out of existence. Analysts will still be needed to assess model results and validate the reasonability of the model predictions. Since data mining software lacks the human experience and intuition to recognize the difference between a relevant and an irrelevant correlation, statistical analysts will remain in high demand.

18.4.3 The Purpose of Data Mining

Data mining helps marketing professionals improve their understanding of customer behavior. In turn, this better understanding allows them to target marketing campaigns more accurately and to align campaigns more closely with the needs, wants and attitudes of customers and prospects.

If the necessary information exists in a database, the data mining process can model virtually any customer activity. The key is to find patterns relevant to current business problems.

Typical questions that data mining answers include:

- Which customers are most likely to drop their cell phone service?
- What is the probability that a customer will purchase at least $100 worth of merchandise from a particular mail-order catalog?
- Which prospects are most likely to respond to a particular offer?

Answers to these questions can help retain customers and increase campaign response rates, which, in turn, increase buying, cross selling and return on investment (ROI).

18.4.4 Scoring the Model

Data mining builds models by using inputs from a database to predict customer behavior. This behavior might be attrition at the end of a magazine subscription, cross-product purchasing, willingness to use an ATM card in place of a more expensive teller transaction, and so on.

The prediction provided by a model is usually called a *score*. A score (typically a numerical value) is assigned to each record in the database and indicates the likelihood that the customer whose record has been scored will exhibit a particular behavior.

For example, if a model predicts customer attrition, a high score indicates that a customer is likely to leave, while a low score indicates the opposite. After scoring a set of customers, these numerical values are used to select the most appropriate prospects for a targeted marketing campaign.

18.4.5 The Role of Campaign Management Software

Database marketing software enables companies to deliver to customers and prospects timely, pertinent, and coordinated messages and value propositions (offers or gifts perceived as valuable). Today's campaign management software goes considerably further. It manages and monitors customer communications across multiple touchpoints, such as direct mail, telemarketing, customer service, point-of-sale, e-mail, and the Web.

Campaign management automates and integrates the planning, execution, assessment, and refinement of possibly tens to hundreds of highly segmented campaigns running monthly, weekly, daily, or intermittently. The software can also run campaigns that are triggered in response to customer behavior or milestones – such as the opening of a new account.

Increasing Customer Lifetime Value

Consider, for example, customers of a bank who only use the institution for a checking account. An analysis reveals that after depositing large annual income bonuses, some customers wait for their funds to clear before moving the money quickly into their stock brokerage or mutual fund accounts outside the bank. This represents a loss of business for the bank.

To persuade these customers to keep their money in the bank, marketing managers can use campaign management software to immediately identify large deposits and trigger a response. The system might automatically schedule a direct mail or telemarketing promotion as soon as a customer's balance exceeds a predetermined amount. Based on the size of the deposit, the triggered promotion can then provide an appropriate incentive that encourages customers to invest their money in the bank's other products. Finally, by tracking responses and following rules for attributing customer behavior, the campaign management software can help measure the profitability and ROI of all ongoing campaigns.

Integrating Data Mining and Campaign Management

The closer integrating data mining and campaign management work together, the more business results. Today, campaign management software uses the scores generated by the data mining model to sharpen the focus of targeted customers or prospects, thereby increasing response rates and campaign effectiveness.

Unfortunately, the use of a model within campaign management today is often a manual, time-intensive process. When someone in marketing wants to run a campaign that uses model scores, he or she usually calls someone in the modeling group to get a file containing the database scores. With the file in hand, the marketer must then solicit the help of someone in the information technology group to merge the scores with the marketing database.

This disjointed process is fraught with problems:

- The large numbers of campaigns that run on a daily or weekly basis can be difficult to schedule and can swamp the available resources.
- The process is error prone; it is easy to score the wrong database or the wrong fields in a database.
- Scoring is typically very inefficient. Entire databases are usually scored, not just the segments defined for the campaign. Not only is effort wasted, but the manual process may also be too slow to keep up with campaigns run weekly or daily.

The solution to these problems is the tight integration of data mining and campaign management technologies. Integration is crucial in two areas:

First, the campaign management software must share the definition of the defined campaign segment with the data mining application to avoid modeling the entire database. For example, a marketer may define a campaign segment of high-income males between the ages of 25 and 35 living in the northeast. Through the integration of the two applications, the data mining application can automatically restrict its analysis to database records containing just those characteristics.

Second, selected scores from the resulting predictive model must flow seamlessly into the campaign segment in order to form targets with the highest profit potential.

18.4.6 The Integrated Data Mining and Campaign Management Process

This section examines how to apply the integration of data mining and campaign management to benefit the organization. The first step creates a model using a data mining tool. The second step takes this model and puts it to use in the production environment of an automated database marketing campaign.

Step 1: Creating the Model

An analyst or user with a background in modeling creates a predictive model using the data mining application. This modeling is usually completely separate from campaign creation. The complexity of the model creation typically depends on many factors, including database size, the number of variables known about each customer, the kind of data mining algorithms used, and the modeler's experience.

Interaction with the campaign management software begins when a model of sufficient quality has been found. At this point, the data mining user exports his or her model to a campaign management application, which can be as simple as dragging and dropping the data from one application to the other. This process of exporting a model tells the campaign management software that the model exists and is available for later use.

Step 2: Dynamically Scoring the Data

Dynamic scoring allows us to score an already-defined customer segment within the campaign management tool rather than in the data mining tool. Dynamic scoring both avoids mundane, repetitive manual chores and eliminates the need to score an entire database. Instead, dynamic scoring marks only relevant customer subsets and only when needed.

Scoring only the relevant customer subset and eliminating the manual process shrinks cycle times. Scoring data only when needed assures "fresh," up-to-date results. Once a model is in the campaign management system, a user (usually someone other than the person who created the model) can start to build marketing campaigns using the predictive models. Models are invoked by the campaign management system.

When a marketing campaign invokes a specific predictive model to perform dynamic scoring, the output is usually stored as a temporary score table. When the score table is available in the data warehouse, the data mining engine notifies the campaign management system and the marketing campaign execution continues.

Here is how a dynamically scored customer segment might be defined:

Where

Length_of_service = 9 And Average balance > 150 And In Model (promo9). score > 0.80

In this example:

Length of service = 9 limits the application of the model to those customers in the ninth month of their 12-month contracts, thus targeting customers only at the most vulnerable time. (In reality, there is likely a variety of contract lengths to consider this when formulating the selection criteria.)

Average balance > 150 selects only customers spending, on average, more than $150 each month. The marketer deemed that it would unprofitable to send the offer to less valuable customers.

Promo9 is the name of a logged predictive model that was created with a data mining application. This criterion includes a threshold score, 0.80, which a customer must surpass to be considered "in the model." This third criterion limits the campaign to just those customers in the model, i.e., those customers most likely to require an inducement to prevent them switching to a competitor.

18.4.7 Data Mining and Campaign Management in the Real World

Ideally, marketers who build campaigns should be able to apply any model logged in the campaign management system to a defined target segment. For example, a marketing manager at a cellular telephone company might be interested in high-value customers likely to switch to another carrier. This segment might be defined as customers who are nine months into a twelve-month contract, and whose average monthly balance is more than $150.

The easiest approach to retain these customers is to offer all of them a new high-tech telephone. However, this is expensive and wasteful since many customers would remain loyal without any incentive. Instead, to reduce costs and improve results, the marketer could use a predictive model to select only those valuable customers who would likely defect to a competitor unless they receive the offer.

18.4.8 The Benefits of Integrating Data Mining and Campaign Management

For Marketers

Improved campaign results through the use of model scores that further refine customer and prospect segments.

Records can be scored when campaigns are ready to run, allowing the use of the most recent data. "Fresh" data and the selection of "high" scores within defined market segments improve direct marketing results.

Accelerated marketing cycle times that reduce costs and increase the likelihood of reaching customers and prospects before competitors.

Scoring takes place only for records defined by the customer segment, eliminating the need to score an entire database. This is important to keep pace with continuously running marketing campaigns with tight cycle times.

Accelerated marketing "velocity" also increases the number of opportunities used to refine and improve campaigns. The end of each campaign cycle presents another chance to assess results and improve future campaigns.

Increased accuracy through the elimination of manually induced errors. The campaign management software determines those records to be scored and when.

For Statisticians

Less time spent on mundane tasks of extracting and importing files, leaving more time for creativity – building and interpreting models. Statisticians have greater impact on corporate bottom line.

18.5 Completing a Solution for Market-Basket Analysis – Case Study

Algorithms for finding rules or affinities between items in a database are well known and well documented in the knowledge discovery community. A prototypical application of such affinity algorithms is in "market basket" analysis – the application of affinity rules to analyzing consumer purchases. Such analyses are of particular importance to the consumer package goods industry. The retailers and wholesalers in the industry generated over 300 billion dollars of sales every year in the United States alone. Despite the economic importance of this industry, data mining solutions to the key business problems are yet to be developed. This section discusses some of the problems of the consumer package goods industry, notes a case study of some of the challengers presented to data miners within this industry, and critiques current knowledge discovery research in these areas.

18.5.1 Business Problem

The consumer package goods industry that exists within a complex economics is informational environment. Mass merchandizing or products is in decline; U.S. consumers are increasing recognized as belonging to fifty (or more) distinct segments, each with its own demographic profile, buying power, product preferences, and media access. The items being sold, consumer products are more diverse than ever before; a single category of food may easily contain hundreds of competing products. Within this highly differentiated environment, strong product brand names continue to offer a strong competitive advantage. By themselves temporary price reductions are not sufficient for establishing consumer loyalty to either a store or product. Consumers are knowledgeable, and mobile, enough to seek out the lowest possible prices for a product. Ultimately consumer value is gained by those retailers able to negotiate favorable terms with their suppliers. Retailers gain the requisite detailed knowledge of customers through the creation of consumer loyalty programs and the use of on-line transaction processing systems; this information about the consumer is a crucial component in retailer–supplier negotiations.

Consumer packaged goods is a mature industry in the United States. Profit is no longer merely a matter of opening more stores, and selling to increasing numbers of consumers; the market is becoming saturated, and the available consumer disposable income largely consumed.

Maintenance of an existing customer base is more important than growing entirely new customers; this new phase of retail growth is based upon selling more and a greater variety of products to pre-existing consumers. The most profitable retailers are those that are able to maintain or reduce their operating costs. Economics of scope, not scale, determine profitability. Data warehousing is one of the foremost technological means of increasing operational efficiency. Efficient consumer response systems, based upon data warehouses, are expected to save the industry \$30 billion a year. Category management, an organizational strategy for enhancing retailer–wholesaler coordination, is another means of increasing operational efficiency. In the following two brief case studies we examine how data warehouses, category management, and data mining techniques show promise for answering the concerns of two large consumer package goods companies.

18.5.2 Case Studies

A major international food manufacturer, with significant equity and a wide variety of manufactured products, is interested in optimizing its product-advertising budget. Like many package goods retailers, this manufacturer has an extensive and rapidly growing advertising budget. Essential to the endeavor is the cooperation of their independent retail outlets in the creation and design of product promotions. The manufacturer sought to create a suite of software tools for the design of promotions, utilizing the newest data mining technology, and to make these tools available in real time to their category managers and to the managers of their retail outlets. The business case suggested that there would be at least three sources of return in the creation of this tool: Improved coordination with retailers, more effective cross sales across product categories, reduced promotional competition from other manufacturers, and enhanced promotional returns. NCR Corporation proposed and designed a state-of-the-art neural network for forecasting and optimizing planned promotions. The network met, or exceeded, industry standards for promotional forecasts (within 15% of actual sales, 85% of the time). Despite the statistical quality of the results the application was never put into production by the manufacturer, the software design necessary to implement the results was too complex. Part of the application complexity stemmed from the hierarchical data types necessitated by the varied products and markets; another component of the complexity was reconciling the different product world views of the manufacturer and the retailer.

A major regional food retailer, a grocer, sought analyses of its consumer transactions within its produce and salad dressing departments. The retailer anticipated improved design of store layouts, improved promotional design, and an insight into the market role of the various highly differentiated products within the category. The retailer clearly anticipated a causal analysis that would reveal the products, which, when purchased by consumers, would lead to additional add-on sales of other products. NCR produced a market basket analysis, which revealed the distinctive purchasing profiles that are associated with each major brand of interest. The NCR analysis revealed that the best selling brands were not those that resulted in the greatest amount of attendant sales. The NCR analysis supported the existing category management plans by the retailer, and also independently confirmed the results of a demographic panel survey. Despite these successes the market basket analysis, by itself, did not produce any new actionable results for the retailer. In the next section on data mining, key data mining algorithms and outputs are examined for their suitability for answering these, and other, consumer package goods questions.

18.5.3 Data Mining Solutions

Affinity algorithms are well understood and well documented by the data mining community. The quintessential application of affinity algorithms is in the area of market basket analysis. For instance, these algorithms when applied to market basket analysis produce rules such as "Those baskets producing product X are also 75% likely to contain product Y." Additional research has focused on optimizing the speed and efficiency with which these rules are found; however additional applied research is needed to the support decision-making needs of the consumer goods industry (and other relevant business groups).

First, affinity algorithms produce individual, isolated rules; associations between groups of products are not revealed. While the analysis can be repeated across all products in a category, or even a store, the number of rules produced grows exponentially. Not only is the computation complex, but the resulting welter of rules is hard to interpret as well.

Second, the output of affinity algorithms seems to suggest causal relationships between products. Yet the algorithms themselves embody no casual assumptions. The nature of product affinities needs to reconsidered, either a new and causal form of affinities analysis needs to be produced, or a thorough understanding of noncausal applications and use of affinity rules needs to be obtained.

Third, affinity algorithms lack robustness. The algorithms produce a point estimate of affinity; yet retailers need to understand how (and if) these rules apply large groups of transaction. A similar issue is the minimum sample size needed to produce robust results.

Fourth, market basket analyses carry implicit information about consumer preferences. Even when consumer identification is missing from transaction data, the data can still be grouped or segmented using data mining techniques to reveal distinct group of consumer preferences. Affinity algorithms imply that samples are taken from homogenous groups of customers; yet business knowledge suggests that consumers are highly varied in taste and expenditure.

Fifth, the market basket analyses, for some set of business questions, may require the rigor of a properly designed statistical experiment. Reasoning from standard to promotional pricing, as well as reasoning from standard display conditions to promotional display conditions, is unwarranted. Yet much of the potential of market basket analysis stems from the capacity of retailers to manipulate product pricing, display, or even attributes to meet consumer need.

Sixth, and finally, standard forecasting tools produce estimates of sales single goods across times. (This is not conventionally the domain of market basket or affinities analysis.) However retailers and manufacturers need to have forecasts for whole groups of products. Producing individual product forecasts, and then aggregating, will not produce optimum forecasts since sales of one product contains information about the potential sales of other products, indeed, the forecasts may not even aggregate correctly. Techniques such as "state space analysis," which combine forecasting with multivariate analysis may prove useful.

18.5.4 Recommendations

The consumer package goods industry is an important, and expansive, industrial segment of the economy. This industry is dependent upon information for its continued economic growth. It is therefore making great progress in collecting large databases of relevant data about its industry. The corresponding questions the industry has about its data are both interesting and economically fruitful. This section considered two case studies of applying standard data mining techniques to industrial questions in the area of consumer package goods. The examples discussed a wholesaler and a retailer who sought better management of product categories and a resulting improved economy of scope. Commercial success of data mining will in part be dependent upon the capacity of algorithms to model complex, hierarchical arrangements of goods and products.

18.6 Data Mining in Finance

Data mining has received much attention as companies and organizations started to ask how they can better utilize the huge data stores they built up over the past two decades. While some interesting progress has been achieved over the past few years, especially when it comes to techniques and scalable algorithms, very few organizations have managed to benefit from the technology. This paradoxical situation of having too much data, yet not be able to utilize it or mine it, arose because of technical and business challenges.

In many cases the desired target variable does not necessarily exist in the database. If the database includes information about customer purchases, a business user might only be interested in customers whose purchases were more than one hundred dollars. Obviously, it would be straightforward to add a new column to the database that contained this information. But this would involve database administrator and IT personnel, complicating a process that is probably complicated already. In addition, the database could become messy as more and more possible targets are added during an exploratory data analysis phase. The solution is to allow the user to interactively create a new target variable. Combing this with an application wizard (it can significantly improve the user's experience, besides simplifying the process, and can help prevent human error by keeping the user on track), it would be relatively simple to allow the user to create computed targets on the fly.

Mining financial data presents special challenges. For one, the rewards for finding successful patterns are potentially enormous, but so are the difficulties and sources of confusions. The efficient market theory states that it is practically impossible to predict long-term financial markets. However, there is good evidence that short-term trends do exist and programs can be written to find them. The data miner's challenge is to find the trends quickly while they are valid, as well as to recognize the time when the trends are number longer effective.

Additional challenges of financial mining are to take into account the abundance of domain knowledge that describes the intricately inter-related world of global financial markets and to deal effectively with time series and calendar effects. The relational data mining (RDM) is a learning method able to learn more expressive rules than other symbolic approaches. RDM is thus better suited for financial mining, because it is able to make better use of application underlying domain knowledge. Relational data mining also has a better ability to explain the discovered rules – ability critical for avoiding spurious patterns, which inevitably arise when the number of variables examined is very large. The earlier algorithms for relational data mining, also known as ILP – inductive logic programming, suffer from a well-known inefficiency. The researchers introduced a new approach, which combines relational data mining with the analysis of statistical significance of discovered rules. This reduces the search space and speeds up the algorithms. They also introduced a set of interactive tools for "mining" the knowledge from the experts. This helps to further reduce the search space.

Data mining does not operate in a vacuum. The results of the data mining process will drive efforts in areas such as marketing, risk management, and credit scoring. Each of these areas is influenced by financial considerations that need to be incorporated in the data mining modeling process. A business user is concerned with maximizing profit, not minimizing RMS error. The information necessary to make these financial decisions (costs, expected revenue, etc.) is often available and should be provided as an input to data mining application.

18.7 Data Mining for Financial Data Analysis

Most banks and financial institutions offer a wide variety of banking services (such as checking, savings, and business and individual customer transactions), credit (such as business, mortgage, and automobile loans), and investment services (such as mutual funds). Some also offer insurance services and stock investment services.

Financial data collected in the banking and financial industries is often relatively complete, reliable, and of high quality, which facilitates systematic data analysis and data mining. Here we present a few typical cases.

Design and construction of data warehouse for multidimensional data analysis and data mining: Like many other applications, data warehouses need to be constructed for banking and financial data. Multidimensional data analysis methods should be used to analyze the general properties of such data. For example, one may like to view the debt and revenue changes by month, region, sector, and other factors, along with maximum, minimum, total, average, trend, and other statistical information. Data warehouses, data cubes, multifeature and discovery-driven data cubes, characteristic and comparative analysis, and outlier analysis all play important roles in financial data analysis and data mining.

Loan payment prediction and customer credit policy analysis: Loan payment prediction and customer credit analysis are critical to the business of a bank. Many factors can strongly or weakly influence loan payment performance and customer credit rating. Data mining methods such as feature selection and attribute relevance ranking may help identify important factors and eliminate irrelevant once. For example factors related to the risk of loan payments include loan-to-value ratio, term of the loan, debt ratio, payment-to-income ratio, customer income level, education level, residence region, credit history, and so on. Analysis of the customer payment history may find that, say, payment-to-income ratio is a dominant factor, while education level and debt ratio are not. The bank may then decide to adjust its loan granting policy so as to grant loans to those whose application was previously denied but whose profile shows relatively low risks according to the critical factor analysis.

Classification and clustering of customers for targeted marketing: Classification and clustering methods can be used for customer group identification and targeted marketing. For example, customers with similar behaviors regarding banking and loan payments may be grouped together by multidimensional clustering techniques. Effective clustering and collaborative filtering methods (i.e., the use various techniques to filter out information, such as nearest neighbor classification, decision trees, and so on) can help identify customer groups, associate a new customer with an appropriate customer group, and facilitate targeted marketing.

Detection of money laundering and other financial crimes: To detect money laundering and other financial crimes, it is important to integrate information from multiple databases (like bank transaction databases and federal or state crime history databases), as long as they are potentially related to the study. Multiple data analysis tools can then be used to detect unusual patterns, such as large amounts of cash flow at certain periods, by certain groups of people, and so on. Useful tools include data visualization tools (to display transactions activities using graphs by time and by groups of people), linkage analysis tools (to identify links among different people and activities), classification tools (to filter unrelated attributes and rank the highly related ones), clustering tools (to group different cases), outlier analysis tools (to detect unusual amounts of fund transfers or other activities), and sequential pattern analysis tools (to characterize unusual access sequences). These tools may identify important relationships and patterns of activities and help investigators focus on suspicious cases for further detailed examination.

18.8 Summary

Database marketing software applications will have a tremendous impact on how business is done in the future. Although the core data mining technology is here today, developers need to take what already exists and turn it into something that business users can work with. The successful database marketing applications will combine data mining technology with a thorough understanding of business problems and present the results in a way that the user can understand. At that point the knowledge contained in a database will be understood by people who can turn what is known into, what can be done.

The section also suggests the implementation of an Internet-based online information retrieval system, which offers agrifood industry sector-specific information, which is directly, linked to management's critical success factors and allows personalization through the implementation of appropriate filtering techniques. The information retrieval system is based on information agents, which search for appropriate documents on distributed information sources. The utilization of information agents has left the early experimental phase and is in the early stages of professional use. Ongoing research on the implementation of agent-based intelligent information retrieval systems for the agrifood sector does focus on the selection of appropriate search directives for the robots, the formulation of taxonomy models that best map users' information needs, and the appropriate retrieval, individualization, and presentation of information from the search results.

In this section the method that integrates data mining and campaign management software to increase the customer value was discussed. The section also considered two case studies of applying standard data mining techniques to industrial questions in the area of consumer package goods. The examples discussed a wholesaler and a retailer who sought better management of product categories and a resulting improved economy of scope. Commercial success of data mining will in part be dependent upon the capacity of algorithms to model complex, hierarchical arrangements of goods and products. Also how data mining is applied in finance and finance data analysis has been described.

18.9 Review Questions

1. Compare data mining and database marketing.
2. Explain in detail data mining for marketing decisions.
3. What is agent-based information retrieval systems?
4. What are the applications of data mining in marketing and explain in detail.
5. Define scoring, campaign management, and customer segment attrition.
6. What is the role of campaign management software?
7. What are the steps involved in implementing integrated DM and campaign management process?
8. What are the benefits of integrating data mining and campaign management?
9. Explain in detail how data mining is used in finance.
10. Describe the design and construction of data warehouse for multidimensional data analysis and DM.
11. Explain data mining for financial data analysis.

19

Banking and Commercial Applications

Objectives:

- The wholesale side of a commercial bank has a significant need to invest in business intelligence and data mining technology because wholesale banking provides a large percentage of a commercial bank's revenue.
- A business intelligence solution, which summarizes this information in the form of query-based reports, augmented by the predictive power of data mining technology, can greatly enhance the corporate decision making process.
- Various groups of analysts at the bank employ data mining software ranging from proprietary solutions (neural networks in stock brokerage, decision trees in credit risk), PC-based single algorithm tools (knowledge Seeker), to SAS on MVS and Intelligent Miner on SP2.
- DM supports a unit view of patterns in data, especially in cases where similar project objectives and data deliver discrepancies result.
- The KDD process used in commercial applications performs data cleaning and data preparation, which has to be done before the actual data mining can take place.
- Decision support is like a symphony of data. It is best to look at decision support from a functional (rather than a tool-based) perspective and to look at available technologies in terms of the functions they provide.
- The model has to be integrated into existing systems, applied on a day-to-day basis, and is shown to meet a target in cost savings.
- To sell products and service performances via Internet in a really permanent and successful manner, this means very special requirements to today's e-commerce platforms.
- The diversification of product and performance portfolios and an improved customer relationship management (CRM) on the basis of asset situations, venturesomeness, payment habits, and consumer behavior will become the driving forces of the use of data mining in e-commerce.

S. Sumathi and S.N. Sivanandam: *Banking and Commercial Applications*, Studies in Computational Intelligence (SCI) **29**, 439–472 (2006)
www.springerlink.com © Springer-Verlag Berlin Heidelberg 2006

- Retail data mining can help identify customer buying behaviors, discover customer shopping patterns and trends, improve the quality of customer service, achieve better customer retention and satisfaction, enhance goods consumption ratios, design more effective goods transportation and distribution policies, and reduce the cost of business.
- The most innovative retailers of today are those who use business intelligence to gain sustained competitive advantage.

Abstract. The applications of data mining in banking, e-commerce, retail and commercial areas with case studies are illustrated in this section. It discusses the application of data mining in the wholesale banking industry, illustrates some of the associated challengers, and recommends the development of a domain-specific knowledge-encoding tool. Although we focus on wholesale banking, these observations have direct parallels elsewhere, including retail banking, pharmaceutical, and manufacturing industries. Section 19.2 includes a case study based on the distributed data mining. The case study discussed was taken from Global Information Technology, Bank of Montreal, Canada. Another case study Dimensional Systems, Cambridge, MA, is discussed in Section 19.4 based on the decision support systems.

There is a tremendous interest in data mining applications in the commercial world. Many companies, after their initial success with their data mining research and pilot projects, are starting to move these projects into real-world business applications. Today most of the data mining development has come from the research (academic) world. As a result, a lot of business issues have not been addressed and so far they have only received cursory attention. The section highlights some of these issues to the KDD research community so that we could bridge the gap between the research (academic) world and the business world.

Data mining is the "non-trivial process of identifying valid, novel, potentially useful and ultimately understandable patterns in data." To sell products and service performances via Internet in a really permanent and successful manner means very special requirements to today's e-commerce platforms. The virtual market in the network of the networks develops faster and, in the future, will be more important than the market of the real world. The offers are more extensive, the purchasing decisions by the customers are made faster, and last but not least: The competitor's product is only a mouse-click click away from the respective Web pages! Section 19.6 outlines the types of data available for mining, describes operation and limitations of the mining algorithms, and discusses the marketing and ethical issues that arise.

Retail data mining can help identify customer buying behaviors, discover customer shopping patterns and trends, improve the quality of customer service, achieve better customer retention and satisfaction, enhance goods consumption ratios, design more effective goods transportation and distribution policies, and reduce the cost of business.

Section 19.8 explores the various applications of business intelligence in the retail industry. Business intelligence refers to a host of technologies like data warehousing, online analytical processing (OLAP), and data mining, which seek to turn data into actionable information.

19.1 Bringing Data Mining to the Forefront of Business Intelligence in Wholesale Banking

Although the basic algorithms of data mining technology have been available for many years, data mining has not yet realized its full potential as an integral part of some business intelligence solutions. As in most industries, the success of a data mining implementation as part of a viable business intelligence solution depends primarily on the accessibility of the data, the level of integration of the data mining software to existing databases, the ease of data manipulation, and the degree of "built-in" domain knowledge. Each of these issues merits careful thought and analysis, although the focus here is on the last issue. The wholesale side of a commercial bank has a significant need to invest in business intelligence and data mining technology because wholesale banking it: provides a large percentage of a commercial bank's revenue, thus making the potential return on investment attractive; needs to manage client relationships across a broad spectrum: from a relatively small corporate entity, to a large multinational corporation with potentially dozens of subsidiaries and closely aligned business partners; and is affected by a large number of factors, including macroeconomic factors such as international economic forces, industry-specific trends, the real estate market, and microeconomic factors such as a particular client's economic health and leadership.

A business intelligence solution that summarizes this information in the form of query-based reports, augmented by the predictive power of data mining technology, can greatly enhance the corporate decision-making process. In a typical data mining project, these data are brought together at an appropriate level of generality that describes product usage as well as client information. Although obtaining data from internal sources and external vendors is not difficult, creating an appropriate data set for mining is challenging for many reasons, a few of which are given here:

"Customer" is not easily defined. A large wholesale bank may work with a "parent" company as well as potentially dozens of subsidiaries, each with its own set of legal and financial constraints.

Product cohorts. These can lead to an array of analytical problem because a corporation's usage of one product set may indirectly their use of another product set because of legal or financial structure.

Patterns of product usage are sporadic, cyclical or inter-related. This perennial problem is found in many industries but is particularly exacerbated in wholesale banking because of product cohorts and complex corporate relationships.

Given that this list is not nearly exhaustive, simply the creation of an appropriate mining database is time consuming and challenging. Furthermore, consideration of these factors in the data mining analysis is crucial for data mining to gain acceptance within wholesale baking marketing channels. Lines of corporate influence, product cohorts, and data quality issues are often known by domain experts but are not directly reflected in the data.

To illustrate, consider a product demand forecasting analysis, which is a typical application of data mining in this industry, where the task is to predict a corporation's demand for a banking product given other product usage information, relevant economic ratios, and other macroeconomic variables. Building a good predictor here requires a clear understanding of the relationship between product groups. A forecasting analysis might indicate the "usage of cash management products imply usage of investment banking products," but this may be information that is implied by bank policies, or is necessary because of other economic considerations. A tool that reflects these internal relationships when delivering results would greatly improve the feasibility of data mining in this industry. The details of such a tool fall outside of the scope of this section, but a few points can be made regarding it. A simple tool may include the following characteristics:

- A straightforward interface to link known concepts together, such as known lines of corporate influence, known product cohorts, etc;
- A data manipulation feature to effectively deal with hierarchical variables. An internal validation of the data mining results against a table of known encoded relationships;
- A graphical display of known relationship, with an overlay of the "predicted" or "discovered" relationships would greatly aid in understanding the model and in the delivery of executive information; and
- A feedback mechanism that gives the analyst the ability to differentially enhance the strength some relationships to enable "what if" analyses.

19.2 Distributed Data Mining Through a Centralized Solution – A Case Study

This is to present a corporate data mining solution, which supports the current and future large-scale analytical needs of most of the Bank of Montreal lines of business. This case study is taken from Global Information Technology, Bank of Montreal, Canada.

19.2.1 Background

Data mining has quickly matured out of isolated, small-scale, PC-based, single algorithm techniques to a robust analytical solution, which utilize a combination of various artificial intelligence algorithms, massively parallel technology, direct both-way access to relational databases and opened systems with published Application Programming Interfaces. In the banking industry, data mining techniques have been accepted by the statisticians' community and utilized side by side with more traditional statistical modeling techniques. Various groups of analysts at the Bank employ data mining software ranging from proprietary solutions (neural networks in stock brokerage, decision trees in credit risk), PC-based single algorithm tools (Knowledge Seeker), to SAS on MVS and Intelligent Miner on SP2. Among these groups the analytical, technological, and statistical skill sets and expertise vary.

Growing interest in Data Mining technology at the Bank is driven by numerous factors:

- *Strategic/political:*
- Growing awareness of DM (at the executive as well as business analyst levels)
- Recognizing need for elastic knowledge-based decision making – gaining competitive advantage by responding quickly to changes
- Growing competition in the industry
- Internal competitiveness among divisions, departments, and their respective leaders
- Influence of various consulting companies
- Increased demand by the business (analysts, product managers, branch managers) to be granted direct access to the information
- *Practical:*
- Growing volumes of relatively clean data stored in data warehouse, data marts, and operational data stores.
- Limitations in standard analytical/statistical approaches in terms of number of variables considered for a model, their scarcity, cardinality, and the (high) number of categorical variables requiring special treatment.
- Increased demand by analysts to work with large data samples or even full data sets (fraud detection, credit risk, etc.)
- The need to analyze data from several or all lines of business (cross selling, credit risk, product cannibalization, customer behavior).

The experience with the current status quo shows that satisfying the above needs locally and on ad hoc bases is neither practically manageable nor cost effective. Therefore, at the Bank they are in the process of implementing a robust *Centralized Data Mining Solution (CDMS)* supported by the creation of *Data Mining Center of Excellence*, an institution responsible for managing the new Hardware/Software (HW/SW) and its utilization by various DM groups, providing data transformation, managing DM metadata, and deploying precanned models to light users via the Intranet.

Benefits of CDMS: The CDMS provides high-speed links/gateways to the major data sources deployed in the organization (bank information warehouse, customer knowledge data mart, credit card, risk management, etc.) totaling more than 3TB of data. The likely power user groups are: database marketing, credit risk, credit card, transfer pricing, and others. These groups will be freed from tedious data transformation and HW/SW maintenance responsibilities.

It is the nature of data mining projects to require initially large space to store data. Contrary to what is said, in the data marts the storage requirements are limited to the initial data crunching and the life of the project. Therefore most of the storage could be freed in several weeks and made available to other projects. Equipped with massively parallel processing and sharing almost 0.5TB of DASD users will benefit from unprecedented processing power allowing them to run on large data sets.

Benefits in Costs: The data mining technology is expensive to acquire (SW and dedicated HW) and difficult to maintain (operating system, underlying database, data mining SW itself, gateways, front ends, etc.). And let us emphasize the support of various high-speed links and gateways, data marts, and other business critical sources of data. It is hard to imagine a LOB being able to justify a purchase of top-notch data mining technology, a solution that could cost an initial $600,000 of investment and $300,000 a year for support and maintenance. Such a system would likely be underutilized by the LOB most of the time.

Obviously, the CDMS is highly scaleable and in the long term is a cost-efficient solution – a win-win for all groups involved and for the organization as a whole. For its growth this point and corporate-managed data mining effort give additional power in negotiations with HW/SW vendors.

Data Mining Meta data: One important responsibility of the Data Mining Center of Excellence, which manages the CDMS, is the creation and management of corporate data mining metadata. This is a relational database outlining all data mining projects, models built, their frequency, contacts to creators, data sources involved, variables, and a models' version control. Details about particular model parameter settings, treatment of NULLs, outliers, etc. will be included. The metadata will be published on the Intranet and accessible by all Bank employees.

This solution will help in interpretation of DM results, especially in cases where similar project objectives and data deliver discrepancies in results. It supports a **unit view of patterns** in data.

19.3 Data Mining in Commercial Applications

19.3.1 Data Cleaning and Data Preparation

An important component of the KDD process is data cleaning and data preparation, which has to be done before the actual data mining can take place. The current state of art of data cleaning is far from being as sophisticated as, say data mining algorithms. Most of the data cleaning is done laboriously be humans. There are few general and automated tools that assist in this process.

The idea is to use some of the data mining techniques in data cleaning. We envision data cleaning to be performed in several stages, ranging from application of simple techniques to increasingly more sophisticated methods. For example, one can use visualization techniques to look for missing data, imperfect data, etc. Or, one can use more sophisticated clustering tools to look for outliers, which may indicate potentially erroneous data. Statistical methods that look for missing data need to be studied in this context. Of course, the entire process will be an iterative one, with periodic evaluation by humans and the reapplication of appropriate cleaning tools.

Let us take an example. Consider a database maintained by a county office. This database contains information on every piece of real estate in the county. Such information might include the address, price owner, size, etc. Such databases may contain lots of errors, missing data, etc. Initially, we could use visualization techniques to locate each house on a map of the county. If a high-priced house appears in a low-income neighborhood, then that is possibly an error. Secondly, we could use multidimensional clustering techniques to look for "outliers" in clusters, which might point to erroneous data. If we have much prior knowledge about the data, we can assume a prior distribution that models the database and use Bayesian methods to compute posterior distributions. Then, we could selectively omit certain data items and re-compute posterior distributions again, thereby allowing us to detect outliers.

19.3.2 Involving Business Users in the KDD Process

Successful data mining applications in business world would require constant interaction and feedback between different users and the data mining process.

User interaction, in different aspects of the KDD process:

Previous knowledge:- The business user may have some underlying (but incomplete) knowledge about the data to be mined. Incorporating this knowledge into the data mining process will allow the data mining process to be more efficient.

On the other hand, the data mining process can also be used as a check against the knowledge that was supplied. For instance, mining may contradict some conventional wisdom. The system should allow for that. Thus we should provide methods to allow data mining algorithms to provide reasonably good results if there is not enough resources for a full-scale job.

Resource constraints:- The need to be able to mine information from a huge resource in real-time (or near real-time) is increasing. The demand of the real world does not allow the luxury of time.

Intermediate feedback:- Each step of the KDD process should provide some intermediate results. For example, results of the data cleaning process, decision of which algorithm to use, etc. The system should be able to provide appropriate feedback to appropriate personnel (users) so that they can make decisions to guide the data mining process.

19.3.3 Business Challenges for the KDD Process

The business issues are many, such as scalability, integration of current systems, data visualization, need for database system support, incremental processing, flexibility and so on. The important issues to be concentrated are:

Data Cleaning and Data Preparation

The research issues are many. How does one design algorithms that can work for heterogeneous datasets? Do the algorithms scale well? Can we perform automatic error "correction" in addition to just error detection?

Knowledge representation and incorporation

How to represent the underlying knowledge that is known? How to incorporate this knowledge into the appropriate KDD algorithm? How to discern whether the underlying knowledge is useful or counterproductive?

Algorithmic issues

To devise algorithms that would take resource constraints into account; to provide algorithms that can provide good intermediate results even if severe resource constraints are imposed. There has been some work on constraint algorithms and incremental algorithms. We need to devise method to incorporate them into the KDD process.

Data visualization and feedback

To provide effective methods for the data mining process and to provide feedback understandable by the user. Different kinds of users may require different kind of visualization techniques. Also, effective means of allowing user feedback to the data mining system need to be devised, ideally coupling these to the data visualization process.

19.4 Decision Support Systems – Case Study

Decision support is like a symphony of data. In the same way we need a variety of musical instruments functioning together for the purpose of playing a symphony, we need a variety of software tools working together for the purpose of doing decision support. It is best to look at decision support from a functional (rather than a tool-based) perspective and to look at available technologies in terms of the functions they provide. In other words, tool features get mapped to DSS (Decision Support System) functions. This case study is taken from Dimensional Systems, Cambridge, MA.

19.4.1 A Functional Perspective

Basic DSS functions

DSS is about synthesizing useful knowledge from large data sets. It is about integration, summarization, and abstraction as well as ratios, trends, and allocations. It is about comparing database generalizations with model-based assumptions and reconciling them when they are different. It is about good, data-facilitated creative thinking and the monitoring of those creative ideas that were implemented. It is about using all types of data wisely and understanding how derived data was calculated. It is about continuously learning and modifying goals and working assumptions based on data-driven models and experience. In short, decision support should function like a virtuous cycle of decision-making improvement. Let us laundry list these concepts to identify the minimum set of basic functions that comprise any DSS framework:

- Data collection,
- Data storage and access organization,
- Dimensional structuring,
- Data synthesis,
- Intuitive representations and access models,
- Predictive models,
- Model verification
- Knowledge sharing,
- Resource allocation strategies,
- Scenario analysis,
- Belief conflict resolution,
- Prescriptions and
- Decision implementation capabilities.

A Cognitive Metaphor for DSS: Beyond Closed Loop Systems

The best metaphor that one can think of for understanding how all these decision support functions fit together is a cognitive one. In contrast, earlier metaphors focused on the unidirectional flow of information from "raw" data to synthesized knowledge. Second-generation metaphors, currently in vogue, focus on bidirectional, closed loop systems wherein the results of DSS analysis are fed back into production systems. The hallmark of a third-generation cognitive metaphor is the interplay of two separate information loops. The first is akin to the closed loop system and one would characterize it as a data-driven loop. But in addition to that loop there exists an inner loop where data driven information meets model-driven goals and beliefs at the moment of decision. Although that inner loop is frequently provided by a living breathing person, it is a function that needs to take place and in automated systems needs to take place in the form of software within the overall decision support system. AI workers have known for a long time that it takes a combination of data-driven and model-driven information to produce high-quality decisions.

Using a cognitive metaphor, the universe of DSS functions is composed of five distinct functional layers within which the two above-mentioned information loops interact: a sensory/motor layer, a primary memory layer, a data-based interpretive and understanding layer, a decision layer, and a model-driven layer of goals and beliefs.

Data-Driven Understanding

From 20,000 feet, data-driven understanding is the process of synthesizing knowledge from large disparate data sets. *Understanding* is a loaded term, so let us break it down into smaller chunks. The key components of understanding are describing, explaining, and predicting. The main obstacles to understanding are lack of tool integration, missing, meaningless, and uncertain data and lack of verification capabilities.

Descriptions form the basis. Examples include "The Cambridge store sold 500 pairs of shoes last week," "Our corporation did 35 million last year," "The Boston stores paid an average of 36 dollars per foot in 1997" and "Boston rent is twice as expensive as the rent in Portland, Maine." Descriptions are more than just measurements. Descriptive processes take whatever raw measurements there are, and through aggregations, ratios, and other certainty-preserving operations, creating a fleshed out multilevel multidimensional description.

In practice, there may be a variety of inferential techniques employed to arrive at a descriptive model of a business or organization. For example, inferential techniques may be used to guess what values may apply to what are otherwise missing cells. When there are a lot of blanks that need to be filled in for a descriptive model the process is akin to data archeology. Good data archeology requires the close integration of OLAP and data mining or statistics tools.

Explanatory modeling starts where descriptive modeling left off. Explanatory models are representations of relationships between descriptions. Such statements as "For every increase of 1% in the prime rate, housing sales decrease by 2%" represent explanations or relationships inferred from descriptions of both housing sales and interest rates. The functionality provided by statistics and data mining tools of all varieties belong in this arena. Regressions (the mother of all analyses), decision trees, neural nets, association rules, and clustering algorithms are examples of explanatory modeling.

Predictive modeling is just an extension of explanatory modeling. One cannot make a prediction without having at least one relationship that we are banking on. And while most all-mining activities aim at building predictive models, the key algorithms are in the discovery of the patterns. Predictions are just the extension of some pattern already discovered. That is why all the data mining algorithm buzzwords, we hear are about pattern discovery techniques, not pattern extension.

OLAP tools do not provide for explanatory or predictive modeling. Data mining does not provide for dimensional structuring. Yet, it is best to perform data mining within an OLAP (multilevel, multidimensional) environment. For example, to design a new promotional campaign based on point-of-sale POS and demographic data we might use

- An OLAP tool to aggregate SKU-level data to product categories,
- The OLAP environment for exploratory analysis and new variable creation,
- Clustering to discover natural segments in the POS data,
- Visualization to interpret the clusters,
- OLAP to incorporate the clusters as higher level aggregation levels,
- More directed mining on the cluster-aggregated data,
- Visualization to interpret the mining results, and
- OLAP to further aggregate the mining-based predictions

all in order to support the data-based brainstorming for a new promotion campaign. In short, we needed to use a combination of OLAP, data mining, and visualization to accomplish a single BI task: promotion development. We can call this kind of integration **DSS fusion**.

The market as a whole is beginning to move in this direction. A number of OLAP companies are adding or claiming to add data mining capabilities, although not all of them are fully integrated with their OLAP products. For example, Holos is adding mining capabilities, Cognos has a simple mining application, MIS AG has a mining application, as does Pilot Software (D&B, Data Intelligence Group). We believe it will be easier for OLAP companies to add data mining capabilities than it will be for data mining companies to add OLAP capabilities.

Although it is good to see so many OLAP vendors offering mining capabilities, these capabilities still need to be better integrated. Mining functions should be as simple to invoke as ratios. It should be possible to perform data transformations from within an OLAP/mining environment. And mining should be fully integrated within the dimensional structure meaning that operations like drill down from an interface to the results of an association rule algorithm should work, and depending on how things were defined either return a set of associations already calculated for lower down in perhaps a product hierarchy, or trigger the calculation of such associations. Thus the same thinking that goes into an OLAP design of what should be precalculated and what can be calculated on demand can apply to data mining as well.

Missing, meaningless, and uncertain data are frequently present in data sets and pose a significant hurdle to understanding. Missing and meaningless data are logically distinct, both need to be distinguished from the value zero (frequently the default value assigned an empty cell), and need to be differentially processed. For example, in an averaging function, where empty cells denoted *missing*, we would need to assign some kind of default value to the empty cells for the purposes of aggregation. In contrast, if the empty cells denoted *meaningless*, we could not assign a default value. Most OLAP and data mining tools lack good empty cell handling techniques.

Unlike missing and meaningless data, uncertain data is present as a data value in a data set. For example, a statement like "We predict that our new brand of ice cream will capture 5% of the market for ice cream products," needs to qualify the uncertainty associated with the estimate of 5%. Typical statistical measures of uncertainty include the variance and bias associated with an estimate. The overall picture of uncertainty can get a little more complicated as derived measures follow from business rules, which have their own sources of uncertainty. For example high-level sales forecasts based on aggregating lower level predicted sales data need to carry forward the uncertainties derived from the predictive models through the aggregation process. What is more, the predictive models themselves may rely on certain "rules of thumb" for their forecasting logic. As more assumptions become embedded in business data, OLAP tools especially will need to provide ways to process uncertainty.

Finally, in the same way as an astronaut's working environment is composed of fabricated living elements (temperature, pressure, oxygen, etc.), a DSS end user/analyst's working environment is composed of fabricated data elements (daily summaries, weekly aggregates, brand reports, and so on). Given the degree to which end users are dependent on derived data as their own inputs, it is crucial that DSS vendors provide better verification capabilities.

19.4.2 Decisions

Deciding is the process or function of combining goals and predictive models. To decide that prices need to be lowered on certain products is the result of a goal to maximize sales and a predictive model that relates sales to product price. To decide that a certain loan applicant should be denied credit is the result of a goal to minimize loan write-offs and a predictive model that relates certain loan applicant attributes with the likelihood of loan default.

If there were no goals, it would be impossible to decide what course of action to take as any action would be as acceptable as any other. Without the goal of maximizing sales, for example, there is no right decision concerning product pricing. And without a predictive model equating product prices with product sales there is no way to know which decision will be most likely to maximize sales.

Decision-making challenges may arise from

- the need to automate certain decision-making functions,
- the need to ensure consistent decisions,

- difficulties analyzing how a decision was made,
- complexities in the predictive models,
- difficulties interpreting stated goals,
- instability in the goals themselves,
- interpersonal dynamics,
- fluctuations in the predictive models, and
- conflicts between data-driven and model-driven beliefs.

Business rule automation tools focus on the first two challenges. Decision analysis tools focus on challenges 3–6. Group decision support tools focus on challenge 7. Challenges 8 and 9 lie a little further down the road.

Looking ahead, we will start to see self-modifying rule systems that continuously monitor the world to see if it behaved like predicted, and when it does not, then changing the predictive models it used to make rules. In the process, systems could try out different scenarios or predictive models and analyze how well the system would have fared under each scenario. We would also like to see rules bases connect to OLAP tools wherein the rules base was the source of cost allocation rules used in the OLAP system. Although OLAP tools provide a sophisticated calculation environment they would benefit from an organized method of defining and managing rules.

We should also be able to deduce rules given goals and predictive models, which brings to the next major category of decision-making software, decision analysis software. The need for decision analysis software kicks in where decisions are based on multiple predictive models with complex measures of uncertainty and where the goals themselves are variable. Typically this appeals to higher up in an organization. Decision analysis is closely related to operations research where there are several mutually exclusive goals and shared scarce resources and the trick is to maximize a global property like profit, stability, or happiness. In summary,

- The BI or DSS space is best looked at from a cognitively based functional approach rather than any tool-centric perspective,
- The center of decision support is at the intersection of data-driven understanding and model-driven goals and beliefs,
- Data collection and storage should be driven by decision-making needs,
- All media should be organized through an integrated semantic model of the enterprise,
- Data mining, OLAP, and data visualization can and ought to work together for the majority of decision support problems,
- The analytical knowledge required for interpreting complex derived data should be disseminated to end user/analysts,
- Descriptive and explanatory modeling should begin leveraging textual data,
- Model-based decision-influencing beliefs should be captured electronically and compared with data-driven versions of the same beliefs,

- Procedures for resolving conflict between data and/or model driven beliefs should be public, and
- Everything should be verifiable.

19.5 Keys to the Commercial Success of Data Mining – Case Studies

Some techniques to the commercial success of data mining are presented based on (i) The experience at BT Laboratories, Data Mining Group, Ipswich and (ii) A service provider's view taken from a global multidisciplinary professional services firm – Arthur Anderson, Zurich, Switzerland.

19.5.1 Case Study 1: Commercial Success Criteria

Success criteria are much more commercially oriented than they were in the early days. It is no longer enough to produce a convincing demonstration system or model. It has to be integrated into existing systems, applied on a day-to-day basis, and be shown to meet a target in cost savings. Defining and estimating the costs and benefits of a proposed project can be difficult, but we have found that the more effort that is spent on this early on, the better.

Reasons for Failure

We have carried out many data mining projects over the years. In our experience, the key to a successful data mining project is not obtaining some data and finding a useful pattern in it. Only a small number of our projects have been unsuccessful in the sense that we could not find anything useful in the data. More usual reasons for failure are:

- inadequately defined objectives,
- an inadequately thought-out exploitation route for results;
- external factors such as changes in the business environment that make the objectives of the project no longer relevant, and re-organizations.

To avoid these problems, we spent a lot of effort in the definition and planning phase of data mining project and have produced our own data mining project guidelines, in conjunction with Syntegra, BT's systems integration arm, for the company to follow.

A commercial data mining application takes more skills and people than just the data miners; the successful projects are those in which data mining is just seen as a component of an overall project or system. This tends to focus the objectives and deliverables of the data mining aspect, and avoids it being overhyped, if that is possible!

Data Extraction and Preprocessing

We still experience delays and difficulties in obtaining the necessary data for projects, particularly if the data is not in a data warehouse. And a lot of effort is spent preprocessing the data prior to analysis. This is not, however, a severe problem if planned for.

Our Rule

We have found that the role of the group has changed in recent years. Up until about a year ago, most of our nonresearch work was data analysis, i.e., doing data mining for another part of the company. We still do a lot of this, but often we find ourselves acting as consultants, advising other parts of the company about data mining. For example:

- Will data mining alone solve our problem?
- What tools do we need?
- What skills do we need?
- Which tool or supplier should we go with?

This changing role is our response to the increase in demand for data mining and also the increasing choice of suppliers and tools.

Areas for Improvement in DM Software

The suggestions for improvements to existing DM software concur with those cited in the literature, for example: integration with relational databases, scalability, etc. Obviously, the onus is not just on the data mining tool developers to achieve integration, but also major database vendors. Databases should support operations often used in data mining efficiently, for example, random sampling.

Several tools suppliers have developed data mining tools, which, it is claimed, can be used by nonspecialists, often via an easy-to-use interface or data visualization. While this is certainly an improvement, we believe such tools will not find a large market of nonspecialists. This is because nonspecialists do not want to have data mining made easy for them – they really do not want to do data mining at all in a general context – what they want is to solve their particular problem, be it, targeting their marketing, highlighting potential fraudsters, etc. The likely outcome will be the development of "vertical" applications for common business problems in particular industry sectors, in which the data mining element is largely hidden from the user. Examples include systems designed to highlight customers likely to leave a mobile telecommunication company for the competition ("churners"), telephony fraud detection, and targeting personalized adverts at Internet users.

Problem Formation

Problem formulation, along with risk assessment and project planning, is very important. We spend a lot of time upfront with the customers of our work working out as well as we can exactly what the business benefit is hoped to be, and how it will be realized.

DM Lifecycle

We have produced our own guidelines for managing data mining projects, called M3, for use by BT. It uses a DM lifecycle similar to those in the literature in Fayyad 1996, but places emphasis on the early stages of a lifecycle: particularly problem formation, data investigation, risk assessment, feasibility check, cost-benefit analysis, project planning, role identification, and defining a clear exploitation route. This can act as a checklist for the analyst where all the relevant aspects have been covered and provide some tips on pitfalls to watch out for. For example, the role identification part of M3 defines several roles that may have to be involved in, and kept committed to, a successful data mining project: the analyst himself/herself, the domain expert, the database designer/administrator, the customer, the end user, the legal expert, the system developer, and the data subjects themselves.

19.5.2 Case Study 2: A Service Provider's View

Arthur Andersen is a global multidisciplinary professional services firm that helps its clients improve their business performance through assurance and business advisory services; business consulting; economic and financial consulting; and tax, legal, and business advisory services. Arthur Andersen Business Consulting in Switzerland offers services and provides implementations in the areas among others of cost management, revenue enhancement, activity-based management, performance management, knowledge management, transaction systems, collaborative systems, data warehousing, and OLAP and data mining technologies.

Since we are both a user of data mining software and provider of data mining solutions, our view includes both aspects: what are the benefits and application areas of data mining technology as well as directions for improvement for data mining software.

- *Interfaces.* As for any information processing system, it is important to get data in and out of the system. The advantage that data mining applications have is that they are usually built on top of a data warehouse. Hence, they do not have to provide access to a variety of operational systems, but just to a database directly. This is to be preferred due to the massive amounts of data that need to be transferred. However, the information chain does not stop with the data mining application. It is merely an intermediate element. The information that comes out of such a system has to go either into an operational system, e.g., a mail targeting or a customer service application, or needs to be passed on to decision takers as part of a knowledge management framework. The amount of data could be small – in the case that only the model and some related information is needed – but more often it will larger – if customer records are to be augmented with credibility scores or customer segment information. In the latter case, this data could be written back to the data warehouse and from there sent to the operational system. By doing this, one could avoid building special interfaces for the various operational systems. Most data mining applications are also not intended for knowledge sharing in contrast to collaborative systems such as electronic mail and Web pages or management information systems. Hence, models that have been created in a data mining process such as decision trees or scorecards need to be further processed electronically to make this information accessible and usable.

- *Model Scoring.* Everyone knows about the things that can happen when one tries to learn a model from large datasets and that can decrease the quality of the model or even make it useless, like, for instance, over learning. Especially, when we try to anticipate the customer's future behavior, e.g., in churn management or credit scoring, we need to know about the quality of the model. Some data mining tools that originate in statistical programs allow model scoring, whereas tools that are more application driven and are geared toward the end user just offer the model at the end. Another point is that the need for a data mining applications originated in a business problem, which was then converted into a data-mining task. Hence, the score needs to be translated back into terms of the original business problem. For example, we need to know about the costs of the errors we will make due to the quality of the model. Data mining works on historical data; but usually new data is constantly flowing into the databases. Rebuilding the model and rescoring the model are two different tasks with different resource consumption. Model scoring could help us to decide when to rebuild the model.
- *Dimensions with Interrelated Values.* Some dimensions are not just a collection of values. First of all, there is the time dimension. The fact that the values in this dimension represent points in time puts them into a certain order and augments them with an adjacency relation. This gives them a completely different meaning compared to unrelated values like customer IDs, company departments, or products. But time is not the only dimension where we have additional information about its values. The geographical location is another dimension where values are linked by an adjacency relationship. We know that certain states are neighbors. This does not give them an order like it is the case in the time dimension but this relationship could help in the explanation of certain phenomena. We also know that countries are composed of smaller administrative districts

and in turn are located in one of the five continents. This knowledge forms a hierarchy on top of the values. Likewise, we put customers into customer segments and products into product groups. Sometimes we could treat this information as an additional attribute, however, this might not always be feasible, especially, if the hierarchies do not have clearly identifiable levels. But this hierarchy puts certain values into a closer relationship than others and it would only make sense to make use of this information.

- *Operational Use vs. Ad hoc Analysis.* We see data mining software as the extension of current decision support systems, such as reporting tools and OLAP technology. We gain more power in two directions: (1) the range of questions we can ask the system to answer and (2) the amount of data that can be processed. Reporting software, especially one that produces paper, is mainly used for regularly, e.g., monthly, quarterly, or yearly, standard reports. There are usually only a few different types of reports and they contain mostly too many details for one user and too few for another. If such a report gives rise to a question that it cannot answer then there is need for an ad hoc analysis, which almost always means calling up someone in IT and have them design and print a special report. OLAP technology tries to cover both: regularly standard reports and ad hoc queries. By providing drill-up and -down as well as drill-through functionality it can serve the users with the right level of detail. Starting from such an electronic report, the user can furthermore adapt it by slicing and dicing to view the data in different ways and perform an ad hoc analysis. We believe that data mining will be used in much like the same ways. There will be standard applications (runs of data mining algorithms and presentations of the results) that will be scheduled regularly. This might be a specific task, which has been designed in a collaborative effort by a business user and a statistician. The configurations and sets of parameters need to be stored. We would like to have a user interface where most if not all of the configuration complexity can be hidden. Another request might be to compare the results of a sequence of such regularly performed applications and to look at changes over time, like for instance review the changes of the criteria for grouping customers into segments. On the other hand, there remains the need for ad hoc analyses. There might be questions arising from standard applications or we need to prepare and monitor the results of nonregular actions. Data mining software will have to provide the means to satisfy both needs.
- *Business Templates and Automated Model Selection.* After having been occupied with implementing all the necessary technology for visualizing multidimensional data and building user interfaces, OLAP software providers are now taking the next step and delivering templates and shells in order to reduce the amount of work required to build a solution to a specific business problem from scratch using a very generic tool. This idea tries to leverage the underlying commonalties of certain classes of business problems and to shift work being done several times by several end users or by

service providers to work being done once by the software producer. Also, now that much of the technology is implemented in all of the different OLAP tools, providing business templates could be a market differentiator for a software company. With suppliers of data mining software trying to add on more and more technology, they still have the focus on the technology as opposed to the focus on the business problem to solve. However, we need specific solutions to specific problems. In addition to the problems of software and user interfaces – something that data mining applications share with OLAP tools – one also needs to select the right algorithm and parameters or the right model from a set of runs of potentially different algorithms. Focusing on a particular business application helps in reducing the interface problem to few sources and target operational systems. By drawing on the experiences of business users of the software and combining them with the expertise that was needed to implement the algorithms for building business templates, software vendors could help reduce costs and efforts of implementing data mining. It could not only differentiate them from other tools that are built on the same technology but also decrease the likelihood of a user blaming the vendor for failures while using its software.

- *Business Applications.* One of the requirements for OLAP software is the ability to build business applications with a number of pages of different standardized views of maybe different data, while hiding most of the complexity of the underlying technology. These pages include settings of which dimensions of the data to show at what level of detail, which have been determined by, for instance, business analysts. This facilitates the use of the software for nontechnical end users and especially for operational, regularly, standardized usage. In our opinion, data mining technology will go in a similar direction. Configurations such as the selection of the algorithm, certain parameters, and the form of representing the results could be developed through the collaboration of business users and statisticians, leaving only a few knobs if at all for the standard user. The page could be extended by other user interface elements such as push buttons or sliders to provide access to preset configurations or standardized views of the results. Being able to build business applications with a user interface and functionality adapted to the end user is, from our point of view, one of the prerequisites to make data mining technology more accessible to a broader range of business users. One wish would also be that data mining software would lay a common ground for communication between business users and statisticians. For instance, being able to demonstrate the effects of algorithm or parameter selections directly on the user's data or, on the other hand, to collectively interpret the results might help the understanding of each other's problems. Also by providing means to translate the business problem into a statistical one and the statistical answer back into a business one, the software could facilitate both communities while building relationship between elements of the business and the statistical world.

- *New Application Areas.* Most of data mining technology is applied to customer data. One of its characteristics is that it is usually stored in a relational database and that there are a huge number of records. However, since in the company we also help the customers in the areas of performance management and activity-based management, we were wondering whether this could also be a possible application area. What if there are a large number of accounts, cost centers, or activities to be analyzed? We think that even a hundred different positions might be too much to be dealt with manually. One could, for instance, look at drastic changes over time, which probably no one would do for one hundred departments. Furthermore, we think that most of the ideas that have been developed regarding customers can also be applied to employees: employee satisfaction instead of customer satisfaction, employee retention instead of customer retention, employee profitability instead of customer profitability. Companies developed bonus programs, as one example, to implement these ideas. With possibly a large number of employees, this would open up a new application area.

Finally, we would propose performing data mining on multidimensional data. Here, the problem is less as regards the number of items per dimension but more with regard to the number of dimensions. With six of seven dimensions and only ten items per dimensions we already have one to ten million data elements. Let us say, we see a drop in revenue. Finding out which product in which configuration in which packaging sold in which location through which channel was mostly responsible for this observation might not be easy and could be time consuming. Furthermore, by only looking at the data on the top level, the business user might not even notice this event, which he should probably act upon at all because it is cancelled out by other data it is aggregated with. Performing multidimensional data mining could again be combined with business knowledge about the different domain that multidimensional OLAP technology has been applied to.

19.6 Data Mining Supports E-Commerce

Many users of the Internet are aware that each time they connect to an on-line shopping server, they leave behind a "footprint" in the site's server logs. The information contained in this footprint is innocuous, but it can be "mined." Data mining is the "non-trivial process of identifying valid, novel, potentially useful and ultimately understandable patterns in data." This section outlines the types of data available for mining, describes operation and limitations of the mining algorithms, and discusses the marketing and ethical issues that arise.

A relatively high product and offer equality of the different offers, a decreasing customer loyalty, a high margin pressure, and the drastically increasing globalization are further factors for the fact that the market chances become more and more narrow.

Web stores that want to guarantee their operators a substantial advantage over their competitors must be able to efficiently support an individualization of the mass marketing so far usual. This individualization and personification in the marketing will become the decisive instrument in the fight for the customers at the beginning of the new millennium.

The more exactly an offer, a price, a bonus, or a tariff is tailored to the customer and the more precise the usage habits, the consumer requirements, the reaction patterns, and the behavioral features of a customer can be predicted, the more positive his purchasing reaction in the store will be. The "tailored" courting and the systematic elimination of "consumer inhibition thresholds" on the basis of previously created customer profiles will open a completely new dimension in the Web marketing. The diversification of product and performance portfolios and an improved customer relationship management (CRM) on the basis of asset situations, venturesomeness, payment habits, and consumer behavior will become the driving forces of this new development.

As no other branch of the business else, the e-commerce is predestined to create the "glassy customer." As with the field of credit and customer card purchases, transaction histories can be examined. The analysis results then flow back to the marketing and form the basis of a selective micro marketing with an enormous degree of efficiency. Through "enriching" the collected data stocks with socio-graphical and microgeographical information (e.g., social background of residential area), the shop operator will allow the shop operator in the future to obtain even more sales-relevant knowledge on his respective customers.

Up selling and cross selling potentials (for example: a person who purchases A will also purchase B in 70% of all cases) can be discovered by means of basket analyses and can be utilized in sales-increasing manner. The hit quote on target-group-optimized offering actions (phone call, mail, or e-mail) can be improved substantially and, thus, unnecessary and cost-intensive "scattering losses" (e.g., stressed customers leave) can be avoided. With that, the "marketing shotgun" belongs to the past. The customer's binding to the product and to the offering enterprise is improved while the sales/distribution costs are reduced. Leaving-endangered customers or customers who have already left can be won back, and new customers can be counted more efficiently. The improvement of sales/traffic ratio must be the objectives of tomorrow's e-commerce solutions. With its data mining solutions, the Prudential System Software is already able to contribute with an essential step in this direction!

19.6.1 Data Mining Application Possibilities in Web Stores

The application possibilities of the data mining in Web stores become particularly clear from those persons' point of view who operate Web stores each day. Just at those points where many decisions are presently still made manually, a decision-supporting data mining system developed by Prudential Systems can contribute to an enormous value increase of Web stores. On principle, two types of administrators operate Web stores: by system administrators and by business administrators.

The task of the system administrator is to make the basic settings immediately after a store has been installed and to change them later, to create the design of the store, to release the store front and, possibly, to block it, to technically maintain the business administration as well as to do database manipulations.

Business administrators are engaged with the actual business. They process orders; supervise the inventory; order with suppliers; administer customer account data; change product offers and descriptions; offer discounts for selected customers; define customer categories; and cause products to be summarized in lists, subcategories, categories, and main catalogs. Moreover, they react to selected, but not ordered baskets with special offers or a new product structure.

During his work, a business administrator has to make many decisions. However, the data that would be necessary for these decisions are present in a form that is not suitable for him. The data mining can fill this gap by the fact that exactly those decisions are automatically supported, which the business administrator presently still has to make manually or cannot make at all.

In the store, there is essentially distinguished between customers and products. Products can be summarized in lists, subcategories, categories, and main catalogs. The summary influences the presentation of the store and, thus, is relevant for the touch-and-buy behavior of the customers with respect to these products. However it is made by the business administrator without the possibility to actually consider the behavior of the customers because he does not have the needed information. Here analyses of the touch-and-buy behavior can give important hints on the classification.

Moreover the business administrator classifies the customers to form customer categories, and customers can assign themselves to profiles. The assignments are always made manually without evaluating the behavior of the customers. So, for example, discounts are allocated manually, and the customers manually determine those catalogs and products that they want to view.

In a big store with many customers, it is surely impossible for the business administrator to perform the assignments completely or even optimally. Moreover, only a few customers will utilize the possibility to assign themselves to profiles or they select the wrong profile. Consequently it is required to cause this assignment process (i.e., the individualization of the store) to be carried out automatically. The data mining is the philosophy that can and will realize this. The PrudSys ECOMMINER developed by Prudential Systems is the tool that puts life into this philosophy!

Letting Business Users Loose: In the commercial world, the term *data mining* has become associated with large, multihundred-thousand dollar projects taking several months and requiring the skills of PhD data miners with years of experience. A business manager wanting to use data mining to gain some competitive advantage might embark upon a large project and work with skilled data miners. But this can be cumbersome: large projects require many rounds of review, assessment of the set of vendors and consultants who offer products and services, project scooping, possibly buying new hardware, working with their IT department, etc. It is quite an obstacle. If this situation is one point on a spectrum, another point at the opposite end would be where applications already exist within the company that would allow a business person to mine their easily. The business manager might just point her Web browser at the site that contains the data and applications for analysis, look around for a while and discover, perhaps, that the jump in her group's expenses this quarter is primarily attributable to travel and hiring expenses, or that the majority of her lost customers were affluent single men who lived close to her competitor's store.

Giving power tools to novice business users can be a recipe for disaster – what if they are pointing the tools at data they do not fully understand? What if they build a predictive model and accidentally include input variables that were not all known before the action the model suggests needs to be taken? What if their data is not clean? And so forth. An experienced data miner is someone who has learned to use considerable caution and knows how to avoid pitfalls and traps.

But does this mean we cannot give "data mining" applications to business users? What if instead of just putting a nice GUI on top of a powerful data mining tool like a neural network and calling it easy-to-use, we built simpler analysis methods, whose results were easy to understand? What if the results were phrased as statements about the existing data, not predictions about future data that could be misconstrued. What if, instead of a raw tool that could be turned on any data they happened to have lying around, instead they had applications that were integrated with a data mart that was constantly updated with sales and other data sources, and analyses were restricted to use only data that was known to be clean? These "analysis methods" might look preposterously simple to expert data miners, yet they might be very useful to business users.

A robotics professor once asked his class to design a robot to wash dishes. The inventions that were turned in stretched the technology to its limits: microsensors and force feedback control to avoid breaking the dishes, vision systems to see the dirt, and joints with many degrees of freedom to allow reaching into the sink, wiping the dish, and setting it into the drying rack. The next day, the professor pointed out that dishwashing machines already exist, are relatively cheap to manufacture, and work quite well.

Large data mining projects certainly have their place – in some cases small improvements in a model can result in huge savings, making the investment easily worthwhile. In other cases, a large project may not be explored due to lack of resources. If no applications exist to allow an executive or manager to look at their business and their data, there may be a huge missed opportunity cost from their lack of understanding of their business.

19.7 Data Mining for the Retail Industry

The retail industry is a major application area for data mining since it collects huge amounts of data on sales, customer shopping history, goods transportation, consumption and service records, and so on. The quantity of data collected continues to expand rapidly, especially due to the increasing ease, availability, and popularity of business conducted on the Web, or e-commerce. Today, many stores also have Web sites where customers can make purchases on-line. Some businesses, such as Amazon.com, exist solely on-line, without any bricks-and-mortar (i.e., physical) store locations. Retails data provide a rich source for data mining.

A few examples of data mining in the retail industry are outlined as follows.

Design and construction of data warehouses based on the benefits of data mining: Since retail data cover a wide spectrum (including sales, customers, employees, goods transportation, consumption and services), there can be many ways to design a data warehouse. The levels of detail to be included may also vary substantially. Since a major usage of a data warehouse is to support effective data analysis and data mining, the outcome of preliminary data mining exercises can be used to help guide the design and development of data warehouse structures. This involves deciding those dimensions and levels that are to be included and what preprocessing to perform in order to facilitate quality and efficient data mining.

Multidimensional analysis of sales, customers, products, time, and region: The retail industry requires timely information regarding customer needs, product sales, trends and fashions, as well as the quality, cost, profit, and service of commodities. It is therefore important to provide powerful multidimensional analysis and visualization tools, including the construction of sophisticated data cubes according to the needs of data analysis. The multifeature data cube is a useful data structure in retail data analysis since it facilitates analysis on aggregates with sophisticated conditions.

Analysis of the effectiveness of sales campaigns: The retail industry conducts sales campaigns using advertisements, coupons, and various kinds of discounts and bonuses to promote products and attract customers. Careful analysis of the effectiveness of sales campaigns can help improve company profits. Multidimensional analysis can be used for this purpose by comparing the amount of sales and the number of transactions containing the sales items during the sales period versus those containing the same items before or after the sales campaign. Moreover, association analysis may disclose which items are likely to be purchased together with the items on sale, especially in comparison with the sales before or after the campaign.

Customer retention-analysis of customer loyalty: With customer loyalty card information, one can register sequences of purchases of particular customers. Customer loyalty and purchase trends can be analyzed in a systematic way. Goods purchased at different periods by the same customers can be grouped into sequences. Sequential pattern mining can then be used to investigate changes in customer consumption or loyalty and suggest adjustments on the pricing and variety of goods in order to help retain customers and attract new customers.

Purchase recommendation and cross-reference of items: By mining associations from sales records, one may discover that a customer who buys a particular brand of perfume is likely to buy another set of items. Such information can be used to form purchase recommendations. Purchase recommendations can be advertised on the Web, in weekly flyers, or on sales receipts to help improve customers in selecting items, and increase sales. Similarly, information such as "hot items this week" or attractive deals can be displayed together with the associative information in order to promote sales.

19.8 Business Intelligence and Retailing

19.8.1 Applications of Data Warehousing and Data Mining in the Retail INDUSTRY

The information economy puts a premium on high-quality actionable information – exactly what business intelligence (BI) tools like data warehousing, data mining, and OLAP can provide to the retailers. A close look at the different retail organizational functions suggests that BI can play a crucial role in almost every function. It can give new and often surprising insights about customer behavior, thereby helping the retailers meet their ever-changing needs and desires.

On the supply side, BI can help retailers identify their best vendors and determine what separates them from not so good vendors. It can give retailers better understanding of inventory and its movement and also help improve storefront operations through better category management. Through a host of analyses and reports, BI can also improve retailers' internal organizational support functions like finance and human resource management.

As quite sensitively portrayed in the movie, large chain superstores have nearly forced small independent retailers to close down. At the same time, these large retailers have gained considerable power in the supply chain. They are increasingly dictating terms to the retailers and inventing new ways of attracting customers. But to hold the customer's imagination for long has remained an elusive dream. Changing tastes and preferences, increasing competition, demographic shifts, and the simple "let's try something new" attitude have all been blamed for customer disloyalty. No wonder retailers today are going that extra mile to reach and understand the customer. They are also getting their act together by streamlining the supply chain, improving storefront operations, and actively exploring alternative channels like the Internet. Technology has played a key role in retailers' effort to compete in this volatile market. Sophisticated retailers have quickly evolved from basic automation to embrace new technologies like CRM, business intelligence, etc. This section explores the various applications of business intelligence in the retail industry. Business intelligence refers to a host of technologies like data warehousing, online analytical processing (OLAP), and data mining, which seek to turn data into actionable information.

19.8.2 Key Trends in the Retail Industry

Rise of superstores: Last two decades has seen the phenomenal rise of the "Chain of superstores" in both the US and Europe. Growing consolidation and globalization in the sector have seen the bargaining power of the retailer increase in the supply chain. We believe that in order to counter saturated domestic markets and increasing competition, leading superstores would continue to expand globally. WalMart acquired Britain's third largest supermarket chain ASDA, to establish itself in Europe. Similarly the grocery giant Safeway has significant presence in both the US and UK.

Customer Relationship Management as a key driver: Smart retailers have reoriented their business around the customer. In the mad rush to acquire new customers, they have realized it is equally important to retain the existing ones. Increased interaction and sophisticated analysis techniques have given retailers unprecedented access to the mind of the customer; and they are using this to develop one-to-one relation with the customer, design marketing and promotion campaigns, optimize store layout, and manage e-commerce operations. For example Safeway uses its ABC loyalty card to record each customer's individual transactions. This coupled with other relevant data has given Safeway tremendous knowledge about customer buying patterns – knowledge that has significantly helped in augmenting customer loyalty.

Supply Chain Management as a key driver: Increasingly retailers are handling their inbound logistics by setting up their own distribution networks. We believe that a vital criterion for success in future would be the ability to harness worldwide distribution and logistics network for purchasing. This global supply chain should ensure high levels of product availability that consumers want to buy.

Rise of Online Retailing: Some say that the Internet will completely change the face of retailing; others believe that the "touch and feel" factor would ultimately dominate and the Net will have only a marginal impact on the shopping behavior. Probably the truth lies somewhere in between. But one thing is sure – online retailing is here to stay. Many retailers realized that and have rushed to start their own e-commerce Web site. We believe that the key to success would be the effectiveness with which retailers integrate the Internet with their existing business model.

19.8.3 Business Intelligence Solutions for the Retail Industry

Business Intelligence (BI) refers to the ability to collect and analyze huge amount of data pertaining to the customers, vendors, markets, internal processes, and the business environment. A data warehouse is the corner stone of an enterprise-wide business intelligence solution; various analytical (OLAP) and data mining tools are used to turn data – stored in the data warehouse – into actionable information.

Delivering Customer Value

Figure 19.1 illustrates the different functions in a typical retail organization. Customer relationship management (CRM) forms the focal point from where the vital insights gained about the customers – using BI tools – are absorbed in the entire organization. BI also plays a critical role in all the other retail functions like supply chain management, store front operations, and channel management. This chapter is an introduction to the various BI applications in the different functions in the retail organization, including support functions like finance and human resources.

Customer Relationship Management

The CRM strategy should include:

(a) Operational CRM: Automating interaction with the customers and sales force, and
(b) Analytical CRM: Sophisticated analysis of the customer data generated by operational CRM and other sources like POS transactions, Web site transactions, and third-party data providers.

A typical retail organization has a huge customer base and often customer's needs are fairly differentiated. Without the means to analyze voluminous customer data, CRM strategy is bound to be a failure. Hence, we believe that analytical CRM forms the core of a retailer's customer relationship strategy.

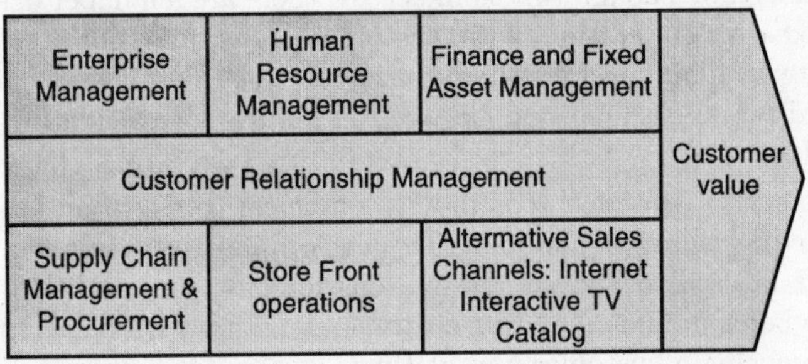

Fig. 19.1. Retail Organization

Marketing and sales functions are the primary beneficiaries of analytical CRM and the main touch points from where the insights gained about the customer is absorbed in the organization.

Analytical CRM uses the key business intelligence tools like data warehousing, data mining, and OLAP to present a unified view of the customer. Following are some of the uses of analytical CRM:

Customer Segmentation: Customer segmentation is a vital ingredient in a retail organization's marketing recipe. It can offer insights into how different segments respond to shifts in demographics, fashions, and trends. For example it can help classify customers in the following segments

- Customers who respond to new promotions
- Customers who respond to new product launches
- Customers who respond to discounts
- Customers who show propensity to purchase specific products

Campaign/Promotion Effectiveness Analysis: Once a campaign is launched its effectiveness can be studied across different media and in terms of costs and benefits; this greatly helps in understanding what goes into a successful marketing campaign. Campaign/promotion effectiveness analysis can answer questions like:

- Which media channels have been most successful in the past for various campaigns?
- Which geographic locations responded well to a particular campaign?
- What were the relative costs and benefits of this campaign?
- Which customer segments responded to the campaign?

Customer Lifetime Value: Not all customers are equally profitable. At the same time customers who are not very profitable today may have the potential of being profitable in future. Hence it is absolutely essential to identify customers with high lifetime value; the idea is to establish long-term relations with these customers. The basic methodology used to calculate customer lifetime value is – deduct the cost of servicing a customer from the expected future revenue generated by the customer, add to this the net value of new customers referred by this customer, and discount the result for the duration of the relationship. Though this sounds easy, there are a number of subjective variables like overall duration of the customer's relation with the retailer, gap between intermediate cash flows, and discount rate. We suggest data mining tools should be used to develop customized models for calculating customer lifetime value.

Customer Loyalty Analysis: It is more economical to retain an existing customer than to acquire a new one. To develop effective customer retention programs it is vital to analyze the reasons for customer attrition. Business intelligence helps in understanding customer attrition with respect to various factors influencing a customer and at times one can drill down to individual transactions, which might have resulted in the change of loyalty.

Cross Selling: Retailers use the vast amount of customer information available with them to cross sell other products at the time of purchase. This effort is largely based on the tastes of a particular customer, which can be analyzed using BI tools based on previous purchases. Retailers can also 'up sell' – sell more profitable products – to the customer at the time of contact.

Product Pricing: Pricing is one of the most crucial marketing decisions taken by retailers. Often an increase in price of a product can result in lower sales and customer adoption of replacement products. Using data warehousing and data mining, retailers can develop sophisticated price models for different products, which can establish price–sales relationships for the product and how changes in prices affect the sales of other products.

Target Marketing: Retailers can optimize the overall marketing and promotion effort by targeting campaigns to specific customers or groups of customers. Target marketing can be based on a very simple analysis of the buying habits of the customer or the customer group; but increasingly data mining tools are being used to define specific customer segments that are likely to respond to particular types of campaigns.

Supply Chain Management and Procurement

Supply chain management (SCM) promises unprecedented efficiencies in inventory control and procurement to the retailers. With cash registers equipped with bar-code scanners, retailers can now automatically manage the flow of products and transmit stock replenishment orders to the vendors. The data collected for this purpose can provide deep insights into the dynamics of the supply chain. However, most of the commercial SCM applications provide only transaction-based functionality for inventory management and procurement; they lack sophisticated analytical capabilities required to provide an integrated view of the supply chain. This is where data warehousing can provide critical information to help managers streamline their supply chain. Some of the applications of BI in supply chain management and procurement are:

Vendor Performance Analysis: Performance of each vendor can be analyzed on the basis of a number of factors like cost, delivery time, quality of products delivered, payment lead time, etc. In addition to this, the role of suppliers in specific product outages can be critically analyzed.

Inventory Control (Inventory levels, safety stock, lot size, and lead time analysis): Both current and historical reports on key inventory indicators like inventory levels, lot size, etc. can be generated from the data warehouse, thereby helping in both operational and strategic decisions relating to the inventory.

Product Movement and the Supply Chain : Some products move much faster off the shelf than others. On-time replenishment orders are very critical for these products. Analyzing the movement of specific products – using BI tools – can help in predicting when there will be need for reorder.

Demand Forecasting : It is one of the key applications of data mining. Complex demand forecasting models can be created using a number of factors like sales figures, basic economic indicators, environmental conditions, etc. If correctly implemented, a data warehouse can significantly help in improving the retailer's relations with suppliers and can complement the existing SCM application.

Storefront Operations

The information needs of the store manager are no longer restricted to the day-to-day operations. Today's consumer is much more sophisticated and she demands a compelling shopping experience. For this the store manager needs to have an in-depth understanding of her tastes and purchasing behavior. Data warehousing and data mining can help the manager gain this insight. Following are some of the uses of BI in storefront operations:

Market Basket Analysis: It is used to study natural affinities between products. One of the classic examples of market basket analysis is the beer–diaper affinity, which states that men who buy diapers are also likely to buy beer. This is an example of "two-product affinity." But in real life, market basket analysis can get extremely complex resulting in hitherto unknown affinities between a number of products. This analysis has various uses in the retail organization. One very common use is for *in-store product placement.* Another popular use is *product bundling*, i.e., grouping products to be sold in a single package deal. Other uses include designing the company's e-commerce Web site and product catalogs.

Category Management: It gives the retailer an insight into the right number of SKUs to stock in a particular category. The objective is to achieve maximum profitability from a category; too few SKUs would mean that the customer is not provided with adequate choice, and too many would mean that the SKUs are cannibalizing each other. It goes without saying that effective category management is vital for a retailer's survival in this market.

Out-Of-Stock Analysis : This analysis probes into the various reasons resulting into an out-of-stock situation. Typically a number of variables are involved and it can get very complicated. An integral part of the analysis is calculating the lost revenue due to product stock out.

Alternative Sales Channels

The success of a retailer in future would depend on how effectively it manages multiple delivery channels like the Internet, interactive TV, catalogs, etc. A single customer is likely to interact with the retailer along multiple channels over a period of time. This calls for an integrated strategy to serve the customer well, which requires smooth flow of information across channels. To ensure smooth flow of information customer data needs to be collected from different channels in one data warehouse. Customer relationship strategy can then be built around this customer-centric data warehouse. We have already seen how analytical CRM can provide analyses over the centralized data warehouse. In this section we explore how data warehousing and data mining can improve the effectiveness of a channel.

E-Business Analysis: The Internet has emerged as a powerful alternative channel for established retailers. Increasing competition from retailers operating purely over the Internet – commonly known as "e-tailers" – has forced the "bricks and mortar" retailers to quickly adopt this channel. Their success would largely depend on how they use the Net to complement their existing channels. Web logs and information forms filled over the Web are very rich sources of data that can provide insightful information about customer's browsing behavior, purchasing patterns, likes and dislikes, etc. Two main types of analysis done on the Web site data are:

Web Log Analysis: This involves analyzing the basic traffic information over the e-commerce Web site. This analysis is primarily required to optimize the operations over the Internet. It typically includes following analyses:

Site Navigation: An analysis of the typical route followed by the user while navigating the Web site. It also includes an analysis of the most popular pages in the Web site. This can significantly help in site optimization by making it more user friendly.

Referrer Analysis: An analysis of the sites, which are very prolific in diverting traffic to the company's Web site.

Error Analysis: An analysis of the errors encountered by the user while navigating the Web site. This can help in solving the errors and making the browsing experience more pleasurable.

Keyword Analysis: An analysis of the most popular keywords used by various users in Internet search engines to reach the retailer's e-commerce Web site.

Web Housing: This involves integration of Web log data with data from other sources like the POS transactions, third-party data vendors etc. Once the data is collected in a single customer-centric data warehouse, often referred to as Web house, all the applications already described under CRM can be implemented. Often a retailer wants to design specific campaigns for users who purchase from the e-commerce Web site. In this case, segmentation and profiling can be done specifically for the e-customers to understand their needs and browsing behavior. It can also be used to personalize the content of the e-commerce Web site for these users.

Channel Profitability: Data warehousing can help analyze channel profitability, and whether it makes sense for the retailer to continue building up expertise in that channel. The decision of continuing with a channel would also include a number of subjective factors like outlook of key enabling technologies for that channel. For example m-commerce, though not a very profitable channel today, has the potential to be a major alternative channel in the years to come.

Product–Channel Affinity: Some product categories sell particularly well on certain channels. Data warehousing can help identify hidden product-channel affinities and help the retailer design better promotion and marketing campaigns.

Enterprise Management

This typically involves the various activities performed by the top management; and the role of data warehousing and data mining is to provide the top management with reports and analyses to meet their decision-making requirements. One possible BI application in this area is:

Dashboard Reporting on KPIs: Key performance indicators like contribution margin, response rate, campaign costs, customer lifetime value can be presented in dashboard reports to the top management to facilitate decision-making process. Also alerts can be triggered if any KPI reaches a predefined threshold level. These reports can incorporate retail industry benchmarks, provided by third-party researchers, which can be used as threshold levels for various KPIs.

Human Resources

Data warehousing can significantly help in aligning the HR strategy to the overall business strategy. It can present an integrated view of the workforce and help in designing retention schemes, improve productivity, and curtail costs. Some BI applications in HR are:

Human Resource Reports/Analytics: Reports and analysis can be generated to support an integrated view of the workforce. Various analyses include staff movement and performance, workforce attrition by store, workforce performance by store, compensation and attrition, and other customized analyses and reports. The HR data can be integrated with benchmark figures for the industry and various reports can be generated to measure performance vis-a-vis industry benchmarks.

Manpower Allocation: This includes allocating manpower based on the demand projections. According to the seasonal variation in demand, temporary manpower can be hired to maintain service levels. The demand levels vary within one working day also, which can be used to allocate resources accordingly.

HR Portal: Employers need to maintain accurate employee data, which can be viewed by the employees for information relating to compensation, benefits, retirement facilities, etc. Payroll data can be integrated with data from other human resource management applications in the HR data warehouse. This data can then be circulated within the organization through the HR portal.

Training and Succession Planning: Accurate data about the skill sets of the workforce can be maintained in the data warehouse. This can be used to design training programs and for effective succession planning.

Finance and Fixed Asset Management

The role of financial reporting has undergone a paradigm shift during the last decade. It is no longer restricted to just financial statements required by the law; increasingly it is being used to help in strategic decision making. Also, many organizations have embraced a free information architecture, whereby financial information is openly available for internal use. Many analytics described till now use financial data. Many companies, across industries, have integrated financial data in their enterprise wide data warehouse or established separate financial data warehouse (FDW). Following are some of the uses of BI in finance:

Budgetary Analysis: Data warehousing facilitates analysis of budgeted versus actual expenditure for various cost heads like promotion campaigns, energy costs, salary, etc. OLAP tools can provide drill down facility whereby the reasons for cost overruns can be analyzed in more detail. It can also be used to allocate budgets for the coming financial period.

Fixed Asset Return Analysis: This is used to analyze financial viability of the fixed assets owned or leased by the company. It would typically involve measures like profitability per square foot of store space, total lease cost vs. profitability, etc.

Financial Ratio Analysis: Various financial ratios like debt–equity, liquidity ratios, etc. can be analyzed over a period of time. The ability to drill down and join inter-related reports and analyses, provided by all major OLAP tool vendors, can make ratio analysis much more intuitive.

Profitability Analysis: This includes profitability of individual stores, departments within the store, product categories, brands, and individual SKUs. A major component of profitability analysis is the costs incurred by stores/ departments and the cost of acquiring, storing, and allocating shelf space to particular product categories, brands, or SKUs. It goes without saying that profitability analysis has an extremely universal appeal and would be required by other groups within the retail organization.

19.9 Summary

In many business settings the comprehensibility of the analyses as well as the degree of validation against known relationships help enhance the perception of the data mining activity, thus creating an environment where this new technology can take root. The success of data mining does not solely depend on the quality of new algorithms, but also on the usability, comprehensibility, and degree of domain-specific knowledge integrated in the tool.

Data mining and analytics in general demand complex set of skills: business, statistics, databases, operations systems. CDMS, the DM Metadata, is a natural platform for sharing those skills and it helps to avoid duplications of effort.

This section discussed so far data mining in the area of customer relationship management for customer segmentation and customer retention analysis as well as for credit scoring. Industries in which this technology is traditionally applied are those that collect customer data for the purpose of billing. They have always been required to store this data and, thus, can make use of it without introducing new operational systems beforehand. We have helped banks, insurances, and Internet service providers. More and more industries are obtaining customer data while implementing information systems for marketing, sales, and customer service. Besides that, every company could benefit from data mining technology applied to the analysis of costs and revenues, activities and processes, and its performance in general.

Retailers are known for innovation. The most innovative retailers of today are those who are using business intelligence to gain sustained competitive advantage. These retailers have also realized that BI can be used strategically only when it is implemented with utmost care and complete support from the top management. We believe that unless all the user groups are consulted and the objectives clearly defined, BI solution cannot be a success. Also, like any other technology solution, BI cannot exist in vacuum. We strongly believe that it is just a means to an end. The wisdom, gathered by analyzing huge amount of data, should reach every corner of the retail organization. The end objective is to convert this wisdom into effective action. And for this the entire organization should be able to leverage the business intelligence network. Thus the chapter involved various case studies and discussed how data mining is suitable for banking and commercial applications.

19.10 Review Questions

1. With a case study, explain how distributed data mining provides a solution in banking.
2. What are the factors that led to data mining technology at the Bank?
3. Explain CDMS – centralized data mining solution.
4. Explain how data mining is used in various commercial applications.
5. State the basics of decision support system.
6. On what basis do the decision-making challenges arise?
7. How does the data mining supports e-commerce and retail industries?
8. What are the applications of data warehousing and mining in the retail industry?
9. What are the key trends in retail industry?
10. Write a short note on enterprise management.
11. Write a short note on finance and fixed asset management.

20

Data Mining for Insurance

Objectives:

- Data mining is a "horizontal" technology, it can be applied in a wide range of enterprises to continuously improve business decision making. Innovation is the order of the day in the insurance industry as providers grapple with a range of business challenges.
- Data mining can be applied to evaluate and refine the business rules and processes that form the basis of their underwriting polices.
- The high-level "pitch" for data mining in insurance underwriting is that consistently having a sound understanding of underlying patterns, trends, and relationships that impact on strategic and operational success ahead of industry peers, and acting on these findings, can enable a sustainable advantage in competitive markets.
- The distinct benefit of data mining is its ability to reveal patterns and relationships in large volumes of data and generate a range of predictors with specified confidence levels – that is of special relevance to those tasked with the underwriting function in insurance businesses.
- The internal data resources used by the leading insurers can be used to streamline, improve, and target their business processes where premium/pricing constraints are real and enduring, lowering their effective delivery cost of their insurance products and their total cost of risk, and enabling the pursuit of specific profitable subcategories of risk in areas of business that were formerly "out of bounds."
- Business intelligence (BI) forms the most critical component of claims management, helping in fraud detection and claims estimation. On the asset management side, it can lower the insurer's risk through sophisticated risk models developed using data mining tolls.
- The insurance industry is totally dependent on the ability to convert raw data into intelligence – intelligence about customers, markets, competitors, and business environment. Over the years data processing technology has progressed phenomenally and tools like data warehousing, OLAP, and

S. Sumathi and S.N. Sivanandam: *Data Mining for Insurance*, Studies in Computational Intelligence (SCI) **29**, 473–498 (2006)
www.springerlink.com
© Springer-Verlag Berlin Heidelberg 2006

data mining, which constitute the cornerstone of an effective business intelligence (BI) environment, have been widely accepted across industries.
- The insurance industry is extremely divided in its adoption of business intelligence environment based on technologies like data warehousing, OLAP, and data mining. Efficiently gathering the information requirements of all the user groups is thus extremely critical for the success of any data warehouse.

Abstract. In this section, the application of business intelligence tools like data warehousing, OLAP, and data mining in insurance and insurance underwriting decision support systems are presented.

In this section we explore background trends and concepts relevant to the insurance underwriting process and overview specific areas where we believe data mining technology can be deployed to improve certainly and efficiency – two important business objectives to insurers at this stage in the industry's evolution.

As customers know, data mining involves the application of technology, relevant business knowledge, and a creative approach to problem solving, to the achievement of defined business objectives. Since data mining has inherently "horizontal" technology, it can be applied in a wide range of enterprises to continuously improve business decision making. Innovation is the order of the day in the insurance industry as providers grapple with a range of business challenges. We suggest that the innovation use of data mining technology in the underwriting process is an area worth exploring.

The insurance industry is totally dependent on the ability to convert raw data into intelligence – intelligence about customers, markets, competitors, and business environment. Over the years data processing technology has progressed phenomenally and tools like data warehousing, OLAP, and data mining, which constitute the cornerstone of an effective business intelligence (BI) environment, have been widely accepted across industries. However, insurance companies have been relatively slow in adopting these tools, primarily because of lack of competition due to protective regulations. But now, they can no longer afford to be complacent as the Internet, deregulation, consolidation, and convergence of insurance with other financial services are fast changing the basic structure of the industry.

20.1 Insurance Underwriting: Data Mining as an Underwriting Decision Support Systems

Persistent overcapacity in the insurance industry has reached the point where more competition on price would be suicidal for carries. This inability to make further concessions on price has shifted the locus of competition to the quality of products and services being offered. Lowering the total cost of risk will require more than simple price cutting. To increase revenues, insurers, reinsurers, and brokers will need to add value to their clients by:

- Enhancing ancillary services such as risks control and claims handling.
- Customizing products for individual industries and individuals clients.

- Finetuning of existing products or developing new ones to address emerging and nontraditional risks.

20.1.1 Data Mining and Insurance: Improving the Underwriting Decision-Making Process

The discussion on business applications of data mining technology in this section is industry specific (insurance). We focus on a core competency and critical business process for insures (underwriting) – a domain with clearly defined "business rules." These business rules are typically developed using traditional statistical tools and methodologies (actuarial analysis), codified for further reference and guidance (an underwriting manual or polices), and administered by highly skilled professionals (underwriters). These professionals apply judgment and discretion, based on their training and experiences, to obtain defined business objectives (an underwriting break-even or profit position at the lowest possible cost to insured and at the lowest possible risk to the insurer).

At ANGOSS Software Corporation they market the data mining solutions as enabling technology that "help businesses make better business decisions every day." This discussion is meant to be informative and illustrative of the broader ways data mining technology can be used to optimize (validate, debunk, and refine) the core business rules that guide decision making in many industries. Many of the underlying concepts and approaches outlined in this section apply equally to the core processes of banking, telecom, retail, and other enterprises. This discussion assists the insurance users in generating dialog, within their companies, about areas where data mining can be applied to evaluate and refine the business rules and processes that form the basis of their underwriting polices.

The approach is not rocket science, but we think it involves more than simply actuarial science; at least as traditionally understood and practiced by actuaries in large insurance organizations. Hopefully, the readers will find some of the themes and ideas presented here useful, regardless of their industry background. We were warned never to watch how sausages and laws are made. Some would say the same thing about decision making in large organizations.

"Gut instinct" and "tribal knowledge" have always played an important role in business decision making for companies large and small. With government oversight and "following the market leader" becoming less important in insurance and elsewhere, most businesses are finding far broader flexibility to chart their own course.

Increasingly, business decision makers face the task of absorbing, analyzing, and acting on the knowledge contained in their own data, as well as in external data sources relevant to their particular business. This intuitively obvious cliche has been overworked by "big ticket" sellers of technology in such areas as "data warehousing," "enterprise relationship management," and "knowledge management," but it is understood by most businesses to be of central and growing importance to their business success.

Properly managing and executing projects revolving around data analysis often seems like a Herculean task – fraught with bureaucracy, political risk, and uncertain business outcomes. This is why advocates of data mining in operational systems – especially "core competency" zones like insurance underwriting – must make sure their course is well charted early on, by engaging senior executives, and proceeding incrementally to define and consistently achieve or exceed stated business objectives.

The high-level "pitch" for data mining in insurance underwriting is that consistently having a sound understanding of underlying patterns, trends, and relationships that impact on strategic and operational success ahead of industry peers, and acting on these findings, can enable a sustainable advantage in competitive markets. In this respect, equipping underwriters with the more flexible, automated decision support systems data mining enables is good business – it is driven by the recognition that the underwriting process itself is undergoing fundamental stress and change as insurers seek to evolve their businesses.

The "fear" is that others may already be doing what the organization's structure prevents from achieving. Despite the industry noise, moving incrementally is a very sound approach – but movement is strongly advised! And do not accept arguments that "data quality" problems exist, that "data" is not available, that "we don't need to do this," or "we are already doing this." The "data" is not the obstacle; organizational complacency is.

Even for many large organizations, underwriting decisions are still driven largely by imperfect knowledge, fixed underwriting policies and procedures that resemble sledgehammers more than scalpels, and ultimately the "gut instinct" of underwriters seeking to maintain existing business and win share in a soft and highly competitive market.

ANGOSS Corporation Customers knows that data mining technology – at its most simplistic level of use – is a flexible and affordable business decision support tool. Data mining technology enables users to explore internal and external data sources, in the context of defined business objectives, to achieve deeper understanding of the underlying drivers of their business success.

When properly understood and used, data mining technology can help enterprises chart their business course aggressively, with higher confidence in more certain outcomes. It is this distinct benefit of data mining – its ability to reveal patterns and relationships in large volumes of data and generate a range of predictors with specified confidence levels – that is of special relevance to those tasked with the underwriting function in insurance businesses.

Insurance companies, like others, face major challenges in their core business as we move through the next decade. The following are some of the relevant trends and developments:

- Increased competition among insurers in both traditional "commodity" lines (automobiles, life and home owners' policies) and more complex insurance products (commercial P&C, workers' compensation, and benefits) for market share.
- Mergers, divestments of entire lines of business, and "demutualization" are preludes to further industry consolidation, rationalization, and specialization.
- Increasing dependence among insures on investment gains (rather than returns from the underwriting process) for bottomline earnings growth.
- Increased capital flows from external sources to reinsurance markets, lowering margins (and profitability) of this business to existing insurers.
- Emergence of technologically savvy "dot com" competitors for certain insurance business lines and insurance functions.
- Repositioning of major insurers around "core competencies" (lines/functions/outsourcing/"back office" services provision)
- Distribution/channel challenges (managing captive versus independent brokers and agents in an "on-line" business environment)
- The shifting of "insurance intermediaries" (third-party claims administrators, consultants, brokers, adjusters, and data source providers) to technologically more advanced on-line delivery and fulfillment systems more closely integrated with insurers own front- and back-office systems.

Despite relatively high levels of profitability, substantial reserves, and limited catastrophic losses in recent years, insurers are collectively fighting – and largely losing – decades old battle to regain lost profitability in their core underwriting business. So where does data mining fit in to all this?

An insurer chooses to accept a given risk because it can find a set of "terms" on which the particular risk can be assumed and the business can be profitably written. These terms comprise four general types: premium rates, policy provisions, the hazard presented in the risk, and the related reinsurance arrangement. As we move forward "risk" is coming to mean not only traditional hazards but broader organizational risks – political, trade, financial, and strategic risks formerly outside of the purview of insurance.

The underwriting process is central to the acceptance of any risk – and the defining and documenting of the terms of that acceptance. Underwriting is typically not a "no" decision but rather a "yes, but" decision. The "but" part of the equation – being the terms of acceptance – is a dynamic and continuously evolving process based – among other things – on consideration of all available relevant (past) data, which may impact on the advisability and cost of assuming a given risk.

In large measure, traditional actuarial analysis – based predominantly on benchmark indicators (such as mortality and morbidity analyses in life and benefits domains) and on the shared information or "collective" tribal knowledge flowing from industry wide statistics (advisory ratings, aggregate loss data and the like) has been used to define "underwriting procedures."

Everyone appreciates how far insurers have come since the early days at Lloyds where risks were formally presented for endorsement and acceptance, but few truly appreciate how far. As the concept of risk broadens to encompass new kinds of risk financing and risk control, an ever-broadening range of data from disparate sources can be found to impact on the entire insurance underwriting process – increasingly in real time! One ANGOSS customer explored a range of social, political, and economic indicators in the context of a dependent variable of "political insurrection" to seek predictors of instability. Lunacy? Or an enlightened attempt to define underwriting parameters for political risk insurance?

And though business in the "front office" at Lloyds may appear to be conducted in much the same fashion, as it was hundreds of years ago, times have definitely changed for tomorrow's leaders in the insurance industry. Leaving aside the technology hype, modern database technologies, as well as "e-business" and supporting infrastructure technologies represent en entirely new kind of "plumbing" – an on-line, real-time "digital nervous system" for the 21st century.

All of these technologies combine to offer exponentially greater interconnectedness and data sharing for insurers and the intermediaries they interface with at an operational level. This connectedness enables a breadth, depth, and speed of analysis – both of internally generated and externally sourced data – that is far more substantial than was considered possible even a decade ago. And good data mining software loves large volumes of data!

Leading insurers (and they are not always – or perhaps even often – the largest ones) know that by making investments in these infrastructure technologies they will be able to achieve substantially greater operational efficiency, while building up substantial internal data resources. These resources can be used to streamline, improve, and target their business processes where premium/pricing constraints are real and enduring, lowering their effective delivery cost of their insurance products and their total cost of risk, and enabling the pursuit of specific profitable sub-categories of risk in areas of business that were formerly "out of bounds."

Examples of enabling technologies that "set the stage" for data mining are rich and varied. They include:

Wireless connectivity to databases, enabling sales agents, claims adjusters, loss control experts and other intermediaries in the field to transmit data to and from their "host" insurer in close to real time. The Internet itself, which enables a range of interactions – "personalization" of information on the customer side, "self-service" investigation of pricing and product options, "self-reporting" by submitting applications and ongoing information, and ultimately, the "binding" of a broader range of insurance contracts to customers in near-real time.

The flexibility, scalability, performance, and declining cost of modem database systems from major vendors like Microsoft, IBM, and Oracle – and the solutions of thousands of application vendors supporting predominantly the

Microsoft database environment – provide unprecedented levels of flexibility in constructing, linking, and analyzing databases quickly and efficiently. Secure payment, funds transfer, outsourced third-party claims administration, and other technology improvements enable more efficient – but still integrated – administration, settlement, and analysis of payments and claims experience.

20.1.2 What does an Insurance Underwriter Do?

Underwriter as Gatekeeper

Few outside the insurance business have a full appreciation of the technical, legal, political, and practical skill set that defines the role of an underwriter in an insurance business. In general underwriters are tasked with the responsibility of assessing – accepting, rejecting, and revising – insurance contracts that the insurer's distribution channels and marketing systems bring forward for consideration.

In this sense they are the "gatekeeper" of the insurer's core business. Their primary goal is to achieve stability of results – particularly the avoidance of wide variations in annual loss ratios – while making adequate protection available to customers at the lowest possible cost. Expressed as a business equation, the revenues derived from insurance premiums should meet or exceed the amount paid out on account of claims, the expenses of securing and administrating their book of business, and enable a profit contribution to the insurer.

Few inside the industry responsible for overseeing the underwriting function fully appreciate how technology is changing and will continue to change the underwriting process. Underwriting decisions traditionally made with "sober reflection" must now be made for competitive reasons in closer to real time with no sacrifice of certainty in outcome. Technology and business imperatives will require that underwriters' reliance on their experience and knowledge of the "rules" must give way to a more flexible but still certain business environment where underwriting rules – permitted risk categories, rating structures, specific, specific terms, and the like – may change rapidly and continuously.

As noted by Marsh & McLennan, within the constraints of government regulation (and subject to the "suicidal" impulses of insurers seeking to expand market share without regard to underwriting costs and losses) traditional "ratings" or sledge hammer approaches to broad classes and categories of risk will increasingly be replaced by customized insurance solutions tailored as to terms to specific customer types and specific industry requirements. Ironically, this objective runs counter to traditional underwriting approaches of diversifying and spreading risk. As a result, the "underwriting procedures manual" of the future is far more likely to be a complex series of nested business rules compiled electronically and continuously optimized and updated for "risk blending" to reflect an insurer's analysis and knowledge of all relevant data to the risk being underwritten and managed across its entire organization.

The traditional approach of basing underwriting decisions on the "law of averages" through time, is being replaced by the recognition that (to paraphrase Keynes) in the long run, insurers operating solely on the "law of averages" will be out of business. The flexibility of modern technology makes far more rapid (and informed) adjustment and optimization of underwriting business rules possible. Within the remaining regulatory and self-regulatory constraints insurers face, the proactive leaders in the industry will seize on these developments for competitive advantages.

General Principles of Underwriting

In opposition to the underwriter's goal of achieving stability and profitability is the fundamental principle of "adverse selection" found throughout the field of insurance – the tendency of insurance applicants in many categories to include a preponderance of "poorer than average for class" risks. Adverse selection arises because of both moral and other hazards – the tendency of those most likely to avail themselves of insurance to seek it out. In performing his or her role and mindful of this fundamental touchstone of risk, an underwriter seeks to achieve fundamental objectives linked to sound underwriting principles. These include:

- Spreading of risk, by committing the insurer to as many different and distinct risks as possible so as to permit the "law of averages" to apply.
- Uniformity of application, by pursuing a coherent enterprise-wide underwriting policy that is designed to achieve predictability and certainty of results across the organization.
- Continuity of risk, by ensuring that the insurer develops a "seasoned" portfolio of business so that the "law of averages" has the time to work.
- Blending of risk, by taking on within each class/type of commitments underwritten by the insurer a sufficient number of commitments of similar type to produce stability of results.

Information Sources

As part of the process of reviewing and defining the terms for any risk, an underwriter, theoretically at least, has available a range of internally and externally sourced data, including applicant-supplied information, through basic application and other information submitted on an ongoing basis.

Insurer-maintained information, obtained through sales representatives (reports and recommendations), claims administration information, information on the covered risk (such as zoning, construction, municipal building code, and similar information on properties), and "site-specific" information (from adjusters, loss prevention experts, brokers, benefits consultants, and other intermediaries involved in the insurance process), third-party-sourced information, from external data sources such as third-party administrators, as well as private reporting sources such as Dun & Bradstreet, Experian, Claritas, Acxiom, and other providers. Industry information provided by insurance bureaus and self-regulatory organizations, including "experience statistics," advisory rates, and other "ratemaking" data such as trends, equilibrium, and expense factors, and permissible loss costs.

Government information, including data made available to the insurance industry as part of the insurance regulatory process and data made more generally available such as census data, economic performance indicators and the like. This has always been the case. However, in practical terms, and in the absence of fully integrated systems enabling rapid access to and linking of all potentially relevant data, insurers have relied quite heavily on imperfect knowledge to complete such analysis and codify procedures in acceptable forms for administration. Underwriting decisions have involved a combination of the "tribal wisdom" codified in writing manuals: constructed with heavy actuarial influences; and the knowledge and experience of their underwriting professionals in interpreting and applying those rules.

In recent years – and particularly with the advent of the Internet – entirely new means exist to acquire, analyze, exchange, and act on a wide range of relevant; useful data. The underwriting function is itself being transformed – because both changing technology and changing business conditions demand this. Moving forward, underwriters (or their decision support professionals) must have access to all of this data and be able to explore, analyze, and respond to the patterns, trends, and relationships that they discover in it.

Although data mining does not hold a monopoly on insight in this area, it can offer unique means of drilling into, exploring, and understanding trends, patterns, and relationships in data. In addition to efficiently handling the garden variety data mining tasks of most interest to an insurer's sales channel (define a "good prospect," what are good predictors of a particular product sell – cross or up, what are the key indicators of attrition so that we can intervene and preserve my customer account) data mining technology can be invoked in a wide range of the areas. The following are examples of the kinds of analysis enabled by well-designed systems, which capture and leverage all available data relevant to risk:

Based on all available data we have access to, what are the key drivers, or predictors an underwriting loss with respect to common customers, categories of risk, classes of risks, or across the entire portfolio of assumed risks? What is the relationship, if any, between the "personalization" attributes of specific customers (defined by their interactions with captive agents or brokers who are submitting information to the database and ultimately by their "self-reporting" of this information on-line)?

What is the relationship, if any, between crime statistics, unemployment levels, changes in property taxes or the indicators or instability (at all levels) and property claims for purposes of pricing a homeowner's policy in a specific area? What is the relationship, if any, between third-party auto claims for theft and vehicle types and how has this changed since last year.

What is the relationship, if any, between urban growth rates and the incidence of auto accident related claims? What is the relationship, if any, between the numbers of PC-based workers in an organization and the cost of an employer benefits plan that offers prescription eye wear?

What is the short-term stress leave claims experience for females in a company with no maternity plan compared with that of a company with a generous maternity and early childhood program? What would be the impact of a change in the deductible threshold for a particular policy or customer type based on claims experience across all classes of risk with differential deductibles? What happens when we experiment with the deductible rules?

These are illustrative examples of a wide range of analytical and exploratory techniques that can be used by underwriters to move far beyond traditional "advisory ratings" or based procedures to accept specific insurance risks.

Being able to automatically explore relevant data – to hypothesize and search for patterns, relationships, and anomalies in the context of defined business objectives – can help guide insurance underwritten in their assessment of risk and in their "yes, but" interactions with their sales channel. They can also act as "probes," guiding actuaries and decision support professionals into areas that would otherwise not have been considered as being related to relevant to risk. Even where traditional statistical measures of confidence may not have been achieved, this process of interactively exploring and understanding patterns and relationships in data can inform and improve the underwriting process because it invariably yields new knowledge to the underwriter.

Should We Go by the Book, Toss the Book, or Write New Chapters?

In virtually all large organizations, underwriting policies are established, codified, signed off on, and implemented after careful consideration and review of a range of business factors. Basic underwriting guidelines and requirements, as well as procedures for varying from them, are carefully developed, documented in manuals (and to some extent, in customized underwriting software applications), and learned through training and continuing education, as well as application in business cases. These codified rules represent the "tribal wisdom" of the specific insurer that has established them.

Data mining helps improve business decision making in large measure by identifying areas where the "tribal wisdom" documented in corporate policies, procedures, manuals, and guidelines ("going by the book") is right, and where it may be wrong – as well as exploring the shades of gray in between. As one area of investigation, then, any insurer can benefit from an exploration of its data in the context of its existing underwriting rules and procedures currently in force.

In areas where data mining reveals that "the book" is clearly right or clearly wrong the business benefit is immediate and obvious; the necessary can be implemented or are unnecessary. In addition, as part of its ongoing relationships with its insurers, information gleaned by the underwriter from its analysis of its own can be shared to build a relationship of more long-lasting value (We know the analysis shows that where customers of the size and type introduce a "wellness" program at work their claims experience improves by X per year in Y% of the cases).

It is in the gray areas, where judgment and "domain experience" is necessary, and perhaps where the margin between profit and loss hangs in the balance, the data mining can assist the underwritten best. An underwriter equipped with the tools at his or her disposal to explore and understand these gray areas can make the judgment calls – the "yes, buts" in choosing to accept specific risks – that are the core skill set of their profession on a more informed basis.

Many insurers, for example, may have concerns over providing workers compensation or other benefits insurance to seasonal employers, industries characterized by high personnel turnover, or work forces with other specific attributes that are defined as being "higher than average" risks. But does their own support this kind of hard and fast approach across entire categories of businesses?

Joining third-party data store information – for example classification, industry information, and customer-level information from a third-party service like Dun & Bradstreet – to claims-related information from an insurer's database may reveal that this is not true, or perhaps true in enough cases to be a strong "general rule" but definitely untrue in others where business may be profitably done (for example, residential construction firms that under x employees and revenues of less than $y operating in a regional market with growth rate above z typically provide a much more stale claims profile regardless of cyclical factors that otherwise impact on risk).

Data mining technology, by rapidly exploring all claims data – whether in any specific class of business or across the entire portfolio – can illuminate for review the historical patterns and relationships between claims and all other selected data elements. This type of analysis may confirm the utility of specific riles, demonstrate that they are incorrect and unnecessary barriers to business, or provide a range of intermediate "but" exceptions that suggest the rules are generally correct but valid (or invalid) in specific instances. These intermediate rules may enable an underwriter to rapidly respond – with specific and appropriate terms – to applicants who may fall within the category of the "general rule" but also are within the "specific exceptions" to it.

Similarly, an insurer may wish to analyze its entire portfolio of residential homeowner policies by captured dimensions – geocoding, household data elements, claims experience, advisory ratings, and application information

– to determine if there are useful predictors that impact on a variable pricing mode for a commodity insurance product. This analysis may be used to implement a set of business rules deployable "on-line" in real time to enable their captive brokers to have much broader "binding authority" over more business lines with some level of control and balances (more sophisticated electronically embedded underwriting rules).

The principle of "utmost fidelity" of the insured in an "e-business" insurance world can also be used to advantage by indicating to customers and prospects that pricing and delivery of insurance products to them is based on the accuracy and completeness of the information they have provided. Over-reliance on misplaced confidence in external data sources is always in "x factor" worth remembering. However, on balance the business benefits of having access to richer data dimensions far outweighs this potential concern.

At the core of the current evolution in the underwriting process is the recognition that traditional underwriting approaches have generated – on balance – a poor record of performance, and that the underwriting process is itself changing as technology and business imperatives force more rapid decision making involving the administration of larger classes of risk, which must be assessed in the context of ever-growing volumes of potentially relevant data.

By adopting and implementing data mining as a decision support function in the underwriting domain insurers are not seeking to displace personnel with "model managers" or "underwriting robots." They are seeking to substantially improve the overall "intelligence" of their decision support system so that today's "tribal knowledge" can be refined (and redefined) tomorrow and dynamically thereafter based on what has happened today and every day prior to today.

What Challenges Does an Underwriter Face?

The underwriter can currently be seen as the "fulcrum" between the sales and risk management functions within an insurance organization. "Sellers," whether captive agents or independent brokers or other intermediaries, are mandated with the responsibility of generating new business (including renewal business). This group seeks confirmation of insurance availability as quickly as possible at the most competitively attractive rates in order to ensure market share growth (and their personal compensation). Risk managers, on the other hand, are mandated with the responsibility of ensuring that the insurer's policies and procedures for the assumption and sharing (through reinsurance) of risks are appropriately implemented.

The underwriter stands between the sales force in the field (give the best pricing; it is war out there) and the actuarial and risk management personnel (have we identified and assessed exposures associated with this piece of business based on claims experience and have we managed this risk?). In the world where the largest insurance firms are consolidating, and where on-line/real-time quotations, pricing, and placing of insurance are becoming a new and viable distribution channel for various insurance products, the time dimension of the underwriting process has its own significance.

20.1.3 How is the Underwriting Function Changing?

The goal of underwriting, expressed as a simplistic formula ignoring such factors as time, the investment P&L and indirect enterprise costs, can be roughly stated as follows:

$$P = A + B + C$$

where P is the premium income from a class of risks, A is the acquisition cost of the business, B is the direct cost of administering the business, and C is the loss experience/payout on the business Looking back at the industry trends highlighted earlier, a sustained (and arguably unsustainable) disequilibrium has existed for many years in this equation (largely offset by a large "G" – the gains delivered from investing premium streams – and small "L" the category of "catastrophic losses"). In increasingly competitive markets there is pressure (tightening) on premium income, which means the addressing of chronic underwriting losses must be addressed through other means.

Unless insurers are content to generate continuing unacceptably high loss ratios on their underwriting business (a result that the "law of averages" suggests would have significant adverse consequences in circumstances where their offsetting investment gains decline, or where catastrophic losses materialize) changes must occur in the cost dimensions of the equation. The corresponding changes in operations this result necessitates are:

- Improving efficiencies in the acquisition of business (acquiring complementary business lines; providing "on-line" channels for attracting new business, and establishing systems to differentiate at a far deeper level potentially good risks versus bad one, are good examples)
- Lowering the administration costs associated with the business (merging operations, outscoring noncore functions of giving up "back office" entirely; as well as improving efficiencies in the administrative of the business by automating process, etc.)
- Improving the loss claims ratio and reducing claims by a range of proactive polices (from "barring" certain types of risks from the underwriting process, through loss prevention and loss reduction, to focused investigation and analysis of claims fraud and abuse)

20.1.4 How can Data Mining Help Underwriters Make Better Business Decisions

Despite the role that data mining – as an analytical solution – can play, we believe it is a mistake to assume that underwriters are going to become or indeed should be "data miners." The nature of their role makes this impractical.

However, underwriters should be direct and substantial consumer of data mining outputs – both descriptive analysis and predictive models; prepaid for their assessment and refinement by decision support professional in insurance organization tasked with this responsibility.

In addition, to facilitate both the process of model generation (a decision support task) and model consumption (an underwriting function) insurance organization should have a long-term action plan in place to exploit key evolutions in technology that are directly relevant to their needs and competitive success. This means not only developing underwriting solutions as "dynamic, expert systems" with a higher level of data mining intelligence built into them, but also implementing complementary technology improvements in other organizational areas to ensure that the data analysis functions central to "next generation" underwriting system across the organization can be preformed continuously, quickly, and efficiently.

In the board sense, an organization commitment to data quality and an understanding of the importance of gathering, analyzing and communicating knowledge derived from data analysis is key to success. Specific examples of areas where the "data capturing" infrastructure can be finetuned to assist include the following:

- Common application forms across the organization for each type of insurance product we offer containing "required filed" data entry to fuel the analysis engine. Emphasis on continuous incremental improvement in the scope and accuracy of data captured of data captured is key. Sales and distribution channels, as well as customers, must be "plugged in" and be given incentives to collaborate in this process.
- Making applications and related processes for the delivery on insurance products and related serious as much as possible an electronic process, with flexible and robust interfaces.
- Enabling channel representatives – captive agents and brokers – to access and submit data with respect to applications electronically.
- Ensuring that supporting personnel – such as adjusters, site inspectors, and loss prevention experts – also are plugged in to insurer database and incentive to collaborate; filing reports electronically and capturing additional supplement data relevant to risk.
- In personal line businesses, ensuring that customers and prospect interactions take place at a "household" level and that these interactions also capture data relevant to both risk and future potential sales or services opportunities.
- In business lines, tracking customers and prospects by multiple data elements, including standard dimensions that permit cross comparative analysis (e.g., number of employees, D&B DUNS identifiers, and "key indicator" scores).

20.2 Business Intelligence and Insurance – Application of Business Intelligence Tools like Data Warehousing, OLAP and Data Mining in Insurance

What is the most important pre-requisite in today's volatile marketplace? The answer is rather straightforward – "information," or more precisely, "actionable information." For no other industry is it as important as for the insurance industry, which is almost totally dependent on current and insightful information. This is exactly what business intelligence (BI) tools like, data warehousing, data mining, and OLAP seek to provide. A close look at the insurance value chain suggests that BI can play a crucial role in almost every aspect of the chain. It can help identify the right customers for target marketing and analyze the reason for customer attraction. It can help the insurer better manage its agents and sales force and improve the effectiveness of actuarial and underwriting functions. BI forms the most critical component of claims management, helping in fraud detection and claims estimation. On the asset management side, it can lower the insurer's risk through sophisticated risk models developed using data mining tolls. And most importantly, BI tools can help insurers provide crucial information to corporate clients, which can go a long way in cementing the insurer's relation with the clients.

Late seventeenth century was an era of growing international trade. New shipping routes were discovered and adventurous sailors brought exotic products from strange and alien lands. But their journeys across the oceans were fraught with danger and unknown risks. This gave rise to a new breed of entrepreneurs – marine underwriters – who agreed to cover the losses in return for a fixed amount of premium. Their business depended on current information about the sea routes, pirates, political condition, weather patterns, conditions aboard the ship, and consumer tastes for exotic products. In order to acquire business information, many marine underwriters began to frequent Edward Lloyd's coffeehouse in London. This was the place where they could share business intelligence with other underwriters and captains of trading to form a society that went on to become the most famous of all insurance companies – Lloyd's of London.

The very "business intelligence" that brought together the marine underwriters at Lloyds' is, if anything, much more crucial for the insurance industry today. It pervades almost every aspect of the value chain and technology has the potential of making it ubiquitous across the organization. Today, an underwriter would not go to a coffeehouse to gain business intelligence, but, probably, to get away from it. And one thing is sure: she cannot afford to stay in the coffeehouse for long.

20.2.1 Insurance Industry Overview and Major Trends

The insurance industry is quite diverse in terms of portfolio of products provided by different companies. The products can be broadly classified into two product lines: property and casually (P&C) and life insurance. Life insurance product line can be further subdivided into life insurance, health insurance, and annuity products.

Growing consolidation and change in the regulatory framework has led many insurers to add new products to their portfolio. This presents its own unique challenge to the insurer in leveraging its greatest asset data. A number of other trends in the insurance industry have also exponentially increased the importance of an effective business intelligence environment; at the same time, these trends are responsible for increasing the complexity of building such an environment.

Growing Consolidation: Consolidation is a major force altering the structure of insurance industry, as insurers seek to create economies of scale and broaden their product portfolios. The aggregated value of mergers and acquisition was $55.7 billion in 2000, up from $41.7 billion in 1999, and a mere $8.5 billion in 1993.

Convergence of Financial Services: Mergers and acquisition of insurance companies with other financial service providers like banks have led to the emergence of integrated financial services companies.

New Distribution Channels: New distribution channels are fast catching up with the traditional insurance agent. Though these channels are not a major threat as yet, they are rapidly changing the way insurers and customers interact with each other.

Focus on Customer Relationship Management: The only viable strategy for insurers today is to focus on the needs of the customers and strive to serve them better. Customers have extremely differentiated needs and, also, the profitability of individual customers differs significantly; hence, an effective CRM strategy becomes the most vital component of an insurer's overall business strategy.

20.2.2 Business Intelligence and the Insurance Value Chain

During the last three decades, insurance companies acquired significant product development capabilities; but they lagged behind in truly understanding the customer. This led most firms to develop products that they could, rather than those required by their customers. But during last few years, deregulation and growing competition has forced insurance companies to move from traditional product-centric operations to customer-centric operations.

To succeed in this market, insurers have to analyze their customer's needs and tailor all the business processes in the value chain to effectively meet their unique requirements. Implicit in this argument is the assumption that insurance companies have the ability and policies to change theminto actionable information. Business intelligence tools like data warehousing, OLAP, and data mining can significantly help in almost all the aspects of the value chain

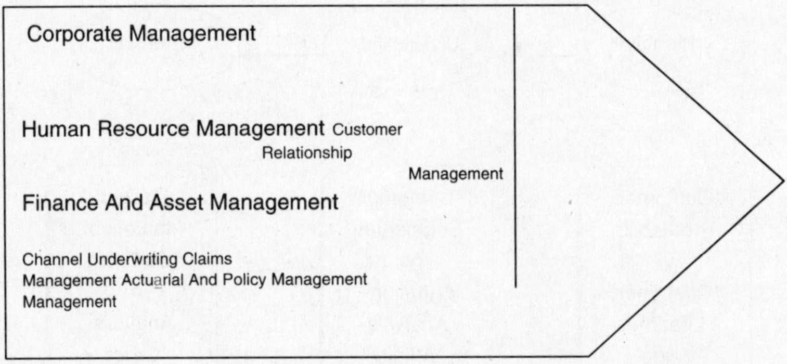

Fig. 20.1. Insurance Value Chain

to achieve this objective. Figure 20.1 illustrates the insurance value chain. In the following discussion we focus on some of the business intelligence applications in each block of the value chain.

20.2.3 Customer Relationship Management

A typical insurance company has a huge customer base, varied product lines with a number of products with each line, many distribution channels, and a market spread across geographies. To effectively interact with customers and design suitable products, the insurer's CRM strategy has to fully utilize the potential of technology. The insurer has to leverage the vast pool of data at each step in the CRM process, and the use the insight gained for developing new products and services to meet the ever-changing needs of the customers.

The CRM process is an insurance company and has three steps:

1. Identify the most profitable or potentially profitable customers for future interaction.
2. Understand their needs and buying patterns, and
3. Interact will them so as to meet all of their expectations.

Figure 20.2 illustrates the role of business intelligence in each step of the CRM process. Right from identifying the most profitable customers to improving the overall quality of customer interaction, BI tools can go a long way in making the insurer's CRM strategy a roaring success.

Customer Profitability: Rather than simply acquiring new customers, it is vitally important to retain and increase the profitability of existing ones. Identifying the most profitable customers is the first step in that direction. To arrive at the overall profitability of a customer, insurers must quantify (a) the costs involved in serving the customer over a period and (b) the revenues realized from the customer during that period. The results of customer profitability analysis can point toward the reasons behind why some customers are not

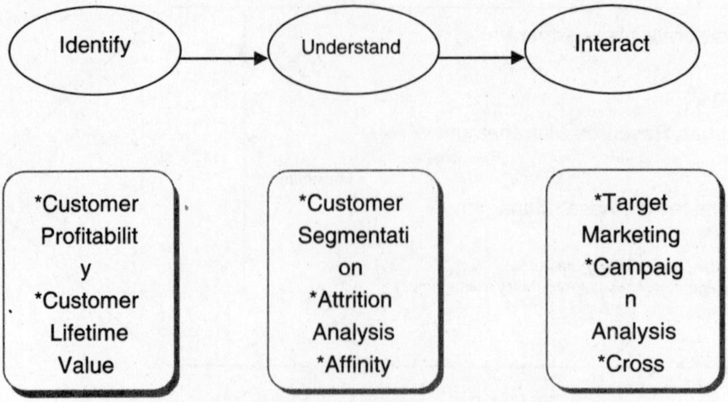

Fig. 20.2. Business Intelligence and CRM

as profitable as others are. For example, a customer might be unprofitable because the products used by her do not match her risk profile. Customer profitability analysis can significantly help in developing new product and customizing existing products for a customer or customer segment.

Customer Lifetime Value: Customer profitability is not the sole measure of a customer's value to the insurance company. A customer may have the potential of buying profitable products in the future; she may also serve as an excellent reference for more profitable customers. Customers lifetime data (LTV) is, hence, a more insightful measure. Often data mining tools are used to model customer lifetime value, taking into account all the factors that have a bearing on the customer's value over the entire course of her relationship with the insurance company.

Customer Segmentation: Segmentation is used to segregate customers, who exhibit common characteristics, in different segments. These segments can then be treated as distinct entities and the future interaction with them can be tailored accordingly. Customer segmentation can save a lot of marketing effort, which would otherwise go waste. Often data mining tools are used for customer segmentation. These tools use "clustering" algorithms for segmenting the entire customer base into clusters, identified on the basis of various demographic, and psychographic factors.

Attrition Analysis: Studies have shown that across industries, acquiring new customers is much more costly than retaining existing ones. This is especially true for insurance. Typically, buying an insurance product is a long-term decision for a customer; and if she decides to switch, it is very likely that she will not come back. Hence retaining the existing customers is of paramount importance; customer attrition analysis is the first step in this direction. It involves analysis of data captured during individual customer contacts at the various touch points. For attrition analysis, customer contact data is coupled with other data sources like claims and policies; the resultant data set is then associated with customers who have switched to analyze the possible reasons behind this decision. The results can also be used to improve the performance of customer touch points.

Affinity Analysis: It is often referred to as market-basket analysis. Certain products show an affinity toward each other and are likely to be bought together. For example, a man in his early thirties who buys a life insurance policy might also be interested in a certain type of annuity. These affinities can be, at times, extremely difficult to unearth and often data mining tools are used for this purpose. These tools use a technique called "association analysis" for arriving at the right combination of products and services for a customer or customer segment.

Target Marketing: Target marketing – marketing to a specific customer group – is a natural outcome of customer segmentation. Once distinct customer segments are identified, BI tools can be used to study the products likely to be bought by the segment. Often data mining is used to develop predictive models to establish the buying propensity of a segment toward various existing or new products. Armed with this knowledge, marketing managers can design specific campaigns targeted at individual segments.

Campaign Analysis: Campaign analysis is used to analyze the effectiveness of marketing or promotion campaigns. The effects of particular campaign on sales of the promoted product can be tracked using BI tools. Often the surge in sales of the promoted product can result in decrease in sales of other related products. BI tools can also help identify such relationships. The campaign data is stored in a data warehouse and can be used to predict the effectiveness of similar campaigns in future.

Cross Selling: Cross selling is a major source of revenue for insurance companies. For effective cross selling, an insurer can leverage the data – housed in the data warehouse – to quickly zero down on the new products that would be required by its existing customers. These can then be offered to them during the next contact.

20.2.4 Channel Management

Traditionally, insurance companies have relied on independent agents, brokers, and direct sales force for distributing their products. These are still the primary distribution channels and are likely to remain so in the near future. Internet is emerging as a popular distribution option for each certain type of insurance products like auto insurance. But most of the insurance products have not been that successful on the Net, primarily because of the complex nature of transactions involved. Fear of channel conflict has also prevented many insurers from fully exploiting its potential. No wonder, most of the traditional insurers currently use the Internet only to provide information about their products. But one thing is sure: insurers will have to quickly integrate the Internet with their existing channels. At the same time they will have to improve the overall distribution effectiveness via the traditional channels. BI tools allow insurers gain insight into the various distribution channels to help them intelligently address the various issues channel management:

Agent and Sales Force Deployment: Based on geography of its customer base, insurers can optimally deploy the right number of agents and sales persons in different locations. This analysis should also include the potential customers for new and existing products offered by the insurer.

Agent Development and Relationship Management: BI tools can leverage the sales data – stored in the data warehouse – to analyze the selling behavior of agents and sales persons. This analysis can identify the best agents and sales people who can then be adequately rewarded. The analysis can be extended to include more subtle behavioral aspects, which can be used to design training programs for agents. For example, some agents might just be selling certain products and totally missing out on others. This may indicate need for product specific training.

Channel Analysis: This involves calculating the performance of various channels. Using BI tools, insurers can compare the performance of various channels and drill down to the level of individual agents and products. The performance should be tracked over time so as to measure the effectiveness of corrective/developmental actions undertaken by the insurer.

E Business Development: A variety of analytics can be done on customer and transaction data captured via the Internet. This data should be integrated with data collected from traditional channels for a more meaningful segmentation of customers who buy policies over the net. This "e-segmentation" can help in designing campaigns specifically for the online customers. There is another source of potential useful data that can significantly help in improving online operations: Web log files. Web log analysis includes

Site Navigation: An analysis of the typical route followed by the user while navigation the Web site. It also includes an analysis of the most popular pages in the Web site. This can help in site optimization by making the Web site user friendly.

Referrer Analysis: An analysis of the sites, which are very prolific in diverting traffic to the insurer's Web site.

Error Analysis: An analysis of the errors encountered by the user while navigating the Web site. This can help in solving the problems and making the browsing experience more pleasurable.

Keyword Analysis: An analysis of the most popular keywords used by various users in Internet search engines to reach the insurer's Web site.

20.2.5 Actuarial

The actuarial function forms the crux of the insurance business. It involves estimating risk pertaining to the asset that is being insured. In case of life and health insurance it involves calculating the probability of accident or death based on various demographic, psychographic, and environmental characteristics. The task of an actuary is extremely complex and has strategic implications for the insurance company; it often takes up to ten years to become a certified actuary. An actuary can use sophisticated mathematical models – developing using data mining tools – to calculate future premiums and to allocate portions of a book of business for reinsurance:

Risk Modeling: Actuaries can develop predictive models using data mining tools, for identifying risk profiles of various customer segments. These models can include risk measures like mean claim amount, claim frequency, and loss ratios. For example, rich men who show propensity toward drinking and drive sports cars constitute a high-risk profile group. Various risk measures can be calculated for this customer segment, which can, then, be used for calculating at the right premium amount.

Reinsurance: A reinsurance company can take up a part of the insurer's risks in return for a part of the premium. In case of a claim, the reinsurance company will pay the corresponding claim amount. Actuaries need to decide the right amount of reinsurance in order to maximize the returns for the risk acceptable to the insurance company. Data mining tools can develop predictive models to arrive at the reinsurance level for the book of business based on the historical claims data residing in the data warehouse. These predictive models can identify suitable policies for reinsurance based on the loss experience of similar policies in the past.

Profitability Analysis: Profitability of the existing products can be tracked along various factors like product line, geographic region, agency, customer segment, etc. This is often that first step in predicting the viability of new products. Based on historical profitability, actuaries can also develop more sophisticated predictive models – using data mining tools – for estimating the marketability of new products and identifying the most profitable customer segments for those products.

20.2.6 Underwriting and Policy Management

An underwriter decides whether the risk undertaken by insuring a client is acceptable or not; and if it is acceptable, she determines the right amount of premium to be charged. Business intelligence tools can leverage the claims, loss, and others stored in a data warehouse – to help improve underwriting and policy management. Following are some of the business intelligence applications in this area:

Premium Analysis: Premium income is the primary source of revenue for an insurance company. Premium analysis allows the tracking of premium performance by a product or product line, a geographical region, an agency or a particular agent, and by a branch office. A variety of reports and analysis can be generated by "slicing and dicing" the data.

Loss Analysis: For some products or product lines, the premium revenue might be less than the cost of serving them. This loss, often termed as *underwriting loss*, may be due to inaccurate initial risk estimate. Insurers need to constantly monitor the loss data to determine the cost of getting new customers and renewing old ones for those products or product lines. This can help in improving profitability of underwriting programs and help insurers salvage their book of business.

20.2.7 Claims Management

The importance of claims management cannot be overemphasized. Speedy and effective claims handling forms the basis of sound customer relationship management. At the same time, the insurer has to guard against the ever-increasing specter of fraudulent claims. Insurance fraud is now very common in America, as a recent study by Insurance Research Council (IRC) pointed out that almost 36% respondents believe that it is acceptable to inflate claim amount to make up for the premium paid over the years. According to another study by the Insurance Information Institute, in the US alone P&C fraudulent claims amounted to a mind-bogging figure of $24 billion or 10% of the total claims in 1999. The figure for health insurance is believed to be almost four times that. Most of this addition cost due to fraud is passed on to the consumer as higher premium.

The opportunity cost of ineffective claims management is extremely high – hasty claim settlement can result in increased fraud-related costs and, at the same time, slow fraud detection can increases the overall claims cycle time, leading to higher customer dissatisfaction. No wonder insurance companies around the world are looking at technology for that "right" claims management solution. Experience of various companies suggests that without sophisticated analytical capabilities such a solution would remain an elusive dream. Following are some BI applications in claims management.

Claims Analysis: Claims analysis is one of the most common BI applications in the insurance industry. It involves analysis of the claims data coupled with other data sources like underwriting and policies. It is primarily used to gauge claims processing efficiency, which has a direct bearing on customer satisfaction.

Claims analysis is also used to understand subtle business trends in claims, which would have been otherwise difficult to spot. Typically OLAP tools are used to analyze and drill down to the detailed level for á better understanding of these trends. For example insurers can use this data to analyze trends in claims and loss patterns, which can help optimize reserve management, leading to lower risk and more available funds for investment. Claims analysis can also help in spotting fraud by analyzing above normal payoffs along different factors like geographical region, agent, and insured party.

In cases of health insurance, claims analysis can reduce abuse by analyzing the behavior of various practitioners. It can spot practitioners who have been consistently prescribing expensive medicines and tests in cases where they are not required. The analysis can also include other procedures and practices like the average length of hospital stay followed by the practitioner.

Fraud Detection: The likelihood of fraud in a claim can be detected by sophisticated analysis of claims data coupled with other internal or external data like payment history, underwriting, and ISO Claim Search Database. Data mining tools are typically used to develop models that can spot patterns in fraudulent claims. A claim, which shows fraud-related symptoms, can be sent to the Special Investigations Unit (SIU) for further investigation before the actual payment is made.

Claims Estimation: The actual value of claims cannot be known beforehand when a new product is launched. In such cases, an estimated value has to be set aside in the form of cash reserves; and such funds cannot be used for long-term investments. The accuracy of these estimates has far-reaching impact on the profitability of the insurance company. OLAP tools can be used to do an analysis on the claims data across geographies and customer segments to arrive at better claim estimates. For higher accuracy, data mining tools can also used to develop sophisticated claim estimation models.

20.2.8 Finance and Asset Management

The role of financial reporting has undergone a paradigm shift during the last decade. It is no longer restricted to just finances required by the law; increasingly it is being used to help in strategic decision making. Historically, interest earned on investments has been a major source of income for insurers and this income has greatly contributed to the overall profitability of insurance industry; at the same time underwriting costs have drastically brought down profitability. To compete in this market, insurers need to increase their return on investments and bring down underwriting costs. This requires ready access to financial data for analysis purposes. Many companies, in an attempt to improve financial reporting and decision making, have integrated their financial data in a financial data warehouse (FDW).

Budgeting: Data warehousing facilitates analysis of budgeted versus actual expenditure for various cost heads like promotion campaigns, underwriting expenses, commissions, etc. OLAP tools can provide drill down facility whereby the reasons for cost overruns can be analyzed in more detail. It can also be used to allocate budgets for the next financial period. Various activity-based costing models can be developed for better costs control and allocation.

Asset Liability Management: Models can be developed using BI tools to measure the insurer's exposure to various risk factors like change in the interest rate structure, share market volatility, etc. These models can be used to predict the performance of portfolio different economic scenarios and predict future liquidity needs of the insurer.

Financial Ratio Analysis: Various financial ratios like debt–equity, liquidity ratios, etc. can be analyzed over a period of time. The ability to drill down and join inter-related reports and analyses – provided by all major OLAP tool vendors – can make ratio analysis much more intuitive.

Profitability Analysis: This includes profitability of individual products, product lines, underwriting activity, and investments. A major component of profitability analysis is a thorough analysis of costs incurred during underwriting, which has been a major factor in bringing down the overall profitability of insurance companies.

Web Reporting and Analysis: Swift decision making requires ready access to financial data via an intuitive interface. Increasingly companies are providing executives concerned Web-based access to financial data – stored in the financial data warehouse. Almost all the standard OLAP tools have a Web interface that can greatly facilitate ad hoc querying report distribution.

20.2.9 Human Resources

Data warehousing can significantly help in aligning the HR strategy to the overall business strategy. It can present an integrated view of the workforce and help in designing retention schemes, improving productivity, and curtailing costs. Some BI applications in HR are:

Human Resource Reports/Analytics: Reports and analysis can be generated to support an integrated view of the workforce. Various analyses include staff movement and performance, workforce attrition by department, workforce performance by department, compensation and attrition, and other customized analyses and reports. The HR data can be integrated with benchmark figures for the insurance industry and various reports can be generated to measure performance vis-à-vis benchmarks.

Manpower Allocation: This includes allocating manpower based on new product launches. According to increased requirement, agent sales people can be deployed in specific regions where demand projections are high or likely to increase.

HR Portal: Employers need to maintain accurate employee data, which can be viewed by the employees the information about compensation, benefits, retirement facilities, etc. Payroll data can be integrated with data from other human resource management applications in the HR data warehouse. This data can then be made visible within the organization through the HR portal.

Training and Succession Planning: Accurate data about the skills sets of the workforce can be maintained in the data warehouse. This can be used to design training programs and for effective succession planning.

20.2.10 Corporate Management

The top management of any insurance company has its own business intelligence requirements. The MIS department is typically responsible for providing all the reports to them. It is also responsible for providing statutory reports to various outside agencies and any other information requirement within or outside the company. This may include information given to its customers in the form of various reports and analysis. A business intelligence environment that leverages data collected across the value chain is possibly the only effective solution for MIS.

Dashboard Reporting: Performance measures like product line profitability, overall underwriting costs, ROI on investment portfolio, etc. can be presented in dashboard reports to the top management to facilitate decision-making process. Also alerts can be triggered if any performance measure reaches a predefined threshold level. These reports can incorporate insurance industry benchmarks, provided by third-party researchers.

Statutory Reporting: Insurers have to provide a number of statutory reports to outside agencies, government bodies, and trade consortia. These reports can be easily generated from the business intelligence environment.

Customer Information Services: Historical claims data can be very vital for institutional clients like those offering workers' compensation against accident. It can help the customers identify major trends in claims and implement suitable corrective actions. Customer information services can not only reduce claims related costs for the insurance company, it can also go a long way in cementing insurance company's relations with its customers.

20.3 Summary

In summary, moving from a traditional "hard copy" underwriting manual with "tribal knowledge" as the guiding force in decision making is inevitable. The long-standing record of sustained industry – wide underwriting losses – expressed in another way – can be seen as a proxy for underinvestment in the technology and personal resources required to improve this core decision-making process. Being able to properly assess and price and ever-growing range of risks requires access to all requires information. In the past this access has been imperfect. It has also been limited within the organization to a relatively small group of users (actuaries) disconnected to the core drivers of insurers businesses and the core consumers of their services – to make better business decision in the underwriting domain is clearly necessary. We think data mining can play a useful role here.

The insurance industry is extremely divided in its adoption of business intelligence environment based on technologies like data warehousing, OLAP, and data mining. Quite a few insurance companies are in an advanced stage of their business intelligence initiative; yet there are many that are oblivious of its benefits. Some insurers have gone for nonscalable temporary solutions, which often fail to leverage the ever-increasing volumes of data. Hence, recognizing the need for an effective business intelligence environment based on the right architecture is vital. But it is just the first step. The real challenge is to make the BI environment an integral part of the decision-making process. Efficiently gathering the information requirements of all the user groups is thus extremely critical for the success of any data warehouse. The belief is that "we build it, they will use it" is wrong. Also a data warehouse cannot be the answer to all the information requirements; hence it is also very important to set clear business objectives for the business intelligence solution with total top management support.

20.4 Review Questions

1. How is data mining used in insurance companies?
2. What does an insurance underwriter do? State the principles of underwriting.
3. What are the various information sources available for mining in insurance sector?
4. How does data mining help underwriters to make better business decisions?
5. Give an overview of mining in insurance industry and about insurance value chain.
6. Discuss in detail channel management, policy management, chain management, finance, and asset management.
7. Explain corporate management technique.

21

Data Mining in Biomedicine and Science

Objectives:

- Data mining can be used to help predict future patient behavior and to improve treatment programs. Data mining has been used in a number of nonmedical domains.
- To turn information into knowledge, health care organizations are implementing data mining technologies to help control costs and improve the efficacy of patient care.
- The data mining algorithms such as rough sets and prediction algorithms significantly reduce patients' risk and diagnosis costs. The proposed algorithms use features extracted from data sets of different origins.
- Sequential pattern analysis and similarity search techniques have been developed in data mining, which has become a powerful tool and contributes substantially to DNA analysis.
- A series of unsupervised neural networks approach to medical data mining were designed and actual data were used to evaluate their performance in identifying natural clusters of patient population.
- The process of diagnosis can be enhanced if the physician can always be up to date with any new information – the kind of information procured through data mining. Data mining is the process of discovering nonobvious and potentially useful patterns in large data repositories such as warehouses.
- An architecture that uses the assistance of data mining to provide decision support for fever diagnosis is proposed. The proposed system also uses concepts of artificial intelligence to carry out the process of diagnosis. The main difference between a usual expert system is that the data mining process enhances the learning process.
- Scientific instruments can easily generate terabytes and even petabytes (a million gigabytes) of information. Scientific data is frequently in the form of images; there is also time series and sequence data such as DNA sequences, which need special algorithms to be dealt with.

S. Sumathi and S.N. Sivanandam: *Data Mining in Biomedicine and Science*, Studies in Computational Intelligence (SCI) **29**, 499–543 (2006)
www.springerlink.com
© Springer-Verlag Berlin Heidelberg 2006

- Databases that used to store millions of bits of useless information can be mined for insights that can greatly profit the miners. Recently, scientific instruments and business systems have been gathering extra information that was apparently useless simply because it was so easy to do.
- Algorithm development and m(ADaM) incorporates algorithms for detecting a variety of geophysical phenomena to address the needs of the earth science community. Its flexible architecture design has made it possible for ADaM to handle the multiple formats, scales, resolutions, and large granule sizes typical of spatial data for many different science problems.

Abstract. The explosive growth in data collection in business and scientific fields has literally forced upon us the need to analyze and mine useful knowledge from it. Data mining refers to the entire process of extracting useful and novel patterns/models form large data sets. Due to the huge size of data and amount of computation involved in data mining, high-performance computing is an essential component for any successful large-scale data mining applications.

The past decade has seen an explosive growth in biomedical research, ranging from the development of new pharmaceuticals and advances in cancer therapies to the identification and study of the human genome by discovering large-scale sequencing patterns and gene functions.

In this section, the applications in biomedicine and DNA data analysis and an unsupervised neural network approach to medical data mining techniques and data mining-assisted decision support for fever diagnosis – case studies are presented. In addition, the applications of data mining in science with case studies are illustrated.

An architecture that uses the assistance of data mining to provide decision support for fever diagnosis is proposed. Today's advanced scientific instruments can easily generate terabytes and even petabytes (a million gigabytes) of information. Although data mining can be invaluable tool in analyzing this data, it faces the additional challenge that scientific data frequently is not in a convenient flat file format. Scientific data is frequently in the form of images, which are relatively easily examined by humans, but which present a myriad of problems for data mining programs. There is also time series and sequence data such as DNA sequences, which need special algorithms to be dealt with effectively. Finally there are categorical values such as protein sequences. The problem with such data is that many algorithms rely on feature vectors allowed by numerical data, so these algorithms cannot be used on categorical data sets. Despite these extra difficulties, scientific data mining has still been making rapid progress.

Data mining is a concept that is taking off in the commercial sector as a means of finding useful information out of gigabytes of data. While products for the commercial environment are starting to become available, tools for a scientific environment are much rarer (or even nonexistent). Yet scientists have long had to search through reams of printouts and rooms full of tapes to find the gems that make up scientific discovery.

This section explores some of the ad hoc methods generally used for data mining in the scientific community, including such things as scientific visualization, and outline how some of the more recently developed products used in the commercial environment can be adapted to scientific data mining.

The Information Technology and Systems Center (ITSC) at the University of Alabama in Huntsville developed the Algorithm Development and Mining (ADaM) system under a research grant from NASA to investigate new methods of processing large volumes of Earth Observing System (EOS) remote-sensing data sets. This system provides knowledge discovery and data mining capabilities for data values as well as for metadata and catalogs the information discovered. ADaM incorporates algorithms for detecting a variety of geophysical phenomena to address the needs of the earth science community. This data mining system has been used for other research studies dealing with topics such as texture classification, image processing, and statistical analysis of earth science data sets. This section provides a detailed description of the ADaM system architecture, design, components, client interface, and the processing environment. It also describes the future directions that ITSC intends to pursue with ADaM.

21.1 Applications in Medicine

With the widespread use of medical information systems that include databases, which have recently featured explosive growth in their sizes, physicians and medical researchers are faced with a problem of making use of the stored data. The traditional manual data analysis has become insufficient, and methods for efficient computer-assisted analysis indispensable, in particular those of data mining and other related techniques of knowledge discovery in databases and intelligent data analysis.

21.1.1 Health Care

The effective use of information and technology is crucial for the health care organizations to stay competitive in today's complex, evolving environment. The challenges faced when trying to make sense of large, diverse, and often complex data source of considerable. In an effort to turn information into knowledge, health care organizations are implementing data mining technologies to help control costs and improve the efficacy of patient care. Data mining can be used to help predict future patient behavior and to improve treatment programs. By identifying high-risk patients, clinicians can better manage the care of patients today so that they do not become the problems of tomorrow.

21.1.2 Data Mining in Clinical Domains

As clinical data becomes larger, either through the proliferation of features to be collected and stored, or the number of records in databases, or both, a concern has been growing as to how to "get the data back out." Even more urgently, researchers, clinicians, and managers have become concerned with getting knowledge out of these large, complex databases. Simple database queries fail to address this concern of several reasons: a query might not retrieve the information desired because of user bias, lack of skill or experience, or limitations of the query software or database platform. In addition, the data in large databases often represents extremely complex concepts that may escape even the most experienced content expert working with a highly competent database developer. As a result, various industries have looked to data mining as a general approach for automatically discovering knowledge "hidden" in these databases. Data mining has been used in a number of nonmedical domains, such as fraud detection and marketing, and is now increasingly being used in a variety of health care database environments, including insurance claims, electronic medical records, epidemiological surveillance, and drug utilization.

21.1.3 Data Mining In Medical Diagnosis Problem

The interest in systems for autonomous decisions in medical and engineering applications is growing, as data is becoming more easily available. Though the two areas – medicine and engineering – appear to be remote in terms of the underlying processes, both face many common challenges. One of the problems of interest of both areas is autonomous prediction. Here one instance of the prediction problem, i.e., the diagnosis problem in medicine is considered.

An SPN (solitary pulmonary nodule) is a lung abnormality that may be cancerous or benign. Over 160,000 people in the US only are affected by lung cancer, and over 90% of them die. It is important that SPNs are diagnosed early and accurately. The clinical diagnosis of SPN using information from noninvasive tests is 40–60% accurate. This figure implies that many patients suspected of malignancy have to undergo biopsy that involves considerable risks (including death) and costs to them. The data mining algorithms such as rough sets (it is one unique theory in data mining) and prediction algorithms as in A. Kusiak *et al.*, 2000, significantly reduce patients' risk and diagnosis costs. In a typical SPN disease occurrence scenario, a nodule is detected on a patient's chest radiograph. As this SPN may be either benign or malignant, further testing is required to determine is exact nature. The diagnosis is perceived to depend on many features, such as the SPN diameter, border character, presence of calcification, patient's age, smoking history, and so on. Multiple medical disciplines are involved collecting a large volume of clinical data at different times and locations, with varying accuracy and consistency. Therefore, an approach that fuses information from different sources and intelligently processes large volumes of data is needed. The proposed algorithms use features extracted from data sets of different origins. The research shows that the number of features (results of noninvasive tests, patient's data, etc.) necessary to diagnose an SPN is smaller than that used in current medical practice. At the same time the decisions made are 100% accurate.

21.2 Data Mining for Biomedical and DNA Data Analysis

Since a great deal of biomedical research has focused on DNA data analysis, we study this application here. Recent research in DNA analysis has led to the discovery of genetic causes for many diseases and disabilities, as well as the discovery of new medicine and approaches for disease diagnosis, prevention, and treatment.

An important focus in genome research is the study of DNA sequences since such sequences form the foundation of the genetic codes of all living organisms. All DNA sequences comprise four basic building blocks (called *nucleotides*): adenine(A), cytosine(C), guanine(G), and thymine(T). These four nucleotides are combined to form long sequences or chains that resemble a twisted ladder.

Human beings have around 100,000 genes. A gene usually comprises hundreds of individual nucleotides arranged in a particular order. There are almost an unlimited number of ways by which the nucleotides can be ordered and sequenced to form distinct genes. It is challenging to identify particular gene sequence patterns that play roles in various diseases. Since many interesting sequential pattern analysis and similarity search techniques have been developed in data mining, data mining has become a powerful tool and contributes substantially to DNA analysis in the following ways.

21.2.1 Semantic Integration of Heterogeneous, Distributed Genome Databases

Due to the highly distributed, uncontrolled generation and the use of a wide variety of DNA data, the semantic integration of such heterogeneous and widely distributed genome databases becomes an important task for systematic and coordinated analysis of DNA databases. This has promoted the development of integrated data warehouses and distributed federated databases to store and manage the primary and derived genetic data. Data cleaning and data integration methods developed in data mining will help the integration of genetic data and the construction of data warehouses for genetic data analysis.

21.2.2 Similarity Search and Comparison Among DNA Sequences

We have studied similarity search methods in time-series data mining. One of the most important search problems in genetic analysis is similarity search and comparison among DNA sequences. Gene sequences isolated from diseased and healthy tissues can be compared to identify critical differences between the two classes of genes. This can be done by first retrieving the gene sequences from the two tissue classes, and then finding and comparing the frequently occurring patterns of each class. Usually, sequences occurring more frequently in the diseased samples than in the healthy samples might indicate the genetic factors of the disease; on the other hand, those occurring only more frequently in the healthy samples might indicate mechanisms that protect the body from the disease. Notice that although genetic analysis requires similarity search, the technique needed here is quite different from that used for time-series data. For example, data transformation methods such as scaling, normalization, and window stitching, which are popularly used in the analysis of time-series data, are ineffective for genetic data since such data are nonnumeric data and the precise interconnections between different kinds of nucleotides play an important role in their function. On the other hand, the analysis of frequent sequential patterns is important in the analysis of similarity and dissimilarity in genetic sequences.

21.2.3 Association Analysis: Identification of Co-occurring Gene Sequences

Currently, many studies are focusing on the comparison of one gene to another. However, most diseases are not triggered by a single gene but by a combination of genes acting together. Association analysis methods can be used to help determine the kinds of genes that are likely to co-occur in target samples. Such analysis would facilitate the discovery of groups of genes and the study of interactions and relationships between them.

21.2.4 Path Analysis: Linking Genes to Different Stages of Disease Development

While a group of genes may contribute to a disease process, different genes may become active at different stages of the disease. If the sequence of genetic activities across the different stages of disease development can be identified, it may be possible to develop pharmaceutical interventions that target the different stages separately, therefore achieving more effective treatment of the disease. Such path analysis is expected to play an important role in genetic studies.

21.2.5 Visualization Tools and Genetic Data Analysis

Complex structures and sequencing patterns of genes are most effectively presented in graphs, trees, cuboids, and chains by various kinds of visualization tools. Such visually appealing structures and patterns facilitate pattern understanding, knowledge discovery, and interactive data exploration. Visualization therefore plays an important role in biomedical data mining.

21.3 An Unsupervised Neural Network Approach to Medical Data Mining Techniques: Case Study

In this section we present the application of unsupervised neural networks with data visualization approach on a set of conventional pathology data. Inherent difficulties in the utilization of such data were overcome by utilizing three data subspaces identified as Drugs, Topography and Morphology. This case study is taken from D.Shalvi et al., Medical Informatics and Computational Intelligence Research Lab, University of Maryland, Baltimore, Maryland.

A series of unsupervised neural networks were designed and actual data were used to evaluate their performance in identifying natural clusters of patient population. Included is a method to examine and validate the underlying reasons for clustering. Preliminary examinations of identified clusters by qualified pathologists have shown promising results, which supports the conclusion that the suggested methodology yields discoveries and medical interpretations that can eliminate or serve as alternatives to special purpose of epidemiological studies.

21.3.1 Knowledge Extraction Through Data Mining

Data collection is often undertaken to monitor individual cases such as a patient in a hospital. This type of data is distinguishable from data sets collected for the purposes of studying a population; for example, determining the television shows that are the most popular. Individual data sets may be gathered and studied collectively for purposes other than those for which the sets were originally created; in such fashion new knowledge may be obtained while simultaneously eliminating one of the largest costs in developing knowledge, data collection. This approach is especially appropriate for medical data, which often exists in vast quantities in an unstructured format. Applying data mining techniques can facilitate systematic analysis.

Data mining is the process of sifting through and analyzing rich sets of domain specific data and then extracting the information and knowledge in the form of new relationships, patterns, or clusters for decision making purposes. Thus data mining is a form of knowledge discovery essential for solving problems in a specific domain. Conventionally, data is gathered to test an existing hypothesis (a top-down search). Alternatively, the existing data is mined and allowed to form natural clusters (a bottom-up finding). Cluster detection may be employed through statistical techniques such as Bayes' theorem or nonstatistical techniques such as unsupervised neural networks, which form clusters on the data set without knowing what the output clusters should model.

In this section Kohonen self-organizing maps (SOMs) are used to cluster a specific medical data set containing information concerning the patient's drugs, topographies (body locations), and morphologies (physiological abnormalities); these categories can be identified as the three input subspaces. Data mining techniques are used to collapse the subspaces into a form suitable for network classification. The goal is to acquire medical knowledge, which may lead to tool formation, automating an assist to medical decisions regarding populations.

21.3.2 Traditional Difficulties in Handling Medical Data

Medical data typically requires a large amount of preprocessing in order to be useful. There is numeric and textual data interspersed. Frequently different symbols are used with the same meaning; "male" may be denoted as "M," "m," 0, or a variety of other formats. One medication or condition may be commonly referred to by a variety of names. There is often a redundancy of data; age may appear in several places. Erroneous data is very common; medical terms are frequently misspelled. Finally, medical data is frequently sparse; when a structure is imposed on medical data much of the structure remains empty for a large portion of the population due to the breadth required of any structure.

A robust data preprocessing system is required in order to draw any kind of knowledge from even medium-sized medical data sets. The data must not only cleaned of errors and redundancy but organized in a fashion that makes sense for the problem; in this today's context, the data must be organized so that the benefits of using unsupervised neural networks may be maximized.

21.3.3 An Illustrative Case Study

For the purposes of this study a confidential real-world medical data set was used. This data set is medical of the many in the medical field in terms of the data difficulties mentioned above; redundant and erroneous data is frequent. The data makes use of and introduces main-specific knowledge organized as three hierarchical trees, identified as Drugs, Topography, and Morphology. Each tree is several layers deep providing sample opportunity for exploiting the trees' structure at various levels of depth. The most significant portion of the Morphology tree is displayed in Fig. 21.1.

After the requisite data preprocessing we can use unsupervised neural networks to organize the data into clusters of patients containing similar features. Data mining is employed in both before and after the neural network; before, to exploit the substructure of the provided trees, and after, to group the finegrain output clusters into larger clusters that perhaps contain more meaningful data. Finally, the original data is mined to explain the output clustering and extract the knowledge of the neural network in terms that the pathologist can understand and qualify. The goal is to discover hidden knowledge of the relationship between drugs, Topography, and Morphology.

21.3.4 Organizing Medical Data

Standard techniques were employed to clean erroneous and redundant data; for example, "GI Bleeding," "Gi Bleed," and "Gastrointestinal Bleeding" were all mapped to "Gastrointestinal Bleeding." If "Albuterol" appeared more than once for a patient, multiple entries were discarded.

The data set used contains only data at the leaves; each leaf node is present in only one or two tuples on average. This poses two problems: any conclusions formed would be statistically insignificant and the level of computation required for such an analysis would be exceedingly high due to the existence of roughly ten thousand different leaf nodes. To alleviate these problems the

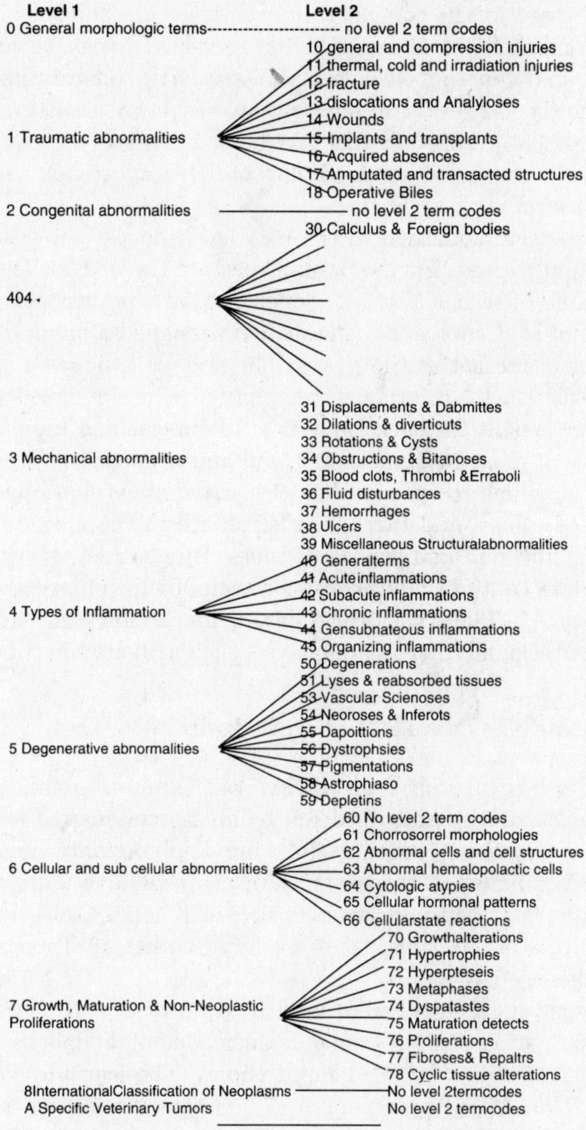

Fig. 21.1. Morphology Tree – Condensed

data was processed at the root level of each tree. At the root level each tree collapses into much fewer nodes; fourteen root-level drugs, sixteen root-level topographies, and ten root-level morphologies. By constraining all data to the root level the degree of differentiation has been greatly reduced from thousands to $40\,(14+16+10)$. The trade-off in cost is a great reduction in precision, while a benefit is the possibility of detecting trends within the data at the general level.

As the trees were collapsed to the root level the per tuple data was converted to bipolar format. For every tuple each of the 40 root level nodes was assigned a value of either 1 or -1 depending on whether any data existed for the leaves of that root node. The node is assigned a value of 1 if at least one data value is present at the leaves. The node is assigned a value of -1 if no data is at the leaves. In other words, only existence is preserved; quantity is lost. The end result for each tuple is a 40-dimensional bipolar array. The original data is at most 18-dimensional, containing one to six drugs, one to six topographies, and one to six morphologies. Each dimension may contain one of thousands of values, and there is no structure to the data due to variable dimensionality and null and repeated values. By contrast, every tuple of the transformed data contains 40 dimensions, each of which may take one of only two values, 1 or -1. The transformed data is much more consistent and lends itself to computationally intensive analysis such as neural networks.

21.3.5 Building the Neural Network Tool

The Kohonen self-organizing map (SOM) was chosen to organize the data in order to make use of a spatially ordered 2-dimensional map of arbitrary granularity. This architecture proved useful when applying data mining and data visualization techniques to the output. An $\mathrm{n} \times \mathrm{n}$ SOM was implemented for $\mathrm{n} = 5$, 7, 10, and 20. The Input layer consists of 40 input nodes, corresponding to the root layers of the three trees: 14 Drug nodes, 16 Topography nodes, and 10 Morphology nodes.

The following parameters were used to train a 10×10 network for trial 0. The training period was set to the default value of 30 epochs, which yields 62430 training iterations for 2081 input tuples. The learning coefficient α is initialized to 0.06. After approximately 7.5 epochs α is halved to 0.03. After another 7.5 epochs it is halved again to 0.015. For the final set of 7.5 epochs it is halved again to become 0.0075. Thus, the network is trained relatively quickly at the beginning; as learning progresses training is reduced to finetuning. After 30 epochs no learning occurs.

Other network parameters are varied over time as well. The frequency estimation parameter β is initialized to 0.0005 and successively halved to 0.0001 for the final set of 7.5 epochs. The conscience parameter γ is initialized to 1.0 and successively halved to 0.125 for the final set of 7.5 epochs. The neighborhood shape is a constant square of variable width. The neighborhood width is reduced over time from 7 nodes to 1 node; this determines the portion of the network that learns after every iteration.

After the network is trained it is used for one final pass through the input data set in which the weights are not adjusted. This network testing provides the final classification of each input data tuple into a single node in the 10×10 grid. The output is taken from the coordinate layer as an (x, y) pair. SOM trial 1 also uses n = 10 and is similar to trial 0 save for random initial conditions. For trial 2 n = 20; correspondingly, the neighborhoods are larger throughout training. In trial 3 n = 5 and in trial 4, n = 7; these trials contain smaller neighborhoods throughout training. For all trials, the learning parameters mentioned above are very similar.

21.3.6 Applying Data Mining and Data Visualization Techniques

The output of the SOM is a population distribution of tuples with spatial significance, which may be decomposed into three subspace distributions representing the decision boundaries determined by the network. The population distribution may be aggregated into larger groups to provide clusters not distinguishable in a high-resolution map. Such clusters can then be analyzed according to the subspace decomposition to determine significant patterns in the data.

Figure 21.2 displays the output data population distribution for the trial 0 network. This grid displays the number of tuples that were classified into each Kohonen layer node (square) during testing. Square (1,1) contains 180 tuples, by far the largest number of any square. Square (2,1) contains only one tuple and square (1,2) contains none. As SOMs tend to shrink or expand the output map as needed to fit the data these valleys suggest that the cluster at (1,1) is very sharp and well defined.

This hypothesis can be confirmed by looking at the raw data, which upon examination reveals that every one of the tuples in square (1,1) contains root level data only for Drug 6, Topography 6 and Morphology 5. The tuple at square (2,1) contains these three root level nodes as well as Drug 7, a difference slight enough for the network to distinguish the tuple by classifying it one square away from (1,1). All of the 34 tuples in square (3,1) contain Drug 6, Topography D and Morphology 5 but only 29 of the 34 tuples contain Topography 6. Clearly the difference between square (3,1) and square (1,1) is greater than that between square (2,1) and square (1,1).

The initial network organization of 100 squares is aggregated to form clusters that are spatially large; the outline of these areas was determined very subjectively. It is represented by any (x, y) coordinate pair, which has a population greater than 10 (an arbitrary threshold) combined to form spatially large clusters.

Cluster A comprises only one square at 1; this is 1% of the output space. However, this cluster of population of 180; or almost 9% of all the tuples! As this tends to distribute the data over the entire 2-dimensional output space,

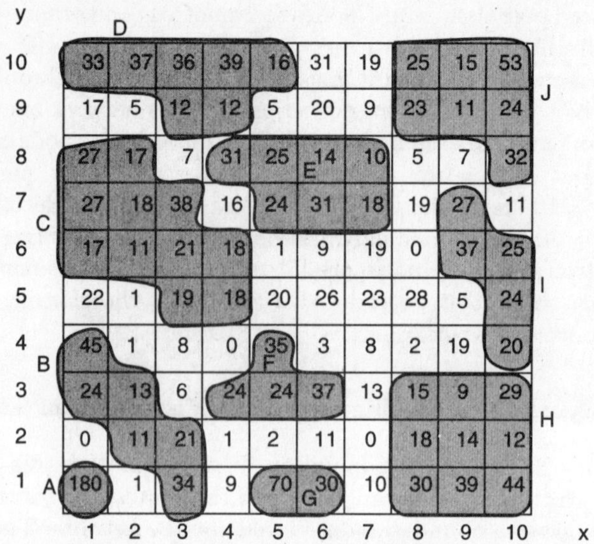

Fig. 21.2. Population of Clusters

this indicates that cluster A is well defined. Population and spatial considerations and on domain-specific knowledge were used to create cluster map in Fig. 21.2. Upon examination of the raw within these clusters one finds similarities between the tuples, which are indicative of medical relationships or dependencies. Numerous hypotheses can be made regarding these relationships, many of which were not a *priori* known.

The SOM groups together tuples in each square regarding to their similarity. The only level at which the SOM can detect similarities between tuples is at the root level of each of the three subspace trees since this was the level of differentiation presented to the SOM's input. Consequently we should expect that tuples in the same are often have the same root level drugs, topographies morphologies. Using data mining principles we can down from the output map into the raw data to over the source of the SOM's clustering.

For instance, square (3,10) belongs to cluster D contains 36 tuples. Drilling down into the data it is covered that of the 36 tuples:
36 tuples (100%) contain Drug 2 (Industrial products) 3 tuples (8%) contain Topography 5 (Digestive Tract) 13 tuples (36%) contain Topog. 6 (Digestive Organs) 3 tuples (8%) contain Topography 7 (Urinary Tract) 12 tuples (33%) contain Topography A (Nervous System & Special Sense Organs)
1 tuple (2%) contain Topog. B (Endocrine System) 5 tuples (13%) contain Topography C (Hematopoietic & Reticuloendothelial Systems) 2 tuples (5%) contain Topog. D (Topog. Regions) 36 tuples (100%) contain Morphology 8 (International Classification of Neoplasms)

Clearly the presence of Drug 2 and Morphology 8 defined are (3,10) since all of the tuples belonging to this are root-level nodes. For most squares, if it is covered that there is one or at most two morphologies each can be found in all or roughly 95% of the tuples. If arbitrary threshold level is dropped to 66% of the tuples there is almost always one or two morphologies that will surpass the threshold. If the threshold level is at 100% the following morphology cluster map may be constructed; this appears as Fig. 21.3.

The number inside each square in the morphology cluster map represents the root-level code of the morphology tree. For example, there are a large group of squares in the bottom of the graph that contain a 5; these square form a cluster for which 100% (the chosen threshold) of the tuples inside the cluster contain Morphology 5, Degenerative Abnormalities. There are at least six distinct clusters in this graph. Clearly the morphology clusters are arranged in a spatially meaningful way by the SOM; one cluster flows into the next.

A comparison between Fig. 21.2 and Fig. 21.3 shows that there appears to be some relation between the Population clusters and the Morphology clusters. Population clusters A (Morph. 5), C(4), F(5), G(5), and J(3) are entirely enclosed by their respective Population clusters. Clusters D and H are almost entirely enclosed, while clusters B, E, and I span two or more clusters, indicating that the Population clusters are not well formed.

The threshold level of 100% was chosen because in most squares, one or two morphologies are found in 100% of the square's tuples. If the thresh-

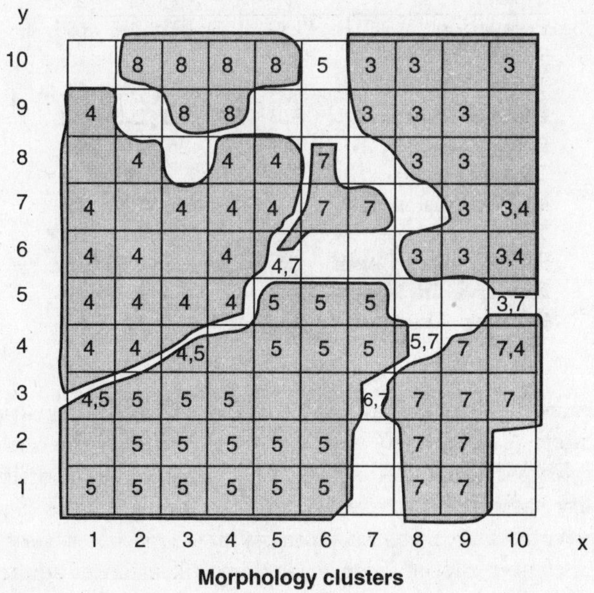

Fig. 21.3. Morphology Clusters

old is dropped to 66%, the clusters somewhat expand in size, and the overlap between the Morphology and Population clusters increases. In particular, cluster H is now completely enclosed by Morphology 7, translating to a cluster in which 66% of the population contains Morphology 7.

A similar drill-down analysis is done for the Drug and Topography subspaces, yielding maps that can be used to explain many of the population clusters. Additionally, trials 1–4 yielded extremely similar results, with identical clusters identifiable to those of trial 0, indicating that the clustering of the data is independent of network size and training parameters. It is now possible to introduce several heuristics to rate the worthiness of the population clusters.

The clusters delineated in the previous section were formed on the basis of population and spatial distribution alone. The worthiness of these clusters can be measured by the following parameters, which make use of square tuple population, cluster square count, and overlap with the three tree subspaces. Subsequently, the population distribution is clustered again with a heavier emphasis placed on tree subspace overlap.

We can estimate how well formed the Population clusters are by looking at the total number of tuples in each cluster compared to the average number of tuples expected for each square of that cluster in a random distribution. Table 21.1 displays this analysis.

Table 21.1. Cluster assessment worksheet

Cluster	A	E	F	G	Av.
Population	180	153	121	106	165
Square Count	1	7	4	2	5.9
Tuples/Square	180	21.9	30.3	53	43.4
Cluster Intensity	**8.6**	**1.1**	**1.5**	**2.5**	**2.1**
Drug Inclusion	1	.43	.25	1	.59
Topo Inclusion	1	.86	1	.5	.69
Morp Inclusion	1	.43	1	1	.85
Total Inclusion	3.0	1.71	2.25	2.5	2.1
Scaled Square Count	1.5	2.4	1.95	1.65	2.2
Relative Inclusion	**4.5**	**4.1**	**4.4**	**4.1**	**4.6**
Cluster Assessment	**38.8**	**4.5**	**6.6**	**10.3**	**9.5**

Tuples/Square reflects the average number of tuples per square for that particular cluster. A completely random distribution of the data would produce 20.81 tuples per square. A well-defined cluster should exhibit more; the cluster Intensity value grossly reflects this; it is simply Tuples/Square divided by 20.81. Cluster E, which has an intensity of 1.1, is not a very well-defined cluster since a cluster chosen from a random collection of squares would on average produce the same intensity. By contrast cluster A, which has an intensity of 8.6, is extremely well defined. Note that the Cluster Intensity parameter does not incorporate the population of neighboring (border) squares. A very small population in the border squares, as is the case with cluster A, should translate to a more well-defined cluster; this factor is not accounted for in the definition and would be a useful parameter modification.

Most clusters exhibit an intensity slightly greater than 1, as would be expected. However, a cluster with an intensity of 1.5 (cluster F) is much more well-defined than a cluster with an intensity of 1.1 (Cluster E); it contains 40% more tuples per square! Clusters A and G overshadow this significant difference.

Cluster assessment becomes more accurate when the Drug, Topography and Morphology cluster maps are included in the analysis. The Inclusion parameters represent the percentage of each population cluster that is encompassed in the most dominant Drug, Topography, and Morphology cluster for that Population cluster, as defined by the 100% threshold. For example, cluster F is composed of 4 squares. One of the 4 squares (25%) is part of Drug cluster A, all 4 squares (100%) are part of Topography cluster 6 and all 4 squares (100%) are part of Morphology cluster 5. The total Inclusion parameter is the sum of the 3 inclusion proportions; it is a relative measure of how well each Population clusters overlaps with the three subspace map clusters. The Scaled Square count parameter resets the Square Count scale range, originally from 1 to 11, to a new range, 1.5 to 3.0. Relative Inclusion is simply Scaled Square Count multiplied by Total Inclusion. Square Count is scaled so that the new range has the same numerical impact ($\sim$1.5 to 3.0) as total inclusion on the derived parameter. In this fashion the dependence on cluster size is reduced.

In terms of Total Inclusion clusters A, B, G and J are notably sharp. This is impressive for large clusters such as B and J. This expectation is realized in the Relative inclusion parameter for which clusters B (5.3) and J(6.2) have significantly higher values than all other clusters. Cluster C (4.6) also has a fairly large Relative Inclusion primarily due to its large size of 11 squares.

Finally, the cluster Assessment parameter is an overall assessment of the quality of the cluster, which combines Cluster Intensity (relative population) and Relative Inclusion (overlap with subspace clusters). As would be expected clusters A and G rate extremely well. Clusters B (6.3) and J (8.0) also rate fairly well, while cluster E rates very poorly. This can be attributed to cluster E's low intensity and inclusion.

These derived parameters are of course very subjective and only crudely attempt to rate a cluster. The sharpness of cluster boundaries is not included in the estimate and there is too great of a reliance on cluster size. Nevertheless, the above numbers are still very useful in rating the worthiness of heuristically determined clusters.

The following clusters were completely enclosed within two or more subspace root nodes:

- A: CNS Drugs; Digestive Organs; Degen. Abnormal.

- B: CNS Drugs; Digestive Organs
- F: Digestive Organs; Degenerative Abnormalities
- G: General Terms, Antihistamines, and Antiinfective Drugs; Degenerative Abnormalities.

These clusters are of particular interest since they relate root level nodes from different trees. Clusters A and F indicate a strong relationship between Digestive Organs and Degenerative Abnormalities; this relationship may simply be symptomatic of necrosis of the liver. A look at raw data indicates that cluster G contains many cases and lipoid degenerations (forms of degenerative Abnormalities) and the patient was often Tetracycline or Isoniazid (penicillins, or antiactive Drugs). This extracted relationship is a data set that has been discovered through data mining.

Knowing that the origin Population cluster map efficient we can recluster the network output map information gleaned from the subspace inclusion parameters. In Figure 21.4 a new cluster map is shown. Each node contains a code representing the root-level nodes for that square at 100%. For example, square (1,1) contains code 665, indicating that 100% of the tuples in that square contain Drug 6, Topography 6, and Morphology 5.

The cluster map was regenerated using clusters the subspace maps as well as spatial configuration of the original SOM feature space to realize nonactive clusters and population (to create larger clusters). Conflicting goals of the cluster formation are the cluster size and dimensionality of cluster definition, given the number of the root level nodes that are contained by the tuples

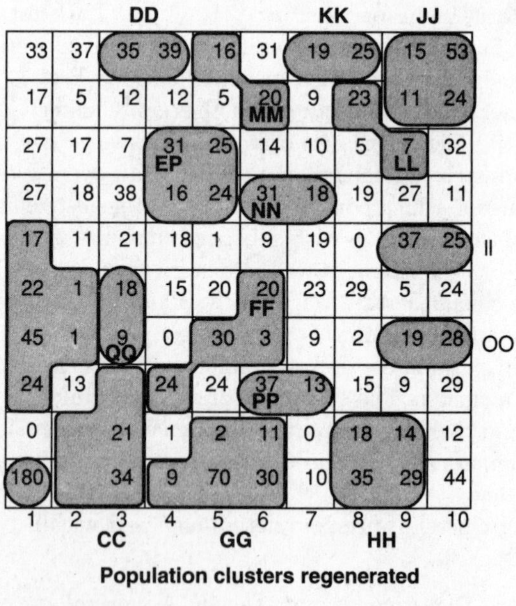

Fig. 21.4. Population clusters regenerated

in the cluster. The increased specificity and enable detection of fine clusters perhaps indicative of known relations on the data that would likely be covered with broader clustered feature space.

All of the multiple square clusters contain at least root level nodes from different trees. Many of the training single squares contain two root level nodes from different trees as well; for example, all 29 tuples in square (10,3) contains Topo. 8 and Morp. 7 as indicated by the code "x87" within the square; the "x" denotes that no single drug was contained by 100% of the tuples. In summary, the root-level node frequency data allows construction of root-level node clusters, individually by the subspace and as an intersection of subspaces. This allows generation more meaningful clusters using spatial, population, and subspace correspondence consideration. Many of the clusters identified using the above method not only comprise multiple squares but often are defined by multiple root-level nodes across the subspace trees. These clusters are a potential gold mine of data nuggets.

21.4 Data Mining – Assisted Decision Support for Fever Diagnosis – Case Study

In this section, we explain the dangers of inaccurate of delayed diagnosis of fever and show that fever diagnosis is an ideal candidate for decision support. An architecture that uses the assistance of data mining to provide decision support for fever diagnosis is proposed. The proposed system also uses concepts of artificial intelligence to carry out the process of diagnosis. We describe the various components of the architecture and the interactions between them. The architecture contains a data warehouse, which is incrementally refreshed with case details. The patterns obtained by mining the warehouse are used to formulate rules and in turn form questions in simple English. An interactive GUI, which poses questions to the physician. Each successive question is chosen based on the responses received previously. We conclude by saying that the results of such a system cannot be expected to occur in any definite time frame information can be discovered only if it is present. This case study is taken from S. Jayakumar et al. S.V. College of Engg, Pennalur, India, 2002.

Apart from being one of the most common "symptoms" encountered by physicians while treating patients, a fever is also a symptom that can have an extremely wide array of causes. Thus, a physician when confronted with a feverish patient, has to ask a number of questions and possibly make a large number of eliminations in order to arrive at the right diagnosis. The causes of fever may vary from neoplastic disease to sunstroke, and may have geographical, age-related, and racial variations. Also there are certain cases, wherein it is absolutely vital that a fever is diagnosed correctly at a very early stage. A classic example would be that of malaria. There are four species of the protozoan Plasmodium, which produce malaria in its various forms. Of these, the malaria caused by *Plasmodium falciparum* is the most dangerous, because, if left untreated, it can cause cerebral malaria, which is fatal. Meningitis is another form of fever that can turn fatal if not treated early.

The factors mentioned above, among others, make diagnosis of fever an ideal candidate for decision support. The process of diagnosis can be enhanced if the physician can always be up to date with any new information – the kind of information procured through data mining. Data mining is the process of discovering nonobvious and potentially useful patterns in large data repositories such as warehouses.

21.4.1 Architecture for Fever Diagnosis

Here we give a brief overview of the proposed architecture. We explain the basic functions carried out by the various components of the architecture and also the interactions between these components. The detailed functioning of each of the components is explained in the sections that follow along with relevant examples.

The main components of the architecture are the Medical Data Definition component (MDD), Physician system interface (PSI), Diagnostic Question Banque (DQBanque), Pattern Extractor and Rule constructor.

The pattern extractor consists of both a data warehouse and a data mining tool set. The data warehouse consists of exhaustive case histories of patients. The patterns that are generated by mining the warehouse are used by the rule constructor in order to construct new rules. These rules are then sent to the DQBanque. The DQBanque performs the task of converting these rules into questions in simple English, while also acting as a storage area for both the rules and the questions. During the process of diagnosis, the DQBanque sends appropriate questions to the PSI, which is a GUI used by the physician. The responses received by the PSI are sent back to the DQBanque, and based on these responses the appropriate follow-up questions are selected and sent to the PSI. The proposed system also contains the MDD that contains the various fever-related terms and their explanations. This component also contains information regarding the form in which data are stored in the warehouse, the facilitating data cleaning.

21.4.2 Medical Data Definition Component

The Medical Data Definition component defines the data and the metadata related to the topic of fever diagnosis. This includes specifics like symptoms, type of fever, diet classification, frequency of symptoms, temperature, duration of symptoms, blood pressure, strain of pathogen, etc. It also includes descriptions of various patient-related information in the form of attributes. This information includes name, age, sex, height, weight, race, and location among others. To be stored in the data warehouse, it is necessary that the data be taken through a data preparation process, or in other words the data must be cleaned. Data cleaning is the process of resolving inconsistencies in format and naming. The MDD helps in specifying the categories under which the data that is got via the PSI will be stored in the data warehouse after cleaning. It has a two-way connection to the pattern generator.

21.4.3 Physician–System Interface

This is basically a GUI that assists the physician in carrying out the process of diagnosis. It serves a dual purpose:

- Displays the appropriate questions, which it acquires from the DQBanque
- Accepts the responses and sends them to the Pattern Extractor and the DQBanque

The questions displayed by the PSI are framed by the DQBanque (in simple English) based on the rules present in it. These questions predominantly require yes/no type responses or at the most, one-phrase answers. The responses thus obtained are sent to both the Pattern Extractor and the DQBanque. The Pattern Extractor will convert the data into the appropriate form (with the help of the MDD) and feed it into the data warehouse. The DQBanque analyzes the responses and determines the next appropriate question to be posed to the physician.

21.4.4 Diagnostic Question Banque

This component serves as a repository for both diagnoses-related rules and the simple English questions derived from those rules. The rules are obtained from the rule constructor as and when they are formed. The DQBanque performs the function of converting these rules into questions, which are in simple English, along with storing these questions. Even after the conversion process, the DQBanque retains the rules in their original form as well. When the diagnosis takes place, at first the DQBanque send over standard patient related and preliminary disease-related questions to the PSI. The responses are sent back to the DQBanque in real time. After a certain stage, the next question to be asked, is decided by the DQBanque on the basis of the responses received so far. Often a particular response would necessitate the asking of a set of questions rather than just one. These questions would be placed in a stack, as would the sets of questions necessitated by each one of these questions. Thus the DQBanque performs the task of traversing all these stacks to whatever extent necessary in order to form a diagnosis.

From the above information we can see that the DQBanque has a one-way connection with the Rule Constructor and a two-way connection with the PSI.

The functioning of the DQBanque is now discussed in the context of diagnosis of bacterial meningitis. The following is a typical set of questions that would be displayed on the PSI during the course of diagnosis of this disease:

What is the name of the patient?

Andrew Miller
What is Andrews' sex?
Male
What is Andrew's age?
25
Country of Residence?
United States
State/Province?
New York
Race
Caucasian

The above questions are examples of standard queries (put to all patients), the responses to which are stored in the hospital records. The individual responses may or may not be fed into the data warehouse.

Temperature?
98.6
Pulse?
100
Respirations?
20
Blood Pressure?
110/70
Recent occurrence of seizures?
Yes

The information collected so far suggests the possibility of both bacterial meningitis and herpes encephalitis headaches?

Yes

If the response had been "No," the next question would have possibly been: Presence of temporal lobe lesions?

A "Yes" response for this question would strongly suggest the presence of herpes encephalitis, and the line of questioning would proceed appropriately. Considering that the response was "yes" to the query about headaches, the questioning continues as follows.

Stiff neck?
Yes
Unusual change in mental status?
Yes

At this point all the classic symptoms of bacterial meningitis have been established.

Thus the typical working of the DQBanque was illustrated in the context of bacterial meningitis.

21.4.5 Pattern Extractor

This component is composed of two parts: the data warehouse and the data mining toolset. The data warehouse basically consists of a data cube and a two-dimensional table. The data cube contains an exhaustive list of case histories with all the possible relevant information such as case number, age, sex, race, geographical location, existing ailments and conditions, hereditary ailments, and current symptoms among others. The 2-D table would contain only information stating the various symptoms and conditions corresponding to each individual fever-causing disease. The data cube is augmented with fresh case details in real time as and when diagnosis takes place. The responses to the questions posed by the PSI (which are either yes/no responses or single phrases) are transferred to the pattern extractor and held in temporary storage. After the necessary data cleaning is done (with the help of the MDD), the data is fed into the data cube. The data cleaning process is done with the assistance of the MDD to resolve any inconsistencies in storage format and naming. Thus we see that the data cube is incrementally refreshed.

The data mining tool set implements a rich set of mining algorithms based on association rules, classification, and time-series analysis. These algorithms are periodically used on the data present in the data cube, in order to generate any kind of "interesting" patterns. The aimed end result of these patterns is basically to find a set of symptoms, conditions, and other factors such as race, age, gender, and location associated with each individual fever-causing disease. The information thus generated is stored in a dynamically expandable 2-D table. A typical row in the table would contain the name of the disease in the first column, and the various associated information in the other columns. So, to put it succinctly, the results of mining the data cube are stored in the 2-D table. The rule constructor receives the information present in the 2-D table and formulates new rules.

So, we see that the pattern extractor has one-way connections to both the PSI and the rule constructor. It also has a two-way connection with the MDD.

When the patient's age is taken as input, he/she is automatically classified as a neonate, a child, or an adult. The data cube present extractor contains an attribute called age group, which stores this information. The data cube also contains attributes specifying the organism, which caused the disease. For example, it is known that H. influenzae or N. meningitis generally causes the occurrence of bacterial meningitis in children. This is the kind of knowledge that can potentially be gained from the data mining process. One can also discover nonobvious racial, geographical, or diet-related patterns in the occurrence of fever. The new knowledge thus found can greatly quicken the process of diagnosis and possibly make it more accurate.

21.4.6 Rule Constructor

The Rule Constructor has a one-way connection to the pattern extractor and another one to the DQBanque. This component gleans the data stored in the

2-D table (present in the pattern extractor) and formulates rules based on this data. Any rule thus formed is first compared to the database of rules present in the DQBanque. If the rule already exists, it is discarded. Else, it is added on to the database of rules in the DQBanque. The following rules illustrate the basic format of rules formed by the Rule Constructor.

IF seizure = yes AND

Headache = yes AND

Stiffneck = yes AND

Mentalstatechange = yes

THEN Bacterial/Meningitis.

IF seizure = yes AND

Nonseasonalfocalneurologicdefect = yes

AND

Temporallobeleisions = yes

THEN Herpesencaphalitis

When a "yes" response is received for the query "seizures?" the DQBanque is searched for rules having an "If" condition "seizures = yes." When such a rule is found, the other conditions of the rule are traversed one by one, and the corresponding questions are chosen from the DQBanque and sent to the PSI. When a certain condition in the rule is not satisfied, that rule is discarded, and the search starts for another rule, which has the condition "seizures-yes." This is basically how questions are chosen and sent to the PSI. Although this explanation basically pertains to a process taking place in the DQBanque, we chose to give it here, as we wanted all rule-related discussions to appear together. These rules have scope for evolving further as the mining process takes place. It would generally be a slow and almost static evolution.

21.5 Data Mining and Science

Although data mining can be invaluable tool in analyzing this data, it faces the additional challenge that scientific data frequently is not in a convenient flat file format. Scientific data is frequently in the form of images, which are relatively easily examined by humans, but which present a myriad of problems for data mining programs. There is also time-series and sequence data such as DNA sequences, which need special algorithms to be dealt with effectively.

An example of mining scientific data was the cataloging of a sky survey. The Second Palomar Observatory Sky Survey took six years to collect three terabytes of image data containing an estimated two billion sky objects. The three thousand photographic images were scanned into 16-bit pixel resolution digital images at $23,040 \times 23,040$ pixels per image. The problem was generating a survey catalog of all these sky objects from this information. Additionally, the attributes and class of each object are needed to be determined and recorded in the catalog. To solve this problem, the Sky Image Cataloging and Analysis Tool (SKI-CAT) system was developed.

The majority of objects in each image were faint, making determination of their class by visual inspection or classical computational approaches in astronomy impossible. SKI-CAT used decision tree-learning algorithms to accurately predict sky object classes. This accuracy was verified by comparison with a set of high-resolution charged-couple device images. SKI-CAT was 94% accurate at predicting the class of sky objects, which increased the number of reliably classified objects by 300%. These results have already helped astronomers discover sixteen new high red-shift quasars. Such quasars are difficult to find and provide clues about the early history of the universe.

The Magellan spacecraft orbited the planet Venus for over five years and used synthetic aperture radar to penetrate the gas and cloud cover to map the surface of the planet. The result is that we have a unique high-resolution map of the entire planet. In fact, we have more of Venus mapped at the 75-m pixel resolution than we do of the Earth because so much of the Earth is covered by water. This data set is valuable because of its completeness and because Venus is the most similar to Earth in size. It is hoped that learning about the geological evolution of Venus will produce valuable lessons about the Earth.

The immense size of this data set prevents planetary geologists from personally examining all the images. To assist geologists in analyzing the Venus map, the Jet Propulsion Laboratory developed the Adaptive Recognition Tool (JARtool). The system seeks to automate the search for small volcanoes by training the system via examples. The geologists would label a small sample of the images and the system would then use these to train itself to recognize small volcanoes. The system would then attempt to locate and measure the planet's estimated one million small volcanoes. It used classification learning to distinguish true detections of volcanoes from false alarms. It performed as well as scientists in identifying common types of small volcanoes, but rarely detected those scientists are not sure about.

The geoscientific data mining system Quakefinder automatically detects and measures tectonic activity in the earth's crust using satellite data. It was used to map the direction and magnitude of group displacements due to the 1992 Landers earthquake in Southern California over a spatial region of several hundred square kilometers at a resolution of 10 m to a subpixel precision of 1 m. Quakefinder is implemented on a 256-node Gray T3D parallel supercomputer so that the gathered data can rapidly produce scientific results. Besides automatically measuring known faults, it also allows automatic knowledge discovery by indicating novel unexplained tectonic activity away from the primary faults never before observed. Future work will focus on the measurement of continuous processes over many images, instead of simply measuring abrupt behavior seen during earthquakes.

In summary, although data mining is still limited in its functionality, its potential is nearly unlimited. Already business, science, and security have derived benefits from its development. Databases that used to store millions of bits of useless information can be mined for insights that can greatly profit the miners. Recently, scientific instruments and business systems have been gathering extra information that was apparently useless simply because it was so easy to do. The creation of data mining makes this excess information useful.

21.6 Knowledge Discovery in Science as Opposed to Business-Case Study

The relatively new discipline of data mining is most often applied to extraction of useful knowledge from business data. However, it is also useful in some scientific applications where this more empirical approach complements traditional data analysis. The example of machine learning from air quality data illustrates this alternative. This case study is taken from Brian J Read, CLRC Rutherford Appleton Lab, Oxon, UK

Data mining is the essential ingredient in the more general process of knowledge discovery in databases (KDD). The idea is that by automatically sifting through large quantities of data it should be possible to extract nuggets of knowledge.

Data mining has become fashionable, not just in computer science (journals and conferences), but particularly in business IT. (An example is its promotion by television advertising.) The emergence is due to the growth in data warehouses and the realization that this mass of operational data has the potential to be exploited as an extension of *business intelligence*.

21.6.1 Why is Data Mining Different?

Data mining is more than just conventional data analysis. It uses traditional analysis tools (like statistics and graphics) plus those associated with artificial intelligence (such as rule induction and neural nets). It is all of these, but different. It is a distinctive approach or attitude to data analysis. The emphasis is not so much on extracting facts, but on generating hypotheses. The aim is more to yield questions rather than answers. Insights gained by data mining can then be verified by conventional analysis.

21.6.2 The Data Management Context

"Information technology" was originally "data processing." Computing in the past gave prominence to the processing algorithms – data were subservient.

Typically, a program processed input data tapes (such as master and detail records) in batches to output a new data tape that incorporated the transaction. The structure of the data on the tapes reflected the requirements of the specific algorithm. It was the era of Jackson Structured Programming. The concept of database broke away from this algorithm-centric view. Data assumed an existence independent of any programs. The data could be structured to reflect semantics of relationships in the real world. One had successively hierarchical, network, relational, and object data models in commercial database management systems, each motivated by the desire to model better the structure of actual entities and their relationships.

A database is extensional, storing many facts. Some information is intentional; that is, it manifests as rules. Some limited success was achieved with deductive databases that stored and manipulated rules, as for example in Prolog-based systems. This encouraged *Expert Systems.* However, it was hard to achieve solid success. The main difficulty was the knowledge elicitation bottleneck: how to convert the thought processes of domain experts into formal rules in computer.

Data mining offers a solution: automatic rule extraction. By searching through large amounts of data, one hopes to find sufficient instances of an association between data value occurrences to suggest a statistically significant rule. However, a domain expert is still needed to guide and evaluate the process and to apply the results.

21.6.3 Business Data Analysis

Popular commercial applications of data mining technology are, for example, in direct mail targeting, credit scoring, churn prediction, stock trading, fraud detection, and customer segmentation. It is closely allied to data warehousing in which large (gigabytes) corporate databases are constructed for decision support applications. Rather than relational database with SQL, these are often multidimensional structures used for the so-called *on-line analytical processing* (OLAP). Data mining is a step further from the directed questioning and reporting of OLAP in that the relevant results cannot be specified in advance.

21.6.4 Scientific Data Analysis

Rules generated by data mining are empirical – they are not physical laws. In most research in the sciences, one compares recorded data with a theory that is founded on an analytical expression of physical laws. The success or otherwise of the comparison is a test of the hypothesis of how nature works expressed as a mathematical formula. This might be something fundamental like an inverse square law. Alternatively, fitting a mathematical model to the data might determine physical parameters (such as a refractive index).

On the other hand, where there are no general theories, data mining techniques are valuable, especially where one has large quantities of data containing noisy patterns. This approach hopes to obtain a theoretical generalization automatically from the data by means of indication, deriving empirical models, and learning from examples. The resultant theory, while may not be not fundamental, can yield a good understanding of the physical process and can have great practical utility.

21.6.5 Scientific Applications

In a growing number of domains, the empirical or black box approach of data mining is good science. Three typical examples are:

Sequence Analysis in Bio Informatics

Genetic data such as the nucleotide sequences in genomic DNA are digital. However, experimental data are inherently noisy, making the search for patterns and the matching of subsequences difficult. Machine-learning algorithms such as artificial neural nets and hidden Markov chains are a very attractive way to tackle this computationally demanding problem.

Classification of Astronomical Objects

The thousands of photographic plates that comprise a large survey of the night sky contain around a billion faint objects. Having measured the attributes of each object, the problem is to classify each object as a particular type of star or galaxy. Given the number of features to consider, as well as the huge number of objects, decision tree-learning algorithms have been found accurate and reliable for this task.

Medical Decision Support

Patient records collected for diagnosis and prognosis include symptoms, bodily measurement, and laboratory test results. Machine-learning methods have been applied to a variety of medical domains to improve decision making. Examples are the induction of rules for early diagnosis of rheumatic diseases and neural nets to recognize the clustered microcalcifications in digitized mammograms that can lead to cancer.

The common technique is the use of data instances or cases to generate an empirical algorithm that makes sense to the scientist and that can be put to practical use for recognition or prediction.

21.6.6 Example of Predicting Air Quality

To illustrate the data mining approach, both advantages and disadvantages, this section describes its application to a prediction of urban air pollution.

Motivation

One needs an understanding of the behavior of air pollution in order to predict it and then to guide any action to ameliorate it. Calculations with dynamical models are based on the relevant physics and chemistry.

An interesting research and development project pursuing this approach is DECAIR (Development of an earth observation data converter with application to air quality forecast- http://www-air.inria.fr/decair/). This concerns a generic system for exploiting urban quality models by incorporating land use and could cover data from remote-sensing satellite images.

To help with the design and validation of such models, a complementary approach is described here. It examines on air quality empirically. Data mining and, in particular, machine-learning techniques are employed with two main objectives:

1. To improve our understanding of the relevant factors and their relationships, including the possible discovery of nonobvious features in the data that may suggest better formulations of the physical models;
2. To induce models solely from the data so that dynamical simulations might be compared to them and that they may also have utility, offering (shortterm) predictive power.

Source Data

The investigation uses urban air quality measurements form the City of Cambridge (UK) [Cambridge City Council Air Quality Monitor: http://www.io-ltd.co.uk/ccc.html]. These are especially useful since contemporary weather data from the same location are also available. The objectives are, for example, to look for and interpret possible correlations between each pollutant (NO, NO_2, NO_x, CO, O_3, and PM_{10} particulates) and

a. the other pollutants;
b. the weather (wind strength and direction, temperature, relative humidity and radiance); looking in particular for lags – that is, one attribute seeming to affect another with a delay of perhaps hours or of days.

Data Preparation

Before trying to apply machine learning and constructing a model, there are three quite important stages of data preparation. The data need to be *cleaned*, *explored*, and *transformed*. In typical applications, this can be most of the overall effort involved.

a) Cleaning

Though not elaborated here, but commonly a major part of the KDD process is data cleaning. In this case, one is concerned with imposing consistent formats for dates and times, allowing for missing data, finding duplicated data, and weeding out bad data – the latter are not always obvious. The treatment of missing or erroneous data needs application dependent judgment.

b) Exploration

Another major preliminary stage is a thorough examination of the data to acquire familiarity and understanding. One starts with basis statistics – means, distributions, ranges, etc – aiming to acquire a feeling for data quality. Other techniques such as sorting, database queries, and especially exploratory graphics help one gain confidence with the data.

c) Transformation

The third preparation step is data set sampling, summarization, transformation, and simplification. Working with only a sample of the full data, or applying a level of aggregation, may well yield insight and result that are quicker (if not even discernible at all) than with the complete data source. In addition, transforming the data by defining new variables to work with can be a crucial step. Thus one might, for instance, calculate ratios of observations, normalize them into bins, bands, or classes.

Initial Analysis

The initial analysis concentrated on the daily averages for the weather measurements and daily maxima of the pollutants. This simplifies the problem, the results providing a guide for a later full analysis. In addition, the peak values were further expressed as bands (e.g., "low," "medium," and "high"). For example, ozone (O_3) values were encoded as

LOW	<50 ppb
MEDIUM	50–90 ppb
HIGH	>90 ppb

The bands relate to standards or targets set by the UK Expert Panel on Air Quality Standards (EPAQS) that the public can appreciate. (For ozone, the recommended limit is 50 ppb as an 8-hour running average).

The data exploration and analysis is guided by domain knowledge and enhanced by it, Examination of the Cambridge air pollution data confirmed initial expectations:

- There is a daily cycle with peaks in the afternoon.
- Sundays have low pollution
- An east wind (from industrial Europe) increases ozone levels.
- Sunlight on nitrogen dioxide (NO_2) produces ozone.
- Particulates (PM_{10}) come from vehicle exhausts.

Cambridge has little industry and within an urban environment traffic is the dominant pollution agent. Its effect depends on the local topography, so mesoscale dynamical models have restricted value.

Modeling

The two principal machine-learning techniques used in this application are neural networks and the induction of decision trees. Expressing their predictions as band values make the results of such models easier to understand.

a) Decision Trees

Applying the C5.0 algorithm to the data to generate a simple decision tree, one gets for ozone bands:

AirTemp = <28.3 → LOW
AirTemp > 28.3
RelHum = <58.1 → HIGH
RelHum > 58.1 → MEDIUM

This suggests how the ozone concentration depends mainly on the air temperature and relative humidity. The same tree, expressed as a rule set is:

Rules for HIGH:

Rule #1 for HIGH:

If AirTemp >28.3

And RelHum = <58.1

Then → HIGH

Rules for LOW:

Rule #1 for LOW:

If AirTemp = <28.3

Then → LOW

Rules for MEDIUM:

Rule #1 for MEDIUM:

If AirTemp > 28.3

And RelHum > 58.1

Then → MEDIUM

Default: → LOW

In fact, the support for these rules is modest. The handicap is that there are too few instances of HIGH ozone days into the data. Reliable predictions would need something more elaborate, but this illustrates the idea.

b) Neural Networks

Alternatively, the daily data can be fitted with an artificial neural network to model the ozone band value. A first attempt yields:

Neural Network "O3band" architecture
Input Layer: 5 neurons
Hidden Layer #1: 4 neurons
Output Layer: 4 neurons
Predicted Accuracy: 96%
Relative Importance of Inputs
Air Temp: 0.29
RelHum: 0.06
Rad: 0.04
Wet: 0.02
Windspeed: 0.004

Again, this shows that air temperature is the dominant predictor. However, given the limited quantity of data summarized to daily values, it is not worth trying to refine the model network.

Software

The air quality data were analyzed using the data mining software package *Clementine* (http://www.spss.com/software/clementine/and http://www.isl.co.uk/-originally from Integral Solutions Ltd. and now from SPSS Inc.) While this provides standard machine-learning algorithms to generate models, its great virtue is the powerful visual environment it offers for data exploration. This case of data exploration and modeling is crucial is in allowing the domain expert to attack the problem and find applicable results.

In summary, work so far supports the common experience in data mining that most of the efforts is in data preparation and exploration. The data must be cleaned to allow for missing and bad measurements. Detailed examination leads to transforming the data into more effective forms. The modeling process is very iterative, using statistics and visualization to guide strategy. The temporal dimension with its lagged correlations adds significantly to the search space for the most relevant parameters.

Investigation that is more extensive is needed to establish under what circumstances data mining might be as effective as dynamical modeling. (For instance urban air quality varies greatly from street depending on buildings and traffic.) A feature of data mining is that it can "short circuit" the postinterpretation of the output of numerical simulations by directly predicting the probability of exceeding pollution thresholds. A drawback is the need for large datasets in order to provide enough high-pollution episodes for reliable rule induction. More generally, data mining analysis is useful to provide a reference model in the validation of physically based simulation calculations.

21.7 Data Mining in a Scientific Environment

The advent of the computer has brought with it the ability to generate and store huge amounts of data. For example, it is not unusual for power users to have the equivalent of three or four encyclopedias worth of data online. When you add the data generated by government and other organizations, such as the recently completed census or the data collected every time you make a purchase at any modern supermarket, the volume of data available is almost incomprehensible. The problem is in how to turn this data into usable information.

However, this is not a new phenomenon. Scientists, especially experimentalists, have long had to tackle this problem. While Isaac Newton may have formulated his theory of gravity when an apple fell on his head, it was still followed by hundreds, if not thousands, of experiments demonstrating, validating, and/or refining the original equation. Taking more recent examples, the volume of data generated by space probes and particle physics, dwarfs anything previously contemplated. Looking closer to home, scientists at ANSTO often analyze data generated over their entire working career of twenty or thirty years.

Over the centuries, various methods have been developed to deal with this volume of data, many of which were seen as major steps forward for mathematics at the time. Some of these methods include fast Fourier transforms, multivariate regression analyses, as well as a whole range of statistical methods. More recently, visualization has been widely adopted by scientists as a means of studying the ever-growing masses of data.

21.7.1 What is Data Mining?

With the current trends in centralization of an organization's data in large databases, particularly in a commercial environment, the process of extracting useful information has become more formalized and the term **Data Mining** has been coined for it. In one of the first papers on commercial data mining, Evangelos Simoudis of IBM defined it as:

The process of extracting previously unknown, comprehensible and actionable information from large databases and using it to make crucial business decisions (Simoudis 1996).

This definition has a definite business flavor and much of IBM's development of data mining has been in this direction. In practice, data mining is a process that can take on different approaches depending on the type of data involved and the objectives desired. As this is still very much an evolving discipline, much work is being undertaken to determine standard processes for the varied environments. Further, as the context in which the data is gathered is often an important component, this must be factored into any analysis.

Data mining consists of three components: the captured data, which must be integrated into organization-wide views, often in a *Data Warehouse*; the *mining* of this warehouse; and the organization and presentation of this *mined* information to enable understanding.

The data capture is a fairly standard process of gathering, organizing, and cleaning up; for example, removing duplicates, deriving missing values where possible, establishing derived attributes and validation of the data. Much of the following processes assume the validity and integrity of the data within the warehouse. These processes are no different to any data gathering and management exercise.

The *Data Mining* process itself is the extraction of valid and previously unknown information, as given in the definition above. There are two approaches: *verification driven*, whose aim is to validate a hypothesis postulated by a user, or *discovery driven*, which is the automatic discovery of information by the use of appropriate tools.

The data mining process is not a simple function, as it often involves a variety of feedback loops since while applying a particular technique, the user may determine that the selected data is of poor quality or that the applied techniques did not produce the results of the expected quality. In such cases, the user has to repeat and refine earlier steps, possibly even restarting the entire process from the beginning. This is best illustrated in Figure 21.5 from Simoudis article.

The *final* step is the presentation of the information in a format suitable for interpretation. This may be anything from a simple tabulation to video production and presentation. The techniques applied depend on the type of data and the audience as, for example, what may be suitable for a knowledgeable group would not be reasonable for a naive audience.

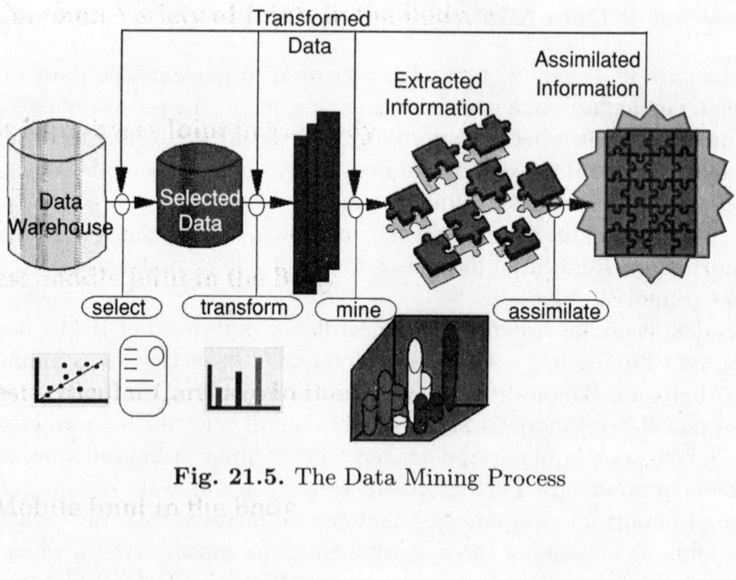

Fig. 21.5. The Data Mining Process

Verification Driven Data Mining

Currently, the most common use of data mining is verification driven and is primarily aimed at confirmation of an idea. Generally, the mechanism is to propose some association or pattern and then to study the data to find support, or otherwise, for the proposal.

There are a number of standard techniques used in verification driven mining; these include the most basic form of query and reporting, presenting the output in graphical, tabular, and textual forms, through to multidimensional analysis and on to statistical analysis.

Discovery-driven Data Mining

The discovery-driven approach depends on a much more sophisticated and structured search of the data for associations, patterns, rules, or functions, and then having the analyst review them for value. The current techniques for performing discovery-driven mining consist of four different approaches: predictive modeling, including neural nets, link-analysis technique that attempts to establish links between records, database segmentation that partitions the data into collections of related records, and finally deviation detection that identifies point that do not fit in a segment.

21.7.2 Traditional Uses of Data Mining

Within the business world, data mining is being seen as a method of tapping into the value of the data with an organization and providing a competitive advantage. An example of this is the analysis of purchase histories, drawn from credit card transactions, preferred customer schemes, frequent shopper schemes, and any other purchasing data that includes customer information. Using a method called *neural segmentation*, a number of different types of purchase patterns can be identified and then customer groupings can be associated with this data.

For instance, such analysis of shopping has identified two groups of people who purchase baking items, the first being older, retired couples, and the second, young couples with large families. The next step may be to look at product linkage; for example, there may be a group of people who purchase men's suits, women's high-fashion shoes, men's ties and expensive chocolates. They do not buy baby clothes, houseware, and greeting cards. This indicates that a store may be able to bring in more customers for a sale of suits if they have chocolates for half-price, or better yet, give away the chocolates.

These procedures can be used further for the analysis of any activity that generates large volumes of data, from specific surveys through to the collection of operational data, such as stock movements, or point-of-sale information. An example of this is *Market Basket Analysis*, which refers to the discovery of patterns within items purchased as is illustrated by such correlations between the purchase of paint and paint brushes or paint thinner. These associations can then be used to determine shelf locations and promotional sales planning.

Such analysis is the main force driving the introduction of data mining within large organizations and, thus, the current interest in such research. It is invariably related to the interrogation of large volumes of data, using high-performance systems and massive amounts of storage. However, there is still the need to apply some common sense to the results as spurious patterns and associations may be found. It is quite possible for an association to be found between the purchase of paint and cat food, which may be caused by other factors that were not part of the original analysis.

Most commonly, data mining is a single step in the entire process of *Decision Support*, and fits into the general process: *Data Warehouse - Data Mining - Decision Support.*

21.7.3 Data Mining in a Scientific Environment

While IBM may be driving data mining in the commercial marketplace, the origins are in fact in scientific computing with considerable work being done at UCLA and the University of Helsinki. Some of the original work was on geophysical databases in an attempt to process some of the large volumes of data they have available.

What is not considered in much of the work on data mining is that most, if not all, of this work is just as applicable to the scientific environment. One of the critical issues with data mining is a *credibility check* being performed by someone who is aware of the field. Most scientists, and in particular experimentalists, have a great respect for their data, being well aware of the dangers of using inapplicable methods for analysis. An excellent example of this is given in Clifford Stoll's new book *Silicon Snake Oil*, in which he describes a study by an astronomer, Professor Li Fang, into the periodic motions of the earth's axis. This study involved the analysis of thousands of years of astronomical measurements. Dr Li had performed all the measurements by hand and Clifford was attempting to show him how easy it would have been with a computer. On presenting his results, Dr Li replied:

When I compare the computer's results to my own, I see that an error has crept in. I suspect it is from the computer's assumption that our data is perfectly sampled throughout history. Such is not the case, especially during the Sung dynasty. And so, it may be that we need to analyze the data in a slightly different manner.

Having a computer, I had naturally cast the problem as simple data analysis.... The real challenge was understanding the data and finding a good way to use it (Stoll 1995).

The underlying principles of the science method, being the cycle *observation-hypothesis-experiment* fits well with the processes of data mining with discovery driven mining working well for the *observation-hypothesis* step and the verification driven mining for the *hypothesis-experiment* step.

As scientists have been working with this principle for centuries, and as most mathematics has been intended to support such scientific endeavors, many, if not all, of the methods are already being used by them. In many cases, the only change is in the terminology, not in the practice.

The final stage in any data mining is the presentation of results and this has both a very long history and is an area of rapid change in scientific work. This stems from simple graphs that scientists have long studied through to the latest techniques in visualization being demonstrated on high-performance graphics workstations.

21.7.4 Examples of Scientific Data Mining

One example of the scientific analysis of such data found in farming and the environment, is optimization of crop yield while minimizing the resources supplied. To minimize the resources, it is necessary to identify what factors affect the crop yield, out of such items as chemical fertilizers and additives; for example, phosphate, the moisture content and type of the soil.

One analysis looked at over 64 separate items measured over a number of years to extract the items that were significant. Initially the analysis was discovery-driven mining to attempt to find what parameters were significant, either by themselves or in conjunction with others. Using such statistical methods as multivariate regression analysis, the parameters that are significant and their relative influence was determined. From this, an equation was developed, which was then further verified through verification driven mining against new datasets.

Of more general interest, global climate change studies, a *hot* research area, are primarily a verification-driven mining exercise. Climate data has been collected for many centuries and is being extended into the more distant past through such activities as analysis of ice core samples from the Antarctic and, at the same time, a number of different predictive models have been proposed for future climatic conditions. The sample data is then used to verify these models by seeing if they accurately predict past conditions, which can then be compared with the sample data. From this, the models are then further refined and used for another round of verification driven mining.

21.7.5 Concluding Remarks

Data mining is a new term and formalism for a process that has been undertaken by scientists for generations. The massive increase in the volume of data collected or generated for analysis with the use of computers has made it an essential tool. However, despite the more *formal* approach, data mining is something that scientists perform on an *ad hoc* basis and can easily adapt to. Many of the methods used for the analysis of the data were originally developed to process scientific data and are used unchanged.

21.8 Flexible Earth Science Data Mining System Architecture

Designing a data mining system for Earth Science applications is complex and challenging. The issues that need to be addressed in the design are (1) variability of data sets, (2) operations for extracting information, and (3) providing the capability to the user to write complex mining plans. Earth Science data sets not only come in different formats, types, and structures, there are also many different states of processing such as raw data, calibrated data, validated data, derived data, or interpreted data. The mining system architecture must be designed to be flexible to handle these variations in data sets. The operations required in the mining system vary for different application areas within Earth Science. Operations could range from general purpose operations such as image processing techniques or statistical analysis to highly specialized, data set-specific science algorithms. The mining system architecture should be flexible in its ability to process new data sets and incorporate new operations without too much effort. The design of the architecture should also allow other users to build new clients to utilize such a system. The Information Technology and Systems Center at the University of Alabama in Huntsville originally developed the Algorithm Development and Mining (ADaM) system under a research grant from NASA Headquarters Research Announcement (NRA) to investigate new methods of processing large volumes of Earth Observing System (EOS) remote-sensing data sets. ADaM is designed to handle the complexity of mining Earth Science data. It can process heterogeneous data sets and allows users to add research problem-specific science algorithms to the system.

This section discusses issues that had to be considered in designing a flexible system architecture. It describes the ADaM system and its user interface as an example of a flexible design. This section also describes the research directions that are evolving from this innovative architecture.

21.8.1 DESIGN ISSUES

As stated in the introduction, several issues had to be considered while designing a flexible mining system for Earth Science. These are:

Data Handling Capabilities

Earth Science data introduces complexity in designing, building, and utilizing a data mining system, because these data sets can be quite varied. They can be point data collected by a meteorological instrument, swath or grid data collected by satellites, or volume scan data collected by weather radar. The formats of these data sets also vary from simple binary or ASCII files to more complex structures such as Hierarchical Data Format for the Earth

Observing System (HDF-EOS). The spatial and the temporal resolutions of these data sets depend upon the measuring instrument and the platform. The spatial resolution could vary from hundreds of kilometers to a few meters. The temporal range of a data file could vary from 15 minutes to a day or longer. Temporal resolution could vary from instantaneous measurements to accumulation of data over some period. To utilize mining techniques over the broad range of data sets, the mining system had to be designed to handle these types of data set variations.

Addition of New Algorithms

In certain circumstances, a known scientific algorithm can be utilized to extract the information needed from data sets. Detecting Mesoscale Convective Systems (MCS) from SSM/I data utilizing the Devlin algorithm is one such example. The data mining system had to be designed to be flexible enough to allow not only data set specific algorithms but also other new algorithms to be added to it without affecting the other operations.

Allow Scientists to Select and Sequence Different Operations

The mining system also needed the capability to allow scientists to create their own mining plans. A mining plan is a sequence of specified steps, where each step is a processing operation. The scientist should be able to piece together different operations/algorithms to reach their goal.

21.8.2 ADaM System Features

The ADaM system was designed using the latest object Oriented techniques to achieve a high degree of portability, accessibility, and modularity. The implementation in standard C++ allows the system to run on multiple operating systems, including IRIX, Linux, and Microsoft Windows NT. One of the design goals was to have ADaM work at both data archive centers or on a user's desktop workstation.

Overview of the Architecture

The ADaM data mining system has been designed to extract content based metadata from large Earth Science data archives. It can detect phenomena or events that are of interest to scientists and then store this information in a way that facilitates the data search and order process. Some mining results are stored in Event/Relationship Search System (E/RSS), an ITSC-developed spatial data search engine used to find coincidences between mining-generated phenomena, climatological events, and static information such as country and river basin boundaries. The data mining engine also provides

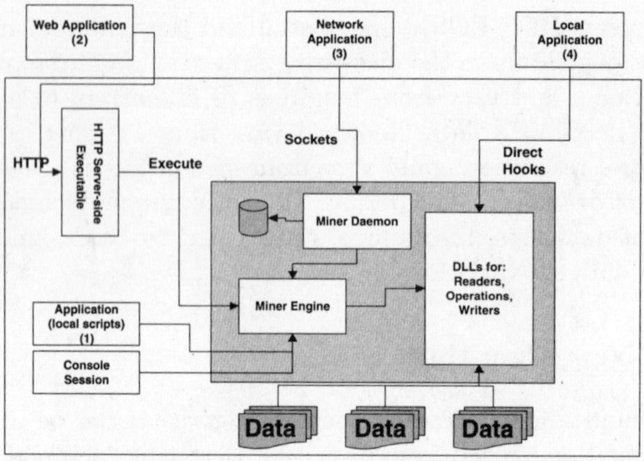

Fig. 21.6. Multiple Process Flows Utilizing the ADaM Data Mining Architecture

other data ordering-related capabilities such as subsetting and custom data product generation through specialized client applications. Custom processing may include gridding, resampling, filtering, format conversion, or other analysis depending on the needs of the customers. For example, ADaM can generate a monthly total rain accumulation image from radar reflectivity data. Both the E/RSS and custom processing client are Web applications, so the clients are capable of running in almost any environment. Figure 21.6 depicts a generalized view of how the ADaM data mining architecture has been utilized. This architecture allows the clients to communicate to the system in a variety of ways such as: (1) The miner engine can be driven directly via local scripts or an interactive console session, (2) A Web application can guide the user in creating mining plans, which execute the mining engine, (3) A network application can submit mining plans via the miner daemon, and (4) The system may also be used as a library with the application directly linking to the individual operations needed.

Processing Flow

The ADaM system architecture is based on a processing stream, in that mining is broken down into a series of steps with results from each step passed to the next one in line. Figure 21.7 illustrates both ADaM's data processing stream, as well as the three basic types of modules: input, processing, and output. The use of data input filters, specialized for a variety of data types, has been instrumental in simplifying the development of the processing and output operations. The selected input filter translates the data into a common internal structure so that the processing operations can all be written for a single data representation.

This allows the addition of new operations to the system without having to address input data format problems. Similarly, the addition of a new input

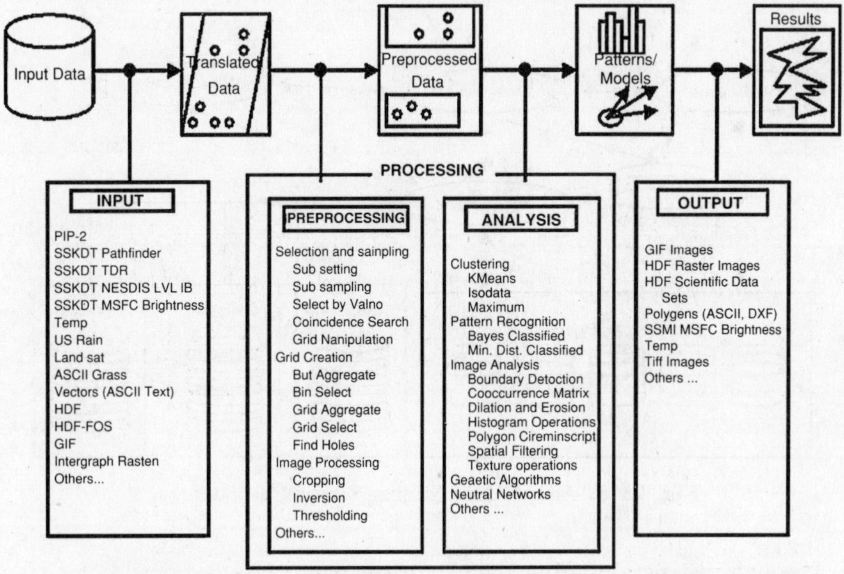

Fig. 21.7. Schematic diagram depicting the stream of a mining system

filter provides access to the entire suite of processing operations for the data type in question. This design feature allows ADaM to handle heterogeneous Earth Science data sets. The mining system currently allows over 120 different operations to be performed on the input data stream. These operations vary from specialized atmospheric science data set specific algorithms to generalized image processing techniques. The last step in the mining process is the selection of the input modules, the output filters effectively insulate the processing operations from having to support all the possible output formats. Since the input data has been converted to ADaM's internal format, the output modules allow the user the option to select either the input format or a different format for the final data product.

Components

In order to allow for the distributed use of the data mining functionality, the ADaM system was designed as a client-server architecture, which supports remote client applications communicating with the data mining server. This allows the server system to be co-located with archived data stores while being driven by either remote or local clients. In support of this architecture, the ADaM data mining system is composed of the mining engine and mining daemon, both located on the server. The daemon supports a specific protocol of messages and listens on a configured port for instructions from client applications.

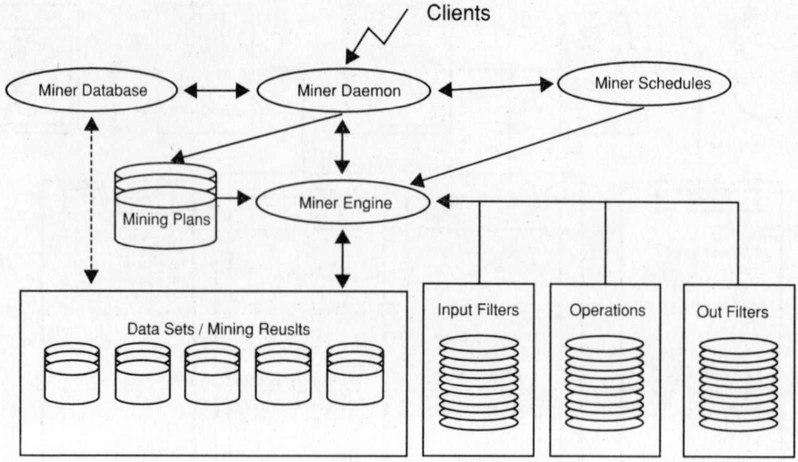

Fig. 21.8. AdaM Data Mining Server Components

Through instructions from the client, the daemon is responsible for managing user access information, file management operations, and job scheduling and management. The daemon ultimately sends the correct information and directions to the engine in the form of a "mining plan" for actual processing. A software interface layer was created providing tools to assist client application developers in communicating with the mining daemon across network sockets. Figure 21.8 depicts the connections between the components of the ADaM client-server architecture. Each component performs a specific well-defined task, and therefore the components themselves may be replaced or updated provided that the new components conform to the same interfaces.

Mining Engine:

The mining engine is the software component that manages the processing of data through a series of specified operations. The input, processing, and output modules are dynamically loaded as needed at execution time, and this allows for the addition of newly developed modules without the need to rebuild the engine. The mining engine interprets a mining plan script that provides the details about each specified operation and the order that they should be executed. Other communication with the mining engine is managed through the mining daemon process.

Mining Daemon:

The mining daemon is the gateway to the mining engine. All network communications with the mining system are handled by the daemon through a message handling protocol. Upon installation the daemon is configured to listen on a specific port for any socket communications. The daemon is capable of handling a fairly rich set of messages that allows it to perform file management duties, command the mining engine, and provide user security screening. The daemon can also determine at run time, which processing modules are available on the server.

Mining Database:

The database component is used to store information that is required for the smooth operation of the system and the interaction of its components. This information includes the names, locations, and related metadata for input data sets available on the server. It also includes information about users, jobs, mining results, and other related information. A relational database is currently used for this task. Access to the database is provided by the daemon.

Mining Scheduler:

The scheduler component examines the list of jobs to be executed on the server and determines which job or jobs to execute at any given time. The scheduling policy used can be unique to each server. The scheduler invokes the mining engine for each job and monitors its progress, updating the job status in the database whenever it changes.

Operations and Data set Input/Output Filters:

Each of the operations and data set filters is implemented as a shared library. The libraries are loaded dynamically by the mining engine, which means that new modules may be added to the system without recompiling or relinking. Each of the operations and filters is completely independent of all the others. All operations and filters either produce or operate on a common format representing scientific data. This design feature allows science specific algorithms to be incorporated into the system with relative ease.

Mining Plan:

The mining plan script conveys the processing instructions to the mining engine. The plan contains the number and sequence of processing steps as well as the detailed parameters (tokens) describing how to perform each step, such as where to find the input data, where to store the output and configuration parameters for all the various operations. Mining plans may be created using the mining plan editor. Since mining plans are text files, they may also be created using any text editor. It is easy for applications to write mining plans. The mining plan begins with a number indicating the number of operations in the plan. The remainder of the plan is a series of token/value pairs where the tokens and values are delimited by newlines.

21.8.3 ADaM Plan Builder Client

In order to allow users to build complex mining plans, ITSC has designed an easy-to-use and functional user interface called the ADaM Plan Builder. This user interface is a client that communicates directly to the mining engine. It makes it easier for the user to select the right operation for the task and to provide values for the parameters for that operation. The individual ADaM operation documentation is written in XML™. Since this standardizes the documentation, the Plan Builder written in Java parses and utilizes the information contained in those XML files. Thus, the Plan Builder Interface utilizes these XML files to provide the user options on the operations available, what parameters each operation requires, the meaning of each parameter, default values for those parameters and finally a sample mining plan. Through the Plan Builder, users can select sample operation steps and modify the values for the parameters according to their needs. The ADaM Plan Builder allows the user to chain together complex mining plans for scientific research.

The Plan Builder also allows the users to edit and modify the mining database. The user can feed the metadata information about the data files to be mined into the database via this client. The database then automatically selects the correct files for mining based on the time range given. The architecture of the Plan Builder is shown in Figure 21.9 and a screen capture of the interface is shown in Figure 21.10.

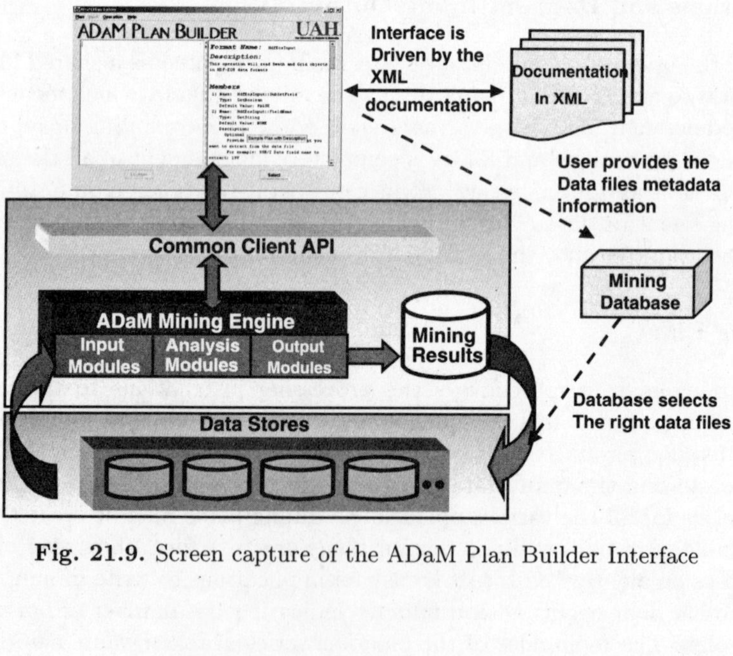

Fig. 21.9. Screen capture of the ADaM Plan Builder Interface

21.8.4 Research Directions

ADaM is currently undergoing a metamorphosis in the sense of becoming uncoupled from an environment that is dependent on centralized processing on a single server platform with the availability of local data. The following sections describe some of the efforts underway to migrate ADaM into a highly distributed environment that will provide broader access to the system and distributed heterogeneous scientific data sets, while addressing improved scalability and flexibility.

Distributed Mining

ITSC is currently investigating and prototyping emerging distributed component technologies. To address the use of distributed mining services and access to distributed data sets, the use of distributed mining services opens the system to greater possibilities of extensibility, performance, scalability, and reliability by distributing the processing burden and lessening the possibility of centralized points of system failure. Current research efforts have been successful with the development of an Earth Science Markup Language [http://esml.itsc.uah.edu] that will make great strides towards realizing generic access to heterogeneous data sets. The integration of ESML technology with planned distributed mining components is expected to result in a virtual processing environment that capitalizes on improved networking bandwidth and underutilized distributed processors.

Grid Mining

Another approach to distributed mining is also being prototyped in the form of grid mining. ITSC researchers, in collaboration with NASA/Ames researchers, have been successful with implementing and testing the ADaM system on the NASA Information Power Grid. The Grid approach employs a sophisticated infrastructure of message passing, scheduling, and security in an effort to utilize large capacity processing and data centers for scientific research. This approach to distributed data mining promises to be of particular benefit to scientific researchers in need of massive processing and data resources.

Mining Onboard Space Craft

ITSC is also investigating and developing an innovative processing system capable of handling the unique constraints and characteristics of the on-board satellite data and information environment. The EnVironmEnt for On-Board Processing (EVE) system will serve as a proof of concept of advanced information systems technology for remote-sensing platforms. EVE's on-board, real-time processing will provide capabilities focused on the areas of autonomous data mining, classification, and feature extraction. These will contribute to

Earth Science research applications, including natural hazard detection and prediction, fusion of multisensor measurements, intelligent sensor control, and the generation of customized data products for direct distribution to users. EVE is being engineered to provide high-performance data processing in a real-time operational environment. A ground-based test bed is being created to provide testing of EVE and associated Earth Science applications in a heterogeneous embedded hardware and software environment.

Concluding Remarks

The ability of the user to create complex mining plans by chaining together different operations is also possible because of the flexibility of the architecture. ITSC plans to utilize its experience in designing ADaM to meet the scientific mining requirements of the next generation of scientists in several other domains. Research efforts are focused toward distributed mining across the Web; mining large volumes of data on the information power grid; and finally designing a system that would be used onboard aircraft or spacecraft to extract features or phenomena as soon as they are sensed by the instrument.

21.9 Summary

An architecture was proposed that uses assistance of data mining to provide decision support for diagnoses of fever. Although we have dealt exclusively with the diagnoses of fever in this section the basic architecture can be implemented for the diagnoses of any other physical condition. The system we proposed is meant to make the process of fever diagnosis less time consuming and possibly more accurate. It must be emphasized that the system would serve only as an accessory and not as a substitute for the human physician. The main difference between a usual expert system is that the data mining process enhances the learning process. It must also be noted that the diagnosing power is never complete. Its knowledge could become better than its current knowledge in several years or months or weeks or days or even in an hour.

Research to expand the types and magnitude of data that data mining systems can effectively mine is well underway. The needs of business, security, and science will provide incentive to invest time and money into such development. Perhaps someday data mining will advance faster than the growth of databases and allow the mining of nearly infinite databases, such as mining the entire World Wide Web.

Is data mining as useful in science as in commerce? Certainly, data mining in science has much in common with that for business data. One difference, though, is that there is a lot of existing scientific theory and knowledge. Hence, there is less chance of knowledge emerging purely from data. However, empirical results can be valuable in science (especially where it borders on engineering) as in suggesting causality relationships or for modeling complex phenomena.

Another difference is that in commerce, rules are soft sociological or cultural and assume consistent behavior. For example, the plausible myth that *"30% of people who buy babies' nappies also buy beer"* is hardly fundamental, but one might profitably apply it as a selling tactic (until perhaps the fashion changes from beer to lager).

On the other hand, scientific rules or laws are, in principle, testable objectively. Any results from data mining techniques must sit within the existing domain knowledge. Hence, the involvement of a domain expert is crucial to the data mining process. Naïve data mining often yields "obvious" results. The challenge is to incorporate rules known a priori into the empirical induction, remembering that the whole KDD process is exploratory and iterative.

ADaM has proven to be an effective and valuable tool to mine Earth Science spatial data. Its flexible architecture design has made it possible for ADaM to handle the multiple formats, scales, resolutions, and large granule sizes typical of spatial data for many different science problems. The design permits the easy addition of new algorithms, especially domain-specific science algorithms. Thus this section discussed the various case studies involved in science and biomedicine with respective illustrations.

As a final point, the biggest of all, the Internet, is becoming more and more important, and while there is useful information, extracting that from the terabytes being added daily is an enormous task. The techniques of data mining are applicable here more than any other domain. However, to make use of it takes time, effort and, above all, people with a knowledge of the field, to differentiate the true solutions from the infeasible.

21.10 Review Questions

1. How does data mining help in clinical and medical diagnosis problems>
2. How does data mining contribute substantially to DNA analysis?
3. With a case study explain the approach of an unsupervised neural network to medical data mining techniques.
4. Explain in detail how data mining is used for fever diagnosis and its application in science.
5. Define data management context as applicable to mining.
6. Explain with example, how data mining is used in scientific applications.
7. What are the traditional uses of data mining?
8. Give the examples of scientific data mining.
9. Explain how data mining is used in scientific environment with an example.
10. Explain the AdaM system features with its architecture and its components.
11. Write a short note on AdaM plan builder client.

22

Text and Web Mining

Objectives:

- Text and Web mining are the technologies analyzing less structured data having the goal to extract knowledge from text databases and from Web log files.
- The data warehouse provides the ideal structure for data mining and knowledge discovery, so the World Wide Web, with its lack of structure, provides the greatest technical challenge for those who would use AI and statistical methods to glean knowledge from data.
- To discover hidden patterns and relationship within the Web data, Web-mining task performs discovery of association rules, discovery of sequential patterns, and discovery of classification rules and data clusters.
- The two domains that pertain to Web mining are Web contents mining and Web usage mining.
- Web mining helps organization determine the lifetime value of customers and cross marketing strategies across product and identify population of potential customers for electronic commerce.
- The major strengths of the WeblogMiner are its scalability, interactivity, and the variety and flexibility of the analyses possible to perform.
- Web servers register a (Web) log entry for every single access they get in which they save the URL requested, the IP address from which the request originated, and a timestamp. To reduce the size of the log files to analyze Weblog analysis tools make assumptions in order to filter out some data like failed requests (i.e., errors) or page graphic requests, or to round off the log entries by combining similar requests.
- In multidimensional data cube, to reduce the data cube construction time, a bridge is built between the Web server and a data cube that would propagate access information directly to the aggregated cells of the data cube.

S. Sumathi and S.N. Sivanandam: *Text and Web Mining*, Studies in Computational Intelligence (SCI) **29**, 545–589 (2006)
www.springerlink.com
© Springer-Verlag Berlin Heidelberg 2006

- Web Usage Mining is the "automatic discovery of user access patterns form Web servers". Examples of tools include Web SIFT, Web Utilization Miner, and Easy Miner.
- In biomedical literature, the first one is an information retrieval task, which retrieves useful MEDLINE documents for biological research with high precision, and the second is a text (data) mining task, which applies association rule mining and term extraction for mining MEDLINE.
- Science & Technology (S&T) text mining is the application of text mining to highly detailed technical material. There are over seventy text mining tools available on the Internet.

Abstract. An Overview on Web mining, WebLogMiner, and Web Usage Mining on Proxy Servers and Text Data Mining in Biomedical Literature Case Studies are described in this section.

The World Wide Web serves as a huge, widely distributed, global information service center for news, advertisements, consumer information, financial management, education, government, e-commerce, and many other information services. The Web also contains a rich and dynamic collection of hyperlink information and Web page access and usage information, providing rich sources for data mining.

Most previous studies of data mining have focused on structured data, such as relational, transactional, and data warehouse data. However, in reality, a substantial portion of the available information is stored in text databases (or document databases), which consist of large collections of documents from various sources, such as news articles, research papers, books, digital libraries, e-mail messages, and Web pages. Text databases are rapidly growing due to the increasing amount of information available in electronic forms, such as electronic publications, e-mail, CD-ROMs, and the World Wide Web (which can also be viewed as a huge, interconnected, dynamic text database).

Data mining and knowledge discovery has become a very useful technique in both research and everyday life. As more and more messages are published on the World Wide Web, the need for extracting important, useful information from the Web is increasing. Data mining in Web, or called Web mining becomes a very hot topic.

Data stored in most text databases are semistructured data in that they are neither completely unstructured nor completely structured. For example, a document may not only contain a few structured fields, such as title, authors, publication date, length, category, and so on, but also contain some largely unstructured text components, such as abstract and contents. There have been a great deal studies on the modeling and implementation of semistructured data in recent database research. Moreover, information retrieval techniques such as text indexing methods have been developed to handle unstructured documents.

Traditional information retrieval techniques become inadequate for the increasingly vast amounts of text data. Typically, only a small fraction of the many available documents will be relevant to a given individual or user. Without knowing what could be in the documents, it is difficult to formulate effective queries for analyzing and extracting useful information from the data. Users need tools to compare different documents, rank the importance and relevance of the documents, or find patterns and trends across multiple documents. Thus, text mining has become an increasingly popular and essential theme in data mining.

Web Usage Mining is an aspect of data mining that has received a lot of attention in recent years. An area that has received much less attention is the investigation of user behavior on proxy servers. Servers of Internet Service Providers (ISPs) log traffic from thousands of users to thousands of Web sites. No doubt that Web server administrators are interested in comparing the performance of their own site with those of competitors. Moreover, this kind of research can give a general overview of user behavior on the Internet or an overview of behavior within a specific sector.

Biological literature databases such as MEDLINE contain rich information required for biological research, currently there is great demand for extracting information from such databases. This section attempts to focus on two issues. The first one is an information retrieval task, which retrieves useful MEDLINE documents for biological research with high precision, and the second is a text (data) mining task, which applies association rule mining and term extraction for mining MEDLINE.

22.1 Data Mining and the Web

Text and Web mining are technologies analyzing less structured data having the goal to extract knowledge from text databases and from Web log files. Good examples of business-related text-mining applications in the domain of efficient handling with the textual information overload use visualization techniques, automatic building of onthologies (subject taxonomies), predicting financial trends from the text from news wires, information extraction from document databases, span filtering, document summarization and text categorization. Typical Web-mining application is in the area of customer profiling from the Web log files analyzing customer trace and historical information with the goals as increase sales of products at Web site. Internet opens wide possibilities to access the remote database. Due to considerable interest of the society to ecology and its vital activity it is reasonable to supply Internet with databases.

If the data warehouse provides the ideal structure for data mining and knowledge discovery, then the World Wide Web, with its lack of structure, provides the greatest technical challenge for those who would use AI and statistical methods to glean knowledge from data. Oren Etzioni in 1996 discussed these challenges as well as some techniques used in overcoming them in several deployed Web based systems. Etzioni in 1997 convincingly argues that the lack of structure that characterizes the Web is only apparent: Large portions of the Web are multilayered sites with data warehouse like structures (on-line catalogs of merchandise); other portion of the Web have very characteristics features (home pages); still other portions are partially labeled by HTML annotations like <title> as well as the linguistic and typographic conventions of files in natural language, postscript, latex, and all Web servers have a domain name that serves to partially limit what they might contain. These last points are most important for systems that learn about the Web by coming to understand common tags and the content associated with certain domain names. In short, Etzioni believes, as this writer certainly does that the Web is more a "gold mine" than a "quagmire."

In these systems, what Etzioni and his colleagues have deployed fall into three board classes that are enumerated as: (a) Resource discovery: locating unfamiliar documents and services on the Web – here the focus is on search: (b) Information extraction: "automatically extracting specific information from the Web resources" – here the focus is on understanding: and (c) Generalization: "uncovering general patterns at individual Web sites and across multiple sites" – here the focus is on learning. All three areas – search, understanding, and learning – are classical AI tasks. And, as for the last of these, learning, the Web-based systems discussed differ from those that merely interact with the *user* to learn his preferences and then search or act on his behalf; *these* systems all learn about the Web *itself* and do so by various methods, including interaction and experience. In that sense they are more genuinely intelligent.

22.1.1 Resource Discovery

Softbots are intelligent software agents – software robots – and one thing any robot wroth the name must be able to do in use tools. METACRAWLER (http://www.metacrawler.com) is a softbot that queries several Web search engines in parallel and produces the different syntax of different search engines and the need to query multiple search engines and prune the results. The main advantage of METACRAWLER is that it uses the multimillion document. Web indices and directories as *tools,* thus eliminating huge storage requirements and acting as essentially a gateway or interface with some intelligent processing. When run on one name, METACRAWLER gives almost number false positives, and also misses a great deal. (In contrast, METAFIND (http://www.metafind.com) – the favorite metasearch engine – and DOGpile (http://www. dogpile.com) are a good deal more comprehensive, but abound with false positives.

22.1.2 Information Extraction

NETBOT JANGO (http://www. Jango.com) will in response to a query visit manufacturer sites for product descriptions, magazine sites for product reviews and vendor sites for price quotes. "JANGO is intelligent enough to create accounts with passwords for vendors and recall them for later use and can also be instructed to make a purchase using the user's billing and shipping, information." As this is written, NETBOT, JANGO is in between versions, with number version available for testing.

22.1.3 Generalization

AHOY! (http://ahoy.cs.washington.edu: 6060) is a softbot that learns what a home page is and where to find them; it can locate home pages with both sensitivity and specificity, given at least some of: e-mail address, affiliation, country. It worked poorly on the test set – the e-address book – we used, failing not only on such typical cases as URLs beginning with ftp://, but also on such obvious cases as http://www.domain.com/-Name/when given name @ domain.com as the e-mail address, but number institutional affiliation. It worked best on home pages for those with academic affiliations. A disclaimer Etzioni in 1996 makes that: None of these systems (except AHOY! which is probably not good enough to market) is today exactly as described by Etzioni and his colleagues in the literature; they have been licensed to Internet companies and significantly changed by their proprietors.

22.2 An Overview on Web Mining

Web mining is the application of data mining technologies to huge Web data repositories. To discover hidden patterns and relationship within the Web data, Web mining task can be divided into 3 general categories that use these data mining technologies: discovery of association rules, discovery of sequential patterns, and discovery of classification rules and data clusters. Basically, there are two domains that pertain to Web mining Web contents mining and Web Usage Mining.

The World Wide Web is continuously growing and "collecting" all kinds of resources. Despite the anarchy in which it is growing, the Web is one of the biggest repositories ever built. As a confluence of data mining and world Web technologies, analyzing and exploring regularities using data mining in Web user behavior can improve systems performance and enhance the quality and delivery of the Internet information services to the end user. It can also help an organization to determine the lifetime value of customers and cross marketing strategies across product and identify population of potential customers for electronic commerce. Analysis of how users are accessing a site is critical for determining effective marketing strategies and optimizing the logical structure of the Web site. For selling advertisements on the World Wide Web, analyzing user access patterns helps in targeting ads to specific groups of users.

Characteristics of Web documents make it not easy for automated discovery of Web-based information:

- The lack of structure of information sources and unique characteristics of the client-server model in the World Wide Web, including
- Differences between the physical topology of Web repositories and user access paths;
- Difficulty in identification of unique users as well as user sessions or transactions.

However, even though the Web is dynamic and unordered, it provides many examples of semistructures as linguistic convention, semistructured documents (e.g., catalogs), Web directories, HTML tags (e.g., <title>) etc.

22.2.1 Taxonomy of Web Mining

Web Content Mining

Web content mining is the process of *extracting knowledge* and *information discovery* from sources across the World Wide Web. Web content mining uses two major approaches: agent-based and database. The goal is to provide structural information, categorize, filter, and interpret documents.

Agent-based Approach

Web agents are intelligent tools for information retrieval and extending data mining techniques to provide a higher level for organization for semistructured data available on the Web.

Intelligent search agents search for relevant information using domain characteristics and user profiles to organize and interpret the discovered information.

Information Filtering/Categorization

A number of Web agents (HyPursuit and Bookmark Organizer) use various information retrieval techniques and characteristics of open hypertext Web documents to automatically retrieve, filter, and categorize them.

Personalized Web agents

Web agents in this category learn user preferences and discover Web information sources based on these preferences, and those of other individuals with similar interests (using collaborative filtering).

22.2.2 Database Approach

Multilevel Databases

At higher level(s) metadata or generalizations are extracted from lower levels, which contain semistructured information stored in various Web repositories such as hypertext documents, and organized in structured collections, i.e., relational or object-oriented databases.

Web Query Systems

Many Web-based query systems and languages utilize standard query language such as SQL, structural information about Web documents, and even natural language processing for the queries that are used in World Wide Web searches.

Characteristics of Web Query Systems

Web Query Systems	Characteristics
W3QL	Combines structure queries, based on the organization of hypertext documents, and content queries, based on information retrieval techniques.
WebLog	Logic-based query language for restructuring extracts information from Web information sources.
Lorel and UnQL	Query heterogeneous and semistructured information on the Web using a labeled graph data model.
TSIMMIS	Extracts data from heterogeneous and semistructured information sources and correlates them to generate an integrated database representation of the extracted information.

Web Usage Mining

Web Usage mining, also known as Web log mining, is the process of *discovering interesting patterns* in Web access logs. Other sources of user information include *referrer logs,* which contain information about the referring pages for each page reference and user registration.

More sophisticated systems and techniques for discovery and analysis of patterns are now emerging. These tools can be placed into two main categories, as discussed below.

Pattern Discovery Tools

The emerging tools for user pattern discovery use sophisticated techniques from AI, data mining, psychology, and information theory, to mine to knowledge from collected data. For example, WEBMINER (Mobasher et al., 2000; Cooley et al., 1999) automatically discovers association rules and sequential patterns from server access logs.

Pattern Analysis Tools

Once access patterns have been discovered, analysts need the appropriate tools and techniques to understand, visualize, and interpret these patterns, such as Web Viz system (Pitkow and Bharat, 1994). Some tools have proposed using OLAP techniques such as data cubes for the purpose of simplifying the analysis of usage statistics from server access mechanism for querying the discovered knowledge.

22.2.3 Web Mining Tasks

- Data preprocessing for Web mining
- Mining patterns and discovering knowledge using association rule mining, time sequential mining, and classification and clustering mining
- Analysis of mined knowledge

Web Usage Mining Architecture

The architecture divides the Web usage mining process into two main parts. The first part includes the domain dependent of transforming the Web data into suitable transaction form. This includes preprocessing, transaction identification, and data integration components. The second part includes the largely domain independent application of generic data mining and pattern matching techniques as part of the system's data mining engine.

Data Pre-processing

Web usage data collected in various logs is at a very fine granularity. There is a need to group individual data collection events into groups, called *Web transactions,* before feeding it to the mining system. Clearly improved data quality can improve the quality of any analysis on it. Portions of Web usage data exist in sources as diverse as *Web server logs, referral logs, registration files,* and index server logs. Intelligent integration and correlation of information from these diverse sources can reveal usage information that may not be evident from any one of them. Hence, there is a need to preprocess the data to make it easier to mine to knowledge.

Pattern/Knowledge Mining

The key component of Web mining is the mining process itself. Web mining has adapted technique from the field of data mining, databases, and information retrieval as well as developing some techniques of its own, e.g., *path analysis.* Web usage mining studies reported to date have mined for *association rules, temporal sequences, clusters,* and *path expressions.* As Web continues to expand, there is a continual need to figure out new kinds of knowledge about user behavior that needs to be mined.

Mined Knowledge Analysis

The output of knowledge mining algorithms is often not in a form suitable for direct human consumption, and hence there is a need to develop tools that incorporate statistical methods, visualization, and human factors to help better understand the mined knowledge. Web mining, in particular, is the creation of intelligent tools that can assist in the interpretation of mined knowledge. Clearly, these tools need to have specific knowledge about the particular problem domain to do any more than filtering based on statistical attributes of the discovered rules or patterns.

22.2.4 Mining Interested Content from Web Document

The World Wide Web is a collection of electronic documents whose size is growing exponentially. This makes it difficult for users to find useful information. The Web provides an enormous resource for many informational topics but does not provide a good means to find the information relevant to specific user interests. The main problem is related to the difficulty of locating and tracking appropriate sources from the enormous number of unstructured documents.

Web Mining Environment (WME)

WME is capable of finding, extracting, and structuring information related to a particular domain from Web documents using general-purpose indices. This approach presents a tool for the automatic translation of a domain specification into a Web miner able to retrieve data in that domain on the Web. Further researches are now focusing on machine learning to improve extraction capability and automatic domain normalization.

To judge the performance, WME was compared to different search engineer such as Alta Vista, Yahoo, and MetaCrawler in calculating the recall measure and precision measure for the extraction task. As a result, WME returns a higher percentage of documents than Yahoo and MetaCrawler with well semistructured domains.

- The WME Architecture

The main phases needed by a WME to satisfy a user query are:

(i) Domain page retrieval
(ii) Retrieved domain page structuring
(iii) Information extraction from a structured page
(iv) User presentation of the results and local database population

The Web engine returns a first approximation of a set of Web pages that might contain the requested information. A WME uses these descriptions, by which Web page addresses are followed, and a set of *weighted string patterns* to sort the addresses in decreasing order with respect to a content-base interest rate. The *weighted string patterns* are defined by the WME designer that creates a set of string pattern templates and assigns to each of them a weight based on his knowledge of the domain. String patterns may also be created at search time by the WME by using user query values. WME designer also defines a threshold used by the WME to extract the most "interesting" pages from the ordered set. The second phase classifies and qualifies text segments in a page based on the structural characteristics of it. Once a page has been structurally analyzed the phase 3 extracts the requested information by first locating the text segments of interest and then extracting the text strings meeting the user query. In the last phase of entity values are presented to the user in the appropriate fields for the query-by-example interface. Moreover, they are inserted into the local database in order to provide fast retrieval to successive queries and to allow data analysis functions to derive useful information.

- Design of a WME Generator (WMEG)

The WMEG processes the Miner Attributes and instantiates the template source codes. The WME developer specifies all the parameters by using a *graphical interface.* These parameters form the *Miner Attributes,* and are the input for the *code generator.* External extraction functions, with some template files, are used as additional inputs.

Relational Fuzzy C-Maximal Density Estimator (RFC-MDE)

RFC-MDE is an approach for automatic discovery of user session profiles in Web log data. It is an extension of Maximal Density Estimator (MDE), which yields a robust estimate of the parameters without assuming that the contamination rate is known. A "user session" is defined as a temporally compact sequence of Web accesses by a user. A new distance measure is defined between two Web sessions that captures the organization of a Web site. The goal of RFC-MDE is to categorize these sessions.

Most data mining applications involve data that is corrupted with noise. Existing robust estimators such as, MDE, suffer from their strong dependence of a known or an assumed amount of noise, or equivalently an estimated scale value. RFC-MDE can perform fuzzy clustering to the sessions with an unknown amount of noise. The RFC-MDE algorithm was successfully used to cluster the sessions extracted from real server access logs into typical user session profiles, and even to identify the noisy sessions and profiles. As a by-product of the clustering processes, associations between different URL addresses on a given site can easily be inferred from the resulting robust profiles.

22.2.5 Mining Pattern from Web Transactions/Logs

Using Web log files, studies have been conduced on analyzing system performance, improving system design, understanding the nature of Web traffic, and understanding user reaction and motivation.

Discovery and analysis of various data relationships is essential in fully utilizing the valuable data gathered in daily transactions. A comprehensive analysis tool must be able to automatically discover such data relationships, including the correlation among Web pages, sequential patterns over time intervals, and classification of users according to their according to their access patterns. This technique must be able to discover relationships in very high traffic servers with very large access logs.

Open Market (e.g., Open Market Inc. 1996), Web Trend (e.g., Software Inc., 1995), and NetGenesis (net. Genesis, net. Analysis desktop., 1996) are Web analysis tools to determine the number of accesses to the server and the individual files within the organization's Web space, the times or time intervals of visits, and domain names and the URLs of users of the Web server. However, these tools are designed to deal with low-to-moderate traffic servers, and they usually provide little or number analysis of data relationships among the accessed files and directories within the Web space.

WEBMINER

WEBMINER is a framework for Web transaction mining to discover association rules and sequential patterns from data collected in World Wide Web transactions. This framework includes formal data transaction models for various Web mining tasks. It can be used in very high-traffic servers with very large access logs. In order to provide a greater degree of user control, a query language has been built on top of the basic engine. The user can specify the type of pattern to look for, and only those are retrieved. The system tends to be extended to incorporate components for clustering and the discovery of classification rules.

- Structure of access log data

According to the Common Log Format specified as part of the HTTP protocol, a log entry contains the client IP address, user id, access time, request method, and the URL of the page accessed, the protocol used for data transmission, an error code, and the number of bytes transmitted. A sample entry from a Web server access log is listed as following:

Moose.cs.umn.edu mobasher – [09/Aug/1996:09:55:50 – 05001] "GET/ ~suharyon/lisa.html HTTP/1.0" 200 654

There are a variety of files accessed as a result of a request by a client to view a particular Web page. Typical examples include:

<filename>.html: the HTML file for the requested page;
<filename>.gif.or <filename>.jpg: image files.
<filename>.map?<x,y>: file mapped to coordinates x and y of an image map file;
<program>.cgi?<arguments>: a server-side executable file.

- Data Cleansing

The primary objective of Web mining is to discover interesting patterns in accesses to various Web pages within the Web space associated with a particular server. According to the task, all irrelevant and redundant log entries such as, image file without a hyper link, image map files, and other multimedia support files can be removed. Data cleansing is performed by checking the suffix of the URL name. All the URL entries with filename suffixes such as, gif, jpeg, JPEG, jpg, JPG, and map are removed from the log. Then the data is formatted appropriately according to the application.

- Data and Transaction Model for Association Rules

Association rule mining is to discover all associations and correlation among data items where the presence of one set of items in a transaction implies (with a certain degree of confidence) the presence of other items. In the context of Web mining, this problem amounts to discovering the correlation among accesses to various files available on the server by a given client. In WEBMINER, each transaction is based on the set of all log entries belonging to the same client (IP address and user id), within a given maximum time gap (provided by the user).

- Data and Transaction Model for Sequential Patterns

Sequential mining is to find intertransaction patterns such that the presence of a set of items is followed by another item in the time-stamp ordered transaction set. Using the temporal characteristics of the data, another important kind of data dependency that can be discovered are similar time sequences. For example, we may be interested in finding common characteristics of all clients that visited a particular file within the time period $[t_1, t_2]$; we may be interested in a time interval (within a day, or within a week, etc.) in which a particular file is most accessed. In WEBMINER, each transaction is defined as a set of all the URL names and their access times for the same client within a user-specified maximum time gap.

- Clustering and Classification

Classification allows one to develop a profile for items belonging to a particular group according to their common attributes. This profile can then be used to classify new data items that are added to database.

In Web mining, classification techniques allow one to develop a profile for clients who access particular server files based on demographic information available on those clients. Much of this information on clients can be obtained by analyzing client requests and the information transmitted by the client browser, including the URL. The relationships can be discovered such as: clients who often access URL_1 tend to be from educational institutions; or clients who placed an on-line order in URL_2, tend to have previously visited the site for Company X, etc.

After obtaining profile information on clients and discovering classifications of data items (server files), it is obviously easy to cluster clients or data items that have similar characteristic together. It can facilitate the development and execution of future marketing strategies.

In a clustering and classification task, the definition of transaction depends on the process. If the process is defined to be clustering based on individual URL names, and *clustering transaction* can be defined to be the same as that of association rules with infinity as a max time gap (in the case of sequential patterns, max. time gap corresponds to the definition of window size).

In classification task, we add additional attributes provided by user to the attributes defined from the clustering task. These additional attributes come from user registration, on-line survey forms, and techniques such as "anonymous ticketing."

22.2.6 Web Access Pattern Tree (WAP tree)

WAP tree is developed for efficient mining of access patterns from pieces of logs. A Web access pattern is a sequential pattern in a large set of pieces of Web logs, which is pursued frequently by users. The novel data structure, WAP trees, stores highly compressed, critical information for access pattern mining and facilitates the development of novel algorithms for mining access patterns in large set of log pieces.

Comparing the scalabilities of WAP mine and GSP, the algorithm proposed in R. Srikant, R. Agrawal, ACM SIGMOD Canada, 1996, with threshold as well as the number of access sequence in the database, the result shows that WAP mine outperforms GSP in quite significant margin, and WAP mine has better scalability than GSP. The success of WAP tree and WAP mine can be credited to the compact structure of WAP tree and the novel *conditional search* strategies.

- Construction of WAP tree

The WAP tree is an effective data structure. It registers all count information for pattern mining, and frees the mining, and frees the mining process from counting candidates by pattern matching. The conditional search strategies narrow the search space efficiently and make best use of WAP-tree structure. It avoids the overwhelming problems of generating explosive candidates in *Apriori*-like algorithms.

It is easy to show that a WAP tree can be partitioned and structured in the form similar to B+ tree, and can be implemented evening pure SQL. Therefore, WAP tree as well as mining use WAP tree is highly scalable.

- WAP mine

Access patterns can be mined sequential pattern mining techniques. Almost all previously proposed methods for sequential pattern mining are based on a sequential pattern version of *Apriori heuristic.* The essential structure of the WAP min algorithm is as follows:

- Mining WAP from WAP tree

The WAP tree structure constructed by Algorithm 1 provides some interesting properties that facilitate mining Web access patterns. WAP tree uses *conditional search,* instead of searching all Web access patterns with same suffix, to mine all Web access patterns. As the suffix becomes longer, the remaining search space becomes smaller potentially.

WebLogMiner

WebLogMiner (Zaiane et al., 1998) is a knowledge discovery tool, which benefits from OLAP and data mining techniques, and multidimensional data cube, to interactively extract implicit knowledge from very large Web log files. Concrete examples using these techniques were given for time-series pattern analysis. The major strengths of this design are its scalability, interactivity, and the variety and flexibility of the analyses possible to perform.

WebLogMiner is not limited by the huge sizes of the log files continuously expanding, the type of data collected in the Web logs, and the techniques used to analyze this data. Despite these strengths, the discovery potential of such a design is still limited due to the current impoverished Web log files. The implementation of a WebLogMiner prototype is still processing. The plan to integrate the technology developed for data mining system DBMiner in WebLogMiner and use DBMiner's sparse data cube capabilities, and develop additional modules for time-series pattern analysis is continuing.

22.3 Text Mining

22.3.1 Definition

Text mining should not be confused with the better known Internet search engine tools or database management capabilities. Analogous to data mining, which extracts useful information from any type of data with large quantities, text mining is a procedure applied to large volumes of free unstructured text. After a traditional search for documents is completed, such as in format of full text, abstracts, or indexed terms, text mining explores the complex relationship among documents.

Science & Technology (S&T) text mining is the application of text mining to highly detailed technical material (Kostoff). There are three major components of S&T text mining.

(1) Information Retrieval, the foundational step of text mining. It is the extraction of relevant records from the source technical literatures or text databases for further processing.
(2) Information Processing, the extraction of patterns from the retrieved data obtained in the previous step. According to Kostoff, it has three components: bibliometrics, computational linguistics, and clustering techniques. This step typically provides ordering, classification, and quantification to the formerly unstructured material.
(3) Information Integration. It is the combination of the information processing computer output with the human cognitive processes.

22.3.2 S&T Text Mining Applications

There are several existing and potential text mining applications.

(1) Retrieving Documents

Text mining can be used to improve the comprehensiveness and relevance of information retrieved from databases. Related research focuses on different techniques for their comprehensiveness and S/N of records retrieved. Most high-quality methods use some type of iterative method with relevance feedback to modify the initial test query for increased comprehensiveness and precision of the records retrieved.

(2) Identify Infrastructure

Text mining can be used to identify the elements of the infrastructure of a technical discipline. These infrastructure elements are the authors, journals, organizations, and other group or facilities that contribute to the advancement and maintenance of the discipline. Additionally, text mining can provide their specific relationships to the total technical discipline or to subdiscipline areas.

(3) Identify Technical Themes/Relationships

Text mining can be used to identify technical themes, their inter-relationships, their relationships with the infrastructure and technical taxonomies through computational linguistics. By categorizing phrases and counting frequencies, S&T text mining can estimate adequacies and deficiencies of S&T in subtechnology areas.

(4) Discovery from Literature

There are different kinds of literature-based discovery: examining relationship between liked, overlapping literatures, and discovering relationships or promising opportunities that would not be found when read separately. Successful performance of literature discoveries can lead to identification of promising new technology opportunities and research directions, such as extrapolation of ideas from on discipline to a disparately related discipline.

(5) Technology Forecasting

In the process of retrieving and relating useful text data, text mining can also provide the time series for trend extrapolation. As an extension of the process, text mining can be used to identify state-of-the-art research & development (R&D) emphases and portend future development.

22.3.3 Text Mining Tools

There are over seventy text mining tools available on the Internet. A thorough documentation written by van Gemert lists 71 text mining tools with brief description of each of the company and its product. One of them is VantagePoint, which was developed by Search Technology, Inc., with the technology assistance of Dr. Alan Porter, the director of Technology Policy and Assessment Center (TPAC) at Georgia Tech.

Innovation Indicators

As mentioned before, bibliometrics is one of the text mining techniques to capture desired information. It uses counts of publications, patents, or citations to measure and interpret the advance of technologies. These counts can then be reasoned as innovation indicators of a certain technology.

Innovation indicators collect information on technology life cycle status, innovation contextual influence, and product market potential concepts. Examples of research and development (R&D) innovation indicators can be obtained by the following bibliometrics measures (TPAC).

(1) Fundamental Research—number of items in databases such as Science Citation Index
(2) Applied Research—number of items in databases such as Engineering Index
(3) Development—number of items in databases such as U.S. Patents
(4) Application—number of items in databases such as Newspaper Abstracts Daily
(5) Societal Impacts—issues raised in the Business and Popular Press abstracts

These innovation indicators can be defined along with the identification of technical themes, where co-occurrence of clustering analysis is required. Therefore, combined with text mining technique, innovation indicators can be generated as aids to show the maturity level of technology, similar in concept to Technology Readiness Levels (TRLs).

Innovation Flow Mapping

Innovation flow mapping is a technique to model the influences on or drivers of technology development in a graphical manner. It can be used as a brainstorming tool in the early stage of planning or examining the prospects for a technology and whether the institutions and the organizational capability exist to complete the development. Similar to an *Interrelationship Diagraph*, an innovation flow map consists of blocks of identified technologies/sources and cause/influence relationship arrows in between blocks. It can depict the location of research domain relative to each other and institutional interest and overlaps.

Competitive Technological Intelligence (CTI)

Competitive technological intelligence (CTI) is another use of text mining techniques. It grew aggressively in the 1990s when companies, universities, and government agencies were in need of knowing what capabilities others have of developing a particular technology. An essential part of CTI comes from topical searches in R&D databases of the text mining process.

CTI analysis can be done between market competitors or countries. An example by Porter and Detampel (1995) is a comparison of the patenting frequency in several areas related to multichip modules for the top two patenters, IBM and AT&T.

22.3.4 Text Data Mining

Data "mining" is not a very good metaphor for what people in the field actually do. Mining implies extracting precious nuggets of ore from otherwise worthless rock. If data mining really followed this metaphor, it would mean that people were discovering new facts within their databases/data warehouses. In practice, data mining applications tend to be (semi-)automated discovery of trends and patterns across very large datasets, usually for the purposes of decision making. But in the case of text, we can take the mining-for-nuggets metaphor seriously.

Text Data Mining vs. Information Retrieval

The goal of IR is to help users find documents that satisfy their information needs. Like looking for needles in a needle stack – the problem is not so much that the desired information is not known, but rather that it co-exists with many other valid pieces of information.

The fact that an IR system can return useful documents does not imply that a new discovery has been made: the information had to have already been known by the author of the text, otherwise she could not have written it down. Text data mining means the extraction of new, never-before encountered information.

How can text data mining help IR?

1. Text clustering to provide thematic overviews of text collections.
2. Automatically generating term associations to aid in query expansion
3. Using co-citation analysis to find general topics within a collection or identify central Web pages.

A second goal of data mining on the Web is to analyze the transactions run in a Web-based system, be it to optimize the system or to find information about the clients in a system.

TDM and Computational Linguistics

Empirical computational linguistics computes statistics over large text collections in order to discover useful patterns. These patterns are used to inform algorithms for natural language processing, such as

a) part-of-speech tagging
b) word sense disambiguation (e.g., prices, prescription, and patent are likely to co-occur with the medical sense of drug; while abuse, paraphernalia, and illicit are likely to occur with the illegal drug sense of the word – Church & Lieberman)
c) bilingual dictionary creation (Armstrong, 1994). See example at the end.

A classification of data mining and text data mining applications

	Finding patterns	**Finding nuggets**	**Finding nuggets**
		Novel	**Non-novel**
Nontextual data	Standard data mining	-	Database queries
Textual data	Computational linguistics	Real text data mining	Information retrieval

Real TDM

1. One body of work uses text **category labels** (associated with Reuters newswire) to find "unexpected patterns" among text articles (Feldman & Dagan, 1995). The main approach is to compare distributions of category assignments within subsets of the document collection. For instance, distributions of commodities in country C1 are compared against those of country C2 to see if interesting or unexpected trends can be found.
2. **On-line new event detection.**
The input is a stream of stories in chronological order, and whose output is a yes/no decision for each story, made at the time the story arrives, indicating whether the story is the first reference to a newly occurring event. In other words, the system must first detect the first instance of what will become a series of reports on some important topic.
3. **Using text to form hypotheses about disease.**
Experts can only read a small subset of what is published in their fields and are often unaware about developments in related fields. It should be possible to find useful linkages between information in related literatures, if the authors of those literatures rarely refer to (are unaware of) each other's work (Swanson, 1987).

For example, when investigating causes of migraine headaches, he extracted various pieces of evidence from titles of articles in the biomedical literature, e.g.

- stress is associated with migraines
- stress can lead to loss of magnesium
- calcium channel blockers prevent some migraines
- magnesium is a natural calcium channel blocker
- spreading cortical depression (SCD) is implicated in some migraines
- high levels of magnesium inhibit SCD
- migraine patients have high platelet aggregability
- magnesium can suppress platelet aggregability

These clues suggest that magnesium deficiency may play a role in some kinds of migraine headache – a hypothesis that did not exist at the time Swanson found these links, but was subsequently supported by medical tests.

This approach has only partly been automated. There is of course a potential for combinatorial explosion of potentially valid links. Sophisticated algorithms are needed for helping with the pruning process, since a good pruning algorithm will want to take into account various kinds of semantic constraints. This is an area of investigation.

4. Using text to uncover social impact.
   Narin et al. 1977 found that the technology industry relies more heavily than ever on government-sponsored research results. They explored relationships among patent text (science references given on the first page) and the original published research literature (looked in the acknowledgements to see who funded the research).
5. The LINDI Project.
   This investigated how researchers can use large text collections in the discovery of new important information.

An important problem in molecular biology is automating the **discovery of the function of newly sequenced genes.**

Human genome researchers perform experiments in which they analyze the co-expression of tens of thousands of novel and known genes simultaneously.

A gene a co-expresses with gene b when both are found to be activated in the same cells at the same time with much more likelihood than chance (*mutual information?*). Given this huge collection of genetic information, the goal is to determine which of the novel genes are medically interesting, meaning that they are co-expressed with already understood genes that are known to be involved in disease.

22.4 Discovering Web Access Patterns and Trends by Applying Olap and Data Mining Technology on Web Logs – Case Study

As a confluence of data mining and WWW technologies, it is now possible to perform data on Web log records collected from the Internet Web page access history. The behavior of the Web page readers is imprinted in the Web server log files. Analyzing and exploring regularities in this behavior can improve system performance, enhance the quality and delivery of Internet information services to the end user, and identify population of potential customers for electronic commerce. Thus, by observing people using collections of data, data mining can bring considerable contribution to digital library designers. We have been developing the knowledge discovery tool, WebLogMiner, for mining Web server log files. This section presents the design of the WebLogMiner, reports the current progress, and outlines the future work in this direction. This case study is taken from Virtual-U Research Laboratory and Intelligent Database Systems Research Laboratory, Simon Fraser University, Canada.

Web servers register a (Web) log entry for every single access they get in which they save the URL requested, the IP address from which the request originated, and a timestamp. With the rapid progress of World Wide Web (WWW) technology, and the ever-growing popularity of the WWW, a huge number of Web access log records are being collected. Popular Web sites can see their Web log growing by hundreds of megabytes every day.

Condensing these colossal files of raw Web log data in order to retrieve significant and useful information is a nontrivial task. It is not easy to perform systematic analysis on such a huge amount of data and therefore, most institutions have not been able to make effective use of Web access history for server performance enhancement, system design improvement.

Using Web log files, studies have been conducted on analyzing system performance, improving system design, understanding the nature of Web traffic, and understanding user reaction and motivation. One innovative study has proposed adaptive sites: Web sites that improve themselves by learning from user access patterns. While it is encouraging and exciting to see the various potential applications of Web log file analysis, it is important to know that the success of such applications depends on what and how much valid and reliable knowledge one can discover from the large raw log data. Currently, there are more than 30 commercially available applications for Web log analysis and many more are available free on the Internet (The University of Illinois maintains a list of Web access analyzers on a Hyper News page accessible at http://union, ncsa.uiuc.edu/HyperNews/get/www/loganalyzers, html).

The most frequent reports predefined by Web log analysis tools are: a summary report of hits and bytes transferred, a list of top requested URLs, a list of top referrers, a list of the most common browsers used, hits per hour/day/week/month reports, hits per domain report, an error report, a directory tree report, etc. Despite the fact that some of the reports can be customized with some of these tools, the majority of the currently available Web log analysis tools have rigid predefined reports. Most if not all of these Web log analysis tools have limitations with regard to the size of the Web log files, whether it is physical in size or practical in time because of the low speed of the analysis. To reduce the size of the log files to analyze Web log analysis tools make assumptions in order to filter out some data like failed requests (i.e., errors) or page graphic requests, or to round off the log entries by combining similar requests. Different assumptions are made for each of the Web log analysis tools resulting in the prospect of different statistics with the same log file. It has been reported that the analysis of the same Web log with different Web log analysis tools ended up with different statistic results.

Overall, the current Web log analysis tools are still limited in their performance, the comprehensiveness and depth of their analyses, and the validity and reliability of their results. The recent progress and development of data mining and data warehousing has made available powerful data mining and data warehousing systems. Many successful data mining systems can handle very large data files like the Web log files. However, we have not seen a systematic study and development of data warehousing and mining systems for mining knowledge from Web access log records. Recent research and development of data mining technology have promoted some studies on efficient data mining for user access patterns in distributed systems, referred to as mining path traversal patterns. Understanding user access patterns in a Web site using these mining techniques not only helps improve Web system design, but also leads to wise marketing decisions (e.g., putting advertisements in proper places, classifying users, etc.). However, mining path traversal patterns is still in its infancy.

In this section we propose to use data mining and data warehousing techniques to analyze Web log records. Based on our experience on the development of relational database and data warehouse-based data mining system, DBMiner, by the Intelligent Database Systems Research Laboratory at Simon Fraser University, and on the development of a Web based collaborative teaching and learning environment, Virtual-U, by the Virtual-U Research Laboratory at Simon Fraser University, we jointly study the challenging issues on data mining in Web log databases, and propose the WebLogMiner system, which performs data mining on Web log records collected from Web page access history.

22.4.1 Design of a Web Log Miner

The most commonly used method to evaluate access to Web resources or user interest in resources is by counting page accesses or "hits." As we will see, this is not sufficient and often not correct. Web server log files of current common Web servers contain insufficient data upon which thorough analysis can be performed. However, they contain useful data from which a well-designed data mining system can discover beneficial information.

Web server log files customarily contain: the domain name (or IP address) of the request; the user name of the user who generated the request (if applicable); the date and time of the request; the method of the request (GET or POST); the name of the file requested; the result of the request (success, failure, error, etc.); the size of the data sent back; the URL of the referring page; and the identification of the client agent.

A log entry is automatically added each time a request for a resource reaches the Web server. While this may reflect the actual use of the resources on a site, it does not record reader behaviors like frequent backtracking or frequent reloading of the same resource when the resource is cached by the browser or a proxy. A cache would store resources and hand them to a client requesting them without leaving a trace in the log files. Frequent backtracking and reload may suggest a deficient design of the site navigation, which can be very informative for a site designer; however, this cannot be measured solely from the server logs. Many have suggested other means of data gathering like client-site log files collected by the browser, or a Java Applet. While these techniques solve problems created by page backtracking and proxy caching, they necessitate the user's collaboration, which is not always available. Until the Web server log files are enriched with more collected data, our data mining process is solely based on the information currently gathered by the Web servers.

Researchers working on Web log analysis discredit the use of Web access counts as indicators of user interest or measure of the interestingness of a Web page. Access counts, when considered alone, can be misleading metrics. For example, if one must go through a sequence of documents to reach a desired document, all documents leading to the final one get their counters incremented even if the user is not interested in them at all. The access counters alone do not account for the user's ability to access the information and the appropriateness of the information to the user. Nonetheless, when access counts are used in conjunction with other metrics, they can help infer interesting findings.

Despite the impoverished state of the server logs, much useful information can be discovered when using data mining techniques. The date and time collected for each successive request can give interesting clues regarding the user interest by evaluating the time spent by users on each resource, and can allow time sequence analysis using different time values: minutes, hours, days, months, years, etc. The domain name collected can allow practical classification of the resources based on countries or type of domain (commercial, education, government, etc.). The sequence of requests can help predict next caching those resources, or by allowing clustering of resources in a site based on user motivation. Notwithstanding, the server logs cannot be used as recorded by the Web server and need to be filtered before data mining can be applied.

In the WebLogMiner project, the data collected in the Web logs goes through four stages. In the first stage, the data is filtered to remove irrelevant information and a relational database is created containing the meaningful remaining data. This database facilitates information extraction and data summarization based on individual attributes like user, resource, user's locality, day, etc. In the second stage, a data cube is constructed using the available dimensions. On-line analytical processing (OLAP) is used in the third stage to drill-down, roll-up, slice, and dice in the Web log data cube. Finally, in the fourth stage, data mining techniques are put to use with the data cube to predict, classify, and discover interesting correlations.

22.4.2 Database Construction from server log Files

Data cleaning and data transformation the data-filtering step is a typical step adopted by many Web log analysis tools. While typically Web log analysis tools may filter out requests for page graphics (as well as sound and video) in order to concentrate on data pertaining to actual page hits, we tend to keep these entries because we believe they can give us interesting clues regarding Web site structure, traffic performance, as well as user motivation. Moreover, one user action can generate multiple server requests, and some of them are requests for page media. Some of these requests are important to deduce the intended action of the user. Another typical cleaning process consists of elimination log entries generated by Web agents like Web spiders, indexes, link checkers, or other intelligent agents that prefetch pages for caching purposes. We chose not to screen out these requests generated by the Web agents. It is often interesting and useful to analyze Web agents' behavior on a site and compare the traffic generated by these automated agents with the rest of the traffic. The data filtering we adopted mainly transforms the data into a more meaningful representation. We tend to consider most of the data are relevant and eliminate a minimal amount of data.

There are two types of data cleaning and data transformation, one that does not necessitate knowledge about the resources at the site and one that does. Cleaning the date and time field of the log entry, for instance, does not need any knowledge about the site itself. The date and time field is simply restructured in a set of fields to specify the day, month, year, hour, minute, and second. Filtering out server requests that failed or transforming server error cods is also generic. Transforming IP address to domain names is independent from the site content as well. However, associating a server request or a set of server requests to an intended action or event clearly necessitates knowledge about the site structure. Moreover, different dynamically generated Web pages can be the result of a single script, thus, an identical server request. Knowledge about the parameters provided to the script to generate the dynamic page, or knowledge about the necessary sequence in the request history before a request for a script, can be essential in disambiguating a server request and associating it to an event.

Metadata provided by the site designers is required for the knowledge-based data cleaning and transformation. The metadata consists of a mapping table between a server requests (URL) with parameters, if available, or a sequence of requests (URLs) and an event with a representative URL. The transformation process replaces the request sequence by the representative URL and adds the event tag to the log entry.

After the cleaning and transformation of the Web log entries, the Web log is loaded into a relational database and new implicit data, like the time spent by event, is calculated. The time spent by event (or page) is approximated from the difference between the time the page for the current event is requested and the time the next page is requested with an upper-bound threshold for the case when the user does not come back to the same server. This notion of time spent is an approximation of the actual perusal duration since it intrinsically includes the time for network transfer, navigation inside the page, etc. It may seem a biased metric but can be very useful comparing pages with the same design.

22.4.3 Multidimensional Web log data cube

Construction and manipulation after the data has been cleaned and transformed, a multidimensional array structure, called a data cube, is built to aggregate the hit counts. The multidimensional data cube has numerous dimensions (i.e., generally more than 3), each dimension representing a field with all possible values described by attributes. For example, the dimension URL may have the attributes: server domain, directory, file name, and extension; or the dimension time may have the attributes: second, minute, hour, day, week, month, quarter, year. Attributes of a dimension may be related by partial order indicating a hierarchical relationship among the dimension attributes. Hierarchies are generally predefined, but in some cases partitioning the dimension in ranges automatically generates these hierarchies. For example the dimension file size can be partitioned into size ranges and later grouped into categories like tiny, small, medium, large, huge.

Examples of dimensions in the Web log data cube include the following. Notice that each dimension is defined on a concept hierarchy to facilitate generalization and specialization along the dimension.

URL of the resource, where the concept hierarchy used is defined on the server directory structure; type of resource, defined on a range hierarchy; size of the resource, defined on a range hierarchy; time at which the resource was requested, defined on a time hierarchy; time spent in page, defined on a range hierarchy of seconds; domain name from which the request originated, defined on a prebuilt domain hierarchy; agent that made the request, defined on a prebuilt hierarchy of known Web agents and browsers; user, defined on a prebuilt user hierarchy, server status, defined on an error code hierarchy.

The multidimensional structure of the data cube provides remarkable flexibility to manipulate the data and view it from different perspectives. The sum cells allow quick summarization at different levels of the concept hierarchies defined on the dimension attributes.

Building this Web log data cube allows the application of OLAP (On-Line Analytical Processing) operations, such as drill-down, roll-up, slice and dice, to view and analyze the Web log data from different angles, derive ratios, and compute measures across many dimensions.

The drill-down operation navigates from generalized data to more details, or specializes an attribute by stepping down the aggregation hierarchy. For example, presenting the number of hits grouped by day from the number of hits grouped by month is a drill-down along the hierarchy time. The roll-up is the reverse operation of the drill-down. It navigates from specific to general, or generalizes an attribute by climbing up the aggregation hierarchy.

For example, the aggregation of total requests from group-by organization by day to group-by country by day is a roll-up by summarization over the server domain hierarchy.

The slice operation defines a subcube by performing a selection on one dimension by selecting one or some values in a dimension. It is a literal cut of a slice (or slices) on the same dimension. For example, the selection domain = ".edu" on the dimension server domain, is a slice on the educational internet domain. The dice operation is a set of consecutive operations on several dimensions. It defines a subcube by performing selections on several dimensions. For example, a subcube can be derived by dicing the Web log data cube on four dimensions using the following clause, country = "Canada" and month = 11/97 and agent = "Mozila" and file type = "cgi." These OLAP operations assist in interactive and quick retrieval of 2D and 3D cross tables and chartable data from the Web log data cube, which allow quick querying and analysis of very large Web access history files.

22.4.4 Data mining on Web log data cube and Web log database

On-line analytical processing and the data cube structure offer analytical modeling capabilities, including a calculation engine for deriving various statistics, and a highly interactive and powerful data retrieval and analysis environment. It is possible to use this environment to discover implicit knowledge in the Web log database by implementing data mining techniques on the Web log data cube. The knowledge that can be discovered is represented in the form of rules, tables, charts, graphs, and other visual presentation forms for characterizing, comparing, associating, predicting, or classifying data from the Web access log.

These data mining functions are briefly explained as follows.

Data Characterization

This function characterizes data in the Web log. It consists of finding rules that summarize general characteristics of a set of user-defined data. The rules are generated from a generalized data cube produced using the Web log data cube and the OLAP operations. For example, the traffic on a Web server for a given type of media in a particular time of day can be summarized by a characteristic rule.

Class Comparison

Comparison plays the role of examining the Web log data to discover discriminant rules, which summarize the features that distinguish the data in the target class from that in the contrasting classes. For example, to compare requests from two different Web browsers (or two Web robots), a discriminant rule summarizes the features that discriminate one agent form the other, like time, file type, etc.

Association

This function mines association rules at multiple levels of abstraction. For example, one may discover the patterns that accesses different resources consistently occurring together, or accesses from a particular place occurring at regular times.

Prediction

Prediction involves predicting values or value distributions of an attribute of interest based on its relevance to other attributes. Both relevance analysis and predictive model construction need statistical analysis techniques. This helps prediction of possible values of missing data or the value distribution of certain attributes in a set of objects. For example, the access to a new resource on a given day can be predicted based on accesses to similar old resources on similar days, or the traffic for a given page can be predicted based on the distribution of traffic on other pages in the server directory.

Classification

Classification consists of building a model for each given class based upon features in the Web log data and generating classification rules from such models. The models are constructed by analyzing a training Web log data set whose class label is known. The classification rules can be used to develop a better understanding of each class in the Web log database, and perhaps restructure a Web site or customize answers to requests (i.e., quality of service) based on classes of requests.

Time-series Analysis

Time-series analysis is to analyze data collected along time sequences, periodicity, and so on. It may also involve attribute relevance analysis, model construction, classification, and prediction. Thus time-series analysis explores most of the techniques developed in the above (data mining functions) plus its own techniques for time-sequence search, similarity analysis, periodicity analysis, and so on. For example, time-series analysis of the Web log data may disclose the patterns and trends of Web page accesses in the last year and suggest the improvement of services of the Web server. Since most data mining functions other than time-series analysis share many commonalities with the work in traditional data mining systems, such as IBM Intelligent Miner (Quest), Silicon Graphics MineSet, DBMiner, this section does not discuss the application of these data mining functions to Web log mining in detail. The focus is on time-series analysis because Web log records are highly time related, and the goals of data mining with Web log records are largely aimed at mining time-related patterns.

The time-series analysis includes network traffic analysis, event sequences and user behavior pattern analysis, transition analysis, and trend analysis. With the availability of data cube technology, such analysis can be performed systematically in the sense that analysis can be performed on multiple dimensions and at multiple systematically in the sense that analysis can be performed on multiple dimensions and at multiple granularities. Moreover, there are major differences in time-series analysis of Web log mining in comparison with other traditional data mining processes.

We take trend analysis as an example to illustrate such a process. In the analysis of the trend of Web accessing in the Virtual-U environment, we would like to see how a user changes his/her Web navigation behavior and focuses his/her attention to interested topics. The analysis takes the following steps.

1. *Data/user selection.* Since a Virtual-U user accesses the Virtual-U Web pages regularly, those who access the Virtual-U Web page only occasionally for curiosity will not be included in the analysis. That is, access regularity will be taken as a standard (or threshold) to filter out scattered accesses.
2. *Cycle detection.* For Web accessing, a regular Virtual-U user usually starts a Virtual-U session, traverses a set of page with possible inputs to some pages, and then leaves the Virtual-U sessions for a while or for a long time before coming back to start another session. Thus the starting or restarting of a Virtual-U Web page, following by a sequence of other local Web page accesses forms a cycle. A data mining task needs to detect such cycles effectively for meaningful analysis. We have developed techniques on how to find such cycles and detect periodically efficiently using data cube structure and OLAP technique.
3. *Trend analysis.* With the accumulation of the discovered sequences and periods, analysis can be performed on them to discover patterns and trends with different interests. One kind of patterns, which can be easily discovered, is the repetition of similar Web page accesses. For the same set of Web pages, many users will access such pages repeatedly in a similar pattern. Number of accesses will be associated with each page node to register such access frequency.

Moreover, with data cube technology, one can roll-up the access history to get general Web page accessing statistics. Such access statistics form a group of Web access trees/graphs. These access trees demonstrate some clear trend along the time axis: taking the structural complexity of the access tree as a focus, one can easily see the trend is the gradually reduced structural complexity of the access tree, which shows more experienced users are more focused on specific topics reduced number of random searches to other scattered topics.

Notice with the data cube technology, such a trend can be demonstrated with different time granularities, such as by week, biweek, month, bimonth, quarter, etc. by simple clicking the button of time axis. However, the trend will be "discovered" by a human with the help of visualization tools instead of a fully automated process.

It is evident now that Web server should collect and enter more information in their logs. It is urgent to study specific needs in Web access pattern analysis and recommend a new structure for Web access logs. This new structure would significantly simplify the data cleaning and data transformation stage. The experience showed us that the data cleaning and data transformation step is not only crucial, but also is the most time consuming. A good data filtering process needs metadata provided by Web site designers.

Due to the important size and the ever-exploding nature of the Web log files, the construction of the multidimensional data cube necessary for the on-line analytical processing and knowledge discovery, is very demanding and time consuming. To reduce the data cube construction time, we plan to build a bridge between the Web server and a data cube that would propagate access information directly to the aggregated cells of the data cube. The incrementally updated Web log data cube would allow real time Web access analysis and data mining.

22.5 Web Usage Mining on Proxy Servers: A Case Study

Web Usage Mining is an aspect of data mining that has received a lot of attention in recent years. Commercial companies as well as academic researchers have developed an extensive array of tools that perform several data mining algorithms on log files coming from Web servers in order to identify user behavior on a particular Web site. Performing this kind of Investigation on the Web site can provide information that can be used to better accommodate the user's needs. An area that has received much less attention is the investigation of user behavior on proxy servers. Servers of Internet Service Providers (ISPs) log traffic from thousands of users to thousands of Web sites. It was a Belgian ISP that showed interest in the subject and consequently provided data from one of their proxy servers for a thesis. This section is a summary of that thesis and lays emphasis on the attained results. The ISP chose to remain anonymous because of privacy issues.

The Internet is generally said to have become available to a large public around 1994-1995. Since that time a great number of companies have thrown themselves on this new medium. In the beginning many entrepreneurs saw great new opportunities to make money by setting up an Internet company. Later on, some of the so-called *brick-and-mortar* companies began to see a need to go on-line. Some of those even changed their business so drastically that not much of the original company was left. Every large company has spent a lot of effort and money to develop a well-established Web site. The ones that have not spent enough effort may find themselves faced with strategic disadvantages in years to come.

In order to have a successful Web site (and especially a successful e-commerce site) it is crucial to know the users of that site. This need has given rise to a whole new field in research, called Web Usage Mining. It is commonly seen as a subdivision of Web Mining, which implies that data mining techniques are applied to data from the World Wide Web. When the data under consideration emerges from Web servers log files, we enter the field out Web usage mining. It is therefore the "automatic discovery of user access patterns form Web servers."

Because it is so important to know one's customers in order to better suit their needs, companies are willing to spend money on the analysis of their log files. As a consequence, apart from tools that were developed by academic researchers, there is simultaneously a significant number of commercial tools that have been developed to meet these needs. Examples of academic tools include WebSIFT (Robert Cooley University of Minnesota, 1999) and Web Utilization Miner (Myra et al., 1998). An example of a commercial Web usage mining tool is EasyMiner, developed by MINEit Software Ltd. All of these tools are designed to understand the most common log file formats so that the process requires very little preprocessing. Unfortunately, when analyzing a log file from a Web server, one can only analyze browsing behavior on a single site. To perform research on a sector or even on general browsing behavior, the log file data of a proxy server is a lot more appropriate because of the many-to-many relationship between sites and users.

22.5.1 Aspects of Web Usage Mining

As in the other form of data mining, in the Web usage mining it is equally important to pay attention to a proper data collection, a thorough preprocessing phase, and the data mining techniques themselves.

22.5.2 Data Collection

Data for Web usage mining can be collected at several levels. We may be faced with data from a single user or a multitude of them on one hand and single site or a multitude of sites. Combining both factors offers four possibilities, as illustrated in Fig. 22.1.

	1 Site	**Multiple sites**
user	Java applets or Javascripts	Modified browser
Multiple users	Server level	Proxy server level

Fig. 22.1. Segments of Web traffic

Data about behavior of a single user on a single site can be collected by means of JavaScript or Java applets. Both methods require user participation in the sense that the user has to enable their functionality. An applet has the additional problem that it may take some time to load the first time. However, it has the advantage that it can capture all clicks, including pressing the back or reload buttons. A script can be loaded faster but cannot capture all clicks.

A modified browser is situated in the second segment. It can capture the behavior of a single user over all visited Web sites. Its advantages over Java applets and JavaScript are that it is much more versatile and will allow data collection about a single user over multiple Web sites. That is why this kind of data collection is used regularly by market research groups, e.g., Nielsen//Netratings, in order to collect information on how certain user groups behave on-line.

The third way of data collection is on the Web server level. These servers explicitly log all user behaviors in a more or less standardized fashion. It generates a chronological stream of requests that come from multiple users visiting a specific site. Since Web servers keep record of these requests anyhow, this information is readily available. Sometimes an analyst will use some additional information to better identify users, such as information from *cookies* or socio-demographic information about the users that may have been collected. This kind of data collection also has a number of drawbacks. Like JavaScripts, it cannot capture page views that were generated by pressing back or reload buttons. Apart from that, it also cannot log page views generated by a cache, either a local cache on the computer of the user, or a cache from an ISP's proxy server.

The fourth level of data collection logs behavior of multiple users visiting multiple Web sites. This kind of information can be found in log files origination from proxy servers. These servers are used by ISPs to give customers access to the World Wide Web. They also function as a cache server. This means that they will keep pages that were recently requested on this server and, if the same request is made by another user shortly after that, they will send the cached page to that user, instead of requesting it once more on the Web server were that page is located.

22.5.3 Preprocessing

Preprocessing is an aspect of data mining of which the importance should not be underestimated. If this phase is not performed adequately, it is not possible for the mining algorithms to provide reliable results.

22.5.4 Data Cleaning

First of all, irrelevant data should be removed to reduce the search space and to skew the result space. Since the intention is to identify user sessions and build up out of page views, not all hits in a log file are necessary. This is true for server logs as well as for proxy logs. A log file generates a hit for every requested file. Since an HTML page may consist of several files (text, pictures, sounds, several frames) it would be useful if we could keep only a single hit for each page view. To get an idea of user behavior, it is only necessary to keep track of the files that the user specifically requested. Very often, all hits with a suffix like .jpg, .gif, .wav, etc. are removed out of the log file. Even though this will also be done in the research that is described later on, it also has a drawback. Sometimes users specifically want to see a picture on a separate page. This page view will be deleted while it should not.

22.5.5 User and Session Identification

After the log file has been cleaned, the next step is to identify users. This very often poses a serious problem. If every computer in the world has its own unique IP-address, there would not be a problem. However, most ISPs make use of *dynamic IP addresses.* This means that every time a user logs on to the Internet, he will be given different addresses. This makes it impossible to distinguish returning users. As a consequences, it is usually simply assumed that every new IP address is the same. This occurs when the agent log shows a change in browser software or operating system. Some sites try to solve the problem of user identification through the use of cookies that contain an identification number. However, users very often delete cookies or disable their use, which makes that this technique is not always reliable either. Other sites try to identify users by asking them for a login and password. It is clear, however, that not every site can do this since it very often scares users away.

Assuming that users have been identified, the next step is to identify sessions. The goal of session identification is to divide the page accesses of each user into individual sessions. A rule of thumb that is commonly used is that when there is an interval of 30 minutes between two page views, the click stream should be divided in two sessions. This rule has been applied since of timeout of 25.5 minutes was established, based on empirical data. This is why, in the further research it is carried out with a timeout of 30 minutes. After users and sessions have been identified, the file under consideration will have additional fields that mention for each line the number of users and of the session.

22.5.6 Data Mining Techniques

For the actual pattern discovery in Web usage mining, mostly the same techniques are employed as in other forms of data mining. The most common ones will be briefly described.

Log file analysis

Even though this is not a data mining technique as such, it is probably the most widely used technique to obtain structured information out of server logs.

There are a large number of tools on the market that will accept the most common log file formats as an input to answer some basic questions that every Web site administrator has. It will provide information such as: the number of hits and page views, the number of unique and returning users, the average length of a page view, an overview of the browsers and operating systems that were used, an overview of keywords that were used in search engines and that led to the Web site, etc. Despite lacking in the depth of its analysis, this type of knowledge can be potentially useful for improving the system performance, enhancing the security of the system, facilitating the site modification task, and providing support for marketing decisions.

Association Rules

In Web usage mining, association rules are used to find out which pages are frequently visited together. In the particular research carried out in this work, they will be used to find out which Web sites and which sectors are frequently visited together. An association rule is usually presented in the following syntax:

$$\text{KeynetBe} \Leftarrow \text{VtmBe} \,\&\, \text{TvlBe}\,(15 : \; 2.788\%, \, 0.27)$$

This rule means that out of the 15 instances (representing 2.788% of the database) that visited the sites of www. vtm.be and www.tv1.be together, 27% also visited www.ketnet.be. The *support* is 15, the *confidence* 27%.

Sequential Patterns

This technique has some similarities with the association rules and is used in this section in addition to those association rules. The difference is that it takes the time dimension into account. The algorithm tries to find sequences in which a certain page (or Web site) usually comes before or after another page (or Web site). In other words, it "attempts to find inter-session patterns such that the presence of a set of items is followed by another item in a time-ordered set of sessions or episodes."

Clustering

In general, clustering is a process of creating a partition so that all the members of each set of the partition are similar according to some metric. In Web usage mining, we can narrow the definition to a *technique to group users in clusters based on their common characteristics.* Clustering algorithms learn in an unsupervised way. They discover their own classes and subsets of related objects in the training set. Then it has to find descriptions that describe each of these subjects.

Classification

Contrary to clustering, classification is a supervised way of learning. The database contains one or more attributes that denote the class of a tuple and these are known as *predicted attributes*, whereas the remaining attributes are called *predicting attributes.* A combination of the predicted attributes defines a class. In the Web domain one is interested in developing a profile of users belonging to a particular class or category. For example, 45% of users who visit two or more sites of television stations in a single session are younger than 21. The algorithms that perform classification include decision tree classifiers, Bayesian classifiers, k-nearest neighbor classifiers, etc.

22.5.7 E-metrics

E-metrics is based on statistics, which is a data mining technique. Therefore, it can be considered to be a Web usage mining method like any other. Moreover, they also try to gain insight into browsing behavior of users and performance of Web sites (NetGenesis- www.netgenesis.com).

E-metrics is measured with which Web sites can be evaluated. They can be compared with regular metrics and ratios as these are used in traditional industry, such as return on investment, net profit, market share, rentability, etc. As Web sites gain a more important position in companies, there emerges a need to evaluate these Web sites – that consume more and more money – and quantify their performance. The intention is to give indications of how well the Web site performs in order to investigate to what extent these measures change over time and how well they perform compared to those of competitors.

Two kinds of e-metrics can be identified, those that can be applied to every Web site and those that were designed for a specific kind of Web site, very often e-commerce sites.

General e-metrics

In this section we briefly describe a few general e-metrics. However, both *stickiness* and *average duration* are explained in more detail because of their importance in the undertaken research.

Stickiness

This is probably one of the most widely used e-metrics. It is a composite metric that indicates the effectiveness with which the content of the page or the Web site can keep the attention of the user. In general it is assumed that sticky sites are better than sites that are less sticky. A possible formula is as follows.

$$\text{Stickiness} = \text{Frequency}^{*}\ \text{Duration}^{*}\ \text{Total site reach}$$

Where

$$\text{Frequency} = \frac{\text{Number of visits in time period T}}{\text{Number of unique users who visited in T}}$$

$$\text{Duration} = \frac{\text{Total amount of time spent viewing all pages}}{\text{Number of visits in time period T}}$$

Where

$$\text{Total site reach} = \frac{\text{Number of unique users who visited in T}}{\text{Total number of unique users}}$$

This formula can be reduced to:

$$\text{Stickiness} = \frac{\text{Total amount of time spent viewing all pages}}{\text{Total number of unique users}}$$

so that one does not need to have all the data for the complete formula to calculate stickiness. Usually stickiness is expressed in minutes per user.

Average duration

This is quite a simple e-metric with which several pages of a Web site (or complete Web sites) can be compared to each other. The metric expresses how long users view a certain page or site on average. In this work the following formula has been used:

$$\text{Average duration} = \frac{\text{Total duration of site (or page) X}}{\text{Total number of page views}}$$

It is impossible to suggest an ideal value for this metric. For entire sites, the value should usually be as high as possible. For individual pages, it depends on the nature of that page. A navigation page should have a low value, which means that users easily find their way, while content pages should have a higher value.

Specific e-metrics

Apart from a vast amount of possible general e-metrics, there are also a great number of metrics specifically for e-commerce sites. These include:

Personalization index: This expresses to what extent data that were asked from the user to fill in on a form, are used to offer a personalized service. This value should be greater than 0.75.

Acquisition cost: This divides promotion costs (in the form of banners) by the number of click-throughs, so that the marketing team can discover to what extent the marketing efforts are effective to acquire users.

Cost per conversion: This divides promotion costs by the number of sales. It is the number that marketing people use to determine the best investment of their promotional budget.

RMF analysis: This is a special analysis in which customers are evaluated on three aspects: Recency (when was the last time they purchased something?), Monetary Value (how much money has he spent on the site already?), and Frequency (how frequent does the user purchase a good on the site?). All users are then placed in this three-dimensional model to find possible user segments.

22.5.8 The Data

The data used for research purposes in this section was offered to us by a Belgian ISP that chose to remain anonymous because of privacy issues. The data used here come from a proxy server that handles request of users that have a broadband Internet connection.

The log file

To offer Internet access, this ISP makes use of six proxy servers, all located in Belgium. Every user is assigned to one of those. As long as this user does not change his installation settings, he will always log on to the same server. This means that the log history of every user can always be found on the same server. This is a clear advantage for data mining research.

Unfortunately, the ISP does not make use of fixed IP addresses. Like most other ISPs, it uses a pool of dynamic IP addresses, which makes it more complicated to identify users. Fortunately, on-the-fly IP addresses—meaning that another address may be used for each file request—are not used here, since it would make this research quite impossible.

At the end of a session, the user can explicitly release his IP address so that it can be used for another user. However, most people do not do this and turn off their computer without doing so. In this case, their address is automatically taken back after 30 minutes, which is convenient because it is also the time interval that will be used to distinguish sessions with the same IP address.

Luckily, there are no time zones in Belgium, because if this were so, it would have to be solved somehow. Also, the server in this research is not load balanced. This, too, would make matters more complicated. As mentioned before, IPs that change within sessions would make it very difficult, if not impossible, to conduct a research. All these problems could arise with other data sets. In the framework of this section, solutions for these problems have not been looked for. It may be an opportunity for future research.

Preprocessing

Before any actual data mining algorithms can be applied on the data, the data needs to be preprocessed so that it can serve as the input of several algorithms. As mentioned in the introduction, most tools that were designed for conventional Web Usage Mining on Web servers perform this preprocessing automatically. In this case, however, this part of the process will have to be executed manually. The advantage of this is that we have more control over the way it is done.

Data Cleaning

First of all, all files with an extension .jpg, .gif, .wav, etc. have been removed from the log file, an action that drastically reduced the size of the file. Secondly, all irrelevant fields were removed. Only three fields were retained: IP address, time stamp, and URL. Finally, only the DNS name in the URLs was kept, in the form of ww.websitename.com (other prefixes and suffixes are possible instead of www and com). This was done in order to facilitate the comparison of different lines in the database, and because the exact pages that were visited are irrelevant in this research. After this process of data cleaning, the original log file that comprised 600MB was reduced to 185MB. The file contains 1,714,813 hits.

User and session identification

The only data that is at disposal to identify users and sessions is the time stamp and the IP address. To start with, every IP address is seen as a different user. However, a different user can represent several sessions. Within the series of lines that are linked to a certain IP address, different sessions are identified by using a *time out* of thirty minutes. This means that if there is a time interval of more than thirty minutes between two consecutive clicks, two sessions will be identified. The problem that we face by doing this is that it is impossible to know whether it was the same person who simply did not touch his computer for half an hour, or effectively a different user that was assigned an IP address because the previous user turned off his computer. However, even if it was the same person, it is still useful to regard his click stream as two (or more) different sessions.

A problem here is that it is impossible to correctly identify the number of unique users, which is needed to calculate stickiness. However, the most important aspect of stickiness will be to compare the results of different sectors with each other. So, if the absolute values of stickiness are somewhat incorrect, that will be crucial. They should be seen relatively to one another. The file contained 7,287 different IP addresses and 10,548 sessions were identified.

22.6 Text Data Mining in Biomedical Literature by Combining with an Information Retrieval Approach – Case Study

This section attempts to focus on two issues. The first one is an information retrieval task, which retrieves useful MEDLINE documents for biological research with high precision, and the second is a text (data) mining task, which applies association rule mining and term extraction for mining MEDLINE. As a result, the obtained association rules may be used for consistency checking and error detection in annotation of MeSH terms in MEDLINE records. This case study is taken from T.N. Tran et al., Tokyo Institute of Technology, Japan.

A large part of the information required for biological research is currently recorded in free-text form such as MEDLINE abstracts. This information is important for many types of analysis, such as classification of proteins into functional groups, discovery of new functional relationship, extraction of protein interaction information, and so on. Recently there has been an increasing deal of research that attempts to extract the abundant information in biological literature databases. Since MEDLINE abstracts are recorded in free-text form, it is natural to apply natural language processing techniques into these kinds of task. To date, a great deal of research has attempted to use information extraction techniques to find gene–gene interactions, protein–protein interactions. In general, most of them are based on natural language processing techniques such as parsing, not from the viewpoint of mining the text sources to discover novel knowledge. The first goal of this research is to address this issue by applying association rules and term extraction for mining MEDLINE.

On the other hand, the task of "preprocessing," i.e., retrieving only MEDLINE documents, which contain relevant biological information prior to the information extraction or text mining from MEDLINE also, plays an important role in biological research. Current systems for accessing MEDLINE such as PubMed (http://www.ncbi.nlm.nih.gov/PubMed/) accept keyword-based queries to text sources and return documents that are hopefully relevant to the query. Since MEDLINE contains an enormous amount of papers and the current MEDLINE search engines is a keyword-based one, the number of returned documents is often large, and many of them in fact are nonrelevant. The approach to solve this is to make use of existing databases of organisms such as *S. cerevisiae* using supervised machine learning techniques.

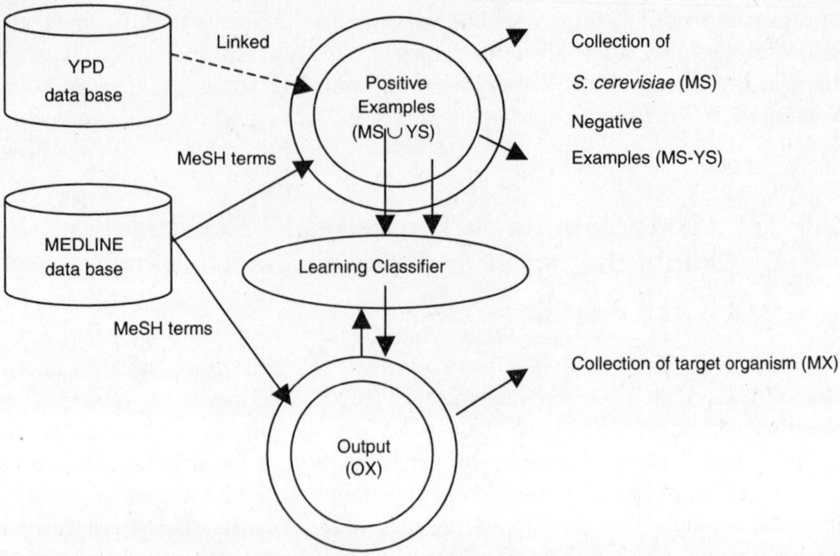

Fig. 22.2. Outline of the Information Retrieval task

22.6.1 Information Retrieval Task – Retrieve Relevant Documents by Making use of Existing Database

Figure 22.2 shows the illustration of the Information Retrieval task. In this figure, YPD database (standing for Yeast Protein Database – http://www.proteome.com/databases/index/html) is a biological database, which contains genetic functions and other characteristics of a kind of *Yeast* called *S. cerevisiae*. Given a certain organism X, the goal of this task is to retrieve its relevant documents, i.e., documents containing useful genetic information for biological research.

Let MX, MS be the sets of documents retrieved from MEDLINE by querying for the target organism X and *S. cerevisiae*, respectively (without any machine learning filtering) and YS be the set of documents found by querying for the YPD terms for *S. cerevisiae* (YS omitted in Figure 22.2 for the reason of simplification). The set of positive and negative examples then is collected as the intersection set and difference set of MS and YS, respectively. Given the training examples, OX is the output set of documents obtained by applying naïve Bayes classifier on MX.

22.6.2 Naïve Bayes Classifier

Naïve Bayes classifiers are among the most successful known algorithms for learning to classify text documents. A naïve Bayes classifier is constructed by using the training data to estimate the probability of each category given the document feature values of a new instance. The probability an instance d belongs to a class c_k is estimates by Bayes theorem as follows:

$$P(C = c_k|d) = \frac{P(d|C = c_k)P(C = c_k)}{P(d)}$$

Since $P(d/C = c_k)$ is often impractical to compute without simplifying assumptions, for the naïve Bayes classifier, it is assumed that the features $X_1, X_2, \ldots X_n$ are conditionally independent, given the category variable C. As a result:= $P(d| C = c_k) = \prod_i P(d_i |C = c_k)$

22.6.3 Experimental results of Information Retrieval task

Two kinds of yeast called *Pombe* and *Candida* have been used as the input organisms. To evaluate the validity of using the machine learning approach, the recall and precision before learning (corresponding to the collection of the input organism in Figure 22.2) and after learning (the output collection in Figure 22.2) have been calculated for both *Pombe* and *Candida*. A certain number of documents (50 in this experiment) in each of dataset are taken randomly, checked by hand whether they are relevant or not, then the number of relevant documents in each dataset is estimated. The recall and precision in each case can be calculated as follows:

$$Precision = \frac{\#documents\, match\, classifier\, and\, relevant}{\#documents\, match\, classifier}$$

$$recall = \frac{\#documents\, match\, classifier\, and\, relevant}{\#documents\, relevant}$$

22.6.4 Text Mining Task – Mining MEDLINE by Combining Term Extraction and Association Rule Mining

We have used the MEDLINE dataset collected from YPD database, which were previously described before as the input for the text-mining task. This dataset consists of 14572 abstracts in HTML form pertaining to *S. cerevisiae.* The text mining task from the collected dataset consists of two main modules: the Term Extraction module and the Association-Rule Generation module. The Term Extraction module itself includes the following stages:

- *XML translation* (the dataset was collected in HTML from): This stage translates the MEDLINE record from HTML form into an XML-like form, conducting some preprocessing dealing with punctuation.
- *Part-of-speech tagging*: Here, the rule-based Brill part-of-speech tagger was used for tagging the title and the abstract part.

- *Term Generation*: Sequences of tagged words are selected as potential term candidates on the basis of relevant morphosyntactic patterns (such as "Non Noun," "Noun Adjective Noun," "Adjective Noun," "Noun Preposition Noun," etc.). For example, "in vivo." "Saccharomyces cerevisiae" are terms extracted from this stage.
- *Stemming*: Stemming algorithm was used to find variations of the same word. Stemming transforms variations of the same word into a single one, reducing vocabulary size
- *Term Filtering*: In order to decrease the number of "bad terms," in the abstract part, only sentences containing verbs listed in the "verbs related to biological events."

After necessary terms have been generated from the Term Extraction module, the Association-Rule Generation module then applies the Apriori algorithm (Agrawal et al. 1994) using the set of generated terms to produce association rules (each line of the input file of Apriori-based program consists of every term extracted from a certain MEDLINE record in the dataset). The goal for text mining in this section is to find relations between the MeSH terms and the substances (location) in a MEDLINE record pertaining to *S. cerevisiae* and to discover the relations between other terms occurred in the title, abstract with the journal name, and so on.

22.6.5 Finding the Relations Between MeSH Terms and Substances

First, we just extracted terms in MeSH terms and Substances fields. We conducted this experiment in order to find the relations between MeSH terms and Substances. Figure 22.3 shows a list of ten rules among obtained rules. Here, those terms begin with "m_" stand for MeSH terms and those terms begin with "s_" stand for Substances. For example, the sixth rule in Figure 22.3 is translated as follows: The rule that in a MEDLINE record if Saccharomyces cerevisiae/genetics and Fungal Proteins/metabolism occur in the MeSH field then Fungal Proteins also occur in the Substances field of this MEDLINE record has a support of 13.5% (i.e., this rule is matched 1971 times in the total of 14572) and a confidence of 99.9%.

- *Freq Verb-Terms*: This variation extracts all terms and also the "frequently-occurring verbs" that occurred in the corresponding sentence (in the stemmed form).
- *None_Freq Verb-Terms*: This variation extracts all terms but does not extract the "frequently occurring verbs."

22.6.6 Finding the Relations Between Other Terms

Next, we attempt to find the relationship between other terms except those that occurred in MeSH terms and Substances fields. For example, association rules concerning the relationship between terms occurred in the title and

1: s_fungal_proteins ← m_fungal_proteins_genetics (26.3%/3829, 99.9%)
2: s_fungal_proteins ← proteins ← m_fungal-proteins_metabolism (21.1%/3070, 99.9%)
3: m_amino_acid_sequence ← m_sequence_homology_ _ amino_acid (13.2%/1922, 94.6%)
4: s_plasmids ← s_plasmids (10.5%/1689, 90.7%)
5: s_plasmids ← m_plasmids (10.6%/1540, 99.5%)
6: s_fungal_proteins ← m_saccharomyces_ cerevisiae_genetics
m_fungal_protein_metabolism (13.5%/1971, 99.9%)
7: m-saccharomyces_ceevisiae_genetics ←amino-acid-sequence
m_base_sequence (23.0%/3343, 80.6%)
8: m_amino_acid_proteins ← m_molecular_sequence_data
s_fungal_proteins (22.4%/3266, 75.8%)
9: s_fungal_proteins ← m_molecular_sequence_data
m_fungal_protein_genetics (16.6%/2417, 100.0%)
10: s_fungal_proteins ← m_molecular_sequence_data
m-fungal_proteins_metabolism (11.2%)/1632, 100.0%)

Fig. 22.3. Ten rules obtained by MeSHterms-Substances (*minimum support* = 0.1, *minimum confidence* = 0.75)

abstract part with the journal name may be interesting. We have used two variations for the reason of comparison as follows:

Figures 22.4 and 22.5 show the list of twenty rules among obtained rules by these two variations with the minimum confidence set to 0.75, respectively. For example, the 12th rule in Figure 22.5 implies that "the rule that in a MEDLINE record if iron sulfur protein occurs this MEDLINE document is published in J Biology chem. Journal has the support of 0.2% (i.e., this rule is matched 19 times in the total of 14572), and the confidence of 84.2%." It can be seen that the relation between journal name and terms extracted from the title and the abstract has been discovered from this example.

It can be seen from Figures 22.4 and 22.5 that making use of terms can produce interesting rules that cannot be obtained using only single-words. The main difference between Freq Verb-Terms (Figure 22.4) and None_Freq Verb-Terms (Figure 22.5) is that the former is used for discovering the relations between "frequently-occurring verbs" and terms, while the latter is used for discovering relations among terms.

22.7 Related Work

There has been a great deal of research aimed at extraction information from biology texts. Despite the attractiveness of automatically extraction of useful information from biomedical text sources, previous works seem to be constrained from extracting factual assertions. Fukuda et al. attempt to identify protein names from biological papers. Andrade and Valencia also concentrate

1: have ← gever_protein (0.2%/25, 80.0%)
2: requir ← gwi4 (0.3%/36, 75.0%)
3: requir ← gdc42 (0.2%/30, 76.7%)
4: requir ← gre-rrna_process (0.3%/36, 77.8%)
5: requir ← groper (0.2%/31, 80.6%)
6: requir ← granslat_activat (0.2%/28, 75.0%)
7: requir ← guclear_fusion (0.2%/29, 75.9%)
8: requir ← gntegrity (0.3%/37, 75.7%)
9: suggest ← gapip (0.2%/27, 77.8%)
10: contain ← gondon (0.3%/32, 84.4%)
11: us ← grobe (0.6%/78, 87.2%)
12: us ← geast_two-hybrid_system (0.2%/27, 85.2%)
13: us ← polymeras_chain_reaction (0.2%/29, 89.7%)
14: yeast ← grf (0.9%/108, 76.9%)
15: yeast ← gpen_read_frame (0.4%/46, 91.3%)
16: identify ← gwo-hybrid_screen (0.3%/34, 94.1%)
17: indicat ← gdrl (0.2%/26, 76.9%)
18: interact ← gwo-hybrid_system (0.6%/76, 82.9%)
19: interact ← gwo-hybrid_screen (0.3%/34, 88.2%)
20: interact ← gwo-hybrid_assay (0.2%/27, 77.8%)

Fig. 22.4. First twenty rules obtained by Freq Verb-Terms (*minimum support* = 0.002, *minimum confidence* = 0.75)

on extraction of keywords, not mining factual assertions. There have been many approaches to the extraction of factual assertion using NLP techniques such as syntactic parsing.

Sekimizu et al. attempt to generate automatic database entries containing relations extracted from MEDLINE abstracts. Their approach is to parse, determine noun phrase, spot the frequently occurring verbs, and choose the most likely subject and object from the candidate NPs in the surrounding text. Rindflesch uses a stochastic part-of-speech tagger to generate an underspecified syntactic parse and then uses semantic and pragmatic information to construct its assertions. This system can only extract mentions of well-characterized genes, drugs cell types, and not the interactions among them. Thomas et al. use an existing Information Extraction system called SRI's Highlight for gathering data on protein interactions. Their work concentrates on finding realizations directly between proteins. Blaschke et al. attempt to generate functional relationship maps from abstracts; however, it requires a predefined list of all named entities and cannot handle syntactically complex sentences.

22.7.1 Future Work: For the Information Retrieval Task

Although using an existing database of *S. cerevisiae* is able to obtain a high precision for other yeasts and organisms, the recall value is still low, especially

1: protein ← oligomer (0.2%/19, 78.9%)
2: protein ← co-immunoprecipitated (0.1%/14, 78.6%)
3: gene ← differently _expression (0.1%/13, 76.9%
4: yeast ← orf (0.9%/108, 76.9%)
5: yeast ← open_read_frame (0.4%/46, 91.3%)
6: yeast ← kb_segment (0.2%/20, 100.0%)
7: yeast ← orf (0.1%/13, 84.6%)
8: yeast ← alyeast (0.1%/18, 94.4%)
9: yeast ← other_orf (0.1%/15, 100.0%)
10: sequ ← other_orf (0.1%/15, 86.7%)
11: essenty ← yeast_cell-viabl (0.1%13, 76.9%)
12: j_biol_chem. ← iron_sulfur_protein (0.2%19, 84.2%)
13: j_boil_chem. ← mitochondry_membran_system (0.1%/16, 75.0%)
14: open_read_frame ← codon (0.3%/32, 81.2%)
15. open_read-frame ← kb_segment (0.2%/20, 85.0%)
16: evidenc ← server_line (0.1%/13, 76.9%)
17: alpha ← a_specific_gene (0.2%/20, 75.0%)
18: nucleu ← nuclear_local_sequ (0.1%/13, 76.9%)
19: er ← secl2p (0.1%/14, 78.6%)
20: assembly ← mitochondry_membran_system (0/1%/16, 100.0%)

Fig. 22.5. First twenty rules obtained by None_Freq Verb_Terms *(minimum support* = 0.001, *minimum confidence* = 0.75)

for the yeasts that are remarkably from *S. cerevisiae.* Since yeasts such as *Candida* might have many unique attributes, we may improve the recall by feeding the documents checked by hand back to the classifier and conduct the learning process again. The negative training set has still contained many positive examples; so we need to reduce the noise by making use of the learning results.

22.7.2 For the Text Mining Task

By combining term extraction and association rule mining, we are able to obtain interesting rules such as the relations among journal names and terms, terms and terms. Particularly, the relations among MeSH terms and "Substances" may be useful for error detection in annotation of MeSH terms in MEDLINE records. However, the current algorithm treats extracted terms such as "cdc37_caryogamy_defect," "cdc37_in_mitosy," "cdc37_mutat" to be mutually independent. It may be necessary to construct semiautomatically term taxonomy, for instance users are able to choose only interesting rules or terms then feedback to the system.

22.7.3 Mutual Benefits between Two Tasks

Gaining mutual benefits between two tasks is also an important issue for future work. First, by applying text mining results, it should be noted that we can decrease the number of documents being "leaked" in the Information Retrieval task. As a result, it is possible to improve the recall. Conversely, since the current text mining algorithm creates many unnecessary rules (from the viewpoint of biological research), it is also possible to apply the information retrieval task first for filtering relevant documents, then apply to the text mining task to decrease the number of unnecessary rules obtained, and to improve the quality of the text mining task.

22.8 Summary

The World Wide Web is continuously growing and "collecting" all kinds of resources, text, multimedia applications, etc. Despite the anarchy in which it is growing, the Web is one of the biggest repositories ever built. Analyzing the Web access logs of different Web sites can help understand the user behavior and the Web structure, thereby improving the design of this colossal collection of resources. Therefore, it is important to build tools to analyze the Web access patterns.

Currently available Web log analysis tools report interesting statistics, but are limited by the huge sizes of the log files continuously expanding, the type of data collected in the Web logs, and the techniques used to analyze this data. Consequently, it is imperative to design a good Web log analysis tool that would overcome the current limitations of the Web log and recommend directives for new Web log standards that would help better analyze the Web access trends and discover useful knowledge from the access records.

We have outlined the design of the system WebLogMiner, which benefits from OLAP and data mining techniques, and multidimensional data cube, to interactively extract implicit knowledge from very large Web log files. Concrete examples using these techniques were given for time-series pattern analysis. The major strengths of this design are its scalability, interactively, and the variety and flexibility of the analyses possible to perform. Despite these strengths, the discovery potential of such a design is still limited due to the current impoverished Web log files.

This section has introduced two tasks concerning information extraction from biological literature databases such as MEDLINE. The first one is an information retrieval, which attempts to retrieve useful documents for biology research with high precision, and the second one is a text mining task which attempts to apply association rule mining and term extractors for mining MEDLINE. It can be seen that making use of the obtained results is useful for consistency checking and error detection in annotation of MeSH terms in MEDLINE records. In future work, combining these two tasks together may be essential to gain mutual benefits for both two tasks. This chapter has also revealed the concepts of text data mining and Web mining.

22.9 Review Questions

1. Give an overview and the taxonomy of Web mining.
2. What are the characteristics of Web mining systems?
3. Explain in detail Web mining environment.
4. Explain relational fuzzy C- Maximal density estimator.
5. Explain Web access pattern tree and Web log miner.
6. Briefly discuss the design methodology of a Web log miner.
7. How is database constructed from server log files?
8. Write a short note on multidimensional Web log data cube.
9. With a case study explain the Web usage mining on proxy servers.
10. What are the various data mining techniques used in Web data mining?
11. Explain in detail naïve Bayes classifier and the information retrieval task.
12. Explain text mining task – Mining MEDLINE.
13. What are the rules obtained by MeSH terms – substances?
14. State the rules obtained by None-Freq Verb-Terms.
15. What are the potential text mining applications and state some of the text mining tools?
16. Compare text mining and information retrieval.
17. Write a note on computational linguistics.

23

Data Mining in Information Analysis and Delivery

Objectives:

- *Information analysis* is an umbrella term that applies to a multitude of techniques for extracting from massive quantities of information various types of important, interesting, or unexpected phenomena.
- The data mining component prepares data for analysis by the automated deduction component. The automated deduction system may discover equivalences that can be used by the data mining component to simplify its search.
- Automated alerts are analytical agents that are designed to automatically find managerially interesting and important information in a database. Alerts can be a powerful analytical tool to keep managers informed as to critical problems and important business opportunities.
- The implications of "alerts run rampant" lead to the ability of the alerts system to provide actionable knowledge to the organization.
- The concept of intelligent information delivery (IID) systems is that it forms a metalayer on top of an alert, or other type of data mining system. The IID monitors the alerts produced and decides which information is most critical to bring to users' attention.
- Information Discovery, Inc. is the leading provider of large-scale data mining oriented decision support software and solutions, introducing pattern management with its breakthrough Pattern Warehouse^{TM} technology.
- **The Data Mining Techniques have been used to discover patterns; the machine-man approaches provide interactive access to the patterns.**
- The modern data mining techniques rely on pattern distillation, rather than data retention. Pattern distillation can be classified into logical, equational, and cross-tabulation methods.

S. Sumathi and S.N. Sivanandam: *Data Mining in Information Analysis and Delivery*, Studies in Computational Intelligence (SCI) **29**, 591–613 (2006)
www.springerlink.com © Springer-Verlag Berlin Heidelberg 2006

Abstract. The Information analysis and delivery, and a characterization of data mining technologies and processes by Information discovery Inc., case study are the subject of this chapter.

Information analysis is an umbrella term that applies to a multitude of techniques for extracting important, interesting, or unexpected phenomena from massive quantities of information various types.

In this chapter, we examine the implications of "alerts run rampant" on the ability of the alerts system to provide actionable knowledge to the organization. We then provide a simple example of an intelligent information delivery (IID) mechanism, that functions as a metalayer to the alerts system. The IID layer evaluates the importance and criticality of alert-created information across all alerts in the system. It then decides on the disposition of specific pieces of information. Finally, we describe how the IID layer can be used as a mechanism to derive knowledge out of analyzed information from data mining systems, in general. This chapter gives information regarding the data mining in information analysis and delivery.

23.1 Information Analysis: Overview

The analysis of information is an area of computer science rapidly growing in importance. Because the information of interest is of a wide variety of nature – and because the type of phenomena, which we seek, varies and often is ill defined, many diverse technologies must be developed and applied in novel ways.

It is often convenient to view information analysis as involving three main steps: data acquisition, information extraction and representation, and analysis.

23.1.1 Data Acquisition

Data of a variety of natures is acquired from a possibly large number of diverse sources. Examples of data and their sources include:

Structured Data

Computer audit trails, financial data, attributes of a complex system, and, generally, data from existing database systems may be subjected to a variety of analyses in an attempt to detect behaviors such as intrusion, fraud, or system malfunction.

Numerical and scientific data from an important subclass of structured data, a subclass whose analysis warrants special consideration. For example, autonomous sensors play an important role in safeguard and nonproliferation applications. While the data produced may be structured (i.e., in a prespecified format with well-defined features), the challenges differ from the above-mentioned sources in several ways. For example, these data may include real-valued vectors of variable length and data that is of a temporal nature (where change of state often is the critical component of the analysis).

It also often is the case that a large amount of complex metadata (e.g., scientific formulas and other types of rules) is required to capture the semantics of this type of data.

Images

Satellite and other types of image data are important for such applications as nonproliferation, climatology, and environmental studies. Such digitized images introduce a variety of new information attributes, such as three-dimensional special and four-dimensional time-space relationships.

Free Text Data

\ Documents, reports, technical articles, and articles from the popular press contain a wealth of information to mine. These sources present a particularly formidable challenge. Though not as difficult as natural language understanding, useful analyses well require context to establish semantics, similarities of topics, patterns of usage, and relevance to target queries.

23.1.2 Extraction and Representation

A crucial aspect of the analysis system is the representation, storage, and retrieval of the information under study. Rather than develop distinct analysis techniques for each type of data we might encounter, the best approach, we argue, is to represent and exploit the salient features of data within a common data model and to develop a uniform analysis methodology that operates upon this common model. Once a suitable representation is chosen, extraction tools are defined for each type of data source to map data from the form gathered into the common representation and to store the resulting data in the underlying database.

The data model devised must be sufficiently rich and flexible to support that variety of data we expect, and it also must be capable of supporting efficiently sophisticated analyses against massively large data sets, including retrieval operations required for data mining. To address these issues, we are interested in customizing and adapting one or more data models well studied in computer science so as to be suitable for the problem. Adaptations may include support for statistical analyses, expert rule bases, and axiom systems; complex hierarchical relationship such as/or relationships; and the identification of data equivalence classes.

23.1.3 Information Analysis

Information analysis requires a suite of sophisticated tools, including:

Data mining tools for discovering and prioritizing potentially interesting information: The research in data mining both explores foundational issues and seeks to apply the results by incorporating the data exploration methodology and algorithmic advances into an experimental software system.

The data mining foundation we have built is based on "information prioritization," a problem model where we are presented with a large number of data points, which must be prioritized. The prioritization produced allows one to pursue items from the highest to lowest ranked until time, money, or interest is exhausted. Another defining characteristic of our work is that we have developed methods, which perform analyses even in the most information-deprived environments (for example, environments lacking labeled training sets, expert rules, and feedback).

The challenges presented by future research include the mining of temporal, special, and textual patterns; the construction of abstract statistical models of undesirable behavior in new domains of interest; and the integration of our statistical data mining techniques with automated reasoning techniques. This last project is considered below.

Automated deduction tools for reasoning about data: An automated deduction tool allows us to make references and draw logical conclusions about retrieved data based on general data based on general rules and relationships. The work in automated deduction focuses on the development of inference rules and strategies needed to reason effectively about problems from mathematics and logic and for application areas such as the analysis of information. In order to develop an effective reasoning component for an information analysis system, we are working on problems such as the following:

Data to be interpreted will be at various levels of abstraction, ranging from raw sensor data to high-level terms that are the output of other data mining and analysis steps. We are working to enhance the inference and search capabilities of the automated deduction system in order to be able to reason effectively at multiple levels of abstraction.

One aspect of the analysis of information is to search for sets of observables that are considered to be evidence for activities or conditions of interest. It often will be the case that several sets of observables will be considered to be "equivalent" evidence for some activity or condition. We are developing strategies to account for equivalence classes in the search for evidence. Specifically, we are attempting to use the automated deduction system to search for evidence using functional rather than strictly syntactic matching criteria.

Integration of data mining and automated reasoning: Data mining and automated reasoning techniques traditionally are applied to quite different types of information analysis problems. We believe each technology can benefit from the other, and together can form the core of powerful information analysis architecture. The directions this integration may take include, for example, the following.

The statistically based data mining system operates as a last line of defense, inspecting data, which the automated deduction system does not flag as violating constraints specified in its rule base.

The data mining component prepares data for analysis by the automated deduction component. For example, the data mining component can provide data at a level of abstraction well suited for analysis by the automated deduction component and can prioritize this information to help guide the automated deduction system's search. Further, clustering of data values suggested by the data mining component can affect the application of inference rules (e.g., whether or not a rule fires).

23.2 Intelligent Information Delivery – Case Study

Recently, users of advanced information systems have begun to realize the value of incorporating automated alerts into their systems. Automated alerts are analytical agents that are designed to automatically find managerially interesting and important information in a database. The agents operate without user intervention, but report important information back to users whenever critical events are found in the database. This case study is taken from Analytic Solutions for Ceres Integrated Solutions, NY.

Alerts can be a powerful analytical tool to keep managers informed as to critical problems and important business opportunities. All it takes, it seems, is having the correct underlying sources of data for the alerts to operate on and then creating the appropriate set of alerts. The problem with automated alerts is that the volume of information automatically returned to the user can quickly become overwhelming. Analysis is easy. Knowledge is hard. A vital component in the development of knowledge is the recognition that an event is something that is important for the recipient to know about; in fact, that it is more important to know about than other significant events.

23.2.1 Alerts Run Rampant

Data mining systems, in general, are geared toward the analysis of vast amounts of data and are designed to produce large quantities of analyzed information that, essentially, have to be sifted through and analyzed before they become useful as business decision making aids. This fact can become a critical issue when applied to automated alert systems. These systems are designed to perform data mining automatically and continuously. An example from the consumer packaged goods (CPG) industry will show the magnitude of the problem.

The CPG industry has, for many years, had the availability of rich sources of data. For most grocery products, vendors such as A.C. Nielsen and IRI sell sales scanner data that tracks all competitive products in a category, by UPC (the individual product, the lowest level information that manufacturers track for sales purposes), in each of fifty or more markets. A typical category can have 1,200 or more UPCs in each market. Most packaged goods manufacturers receive updates weekly. This means that *a single alert measure* can track 60,000 possible events each week.

There are, however, many more than a single important alert measure that packaged goods manufacturers need to track. Some key alert measures for the packaged goods industry include short-term market share changes for all UPCs in the market, trends in market share changes, introductions of new competitive items, competitor price changes, and changes in competitive promotional activity. Competitive activity is inferred by observing such things as retailer promotion pricing actions, increased levels of distribution for a competitor's UPCs, and retailer promotions, such as increases in point of purchase displays, major ads, and coupon activity. Since packaged goods marketers typically micromarket, each UPC has to be tracked, by market, for each alert measure.

There can be easily hundreds of thousands (or many more) events being tracked automatically. Because the CPG marketing environment is highly competitive and dynamic, there can easily be thousands of events that set off trigger conditions to alert a user. The situation gets even more overwhelming when we consider the fact that advanced marketing analysis systems in the CPG industry sometimes also embed sophisticated data mining technology that automatically analyzes causal factors associated with some alert conditions. The ensuing report, then, includes not only alert information, but also details of an analysis. The information overload that results can set up a condition where the user has to either spend all his or her time on reviewing the results of alerts, or ends up just ignoring the output of the alerts system.

23.2.2 What an Intelligent Information Delivery System is

An intelligent information delivery system is essentially a knowledge-based system that:

1. Monitors and intercepts all outputs from the alert system,
2. Performs some analyses on the set of alerts that evaluates alert output based on the totality of what is known, and
3. Applies business rules to the output of the process to determine which outputs from the alert system are critically important for a user to know about.

The IID system functions as a meta-analysis layer for the alerts system. It evaluates alert-created information across all alerts in the system. Based on results of analysis and the rules contained in the system, it decides on the relative importance of the various alerts. The IID system also decides whom, that is, which users should receive specific pieces of information. The IID knowledge base contains rules related to managerial objectives that guide the selection of output for individual users. Development of this knowledge base is based on conducting knowledge engineering sessions with key business users to determine specific business rules to incorporate in the system. Actual application to individual users is based on creating user settings stored in a database table and accessed by the meta-analysis layer.

23.2.3 Simple Example of an Intelligent Information Delivery Mechanism

As an illustration, we provide a simple example of an IID system. The system has all three components:

1) An alert monitor
2) Meta-analysis capabilities, and
3) A business rule knowledge base.

The IID system is designed to support a consumer packaged goods alert system.

Alert Monitor

The alert system in the example polls the database and performs its analyses weekly to coincide with database updates based on marketplace scanner data purchased from IRI or A.C. Nielsen. The focus of the system is on information contained in this data. The IID Alert Monitor intercepts all alerts that are in its domain of knowledge. No alerts are passed onto users at this time. The Monitor holds the alert information until all the alerts have finished processing the updated information in the database.

Meta-Analysis Layer

The Meta-Analysis Layer synthesizes results of the alert process and performs further analysis. For example, it will do cross-market analysis of alerts to discover whether or not an alert condition is specific to a single market, or whether it reflects a more general condition. It will also check if the alert is a one-time occurrence or whether there has been a pattern of these conditions over time.

The Meta-Analysis Layer also makes an assessment of the overall volatility in the marketplace. Highly volatile markets can be expected to have many fluctuations in market share, retailer promotional activity, and competitor product introductions. Some alerts that might be considered significant in a nonvolatile market, after this assessment, may no longer be important enough to report.

Business Rules Knowledge Base

The Business Rules Knowledge Base contains rules developed by conducting in-depth interviews with key business managers in the organization responsible for taking action based on the results of the alert system. The business rules are mapped against the meta-analyzed alerts to determine which alerts are really important to know about, and who receives which alerts.

Consumer packaged goods marketers often focus on Brand Development Index (BDI) and Category Development Index (CDI) measures when running their business. BDI ranks markets as to the strength of the brand in that market. Markets where the brand has a high market share, high BDI markets, are ranked ahead of markets where the brand has a low market share. CDI ranks markets as to the strength of the overall category in the market. Markets where category sales are high (high CDI markets) are ranked ahead of markets where category sales are low.

Brand strategies often incorporate the relative importance of these measures and how to use them. For example, a brand strategy that focuses on increasing market share may often focus on high opportunity markets – those with high CDI, but low BDI. A brand strategy that focuses on maintaining current brand strength may focus on high BDI markets. An important element, then, of the business rules knowledge base may be the incorporation of rules related to BDI and CDI.

An emphasis on BDI and CDI could lead to the following rules:

Rule 1:
IF Brand Strategy is to focus on High Opportunity Markets
THEN Alerts should be ranked by the CDI of the market they relate to
Rule 2:
IF Brand Strategy is to focus on High Brand Strength Markets
THEN Alerts should be ranked by the BDI of the market they relate to

There will be other rules relating to other strategies that incorporate additional factors. Other rules in the knowledge base may relate to results of cross-market analysis, prioritization of negative information about the marketers brand, prioritization of positive information about competitors' brands, priority given to trends versus one-time events, and thresholds related to when to consider competitive activity important.

Rule 3:
IF Cross-market analysis shows an over all strong pattern THEN this is an important alert
Rule 4:
IF UPC is for our Brand
AND there is a downward trend in market share THEN this is an important alert
Rule 5:
IF UPC is for Key Competitors' Brand AND there has been a Highly Significant increase in market share THEN this is an important alert
Rule 6:
IF Market share change for a UPC is $>$ twice the average Market Share change THEN This is a Highly Significant increase in market share
Rule 7:
IF UPC is for Key Competitors' Brand AND there has been at least a three-month trend in price decreases THEN this is an important alert

The above is just a small sample of the business rules knowledge base that would be developed for even a simple intelligent Information delivery system. However, even a simple IID system can reduce the volume of output of alerts from hundreds of pages containing thousands of analyses to just the few most important findings.

23.3 A Characterization of Data Mining Technologies and Processes – Case Study

While myriad of approaches to data mining has been proposed, just a few fundamental techniques form the basis of most systems. Here we provide a characterization of these fundamental technologies and outline a framework for viewing data mining processes in terms of three families of techniques. Information Discovery, Inc. is the leading provider of large-scale data mining-oriented decision support software and solutions, introducing pattern management with its breakthrough Pattern Warehouse^{TM} technology and offering two comprehensive product suites. The Data Mining Suite^{TM} products directly access very large multitable SQL repositories to find powerful multiform patterns. The Knowledge Access Suite^{TM} incrementally stores these premined patterns in a Pattern Warehouse^{TM} for access by business users. The company also offers a wide range of discovery and data mining solutions, strategic consulting, and warehouse architecture design, as well as customized solutions for banking, financial services, retail, customer packaged goods, manufacturing, and web log analysis. This case study is taken from Information Discovery, Inc., Marketing Communications, CA.

The past year has seen a dramatic surge in the level of interest in data mining, with business users wanting to take advantage of the technology for a competitive edge. The IT departments in most Fortune 500 companies are suddenly tasked to respond to deployment questions relating to data mining. The growing interest in data mining has also resulted in the introduction of a myriad of commercial products, each described with a set of terms that sound similar, but in fact refer to very different functionality and based on distinct technical approaches.

The IT managers charged with the task of selecting a decision support system often face a challenge in responding to the needs of the business users because the underlying concept of data mining are far more complex than traditional query and reporting, and to add to the pressure the needs of the business users are usually urgent, requiring decisions that need to be made quickly.

However, while various approaches to data mining seem to offer distinct features and benefits, in fact just a few fundamental techniques form the basis of most data mining systems. Here we provide a characterization of these fundamental technologies, outline a framework for viewing data mining processes, and provide suggestions for the suitability of each technology for various business applications.

We define data mining as "a decision support process in which we search for patterns of information in data." This search may be done just by the user (i.e., by performing queries; in which case it is quite hard) or may be assisted by a smart program that automatically searches the database for the user and finds significant patterns. Once found, the information needs to be presented in a suitable form, with graphs, reports etc. The approaches we discuss here characterize the data mining paradigms from the point of view of discovery and prediction, not from the view point of machine–man pattern access. As in Parsaye, 1997, after the techniques discussed here have been used to discover patterns, the machine–man approaches provide interactive access to the patterns.

23.3.1 Data Mining Processes

Traditionally, there have been two types of statistical analyses: confirmatory analysis and exploratory analysis. In confirmatory analysis, one has a hypothesis and either confirms or refutes it. However, the bottleneck for confirmatory analysis is the shortage of hypotheses on the part of the analyst. In "exploratory analysis," one finds suitable hypothesis to confirm or refute. Here the system takes the initiative in data analysis, not the user.

The concept of "initiative" also applies to multidimensional spaces. In a simple OLAP access system, the user may have to think of a hypothesis and generate a graph. But in OLAP data mining, the system thinks of the questions by itself. We use the term *data mining* to refer to the automated process of data analysis in which the system takes the initiative to generate patterns by itself. From a process-oriented view, there are three classes of data mining activity: *discovery, predictive modeling, and forensic analysis.*

- *Discovery* is the process of looking in a database to find hidden patterns without a predetermined idea or hypothesis about what the patterns may be. In other words, the program takes the initiative in finding what the interesting patterns are, without the user thinking of the relevant questions first. In large databases, there are so many patterns that the user can never practically think of the right questions to ask. The key issue here is the richness of the patterns that can be expressed and discovered and the quality of the information delivered – determining the power and usefulness of the discovery technique.

As a simple example of discovery with system initiative, suppose we have a demographic database of the US. The user may take the initiative to ask a question from the database, such as "what is the average age of bakers?" The system may then print 47 as the average age. The user may then ask the system to take the initiative and find something interesting about "age" by itself. The system will then act as a human analyst would. It will look at some data characteristics, distributions, etc. and try to find some data densities that might be away from ordinary. In this case the system may print the rule: "IF Profession = Athlete THEN Age <30, with a 71% confidence."

This rule means that if we pick 100 athletes from the database, 71 of them are likely to be under 30. The system may also print: "IF Profession = Athlete THEN Age, <60, with a 97% confidence." This rule means that if we pick 100 athletes from the database, 97 of them are likely to be under 60. This delivers information to the user by distilling pattern from data.

- In *predictive modeling* patterns discovered from the database are used to predict the future. Predictive modeling thus allows the user to submit records with some unknown field values, and the system will guess the unknown values based on previous patterns discovered from the database. While discovery finds patterns in data, predictive modeling applies the patterns to guess values for new data items.

To use the example above, once we know that athletes are usually under 30, we can guess someone's age if we know that they are an athlete. For instance, if we are shown a record for John Smith whose profession is athlete by applying the rules we found above, we can be over 70% sure that he is under 30 years old, and we can be almost certain that he is under 60. Note that discovery helps us find "general knowledge," but prediction just guesses that value for the age of a specific individual. Also note that in this case the prediction is "transparent" (i.e., we know why we guess the age as under 30). In some systems the age is guessed, but the reason for the guess is not provided, making the system "opaque."

- *Forensic analysis* is the process of applying the extracted patterns to find anomalous or unusual data elements. To discover the unusual, we first find what is the norm, then we detect those items that deviate from the usual within a given threshold. Again, to use the example above, once we notice that 97% of athletes are fewer than 60, we can wonder about the 3% who are over 60 and still listed as athletes. These are unusual, but we still do not know why. They may be unusually healthy or play sports where age is less important (e.g., golf) or the database may contain errors, etc. Note that discovery helps us find "usual knowledge," but forensic analysis looks for unusual and specific cases.

Each of these processes can be further classified. There are several types of pattern discovery such as If/Then rules, associations, etc. While the rules discussed above have an IF-THEN nature, association rules refer to item groupings (e.g., when someone buys one product at a store, they may buy another product at the same time – a process usually called *market basket analysis*). The power of a discovery system is measured by the types and generality of the patterns it can find and express in a suitable language.

23.3.2 Data Mining Users and Activities

It is necessary to distinguish the data mining processes discussed above from the data mining activities in which the processes may be performed, and the users who perform them. First, the users. Data mining activities are usually performed by three different classes of users: *executives, end users, and analysts.*

- *Executives* need top-level insights and spend far less time with computers than the other groups – their attention span is usually less than 30 minutes. They may want information beyond what is available in their executive information system (EIS). End users and analysts usually assist executives.
- *End users* know how to use a spreadsheet, but they do not program – they can spend several hours a day with computers. Examples of end users are sales people, market researchers, scientists, engineers, physicians, etc. At times, managers assume the role of both executive and end user.
- *Analysts* know how to interpret data and do occasional computing but are not programmers. They may be financial analysts, statisticians, consultants, or database designers. Analysts usually know some statistics and SQL.

These users usually perform three types of data mining activity within a corporate environment: *episodic, strategic, and continuous data mining.*

- In episodic mining we look at data from one specific episode such as a specific direct marketing campaign. We may try to understand this data set, or use it for prediction on new marketing campaigns. Episodic mining is usually performed by analysts.
- In strategic mining we look at larger sets of corporate data with the intention of gaining an overall understanding of specific measures such as profitability. Hence, a strategic mining exercise may look to answer questions such as: "where do our profits come from?" or "how do our customer segments and product usage patterns relate to each other?"
- In continuous mining we try to understand how the world has changed within a given time period and try to gain an understanding of the factors that influence change. For instance, we may ask: "how have sales patterns changed this month?" or "what were the changing sources of customer attrition last quarter?"

Obviously continuous mining is an on-going activity and usually takes place once strategic mining has been performed to provide a first understanding of the issues. Continuous and strategic mining are often directed toward executives and managers, although analysts may help them here. As we see later, different technologies are best suited to each of these types of data mining activity.

23.3.3 The Technology Tree

The top-level dichotomization of the data mining technologies can be based on the retention of data; that is, do we still keep or need the data after we have mined it? In most cases, not. However, in some early approaches much of the data set was still maintained for future pattern matching. Obviously, these retention-based techniques only apply to the tasks of predictive modeling and forensic analysis, and not knowledge discovery since they do not distill any patterns.

As one would expect, approaches based on data retention quickly run into problems because of large data sets. However, in some cases predictive results can be obtained with these techniques and for the sake of completeness. The approaches based on pattern distillation fall into three categories: logical, cross-tabulatioal, and equational.

Not all approaches based on pattern distillation provide knowledge, since the patterns may be distilled into an "opaque" language or formalism not easily readable by humans such as very complex equations. Hence, some of these approaches produce "transparent" and understandable patterns of knowledge, others just produce patterns used for opaque prediction.

Data Retention

While in pattern distillation we analyze data, extract patterns and then leave the data behind, in the retention approaches the data is kept for pattern matching. When new data items are presented, they are matched against the previous data set.

A well-known example of an approach based on data retention is the "nearest neighbor" method. Here, a data set is kept (usually in memory) for comparison with new data items. When a new record is presented for prediction, the "distance" between it and similar records in the data set is found, and the most similar (or nearest neighbors) are identified.

For instance, given a prospective customer for banking services, the attributes of the prospect are compared with all existing bank customers (e.g., the age and income of the prospect are compared with the age and income of existing customers). Then a set of closest "neighbors" for the prospect are selected (based on closest income, age, etc.).

The term *K-nearest neighbor* is used to mean that we select the top K (e.g. top 10) neighbors for the prospect. Next, a closer comparison is performed to select which new product is most suited to the prospect, based on the products used by the Top K (e.g., top 10) neighbors. Of course, it is quite expensive to keep all the data, and hence sometimes just a set of "typical cases" is retained. We may select a set of 100 "typical customers" as the basis for comparison. This is often called *case-based reasoning.*

Obviously, the key problem here is that of selecting the "typical" customers as cases. If we do not really understand the customers, how can we expect to select the typical cases, and if the customer base changes, how do we change the typical customers?

Another usually fatal problem for these approaches has to do with databases with a large number of non-numeric values (e.g., many supermarket products or car parts). Since distances between these non-numeric values are not easily computed, some measures of approximation need to be used – and this is often hard to come by. And if there are many non-numeric values, there will be too many cases to manage.

Pattern Distillation

These technologies extract patterns from a data set then use the patterns for various purposes. Naturally, the first two questions to ask here are: What types of patterns can be extracted and how are they represented?

Obviously, patterns need to be expressed within formalism and a language. This choice given rise to three distinct approaches: logic, equations, or cross-tabulations. Each of these approaches traces its historical roots to a distinct mathematical origin.

For instance, let us consider the distinction between equations and logic. In an equational system operators such as "plus" and "times" may be used to relate variables together, e.g., $(a * X) + b)$ while in a logical system the key operators are conditional (e.g., IF $6 < X < 7$ THEN $1 < y < 2$).

Logic can deal with both numeric and non-numeric data. Equations require all data to be numeric, while cross-tabulations are the reverse and only work on non-numeric data; a key source of problems. But more importantly, equations compute distances form surfaces (such as lines) while cross-tabs focus on co-occurrences.

Neural networks are opaque equational techniques since internally they compute surfaces within a numeric space. As data is repeatedly fed into the network, the parameters are changed so that the surface becomes closer to the data point.

When discussing data mining, it is necessary to distinguish between "directed analysis" and free form "roams" through the database. In directed analysis, also called supervised learning, there is a "teacher" who teaches the system, by saying when prediction was correct or incorrect. Here the data has a specific column that is used as the goal for discovery or prediction.

In unsupervised learning, the system has number teacher, but simply tries to find interesting clusters of patterns within the data set. Most of the business applications of data mining involve directed data mining, while unsupervised discovery can sometimes be used for data segmentation or clustering (e.g., finding classes of customers that group together).

Logical Approaches

Logic forms the basis of most written languages and is essential for "left-brain" thinking. Patterns expressed in logical languages are distinguished by two main features: on one hand they are readable and understandable, on the other hand they are excellent for representing crisp boxes and groupings of data elements.

The central operator in a logical language is usually a variation on the well known If/Then statements (e.g., "If it is raining, then it is cloudy"). However, let us note that while the most common form of logic is conditional logic, often we may need to use other logical forms such as association logic with When/Also rules, (e.g., When paint is purchased, also a paint brush is purchased. While the propositional and predicate logic (i.e., conditional logics) are best known, other forms of logic (e.g., variational and trend logics) are also useful in business data analysis.

Conditional logic systems can be separated into two distinct groups: rules and decision trees. Conditional rules may be implemented by induction or genetic algorithms and there are several approaches for generating decision trees (e.g., CART, CHAID, C4.5)

Rules

Logical relationships are usually represented as rules. The simplest types of rules express conditional or association relationships. A conditional rule is a statement of the form:

If Condition1
Then Condition2

For instance, in a demographic database we may have a rule: If Profession = Athlete Then Age <30. Here we compare the values within fields of a given table (i.e., we have an "attribute-value" representation). Here Profession is the attribute and Athlete the value. Another example of an attribute-value expression is "State = Arizona", where State is the attribute and Arizona the value.

Conditional rules usually work on tables with attributes (i.e. fields) and values, such as below.

Name	Profession	Age	John	Smith	Athlete	27	...	...	...

Rules may easily go beyond attribute-value representations. They may have statements such as "Shipping_State = Receiving_State". Here, in attribute logic, we compare the values of two fields, without explicitly naming any values. This relationship cannot be stated by decision trees or cross-tabs.

Affinity logic is distinct from conditional logic both in terms of the language of expression and the data structures it uses. Affinity analysis (or association analysis) is the search for patterns and conditions that describe how various items "group together" or "happen together" within a series of events or transactions. An affinity rule has the form:

When Item1 Also Item2

An example of this is, When Paint, Also Paint Brush. A simple affinity analysis uses a transaction table such as:

Transaction #	Item
123	Paint
123	Paint Brush
123	Nails
124	Paint
124	Paint Brush
124	Wood
125	–

to identify items that group together within transactions. Here, the transaction# field is used to group items together, while the item# field includes the entities being grouped. In this example, the affinity for transactions 123 and 124 is the pair (Paint, Paint Brush). Please note that this is a distinct data structure from the conditional logic rule above.

As pointed out in "Data Mining with OLAP Affinities," "flat affinities" need to be extended to dimensional or OLAP affinities for better results. A dimensional affinity has the form:

Confidence = 95%
IF
Day = Saturday
WHEN
Item = Paint Brush
ALSO
Item = Paint

Here logical conditions and associations are combined. This form of hybrid structure delivers the real power of transparent logic. Rules have the advantage of being able to deal with numeric and non-numeric data in a uniform manner. When dealing with numeric data, some approaches have to break numeric fields into "code" or specific values. This may effectively remove all numeric consideration form the codes, thus resulting in the loss of patterns. For instance, the field Age may need to be broken into 3 ranges (1–30), (31–60), (61–100), corresponding to young, middle-aged, and old. Of course, the data may hold patterns that overlap any of these ranges (e.g., the range 27–34) may be very significant for some patterns and any approach based on code assignment will miss these.

Rules can also work well on multidimensional and OLAP data because they can deal with ranges of numeric data and their logical formats allows their patterns to be merged along multiple dimensions.

Rules do at times look like decision trees, but despite the surface level similarity they are a distinct and different technique. This is easy to see when we consider the fact that decision trees do not express associations, or attribute-based patterns such as "Shipping_State = Receiving_State" where the values of two fields are compared, without explicitly naming any values.

The main weakness of rules stems from their inability to deal with smooth surfaces that typically occur in nature (e.g., finger-print identification, facial recognition). These "naturally smooth" surfaces are often best approximated by equational approaches such as neural nets.

Below we review two approaches to rule generation, namely induction and genetic algorithms. However, these are not the only approaches to data mining with rules. Some approaches try to precompute every possible rule that a data set could include. In these cases, only a few columns of data may be used because the logical space is so large. Hence we do not review these since they are not practical for large-scale applications.

Rule Induction

Rule induction is the process of looking at a data set and generating patterns. By automatically exploring the data set, the induction system forms hypotheses that lead to patterns.

The process is in essence similar to what a human analyst would do in exploratory analysis. For example, given a database of demographic information, the induction system may first look at how ages are distributed, and it may notice an interesting variation for those people whose profession is listed as professional athlete. This hypothesis is then found to be relevant and the system will print a rule such as.

IF Profession = Athlete
THEN Age <30.

This rule may have "a confidence" of 70% attached to it. However, this pattern may not hold for the ages of bankers or teachers in the same database.

We must also distinguish between fuzzy and inexact rules. Inexact rules often have a "fixed" confidence factor attached to them, i.e., each rule has a specific integer or percentage (such as 70%) representing its validity. However, the confidence in fuzzy rules can vary in terms of the numeric values in the body of the rule; for instance the confidence may be proportional to the age of a person and as the age varies so does the confidence. In this way fuzzy rules can produce much more compact expression of knowledge and lead to stable to behavior.

Rule induction can discover very general rules, which deal with both numeric and non-numeric data. And rules can combine conditional and affinity statements into hybrid patterns. A key issue here is the ability to go beyond flat databases and deal with OLAP patterns.

Genetic Algorithms

Genetic algorithms also generate rules from data sets but do not follow the exploration-oriented protocol of rule induction. Instead, they rely on the idea of "mutation" to make changes in patterns until a suitable form of pattern emerges via selective breeding. The genetic crossover operation is in fact very similar to the operation breeders use when they cross-breed plants and/or animals. The exchange of genetic material by chromosomes is also based on the same method. In the case of rules, the material exchange is a part of the pattern the rule describes.

Let us note that this is different from rule induction since the main focus in genetic algorithms is the combination of patterns from rules that have been discovered so far, while in rule induction the main focus of the activity is the dataset. Genetic algorithms are not just for rule generation and may be applied to a variety of other tasks to which rules do not immediately apply, such as the discovery of patterns in text, planning and control, system optimization, etc.

Decision Trees

Decision trees express a simple form of conditional logic. A decision tree system simply partitions a table into smaller tables by selecting subsets based on values for a given attribute. Based on how the table is partitioned, we get a different decision tree algorithm such as CART, CHAID, and C4.5.

For example, consider the table:

Manufacturer	**State**	**City**	**Product Color**	**Profit**
Smith	CA	Los Angeles	Blue	High
Smith	AZ	Flagstaff	Green	Low
Adams	NY	NYC	Blue	High
Adams	AZ	Flagstaff	Red	Low
Johnson	NY	NYC	Green	Avg
Johnson	CA	Los Angeles	Red	Avg

A decision tree from this table can be drawn. This decision tree first selected the attribute State to start the partitioning operation, then the attribute Manufacturer. Of course, if there are 100 columns in the table, the question of which attribute to select first becomes crucial. In fact, in many cases, including the table above, there is no best attribute, and whichever attribute the tree chooses there will be information loss. For example the two facts:

(a). "Blue products are high profit."
(b). "Arizona is low profit."

can never be obtained from the table above with a decision tree. We can either get fact (a) or fact (b) from the tree, not both, because a decision tree selects one specific attribute for partitioning at each stage. Rules and cross-tabs, on the other hand, can discover both of these facts.

23.3.4 Cross-Tabulation

Cross-tabulation is a very basic and simple form of data analysis, well known in statistics, and widely used for reporting. A two dimensional cross-tab is similar to a spreadsheet, with both row and column headings as attribute values. The cells in the spreadsheet represent an aggregate operation, usually the number of co-occurrences of the attribute values together. Many cross-tabs are effectively equivalent to a 3D bar graph that displays co-occurrence counts. Consider the table in the previous section. A cross-tab for the profit level could look as follows:

CA	AZ	NY	Blue	Green	Red	Profit
1	0	1	2	0	0	High
1	0	1	0	1	1	Avg
0	2	0	0	1	1	Low

Here we have not included the fields Manufacturer and City because the cross-tab would look too large. However, as is readily seen here, the fact that the count of co-occurrence of Blue and High is above the others indicates a stronger relationship.

When dealing with a small number of non-numeric values, cross-tabs are simple enough to use and find some conditional logic relationships (but not attribute logic, affinities, or other forms of logic). Cross-tabs usually run into four classes of problems: first when the number of non-numeric values goes up, second when one has to deal with numeric values, goes up, second when one has to deal with numeric values, third when several conjunction are involved, and fourth when the relationships are not just based on counts.

Agents and belief networks are variations on the cross-tab theme and are discussed next.

Agents

The term *agent* is sometimes used (among its other uses) to refer to cross-tabs that are graphically displayed in a network and allow some conjunctions (i.e., ANDs). In this context the term *agent* is effectively equivalent to the term *field-value pair.*

Like other cross-tab techniques, when dealing with numeric values, agents have to break the numbers into fixed "codes" (e.g., break Age into three age classes: (1–30), (31–60), (61–100)). Of course, the data may hold patterns that overlap any of these ranges (e.g., the range (27–34) and these will not be detected by the agent). And if the ranges selected are too small, there will be too many of them and larger patterns will be missed. Moreover, this inability to deal with numeric data causes problems with multidimensional data.

Belief Networks

Belief networks (sometimes called *causal networks*) also rely on co-occurrence counts, but both the graphic rendering and the probabilistic representation are slightly different from agents. Belief networks usually illustrated using a graphical representation of probability distributions (derived from counts). A belief network is thus a directed graph, consisting of nodes (representing variables) and arcs (representing probabilities dependencies) between the node variables. Each node contains a conditional probability distribution that describes the relationship between the node and the parents of that node. The belief network graph is acyclic, meaning that there are no cycles.

Equational Approaches

The underlying method of pattern expression in these systems is "surface construction" rather than logical expression or co-occurrence counts. Such systems usually use a set of equations to define a "surface" within a numeric space, then measure distances from this surface for a prediction.

The best known example of such a surface is a straight line in a two-dimensional space with the simple equation $Y = (a * X) + b$ and leads to the well-known approach of linear regression in statistics. As the parameter "a" varies in this equation, the slope of the line changes. Regression works well when the points to be approximated lie on a straight line; it is also possible to use nonlinear equations to approximate smoother surfaces.

When the surfaces are even more complex (e.g., $Y = (X2 + X + (1/X)))$, or when there are several dimensions, the ability of humans to understand the equations and surfaces decreases rather quickly. The system becomes opaque or "black box." However, it is still possible to construct such surfaces. In fact, neural nets are known to be "universal approximators" in theory. They can come close to any function. However, the present theory does not specify the practical limits of nets for achieving such approximation on large data sets and most neural net implementations rely on sampling.

The equational approaches almost always require the data set to be all numeric. Non-numeric data needs to be "coded" into numbers (the reverse of what cross-tabs do). This often causes a number of problems, as discussed below.

23.3.5 Neural Nets

Neural nets are a class of predictive modeling system that work by iterative parameter adjustment. Structurally, a neural network consists of a number of interconnected elements (called neurons) organized in layers, which learn by modifying the connection strengths (i.e., the parameters) connecting the layers.

Neural nets usually construct complex equational surfaces through repeated iterations, each time adjusting the parameters that define the surface. After many iterations, a surface may be "internally" defined that approximates many of the points within the data set.

The basic function of each neuron is to: (a) evaluate input values, (b) calculate a total for the combined input values, (c) compare the total with a threshold value, and (d) determine what its own output will be. While the operation of each neuron is fairly simple, complex behavior can be created by connecting a number of neurons together. Typically, the input neurons are connected to a middle layer (or several intermediate layers), which is then connected to an outer layer.

To build a neural model, we first train the net on a "training dataset," then use the trained net to make predictions. We may, at times, also use a "monitoring data set" during the training phase to check on the progress of the training. Each neuron usually has a set of weights that determine how it evaluates the combined strength of the input signals. Inputs coming into a neuron can be either positive (excitatory) or negative (inhibitory). Learning takes place by changing the weights used by the neuron in accordance with classification errors that were made by the net as a whole.

The inputs are usually scaled and normalized to produce a smooth behavior. During the training phase, the net sets the weights that determine the behavior of the intermediate layer. A popular approach is called *backpropagation* in which the weights are adjusted based on how closely the network has made guesses. Incorrect guesses reduce the thresholds for the appropriate connections. Neural nets can be trained to reasonably approximate the behavior of functions on small- and medium-sized data sets since they are universal approximate. However, in practice they work only on subsets and samples of data and at times run into problems when dealing with larger data sets (e.g., failure to converge or being stuck in a local minimum).

It is well known that backpropagation networks are similar to regression. There are several other network training paradigms that go beyond backpropagation, but still have problems in dealing with large data sets. One key problem for applying neural nets to large data sets is the preparation problem. The data in the warehouse has to be mapped into real numbers before the net can use it. This is a difficult task for commercial data with non-numerous values.

Since input to a neural net has to be numeric (and scaled), interfacing to a large data warehouse may become a problem. For each data field used in a neural net, we need to perform scaling and coding. The numeric (and date) fields are scaled. They are mapped into a scale that makes them uniform (i.e., if ages range between 1 and 100 and number of children between 1 and 5, then we scale these into the same interval, such as -1 to $+1$). This is not a very difficult task.

However, non-numeric values cannot easily be mapped to numbers in a direct manner since this will introduce "unexpected relationship" into the data, leading to errors later. For instance, if we have 100 cities, and assign 100 numbers to them, cities with values 98 and 99 will seem more related together than those with numbers 21 and 77. The net will think these cities are somehow related, and this may not be so.

To be used in a neural net, values for nonscalar fields such as City, State, or Product need to be coded and mapped into "new fields", taking the values 0 and 1. This means that the field State, which may have the 7 values: {CA, NY, AZ, GA, MI, TX, VA), is no longer used. Instead, we have 7 new fields, called CA, NY, AZ, CA, MI, TX, VA each taking the value 0 to 1, depending on the value in the record. For each record, only of these fields has the value 1, and the others have the value 0. In practice, there are often 50 states requiring 50 new inputs.

Now the problem should be obvious. "What if the field City has 1,000 values?" Do we need to introduce 1,000 new input elements for the net? In the strict sense, yes, we have to. But in practice this is not easy, since the internal matrix representation for the net will become astronomically large and totally unmanageable. Hence, by-pass approaches are often used.

Some systems try to overcome this problem by grouping the 1,000 cities into 10 groups of 100 cities each. Yet, this often introduces bias into the system, since in practice it is hard to know what the optimal groups are, and for large warehouses this requires too much human intervention. In fact, the whole purpose of data mining is to find these clusters, not ask the human analysts to construct them.

The distinguishing power of neural nets comes from their ability to deal with smooth surfaces that can be expressed in equations. These suitable application areas are varied and include fingerpoint identification and facial pattern recognition. However, with suitable analytical effort neural net works can also succeed in many other areas such as financial analysis and adaptive control. Eventually, the best way to use neural nets on large data sets will be to combine them with rules, allowing them to make predictions within hybrid architecture.

23.4 Summary

Automated deduction can be used to identify a subset of data for further interpretation using data mining strategies. Information prioritization results obtained by mining a database of successful proofs can be used to develop new search strategies for automated deduction.

The automated deduction system may discover equivalences that can be used by the data mining component to simplify its search. Eventually, the two components may be even more tightly coupled, iterating to extract information at successively higher levels of abstraction, and interacting in a hypothesize-test mode.

The concept of intelligent information delivery systems – systems that form a metalayer on top of an alert, or other type of data mining system was introduced. The IID monitors the alerts produced and decide which information is most critical to bring to users' attention. We have also given a brief example of an IID system.

As data mining systems and systems of alerts that repetitively and automatically analyze information in databases become more prevalent, the problem of what to do with all the answers that come out will become increasingly important. The alternative is that over time, users of these systems will find that the more analysis they receive, the less they end up knowing. Also a case study based on the characterization of data mining technologies and processes is given.

The fundamental techniques used for data mining can be classified into distinct groups, each offering advantages and trade-offs. The modern techniques rely on pattern distillation, rather than data retention. Pattern distillation can be classified into logical, equational, and cross-tabulation methods. The underlying structure of these approaches was discussed and compared. Hybrid approaches are likely to succeed best, merging logic and equations with multi-dimensional analysis. However, the overstructure of how these techniques are used should be viewed in the context of machine–man interaction.

23.5 Review Questions

1. Define extraction and representation.
2. What is information analysis?
3. Explain with case study intelligent information delivery with its components.
4. How is data mining process characterized?
5. State the data mining users and their activities.
6. What is data retention and pattern distillation?
7. How are logical approaches formed in the information analysis?
8. How are decision tree and cross-tabulation formed in the intelligent information delivery model?

Data Mining in Telecommunications and Control

Objectives:

- Data mining in telecommunication industry helps to understand the business involved, identify telecommunication patterns, catch fraudulent activities, make better use of resources, and improve the quality of service.
- A large class of data mining algorithms developed for this purpose includes CART, C4.5, neural networks, and Bayesian classifiers, among others. One of the assumptions made by these algorithms, which are carried over into data mining applications, is that of clean data.
- The ability to handle noise in this case is obviously critical to the successful application of data mining algorithms; the treatment of noise typically falls short of handling the complete problem of data error.
- The companies in the telecommunications industry face the *problem of churning.* Churning is the process of customer turnover. This is a major concern for the companies having many customers who can easily switch to other competitors.
- Data mining is one solution to do appropriate credit scoring and to combat churns in the telecom industry.
- Data mining may be used in churn analysis to perform two key tasks: Predict whether a particular customer will churn and when it will happen; Understand why particular customers churn.
- Decision support in telecommunications forms the rules that can be used as decision support rules (for the exchange operator) or directly to automate the operation of the exchange.
- In control systems the purpose is to present a real-time knowledge processing (RTKP) procedure based on conjunctive and disjunctive matrices and operators.
- The field of knowledge discovery in databases (KDD) has delivered a variety of techniques to discover patterns from vast amount of data, which helps in mining for complex data.

S. Sumathi and S.N. Sivanandam: *Data Mining in Telecommunications and Control*, Studies in Computational Intelligence (SCI) **29**, 615–627 (2006)
www.springerlink.com © Springer-Verlag Berlin Heidelberg 2006

Abstract. The data mining applications in telecommunications industry, and a learning system for decision support in telecommunications case study, knowledge processing in control systems, and aircraft control case study are discussed in this section. A few scenarios where data mining may improve telecommunication services are discussed.

The deregulation of the telecommunications industry in many countries and the development of new computer and communication technologies and the telecommunication market are rapidly expanding and highly competitive. This creates a great demand for data mining in order to help understand the business involved, identify telecommunication patterns, catch fraudulent activities, make better use of resources, and improve the quality of service.

In order to determine the needs of the telecommunication industry with respect to the data mining, an extensive literature survey analysis was performed at Telkom. They expressed a need for mining the data stored in the Telkom data warehouse. Almost all areas of Telkom's business can benefit from data mining, but in particular marketing and sales department. A serious problem for Telkcom, and for most companies in the telecommunications industry, is the *problem of churning.* Churning is the process of customer turnover.

A case study for decision support in telecommunications has been described. History data describing the operation of a telephone exchange is analyzed by the system to reconstruct understandable event descriptions. This case study is taken from Gerstner Laboratory, Czech Technical University, Czech Republic.

Real-time knowledge-based or knowledge-processing systems are playing an increasingly important role in transportation, manufacturing, control, and robotic and aerospace systems. They are no longer limited to low-level control functions. Control, supervision, and monitoring of complex hierarchical systems in dynamic and sometimes unpredictable or hazardous environments are typical tasks of current man-made systems.

Current development in real-time artificial intelligence is driven by a need to make knowledge-based systems work in real-time and a need to integrate knowledge-based approaches to handle the complexities of problem-solving behavior in control systems. The purpose is to present a real-time knowledge processing (RTKP) procedure based on conjunctive and disjunctive matrices and operators.

The case study taken from Sylvain Letourneau, University of Ottawa, Canada, is discussed to explain the how data mining is used for maintenance of complex systems. The anticipated contributions of this study were related to two fundamental problems in the fi eld of knowledge discovery in databases: i) automatic preparation of the data prior to model development and ii) use of diverse sources of information.

24.1 Data Mining for the Telecommunication Industry

The telecommunication industry has quickly evolved from offering local and long-distance telephone services to providing many other comprehensive communication services, including voice, fax, pager, cellular phone, images, e-mail, computer, and Web data transmission, and data traffic. The integration of telecommunication, computing network, Internet, and numerous other means of communications and computing is also underway.

The following are a few scenarios where data mining may improve telecommunication services.

24.1.1 Multidimensional Analysis of Telecommunication Data

Telecommunication data are intrinsically with dimensions such as calling time, duration, location of caller, and type of call. The multidimensional analysis of such data can be used to identify and compare the data traffic, system work load, resource usage, user group behavior, profit, and so on. For example, an analyst in the industry may wish to regularly view charts regarding calling source, destination, volume, and time-of-day usage patterns. Therefore, it is often useful to consolidate telecommunication data into large data warehouse and routinely perform multidimensional analysis using OLAP and visualization tools.

24.1.2 Fraudulent Pattern Analysis and the Identification of Unusual Patterns

Fraudulent activity costs the telecommunication industry millions of dollars a year. It is important to identify potentially fraudulent users and their atypical usage patterns; detect attempts to gain fraudulent entry to customer accounts; and discover unusual patterns that may need special attention, such as busy-hour, frustrated call attempts, switch and route congestion patterns, and periodic calls from automatic dial-out equipment (like fax machine) that have been improperly programmed. Many of these types of patterns can be discovered by multidimensional analysis, cluster analysis, and outlier analysis.

24.1.3 Multidimensional Association and Sequential Pattern Analysis

The discovery of association and sequential patterns in multidimensional analysis can be used to promote telecommunication services. For example, suppose we would like to find usage patterns for a set of communication services by customer group, by month, and by time of day. The calling records may be grouped by customer in the following form:

(Customer_id, residence, office, time, date, service_1, service_2,...)

A sequential pattern like "If a customer in the Los Angeles area works in a city different from her residence, she is likely to first use long-distance service between two cities around 5 pm and then use a cellular phone for at least 30 minutes in the subsequent hour every weekday" can be further probed by drilling up and down in order to determine whether it holds for particular pairs of cities and particular groups of persons (e.g., engineers, doctors, etc.). This can help promote the sales of specific long-distance and cellular phone combinations, and improve the availability of particular services in the region.

24.1.4 Use of Visualization Tools in Telecommunication Data Analysis

Tools for OLAP visualization, linkage visualization, association visualization, clustering, and outlier visualization have been shown to be very useful for telecommunication data analysis.

24.2 Data Mining Focus Areas in Telecommunication

The experience with applications of interest to the telecommunications business was carried out at Bell Atlantic STC, NY, where we focus on machine learning algorithms. A large class of data mining algorithms has developed out of ideas investigated earlier, by researchers and developers of machine learning algorithms. Notable examples include CART, C4.5, neural networks, and Bayesian classifiers, among others. One of the assumptions made by these algorithms, which are carried over into data mining applications, is that of clean data.

All of these algorithms, and other like them, do relax the assumption from its strictest terms. They do not assume perfectly clean data, but rather assume that the data might be noisy. While the ability to handle noise is obviously critical to the successful application of data mining algorithms, the treatment of noise typically falls short of handling the complete problem of data error.

24.2.1 Systematic Error

Systematic errors arise in many applications, and they may be due to any of the following:

- Errors of calibration of instruments.
- Personal errors. These are errors caused by habits of individual observes.
- Imperfect technique.

We have found many examples of these in some of the telecommunications applications, investigated at Bell Atlantic Science and Technology Center, NY. One of these applications is classification of customer-reported telephone problems in the local loop of the telephone network. Problem diagnoses are high level, describing roughly that segment of the local loop where the trouble might be found, so that an appropriate technician might be dispatched to repair the trouble. The diagnoses are dispatched to the customer's premise: dispatch to the cable, dispatch to the central office, hold for further testing. The data describing the troubles includes information about the type of switch to which the customer's line is connected and electrical readings such as voltages and resistance, among others. The data mining problem here is to consider a large database of past troubles and their resolutions, and to develop rules for sending the appropriate technicians out to fix problems that have a certain profile. The electrical readings that are a large component of the data are obtained via an automated line testing system. The line testing must be calibrated regularly, but in practice this rarely occurs. As a result, the system becomes miscalibrated, and all readings reported for a set of lines on a given day might be off by a systematic amount. Furthermore, the system's baseline readings can differ from day to day.

This source of systematic error is known, but there are no mechanisms in place to handle so that it can be eliminated from the data. Given the heavy load handled by the company, it is not clear that careful calibration can become a high priority item. Thus we can expect that the problem will persist. People can also affect the data in a systematic way. In particular, one source of the diagnoses for troubles is the technicians who fix the problems. They report results using a complex coding system. If a technician has memorized the wrong code to represent the outcome of a repair, it will be wrong consistently. Again, we have a good sense of the source of the problem, but it is not clear that it can be controlled. Also, aside from maintaining a profile of each technician, it is not clear that there is a mechanism that could automatically correct for these errors.

There are a number of different scenarios that arise with respect to systematic data error.

(1) The systematic error is well understood. In these cases, the data can be "cleaned" and data mining algorithms can be applied to the clean data.
(2) The errors can be reconciled. There are applications in which data may be obtained from several sources. In these cases, it may be possible to retain data that are consistent over the sources. This has the effect of cleaning the data by making the assumption that the data might have errors but that the errors will not be consistent over the various sources. We found that with the local-loop diagnosis application, we were able to use a variety of data sources to reconcile diagnostic error (through we were not able to account for calibration error).
(3) The data cannot be cleaned. These are cases where the error exists, but cannot be removed from the data. It is important to note that in these cases, the sources of the error might, in fact, be quite well known, but that additional complications make it difficult to pull the error out of the data.

One obvious reaction to these situations is to throw up our hands and assume that the application of data mining techniques will provide no useful results. But this reaction is unreasonable.

(1) If the amount of systematic error is small, or if the right algorithm is applied, the impact of the error might be small relative to other gains of the data mining.

(2) Data mining techniques might be useful for helping to identify systematic error, making the process of cleaning one's data a possibility.
(3) There are applications for which only a small amount of mined information can go long way to benefiting a company. In these cases, it is not in our best interest as data miners to simply dismiss an application as being "too hard." In the application described above, an improvement of only 1% over the current dispatch procedure could save the company over $3,000,000 annually.

More work needs to be done on:

(1) Developing data mining algorithms for cleaning systematic error out of data.
(2) Analyzing the tools we have so that we can determine how they are actually affected by different types of error.

24.2.2 Data Mining in Churn Analysis

Data mining is the discovery of knowledge from data, and uses a variety of tools ranging from classical statistical methods to neural networks and other new techniques originating from machine learning and artificial intelligence. Recently, data mining has been used with substantial results in enabling and improving database marketing, process optimization, and detecting fraud.

Almost all areas of Telkom's business can benefit from data mining, but in particular marketing and sales department. A serious problem for Telkcom and for most companies in the telecommunications industry is the *problem of churning.*

Churning is the process of customer turnover. This is a major concern for the companies having many customers who can easily switch to other competitors. Competition will become fierce in the next years as new companies enter the South African market.

Given the increase in the customer choice, there will be increase in the churn rate. Anderson Consulting recently estimated customer churn level of 30% per year in the cellular phone markets. The cost of churn in the telecomm industry is large.

A 1995 report from the Digital Equipment Corporation estimated the cost of churning in the wireless communication to be around $400 per new subscriber. It is clear that spending money holding on to existing customer is more efficient than acquiring the new customer.

In South Africa a further churn-related problem has been identified. Telkom, as requested by the South African Government, should install many phones in previously disadvantaged communities and homes. Because the Client in these areas is not financially self-sufficient, the churn problem aggravated. This is also an example of the closely related to the credit-scoring problem, i.e., the decision to accept a client in the face of the associated risk involved. Data mining is one solution to do appropriate credit scoring and to combating churns in the telecom industry.

Data Mining may be Used in Churn Analysis to Perform Two Key Tasks

Predict whether a particular customer will churn and when it will happen;
Understand why particular customers churn.

These prediction and understanding tasks represent the two most important aspects of data mining in use today. By predicting which customers are likely to churn, the company can reduce the rate of churn by offering customers new incentives to stay. By understanding why customers churn the company can also work on changing their service so as to satisfy these customers pro-actively. In addition the data mining tools so as to choose the best strategy in terms of cost and effort can quantify the chance of the customer churning after action is taken.

24.3 A Learning System for Decision Support in Telecommunications – Case Study

We present a system for decision support in telecommunications. History data describing the operation of a telephone exchange is analyzed by the system to reconstruct understandable event descriptions. The event descriptions are processed by an algorithm inducing rules describing regularities in the events. The rules can be used as decision support rules (for the exchange operator) or directly to automate the operation of the exchange. This case study is taken from Gerstner Laboratory, Czech Technical University, Czech Republic.

In spite of the explosion of information technologies based on written communication, the most common and most frequently used tool is the telephone. Up-to-date private branch exchanges (PBX) provide comfort in managing the telephone traffic, namely regarding calls coming into an enterprise from the outside world. Communication proceeds smoothly provided that the caller knows with whom she wants to communicate and the person is available. In the opposite case, there is a secretary, receptionist, operator, or colleague that can for instance help to find a substituting person.

The operator is a person with no direct product, but with strong impact on productivity of other people. Despite that, a wide range of companies cancels the post of the telephone operator. The reason is that it is not easy to find a person who is intelligent enough to be good operator and to be modest enough to be just an operator. This opens area for computers – the computer is paid for only once so numbers fix costs set in. Moreover, the machine can work nonstop and provide additional data suitable for analysis allowing for improvements of the telecommunication traffic. Currently there are several domains where computers are used in the PBX area (neglecting the fact that PBX itself is a kind of computer):

- *Automated attendant* – a device that welcomes a caller in a unified manner and allows him usually to reach a person, or choose a person from a spoken list; in both cases the calling party is required to co-operate.
- *Voice mail* – a device allowing to leave a spoken message to an unavailable person and some rather sophisticated methods of delivering the messages are available.
- *Information service* – the machine substitutes a person in providing some basic information usually organized into an information tree; the calling party is required to co-operate.

The aim of the above listed tools is to satisfy a caller even if there is no human service available at the moment. But all such devices are designed in a static, simple manner – they act always the same way. The reason is simple – they do not consider who is calling nor what they usually want – as opposed to the human operator. Comparing a human operator/receptionist to a computer, we can imagine the following the improvements of the automated telephony:

1. Considering who is calling (by the identified calling party number) and what number was dialed by the caller, the system can learn to determine the person most probably desired by the caller; knowledge can be obtained either from previous cases (taking into account other data like daytime, explicit information – long absence of some of the company's employee, etc.) or by "observing" the way how the caller was handled by humans before; this could shorten the caller's way to get the information she needs.
2. The caller can be informed by a machine in a spoken language about the state of the call and suggested most likely alternatives; messages should be "context sensitive."

Naturally, the finite goal of computerized telephony is a fully "duplex" machine that can both speak and comprehend spoken language so that the feedback with the caller can proceed in a natural dialog. We present a methodology where the goal is to satisfy the goal 1.

The task was defined by a telecommunication company that installs PBX switchboards in various enterprises. The experiments are based on the PBX logging data coming from one of the enterprises. The methodology is reflected in a unified system with inductive (learning) capabilities to produce decision support rules based on the data describing the previous PBX switching traffic. The system can be naturally adapted to the condition of a specific company (by including a formally defined enterprise-related background knowledge) as well as in the case of a change in the PBX firmware (again via an inductive process).

The language of Prolog (a subset of the language of first-order logic) is employed as a unified formalism to represent the input data, the background knowledge, the reasoning mechanism, and the output decision support rules. The reason for this is the structured nature of the data with important dependence between individual records, and the fact that sophisticated paradigms are available for learning in first order logic. These paradigms are known as *inductive logic programming (ILP)*. The fundamental goal of ILP is the induction of first-order logic theories from logic and background knowledge. In the recent years, two streams of ILP developed, called the *normal setting* (where roughly theories with a "predictive" nature are sought) and the *nonmonotonic setting* (where the theories have a "descriptive" character).

24.4 Knowledge Processing in Control Systems

Several large real-time applications are required to operate in environments that are not fully structured. The lack of information and uncertainty of the environment requires the use of problems-solving techniques. Elevator group control is one such application. There are many possible situations comprising the state of all elevators, existing calls in the building, completion of previously scheduled cars, and combining new hall calls with performance criteria. There are several possible corresponding schedules and as new hall calls appear, the scheduled cars must be revised frequently. Since entering all the possible responses (schedules) into the computer is unfeasible, automating the response construction process will be required. Factory scheduling is another such application.

Looney proposed a matrix procedure for real-time knowledge processing considering only production rules with one antecedent. His procedure however does not preserve its matrix form when several antecedents are present in a rule. Furthermore, when handling multiple antecedents, the procedure is difficult to analyze and to predict if it could meet deadlines due to the chaining scheme adopted. This is a critical issue in real-time situations.

Alternate schemes based on a network type of representation for production rules are the RETE (C.L.Forgy, AI, in 1982 and EUREK (M.Funabashi et al., in 1988) procedures. However, both procedures lack the predictability property, which is essential in real-time applications.

More recently, Paul et al. developed an approach, which integrates problem-solving methodology and architectural primitives to reduce the variance at methodology level and at problem-solving level. Using this approach they have shown that problem-solving and real-time task coexist within a readily analyzable framework.

The purpose is to present a real-time knowledge processing (RTKP) procedure based on conjunctive and disjunctive matrices and operators. The proposed procedure affords the setting up of the focus of attention mechanisms and guarantees its respond time. Those are important characteristics that real-time knowledge-based systems should have.

24.4.1 Preliminaries and General Definitions

A typical RTKP system acting as a direct digital control system is shown in Fig. 24.1. The RTKP module is connected to information sources and receivers. Sources may be sensors connected to a process, human users, or even computer programs in large integrated systems. Receivers can be either actuators, human users, or again computer programs. The main idea behind this scheme is that RTKP takes information from a system, processes this information with the knowledge stored in it and then outputs new information to the system. Outputs are the control decisions. Figure 24.2 shows an RTKP system performing supervisory control tasks. The knowledge-processing task is encapsulated within a server to guarantee temporal isolation between it and conventional real-time tasks.

Internally, RTKP is divided into four basic parts. First is a preprocessor module responsible for the transformation of input information into the internal representation model is used. This module is also responsible for any mathematical treatment (by making transformation of variables, for example), as well as to preprocess task-dependent knowledge. The postprocessor module translates the internal representation model into output information in a former as required by the process. Between those two modules, there are the inference engine and the knowledge base. The last is the internal knowledge repository, coded in a usable format. Information provided by the preprocessor module and knowledge base is processed by the inference engine to generate the desired outputs.

The RTKP is defined by:

- Internal representations of information provided by the preprocessor module and information to be converted by the postprocessor module,
- An internal representation of the knowledge base,
- A procedure for the inference engine.

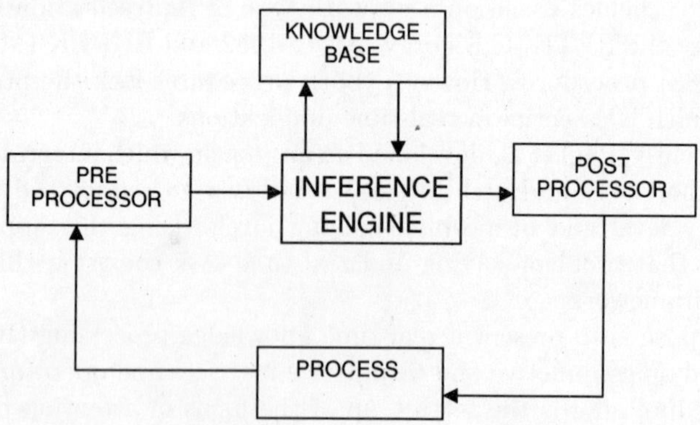

Fig. 24.1. Typical RTKP structure in direct digital control

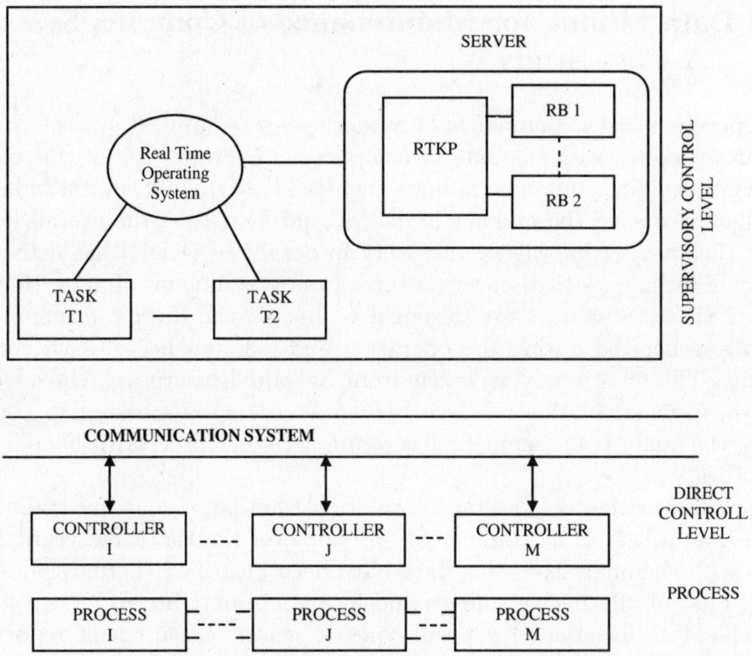

Fig. 24.2. RTKP in supervisory control systems

The knowledge base is composed of two parts: rule base and fact base. A fact is the truth value associated to particular proposition (or a term) used to store knowledge. A fact base is a set of terms, where each term has a meaning related to the process (an associated proposition). Propositions are considered within the prepositional logic framework, although they could also represent fuzzy propositions. In the proposed procedure, the fact base is represented by a fact vector where each component is related to term and contains its truth value.

The rule base can be viewed in two different representations. The first representation is for purposes of knowledge acquisition and analysis (called *the virtual representation*). It is presented as a collection of rules like: if <antecedent> then <consequent>, where <antecedent> represents a disjunctive association of terms. The second representation is coded version of the knowledge for processing purposes.

It is relevant now to review the main requirements that real-time knowledge processing systems should fulfill integration of numeric/symbolic processing, continuous operation, focus of attention mechanisms, interrupt manager services, optimum use of environment, response time warranty, temporal data processing, truth data maintenance, and the fact it dispenses explanatory modules.

24.5 Data Mining for Maintenance of Complex Systems – A Case Study

The operation and maintenance of modern sensor-equipped systems such as aircraft generates vast amounts of complex data. Proper use of this data to predict or explain component failures may lead to savings of several thousands of dollars, reducing the number of delays, and increasing the overall level of safety. The field of knowledge discovery in databases (KDD) has delivered a variety of techniques to discover patterns from vast amount of data. However, none of these techniques are designed to handle the diverse forms of data typically generated during the operation and maintenance of such complex systems. This case study is taken from Sylvain Letourneau, University of Ottawa, Canada. In this research, we study the specific issues to consider during the analysis of commercial aircraft data and process to handle these difficulties.

We aim at extracting useful information from large amounts of data collected from a fleet of 34 commercial aircraft over the last three years. Many issues with the analysis of this data have been identified (Letourneau et al., 1997). First of all, diverse sources and formats of data are to be considered. Available data includes: i) various types of sensor measurement reports describing the status of the aircraft in different phases of operation, ii) warning and failure messages generated when particular conditions occur, and iii) descriptions of aircraft problems along with the maintenance actions taken for each of them. Various sources of background knowledge are also available such as troubleshooting guides, training manuals, and empirical studies. A second difficulty comes from the complexity and the quality of the data. The number of parameters is high (i.e., often more than 100), several parameters are expected to have time-series relationships, and problems such as missing values, improper data types, and out-of-range data are frequently observed. Some sensor measurements must also be normalized due to the influence of variations in the environment.

We propose two major research directions to address these difficulties: i) development of a powerful data preprocessing approach to handle data complexity and data quality problems and ii) integration of the multiple sources of information described above to improve the quality of models learned from the data.

The data preprocessing approach proposed in the research should address the following tasks: i) normalization of the data according to the effects of contextual variations in the environment, ii) cleaning of the data, and iii) labeling of the instances so that supervised machine learning techniques can be used. We have developed the core of a novel domain-independent normalization technique that makes use of the analysis of variance (Letourneau.Matwin & Famili, 1998). Results from experiments with a large data set in the aircraft domain have shown that the proposed approach is powerful in reducing the number of false alarms caused by random fluctuations in the environment. The current focus is on cleaning of the data and labeling of the instances.

The second major aspect of this work will address the use of different sources of information available during the operation and maintenance of complex systems. We studied the use of the domain information regarding the following aspects: i) extraction of the most appropriate features ii) analysis of meaningfulness of obtained models, and iii) improvement of the accuracy. This research is expected to enhance the process of knowledge discovery in databases so that it could be successfully applied to the maintenance and operation of complex systems.

24.6 Summary

This section has described data mining applications in telecommunications industry, and a learning system for decision support in telecommunications case study, knowledge processing in control systems and aircraft control case study.

Data mining can be applied fruitfully, as in network capacity utilization. In network capacity utilization, planning for telecommunication markets that expand, and the ability to service customers is highly affected by the capacity planning that has taken place long before. Data mining provides understanding of the underlying patterns and structures of service usage by customer groups. This insight allows capacity planners to optimize the investments in network facilities to better serve customers, while avoiding costly overexpansions, i.e., having enough capacity to deal with growing markets just at the right time.

24.7 Review Questions

1. How can data mining improve telecommunication services?
2. Write a short note on systematic error observed in mining.
3. How is data mining used in churn analysis?
4. Explain how data mining is used in PBX areas.
5. With typical structure explain real-time knowledge processing (RTKP) in direct digital control and supervisory control systems.

25

Data Mining in Security

Objectives:

- Data mining offers a convenient way to monitor the large computer networks, thus providing the security. The data mining system does this by developing a profile of the typical activities of each user in the network.
- To improve the accuracy, data mining programs are used to analyze audit data and extract features that can distinguish normal activities from intrusions; we use artificial anomalies along with the normal and/or intrusion data to produce more effective misuse and anomaly detection models.
- To improve efficiency, the computational costs of features are analyzed and a multiple-model cost-based approach is used to produce detection models with low cost and high accuracy.
- To improve usability, adaptive learning algorithms are used to facilitate model construction and incremental updates; unsupervised anomaly detection algorithms are used to reduce the reliance on labeled data.
- Security of network systems is becoming increasingly important; intrusion detection systems (IDSs) have thus become a critical technology.
- IDSs models generalize from both known attacks and normal behavior in order to detect unknown attacks.
- Data mining-based IDSs (especially anomaly detection systems) have higher false positive rates than traditional handcrafted signature-based methods, making them unusable in real environments.
- The anomaly detection algorithms explore the use of information-theoretic measures, i.e., entropy, conditional entropy, relative entropy, information gain, and information cost to capture intrinsic characteristics of normal data and use such measures to guide the process of building and evaluating anomaly detection models.

Abstract. Data mining can be used to improve efficiency, quality of data, marketing and sales, and has many more benefits. Furthermore, even in the case of security problems, we have addressed the case where data mining tools could be used to

S. Sumathi and S.N. Sivanandam: *Data Mining in Security*, Studies in Computational Intelligence (SCI) **29**, 629–648 (2006)
www.springerlink.com © Springer-Verlag Berlin Heidelberg 2006

detect abnormal behavior and intrusions in the system. Data mining also has many applications in detecting fraudulent behavior. While all of these applications of data mining can benefit humans, there is also a dangerous side to mining, since it could be a serious threat to the security and privacy of individuals. An overview on data mining in security and real-time data mining-based intrusion detection system case study is illustrated in this chapter.

Security of network systems is becoming increasingly important, as more and more sensitive information is being stored and manipulated online. Intrusion detection systems (IDSs) have thus become a critical technology to help protect these systems.

Most IDSs are based on handcrafted signatures that are developed by manual encoding of expert knowledge. These systems match activity on the system being monitored to known signatures of attacks. The major problem with this approach is that these IDSs fail to generalize to detect new attacks or attacks without known signatures. Recently, there has been an increased interest in data mining-based approaches to building detection models for IDSs. These models generalize from both known attacks and normal behavior in order to detect unknown attacks. Domain experts can also generate them in a quicker and more automated method than manually encoded models that require difficult analysis of audit data. Several effective data mining techniques for detecting intrusions have been developed, many of which perform close to or better than systems engineered by domain experts.

This chapter discusses several problems inherent in developing and deploying real-time data mining-based IDS and presents an overview of the research, which addresses these problems.

25.1 Data Mining in Security Systems

In June 1994, a computer expert in St. Petersburg, Russia, Vladmir Leonidovich Levin, penetrated the Citibank electronic funds transfer network. Over the course of five months, he funneled 10 million dollars into accounts in California, Israel, Germany, Finland, the Netherlands and Switzerland. He was eventually apprehended, and most of the money was recovered, but the incident revealed the vulnerability of large databases to computer hackers.

Incidents such as the one described above make the security of a company's computer system a serious issue in today's business world. System administrators and security officers monitor these computer networks, often comprising thousands of computers and terabytes of storage space. Their job is daunting; especially since a security violation on one workstation could become a multimillion-dollar incident. The Computer Emergency Response Team, an organization of computer security professionals, estimates that only five percent of companies whose security has been compromised are even aware that they have been infiltrated. Although the raw information needed to detect an intrusion is often available in the audit data recorded by each computer, there is far too much of it generated each day for the system administrators and security officers to inspect it. Even if they tried, the vast majority of the audit record would be completely mundane and innocuous actions.

Data mining offers a convenient way to monitor these large computer networks. By detecting anomalous activities in the logs of computers, a data mining system could flag suspicious events for later inspection by system administrators, allowing them to avoid checking all the normal daily activities. The data mining system does this by developing a profile of the typical activities of each user in the network. Deviations from the expected pattern could be harmful or abusive behavior and would therefore be flagged. The system would have to be flexible enough to compensate for normal deviations from expected behavior like users learning new programs or doing new tasks. One study done at the Purdue University found that a data mining system was able to identify a profiled user 99% of the time and differentiated between a profiled user and another user with almost 94% accuracy.

25.2 Real Time Data Mining-Based Intrusion Detection Systems – Case Study

In this section, we present an overview of the research in real-time data mining-based intrusion detection systems (IDSs) taken from W. Lee (1998). North Carolina State University, Raleigh, NC. We focus on issues related to deploying a data mining-based IDS in a real-time environment. We describe the approaches to address the three types of issues: accuracy, efficiency, and usability. To improve the accuracy, data mining programs are used to analyze audit data and extract features that can distinguish normal activities from intrusions; we use artificial anomalies along with the normal and/or intrusion data to produce more effective misuse and anomaly detection models. To improve efficiency, the computational costs of features are analyzed and a multiple-model cost-based approach is used to produce detection models with low cost and high accuracy. We also present a distributed architecture for evaluating cost-sensitive models in real time. To improve usability, adaptive learning algorithms are used to facilitate model construction and incremental updates; unsupervised anomaly detection algorithms are used to reduce the reliance on labeled data. We also present architecture consisting of sensors, detectors, a data warehouse, and model generation components. This architecture facilitates sharing and storage of audit data and the distribution of new and update models. This architecture also improves the efficiency and scalability of the IDS.

However, successful data mining techniques are themselves not enough to create deployable IDSs. Despite the promise of better detection performance and generalization ability of data mining-based IDSs, there are some inherent difficulties in the implementation and deployment of these systems. We can group these difficulties into three general categories: accuracy (i.e., detection performance), efficiency, and usability. Typically, data mining-based IDSs (especially anomaly detection systems) have higher false positive rates than traditional handcrafted signature-based methods, making them unusable in real environments. Also, these systems tend to be inefficient (i.e., computationally expensive) during both training and evaluation. This prevents them from being able to process audit data and detect intrusions in real time. Finally, these systems require large amounts of training data and are significantly more complex than traditional systems. In order to be able to deploy real-time data mining-based IDSs, these issues must be addressed.

In this section, we discuss several problems inherent in developing and deploying a real-time data mining-based IDS and present an overview of the research, which addresses these problems. These problems are independent of the actual learning algorithms or models used by an IDS and must be overcome in order to implement data mining methods in a deployable system. An effective data mining-based IDS must address each of these three groups of issues. Although there are tradeoffs between these groups, each can generally be handled separately. We present the key design elements and group them into which general issues they address.

25.2.1 Accuracy

Crucial to the design and implementation of effective data mining-based IDS is defining specifically how detection performance, or accuracy, of these systems is measured. Because of the difference in nature between a data mining-based system and a typical IDS the evaluation metrics must take into account factors that are not important for traditional IDSs.

At the most basic level, accuracy measures how well an IDS detects attacks. There are several key components of an accuracy measurement. One important component is detection rate, which is the percentage of attacks that a system detects. Another component is the false positive rate, which is the percentage of normal data that the system falsely determines to be intrusive. These quantities are typically measured by testing the system on a set of data (normal and intrusions) that are not seen during the training of the system in order to simulate an actual deployment.

There is an inherent tradeoff between detection rate and false positive rate. One way to represent this tradeoff is by plotting the detection rate versus false positive rate on a curve under different parameter values creating an ROC curve. [Receiver Operating Characteristics (ROC) graphs are used in many detection problems because they depict the tradeoffs between detection rate and false positive rate.] A method to compare accuracy between two IDSs is to examine their ROC curves.

In practice, only the small portion of an ROC curve corresponding to acceptably low false positives is of interest, as in a deployable system, only a low false positive rate can be tolerated. Handcrafted methods typically have a fixed detection threshold they perform at a constant detection rate across different false positive rates. In an ROC curve, we can assume that their curve is a straight line at each detection level. Data mining-based systems have the advantage of potentially being able to detect new attacks that handcrafted methods tends to miss. Data mining-based IDSs are only useful if their detection rate is higher than a handcrafted method's detection rate with an acceptably low false positive rate. Given this framework, the goal is to develop a data mining-based IDS that is capable of outperforming handcrafted signature-based systems at the tolerated false positive rate.

We have developed and applied a number of algorithm-independent techniques to improve the performance of data mining-based IDSs. In this section, we focus on a few particular techniques that have been proven to be empirically successful. We first present a generic framework for extracting features from audit data, which help the discriminate attacks from the normal data. These features can then be used by any detection model building algorithm. We then describe a method for generating artificial anomalies in order to decrease the false positive rate of anomaly detection algorithms. The research has shown that by generating artificial anomalies, we can improve the accuracy of these ID models. Finally, we present a method of combining anomaly and misuse (or signature) detection models. Typically misuse and anomaly detection models are trained and used in complete isolation from each other. The research has shown that by combining the two types of models, we can improve the overall detection rate of the system without compromising the benefits of either detection method.

25.2.2 Feature Extraction for IDS

Two basic premises of intrusion detection are that system activities are observable, e.g., via auditing, and there is distinct evidence that can distinguish normal and intrusive activities. We call the evidence extracted from raw audit data *features*, and use these features for building and evaluating intrusion detection models. Feature extraction (or construction) is the process of determining what evidence that can be taken from raw audit data is most useful for analysis. Feature extraction is thus a critical step in building an IDS. That is, having a set of features whose values in normal audit records differ significantly from the values in intrusion records is essential for having good detection performance.

We have developed a set of data mining algorithms for selecting and constructing the features from audit data. First the raw (binary) audit data is processed and summarized into discrete records containing a number of basic features such as in the case of network traffic: timestamp, duration, source and destination IP addresses and ports, and error condition flags. Specialized data mining programs are then applied to these records to compute the frequent patterns describing correlations among the features and frequently co-occurring events across many records. A pattern is typically from $A, B \rightarrow C, D[\textit{confidence, support}]$, which translates to events A and B are followed by events C and D with a certain confidence and occur with a certain frequency in the data (the patterns' support). The consistent patterns are of normal activities and the "unique" patterns associated with an intrusion are then identified and analyzed to construct additional features for connection records. It can be shown that the constructed features can indeed clearly separate intrusion from normal ones. Using this approach, the constructed features are more grounded on empirical data, and thus more objective than an expert knowledge. Results from the 1998 DARPA Intrusion Detection Evaluation showed that an IDS model constructed using these algorithms was one of the best performing of all the participating systems.

As an example, let us consider the SYN flood attack. When launching this attack, an attacker uses many spoofed source addresses to open many connections, which never become completely established (i.e., only the first SYN packet is sent, and connection remains in the "S0" state) to some port on a victim host (e.g., *http*). We compared the patterns from the 1998 DARPA data set that contain SYN flood attacks with the patterns from a "baseline" normal data set (of the same network), by first encoding the patterns into numbers and then computing "difference" scores. The following pattern, a frequent episode, has the highest "intrusion-only" (i.e., unique for the intrusion) score: *"(flag = S0, service = http, dst_host = victim), (flag = S0, service = http, dst_host = victim) → (flag = S0, service = http, dst_host = victim)* [0.93, 0.03, 2]." This means that 93% of the time, after two http connections with S0 flag are made to host victim, within 2 seconds from the first of these two, the third similar connection is made; and this pattern occurs in 3% of the data. Accordingly, the feature construction algorithm parses the pattern features: "a count of connections to the same dst_host in the past 2 seconds," and among these connections, "the percentage of those that have the same service, and the percentage of those that have the S0 flag." For the two "percentage" features, the normal connection records have values close to 0, but the connection records belonging to syn_flood have values above 80%. Once these discriminative features are constructed, it is easy to generate the detection rules via either manual (i.e., hand-coding) or automated (i.e., machine learning) techniques. For example, we use RIPPER, W.W. Cohen et al., in 1995, an inductive rule learner, to compute a detection rule for syn_flood using these extracted features: if for the past 2 seconds, the count of connections to the same dst_host is greater than 4; and the percentage of those that have the same service is greater than 75%; and the percentage of those that have the "S0" flag is greater than 75%, then there is a syn_flood attack.

25.2.3 Artificial Anomaly Generation

A major difficulty in using machine learning methods for anomaly detection lies in making the learner discover boundaries between known and unknown classes. Since there are no examples of anomalies in the training data (by definition of anomaly), a machine-learning algorithm will only uncover boundaries that separate different known classes in training data. A machine-learning algorithm will not specify a boundary between the known data and unseen data (anomalies). We present the technique of *artificial anomaly generation* to enable the traditional learners to detect anomalies. Artificial anomalies are injected into the training data to help the learner discover a boundary around the original data. All artificial anomalies are given the class label *anomaly*. The approach to generating artificial anomalies focuses on "near misses," instances that are close to the known data, but are not in the training data. We assume the training data are representative, hence near misses can be safely assumed to be anomalous.

Since we do not know where the exact decision boundary is between the known and anomalous instances, we assume that the boundary may be very close to the existing data. To generate the artificial anomalies close to the known data, a useful heuristic is to randomly change the value of one feature of an example to a value that does not occur in the data while leaving the other features unaltered.

Some regions of known data in the instance space may be sparsely populated. We can think of the sparse regions as small islands and dense regions as large islands in an ocean. To avoid overfitting, learning algorithms are usually biased toward discovering more general models. Since we only have known data, we want to prevent models from being overly general when predicting these known classes. That is, we want to avoid the situation where sparse regions may be grouped into dense regions to produce singularly large regions covered by overly general models. It is possible to produce artificial anomalies around the edges of these sparse regions and coerce the learning algorithm to discover the specific boundaries that distinguish the regions from the rest of the instance space. In other words, we want to generate data that will amplify these sparse regions.

Sparse regions are characterized by infrequent values of individual features. To amplify sparse regions, we proportionally generate more artificial anomalies around sparse regions depending on their sparsity using *Distribution Based Artificial Anomaly* algorithm (DBA2) in W. Fan, PhD thesis, 2001. The algorithm uses the frequency distribution of each feature's values to proportionally generate a sufficient amount of anomalies.

25.2.4 Combined Misuse and Anomaly Detection

Traditionally, anomaly detection and misuse detection are considered separate problems. Anomaly detection algorithms typically train over normal data while misuse algorithms typically train over labeled normal and intrusion data. Intuitively a hybrid approach should perform better; in addition, it has the obvious efficiency advantages in both model training and deployment than using two different models. We use the artificial anomaly generation method to create a single model that is both a misuse and anomaly detection method. This allows us to use traditional supervised inductive learning methods for both anomaly detection and misuse detection at the same time. We train a single model from a set of data that contains both normal records and records corresponding to intrusions. In addition, we also generate artificial anomalies using the DBA2 algorithm and train the algorithm over the combined data set. The learned model can detect anomalies and intrusions concurrently.

We learn a single rule set for combined misuse and anomaly detection. The rule set has rules to classify a connection to be normal, one of the known intrusion classes, or anomaly. In order to evaluate this combined approach, we group intrusions together into a number of small clusters. We create multiple data sets by incrementally adding each cluster into data set and regenerating artificial anomalies. This is to simulate the real-world process of developing and discovering new intrusions and incorporating them into the training set. We learn models that contain misuse rules for the intrusions that are known in the training data, anomaly detection rules for unknown intrusions in left-out clusters, and rules that characterize normal behavior.

We have seen that the detection rates of known intrusions classified by misuse rules in models learned with and without artificial anomalies are indistinguishable or completely identical. This observation shows that the proposed DBA2 method does not diminish the effectiveness of misuse rules in detecting particular categories of known intrusions.

25.2.5 Efficiency

In typical applications of data mining to intrusion detection, detection models are produced off-line because the learning algorithms must process tremendous amounts of archived audit data. These models can naturally be used for off-line intrusion detection (i.e., analyzing audit data off-line after intrusion detection should happen in real time, as intrusions take place, to minimize security compromises). In this section, we discuss the approaches to make data mining-based ID models work efficiently for real-time intrusion detection.

In contrast to off-line IDSs, a key objective of real-time IDS is to detect intrusions as early as possible. Therefore, the efficiency of the detection model is a very important consideration. Because the data mining-based models are computed using off-line data, they implicitly assume that when an event is being inspected (i.e., classified using an ID model), all activities related to the event have completed so that all features have meaningful values available for model checking. As a consequence, if we use these models in real time without any modification, then an event is not inspected until complete information about that event has arrived and been summarized, and all temporal and statistical features are computed. This scheme can fail miserably under real-time constraints. When the volume of an event stream is high, the amount of time taken to process the event records within the past n seconds and calculate statistical features is also very high. Many subsequent events may have terminated (and thus completed with attacks actions) when the "current" event is finally inspected by model. That is, the detection of intrusions is severely delayed. Unfortunately, DoS attacks, which typically generate a large amount of traffic in a very short period time, are often used by intruders to first overload an IDS and use the detection delay as a window of opportunity to quickly perform their malicious intent. For example, they can even seize control of the host on which the IDS lives, thus eliminating the effectiveness of intrusion detection altogether.

It is necessary to examine the time delay associated with computing each feature in order to speed up model evaluation. The time delay of a feature includes not only the time spent for its computation, but also the time spent waiting for its readiness (i.e., when it can be computed). For example, in the case of network auditing, *the total duration* of a network connection can only be computed after the last packet of the connection has arrived, whereas the *destination host* of a connection can be obtained by checking the header of the first packet.

From the perspective of cost analysis, the efficiency of an intrusion detection model is its *computational cost,* which is the sum of the time delay of the features used in the model. Based on the feature construction approaches discussed in previous section, we can categorize features used for network intrusion detection into 4 cost levels:

- Level 1 features can be computed from this packet, e.g., the *service.*
- Level 2 features can be computed at any point during the life of the connection state (*SYN_WAIT*, *CONNECTED*, *FIN_WAIT*, etc.).
- Level 3 features can be computed at the end of the connection, using only the information about the connection to be examined, e.g., the total number of bytes sent *from source to the destination.*
- Level 4 features can be computed at the end of the connection, but require access to data of potentially many other prior connections. These are temporal and statistical features and are the most costly to compute.

In order to conveniently estimate the cost of a rule, we assign a cost of 1 to the level 1 features, 5 to the level 2 features, 10 to level 3, and 100 to level 4. These cost assignments are very close to the actual measurements we have obtained via extensive real-time experiments. However, we have found that the cost of computing the level 4 features is linearly dependent on the amount of connections being monitored by the IDS within the time window used for computation, as they require iteration of the complete set of recent connections.

In this section, we discuss approaches to reduce computational cost and improve the efficiency of the real-time intrusion detection models. Next, we describe the techniques for *cost-sensitive modeling* and real-time system for implementing *distributed feature computation* for evaluating multiple cost-sensitive models.

25.2.6 Cost-Sensitive Modeling

In order to reduce the computational cost of IDS, detection rules need to use low-cost features as often as possible while maintaining a desired accuracy level. We propose a multiple rule set approach in which each rule set uses the features from different cost levels. Low-cost rules are always evaluated first by IDS, and high-cost rules are used only when low-cost rules cannot predict with sufficient accuracy.

In the domain of network intrusion detection, we use four different levels of costs to compute features, as discussed in the pervious method. Features of costs 1, 5, and 10 are computed individually and features of cost 100 can be computed in a single lookup of all the connections in the past n seconds. With the above costs and goals in mind, we use the following multiple rule set approach:

- We first generate multiple training sets T_{1-4} using different feature subsets. T_1 uses only cost 1 features. T_2 uses the features of costs 1 and 5, and so forth, upto T_4, which uses all available features.
- Rule sets R_{1-4} are learned using their respective training sets. R_4 is learned as an ordered rule set [An ordered rule set is in the form of "**If** condition$_1$ **then** action$_1$ **else if** condition$_2$ **then** action$_2$....**else** action$_n$". The rules are checked sequentially. We typically place rule for the most prevalent class, i.e., *normal,* as the first rule.] for its efficiency, as it may contain the mostly costly features. R_{1-3} are learned as unordered rule sets [An unordered rule set is in the form of "**If** condition$_1$ **then** action$_1$; **if** condition$_2$ **then** action$_2$;......**If** condition$_n$ **then** action$_n$"the rules can be checked in parallel], as they will contain accurate rules for classifying *normal* connections.
- A precision measurement p_r [Precision describes how accurate a prediction is. Precision is defined as $p = |P \cap W||P|$, where P is the set of predictions with label i, and W is the set of all instances with label i in the data set] is computed for *every rule r*, expect for rules in R_4.
- A threshold value τ_I is obtained for every single class, which determines the tolerable precision required in order for a classification to be made by any rule set except for R_4.

In real-time execution, the feature computation and rule evaluation proceed as follows:

- All cost 1 features used in R_1 are computed for the connection being examined. R_1 is then evaluated and prediction time i is made.
- If $p_r > \tau_I$, the prediction i is fired. In this case, no more features are computed and the system examines the next connection. Otherwise, additional features required by R_2 are computed and R_2 is evaluated in the same manner as R_1.
- Evaluation continues with R_3, followed by R_4, until the prediction is made.
- When R_4 (an ordered rule set) is reached, features are computed as needed while evaluation proceeds from the top of the rule set to the bottom. The evaluation of R_4 does not require any firing condition and will always generate a prediction.

In the experiments, they used data from the 1998 DARPA evaluation. The detailed experimental set up and results can be found in W. Lee et al., in 2000. In summary, the multiple model approach can reduce the computational cost by as much as 97% without compromising predictive accuracy, where the cost for inspecting a connection is the total computational cost of all unique features used before a prediction is made. If multiple features of cost 100 are used, the cost is counted only once since they can all be calculated in a single iteration through the table of recent connections.

25.2.7 Distributed Feature Computation

We have implemented a system that is capable of evaluating a set of cost-sensitive models in real time. This system uses a sensor for extracting light-weight, or "primitive" features from raw network traffic data to produce connection records, and then offloads model evaluation and higher level feature computation to a separate entity, called JUDGE. The motivation for offloading this computation and evaluation is that it is quite costly and we do not wish to overburden the sensor (in this case a packet-sniffing engine).

JUDGE uses models that have been learned using the techniques described previously. That is, there exists a sequence of models, each of which uses increasingly more costly features than the previous model. Models are evaluated and higher level features are computed at different points in a connection by JUDGE as more primitive features become available.

The sensor informs JUDGE of new feature values, or updates to feature values that are maintained throughout the connection's life, whenever there is a change in the connection's state (e.g., a connection has gone from SYN_WAIT to CONNECTED). Sensors also update certain feature values whenever there is an "exception" event. Exceptions are certain occurrences, which should immediately update the value of a specific feature. For example, if two fragmented packets come in and the offsets for defragmentation are correct, the *bad_frag_offset* feature must be updated immediately.

Upon each state change and exception event, JUDGE computes the set of features that are available for the given connection. If the set of features is a proper subset of set of light-weighted features, (the level 1 and 2 features described earlier) used by one of the ID models, then higher level features are computed and that model is evaluated. The logic for determining when a prediction is made is the same as described before.

Once a prediction is made by JUDGE a complete connection record, with the label, is inserted into a data warehouse as described in the system architecture outlined previously. We have currently implemented this system using NFR's Network Flight Recorder as the sensor, although the protocol for communication between the sensor and JUDGE would allow any sensor, which extracts features from a data stream to be used.

Usability

A data mining-based IDS is significantly more complex than a traditional system. The main cause for this is that data mining systems require large sets of data from which to train. The hope to reduce the complexity of data mining systems has led to many active research areas.

First, management of both training and historical data sets is a difficult task, especially if the system handles many different kinds of data. Second, once new data has been analyzed, models need to be updated. It is impractical to update models by retraining over all available data, as retraining can take weeks, or even months, and updated models are required immediately to ensure the protection of the systems. Some mechanism is needed to adapt a model to incorporate new information. Third, many data mining-based IDSs are difficult to deploy because they need a large set of clean (i.e., not noisy) labeled training data. Typically the attacks within the data must either be manually labeled for training signature detection models, or removed for training anomaly detection models. Manually cleaning training data is expensive, especially in the context of large networks. In order to reduce the cost of deploying a system, we must be able to minimize the amount of clean data that is required by the data mining process.

We present an approach to each of these problems. We use the technique *adaptive learning,* which is a generic mechanism for adding new information to a model with out retraining. We employ *unsupervised anomaly detection*, which is a new class of intrusion detection algorithms that do not rely on labeled data. In the next section, we present a system architecture, which automates model and data management.

Adaptive Learning

We propose to use ensembles of classification models ($R_{1,...,}R_4$ described earlier is an example of an ensemble of classification models) as a general and algorithm-independent method to adapt the existing models in order to detect newly established patterns. The goal is to improve the efficiency of both learning and deployment. In reality, when a new type of intrusion is discovered, it is very desirable to be able to quickly adjust an existing detection system to detect the new attack, even if the adjustment is temporary and may not detect the new attack.

At the same time, after we have at least some method of defense, we can look for possibly better ways to detect the attack, which involves recomputing the detection model and may take much longer period of time to compute. When a better model is computed, we may choose to replace the temporary model. For such purposes, we seek a "plug-in" method, i.e., we efficiently generate a simple model that is only good at detecting the new intrusion, and plug or attach it to the existing models to enable detection of new intrusions. Essentially, we efficiently generate a lightweight classifier (i.e., classification

- **If** (H_1 (x) = *normal*) V (H_1 (x) = *anomaly*) **then**
  - **if** H_2 (x) = *normal*
    then *output* ← H_1 (x) (*normal or anomaly*)
  - **else** *output* ← *new_intrusion*
- **else** *output* ← H_1**(x)**

Fig. 25.1. Ensemble-based Adaptive Learning Configuration

model) for the new pattern. The existing main detection model remains the same. When the old model detects an anomaly, this data record is sent to the new classifier for further classification. The final prediction is the function of both the old classifier and the new classifier. Computing the new classifier is significantly faster than generating a monolithic model for all established patterns and anomalies.

In one such configuration, given an existing classifier H_1, an additional classifier, H_2, is trained from data containing normal records and records corresponding to the new intrusion. We refer to H_1 as the existing IDS model, while H_2 refers to a new model trained specifically for a new or recently discovered attack. The decision rules in Fig. 25.1 are evaluated to compute the final outcome. This method is independent of the actual model building algorithm. Each classifier can be anything from a decision tree, a rule-based learner, to a neural network, etc.

We have experimented with different configurations to test the effectiveness of this approach. The cost of training of the proposed method (as measured in cost-sensitive modeling) is almost 150 times less expensive than learning a monolithic classifier trained from all available data, and the accuracy of both essentially equivalent.

Unsupervised Learning

Traditional model building algorithms typically require a large amount of labeled data in order to create effective detection models. One major difficulty in deploying a data mining-based IDS is the need for labeling system audit data for the use by these algorithms. For misuse detection systems, the data needs to be accurately labeled as either normal or attack. For anomaly detection system, the data must be verified to ensure it is completely normal, which requires the same effort. Since models (and data) are specific to the environment on which the training data was gathered, this cost of labeling the data must be incurred for each deployment of the system.

Ideally, we would like to build detection models from the collected data without needing to manually label it. In this case, the deployment cost would greatly be decreased because the data would not need to be labeled. In order to build these detection models, we need a new class of model building algorithm. These model-building algorithms can take as input unlabeled data and create a detection model. We call these algorithms unsupervised anomaly detection algorithms.

In this section, we present the problem of unsupervised anomaly detection and relate it to the problem of outlier detection in statistics. We present an overview of two unsupervised anomaly detection algorithms that have been applied to intrusion detection.

These algorithms can be also referred to as anomaly detection over noisy data. The reason the algorithm must be able to handle noise in the data is that we do not want to manually verify that the audit data collected is absolutely clean (i.e., contains no intrusions).

Unsupervised anomaly detection algorithms are motivated by two major assumptions about the data, which are reasonable for intrusion detection. The first assumption is that anomalies are very rare. This corresponds to the fact that normal use of the system greatly outnumbers the occurrence of intrusions. This means that the attacks compose relatively small proportion of the total data. The second assumption is that the anomalies are quantitatively different from the normal elements. In intrusion detection this corresponds to the fact that attacks are drastically different from the normal usage.

Since anomalies are very rare and quantitatively different from the normal data, they stand out as outliers in the data set. Thus, we can cast the problem of detecting the attacks into an outlier detection problem. Outlier detection is the focus of much literature in the field of statistics.

In intrusion detection, intuitively, if the ratios of attacks are different to normal data is small enough, then because the attacks are different, the attacks stand out against the background of normal data. We can thus detect the attack within the data set.

We have performed experiments with two types of unsupervised anomaly detection algorithms, each for a different type of data. We applied probabilistic-based unsupervised anomaly detection algorithm to build detection models over the system calls and a clustering-based unsupervised anomaly detection algorithm for network traffic.

The probabilistic approaches to detect outliers are by estimating the likelihood of each element in the data. We partition the data into two sets, normal elements and anomalous elements. Using a probability-modeling algorithm over the data, we compute the most likely partition of the data.

The clustering approach detects the outliers by clustering the data. The intuition is that the normal data will cluster together because there is a lot of it. Because anomalous data and normal data are very different from each other, they do not cluster together. Since there is very little anomalous data relative normal data, after clustering, the anomalous data will be in the small clusters. The algorithm first clusters the data and then labels the smallest clusters as anomalies.

25.2.8 System Architecture

The overall system architecture is designed to support a data mining-based IDS with the properties described throughout this section. As shown in Fig. 25.2, the architecture consists of sensors, detectors, a data warehouse, and a model generation component. This architecture is capable of supporting data gathering, sharing, and analysis, and also data archiving and models generation and distribution.

The system is designed to be independent of the sensor data format and model representation. A piece of sensor data can contain an arbitrary number of features. Each feature can be continuous or discrete, numerical, or symbolic. In this framework, a model can be anything from a neural network, to a set of rules, to a probabilistic model. To deal with this heterogeneity, an XML encoding is used so each component can easily exchange data and/or models.

The design was influenced by the work in standardizing the message formats and protocols for IDS communication and collaboration: the Common Intrusion Detection Framework (CIDF, funded by DARPA) and the more recent Intrusion Detection Message Exchange Format (IDMEF by the Intrusion Detection Working Group of IETF, the Internet Engineering Task Force). Using CIDF or IDMEF, IDSs can securely exchange attack information, encoded in the standard formats, to collaboratively detect distributed intrusions. In the architecture, data and model exchanged between the components are

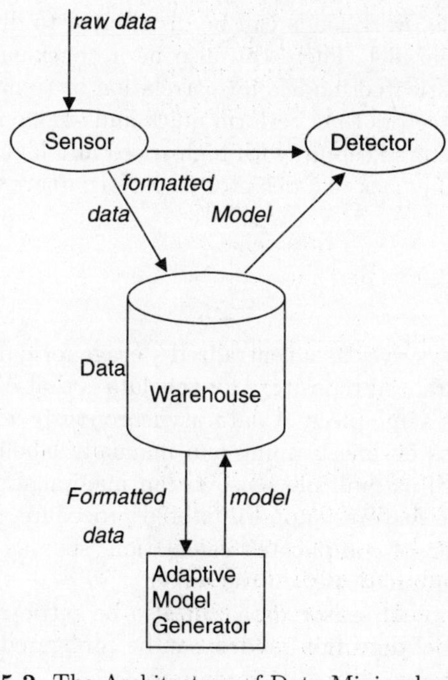

Fig. 25.2. The Architecture of Data Mining-based IDS

encoded in our standard message format, which can be trivially mapped to either CIDF or IDMEF formats. The key advantage of the architecture is its high performance and scalability. That is, all components can reside in the same local network, in which case, the work load is distributed among the components; or the components can be in different networks, in which case, they can also participate in the collaboration with other IDSs in the Internet. In the following sections we describe the components depicted in Fig. 25.2 in more detail.

Sensor

Sensors observe raw data on a monitored system and compute feature for use in model evaluation. Sensors insulate the rest of the IDS from the specific low-level properties of the target system being monitored. This is done by having all of the sensors implement a Basic Auditing Module (BAM) framework. In a BAM, features are computed from the raw data and encoded in XML.

Detectors

Detectors take processed data from sensors and use a detection model to evaluate the data and determine if it is an attack. The detectors also send back the result to the data warehouse for further analysis and report.

There can be several (or multiple layers of) detectors monitoring the same system. For example, work loads can be distributed to different detectors to analyze events in parallel. There can also be a "back-end" detector, which employs very sophisticated models for correlation or trend analysis, and several "front-end" detectors that perform quick and simple intrusion detection. The front-end detectors keep up with high-speed and high-volume traffic and must pass data to the back-end detector to perform more thorough and time-consuming analysis.

Data warehouse

The data warehouse serves as a centralized storage for data and models. One advantage of a centralized repository for the data is that different components can manipulate the same piece of data asynchronously with the existence of a data base, such as off-line training and manually labeling. The same type of components, such as multiple sensors, can manipulate data concurrently. Relational database features support "stored procedure calls," which enable easy implementation of complicated calculation, such as efficient data sampling carried out automatically on the server.

Arbitrary amount of sensor data can also be retrieved by a single SQL query. Distribution of detection models can be configured to push or pull.

The data warehouse also facilitates the integration of data from multiple sensors. By correlating data/results from different IDSs or data collected over a longer period of time, the detection of complicated and large scale attacks becomes possible.

Model Generator

The main purpose of the model generator is to facilitate the rapid development and distribution of new intrusion detection models. In this architecture, an attack detected first as an anomaly may have its exemplary data processed by the model generator, which in turn, using the archived normal and intrusion data sets from the data warehouse, automatically generates a model that can detect the new intrusion and distributes it to the detectors (or any other IDSs that may use these models). Especially useful are unsupervised anomaly detection algorithms because they can operate on unlabeled data that can be directly collected by the sensors.

We have successfully completed a prototype implementation of a data mining and CIDF-based IDS. In this system, a data mining engine equipped with feature extraction programs and machine learning programs serves as the model generator for several detectors. It receives audit data for anomalous events from a detector, computes patterns from the data, compares them with historical normal patterns to identify the "unique" intrusion patterns, and constructs features accordingly. Machine learning algorithms are then applied to compute the detection model, which is encoded as GIDO and sent to all the detectors. Much of the design and implementation efforts had been on extending the common intrusion detection models (CISL). In preliminary experiments the generator is able to produce and distribute new effective models upon receiving audit data.

Related Work

The research encompasses many areas of intrusion detection, data mining, and machine learning. In this section, we briefly compare our approaches with related efforts.

In terms of feature construction for detection models, DC-1 (Detector Constructor, T. Fawcett et al., 1997) first invokes a sequence of operation for constructing features (indicators) before constructing a cellular phone fraud detector (a classifier). We are faced with a more difficult problem here because there is no standard record format for connection or session records (we had to invent our own). We also need to construct temporal and statistical features not just for individual records, but also different connections and services. That is, we are modeling different logical entities that take on different roles and whose behavior is recorded in great detail. Extracting these from a vast and overwhelming stream of data adds considerable complexity to the problem.

The work most similar to unsupervised model generation is a technique developed at SRI in the Emerald system. Emerald uses the historical records to build normal detection models and compares the distributions of new instances to historical distributions. Discrepancies between the distributions signify an intrusion. One problem with this approach is that intrusions present in the historical distributions may cause the system to not detect similar intrusions in unseen data.

Related to automatic model generation is adaptive intrusion detection. Teng et al. (1990) perform adaptive real-time anomaly detection using inductively generated sequential patterns. Also relevant is Sobirey's work on adaptive intrusion detection using an expert system to collect data from audit sources.

Many different approaches to building anomaly detection models have been proposed. Stephanie Forrest presents an approach for modeling normal sequences using look-ahead pairs and contiguous sequences. Helman and Bhangoo present a statistical method to determine sequences, which occur more frequently in intrusion data as opposed normal data. Lee et al. use a prediction model trained by a decision tree applied over the normal data. Ghosh and Schwarzbard use the neural networks to model normal data. Lane and Brodley examine unlabeled data for anomaly detection by looking at user profiles and comparing the activity during an intrusion to the activity under normal use.

Cost-sensitive modeling is an active research area in data mining and machine learning communities because of the demand from application domains such as medical diagnosis and fraud and intrusion detection. Several techniques have been proposed for building optimized for given cost metrics. In our research we study the principles behind these general techniques and develop new approaches according to the cost models specific to IDSs. In intrusion data representation, related work is the IETF Intrusion Detection Exchange Format project and CIDF effort.

25.3 Summary

Although only a few companies currently have data mining systems checking their audit records, the number is expected to dramatically increase in the near future as companies desperately try to ensure that their computer systems are secure from intrusion.

In this section, we have outlined the breadth of the research efforts to address important and challenging issues of accuracy, efficiency, and usability of real-time IDSs.

We have implemented feature extraction and construction algorithms for labeled audit data (i.e., when both normal and intrusion data sets are given). We are implementing algorithms for unlabeled data (which can be purely normal or possibly containing unknown intrusions).

We have developed several anomaly detection algorithms. In particular, we have completed the implementation of and extensive experimentation with "artificial anomaly generation" approaches. We are exploring the use of information-theoretic measures, i.e., entropy, conditional entropy, relative entropy, information gain, and information cost to capture intrinsic characteristics of normal data and use such measures to guide the process of building and evaluating anomaly detection models. We are also developing efficient approaches that use statistics on packet header values for network anomaly detection.

We studied the computational costs of features and models and have implemented a multiple model-based approach for building models that incurs minimal computational cost while maintaining accuracy. We have also developed a real-time system, "Judge," for evaluating models learned using this method.

We are developing adaptive learning algorithms to facilitate model construction and incremental updates. We are also developing unsupervised anomaly detection algorithms to reduce the reliance on labeled training data. We have completed the design and specification of our system architecture with sensor, detector, data warehouse, and modeler components. A prototype system has been implemented and we will continue to build on and experiment with this system.

We are developing algorithms for data mining over the output of multiple sensors. This is strongly motivated by the fact that single sensors do not typically observe entire attack scenarios. By combining the information from multiple sensors we hope to improve detection accuracy.

The ultimate goal of the research is to not only to demonstrate the advantages of our approaches but also to provide useful architectures, algorithms, and tool sets to the community to build better IDSs in less time and with greater ease. Toward this end, we are integrating the feature construction and unsupervised anomaly detection algorithms into the model generator, and building detectors that are equipped with misuse and anomaly detection algorithms. We are deploying the prototype IDS on real-world networks in order to improve the techniques.

A serious limitation of the current approaches (as well as with most existing IDSs) is that we only do intrusion detection at the network or system level. However, with the advent and rapid growth of e-commerce (or e-business) and e-government (or digital government) applications, there is an urgent need to do intrusion and fraud detection at the application level. This is because many attacks may focus on applications that have number effect on the underlying network or system activities. We have previously successfully developed data mining approaches for credit card fraud detection. We plan to start research efforts on IDSs for e-commerce and e-government applications in the near future. We anticipate that we will be able to extend our current approaches to develop application-level IDSs because the system architecture and many of our data mining algorithms are generic (i.e., data format independent). For example, we can develop (and deploy) a sensor for a specific application, and extend the correlation algorithms, with application domain knowledge, in the detectors to combine evidence from the application and the underlying system in order to detect intrusion and frauds. The relevant reports detailing the results described in this case study can be found at http://www.csc.ncsu.edu/faculty/lee/project/id.html, http://www.cs.columbia.edu/ids, and http://www.cs.fit.edu/~pkc/id.

Thus this section has given an overview on data mining in security and real-time data mining-based intrusion detection system case study.

Review Questions

1. How is data mining used in security systems?
2. With a case study explain data mining intrusion detection systems (IDSs).
3. Draw the architecture of data mining based IDS and explain.

APPENDIX-I

Data Mining Research Projects

Abstract. We all live in the information age. The amount of information being collected by businesses, companies, and agencies is large. Recent advances in technologies to automate and improve data collection have only increased the volumes of data. The need for collecting data is to extract useful information. Data mining is primarily the process of knowledge discovery in databases. The data of interest is the previously unknown and potentially useful information contained in the database. Data mining therefore has become a research area of interest to effectively answer to this need. We present the data mining research projects collected from various universities and laboratories.

A.1 National University of Singapore: Data Mining Research Projects

A.1.1 Cleaning Data for Warehousing and Mining

Data from real-world sources are often erroneous, incomplete, and inconsistent, which can result from operator error, system implementation flaws, etc. Such low-quality data is not suitable for effective data mining. Data cleaning has been identified as an important problem. However, little progress has been made thus far. In this project, we study the issues related to data cleaning with the aim of developing an engineering approach that can be useful to the user. The project consists of three phases:

- Identify and categorize the possible errors in data from multiple sources;
- Survey the available and potentially usable techniques to address the problem; and
- Develop a system that can identify and resolve some of the errors. (Contact person. Dr. Lu. Hongjun.)

A.1.2 Data Mining in Multiple Databases

It is common that many databases are kept in an organization. They are collected to serve different purposes. Data mining in individual databases has attracted a lot of attention. Some encouraging results have been achieved. It is time now to consider how we can make use of all the databases in an organization for data mining. Many issues remain unresolved. Significant ones are:

- Will one database help in the data mining of another database?
- How can we consider multiple databases simultaneously for data mining?
- Can current data mining techniques help in this new situation or should we develop new techniques?
- What is so special about mining of multiple databases? (Contact person Dr. Liu Huan.)

A.1.3 Intelligent WEB Document Management Using Data Mining Techniques

With the development of Internet/Web technology, the volume of Web documents increases dramatically. Effective management of the documents is becoming an important issue. The objective of this project is to build an intelligent system that can help Web masters manage Web documents so that they can serve the users better. We first survey the available and potentially useful techniques for discovering access patterns of Web documents stored in an information provider's Web server. The major issues include establishing measurements and heuristics on user access patterns and developing techniques to discover and maintain such discovered patterns. The results are then expanded in the direction of using the discovered user access patterns to manage Web documents so that information subscribers can access information of interest more efficiently. Techniques to be investigated include clustering of web documents, prefetching and caching, and customized linkage of Web documents. (Contact person: Dr. Lu. Hongjun.)

A.1.4 Data Mining with Neural Networks

While the use of neural networks for pattern classification has been common in practice, neural networks have not been widely applied in data mining applications. The reason is that neural network decision process is not easily explainable in terms of rules that human experts can verify. In the past two years, we have investigated the problem of extracting rules from trained neural networks. Our results have been encouraging. The rules extracted by our algorithms are not only more concise than those generated from decision trees, but are, in general, more accurate. (Contact person: Dr. Rudy Setiono.)

A.1.5 Data Mining in Semistructured Data

As the amount of data available on-line grows rapidly, more and more data is semistructured and hierarchical, i.e., the data has no absolute schema fixed in advance, and whose structure may be irregular or incomplete. Semistructured data arises when the source does not impose a rigid structure and when the data is obtained by combining several heterogeneous data sources. As an example of semistructured, hierarchical data source is the Web. Other example of semistrucutured data include: the result of integrating heterogeneous data sources, Bib Tex files, genome databases, drug and chemical structures, libraries of programs, and more generally, digital libraries, and on-line documentation. The goal of this project is to develop a general framework of mining associative patterns from semistructured and hierarchical data. (Contact person: Dr. Wang Ke.)

A.1.6 A Data Mining Application – Customer Retention in the Port of Singapore Authority (PSA)

"Consumer retention" is an increasingly pressing issue in today's ever-competitive commercial arena. This is especially relevant and important for all sales and services related industries. Motivated by this real-world problem in the context of the Port of Singapore Authority (PSA), our work proposed a solution, which integrates the various techniques of data mining, such as decision-tree induction, deviation analysis, and mining multiple concept-level association rules to from an intuitive and novel approach to gauging customer's loyalty and predicting their likelihood of defection. Immediate actions triggered by these "early warnings" are often the keys to the eventual retention of the customer's involved. (Contact person: Dr. Liu Huan, Md. Farhad Hussain, Ng Kian Sing.)

A.1.7 A Belief-Based Approach to Data Mining

The work focused on improving the existing "association rule" discovery technique for practical applications. This was motivated by a collaborative effort with The Ministry of Education (MOE) to discover "interesting" knowledge in the Gifted Education Program (GEP). Although the "Apriori" algorithm can extract a large number of associative patterns, their sole reliance on the objective criteria of "support" and "confidence" for the generation of strong rules has deprived the user of flexibility and control over what kind of rules is "interesting" and specifically required. To overcome the deficiencies, we introduced the subjective criterion of "payoff" and proposed a belief-based data mining framework to incorporate the user's preferences (transient requirements) and beliefs (firm knowledge) into the process of association rule mining. With this, we are also able to find interesting "reliable exceptions," as opposed to only patterns having high predictive accuracy. This approach is further applied into solving the "interestingness" problem for the other tasks of data mining, such as "classification" and "correlation" rule mining. (Contact person: Dr. Liu Huan, Md. Farhad Hussain, Ng Kian Sing.)

A.1.8 Discovering Interesting Knowledge in Database

Knowledge discovery is the extraction of implicit, previously unknown, and potentially useful information from data. There is a growing realization and expectation that data, intelligently analyzed and presented, will be a valuable resource to be used for a competitive advantage. Over the past few years, many techniques have been developed for the discovering of hidden knowledge (which is normally expressed as patterns) in databases. While these techniques were shown to be successful in numerous industrial applications, new problems have also emerged. One of the problems is that it is all too easy to discover a huge number of patterns in a database, and most of these patterns are actually useless or uninteresting to the user. But because of the huge number, it is difficult for the user to comprehend and to identify those patterns that are useful to him/her. To prevent the user from being overwhelmed by the large number of patterns, techniques are needed to identify those interesting ones and/or to rank all the patterns according to their interestingness. In this project, we study this problem and design techniques to perform the identification and/or the ranking tasks. (Contact person: Dr. Liu Bing.)

A.1.9 Data Mining for Market Research

Data mining is the extraction of implicit, previously unknown, and potentially useful information from data. It is now well realized that data, intelligently analyzed and presented, will be a valuable resource to be used for a competitive advantage. In marketing research, a marketing manager typically analyzes customer data in order to target the right customers for his company's products, to retain the profitable customers, and to attract new customers. There is a growing field of research called *database marketing* that makes use of data mining techniques to help marketing managers, to analyze the data to segment the market, to position their products, and to study customers buying behaviors and patterns. (Contact person: Dr. Liu Bing.)

A.1.10 Data Mining in Electronic Commerce

Electronic commerce is the cutting edge of business today. It allows consumers and companies to buy and sell on the computer network. Clearly, with the electronic commerce, a huge amount of information can be generated and collected, e.g., information about how customers search for products, what they buy, how they make purchase decisions, etc. Such information can be analyzed and mined to discover interesting patterns. These patterns can then be used to better serve customers and to help marketing and advertising activities. (Contact person: Dr. Liu Bing.)

A.1.11 Multidimensional Data Visualization Tool

The amount of data stored in databases is growing very rapidly. It has been estimated that the amount of information in the world doubles every 20 months. With this flood of data, there is a growing realization and expectation that data, intelligently analyzed and presented, will be a valuable resource to gain a competitive advantage. Many approaches have been proposed ranging from artificial intelligence techniques to visualization techniques. In essence, data visualization techniques extend the capability of traditional business graphics to support the visualization of multidimensional data. It uses graphical methods to transform multidimensional databases into images, which allow patterns inherent in the data to reveal themselves. In this project, the student is first to survey the existing data visualization techniques. A subset of the techniques will be selected for implementation. These techniques will be tested against a few real-life databases. As an extension of this project, the student is expected to derive and implement an improved visualization tool for the real-life databases. (Contact person: Dr. Wynee Hsu.)

A.1.12 Clustering Algorithms for Data Mining

Knowledge discovery is the extraction of implicit, previously unknown, and potentially useful information from data. Databases have now been used almost in every area. The amount of data stored in databases is growing very rapidly. With this flood of data, clearly little of it will be seen by human eye. If it will be understood at all, it will have to be analyzed by computers. One interesting aspect that can be gathered from data is the clustering information. Clustering tries to identify a finite set of categories or clusters in the data. It groups data tuples together according to some similarity measures. Clustering is seen as the discovery of "natural" classes in data. These classes reflect basic causal mechanisms that make some tuples look more like each other than the rest of the data tuples. In this project, the student is first to study and implement several well-known clustering algorithms. He/she will then conduct experiments to determine how well the algorithms performed on different databases. As an extension of this project, the student will investigate how to improve the efficiency and accuracy of the clustering algorithms. (Contact person: Dr. Wynee Hsu.)

A.1.13 Web Page Design for Electronic Commerce

Hon-part: In electronic commerce, well-designed Web pages will attract more customer and business, and shorten on-line search time. In this project, we propose a framework for designing Web pages of directories/catalogs of products. In the first stage, we propose an algorithm that, given a natural categorization of products/services and frequencies of accessing product, finds the optimal design of the Web pages for the products. We try to make this a general framework. In the second stage, we implement this algorithm and test its performance. Finally, we evaluate the framework. (Contact person: Dr. Wang Ke.)

A.1.14 Data Mining Application on Web Information Sources

Hon-part: Data mining was ranked one of the top ten technologies in a recent survey. Though most information are in the form of Web documents, especially those for electronic commerce, most data mining tools are only designed for well-structured databases. This project looks into how the traditional data mining tools can be used for Web information sources, and what extension is needed. In other words, we address the issue of enlarging the application domain of data mining technologies to those generated by electronic market place and commerce. In some cases, existing data mining tools can be used. In some other cases, data has to be preprocessed first. There are some initial observations and ideas on how these can be done. As an extension of this project, this part enhances the flexibility and application of the proposed framework. For example, by adding more components or removing assumptions, we can apply the data mining tool designed to more diversifying information sources, thus, make the tool more practically useful. (Contact person: Dr. Wang Ke.)

A.1.15 Data Mining in Finance

Data mining was ranked one of the top ten technologies in a recent survey. Today, data mining is being increasingly applied in finance to support financial asset management and risk management. Its use in computational finance will have a major impact in the modeling of currency markets, in tactical asset allocation, bound and stock valuation and portfolio optimization.

We start with a survey in data mining technologies and the background of finance. We then identify the areas of finance in which data mining is particularly useful. Furthermore, we also look into new requirements and issues of data mining in finance that are not typical of data mining in other fields, such as time series requirements. Through this study, the student should be able to argue whether price changes is predictable, and know how to apply a suitable data mining tool in a certain situation, how to evaluate the result. This will be done by one case study. (Contact person: Dr. Wang Ke.)

A.1.16 Document Summarization

In the electronic marketplace, one of the daily operations is summarizing text documents according to different requirements. For example, a progress report highlighting achievements may be read by the department head, a detailed version stressing research plans and potential applications may be read by the research grant committee, a condensed statement stressing ongoing activities and images may be released to visitors and the press.

We study a framework in which summarization can be automatic. A starting point could be some requirements on the raw text from which the summarization is performed. As new cases are presented, the system should learn new patterns of different types of summarization. This can be done by involving some user interaction. An interface will be built for presenting the summarized documents and user instructors. We apply to summarize some real documents and to verify its effectiveness. (Contact person: Dr. Wang Ke.)

A.1.17 Data Mining and Intelligent Data Analysis

Since data mining is still at relatively new research field, a great deal of research and applications remains to be done. In this project, we aim to push this technology further, and make National University of Singapore one of the key contributors of this technology and Singapore industry one of the main users.

Data mining has been recognized as an important technology for businesses internationally. Locally, there are many companies in Singapore that are interested in this technology. Few, however, have made much progress. One key reason is that initial cost is high and no expertise is available. With this project, we will have a pool of experts (with more to be trained) readily available to help the local industry. In the past month, we conducted a survey among some of our major local companies. The survey results show that research done at the National University of Singapore. Half of the companies are also willing to support and to participate in this proposed project as collaborators. Thus, the main purposes of this project are:

- To develop new data mining techniques, and to improve upon existing techniques.
- To build data mining tools (both generic and industry specific ones) that can be readily used by industry users.
- To establish ourselves as a center of expertise for data mining research and applications both locally and internationally.

Proposed Research Program

This proposed project consists of three main parts: basic and applied research, data mining tool developments, and seminars/workshops for local industry. Each of these parts will not be worked on in isolation. Instead each part will complement the others. Below, we briefly describe the three parts.

Basic and Applied Research

In research, the group will investigate the following main projects simultaneously.

Data Cleaning

Data from real-world sources are often erroneous, incomplete, and inconsistent, which can result from operator error, system implementation flaws, etc. Such low-quality data is not suitable for effective data mining. Data cleaning has been identified as an important problem. However, little progress has been made thus far. In this project, we study the issues related to data cleaning with the aim of developing an engineering approach that can be useful to the user. The project consists of three phases: (1) identify and categorize the possible errors in data from multiple sources, (2) survey the available and potentially usable techniques to address the problem; and (3) develop a system that can identify and resolve some of the errors.

Data Mining in Multiple Databases

It is common that many databases are kept in an organization. They are collected to serve different purposes. Data mining in individual databases has attracted a lot of attention. Some encouraging results have been achieved. It is time now to consider how we can make use of all the databases in an organization for data mining. Many issues remain unresolved. Significant ones are:

(1) Will one database help in the data mining of another database?
(2) How can we consider multiple databases simultaneously for data mining?
(3) Can current data mining techniques help in this new situation or should we develop new techniques?
(4) What is so special about mining of multiple database?

Intelligent WEB Document Management Using Data Mining Techniques

With the development of Internet/Web technology, the volume of web documents has increased dramatically. Effective management of the documents is becoming an important issue. The objective of this project is to build an intelligent system that can help Web masters manage Web documents so that they can serve the users better. We first survey the available and potentially useful techniques for discovering access patterns of Web documents stored in an information provider's Web server. The major issues include establishing measurements and heuristics on user access patterns and developing techniques of using the discovered user access patterns and developing techniques to discover and maintain such discovered patterns. The results are then expanded in the direction of using the discovered user access patterns to manage Web documents so that information subscribers can access information of interest more efficiently. Techniques to be investigated include clustering of Web documents, prefetching and caching, and customized linkage of Web documents.

Data Mining with Neural Networks

While the use of neural networks for pattern classification has been common practice, neural networks processes have not been widely applied in data mining applications. The reason is that neural network decision process is not easily explainable in terms of rules that human experts can verify. In the past two years, we have investigated the problem of extracting rules from trained neural networks. Our results have been encouraging. The rules extracted by our algorithms are not only more concise than those generated from decision trees, but are, in general, more accurate. Our algorithms can extract rules of the following forms:

Symbolic rules, e.g., if (married = yes) and (sex = male), then .....
MofN rules, e.g., if 2 of the 3 conditions {number of children is not more than 3, married more than 10 years, owns private property} are satisfied, then .....
Oblique rules e.g., if (monthly salary – 1.5* monthly mortgage), then.....

In this project, we plan to implement these algorithms on a 32 processor Fujitsu AP3000 parallel computer to speed up the training process.

Data Mining in Semistructured Data

As the amount of data available on-line grows rapidly, more and more data is semistructured and hierarchical, i.e., the data has no absolute schema fixed in advance, and whose structure may be irregular or incomplete. Semistructured data arises when the source does not impose a rigid structure and when the data is obtained by combining several heterogeneous data sources. An example of semistructured, hierarchical data source is the Web. Other examples of semistructured data include: the result of integrating heterogeneous data sources, Bib Tex files, genome databases, drug and chemical structures, libraries of programs, and more generally, digital libraries, and on-line documentation. Since there is no fixed schema in semistructured data, the conventional data mining techniques that work with feature vector representation will not be applicable. The goal of this project is to develop a general framework of mining associative patterns from semistructured and hierarchical data.

Subjective Interestingness of the Discovered Patterns

Current research in data mining mainly focuses on the discovery algorithms and visualization techniques. There is a growing awareness that, in practice, it is easy to discover a huge number of patterns in a database where most of these patterns area actually obvious, redundant, and useless or uninteresting to the user. To prevent the user from being overwhelmed by a large number of uninteresting patterns, techniques are needed to identify only the useful/interesting patterns and present them to the user. We have been working on this problem for the past two years. In this project, we consolidate our findings and carry the research further to produce more effective techniques to discover or to identify interesting patterns.

Data Mining Tool Developments

Parallel to our research effort, we have a group of programmers to develop a set of data mining tools. The objective of this part of the program is to put research into practice, i.e., building the next generation data mining systems that can be readily used by industry users.

Mining from Image Databases

Images are a powerful means of conveying information to human. Hence, many real-life applications involve processing and analyzing a large number of images. For example, in the terrain-matching applications, we have thousands of images that are returned by the satellite, which need to be processed and mapped; in the archeology domain, all ancient artifacts are photographed and stored for efficient retrieval later on; in the medical domain, images such as mammograms, ultrasound images, X-ray images, MRI images are already a standard part of health care industry. In spite of the widespread use of images, there is no effective query/retrieval technique for general image databases – the main reason being that we do not understand the underlying relationships between the low-level image features (that can be extracted using our current image processing techniques) and the high-level image constructs that a human typically recognizes. In this project, we explore the feasibility of adapting data mining techniques on image databases.

A.2 HP Labs Research: Software Technology Laboratory

A.2.1 Data Mining Research

Data mining is the process of finding interesting and valuable patterns in large datasets. The discovery of these patterns draws upon such research fields as machine learning, statistics, databases, information retrieval, and information visualization, to name a few.

Emerging e-commerce activities have created an unprecedented need for capabilities to discover, understand, and organize information and services on the Internet. We are developing capabilities to intelligently manage structured and unstructured information and distill nuggets of knowledge from them, so as to revolutionize customer relationship management and enable innovative supply chain structures and practices. We work with HP business units as well as leading-edge external customers to identify exciting research problems and high-potential application opportunities. With a focus on building technologies to enable Smart Enterprises in the Internet age, we develop tools, technologies, and applications, which become accessible as e-services.

Let us illustrate with some of the areas we are working on:

Data Mining and Machine Learning

- Cost and distribution sensitive learning

In real-world applications of data mining, we often face domains in which class distributions are greatly skewed and classification error costs are unequal. In these situations, evaluating classifiers is difficult because classification accuracy, the metric by which most evaluation is currently done, is misleading and inadequate. To make matters worse, class distributions in these domains may drift over time, and error costs may be known only approximately. We have developed a framework, based on ROC analysis, which enables us to analyze and visualize performance separately from assumptions about class distributions and error costs. This allows us to create robust machine learning systems that can operate in changing, uncertain environments.

- Genetic Programming

Genetic programming is a machine-learning techniques that uses the metaphor of natural selection to evolve arbitrarily complex programs that solve specific problems. Here at Labs, we have developed (and are continuing to refine) GPLab, a robust, powerful, and, above all, flexible platform for genetic programming research, experimentation, and application. We are also conducting fundamental research aimed at expanding the power of the genetic programming paradigm as well as applying the technique to problems such as the customer-specific optimization of Web-based interactions.

- Clustering Algorithms

Clustering algorithms are useful in application such as customer segmentation and text mining. Some of our research efforts are aimed at designing new clustering algorithms. For example, we have developed K-harmonic means, which has been shown to be more robust when compared with the industry-leading K-harmonic means robustness is due to the discovery of an objective function that enables the search algorithm to gracefully break out of local optimums that conventional algorithms are not capable of doing.

Most clustering algorithms are not designed to work with data that exceeds the size of main memory. Since we tend to apply these algorithms to very large datasets, we have studied scalability issues. Some of our research investigates parallelizing clustering algorithms. Other work involves divide-and-conquer style clustering in the data stream model of computation, as well as investigating the sample complexity of clustering.

- Uncertainty and Difficulty of Learning

In most classification, diagnosis, and prediction problems, uncertainty plays a significant role. We are interested in techniques for modeling this uncertainty explicitly, for example, using Bayesian networks. We have extended techniques for learning probabilistic models from operational (and therefore dirty) data, and we have developed validation and explanation methods for Bayesian networks. In additional, we are investigating models of cost or utility in our solutions to these problems. Cost-sensitive learning, ROC analysis, multiattribute utility theory, and decision-theoretic predictive models are among some of our areas of interest.

We are also investigating the question of "difficulty" of learning, i.e., to what extent we can be sure that there are significant patterns hidden in the dataset at all, such that we are presenting a meaningful task to any learning algorithm.

Visualization, OLAP and Commerce Automation

- Visualization Techniques

We rely heavily on visualization techniques to guide researchers' intuition, and to provide much more intuitive ways for users to understand the mining results. We are developing an information visualization platform, on which templates for visualizing large amount of data can be plugged in and assembled quickly to build visualization solutions.

We use several innovative techniques on this platform: (1) hidden visual relationship and structures for un-cluttering displays; (2) simultaneous visual presentations for high-dimensional knowledge discovery; and (3) self-organized directed association visualization 3D system for items associations. These techniques have been applied to several data mining visualization applications, such as topic hierarchy for document navigation, telecom switch mining, Web-based trouble shooting, and e-business market basket analysis.

- OLAP (On-Line Analytical Processing)

OLAP is a technology for large-scale, multilevel, and multidimensional data analysis. At HP Labs we are developing a data warehouse/OLAP framework. Based on this framework, we are tackling issues like scaling the whole operation chain, including data cleaning, loading, maintenance, access, and analysis. Our framework is characterized by using OLAP servers as scalable computation engines rather than using them purely as front-end analytical tools. Special emphasis is placed on the scalability, incremental analysis, OLAP programming, and OLAP-based data mining such as generating extended multilevel and multidimensional association rules. Applications of this work include telecommunication traffic analysis, fraud detection, and calling pattern generation; e-commerce service and customer behavior analysis and profiling; and Web access analysis.

- Commerce Automation

E-commerce applications involve parallel, heuristic, and dynamically varied tasks that are orchestrated to facilitate automated, dynamic interenterprise collaboration. To support such applications, we are developing a dynamic collaboration infrastructure called *Common Business Automation* (CBA) based on a dynamic software agent framework.

Text Mining and Feature Engineering

- We are developing a complete framework for text mining, which takes into account text cleaning and feature engineering. For example, we are investigating tools that help build a thesaurus of synonyms of terms specific to a technical domain. We are learning the correlation between short, cryptic descriptions of symptoms captured by customer support engineers and the final diagnoses of the parts replaced. We are developing methods to enable an enterprise to profile their customers better by categorizing and learning form their dialogs on the Web or with service agents. Finally, we are also investigating methods for categorizing and clustering technical and operational documents.

A.3 CRISP-DM: An Overview

A.3.1 Moving from Technology to Business

Despite the development of technology to support a huge database, the rapid spread of computerization in all industries presents users with the problem of interpreting vast amounts of data. Data mining provides a way of unlocking the value hidden within. Although algorithms and tools abound, data mining at present is more of an art than a well-understood, reliable process. There exists no generally available, practical data mining process. This particularly hinders data mining projects involving huge databases and is a major barrier to infrastructure adoption of data mining by large corporate users.

CRISP-DM – "CRoss-Industry Standard Process for Data Mining" – moves away from this focus on technology by addressing the needs of all levels of users in developing data mining technology to solve business problems. Starting from the embryonic knowledge discovery processes used in industry today and responding directly to user requirements, this project defined and validated a data mining process that is generally applicable in diverse industry sectors. This will make large data mining projects faster, more efficient, more reliable, more manageable, and less costly. CRISP-DM has been kept sufficiently lightweight, however, to benefit even small-scale data mining investigations. Issues addressed include:

- Mapping from business issues to data mining problems
- Capturing and understanding data

- Identifying and solving problems within the data
- Applying data mining techniques
- Interpretation of data mining results within the business context
- Deployment and maintenance of data mining results
- Capture and transfer of expertise to ensure future projects benefit from experience

As well as providing a process structure for carrying out data mining, the project also aims to provide guidance on potential problems and solutions, which can occur in mining projects.

A.3.2 Process Model

The current process model for data mining provides an overview of the life cycle of a data mining project. It contains the corresponding phases of a project, their respective tasks, and relationships between these tasks. At this description level, it is not possible to identify all relationships. There possibly exist relationships between all data mining tasks depending on goals, background, and interest of the user, and most importantly depending on the data. An electronic copy of the CRISP-DM version 1.0 process guide and user manual is available free of charge. This contains step-by-step directions, tasks, and objectives for each phase of the data mining process.

The life cycle of a data mining project consists of six phases. The sequence of the phases is not strict. Moving back and forth between different phases is always required. It depends on the outcome of each phase, or which particular task of a phase, that has to be performed next.

Below follows a brief outline of the phases:

- Business understanding

This initial phase focuses on understanding the project objectives and requirements from a business perspective, and then converting this knowledge into a data mining problem definition, and a preliminary plan designed to achieve the objectives.

- Data understanding

The data preparation phase starts with an initial data collection and proceeds with activities in order to get familiar with the data, to identify data quantity problems, to discover first insight into the data, or to detect interesting subsets to form hypotheses for hidden information.

- Data preparation

The data preparation phase covers all activities to construct the final dataset (data that will be fed into the modeling tools(s)) from the initial raw data.

Data preparation tasks are likely to be performed multiple times, and not in any prescribed order. Tasks include table, record, and attribute selection, as well as transformation and cleaning of data modeling tools.

- Modeling

In this phase, various modeling techniques are selected and applied, and their parameters are calibrated to optimal values. Typically, there are several techniques for the same data mining problem type. Some techniques have specified requirements on the form of data. Therefore, stepping back to the data preparation phase is often needed.

- Evaluation

At this stage in the project you have built a model (or models) that appears to have high quality, from a data analysis perspective. Before proceeding to final deployment of the model, it is important to more thoroughly evaluate the model, and review the steps executed to construct the model, to be certain it properly achieves the business objectives. A key objective is to determine if there are some important business issues that have not been sufficiently considered. At the end of this phase, a decision on the use of the data mining results should be reached.

- Deployment

Creation of the model is generally not the end of the project. Even if the purpose of the model is to increase knowledge of the data the knowledge gained will need to be organized and presented in a way that the customer can use it. Depending on the requirements, the deployment phase can be as simple as generating a report or as complex as implementing a repeatable data mining process. In many cases it will be the customer, not the data analyst, who will carry out the deployment steps. However, even if the analyst will not carry out the deployment effort it is important for the customer to understand up front what actions will need to be carried out in order to actually make use of the created models.

A.4 Data Mining SuiteTM

The Data Mining SuiteTM is truly unique, providing the most powerful, complete, and comprehensive solution for enterprise-wide, large-scale decision support. It leads to the world of discovery with the exceptional ability to directly mine large multitable SQL databases.

The Data Mining Suite works directly on large SQL repositories with number need for sampling or extract files. It accesses large volumes of multi-table relational data on the server, incrementally discovers powerful patterns, and delivers automatically generated English text and graphs as explainable documents on the intranet.

The Data Mining Suite is based on a solid foundation with a total vision for decision support. The three-tiered, server-based implementation provides highly scalable discovery on huge SQL databases with well over 90% of the computations performed directly on the server, in parallel if desired.

The Data Mining Suite relies on the genuinely unique mathematical foundation we pioneered to usher in a new level of functionality for decision support. This mathematical foundation has given rise to novel algorithms that work directly on very large datasets, delivering unprecedented power and functionality. The power of these algorithms allows us to discover rich patterns of knowledge in huge databases that could have never been found before.

With server-based discovery, the Data Mining Suite performs over 90% of the analyses on the server, with SQL, C programs, and Java. Discovery takes place simultaneously along multiple dimensions on the server and is not limited by the power of the client. The system analyzes both relational and multidimensional data, discovering highly refined patterns that reveal the real nature of the dataset. Using built-in advanced mathematical techniques, these findings are carefully merged by the system and the results are delivered to the user in plain English, accompanied by tables and graphs that highlight the key patterns.

The Data Mining Suite pioneered multidimensional data mining. Before this, OLAP had usually been a multidimensional manual endeavor while data mining had been a single dimensional automated activity. The Rule-based Influences Discovery SystemTM bridged the gap between OLAP and data mining. This dramatic new approach forever changed the way corporations use decision support. No longer are OLAP and data mining viewed as separate activities, but are fused to deliver maximum benefit. The patterns discovered by the system include multidimensional influences and contributions, OLAP affinities and associations, comparisons, trends, and variations. The richness of these patterns delivers unparalleled business benefits to users, allowing them to make better decisions than ever before.

The Data Mining Suite also pioneered the use of incremental pattern-base population. With incremental data mining the system automatically discovers changes in patterns as well as the patterns of change. For instance, each month sales data is mined and the changes in the sales trends as well as the trends of change in how products sell together are added to the pattern base. Over time, this knowledge becomes a key strategic asset to the corporation.

The Data Mining Suite currently consists of these modules:

- Rule-based Influence Discovery
- Dimensional Affinity Discovery
- Trend Discovery Module
- Incremental Pattern Discovery
- Forensic Discovery
- The Predictive Modeler

These truly unique products are all designed to work together, and in concert with the Knowledge Access SuiteTM.

A.4.1 Rule-based Influence Discovery

The Rule-based Influence Discovery System is aware of both influences and contributions along multiple dimensions and merges them in an intelligent manner to produce very rich and powerful patterns that cannot be obtained by either OLAP or data mining alone. The system performs multitable, dimensional data mining at the server level, providing the best possible results. The Rule-based Influence Discovery System is not a multidimensional repository, but a data mining system. It accesses granular data in a large database via standard SQL and reaches for multidimensional data via a ROLAP approach of the user's choosing.

A.4.2 Dimensional Affinity Discovery

The Affinity Discovery System automatically analyzes large datasets and finds association patterns that describe how various items "group together" or "happen together." Flat affinity just tells us how items group together, without providing logical conditions for the association. Dimensional (OLAP) affinity is more powerful and describes the dimensional conditions under which stronger item groupings take place. The Affinity Discovery System includes a number of useful features that make it a unique industrial strength product. These features include hierarchy and cluster definitions, exclusion lists, unknown-value management, among others.

A.4.3 The OLAP Discovery System

The OLPA Discovery System is aware of both influences and contributions along multiple dimensions and merges them in an intelligent manner to produce very rich and powerful patterns that cannot be obtained by either OLAP or data mining alone. The system merges OLAP and data mining at the server level, providing the best possible results. The OLAP Discovery System is not an OLAP engine or a multidimensional repository, but a data mining system. It accesses granular data in a large database via standard SQL and reaches for multidimensional data via an OLAP/ROLAP engine of the user's choosing.

A.4.4 Incremental Pattern Discovery

Incremental pattern Discovery deals with temporal data segments that gradually become available over time, e.g., once a week, once a month, etc. Data is periodically supplied to the Incremental Discovery System in terms of a "data snap-shot," which corresponds to a given time-segment, e.g., monthly sales figures. Patterns in the data snapshot are found on a monthly basis and are added to the pattern base. As new data becomes available (say once a month) the system automatically finds new patterns, merges them with the previous patterns, stores them in the pattern base, and notes the differences from the previous time periods.

A.4.5 Trend Discovery

Trend Discovery with the Data Mining Suite uncovers time-related patterns that deal with change and variation of quantities and measures. The system expresses trends in terms of time-grains, time-windows, slopes, and shapes. The time-grain defines the smallest grain of time to be considered, e.g., a day, a week, or a month. Time-windows define how time grains are grouped together, e.g., we may look at daily trends with weekly windows, or we may look at weekly grains with monthly windows. Slopes define how quickly a measure is increasing or decreasing, while shapes give us various categories of trend behavior, e.g., smoothly increasing vs. erratically changing.

A.4.6 Forensic Discovery

Forensic Discovery with the Data Mining Suite relies on automatic anomaly detection. The system first identifies what is usual and establishes a set of norms through pattern discovery. The transactions or activities that deviate from the norm are then identified as unusual. Business users can discover where unusual activities may be originating and the proper steps can be taken to remedy and control the problems. The automatic discovery of anomalies is essential in that ingenious tactics used to spread activities within multiple transactions usually cannot be guessed beforehand.

A.4.7 Predictive Modeler

The Data Mining Suite Predictive Modeler makes predictions and forecasts by using the rules and patterns, which the data mining process generates. While induction performs pattern discovery to generate rules, the Predictive Modeler performs pattern matching to make predictions based on the application of these rules. The predictive models produced by the system have higher accuracy because the discovery process works on the entire dataset and need not rely on sampling.

The output from the seven-component products of the Data Mining Suite is stored within the pattern base and is accessible with PQL: The Pattern Query LanguageTM. Readable English text and graphs are automatically generated in ASCII and HTML formats for the delivery on the inter/intranet.

The Data Mining Suite is distinguished by the following unique capabilities:

- *Direct access to very large SQL databases*

The Data Mining Suite works directly on very large SQL databases and does not require samples, extracts, and/ or flat files. This alleviates the problems associated with flat files which lose the SQL engine's power (e.g., parallel execution) and which provide marginal results. Another advantage of working on an SQL database is that the Data Mining Suite has the ability to deal with both numeric and non-numeric data uniformly. The Data Mining Suite does not fix the ranges in numerical data beforehand, but finds ranges in the data dynamically by itself.

- *Multi-Table Discovery*

The Data Mining Suite discovers patterns in multitable SQL databases without having to join and build an extract file. This is a key issue in mining large databases. The world is full of multitable databases, which cannot be joined and meshed into a single view. In fact, the theory of normalization came about because data needs to be in more than one table. Using single tables is an affront to all the work of E.F. Codd on database design. If you challenge the DBA in a really large database to put things in a single table you will either get a laugh or a blank stare – in many cases the database size will balloon beyond control. In fact, there are many cases where no single view can correctly represent the semantics of influence because the ratios will always be off regardless of how you join. The Data Mining Suite leads the world of discovery with the unique ability to mine large multitable databases.

- *No Sampling or Extracts*

Sampling theory was invented because one could not have access to the underlying population being analyzed. But a warehouse is there to provide such access. As pointed out in the paper "Small Data, Small Knowledge" sampling may cause various problems for data mining. The Data Mining Suite prefers to work on the whole dataset, AND can also work on samples based on user insistence – which is not recommended. After all, the data is there, let us use it.

- *General and Powerful Patterns*

The format of the patterns discovered by the Data Mining Suite is very general and goes far beyond decision trees or simple affinities. The advantage to this is that the general rules discovered are far more powerful than decision trees. Decision trees are very limited in that they cannot find all the information in the database. Being rule-based keeps the Data Mining Suite from being constrained to one part of a search space and makes sure that many more clusters and patterns are found – allowing the Data Mining Suite to provide more information and better predictions. (Also see the paper "Rules are Much More than Decision Trees".)

- *Language of Expression*

The Data Mining Suite has a powerful language of expression, going several times beyond what most other systems can handle. For instance, for logical statements it can express statements such as "IF Destination State = Departure State THEN..." or "IF State is not Arizona THEN..." Surprisingly most other data mining systems cannot express these simple patterns. And the Data Mining Suite pioneered dimensional affinities such as IF Day = Saturday WHEN Paint Brush is purchased ALSO Paint is purchased." Again most other systems cannot handle this obvious logic. (See the paper "Data Mining with OLAP Affinities" for more details.)

- *Uniform Treatment of Numeric and Non-numeric Data*

The Data Mining Suite is unique in its ability to deal with various data types in a uniform manner. It can smoothly deal with a large number of non-numeric values and also automatically discovers ranges within numeric data. Moreover, the Data Mining Suite does not fix the ranges in numerical data but discovers interesting ranges by itself. For example, given the field Age, the Data Mining Suite does not expect this to be broken into 3 segments of (1–30), (31–60), (61 and above). Instead it may find two ranges such as (27–34) and (48–61) as important in the dataset and will use these in addition to the other ranges.

- *Use of Data Dependencies*

Should a data mining system be aware of the functional (and other dependence) that exist in a database? "Yes" and very much so. The use of these dependencies can significantly enhance the power of a discovery system – in fact ignoring them can lead to confusion. The Data Mining Suite takes advantage of data dependencies.

- *Server-based Architectures*

The Data Mining Suite has a three-level client server architecture whereby the user interface runs on a thin intranet client and the back-end process for analysis is done on a Unix server. The majority of the processing time is spent on the server and these computations run both by using parallel SQL and non-SQL calls managed by the Data Mining Suite itself. Only about 50% of the computations on the server are SQL based and the other statistical computations are already managed by the Data Mining Suite program itself at time by starting separate processes on different nodes of the server.

- *System Initiative*

The Data Mining Suite uses system initiative in the data mining process. It forms hypothesis automatically based on the character of the data and converts the hypothesis into SQL statements forwarded to the RDBMS for execution. The Data Mining Suite then selects the significant patterns filter the unimportant trends.

- *Transparent Discovery and Predictions*

The Data Mining Suite provides explanations as to how the patterns are being derived. This is unlike neural nets and other opaque techniques in which the mining process is a mystery. Also, when performing predictions, the results are transparent. Many business users insist on understandable and transparent results.

- *Not Noise Sensitive*

The Data Mining Suite is not sensitive to noise because internally it uses fuzzy logic analysis. As the data gathers noise, the Data Mining Suite will only reduce the level on confidence associated with the results provided. However, it will still produce the most significant findings from the dataset.

- *Analysis of Large Databases*

The Data Mining Suite has been specifically tuned to work on databases with an extremely large number of rows. It can deal with datasets of 50 to 100 million records on parallel machines. It derives its capabilities from the fact that it does not need to write extracts and uses SQL statements to perform its process. Generally the analyses performed in the Data Mining Suite are performed on about 50 to 120 variables and 30 to 100 million records directly. It is, however, easier to increase the number of records based on the specific optimization options with the Data Mining Suite to deal with very large databases. These unique features and benefits make the Data Mining Suite the ideal solution for large-scale data mining in business and industry.

A.5 The Quest Data Mining System, IBM Almaden Research Center, CA, USA

A.5.1 Introduction

The goal of the quest project at the IBM Almaden Research Center is to develop technology to enable a new breed to data-intensive decision-support applications. This section is a capsule summary of the current functionally and architecture of the quest data mining systems.

Our overall approach has been to identify basic data mining operations that cut across applications and develop fast, scalable algorithms for their execution. We wanted our algorithms to:

- *Discover* patterns in very large databases, rather than simplify verify that a pattern exists:
- Have a *completeness* property that guarantees that all patterns of certain types have been discovered:
- Have high performance and near-linear scaling on very large (multiple gigabytes) real-life database.

We discuss the operations of discovering association rules, sequential patterns, time-series clustering, classification, and incremental mining. Further information about Quest can be obtained from http://www.almaden.ibm.com/cs/quest. IBM is making the Quest technology commercially available through the data mining product, IBM intelligent Miner.

A.5.2 Association Rules

We introduced the problem of discovering association rules in Agrawal, Imicelinski, & Swami 1993b. Given a set of transactions, where each transaction is a set of literals (called items), an association rule is an expression of the form X ⇒ Y, where X and Y are sets of items. The intuitive meaning of such a rule is that transactions of the database, which contain X tend to contain Y. An example of an association rule is: "30% of transactions that contain beer also contain diapers; 2% of all transactions contain both of these items." Here 30% is called the *confidence* of the rule, and 2% the *support* of the rule. The problem is to find all association rules that satisfy user-specified minimum support and minimum confidence constraints. Applications include discovering affinities for market basket analysis, store layout, customer segmentation based on buying patterns, etc.

A.5.3 Apriori Algorithm

The problem of mining association rules is decomposed into two subproblems.

- Find all combinations of items that have transaction support above minimum support. Call those combinations *frequent* itemsets.
- Use the frequent itemsets to generate the desired rules. The general idea is that if, say $ABCD$ and AB are frequent itemsets, then we can determine if the rule $AB \Rightarrow CD$ holds by computing the ratio r = support $(ABCD)$/support (AB). The rule holds only if $r \geq$ minimum confidence. Note that the rule will have minimum support because $ABCD$ is frequent.

The Apriori algorithm (Agrawal & Srikant 1994) used in Quest for finding all frequent itemsets is given in Fig. A.8.

It makes multiple passes over the database. In the first pass, the algorithm simply counts item occurrence to determine the frequent 1-itemsets (item sets with 1 item). A subsequent pass, say pass k_1, consists of two phases. First, the frequent itemsets $\mathrm{L}_{\mathrm{k}-1}$ (the set of all frequent (k–1)-itemsets found in the (k–1) the pass are used to generate the candidate itemsets C_{k} using the *apriori-gen ()* function. This function first joins $L_{k-1}$with L_{k-1}, the joining condition being that the lexicographically ordered first $k-2$ items are the same. Next, it deletes all those itemsets from the join result that have some (k–1)-subset that is not in $L_{\mathrm{k}-1}$, yielding C_{k}. For example, let L_3 be {{1 2 3}, {1 2 4}, {1 3 4}, {1 3 5}, {2 3 4}}. After the join step, C_4 will be {{1 2 3 4}, {1 3 4 5}. The prune step will delete the itemset {1 3 4 5} because the itemset {1 4 5} in not in L_3. We will then be left with only {1 2 3 4} in C_4.

```
Procedure AprioriAlg()
Begin
L1 = {frequent 1-itemsets);
For (k: = 2; L_{k-1} ≠ φk ++) do {
Ck: = apriori-gen (L_{k-1}); // New candidates
for all transactions t in the dataset do {
for all candidates c ∈ Ck contained in t do
c.count ++;
}
Lk: = {c ∈ Ck | c. count ≥ min-support}
}
Answer: = ∪k Lk;
end
```

Fig. A.8. Apriori Algorithm

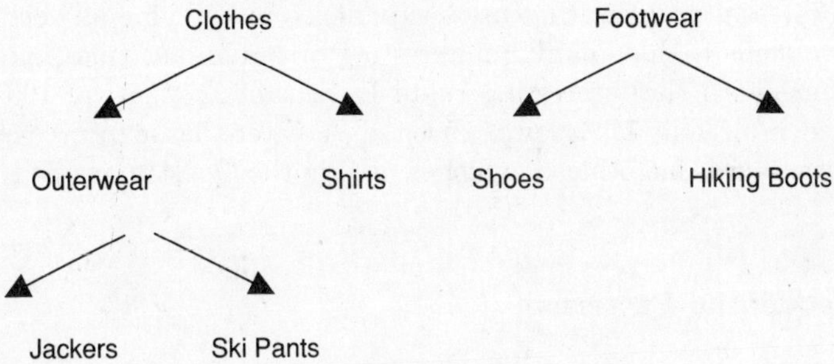

Fig. A.9. Example of a Taxonomy

The algorithm now scans the database. For each transaction, it determines which of the candidates in C_k are contained in the transaction using a hash-tree data structure and increments the count of those candidates. At the end of the pass, C_k is examined to determine which of the candidates are frequent, yielding L_k. The algorithm terminates when L_k becomes empty.

Generalizations

Very often, taxonomies (*is* – as hierarchies) over the items are available. An example of a taxonomy is shown in Fig. A.9: this taxonomy says that Jacket *is* – a Outerwear, ski pants *is* – a Outerwear, Outerwear *is* – a clothes, etc. users are often interested in generating rules that span different levels of the taxonomy. For example, we may infer a rule that people who buy Outerwear tend to buy Hiking Boots from the fact that people bought Jackets with Hiking Boots and ski pants with hiking boots. However, the support for the rule "Outerwear ⇒ Hiking Boots" may not be the sum of the supports for the rules "Jackets ⇒ Hiking Boots" and "Ski pants ⇒ Hiking Boots" since some people may have bought Jackets, Ski Pants, and Hiking Boots in the same transaction. Also, "Outerwear ⇒ Hiking Boots" may be a valid rule, while "Jackets ⇒ Hiking Boots" and "clothes ⇒ Hiking Boots" may not. The former may not have minimum confidence and the latter may not have minimum confidence. This generalization of association rules and the algorithm used in Quest for finding such rules are described in Srikant & Agrawal 1995.

Another generalization of the problem of mining association rules is to discover rules in data containing both quantitative and categorical attributes. An example of such a "quantitative" association rule might be that "10% of married people between age 50 and 60 have at least 2 cars." We deal with quantitative attributes by fine partitioning the values of the attribute and then combining adjacent partitions as necessary. We also have measures of partial completeness that quantify the information loss due to partitioning. This generalization and the algorithm for finding such rules used in quest are presented in Srikant & Agrawal 1996a.

One potential problem that users experience in applying association rules to real problems is that many uninteresting or redundant rules may be generated along with the interesting rules. In Srikant & Agarwal 1995 further generalized in Srikant & Agrawal 1996a, a "greater-than-expected-value" interest measure was introduced, which is used in the Quest to prune redundant rules.

A.5.4 Sequential Patterns

We introduced the problem of discovering *sequential patterns* in Agrawal & Srikant 1995. The input data is a set of sequences, called data sequences. Each data sequence is a list of transactions, where each transaction is a set of items (literals). Typically there is a transaction-time associated with each transaction. A *sequential pattern* also consists of a list of sets of with a user-specified minimum *support*, where the support of a sequential pattern is the percentage of data sequences that contain the pattern.

For example, in the database of a book club, each data sequence may correspond to all book selections of a customer, and each transaction to the books selected by the customer in one order. A sequential pattern might be 5% of customers bought "Foundation," then "Foundation and Empire," and then "Second Foundation." The data sequence corresponding to a customer who bought some other books in between these books still contains this sequential pattern; the data sequence may also have other books in the same transaction as one of the books in the pattern. Elements of a sequential pattern can be sets of items, for example, "Foundation" and "Ringworld Engineers," followed by "Foundation and Empire" and "Ringworld Engineers," followed by "Second Foundation." However, all the items in an element of a sequential pattern must be present in a single transaction for the data sequence to support the pattern.

This problem was initially motivated by applications in the retailing industry, including attached mailing, add-on sales, and customer satisfaction. But the results apply to many scientific and business domains. For instance, in the medical domain, a data sequence may correspond to the symptoms or diseases of a patient, with a transaction corresponding to the symptoms exhibited or diseases diagnosed during a visit to the doctor. The patterns discovered using the data could be used in disease research to help identify symptoms/diseases that precede certain diseases.

Generalizations

The basic definition of sequential patterns was generalized in Srikant & Agrawal 1996b to incorporate following features:

- *Introduction of time constraints.* Users often want to specify maximum and/or minimum time gaps between adjacent elements of the sequential pattern. For example, a book club probably does not care if someone bought "Foundation," followed by "Foundation and Empire" three years later; they may want to specify that a customer should support a sequential pattern only if adjacent elements occur within a specified time interval, say three months.
- *Flexible definition of a transaction.* For many applications, it is immaterial if items in an element of a sequential pattern were present in two different transactions, a long as the transactions times of those transactions are within some small time window. That is, each element of the pattern can be contained in the union of the items bought in a set of transactions, as long as the difference between the maximum and minimum transactions times is less than the size of a sliding time window. For example, if the book club specifies a time window of a week, a customer who ordered the "Foundation" on Monday, "ring world" on Saturday, and then "Foundation and Empire" and "Ringworld Engineers" in a single order a few weeks later would still support the pattern "'Foundation' and 'Ringworld,' followed by 'Foundation and Empire' and 'Ringworld Engineers."'

In addition, if there were taxonomies (*is-a* hierarchies) over the items in the data, the sequential patterns could now include items across different levels of taxonomy. (See Srikant & Agrawal 1996b for a description of the GSP algorithm used in quest for finding such generalized sequential patterns.)

A.5.5 Time-series Clustering

Time-series data constitute a large portion of data stored in computers. The capability to find time series (or portions thereof) that are "similar" to a given time series or to be able to find groups of similar time series has several applications. Examples include identifying companies with similar pattern of growth, finding products with similar selling patterns, discovering stocks with similar price movements, determining portions of seismic waves that are not similar to spot geological irregularities, etc.

We introduced a model of time-series similarity in Agrawal et al., 1995a. In this model, two time series are considered to be similar if they have enough nonoverlapping time-ordered pieces (subseries) that are similar. The amplitude of one of the two time-series is allowed to be scaled by any suitable amount and its offset adjusted appropriately before matching the other series. Two subsequences are considered similar if one lies within an envelope of $\in$ width around the other, ignoring outliers. The matching subseries need not be aligned along the time axis.

The matching system used in quest is described in Agrawal et al., 1995a. It consists of three main parts: (i) "atomic" subseries matching, (ii) long subseries matching, and (iii) series matching. The basic idea is to create a fast, indexable data structure using small, atomic subseries that represents all the series up to amplitude scaling and offset, and find atomic matches by doing a self-join on this structure. The initial prototype used the R+tree for this representation. A faster data structure described in Shim, Srikant, & Agrawal 1996 is now used in its place. The second stage employs fast algorithm for stitching atomic matches to form long subseries matches, allowing nonmatching gaps to exist between the atomic matches. The third stage linearly orders the subseries matches found in the second stage to determine if enough similar pieces exist in the two time series. In every stage, the system allows for the flexibility of user/system-defined matching parameters without sacrificing efficiency.

Classification

Classification is a well-recognized data mining operation and it has been studied extensively in statistics and machine learning literature. However, most of the current classification algorithms have the restriction that the training data should fit in memory. In data mining applications, very large training sets with several million examples are common. We therefore wanted to design a classifier with magnitude (without resorting to sampling/partitioning). The ability to classify larger training data can also lead to improved classification accuracy.

SLIQ (Supervised Learning In Quest), described in Mehta, Agrawal, & Rissanen 1996, is a decision tree classifier, designed to classify large training data. It uses a presorting technique in the tree-growth phase. This sorting procedure is integrated with a breath-first tree-growing strategy to enable classification of disk-resident datasets. In the pruning phase, it use a pruning strategy based on the Minimum Description Length (MDL) principle. The net result of these techniques is that, given training data that can be handled by another decision tree classifier, SLIQ exhibits the same accuracy characteristics, but executes much faster and produces smaller trees. Moreover SLIQ can potentially obtain higher accuracies by classifying larger (disk-resident) training datasets, which cannot be handled by other classifiers.

While SLIQ was the first classifier to address several issues in building a fast scalable classifier and it gracefully handled disk-resident data that are too large to fit in memory, it still requires some information to stay memory resident. Furthermore, this information grows in direct proportion to the number of input records, putting a hard limit on the size of training data. We have recently designed a new decision-tree-based classification algorithm, called SPRINT (Scalable PaRallelizable Induction of decision Trees) that for the first time removes all of the memory restrictions, and is fast, scalable, and easily parallelizable. The algorithm presented in Shafer, Agrawal, & Mehta 1996, can classify datasets irrespective of the number of classes, attributes, and examples (records), making it an attractive tool for data mining.

A.5.6 Incremental Mining

As the data mining technology is applied in the production mode, the need for incremental/active mining arises (Agrawal & Psaila 1995). Rather than applying a mining algorithm to the whole data, the data is first partitioned according to time periods. The granularity of the time period is application dependent. The mining algorithm is now applied to each of the partitioned datasets and patterns are obtained for each time period. These patterns are collected into a database. In this database, each statistical parameter of a pattern will have a sequence of values, called the history of the parameter for the pattern. We can now query the database using predicates that select patterns based on the shapes of the history of some or all parameters. A shape query language is presented for this purpose in Agrawal et al., 1995b.

The user can specify triggers over the database in which the triggering condition is a query on the shape of the history. As fresh data comes in for the current time period, the mining algorithm is run over this data, and the database is updated with the generated patterns. This update causes the histories of the patterns to be extended. This, in turn, may cause the triggering condition to be satisfied for some patterns and the corresponding actions to be executed.

Such active systems can be used, for instance, to build easily warning systems for spotting trends in the retail industry. For example, if we were mining association rules, we would have histories for the support and confidence of each rule. Following the promotion for an item X, the user may specify a notification trigger on the rule X $\Rightarrow$ Y, the triggering condition being that the support history remains stable, but the confidence history takes the shape of a downward ramp. Firing of this trigger will signify that if the goal of promoting X was to drag the sale of Y, it was not fulfilled. The loyalists continued to buy X and Y together, but the new buyers cherry picked X.

A.5.7 Parallelism

Given that mining can involve very large amounts of data, parallel algorithms are needed. Quest algorithms have been parallelized to run on IBM's shared-nothing multiprocessor SP2. The parallel implementation of the mining of association rules is described in Agrawal & Shafer 1996. The implementation shows linear scale-up for association rules. Mining of sequential patterns is also parallellized using similar techniques. We have also parallelized the SPRINT classification algorithm Shafer, Agrawal & Mehta 1996, where all processors work together to build a single classification model. Measurements from these implementations show excellent scaleup, speedup, and sizeup characteristics.

A.5.8 System Architecture

Figure A.10 shows the system architecture of the Quest system. The mining algorithms run on the server close to the data source. Users interact with the system through a GUI that can run on the same workstation or on a different client machine. There is an open API using which the user can optionally import results of any mining operation into software of choice. An interesting aspect of the Quest architecture is its I/O architecture. There is a standard stream interface defined for all accesses to input, insulating the algorithm code from data repository details, which are encapsulated in a data access API. Thus, it is easy to add new data repository types of the Quest system.

The Quest system runs both the AIX and MVS platforms, against data in flat files as well as DB2 family of database products. Databases can be accessed in a loosely coupled mode using dynamic SQL. However, for better performance, it is possible to run the mining algorithm in a tightly coupled mode as described in Agrawal & Shim 1996.

A.5.9 Future Directions

We plan to continue on the current path of identifying new data mining operations and developing fast algorithms for their execution. Two operations that we are currently focusing on are deviation detection (Arning & Agrawal 1996) and segmentation of high-dimensional data. We are also interested in mining data types other than structured data, such as text, particularly in the context of the worldwide web. Finally, we are also exploring the interaction between discovery-driven and verification-driven data mining, especially in QLAP databases.

A.6 The Australian National University Research Projects

A.6.1 Applications of Inductive Learning

Supervisor: Professor John Lloyd (jwl@csl.anu.edu.au)

In the computer sciences laboratory, we are developing a machine learning system, which includes classification functions from examples. Applications

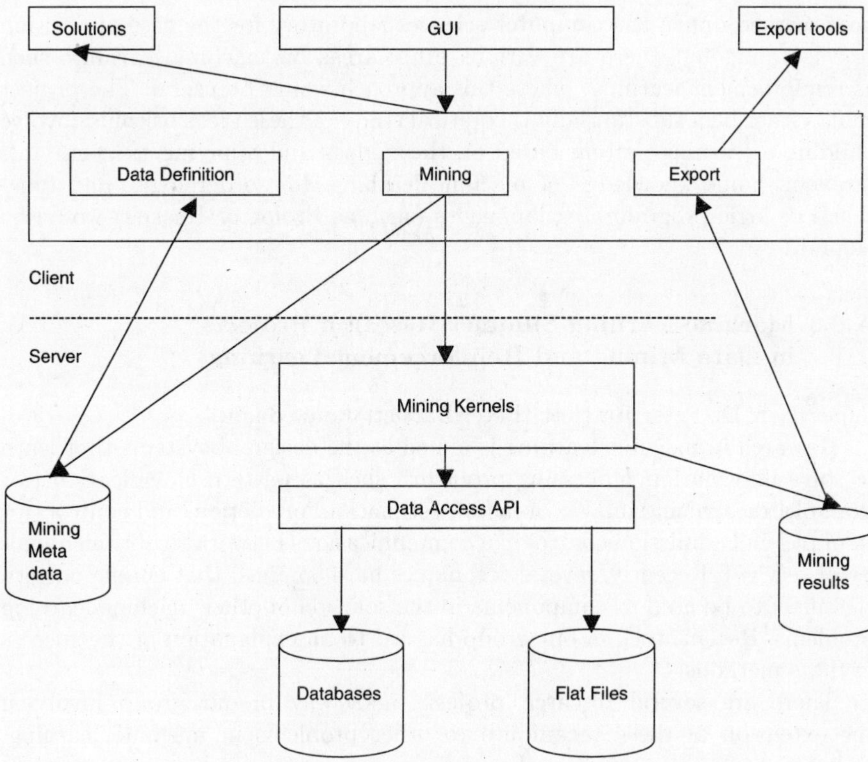

Fig. A.10. The Quest System Architecture

of such a system are ubiquitous and important. For example, it is possible to classify chemicals as carcinogenic or not from a number of examples of chemicals whose carcinogenicity is known.

This project will involve applying the system in some new application areas. This will include identifying the applications, modeling them in a logical representation language, and experimenting with various parameters for the learning system to obtain acceptable classifications. Some knowledge of machine learning and functional or logical programming languages (such as prolog or Haskell) would be helpful.

A.6.2 Logic in Machine Learning

Supervisor: Professor John Lloyd (jwl@csl.anu.edu.au)

Traditionally, machine-learning systems have employed simple languages, such as the attribute-value language, for knowledge representation. However, there are many applications where these simple knowledge representation languages are inadequate and so there is good motivation to consider more sophisticated languages, such as first-or higher-order logic.

This project will involve studying the use of higher-order logic for knowledge representation in machine learning. Much of the detail of this has already been worked out in the computer sciences laboratory for the case of decision tree learning, but there are various other areas on machine learning, such as reinforcement learning, where this approach would be useful. The project could either be a substantially theoretical study of these ideas or could involve building a learning system based on these ideas and applying it to suitable problem. Some knowledge of machine learning, first-order logic, and functional or logic programming languages (such as Prolog or Haskell) would be helpful.

A.6.3 Machine-learning Summer Research Projects in Data Mining and Reinforcement Learning

Supervisor: Dr Peter Bartlett (Peter.Bartlett@anu.edu.au)

Research in machine learning is aimed at the design of systems that learn to solve information processing problems, such as pattern classification (for example, recognizing images of faces), estimation, prediction, and control (for example, call admission control in a communications network to optimist quality of service). Recently, several techniques have emerged that enable pattern classifiers to be used as components in the solution of other machine learning problems. Recent work in our group has led to an explanation of the success of these methods.

There are several research projects underway in our group involving the extension of these techniques to other problems in machine learning, including

- Data segmentation problems, which arise in data mining (for instance, identifying fraudulent mobile phone use from calling behavior), and
- Reinforcement learning (learning to control complex systems that are difficult to model accurately, but for which some performance measure is available).

A.6.4 Computational Aspects of Data Mining (3 Projects)

Contact Person: Dr. Stephen Roberts

Data mining is an attractive new area of research, applying methods from statistics, machine learning, and parallel computing to the analysis of very large datasets. Some of the main challenges in data mining include functions with many variables (not just 3 or 4) a large number of data points (trillions), ill-defined goals of searches.

The Advanced Computational CRC (ACSys) has access to very large datasets and collaborates with several institutions like NRMA and the Taxation Office. Summer Scholars will explore the computational challenges, which are connected to finding information in such large datasets. The scholars will have regular interaction with senior researchers, postdoctoral fellows, and graduate students within the data mining group within the CSL and ACSys.

The aim of the projects is to develop, analyze, and implement data mining algorithms, which can extract useful information from very large data sizes. The data size may be of the order of gigabytes or even terabytes. Particular requirements for the algorithms are scalability with respect to the number of data points even in the event of high dimensionality. Depending on the interests of the students the work can have components of experimental exploration of computational performance using both real and simulated data development and analysis of algorithms and software mathematical error and performance analysis.

As examples, the following projects are possible:

(1) Clustering techniques based on density estimators

Clustering, i.e., finding groups of similar objects is a central theme in data mining. While the k-means algorithm is one of the most popular at the moment, strong contenders are based on the estimation of density. Following a review of the literature the various methods should be compared and a new method based on finite element techniques investigated.

(2) Parallel high-dimensional density estimation

The modeling of high-dimensional datasets poses a particular challenge both for approximation and the parallel implementation. In this project we investigate the parallelization of a density estimation algorithm.

(3) Adaptive methods for high-dimensional problems

Various ways have been suggested to adaptively approximate functions of many variables. We investigate the use of adaptive methods of approximate high-dimensional data.

A.6.5 Data Mining the MACHO Database

Supervisor: Bill Clarke (llib@compter.org)

The MACHO database is an astronomical database of the intensities of about 20 million stars, recorded every night for several years. The data mining researchers here have been primarily using in-house developed algorithms for investigating the variable stars within the database, and we would like to compare our results with those from a commercial data mining package.

The aim of this project will be to investigate (a subset of) the variable stars recorded within the MACHO database using a commercial data mining package - such as SGI's MineSet, or Clementine - to identify interesting classes of stars.

A.6.6 Artificial Stereophonic Processing

Supervisor: Ms.Biljana.Rdlovic (Biljana.Radlovic@syseng.anu.edu.au)

The sound recordings made in a closed room are less successful when the microphone cannot be placed close to the speaker. The quality of the sound reproduced through the earphones to the listener in a remote room will be degraded due to the sound reflections picked up by the microphone. Experiments indicate that the quality and the fidelity of the sound reproduced to the listener are improved by applying to the ears two different versions of the signal recorded by the microphone in a room. This effect can be attributed to an artificial stereophonic effect.

The main task of this project will be to design a bank of band-pass filters, which can be used to "split" the frequency domain between the two ears in a satisfactory fashion. Prior to implementation of the filtering system in C code or MATLAB, the student will be involved in digital recording of the echoed sound and provided with some understanding of the laboratory computer audio system. The project is aimed to shed more light on possible application of the developed technique in practical situations in which the standard techniques are ineffective and lack robustness.

A.6.7 Real-time Active Vision

Supervision: Dr. Sebastien Rugeaux and Dr. Alex Zelinksky (Alex.Zelinsky@anu. edu.au)

The aim of this project is to develop real-time control and vision algorithms for an active vision head. The Robotics Systems Lab has just completed the design and manufacture of a new generation active head that is capable of reproducing some of the motion of the human eyes with very fast velocities and acceleration.

Algorithms for position and velocity control of two degrees of freedom pan/tilt device have already been implemented on a Motion Controller card that runs on a standard PC with the Linux operating system. Real-time vision algorithms, such as optical flow and normalized correlation, have also been implemented using the MMX extension available on Pentium-based computers. A high-performance vision board (Hitachi IP5005), that is capable of doing real-time image processing, is also available. The goal of the project will be to integrate the vision and control algorithms into a single application in order to implement real-time visual behaviors for the active head. Possible behaviors include target detection based on velocity cues and/or object recognition, fast acquisition using saccadic motion, and smooth tracking of moving targets in cluttered environments.

A.6.8 Web Teleoperation of a Mobile Robot

Supervision: Dr. Ken Taylor and Dr. Alex Zelinsky

A robot arm was made controllable using Web browser technology (http://telerobot.mech.uwa.edu.au) at the University of Western Australia in 1994 and has been under development ever since using an interactive approach based on analysis of operator behavior. This is a forerunner for what is likely to be a wide range of future teleoperated applications, including remote operation of mining equipment and Web control of machine tools for remote manufacturing.

This project is to set up the mobile robot owned by Robotic Systems for Web-based teleoperation using the software developed for the arm at UWA. The project involves adapting existing software for a mobile application, devising strategies for transmitting robot state information effectively to operators, developing strategies for keeping the robot online, and monitoring operator behavior.

Tele telerobot arm at UWA is continually in use attracting more than 5000 operators a month and the developer of the mobile teleoperated robot will also gain the satisfaction of seeing their work in use by the Web community.

A.6.9 Autonomous Submersible Robot

Supervision: Dr. David Wettergreen and Dr. Alex Zelinsky (Alex.Zelinsky @anu.edu.au)

Australia has one of the longest coastlines in the world. Its continental shelf contains enormous natural resources, including coral reefs, marine fisheries, and oil and gas reserves. Such a resource must be managed and protected. The efficient management of our coastal resources will require autonomous machines that are capable of routine monitoring, inspection, and maintenance tasks like inspecting and repairing weld joints of oil rigs.

The ANU is building a submersible robot, called Kambara, which will serve as a test bed for experimental research in autonomous subsea operations. This robot will allow researchers to study the fundamental problems of controlling the operations of an underwater vehicle. Most present-day submersible robots work via teleoperation where a human operator (on a remote ship) controls every aspect of the robot's operations. This is inefficient both in terms of human resources and the execution speed of tasks. Our aim is to incorporate on-board sensors such as sonar, gyros, depth sensors, and computer vision that will allow the robot to make its own on-board decisions. The ANU submersible has: a max operational depth of 20m, a mass of 100 kg, and a volume of less than one cubic meter. The five thrusters produce five independent degrees of freedom controllable by on-board systems.

Projects include working on the development of a vision system for the submersible, building a system of sensors, models, and algorithms that estimate the robots, position, orientation, and velocity in the water. And the development of a graphical operator interface that receives state information from the robot and presents it in a graphically intuitive form to the operator. Testing of the submersible is done in an on-campus tank.

A.6.10 The SIT Project

The Science and information Technology (SIT) project is aimed at decision makers and policy makers in corporate and government environments. Besides generating visions of the future we aim to identify chances and pitfalls and do recommendations based on what we find. In the project we will have three focus areas, data mining in science, data mining in corporate and government environments, and data mining for the individual user. In each of these application areas we will chart possible consequences. First, we will list the developments of methods and technology for data mining, based on assumptions on ICT developments of methods and technology for data mining, based on assumptions on ICT developments in the next 20 years. Societal needs and threats will pose accents on further exploration of the three application areas.

Some examples of data mining in science have already been mentioned. In commerce, data mining is use for predicting customer behavior, product data analysis and for risk assessment. Governments can use data mining to detect and prevent crime. But also the civilian can profit from data mining: by helping him find information on the Internet, by making digital libraries accessible, or by enabling multimedia searches. On the other hand, that same civilian will increasingly by confronted personally with the knowledge from science, corporations, and governments that data mining helped generate.

A.7 Data Mining Research Group, Monash University Australia

The Monash University is currently providing advanced research and training in data mining technologies for business, government, and industry.

A.7.1 Current Projects

- "Marketing to Individuals using Data Mining," research project sponsored by Coles-Myer.
- "Data Mining in the Insurance Industry," Ph.D research project sponsored by AAMI.
- "Increasing Wagering Turnover Through Data Mining," research project sponsored by TAB Limited.
- "The Transition from Data Warehousing to Data Mining," Masters research project sponsored by FITR '99 funding.
- "Rule Extraction from Trained Neural Networks," Masters research project sponsored by FITR '99 funding.
- "Neural Networks for Market Basket Analysis," Masters research project.
- "Data Mining for Customer Relationship Management," Masters research project sponsored by FITR '99 funding.

- "Web Usage Mining using Neural Networks," Masters research project sponsored by FITR '99 funding.
- "Data Visualization Techniques and their Role in Data Mining," Masters research project.
- "Evolutionary rule generation for credit scoring," PhD research project.
- "Improved decision trees for data mining," PhD research project.
- "Data Mining in the Pharmaceutical Industry," Masters research project.
- "Data Mining the Malaria Genome," research project in collaboration with Faculty of Medicine, sponsored by Monash University Research Fund.
- "Neural network modelling of water-related gastroenteritis outbreaks," PhD research project in collaboration with Faculty of Medicine.
- "Data Mining with Self Generating Neural Networks," PhD research project.
- "Development of Intelligent OLAP Tools," research project.
- Mining Stock Market Data," research project.
- "Data Mining by Structure Adapting Neural Networks," research project.
- "Rule Discovery with Neural Networks in Different Concept Levels," research project.
- "Cash flow forecasting using neural networks," research project in collaboration with AXA Australia.
- "Neural Networks for market response modeling," PhD research project.
- "Advertising effectiveness modeling," research project.
- "Temporal effects in direct marketing response," Masters research project.

A.7.2 ADELFI – A Model for the Deployment of High-Performance Solutions on the Internet and Intranets

Objective

The main objective of the ADELFI project is to develop generic software tools for the deployment of Internet-based, computationally intensive (CFD) and data intensive (data mining) applications.

The specific measurable objectives of the project are:

- To develop and implement distributed software tools for the deployment of high-performance solutions on the Internet and/or corporate Intranets.
- To exemplify the generic nature of the software tools by demonstrating two diverse applications – computational fluid dynamics (CFD) and data mining.
- To develop universal clients, so that end users can access the high-performance solutions offered to them using Java-enabled Web browsers.
- To test, validate, and benchmark the new software infrastructure by running real applications from the Web using the Internet and corporate Intranets.

The new architecture will be based on a three-tier client/server application model:

- The first tier is a universal client (Web browser) running Java applets. The client software will be downloaded from a Web server to a machine requiring only to have a Java-enabled Web browser. The complete environment for setting up CFD and data mining applications will be supported by the Java applets.
- The middle tier is the application server, which integrates and synchronizes back-end components to fulfill client requests. The application server is designed and implemented using Enterprise Java beans.
- The third tier runs data and computationally intensive applications on high-performance parallel serves. The ADELFI architecture is using CORBA to access and execute the HPC applications on parallel computers.

The foreseen business and industrial benefits are as follows.

- For the industrial users, to increase productivity by accessing high-performance systems, without the need to invest on expensive parallel computers.
- For the industrial users to access powerful applications over the Internet, or corporate intranets, and dynamically control the application from a Java-based user interface running in their Web browser.

Across a wide variety of fields, data is being collected and accumulated at a dramatic pace. There is an urgent need for a new generation of computational techniques and tools to assist humans in extracting useful information (knowledge) from the rapidly growing volumes of data. These techniques and tools are the subject of the emerging field of data mining, that is, knowledge discovery in databases (KDD).

Mathematics Projects on Data Mining:

Classification of Time Series in Data Mining Applications

Many real-world databases possess time components. Temporal (having time components) databases arise in science and engineering domains such as metrological and hydrological databases, as well as in business such as stock exchange databases and the databases storing various kinds of business transactions. Temporal data mining deals with problems of knowledge discovery from these large temporal databases. One objective of temporal data mining is to find and characterize interesting sequential patterns from temporal datasets. For example, some sequential patterns of patients visiting doctors may indicate fraud committed in claiming benefit from health insurance companies. Our current focus is on classification of time series. We study the integrated use of techniques from signal processing (e.g., fast Fourier transforms) and pattern recognition (e.g., K-means) to solve this large temporal database mining problems.

Subspace Clustering of High-dimensional Data for Data Mining Applications

The clustering problem is well known in the database literature for its numerous applications in problems such as customer segmentation, classification, and trend analysis. In high-dimensional spaces not all dimensions may be relevant to a given cluster. One way of handling this is to pick the closely correlated dimensions and find clusters in the corresponding subspace. The number of dimensions in each such cluster-specific subspace may also vary. Hence it may be impossible to find a single small subset of dimensions for all the clusters. We use a bottom-up algorithm, exploiting the monotonicity of our clustering criterion with respect to dimensionality. Remark: both projects involve computer programming.

Potential Postgraduate Projects in Data Mining

With recent advances in data acquisition and storage, extremely large data collections (in the gigabytes to petabytes range) have become very common. About 10 years ago, the systematic study of techniques for discoveries of information in such large collections was initiated and such techniques are commonly referred to as data mining algorithms. Our group of researchers at the Australian National University has been active since 1997 in the development and analysis of data mining techniques using predictive modeling and clustering.

The main focus of our work is on the computational aspects relating to the large number of data records available, the complexity of the data and parallel algorithms. Techniques studied include scalable smoothing techniques, wavelet-based methods, and scalable parallel algorithms. Much of the research is done jointly with government and commercial organizations and our students participate actively in the interaction with industry and government. The proposed research will make use of the national high-performance computing facilities at the ANU APAC (Australian Partnership for Advanced Computing). The main aim is to get better and faster algorithms, which ultimately bring the data to the desktop such that end users can do analysis of very large datasets in real time.

Database Record Linkage

Contact: Peter Christen and Ole Nielsen

Linking records from different data sources is a complex and computational-intensive task. The aim is to combine records with information about certain entities (e.g., patients, companies, families) from different data sources. As an example, a health department might be interested what kind of injuries people have in car accidents. Thus, data from an emergency database and from police or insurance databases (or both) have to be linked.

If a common identifier is present in both databases, the linkage can be done deterministic, but if such an identifier is missing probabilistic linkage methods have to be applied. Attributes and values in a database are given certain weights, and each linked pair of records gets an overall score or probability. Other issues involved in the linkage process are data cleaning, and dealing with missing or faulty data.

This project aims in integrating a linkage into the DM tools, a data mining toolbox based on the scripting language Python currently under development at the ANU Data Mining group. Aspects of this project are literature research, comparison of different linkage algorithms, and finally the implementation (first as a Python prototype and later as a parallel C/MPI code if time permits).

Data Mining with Additive Models

Contact: Markus Hegland

This particular study investigates the application of additive models, which are functions of the form:

$$f(x1, \ldots, xn) = fl(x1)+, \ldots, +fn(xn),$$

where the attributes x1,...,xn can be simple (numbers or categories) or composite (e.g., sets, arrays or graphs). Our earlier work has showed that additive models are effective data mining tools as they handle very large numbers of attributes effectively and they also have good approximation properties in this case. However, very little work has been done on composite attributes and this is where this project shall contribute new insights. The focus of the project depends on the interests of the student and could include the development and analysis of new algorithms, the study of approximation properties and the application in data mining projects.

Depending on the actual direction taken the work is suitable for studies in mathematics or computer science. Preferably this research should ultimately lead to a Ph.D. but initial work, including implementation and literature studies, could also be done in an honors project.

Parallel Algorithms for Predictive Modelling

Contact: Peter Christen and Markus Hegland

Data mining applications have to deal with increasingly large datasets and complexity. Only algorithms, which scale linearly with data size, are feasible for successful data mining applications.

In our group we are developing algorithms for predictive modeling of high-dimensional and very large datasets that are both scalable with the number of data records as well as number of processors if implemented in parallel. These algorithms are based on techniques like finite elements, thin plate splines, wavelets, additive models, and clustering. Prototype implementations have been developed in Matlab, Python, C, and MPI, and we have used them successfully in several real-world data mining projects.

This research project will involve the further development of parallel data mining algorithms based on the available mathematical algorithms and prototypes. Besides scalability other important aspects are data distribution, load balancing, and integration into a larger data mining framework currently being developed in our group.

Wavelet-based Methods for Predictive Modeling

Contact: Ole Nielsen, Markus Hegland and Zuowei Shen

A fundamental issue in data mining is the development of algorithms to extract useful information from very large databases. One important technique is to estimate a smooth surface approximating the data. Such an approximation can be used for visualization, prediction, or classification purposes. However, the size of datasets tends to grow steadily both in terms of the number of records and the number of attributes. While the former issues requires that algorithms scale linearly with the number of records in order to be feasible, the latter require the so-called curse of dimensionality to be addressed. The complexity of the smoothing problem grows exponentially with the dimension so any algorithm computing and storing a smooth surface exactly becomes infeasible for dimensions higher than 4 or 5. Hence, approximative methods that balance accuracy against complexity are needed.

Existing technologies for dealing with high-dimensional data include neural nets, classification and regression trees, and regression splines. Our group has developed an alternative approach using wavelets and a prototype written in Matlab has successful addressed smoothing problems in six dimensions. It works by computing a projection onto spaces of low-density formed as a sum of tensor products of multilevel spaces carefully chosen such that the approximation properties are good for reasonable smooth functions and the algorithmic complex city is reduced significantly.

This project will involve the further development of these ideas based on the available theory and prototypes. Issues to be addressed include (but are not limited to) the following:

- The current implementation is based on one of the simplest refinable function – the "hat" function, which can locally reproduce linear polynomials exactly. Using refinable basis functions that are smoother will lead to even better approximation properties of the method and may improve its predictive power significantly.
- However, this will involve defining suitable corresponding boundary wavelet functions, which is not a trivial thing to do.
- The method can be improved by introducing adaptability: The surface can be compressed further depending on its actual shape using standard wavelet compression methods.

- The wavelet smoother should be integrated into larger data mining framework currently being developed in our group and applied to real-life classification and regression problems in order to compare it to other state-of-the art technologies.

A.8 Current Projects, University of Alabama in Huntsville, AL

Nakajima (1996) explained actual applications of data mining utilizing fuzzy logic both from development point of view and also from a conceptual point of view. The two projects described by him and presented here are projects, which show the application of this technology.

A.8.1 Direct Mailing System

The problem is to identify customers from a customer database who can be targeted for a sale under the assumption that these customers responded positively to advertisements mailed to them. The additional constraint is that the mailing list budget is limited and the number of advertisements to be mailed is to be controlled to increase profit. The first step involves analyzing the database for attributes like "frequency of visits to the store," "sum of purchases," etc. Analysis and plots of the data then determine the cluster of good customers. Next, one has to find the attribute relationships to define a query condition, which is represented by a pair of attributes and a fuzzy linguistic value. One then verifies and refines the query condition by using another customer database. Thus the customer database is ranked and sorted by degree values based on a given fuzzy query condition. The customers retrieved by the query determine the list of potential of good customers.

A.8.2 A Vibration Sensor

A product that was used to sense vibrations and predict the causes of these vibrations (i.e., earthquakes, etc.) was improved by utilizing fuzzy rules. The original sensor was based on simple threshold rule. The error rate for this sensor was around 12%. The fuzzy rules were created by analyzing the actual data in specified cases of earthquakes, automobiles, etc. A feature extraction was done on the dataset to identify each kind of causes. Relationships between the feature parameters and the kind of vibration were discovered to develop the fuzzy rules. These rules were then tested and refined. The accuracy of sensor improved dramatically with the error rate falling to within1%.

A.8.3 Current Status

Data mining and application of fuzzy logic in data mining is still a relatively new field. The possibilities of exploiting these techniques on different datasets to extract information are still immense. To determine the rules of knowledge extraction using fuzzy logic to form a subset of data mining, any changes in approaches to the techniques in data mining would also directly impact how fuzzy logic is applied.

Data mining research is now being geared toward answering the needs of the ever-changing technology. Han 1997 lists possible areas of research in data mining. They are:

- Integrating data warehousing with on-line analytical processing technology, which would allow the data miner the freedom to find interesting patterns in any dimensions of the data cube.
- Allow the user to have the ability of mining multiple kinds of knowledge.
- Devising the data mining query language, which would be similar to relational languages and allow the user to specify data mining tasks.
- Have the capability of handling complex data, not only from datasets, but also the Internet.
- Algorithms, which increase the performance of current data mining algorithms.
- Creative visualization of the dataset to allow interactive data mining.
- Integrate an expert system with the mined knowledge for decision making.

A.8.4 Data Mining Using Classification

The description below taken from B. Moxon's article "Defining Data Mining," which appeared in the August issue of *DBMS Online.*

Classification: Classification, perhaps the most commonly applied data mining technique, employs a set of preclassified examples to develop a model that can classify the population of records at large. Fraud detection and credit-risk applications are particularly well suited to this type of analysis. This approach frequently employs decision tree or neural network-based classification algorithms. The use of classification algorithms begins with a training set of preclassified example transactions. For a fraud detection application, this would include complete records of both fraudulent and valid activities, determined on a record-by-record basis. The classifier-training algorithm uses these preclassified examples to determine the set of parameters required for proper discrimination. The algorithm then encodes these parameters into a model called a classifier.

The approach affects the explanation capability of the system. Once an effective classifier is developed, it is used in a predictive mode to classify new records into these same predefined classes. For example, a classifier capable of identifying risky loans could be used to aid in the decision of whether to grant a loan to an individual.

A.8.5 Email Classification, Mining

Project Group: Shaofeng, Zhu, Xiangfei Zhu, Wenzhi Lai, Huimin Zhao {szhu, zhuzx2, lwzhi}@cs. utexas.edu, hzhao@mail.utexas.edu

Description of Project

Our project intends to build a powerful, efficient email application that have following salient features on classification:

1. Classification based on email address
2. Classification based on email subject (title)
3. Classification based on the content of an email

Upon the above achievements, such an email application can easily implement a filter that throw out junk mails. In case of some mailing lists, the user can recover it from the junk by setting the mail list's address in the application, then it will not be discarded.

A.8.6 Data-based Decision Making

The major component of data based decision making is data mining a multidisciplinary approach to the extraction of hidden, predictive relationships from large databases. It is used in target marketing, risk management, and product development.

The challenge to today's manager is to understand both the power of data mining and its limitations Rewards have been substantial for companies that have tied data mining projects to strategic initiatives; many more companies have squandered resources on poorly planned projects.

As a field, data mining has evolved from earlier computer-driven disciplines: first, in the 1960s, data collection by dedicated, large-scale computer systems; next, beginning in the 1980s, dynamic access to data through relational databases on mainframe, midrange, mini, and desktop computers; and, finally, in the 1990s, multidimensional databases and OLAP (on-line analytical processing) coming together to create data warehousing and decision support systems.

Although the discipline of data mining is an outgrowth of data warehousing and OLAP, data mining refers specifically to the automated extraction of hidden relationships within an organization's data. The development of techniques for massive parallel processing, the steady decrease in data storage costs, and the automated collection of detailed, organization-specific information are the environmental factors behind data mining; demand for higher returns on marketing activity, for minimization of lending risk, and for maximization of the return on new investments are the drivers behind the current interest in data mining. While data mining is, in the strictest sense, simply another application of computer science to a business problem, its dependence on and proximity to emerging software and hardware frontiers makes it an appealing area for study.

From the software side, data mining involves the application of leading-edge technologies like machine learning and artificial intelligence to statistical modeling and neural networks. From the hardware side, data mining is independent of recent developments in multiprocessor computing and redundant storage technology.

Finally, because data mining relies on these emerging technologies, specific implementations of the discipline can be both risky and costly. Therefore, these projects demand the informed awareness of general managers.

A.8.7 Data Mining in Relational Databases

Relational data mining, whose main part is Inductive Logic Programming (ILP), aims at developing practical algorithms for learning relational descriptions from relational databases in a declarative knowledge representation formalism, which is more expressive than the attribute-value representation of more traditional data mining approaches.

A.8.8 Environmental Applications and Machine Learning

Environmental data often need to be analyzed in order to obtain information necessary for environmental management decisions. Sometimes it is necessary to use the data to build a model of the environmental process that we want to manage. In other cases the analysis is necessary to understand the environmental processes studied, by identifying and understanding the interrelationships of different parameters. Machine-learning techniques that elicit regularities from data can be used to this end.

Text and Web Mining

Text and Web mining are technologies analyzing less structured data having the goal to extract knowledge from text databases and from Web log files. Good examples of business-related text mining applications in the domain of efficient handling with the textual information overload using:- visualization techniques,- automatic building of onthologies (subject taxonomies),- predicting financial trends from the text from news wires,- information extraction from document databases,-spam filtering, document summarization and text categorization. Typical Web mining is in the area of customer profiling from the Web log files analyzing customer trace and historical information with the goals as:- increase sales of products at the web site.

Recognizing Client's Behavior to Deliver Personalized Services

On the Internet, businesses need to provide a more proactive interface to their customer than what they have done to date. The effectiveness of businesses on the Internet will depend on their ability to utilize it as a two-way communication channel that anticipates the needs of the customers within the context of their visit and provides content and services that meet these needs. Achieving this goal involves, -analyzing multiple channel interactions and profiling behavior trends;- using all available historical and real time data to update customer profiles continually, a process called *context assembly*;- employing multiple paradigms to be reactive to changing customer context;- providing content morphing and orchestration engines to drive messages to customers at the right time and in the right context. Real case studies will support a discussion on the technologies available to support these business requirements.

Gaining Insights into Customer Needs using Data Mining

In today's economy, there is a rapidly increasing opportunity for global consumers to utilize the convenience of multiple channels to conduct their browsing for new product information, their value comparison with competition and ultimately their purchasing. The data that is gleaned during these processes is a critical resource for analysis with the aim of learning about customer spending patterns, loyalty, and behavioral trends. Knowledge derived from the analysis allows businesses to make more informed, contextualized marketing programs to retain their customers, increase their spend and reduce operating costs – a route to greater profitability.

A.8.9 Current Research Projects

Data mining and data warehousing, including mining knowledge in data warehouses, relational, objects, relational and object-oriented databases.

A data warehouse and relational data mining system, DBMiner, has been designed, developed, and demonstrated in major international conferences (including ACM-SIGMOD '96), licensed to industry, being beta-tested by some companies, and its educational version has been available on the market since June 1998.

Spatial databases, including object-oriented modeling, indexing, query processing, and data mining in spatial databases and geographic information systems. A spatial data mining system, GeoMiner, has been designed, constructed, and demonstrated in major international conferences, including ACM-SIGMOD '97.

Multimedia databases, including multimedia database indexing, buffer management, and data mining in multimedia database systems.

A multimedia data mining system, *MultiMeidaMiner,* has been designed and demonstrated in major international conferences, including ACM-SIGMOD '98.

Deductive and object-oriented databases, including rule compilation, deductive query processing, and integration of deductive and object-oriented database systems.

A deductive database system, LogicBase, has been constructed and demonstrated in major international conferences (including ACM-SIGMOD '94) and made available on the Internet.

Internet-based information technology, including text data mining, Web-based data warehousing, and Web-based data mining. A Web-log data mining system, *WebLogMiner,* has been designed and is under construction.

A.8.10 Web Mining

In today's society there is an increasing need for automated systems providing personalized recommendations to a user faced with a large number of choices. The products customization trend coupled with e-commerce where customers were not provided with an option to examine the products – "off-shelf" in a traditional sense, make the problem of providing accurate personalized recommendations very important. Automated methods are needed to provide a large number of users with the ability to efficiently locate and retrieve information according to their preferences. The task of collaborative filtering is to predict preferences of an active user on unseen items given preferences of other users, typically expressed as numerical ratings.

A.8.11 Neural Networks Applications to ATM Networks Control

Accurate traffic prediction can be used to optimally smooth delay sensitive traffic and increase multiplexing gain in asynchronous transfer mode (ATM) networks.

The objectives of this work are to investigate applicability of neural network techniques for prediction of either the following or several following frame sizes using the information of previous frame sizes.

Integrating Dynamic Learning and Expert Knowledge

Traditional neural network learning involves optimization of the interconnection weights between neurons on a prespecified network architecture. Determination of an appropriate architecture is a challenging problem, which is typically approached through an expensive trial-and-error process.

Applying Neural Networks in Marketing

The neural network is emerging as an important new tool for data mining in marketing. Used in conjunction with conventional statistical tools the neural network can be applied to a wide range of marketing problems from prospect scoring and retention analysis, to data imputation, retail modeling and time series prediction. This presentation will introduce neural networks from a marketing perspective and illustrate by means of case studies from the financial services and FMCG sectors how they can be used to complement, and in some cases supplant more traditional techniques. Two particular applications of neural networks will be considered in detail: prospect scoring and data imputation.

Two Ways of Extracting Meaning from Neural Models

Neural networks can model nonlinear relationship in our data. After describing what this means, two examples of ways in which nonlinear relationships can be analyzed to generate knowledge will be presented. The first of these looks at predicting tourist movements using a multilayer perceptron, while the second examines the generation of customer profiles and the extraction of a fuzzy rule set from a credit risk database using Kohonen's self-organizing map.

Applying Case-based Reasoning to Data Mining

Case-based reasoning is one of the newest techniques in artificial intelligence. CBR uses artificial intelligence inductive learning techniques with a number of enhancements to add further knowledge to the learning process. The application of this approach to the data mining of a financial database and the production of induction trees for inspection and verification will be presented, along with techniques for including domain knowledge for improved learning.

A.8.12 Scientific Topics

- Bayesian Data Mining in Marketing and Customer Satisfaction
- Categorical Data Mining and Knowledge Discovery on the Web
- Data Mining and Experimental Surveys on the Web
- Data Mining and Knowledge Discovery in Medicine and Medical DataBases
- Data Mining in Computational Physics and Historical Artifacts
- Data Mining in Default Risk Analysis and Credit Scoring
- Data Mining in Econometrics and Forecasting
- Data Mining in Finance and Financial Engineering
- Data Mining in Manufacturing
- Data Mining in Market Segmentation and Consumer Preferences
- Data Mining Massive Data Sets
- Data Mining Tools Workshop and Tutorial
- Data Mining with GMDH (Group Method of Data Handling)
- Data Mining with Statistical Pattern Recognition and Self-Organizing Neural Nets

- Dealing with Databases, Data Warehouses, Scalability, and On-line Analytical Processing (OLAP)
- Distributed Data Mining
- Economy of E-Business and Taxation
- High-dimensional Data and Computational Feasibility
- High-performance Computing, Parallel Processing, and Software Tools in Data Mining
- Information and Data Quality, Data Transformation, and Cleaning
- Intelligent Data Mining with Information Complexity, Statistical Modeling, and Genetic Algorithms
- Knowledge Management Issues
- Kohonen Self-Organizing Maps and Spatial Data Mining
- Mixed-Data Mining with Gifi System
- Present and Future E-Business
- Principles of Data Mining; Data Mining Tasks, and Techniques
- The Role of Classification and Clustering Techniques in Data Mining
- The Role of Genetic Algorithms in Data Mining
- The Role of Minimum Message Length Technology in Data Mining with Applications in Finance and Pattern Recognition
- Understanding Search Engines and Text Mining
- Visualization and Visual Data Mining
- Web-Based Data Mining
- Association Rules and Market Basket Analysis,
- Text Data Analysis,
- Data Mining Techniques for handling Numeric Prediction Problems
- Modeling Techniques for Decision Support
- Statistics and Machine Learning for Data Analysis in the Life Sciences
- Data mining techniques and applications in medicine
- Decision support in the Implementation of Environment Policy
- Why to use data mining and decision support in marketing
- Data mining for brand equity management
- Improving Direct Mail Targeting with Neural Networks

A.8.13 Application Areas

- Web Data Mining
- Financial, Accounting, and E-Business Data
- E-business Developments Affecting Supply Chain Design and Management
- Pharmaceutical/Medical/Health Industry
- Genome Data Mining
- Security, Privacy, and Intellectual Ownership
- Intelligent Data Mining Algorithms and Software
- Data Mining of Scanner Data
- Architectural Support for Data Mining
- Performance Comparison of Different Learning Algorithms

A.9 Kensington Approach Toward Enterprise Data Mining Data Mining Group, Imperial College Parallel Computing Centre. Jaturon Chattratichat, Yike Guo, Stefan Hedvaill, and Martin Kohler,

The Kensington system, which is being developed at the Imperial College Parallel Computing Centre in University of London, aims to provide an enterprise solution for large-scale data mining in environments where data is logically and geographically distributed over multiple databases. Supported by an integrated visual programming environment, the system allows an analyst to explore remote databases and visually define and execute procedures that model in the entire data mining process. It also provides learning algorithms, optimized of high-performance platforms, for the most common data mining tasks. Decision models generated by the system design are to provide an integrated, flexible and powerful data mining environment as the basis for customized domain-specific applications. The main features of the system design are discussed in turn below.

A.9.1 Distributed Database Support

Today, many companies store large quantities of data in data warehouses. The data is potentially rich and useful for data mining. A data mining system should allow seamless integration of both local files and remote databases. The Kensington system enables database integration in preparation of data mining by providing remote database access via JDBC. Analysts can query and retrieve data from their remote and distributed databases across the Internet. The ability of query several remote databases concurrently means that an analyst can now efficiently combine and enrich the data for mining.

A.9.2 Distributed Object Management

The Kensington system adopts a three-tier approach based on the Enterprise JavaBeans (EJB) component architecture, to support data mining in an enterprise environment. The component-based middleware is designed to support scalability and extensibility. Application servers can be transparently distributed for scalability or replicated for increased availability. The system also supports efficient management of resources and multitasking capabilities. In an enterprise where resources such as databases and high-performance servers are shared, the Kensington system enables efficient resource management and scheduling.

The data mining procedures that are defined and customized with the Kensington system can be flexibly deployed in the enterprise.

Because a data mining procedure is treated as a graph of components, each of them can be scheduled to use appropriate resources. The middleware's management strategy of the logical component and physical resources ensures that all facilities are used efficiently.

A.9.3 Groupware, Security, and Persistent Objects

In an enterprise where information is often shared within a workgroup, it is important that a data mining system supports the exchange of information in order to enhance productivity. Therefore, the Kensington system enables persistent storage of components so that they may be transparently shared and reused. Important information such as data, defined data mining procedures/templates, or generated decision models are managed as persistent objects, which can easily be exchanged between group members. The system provides on strong security for data transfer and model distribution through secure socket communications. Access control mechanisms protect users' or group's private resources from unauthorized access.

A.9.4 Universal Clients – User-friendly Data Mining

For maximum flexibility and easy deployment, client tools are Java applets that run securely in Web browsers anywhere on the Internet. A data analyst is therefore not bound to any specific location or computer.

Effective human–computer interaction is a strong feature of the Kensington system. Based on the visual programming paradigm, the Kensington client provides an integrated workspace for the visual construction of data mining procedures. The workspace includes wizards and templates for database connection, shows the user's view of the persistence object store, and provides the data mining task construction area. A data mining procedure is built visually as a connected graph and executed on request. The model or models returned by the mining components can be viewed with appropriate visualization applets in the client. Besides various data mining algorithms and data manipulation tools, the client interface also provides Java-based visualization tools for data and model analysis. Data visualization allows users to view and manipulate data before it is mined. Complex models, produced from data mining algorithms, are presented as interactive visual objects. The Kensington system provides various 2D graphing tools and a 3D scatter visualizer for data visualization. A decision tree visualizer, association rule visualizer, and a cluster visualizer are examples of tools used to present mining models to the user.

A.9.5 High-Performance Server

An important issue in data mining is the speed and performance of the task. In a competitive business environment where quick and precise decisions are needed, it is essential that a data mining task is performed within a reasonable time. Given the enormous size of data accumulated today, many analysts have turned to high-performance computers for a solution. Kensington's middleware serves as a gateway for connecting high-performance servers to thin clients and distributed databases. In addition, the system provides several optimized parallel algorithms for classification, clustering, association rule analysis, and neural networks.

APPENDIX-II

Data Mining Standards

Abstract. The information presented in this appendix is collected from Arati Kadav Jaya Kawale Pabitra Mitra, Department of Computer Science and Engineering, IIT, Kanpur. In this section, we have consolidated all the current data mining standards. We have categorized them into process standards, XML standards, standard APIs, Web standards and grid standards, and discussed them in considerable detail. We have also designed an application using these standards. We later also analyze the standards and their influence on data mining application development, and later point out areas in the data mining application development that need to be standardized. We also talk about the trend in the focus areas addressed by these standards.

Researchers in data mining and knowledge discovery are creating new, more automated methods for discovering knowledge to meet the needs of the 21st century. This need for analysis will keep growing, driven by the business trends of one-to-one marketing, customer-relationship management, enterprise resource planning, risk management, intrusion detection, and Web personalization – all of which require customer-information analysis and customer-preferences prediction.

Deploying a data mining solution requires collecting data to be mined, cleaning, and transforming its attributes to provide the inputs for data mining models. Also these models need to be built, used, and integrated with different applications. Moreover it is required that currently deployed data management software be able to interact with the data mining models using standards APIs. The scalability aspect calls for collecting data to be mined from distributed and remote locations. Employing common data mining standards greatly simplifies the integration, updating, and maintenance of the applications and systems containing the models.

Over the past several years, various data mining standards have matured and today are used by many of the data mining vendors, as well as by others building data mining applications. With the maturity of data mining standards, a variety of standards-based data mining services and platforms can now be much more easily developed and deployed. Related fields such as data grids, Web services, and the Semantic Web have also developed standards-based infrastructures and services relevant to KDD. These new standards and standards-based services and platforms have the potential for changing the way the data mining is used.

The data mining standards are concerned with one or more of the following issues:

1. The overall process by which data mining models are produced, used, and deployed: This includes, for example, a description of the business interpretation of the output of a classification tree.
2. A standard representation for data mining and statistical models: This includes, for example, the parameters defining a classification tree.
3. A standard representation for cleaning, transforming, and aggregating attributes to provide the inputs for data mining models: This includes, for example, the parameters defining how zip codes are mapped to three-digit codes prior to their use as a categorical variable in a classification tree.
4. A standard representation for specifying the settings required to build models and to use the outputs of models in other systems: This includes, for example, specifying the name of the training set used to build a classification tree.
5. Interfaces and Application Programming Interfaces (APIs) to other languages and systems: There are standard data mining APIs for Java and SQL. This includes, for example, a description of the API so that a classification tree can be built on data in a SQL database.
6. Standards for viewing, analyzing, and mining remote and distributed data: This includes, for example, standards for the format of the data and metadata so that a classification tree can be built on distributed Web-based data.

The current established standards address these different aspects or dimensions of data mining application development. They are summarized in Table II.1.

Section 2 describes the above standards in detail. In section 3 we design and develop a data mining application using the above standards. Section 4 analyzes the standards and their relationship with each other and proposes the areas where standards are needed. Section 5 illustrates examples on the standards and Section 6 concludes this chapter

II.1 Data Mining Standards

II.1.1 Process Standards

CRISP-DM

CRISP-DM stands for CRoss Industry Standard Process for Data Mining. It is industry, tool, and application neutral standard for defining and validating data mining process. It was conceived in late 1996 by DaimlerChrysler, SPSS, and NCR. The latest version is CRISP-DM1.0.

Table II.1. Summary of Data Mining Standards

Areas	Data Mining Standard	Description
Process Standards	Cross Industry Standard Process for Data Mining (CRISP-DM)	Captures data mining process: Begins with business problem and ends with the deployment of knowledge gained in the process.
XML Standards	Predictive Model Markup Language (PMML)	Model for representing data mining and statistical data.
	Common Warehouse Model for Data Mining (CWM-DM)	Model for metadata that specifies metadata for building settings, model representations, and results from model operations Models are defined through the Unified Modeling Language.
Standard APIs	SQL/MM, Java API (JSR-73), Microsoft OLE-DB	API for Data Mining applications
Protocol for transport of remote and distributed data.	Data Space Transport Protocol (DSTP)	DSTP is used for distribution, enquiry, and retrieval of data in a data space.
Model Scoring Standard	Predictive scoring and update protocol (PSUP)	PSUP can be used for both on-line real-time scoring and updates as well as scoring in an off-line batch environment (Scoring is the process of using statistical models to make decisions.)
Web Standards	XML for analysis (XMLA)	Standard Web service interface designed specifically for on-line analytical processing and data-mining functions (uses Simple Object Access Protocol (SOAP))
	Semantic Web	Semantic Web provides a framework to represent information in machine processable form and can be used to extract knowledge from data mining systems.
	Data Space	Provides an infrastructure for creating a Web of data. Is built around standards like XML, DSTP, PSUP. Helps handle large data sets that are present on remote and distributed locations.
Grid Standards	Open Grid Service Architecture	Developed by Globus – this standard talks about service-based open architecture for distributed virtual organizations. It will provide data mining engine with secure, reliable, and scaleable high bandwidth access to the various distributed data sources and formats across various administrative domains.

Motivation

As the market interest in data mining was resulting into its widespread uptake every new adopter of data mining was required to come up with his own approach of incorporating data mining in his current set up. There was also a requirement of demonstrating that data mining was sufficiently mature to be adopted as a key part of any customer's business process. CRISP-DM provided the standard process model for conceiving, developing, and deploying a data mining project, which is nonpropriety and freely distributed.

Standard Description

The CRISP-DM organizes the process model into a hierarchical process model. At the top level the task is divided into phases. Each phase consists of several second-level generic tasks. These tasks are complete (covering the phase and all possible data mining applications) and stable (valid for yet unforeseen developments). These generic tasks are mapped to specialized tasks. Finally these specialized tasks contain several process instances, which are record of the actions, decisions, and results of an actual data mining engagement process. This is depicted in Fig. II.1.

Mapping of the generic tasks (e.g., task for cleaning data) to specialized task (e.g., cleaning numerical or categorical value) depends on the data mining context. CRISP-DM distinguishes between four different dimensions of data mining contexts. These are:

Application domain (areas of the project, e.g., response modeling)
Data mining problem type (e.g., clustering or segmentation problem)

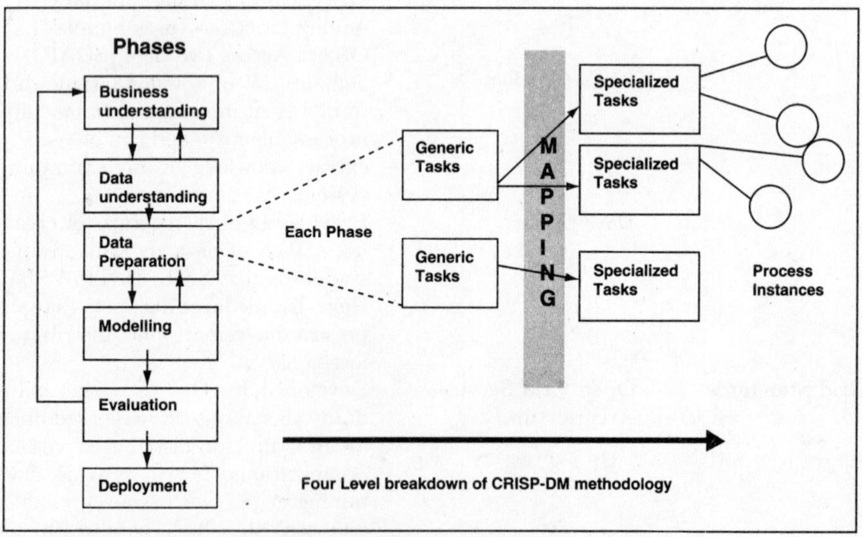

Fig. II.1. CRISP-DM process Model

Technical aspect (issues like outliers or missing values)
Tool and technique (e.g., Clementine or decision trees).

The more value for these different context domains are fixed, the more concrete is the data mining context. The mappings can be done for the current single data mining project in hand or for the future.

The process reference model consists of phases shown in Fig. II.1 and summarized in Table II.2. The sequence of the phases is not rigid. Depending on

Table II.2. Phases in CRISP-DM Process Reference Model

Phases	Description
Business understanding	Focuses on assessing and understanding the project objectives and requirements from a business perspective, then converting this knowledge into a data mining problem definition and a preliminary plan designed to achieve the objectives.
Data understanding	-Starts with an initial data collection. -The data collected is then described and explored (e.g., target attribute of a prediction task is identified). - Then the data quality is verified (e.g., noise or missing values).
Data preparation	Covers all activities to construct the final dataset (data that will be fed into the modeling tool(s)) from the initial raw data. Data preparation tasks are likely to be performed multiple times and not in any prescribed order. The data to be used for analysis is -Selected -Cleaned (their data quality is raised to the level required by the analysis technique) -Constructed (e.g., derived attributes like area = length * breadth are created) -Integrated (information from multiple tables is combined to create new labels) and formatted.
Modeling	-Specialized Modeling techniques are selected (e.g., decision tree with C4.5 algorithm) -Test design is generate to test model's quality and validity. -The modeling tool is run on created data set. -The model is assessed and evaluated. (accuracy tested)
Evaluation	-The degree to which the model meets the business objectives is assessed. -The model undergoes a review process identifying the objectives missed or accomplished based on this whether the project should be deployed or not is determined.
Deployment	Depending on the requirements, the deployment phase can be as simple as generating a report or as complex as implementing a repeatable data mining process across the enterprise. A deployment plan is chalked out before actually carrying out the deployment.

the outcome of each phase, which phase or which particular task of a phase to be performed next is determined.

Interoperability with Other Standards

CRISP-DM provides a reference model, which is completely neutral to other tools, vendors, applications, or existing standards.

II.1.2 XML Standards/ OR Model Defining Standards<TODO>

PMML

PMML stands for The Predictive Model Markup Language. It is being developed by the Data Mining Group [DMG], a vendor-led consortium, which currently includes over a dozen vendors, including Angoss, IBM, Magnify, MINEit, Microsoft, National Center for Data Mining at the University of Illinois (Chicago), Oracle, NCR, Salford Systems, SPSS, SAS, and Xchange. PMML is used to specify the models. The latest version of PMML Version 2.1 was released in March 2003. There have been 6 releases so far.

Motivation

A standard representation for data mining and statistical models was required. Apart from this it was required that it be relatively narrow so that it could serve as common ground for several subsequent standards so that these standards could interoperate.

Standard Description

PMML is an XML mark up language, which provides a way for applications to define statistical and data mining models and to share models between PMML-compliant applications.

It allows users to develop models within one vendor's application, and use other vendors' applications to visualize, analyze, evaluate, or otherwise use the models.

It describes the inputs to data mining models, the transformations used prior to prepare data for data mining, and the parameters that define the models themselves. PMML consists of the components summarized in Table II.3.

Since PMML is an XML-based standard, the specification comes in the form of an XML Document Type Definition (DTD). A PMML document can contain more than one model. If the application system provides a means of selecting models by name and if the PMML consumer specifies a model name, then that model is used; otherwise the first model is used. (Refer to sect. 5.1 for an example of PMML.)

Table II.3. PMML Components of Data Mining Model

PMML Component	Description
Data Dictionary	Data dictionary contains data definitions that do not vary with the model. -Defines the attributes input to models -Specifies the type and value range for each attribute.
Mining Schema	The mining schema contains information that is specific to a certain model and varies with the model. Each model contains one mining schema that lists the fields used in the model. These fields are a subset of the fields in the data dictionary, e.g., the mining schema specifies the usage type of an attribute, which may be active (an input of the model), predicted (an output of the model), or supplementary (holding descriptive information and ignored by the model).
Transformation Dictionary	Defines derived fields. Derived fields may be defined by: -Normalization that maps continuous or discrete values to numbers -Discretization that maps continuous values to discrete values -Value mapping that maps discrete values to discrete values -Aggregation that summarizes or collects groups of values, e.g., by computing averages.
Model Statistics	The model statistics component contains basic univariate statistics about the model, such as the minimum, maximum, mean, standard deviation, median, etc. of numerical attributes.
Model Parameters	PMML also specifies the actual parameters defining the statistical and data mining models per se. The different models supported in version 2.1 are: regression models, clusters models, trees, neural networks, Bayesian models, association rules, sequence models.
Mining Functions	Since different models like neural networks and logistic reasoning can be used for different purposes, e.g., some instances implement prediction of numeric values, while others can be used for classification. Therefore, PMML version 2.1 defines five different mining functions, which are association rules, sequences, classifications, regression, and clustering.

Interoperability with Other Standards

PMML is complementary to many other data mining standards. Its XML interchange format is supported by several other standards, such as XML for Analysis, JSR 73, and SQL/MM Part 6: Data Mining. PMML provides applications a vendor-independent method of defining models so that proprietary issues and incompatibilities are no longer a barrier to the exchange of models between applications.

CWM-DM

CWM-DM stands for Common Warehouse Model for Data Mining. It was specified by members of the JDM expert group and has many common elements with JDM. It is a new specification for data mining metadata and has recently been defined using the Common Warehouse Metadata (CWM) specification from Object Management Group.

Motivation

Different data warehousing solutions, including data mining solutions, should be provided transparently to applications through a unified metadata management environment. Metadata not only links individual software components provided by one software vendor, but it also has the potential to open a data warehousing platform from one provider to third-party analytic tools and applications. The Common Warehouse Metamodel is a specification that describes metadata interchange among data warehousing, business intelligence, knowledge management and portal technologies. The OMG Meta-Object Facility bridges the gap between dissimilar metamodels by providing a common basis for metamodels. If two different metamodels are both MOF-conformant, then the models based on them can reside in the same repository.

Standard Description

The CWM-DM consists of the following conceptual areas that are summarized in Table II.4. CWM DM also defines tasks that associate the inputs to mining operations, such as build, test, and apply (score).

Table II.4. CWM-DM conceptual areas

CWM-DM areas	**Description**
Model description	This consists of: -MiningModel, a representation of the mining model itself - MiningSettings, which drive the construction of the model - ApplicationInputSpecification, which specifies the set of input attributes for the model -MiningModelResult, which represents the result set produced by the testing or application of a generated model.
Settings	Mining Settings has four subclasses representing settings for • Statistics Settings • Clustering Settings • Supervised Mining Settings • Association Rules Settings.

The Settings represents the mining settings of the data mining algorithms on the function level, including specific mining attributes.

The Attributes defines the data mining attributes and has Mining Attribute as its basic class.

Interoperability with Other Standards

CWM supports interoperability among data warehouse vendors by defining Document Type Definitions (DTDs) that standardize the XML metadata interchanged between data warehouses.

The CWM standard generates the DTDs using the following three steps: First, a model using the Unified Modeling Language is created. Second the UML model is used to generate a CWM interchange format called the MetaObject Facility/XML Metadata Interchange. Third, the MOF/XML is converted automatically to DTDs.

II.1.3 Web Standards

With the expansion of the World Wide Web, it has become one of the largest repositories of data. Hence it is possible that data to be mined is distributed and needs to be accessed via Web.

XMLA

Microsoft and Hyperion had introduced XML for Analysis, which is a Simple Object Access Protocol (SOAP)-based XML API designed for standardizing data access between a Web client application and an analytic data provider, such as an OLAP or data mining application. XMLA APIs support the exchange of analytical data between clients and servers on any platform and with any language.

Motivation

Under traditional data access techniques, such as OLE DB and ODBC, a client component that is tightly coupled to the data provider server must be installed on the client machine in order for an application to be able to access data from a data provider. Tightly coupled client components can create dependencies on a specific hardware platform, a specific operating system, a specific interface model, a specific programming language, and a specific match between versions of client and server components. The requirement to install client components and the dependencies associated with tightly coupled architectures are unsuitable for the loosely coupled, stateless, cross-platform, and language independent environment of the Internet. To provide reliable data access to Web applications the Internet, mobile devices, and cross-platform desktops need a standard methodology that does not require component downloads to the client. Extensible Markup Language (XML) is generic and can be universally accessed.

XML for Analysis advances the concepts of OLE DB by providing standardized universal data access to any standard data source residing over the Web without the need to deploy a client component that exposes COM interfaces. XML for Analysis is optimized for the Web by minimizing roundtrips to the server and targeting stateless client requests to maximize the scalability and robustness of a data source.

Standard Description

XMLA – XML-based communication API – defines two methods, Discover and Execute, which consume and send XML for stateless data discovery and manipulation.

The two APIs are summarized in Table II.5.

ResultSet: This required parameter contains the result set returned by the provider as a Rowset object. The Discover and Execute methods enable users to determine what can be queried on a particular server and, based on this, submit commands to be executed.

The client having the URL for a server hosting a Web service sends Discover and Execute calls using the SOAP and HTTP protocols to the server. The server instantiates the XMLA provider, which handles the Discover and Execute calls. The XMLA provider fetches the data, packages it into XML, and then sends the requested data as XML to the client. (Refer to Sect. 5.2 for a detailed example of XMLA.)

Interoperability with Other Standards

XMLA specification is built upon the open Internet standards of HTTP, XML, and SOAP, and is not bound to any specific language or technology.

Semantic Web

The World Wide Web Consortium (W3C) standards for the Semantic Web defines a general structure for knowledge using XML, RDF, and ontologies [W3C SW]. The Semantic Web approach develops languages for expressing information in machine-processable form. The Semantic Web provides a common framework that allows data to be shared and reused across application, enterprise, and community boundaries. It is a collaborative effort led by W3C with participation from a large number of researchers and industrial partners and is based on the Resource Description Framework (RDF), which integrates a variety of applications using XML for syntax and URIs for naming.

This infrastructure in principle can be used to store the knowledge extracted from data using data mining systems, although at present, one could

Table II.5. XMLA APIs

XMLA APIS	Description
Discover	It is used to obtain information (e.g., a list of available data sources) and metadata from Web services. The data retrieved with the Discover method depends on the values of the parameters passed to it. Syntax: Discover ([in] RequestType As EnumString, [in] Restrictions As Restrictions [in] Properties As Properties,// [out] Resultset As Rowset) *RequestType:* Determines the type of information to be returned *Restrictions:* Enables the user to restrict the data returned in Resultset *Properties:* Enables the user to control some aspect of the Discover method, such as defining the connection string, specifying the return format of the result set, and specifying the locale in which the data should be formatted. The available properties and their values can be obtained by using the DISCOVER_PROPERTIES request type with the Discover methodResultSet. *ResultSet:* This required parameter contains the result set returned by the provider as a Rowset object.
Execute	The Execute method is used for sending action requests to the server. This includes requests involving data transfer, such as retrieving or updating data on the server. Syntax: Execute ([in] Command As Command, [in] Properties As Properties, [out] ResultSet As ResultSet) *Command*: It consists of a provider-specific statement to be executed. For example, this parameter contains a <Statement> tag that contains an SQL command or query. *Properties*: Each property allows the user to control some aspect of the Execute method, such as defining the connection string, specifying the return format of the result set, or specifying the locale in which the data should be formatted.

argue that this is more of a goal than an achievement. As an example of the type of knowledge that can be stored in the Semantic Web, RDF can be used to code assertions such as "credit transactions with a dollar amount of $1 at merchants with a MCC code of 542 have a 30% likelihood of being fraudulent."

Data Space

Data space is an infrastructure for creating a web of data or data webs. The general operations in the Web involve browsing remote pages or documents whereas the main purpose of having a data space is to explore and mine remote columns of distributed data. Data webs are similar to Semantic Webs except that they house data instead of documents.

Motivation

The Web today contains a large amount of data. Although the amount of scientific, health care, and business data is exploding, we do not have the technology today to casually explore remote data either to mine distributed data. The size of individual data sets has also increased. There are a certain issues involved in the process of analyzing such a data. The multimedia documents on the Web cannot be directly used for the process of mining and analyzing. Another issue is that the current Web structure does not optimally support handling of large data sets and is best suited only for browsing hypertext documents. Hence there is a need to have a standard support to this data. The concept of a data space helps explore, analyze, and mine such data.

Standard Description

The DataSpace project is supported by the National Science Foundation and has Robert Grossman as its director. DataSpace is built around standards developed by the Data Mining Group and W3C. The concept of a DataSpace is based upon XML and Web services, which are W3C-maintained standards. DataSpace defines a protocol DSTP (DataSpace Transfer Protocol) for distribution, enquiry, and retrieval of data in a DataSpace. It also works with the real-time scoring standard PSUP (Predictive Scoring and Update Protocol).

The DataSpace consists of the components as shown in Fig. II.2: DSTP is the protocol for the distribution, enquiry and retrieval of data in a DataSpace. The data could be stored in files, databases or distributed databases. It has a corresponding XML file, which contains Universal Correlation Key tags (UCK) that act as identification keys.

The UCK is similar to a primary key in a database. A join can be performed by merging data from different servers on the basis of UCKs.[DSTP]

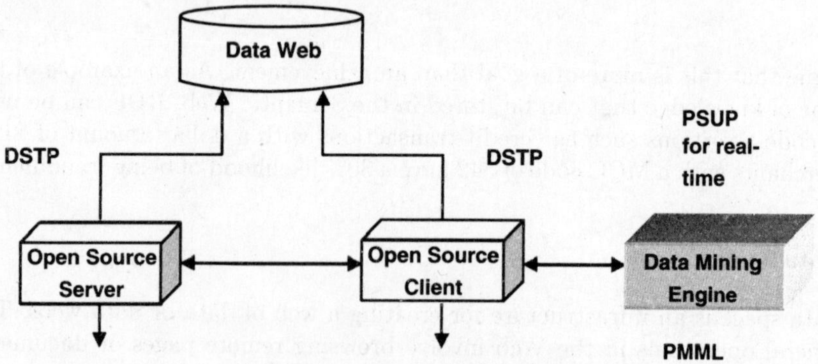

Fig. II.2. DataSpace Architecture

Table II.6. Summary of Data Space Standards

Standard	Description
DSTP	Provides direct support for attributes, keys and meta data. Also supports: Attribute Selection Range Queries Sampling Other functions for accessing and analyzing remote data
PSUP	Is a protocol is a protocol for event driven, real time scoring. PSUP provides the ability to use PMML in real time applications.

The Predictive Scoring and Update Protocol is a protocol for event driven, real time scoring. Real time applications are becoming increasing important in business, e-business, and health care. PSUP provides the ability to use PMML applications in real time and near real time applications.

For the purpose of data mining a DSTP client is used to access the remote data. The data is retrieved from the required sites and DataSpace is designed to interoperate with proprietary and open source data mining tools. In particular the open source statistical package R has been integrated into Version 1.1 of DataSpace and is currently being integrated into Version 2.0. DataSpace also works with predictive models in PMML, the XML markup language for statistical and data mining models. The above mentioned standards are briefed in Table II.6.

II.1.4 Application Programming Interfaces (APIs)

Earlier application developers wrote their own data mining algorithms for applications, or used sophisticated end-user GUIs. The GUI package for data mining included a complete range of methods for data transformation, model building, testing and scoring. But it remained challenging to integrate data mining and the application code due to lack of proper APIs to do the task. APIs were vendor specific and hence proprietary. Thus the product developed would become dependent and hence risky to market. To switch to a different vendor's solution the entire code had to be re-written, which made the process costly.

In short it was realized that data-mining solutions must co-exist. Hence the need arose to have a common standard for the APIs. The ability to leverage data mining functionality via a standard API greatly reduces risk and potential cost. With a standard API customers can use multiple products for solving business problems by applying the most appropriate algorithm implementation without investing resources to learn each vendor's proprietary API. Moreover, a standard API makes data mining more accessible to developers while making developer skills more transferable. Vendors can now differentiate themselves on price, performance, accuracy, and features.

SQL/ MM DM

SQL/MM is an ISO/IEC international standardization project. The SQL/MM suite of standards includes parts used to manage full-text data, spatial data, and still images. Part 6 of the standard addresses data mining.

Motivation

Database systems should be able to integrate data mining applications in a standard way so as to enable the end user to perform data mining with ease. Data mining has become a part of modern data management and could be said to be a sophisticated tool to extract information or to aggregate the original data. SQL is a language widely used by database users today and provides basic operations of aggregate, etc. Thus data mining could be said to be a natural extension to the primitive functionalities provided by SQL. Hence it becomes obvious to standardize data mining through SQL.

Standard Description

The SQL/MM Part 6: Data mining standard provides an API for data mining applications to access data from SQL-MM compliant relational databases. It defines structured user-defined types, including associated methods to support data mining. It attempts to provide a standardized interface to data mining algorithms that can be layered atop of any object-relational database system and even deployed as a middleware when required. Table II.7 below provides a brief description of the standard:

Java API's

Java Specification Request -73 (JSR-73), also known as Java Data Mining (JDM), defines a pure Java API to support data mining operations. The JDM development team was led by Oracle and included other members like Hyperion, IBM, Sun Microsystems, and others.

Motivation

Java has become a language that is widely used by application developers. The Java 2 Platform, Enterprise Edition (J2EE), provides a standard development and deployment environment for enterprise applications. It reduces the cost and complexity of developing multitier enterprise services by defining a standard, platform-independent architecture for building enterprise components.

JSR-73 provides a standard way to create, store, access and maintain data and metadata supporting data mining models, data scoring and data mining

Table II.7. Summary of SQL/MM DM Standard

	Description	
Data Mining Techniques	4 Different data mining techniques supported by this:	
	Row Model	Allows to search for patterns and relationships between different parts of your data
	Clustering Model	Helps grouping of Clusters
	Regression Model	Helps predict the ranking of new data base upon the analysis of existing data
	Classification Model	Helps predicting the grouping or class of the new data
Data Mining Stages	3 distinct stages through which data can be mined	
	Train	Choose technique most appropriate Set parameters to orient the model Train by applying reasonably sized data
	Test	For classification and regression test with known data and compare the model's predictions
	Apply	Apply the model to the business data
Supporting Data Types	**DM_*Model,** Defines the model that you want to use when mining your data **DM_*Settings** Stores various parameters of the data mining model, e.g. - Depth of a decision tree - Maximum number of clusters **DM_*Result** Created by running data mining model against real data **DM_*TestResult** Holds the results of testing during the training phase of the data mining models **DM_*Task** Stores the metadata that describe the process and control of the testing and of the actual runnings. where * could be "Clas" – Classification Model "Rule" – Rule Model "Clustering" – Clustering Model "Regression" – Regression Model	

results serving J2EE compliant application servers. It provides a single standard API or data mining system that will be understood by a wide variety of client applications and components running on the J2EE platform. This specification does not preclude, however, the use of JDM services outside of the J2EE environment.

Standard Description

Defining compliance for vendor specification asks for addressing several issues. In JDM, data mining includes the functional areas of classification, regression, attribute importance, clustering, and association. These are supported by supervised and unsupervised algorithms as decision trees, neural networks, naïve Bayes, support vector machines, K-means on structured data. A particular implementation of this specification may not necessarily support all interfaces and services provided by JVM. JDM is based on a generalized, object-oriented, data mining conceptual model leveraging emerging data mining standards such the Object Management Group's Common Warehouse Metadata (CWM), ISO's SQL/MM for Data Mining, and the Data Mining Group's Predictive Model Markup Language (PMML), as appropriate implementation details of JDM are delegated to each vendor.

A vendor may decide to implement JDM as a native API of its data mining product. Others may opt to develop a driver/adapter that mediates between a core JDM layer and multiple vendor products. The JDM specification does not prescribe a particular implementation strategy, nor does it prescribe performance or accuracy of a given capability or algorithm. To ensure J2EE compatibility and eliminate duplication of effort, JDM leverages existing specifications. In particular, JDM leverages the Java Connection Architecture to provide communication and resource management between applications and the services that implement the JDM API. JDM also reflects aspects the Java Metadata Interface. [JDM]

Microsoft OLEDB-DM

In July 2001 Microsoft released specification document for first real industrial standard for data mining called OLEDB for Data Mining. This API is supported by Microsoft and in part of release of Microsoft SQL Server 2000 (Analysis Server component). (Refer to Sect. 5.3 for an overview of OLEDB.)

Motivation

An industry standard was required for data mining so that different data mining algorithms from various data mining ISVs can be easily plugged into user applications. OLEDB-DM addressed the problem of deploying models (once the model is generated, how to store, maintain, and refresh it as data in the warehouse is updated, how to programmatically use the model to do

Table II.8. Summary of Java Data Model Standards

Architectural Components	JDM has 3 logical components **Application Programming Interface**: Is the end user visible component of a JDM implementation that allows access to the services provided by the data mining engine. An application developer would require the knowledge of only this librar **Data Mining Engine**: Provides the infrastructure that offers a set of data mining services to the API clients **Metadata repository**: Serves to persistent data mining objects. The repository can be based on the CWM framework.
Data Mining Functions	JDM specifies the following data mining functions: **Classification**: Classification analyzes the input or the *build data* and predicts to which class a given case belongs. **Regression**: Regression involves predicting a continuous, numerical valued target attribute given a set of predictors. **Attribute Importance**: Determines which attributes are most important for building a model. Helps users to reduce the model build time, scoring time, etc. Similar to feature selection. **Clustering**: Clustering Analysis finds out clusters embedded in the data, where a cluster is a collection of data objects similar to one another. **Association**: Has been used in market basket analysis and analysis of customer behavior for the discovery of relationships or correlations among a set of items.
Data Mining Tasks	Data mining revolves around a few common data mining tasks: **Building a Model**: Users define input tasks specifying the parameters model name, mining data, and mining settings. JDM enables users to build models in the functional areas – classification, regression, attribute importance, clustering and association.
	Testing a Model: Gives an estimate of the accuracy a model has in predicting the target. Follows model building to compute the accuracy of a model's predictions when the model is applied to a previously unseen data set. Input consists of model and data for testing the model. Test results could be confusion matrix, error estimates, etc. *Lift* is a measure of effectiveness of a predictive model. A user may specify to compute lift. **Applying a Model**: Model is finally applied to a case. Produces one or more predictions or assignments. JDM enables **Object Import and Export**: Could be useful in • Interchange with other DMEs • Persistent storage outside the DME • Object inspection or manipulation • To enable import and export of system metadata JDM specifies 2 standards for defining metadata in XML • PMML for mining models • CWM • **Computing statistics** on data: Provides to compute various statistics on a given physical data set. • **Verifying task correctness**
Extension Packages	javax.datamining javax.datamining.settings javax.datamining.models javax.datamining.transformations javax.datamining.results
Conformance Statement	JDM API standard is flexible and allows vendors to implement only specific functions that they want their product to support. Packages are divided into 2 categories -Required: Vendors must provide an implementation for this. -Optional: A vendor may or may not implement these.

predictions on other data sets, and how to browse models over the life cycle of an enterprise). Another motivation to introduce OLE DB DM was to enable enterprise application developers to participate in building data mining solutions. For this it was required that the infrastructure for supporting data mining solution is aligned with traditional database development environment and with APIs for database access.

Standard Description

OLE DB for DM is an OLE DB extension that supports data mining operations over OLE DB data providers. It has a concept of data mining providers: Software packages that provide data mining algorithms. Data mining consumers: Those applications that use data mining features. OLE DB for DM specifies the API between data mining consumers and data mining providers. It introduces two new concepts of cases and models in the current semantics of OLEDB.

CaseSets: Input data is in the form of a set of cases (caseset). A case captures the traditional view of an "observation" by machine learning algorithms as consisting of all information known about a basic entity being analyzed for mining as opposed to the normalized tables stored in databases. It makes use of the concept of nested tables for this.

Data mining model (DMM): It is treated as if it were a special type of "table": A caseset is associated with a DMM and additional metainformation while creating (defining) a DMM. When data (in the form of cases) is inserted into the data mining model, a mining algorithm processes it and the resulting abstraction (or DMM) is saved instead of the data itself. Once a DMM is populated, it can be used for prediction, or its content can be browsed for reporting.

The key operations to support on a data mining model are shown in Table II.9.

This model has an advantage of having a low cost of deployment. (Refer Sect. 5.3 for an example.)

Interoperability with Other Standards

OLE DB for DM is independent of any particular provider or software and is meant to establish a uniform API. It is not specialized to any specific mining model but is structured to cater to all well-known mining models.

II.1.5 Grid Services

Grids are collections of computers or computer networks, connected in a way that allows for sharing of processing power and storage as well as applications and data. Grid technologies and infrastructures are hence defined as supporting the sharing and coordinated use of diverse resources in dynamic, distributed "virtual organizations."

Table II.9. DMM Operations

Operations on DMM	Description	Syntax
Define	Identifying the set of attributes of data – to be predicted – to be used for prediction and the algorithm used to build the mining model	CREATE statement
Populate	Populating a mining model from training data using the algorithm specified in its definition above	Repeatedly via the INSERT INTO statement (used to add rows in a SQL table), and emptied (reset) via the DELETE statement.
Predict	Predicting attributes for new data using a mining model that has been populated	Prediction on a dataset made by making a PREDICTION JOIN between the mining model and the data set.
Browse	Browsing a mining model for reporting and visualization applications	Using SELECT statement

OGSA and data mining

The Open Grid Services Architecture (OGSA) represents an evolution toward a Grid architecture based on Web services concepts and technologies. It consists of a well-defined set of basic interfaces used to communicate extensibility, vendor neutrality, and commitment to a community standardization process. It uses the Web Services Description Language (WSDL) to achieve self-describing, discoverable services and interoperable protocols, with extensions to support multiple coordinated interfaces and change management.

Motivation

In a distributed environment, it is important to employ mechanisms that help in communicating interoperably. A service-oriented view partitions this interoperability problem into two subproblems:

Definition of service interfaces and the identification of the protocol(s) that can be used to invoke a particular interface

Agreement on a standard set of such protocols. A service-oriented view allows local/remote transparency, adaptation to local OS services, and uniform service semantics.

A service-oriented view also simplifies encapsulation behind a common interface of diverse implementations that allows for consistent resource access across multiple heterogeneous platforms with local or remote location transparency, and enables mapping of multiple logical resource instances onto the same physical resource and management of resources. Thus service definition is decoupled from service invocation. OGSA describes and defines a service-oriented architecture composed of a set of interfaces and their corresponding behaviors to facilitate distributed resource sharing and accessing in heterogeneous dynamic environments. Data is inherently distributed and hence the data mining task needs to be performed keeping this distributed environment in mind. Also it is required to provide data mining as a service. Grid technology provides secure, reliable, and scaleable high-bandwidth access to distributed data sources across various administrative domains that can be exploited.

Standard Description

Figure II.3 shows the individual components of the service-oriented architecture (SOA). The service directory is the location where all information about all available grid services is maintained. A service provider that wants to offer services publishes its services by putting appropriate entries into the service directory. A service requestor uses the service directory to find an appropriate service that matches its requirements. An example of data mining scenario using this architecture is as follows. When a service requestor locates a suitable data mining service, it binds to the service provider, using binding information maintained in the service directory.

The binding information contains the specification of the protocol that the service requestor must use as well as the structure of the request messages and the resulting responses. The communication between the various agents occurs via an appropriate transport mechanism. Grid offers basic services that include resource allocation and process management, unicast and

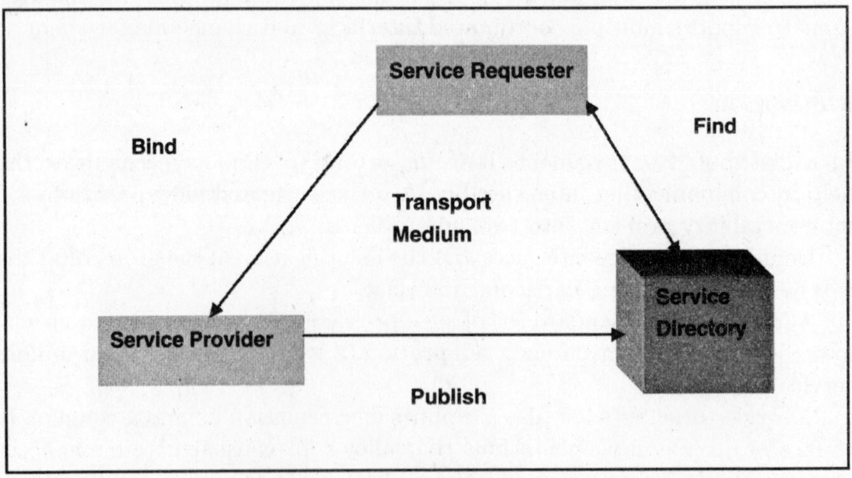

Fig. II.3. Service-oriented architecture

multicast communication services, security services, status monitoring, remote data access, etc.

Apart from this there is data grid that provides Grid FTP (a secure, robust, and efficient data transfer protocol) and metadata information management system. Hence, the grid-provided functions do not have to be re-implemented for each new mining system, e.g., single sign-on security, ability to execute jobs at multiple remote sites, ability to securely move data between sites, broker to determine best place to execute mining job, job manager to control mining jobs, etc. Therefore, mining system developers can focus on the mining applications and not the issues associated with distributed processing. However, the standards for these are yet to be developed.

Interoperability with other standards:

The standard for Grid Services is yet to emerge.

II.2 Developing Data Mining Application Using Data Mining Standards

In this section we describe a data mining application. We then describe its architecture using data mining standards. However we see that not all the architecture constructs can be standardized as no standards are available for them. We point this out in more detail in Sect. II.4 below.

II.2.1 Application Requirement Specification

A multinational food chain has its outlets in several countries, e.g., India, USA, and China. The outlets in each of these want information regarding:

- Combinations of food items that constitute their happy meal.
- Most preferred food items they need to target for their advertisements in the respective country.
- Preferred seasonal food items.
- Information about the food item, their prices and their popularity and coming up with patterns that reveal the relationship between the pricing and the popularity.

The above information must be obtained from these transactions solely as the food chain company does not want to indulge in any surveys. All the customer transactions of each outlet are recorded. The transactions contain along with customer id, the food items, their prices, and the time at which the order was placed. However each outlet could store transactions in different databases like Oracle, Sybase for the same. As we see this is a typical data mining application. In the next section we describe the run-time architecture of the data mining system. We also see how application of standards makes the components of this architecture independent of each other as well as of the underlying platform or technology.

II.2.2 Design and Deployment

Architecture Overview

In the architecture shown in Fig. II.4, the outlets (data sources) are spread in multiple locations (Location A, B, C) henceforth referred to as *remote data sources.* The data before being mined has to be aggregated in a single location. For this we use a client server architecture. Each of the remote data sources have a data server which might connect to the respective database using any standard drivers. A client is deployed in the location where data to be mined is collected. This client contacts these servers for browsing or retrieving data. As mentioned in the figure we need a standard for data transport over the Web so that this entire client server architecture can be independently developed and deployed.

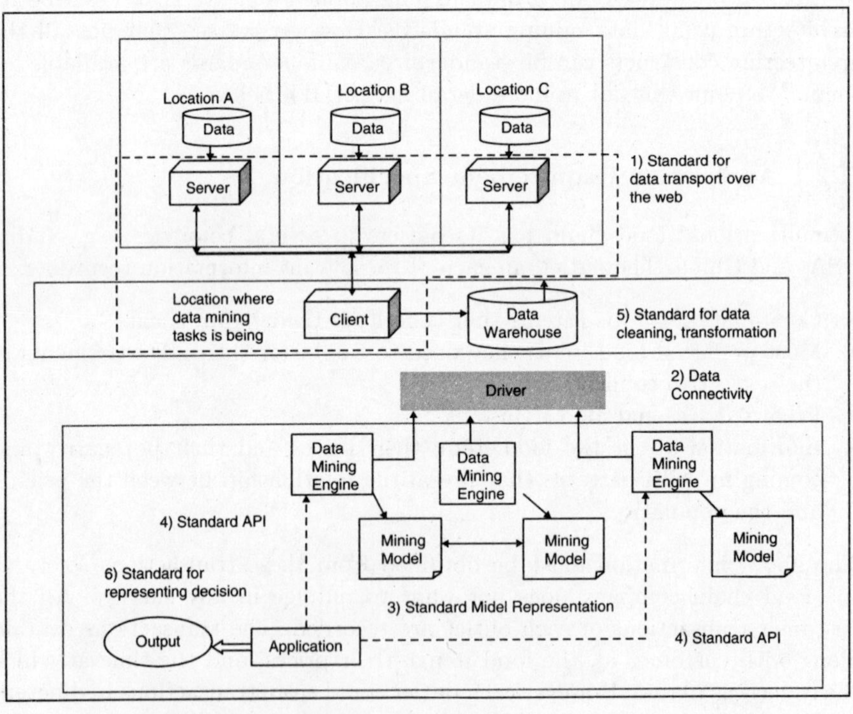

Fig. II.4. Architecture of Data Mining Application

The client stores the data in a data warehouse so that data mining operations can be performed on it. But before the data is to be mined it needs to be cleaned and transformed. Some standards should be present for this purpose.

The DataMining Engine accesses the data in the warehouse with the help of standard data connectivity mechanisms. It produces a mining model such a decision tree, etc. This model is then used to discover patterns in the data. It is required that the model produced be represented in a standard format so as to allow interoperability across vendors as well as different data mining engine. Hence a standard is required for the same.

The data mining engine is accessible to the end user via an application programming interface. The application requiring data mining contains the calls of the API. This set of APIs should be standardized so as to allow the application to switch to a different vendors solution without being concerned about changing his entire code.

Also, once the data mining task is performed the output produced needs to be incorporated into the existing business model. Hence the decisions or suggestions recommended by the data mining model need to be stored. For this a standardized decision model is required that incorporates this decision model with the current business model.

Standards Employed in the Architecture

For the data transport over the Web the standard DSTP [Sect. 2.3.3] is employed. The mining model produced by the data mining engine is PMML [Sect. 2.2.1] compliant so as to enable interoperability. If not PMML then any model that confirms to metamodel specifications of CWM-DM [Sect. 2.2.2] must be used. However the most widely used model currently is PMML. The data mining engine connects to the data warehouse using any of the JDBC or ODBC drivers. Here we use JDBC driver for it. The application uses the data mining services with the help of the standard API –JSR-73. [Sect. 2.4.2].

The entire system should be developed using the Process Standard CRISP-DM [Sect. 2.1.1].

If we want this data mining application to be deployed as a Web service then we can use a provider server at this end that supports XMLA's Execute and Discover APIs [Sect. 2.3.1]. Thus any third party can fire queries without having any software installed at its end.

Standards Not Yet Defined

As we see there are no current standards that can be used for data transformation. Also there is no standard decision model that could incorporate the output of a data mining task into our engine. We discuss this further in sect. II.4.

Scoring should also be integrated with the mining applications via published standard API's and run-time library scoring engines. Automation of the scoring process will reduce processing time, allow for the most up-to-date data to be used, and reduce error.

II.3 Analysis

Earlier data mining comprised algorithms working on flat files with no standards. Industry interest led to development of standards that enabled representation of these algorithms in a model and separation of online development of these models with their deployment. These standards are maturing and becoming rich enough to cover both the data preparation as well as the scoring. In parallel, standards are being developed for integrating data mining standards with the standards in near-by communities such as grid computing and Web-services based computing.

We saw that the standards introduce high requirements on data mining products mainly demanded by users of data mining technology. From the standards we have studied we conclude that they achieve the following two goals:

- Unify existing products with well-known functionality (Partially) design the functionality such that future products match real-world requirements. Therefore these early standards will drive the data mining products and not vice versa.

Also the narrow nature of earlier standards like PMML serves as a common ground for several emerging standards. Adoption of this standard helps in exporting models across different applications. Together PMML and CWM-DM also provide mechanism for abstracting specific data mining technologies by hiding the complexity of the underlying data mining algorithm. Also specific APIs enable seamless integration of data mining solutions into existing infrastructure. We also see that different standards address disjoint areas of data mining application design and development, so they easily interoperate. However, we are yet to have mature standards in the following area:

- Input data transformation, cleaning
- Standard for having a decision model of the output data mining applications that can be seamlessly incorporated in the existing business model.
- Standard for model scoring, which can be integrated with the driving applications via standardized published API's and run-time library scoring engines.

Some standards need to have additional information where for each data mining task performance versus accuracy can be queried. This is required since data mining API users may not know the cost performance trade off of a particular algorithm chosen. This standard's utility will be uncovered more if more than one implementation of a given data mining algorithm example decision trees is provided.

In this survey it is found that integrating data mining applications is still a challenge as all the standards have not yet been completely adopted by the vendors. However we see promising trends as now vendors (for example Oracle) are providing support for data mining applications.

II.4 Application Examples

II.4.1 PMML Example

Data dictionary for Fisher's Iris data set:

<DataDictionary numberOfFields = "5">

<DataField name = "Petal_length" optype = "continuous"/>

<DataField name = "Petal_width" optype = "continuous"/>

<DataField name = "Sepal_length" optype = "continuous"/>

<DataField name = "Sepal_width" optype = "continuous"/>

<DataField name = "Species_name" optype = "categorical">

<Value value = "Setosa"/>

<Value value = "Verginica"/>

<Value value = "Versicolor"/>

< /DataField>

< /DataDictionary>

Corresponding mining schema:

<MiningSchema>

<MiningField name = "Petal_length" usageType = "active"/>

<MiningField name = "Petal_width" usageType = "active"/>

<MiningField name = "Sepal_length" usageType = "supplementary"/>

<MiningField name = "Sepal_width" usageType = "supplementary"/>

<MiningField name = "Species_name" usageType = "predicted"/>

< /MiningSchema>

Node of a decision tree built from the data:

<Node score = "Setosa" recordCount = "50">

<SimplePredicate field = "Petal_length" operator = "lessThan" value = "24.5"/>

<ScoreDistribution value = “Setosa” recordCount = “50”/>

<ScoreDistribution value = “Verginica” recordCount = “0”/>

<ScoreDistribution value = “Versicolor” recordCount = “0”/>

< /Node>

Association Rule Example:

<PMML>...

<!– We have three items in our input data –>

<Item id = “1” value = “Cracker” /> <Item id = “2” value = “Coke” />

<Item id = “3” value = “Water” />

<!– and two frequent itemsets with a single item –>

<**Itemset id = “1”** support = “1.0” numberOfItems = “1”>

<ItemRef itemRef = “1” /> </Itemset>

<**Itemset id = “2”** support = “1.0” numberOfItems = “1”>

<ItemRef itemRef = “3” /> </Itemset>

<!– and one frequent itemset with two items. –>

<Itemset id = “3” support = “1.0” numberOfItems = “2”>

<ItemRef itemRef = “1” /> <ItemRef itemRef = “3” />

< /Itemset>

<!– Two rules satisfy the requirements –>

<**AssociationRule support = “1.0” confidence = “1.0” antecedent = “1” consequent = “2”**/>

<AssociationRule support = “1.0” confidence = “1.0” antecedent = “2” consequent = “1” />

< /AssociationModel>

< /PMML>

II.4.2 XMLA Example

Source: XML for Analysis Specification Version 0.90

Sports Statistics Data Provider

A major sports provider makes its sports statistics available for interactive analysis over the Internet as a Microsoft.NET service called Sports Web Pages.

The service uses the XML for Analysis specification to enable access to both data and analytical models. The Sports Web Site Web pages create a platform-neutral thin analysis client application that speaks to an XML for Analysis provider on the Web server. Client users can use any of several different ways to access this information from any device to find interesting information about their favorite sports.

For example, in one scenario a user is at a basketball stadium watching his home team. While discussing the game with his friends, he wants to know the top winning margins for his team. To get this information, he uses his Internet-enabled cell phone to connect to the mobile version of Sports Web Site, and then he uses the analysis client to retrieve a list of available sports databases. He chooses a database for basketball statistics, reviews predefined queries offered by the client application (player statistics, team wins, and so on), and finds exactly the query he needs. He then executes this query and learns that his team is about to break a 40-year-old record!

Another user goes through similar steps, but instead of using a cell phone, he uses a traditional PC or the Web browser in his interactive TV box. In the above scenarios the two users interact with the Sports Web Site client, which in turn sends a sequence of Discover and Execute methods to fetch the users' selections. A Discover method returns metadata that lists the available databases and details about a given database. Preformulated queries in the client run an Execute method when chosen by the user. In the cell phone scenario, only one query is submitted using the Execute method. The results of the Execute methods return a dataset that provides the requested data, which the client Web page formats for the user.

II.4.3 OLEDB

A vast amount of the critical information necessary for conducting day-to-day business is found outside of traditional, corporate production databases. Instead, this information is found in file systems, in indexed-sequential files (e.g., Btrieve), and in personal databases such as Microsoft Access and Microsoft Visual FoxPro; it is found in productivity tools such as spreadsheets, project management planners, and electronic mail; and more and more frequently, it is found on the World Wide Web.

To take advantage of the benefits of database technology, such as declarative queries, transactions and security businesses had to move the data from its original containing system into some type of database management system (DBMS). This process is expensive and redundant. Furthermore, businesses need to be able to exploit the advantages of database technology not only when accessing data within a DBMS but also when accessing data from any other type of information container. To address this need, Microsoft created OLE DB.

OLE DB is a set of component object model (COM) interfaces that provide applications with uniform access to data stored in diverse information sources and that also provide the ability to implement additional database services.

These interfaces support the amount of DBMS functionality appropriate to the data store, enabling it to share its data.

II.4.4 OLEDB-DM Example

<todo Refine>

Create an OLE DB Data Source Object and Obtain an OLE DB Session Object

This is the standard mechanism of connecting to data stores via OLE DB.

Create the Data Mining Model Object

Using an OLE DB command object, the client executes a CREATE statement that is similar to a CREATE TABLE statement.

```
CREATE MINING MODEL [Age Prediction](
[Customer ID] LONG KEY,
[Gender] TEXT DISCRETE,
[Age] DOUBLE DISCRETIZED() PREDICT,
[Product Purchases] TABLE(
[Product Name] TEXT KEY,
[Quantity] DOUBLE NORMAL CONTINUOUS,
[Product Type] TEXT DISCRETE RELATED TO [Product Name]
)
)
USING [Decision Trees]
```

Insert Training Data into the Model

In a manner similar to populating an ordinary table, the client uses a form of the INSERT INTO statement. Note the use of the SHAPE statement to create the nested table.

```
INSERT INTO [Age Prediction](
[Customer ID], [Gender], [Age],
[Product Purchases](SKIP, [Product Name], [Quantity],
[Product Type])
)
```

```
SHAPE {
SELECT [Customer ID], [Gender], [Age] FROM Customers
ORDER BY [Customer ID]
}
APPEND(
{SELECT [CustID], [Product Name], [Quantity],
[Product Type]
FROM Sales
ORDER BY [CustID]}
RELATE [Customer ID] To [CustID])
AS [Product Purchases]
```

Use the Data Mining Model to make some Predictions

Predictions are made with a SELECT statement that joins the model's set of all possible cases with another set of actual cases. The actual cases can be incomplete. In this example, the value for "Age" is not known. Joining these incomplete cases to the model and selecting the "Age" column from the model will return a predicted "age" for each of the actual cases.

```
SELECT t.[Customer ID], [Age Prediction].[Age]
FROM [Age Prediction]
PREDICTION JOIN(
SHAPE {
SELECT [Customer ID], [Gender], FROM Customers
ORDER BY [Customer ID]
}
APPEND (
{SELECT [CustID], [Product Name], [Quantity]
FROM Sales ORDER BY [CustID]}
RELATE [Customer ID] To [CustID]
)
AS [Product Purchases]
) as t
```

```
ON [Age Prediction].Gender = t.Gender and
[Age Prediction].[Product Purchases].[Product Name] =
t.[Product Purchases].[Product Name] and
[Age Prediction].[Product Purchases].[Quantity] =
t.[Product Purchases].[Quantity]
```

II.4.5 SQL/MM Example

DM_RuleModel type represents models that are the result of the search for assoc. rules

```
<!–definition–>
CREATE TYPE DM_RuleModel AS
(
DM_content CHARACTER LARGE OBJECT(DM_MaxContentLength)
)
<!–public members –>
STATIC METHOD DM_impRuleModel
(input CHARACTER LARGE OBJECT(DM_MaxContentLength))
RETURNS DM_RuleModel
METHOD DM_expRuleModel ()
RETURNS CHARACTER LARGE OBJECT(DM_MaxContentLength
METHOD DM_getNORules ()
RETURNS INTEGER
METHOD DM_getRuleTask ()
RETURNS DM_RuleTask
```

II.4.6 Java Data Mining Model Example

The following code illustrates how to build a clustering model on a table stored in a location that is expressed as a URI (uniform resource identifier). Vendors can design their own URIs and thus we do not use a practical URI in this example. It is assumed that a URI is available in this example.

```
// Create the physical representation of the data
(1) PhysicalDataSetFactory pdsFactory = (PhysicalDataSetFactory)
dmeConn.getFactory("javax.datamining.data.PhysicalDataSet");
(2) PhysicalDataSet buildData = pdsFactory.create(uri);
(3) dmeConn.saveObject("myBuildData", buildData, false)
// Create the logical representation of the data from physical data
(4) LogicalDataFactory ldFactory = (LogicalDataFactory)
dmeConn.getFactory("javax.datamining.data.LogicalData");
(5) LogicalData ld = ldFactory.create(buildData);
(6) dmeConn.saveObject("myLogicalData", ld, false);
// Create the settings to build a clustering model
(7) ClusteringSettingsFactory csFactory = (ClusteringSettingsFactory)
dmeConn.getFactory("javax.datamining.clustering.ClusteringSettings");
(8) ClusteringSettings clusteringSettings = csFactory.create();
(9) clusteringSettings.setLogicalDataName("myLogicalData");
(10) clusteringSettings.setMaxNumberOfClusters(20);
(11) clusteringSettings.setMinClusterCaseCount(5);
(12) dmeConn.saveObject("myClusteringBS", clusteringSettings, false);
// Create a task to build a clustering model with data and settings
(13) BuildTaskFactory btFactory = (BuildTaskFactory)
dmeConn.getFactory("javax.datamining.task.BuildTask");
(14) BuildTask task = btFactory.create("myBuildData",
"myClusteringBS","myClusteringModel");
(15) dmeConn.saveObject("myClusteringTask", task, false);
// Execute the task and check the status
(16) ExecutionHandle handle = dmeConn.execute("myClusteringTask");
(17) handle.waitForCompletion(Integer.MAX_VALUE); // wait until done
(18) ExecutionStatus status = handle.getLatestStatus();
(19) if(ExecutionState.success.isEqual(status.getState()))
(20) // task completed successfully...
```

II.4.7 Web Services

The term *Web services* describes an important emerging distributed computing paradigm that focuses on simple, Internet-based standards (e.g., eXtensible Markup Language: XML) to address heterogeneous distributed computing. Web services define a technique for describing software components to be accessed, methods for accessing these components, and discovery methods that enable the identification of relevant service providers. Web services are programming language, programming model, and system software-neutral. Web services standards are being defined within the W3C and other standards bodies and form the basis for major new industry initiatives such as Microsoft (.NET), IBM (Dynamic e- Business), and Sun (Sun ONE). We are particularly concerned with three of these standards: SOAP, WSDL, and WS-Inspection.

II.5 Conclusion

We could say that the main parts of the systems needed to be standardized are: input data formats, output models, and integration of the data mining systems into the other systems and vice versa.

Currently some maturing standards exist for data mining namely PMML, XMLA, SQL/MM, OLEDB-DM, JDM (JSR-73), CRISP-DM, and CWM-DM.

However these standards are not sufficient and we see that collaboration is required from standards from related areas like Web and grid and we see efforts being put in merging the data mining standards with the Web services, grid services, and Semantic Web standards. There are emerging standards KDD workflow, data transformations, real-time data mining, and data webs. Current solutions for the various parts of the data mining process needed to be standardized are more or less hidden in the typically closed architectures of the data mining products. This is because there is lack of documentation in this area. We have tried to come up with consolidated documents that talk about various standards in reasonable details.

Appendix 3A

Intelligent Miner

Abstract. The information presented in this appendix is based on IBM Intelligent Miner for Data Corporation.

Information technology has developed rapidly over the last three decades. Many organizations store increasingly large volumes of data on their computer systems. Useful information might be hidden in the data in the form of implicit patterns and connections that are not easy to discern using conventional data queries and statistical calculations.

Data mining is the process of discovering valid, previously unknown, and ultimately comprehensible information from large stores of data. We can use extracted information to form a prediction or classification model, or to identify similarities between database records. The resulting information can help we make more informed decisions.

The Intelligent Miner supports a variety of data mining tasks. For example, a retail store might use the Intelligent Miner to identify groups of customers that are most likely to respond to new products and services or to identify new opportunities for cross-selling. An insurance company might use the Intelligent Miner with claims data to isolate likely fraud indicators.

3A.1 Data Mining Process

Data mining is an iterative process that typically involves selecting input data, transforming it, running a mining function, and interpreting the results. The Intelligent Miner assists with all the steps in this process. We can apply the functions of the Intelligent Miner independently, iteratively, or in combination. Mining functions use elaborate mathematical techniques to discover hidden patterns in our data. After interpreting the results of our data mining process, we can modify our selection of data, data processing and statistical functions, or mining parameters to improve and extend our results. Figure 3A.1 shows the basic data mining process.

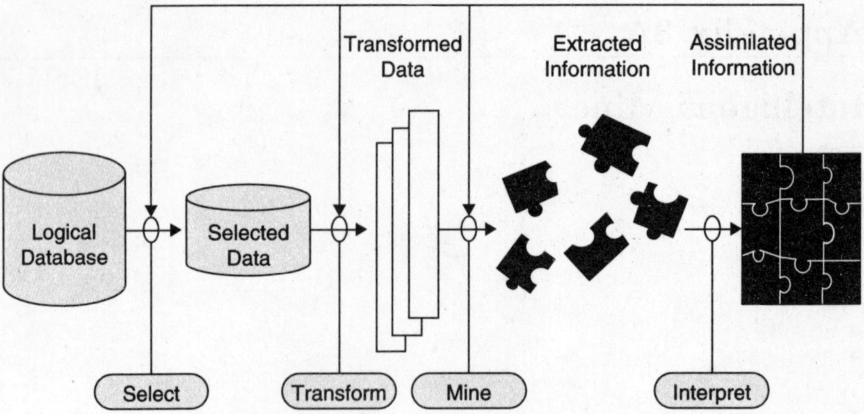

Fig. 3A.2. The data mining process

3A.1.1 Selecting the Input Data

The first step in data mining is to specify the input data that we want to mine and analyze. A data source might not contain all the data that we want to use for a specific data mining objective, or it might include irrelevant data. The data that we want to mine might be in one or more database tables, views, or flat files.

The Intelligent Miner helps us select specific data from a variety of data types and sources to create input data to which we can apply preprocessing and mining functions. For example, we can omit data fields or select a subset of records in one table and then combine them with data selected from another table.

3A.1.2 Exploring the Data

At any point in the process, we can use statistical functions to explore and analyze the data. We might want to apply statistical analysis as we consider input data variables for a mining function. We can also use statistics functions to transform data to create input fields for mining. In addition, these functions are useful for evaluating the output data generated by the mining functions.

3A.1.3 Transforming the Data

After we specify the input data, we can transform it using the Intelligent Miner preprocessing functions. Processing functions such as discretization, filtering, and joining help we organize our data so that we can mine it effectively. For example, if our data contains the fields Salary and Commission, we might aggregate the values of these fields and create a data field named Total Salary. We might also use an Intelligent Miner function to remove Null values from our input data so that they do not affect the results of the data mining process.

3A.1.4 Mining the Data

Transformed data is subsequently mined using one or more mining functions. The Intelligent Miner has the following types of mining functions:

- Associations
- Neural Classification
- Tree Classification
- Demographic Clustering
- Neural Clustering
- Sequential Patterns
- Similar Sequences
- Neural Prediction
- Radial Basis Function (RBF) Prediction

3A.2 Interpreting the Results

We can analyze the results of the data mining process with respect to our decision-support objectives. Visualization tools allow us to view the results and identify important information laid bare by the mining process. We can export the results to a remote workstation so that they can be viewed at a different location. We can also copy certain results to the clipboard to make them available for other tools, such as spreadsheets or statistical applications. Moreover, we can print the results.

Data mining can be an iterative process. When we look at previous results, we might want to adjust the mining settings for a further mining run to improve the result quality. Suppose that we want to use direct marketing to offer new services to our customers and want to target only those customers most likely to be interested.

Example: The first step might be to use the Intelligent Miner to discover categories or clusters in our customer database. We can then look for group characteristics that might make customers more or less likely to accept the new services. The process of clustering our customer database might include the following steps:

1. Select relevant customer data as input data.
2. Transform the data by reorganizing it, eliminating duplicate records, or converting it from one form to another.
3. Specify the location of resulting output data.
4. Specify the parameters for the Demographic Clustering mining function.
5. Run the Demographic Clustering mining function (clustering mode).
6. Visualize the resulting data.
7. Analyze and interpret the results.
8. Repeat steps 3 to 6 until the results satisfy our needs.

3A.3 Overview of the Intelligent Miner Components

This section provides a high-level overview of the Intelligent Miner architecture

The Intelligent Miner communicates between mining and preprocessing functions on the server, as well as between the administrative and visualization tools on the client. The client component includes a user interface from which we can invoke functions on an Intelligent Miner server. The results are returned to the client where we can visualize and analyze them. The client components are available for AIX, OS/2, Windows 95, Windows 98, and Windows NT operating systems.

The server software is available for AIX, OS/390, AS/400, Sun Solaris, and Windows NT systems. In addition, on AIX, Sun Solaris, and Windows NT systems, the server software supports parallel mining with multiple processors. We can have client and server components on the same machine.

Figure 3A.2 shows the client and server components of the Intelligent Miner:

3A.3.1 User interface

A program that allows us to define data mining functions in a graphical environment. We can define preferences for the user interface, which are stored on the client.

3A.3.2 Environment Layer API

A set of API functions that control the execution of mining runs and results. Sequences of functions and mining operations can be defined and executed using the user interface through the environment layer API. The environment layer API is available on all server operating systems.

3A.3.3 Visualizer

A tool that displays the results produced by a mining or statistical function. The Intelligent Miner provides a rich set of visualization tools. We can also use other visualization tools.

3A.3.4 Data Access

Data access to flat files, database tables, and database views.

3A.4 Running Intelligent Miner Servers

This section describes how to run Intelligent Miner servers in the supported operating environments. It also describes how to set up the parallel operating environment for AIX, Sun Solaris, and Windows NT systems.

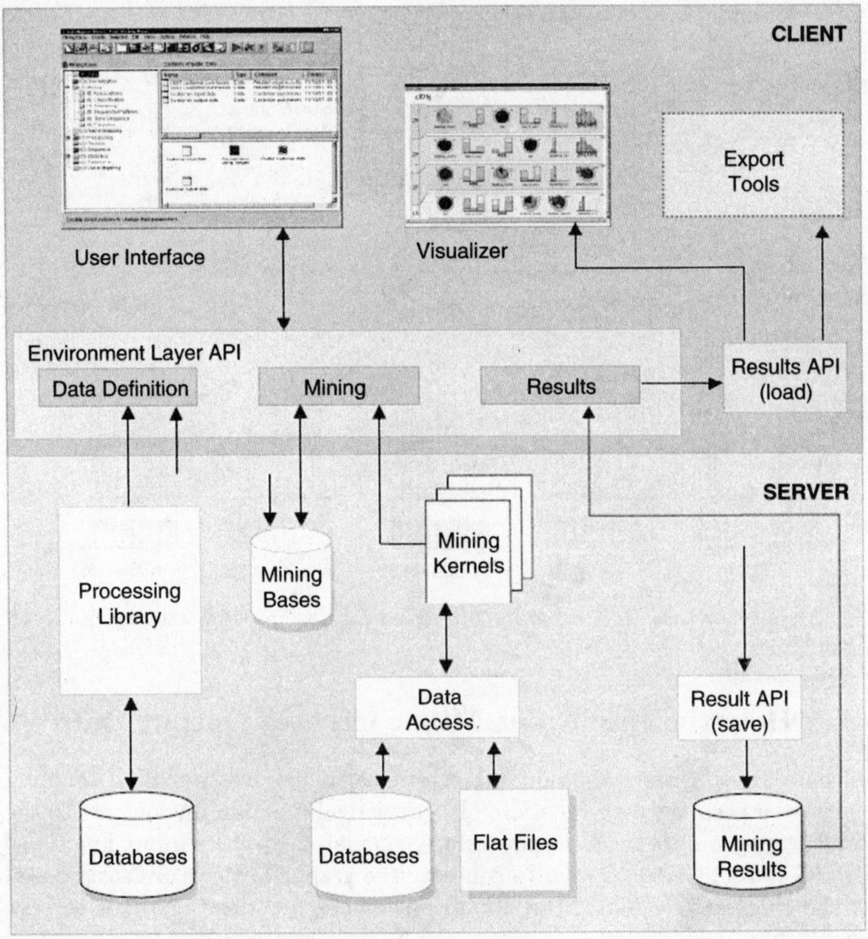

Fig. 3A.4. The Intelligent Miner architecture

We can run the Intelligent Miner server on AIX, AS/400, OS/390, Sun Solaris, and Windows NT systems.

- On AIX and Windows NT, we can run the Intelligent Miner in local mode, that is, we can run the client and the server software on the same workstation. The Windows NT version also provides a standalone mode, which means that we run the Intelligent Miner on a single workstation without a server process. However, we must install the client and the server software on that workstation. In client/server mode, we can connect a client to the server.
- On AS/400, OS/390, and Sun Solaris, we must connect one of the supported clients to the server.

Alternatively, we can write applications of our own using the Intelligent Miner application-programming interface (API).

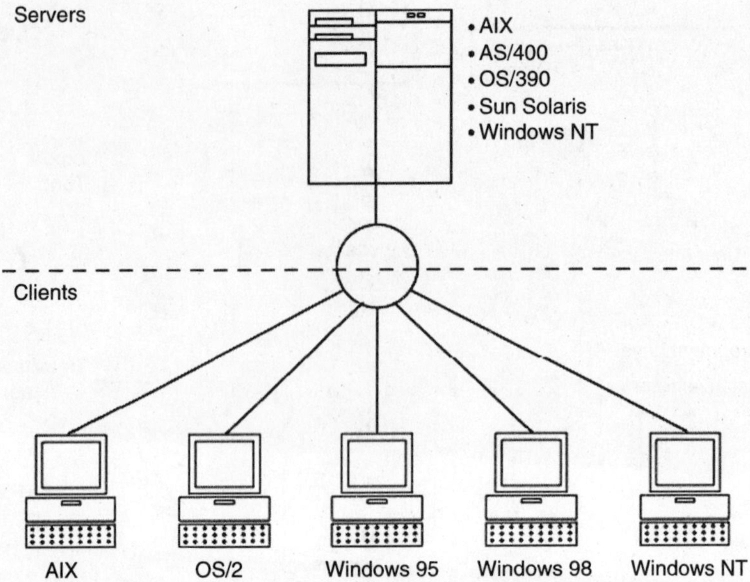

Fig. 3A.5. shows the different client/server configurations supported by the Intelligent Miner.

3A.5 How the Intelligent Miner Creates Output Data

Output data consists of the values that were calculated when we used a mining function, a preprocessing function, or a statistical function. In contrast to the results, output data contains the values calculated by the various functions. Results contain the data used to generate the graphical views or charts based on the calculated values. Most Intelligent Miner functions allow us to save output data.

The mining and statistical functions create output data as flat files or database tables. Preprocessing functions create output data as database tables or database views.

The following rules apply to all types of output data:

- If we select a data object that is marked as **read only** on the output data page, the attempt to create output data fails.
- If the output data does not exist, an attempt is made to create it even if we select **Output should be appended to the specified table** on the database table or view page of the Data wizard.

3A.5.1 Partitioned Output Tables

If our database management system supports partitioned tables and we specify output data of the type **Table**, the Intelligent Miner tries to create a partitioned output table provided that any of the following conditions or set of conditions applies:

- We specify a partitioned table space for the output data.
- We specify **Read and write** or **Overwrite existing records** for our output data, but no table space. The output data already exists as a partitioned table.
- We do not specify a table space for the output data and the input data is a partitioned table.

3A.5.2 How the Partitioning Key is Created

Partitioned tables require a partitioning key. Normally, this is not a problem. For all mining and statistics functions, the Intelligent Miner adds the appropriate fields automatically if we did not specify them in the lists of input and output fields. However, if a partitioning key cannot be determined for some reason, the attempt to create output data fails. Which partitioning key the Intelligent Miner uses is determined as follows:

- The Intelligent Miner uses the partitioning key of the input data if the following conditions apply:
  - The input data is a partitioned table.
  - All of the input table's partitioning key fields are included in the output data.
  - In the case of DB2 for OS/390, the number of partitions in table space of the input table is the same as the number of partitions in the table space of the output table.
- If the partitioning key of the input data cannot be used, the Intelligent Miner uses the partitioning key of an existing table if the following conditions apply:
  - The output data already exists as a partitioned table.
  - We specified the **Read and write** or **Overwrite existing records** option for the output data.
  - All of the existing table's partitioning key fields are included in the output data.
  - For DB2 for OS/390, the number of partitions in the table space of the existing table is the same as the number of partitions in the table space of the output table.
- If the partitioning key of an existing table cannot be used and the database management system creates default partitioning keys, the Intelligent Miner allows the database management system to create a default partitioning key for the partitioned output table.

3A.6 Performing Common Tasks

This section describes the basic concepts and features of the Intelligent Miner and explains how to perform tasks necessary to achieve most data mining goals.

3A.7 Understanding Basic Concepts

In general, data mining in the Intelligent Miner is accomplished through the creation of interrelated objects. The objects are displayed as icons and represent the collection of attributes or settings that define the data or function. We create *settings objects* to perform a particular task. The Intelligent Miner, for example, creates other objects that contain the findings of a mining run.

We save the objects of a particular data mining project as a group called *mining base*. We can create a mining base for each mining objective or project that we put together.

3A.7.1 Getting Familiar with the Intelligent Miner Main Window

3A.8 Main Window Areas

The main window (shown in Fig. 3A.8.1) has the following areas:

3A.8.1 Mining Base Container

The mining base container on the left side of the main window shows the types of objects that are stored in a mining base. (See Fig. 3A.4.) There is a folder or subfolder for each type of Intelligent Miner object that we can create. We can expand folders by clicking on the plus sign next to the folder. Similarly,

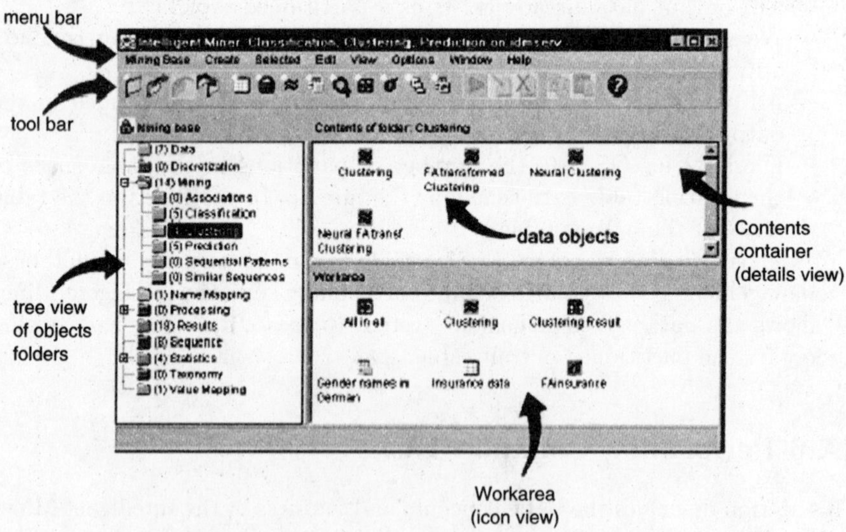

Fig. 3A.8.1. Intelligent Miner main window

we can collapse folders by clicking on the minus sign next to an expanded folder. The number in parentheses in front of each folder displays the number of objects in that folder or in any of its subfolders. When we click a folder, the objects or subfolders are displayed in the contents container.

3A.8.2 Contents Container

We can work with settings objects of a given type in the contents container. The contents container is in the upper right corner of the main window. Use the **View** menu to indicate how we want the objects displayed, for example, by icon or with details. We can select an object and click **Selected Ã Open** on the menu bar. This opens the settings notebook of the object or, if we selected a result object, the corresponding visualizer. We can also double click the object.

3A.8.3 Work Area

The work area (lower right corner) allows us to work with settings objects of different types. (See Fig. 3A.4 on page 65.) We can drag and drop objects from the contents container into the work area to create shortcuts to our objects. This way, we can work with different types of objects without having to click folders and find objects. The work area contents of a mining base are saved as part of our preferences data. It is restored when we restart the client.

When we position the mouse pointer over the toolbar icons, brief explanations are displayed. More help is displayed at the bottom of the main window when we position the mouse pointer over some of the interface elements.

3A.8.4 Creating and Using Mining Bases

Mining bases are displayed in the main window of the Intelligent Miner, as shown in Fig. 3A.4. In the main window, we also work with the objects that belong to the mining bases. The mining base holds descriptive information about the objects. It does not contain the data that is being analyzed. Creating objects in the mining base is a task that we must perform frequently when using the Intelligent Miner. The task of creating an object is essentially the same, no matter which type of object we create. A wizard guides us through the process.

Before we can manipulate objects in a mining base, we must open it. If our client runs on a system other than the Intelligent Miner server, we must first select a server and provide a user ID and password. These and other preferences can be set in the Preferences notebook. See "Setting preferences for the Intelligent Miner."

However, when we run the Intelligent Miner in local mode or in stand-alone mode, that is, client and server software are installed on the same machine, we can open mining bases directly.

The main window provides a standard menu bar and taskbar for opening mining bases and creating objects. Use the **Mining Base** menu or taskbar icon to open an existing mining base or create a new one. With a new mining base we can begin creating objects that can be stored as part of the new mining base.

3A.9 Conclusion

Data mining is the process of discovering valid, previously unknown, and ultimately comprehensible information from large stores of data. We can use extracted information to form a prediction or classification model, or to identify similarities between database records. The resulting information can help us make more informed decisions.

The Intelligent Miner supports a variety of data mining tasks. For example, a retail store might use the Intelligent Miner to identify groups of customers that are most likely to respond to new products and services or to identify new opportunities for cross-selling. An insurance company might use the Intelligent Miner with claims data to isolate likely fraud indicators.

Appendix 3B

Clementine

Abstract. The information presented in this appendix is based on Clementine Data Mining Workbench from SPSS Corporation.

Clementine is an integrated data mining workbench. That is, it provides a wide array of data mining techniques, along with prebuilt vertical solutions, in an integrated and comprehensive manner, with a special focus on visualization and ease of use.

The latest release (Clementine 9.0) incorporates many enhancements, which fall into a number of areas. Productivity is, perhaps, the most important of these. One of the key concerns within the business community about analytic activity is the time that is taken. In Clementine 9.0, there are many improvements that are targeted specifically at enhancing the productivity of the analyst. In addition, data mining is no longer a backroom activity – its results are now widely deployed throughout the enterprise and are leveraged by many users – this new release includes a number of significant features designed to assist in this enterprise-wide deployment.

3B.1 Key Findings

The following represent the key facts of which prospective users should be aware:

- Clementine supports the CRISP-DM methodology for data mining.
- Clementine is not restricted to one, or even a few, data mining algorithms and offers a wide range of decision tree, neural network, and other algorithms, which may be used either individually or in combination.

We found the act of using Clementine highly intuitive, using visual programming techniques to define processes and try out ideas. In addition, the CLEM Expression Builder is relatively simple to use even for novices, while the statistical and graphical displays are easy to understand and interpret.

Users do not have to know how different algorithms work, just what they do. However, beginners may find it easy to forget which algorithm does what. We can refer to the help system if necessary, but there is also a "ready reference" for each operation, which appears in the form of pop-up text when we move our mouse pointer over the icon representing that operation.

Following the acquisition of Data Distilleries by SPSS, the process of deploying models into mainstream applications that can benefit from the addition of analytical power has been considerably enhanced.

Front-end connectivity for the users of databases that offer kernel support for data mining processes (SQL Server, DB2 and Oracle) is supported by the fact that we can now generate native SQL for the appropriate database, directly from Clementine, in a fashion that is transparent to the user.

Traditionally, data mining has usually been regarded as a relatively complex discipline. However, SPSS would argue with this view. In its opinion any complexity or difficulty has primarily been put about by those trying to sell consulting. While it would admit that there is a skill involved in translating between business problems and the data mining environment, the company does not believe that it is necessary for users to understand what is going on in the engine.

3B.2 Background Information

Clementine was originally developed by Integral Solutions Ltd (ISL), which was formed in 1989 by Dr Alan Montgomery and five colleagues from a management buyout of the AI Products Division of SD-Scicon, an IT consultancy that subsequently became a part of EDS.

From the outset, the aim of ISL was to build an integrated environment for data mining operations that could be used and understood by business people, without the need to rely on technical experts. Clementine itself began as a research product in 1992, which was part-funded by the UK's Department of Trade and Industry, and the product was first brought to market in 1994.

Clementine was one of the very first products to bring machine learning to business intelligence, while providing a user interface that was intelligible to nonexperts. As a result it rapidly gained a leadership position within the market, which it has retained to this day.

However, the success of Clementine was always inherently limited by the fact that ISL was a relatively small organization (no more than about 40 staff), which was always likely to struggle against the giants of the IT industry. The takeover of ISL by SPSS in 1998 therefore opened up new opportunities for Clementine that might otherwise not have been available.

SPSS itself was founded in 1968 and earned its reputation primarily as a provider of statistical software, plus graphical applications to represent those statistics. In fact, prior to its acquisition of ISL, SPSS had two data mining products of its own: AnswerTree and Neural Connection. As might be deduced from their names, the former offered support for decision tree algorithms while the latter supported neural networking. AnswerTree is still sold by SPSS as an algorithmic tool rather than as a generalized workbench but Neural Connection is now defunct.

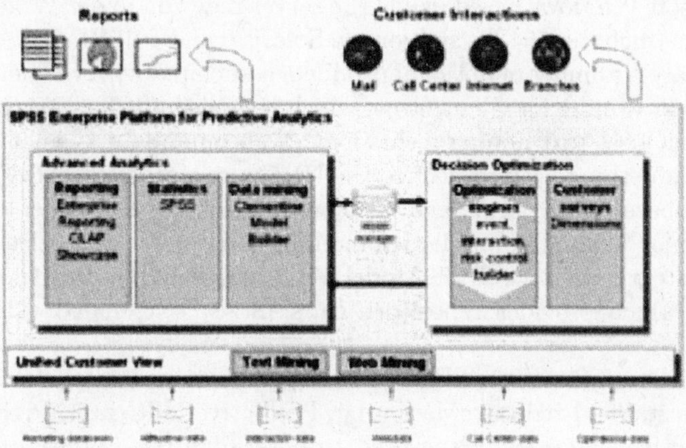

Fig. 3B.8.1. Enterprise Platform for Predictive Analytics

In recent years the company has expanded significantly through a process of acquisition. Most recently, SPSS acquired Data Distilleries, a move that, at first sight, might have led one to suspect a significant overlap with Clementine, but which has actually resulted in a very synergistic leap forward in capability. Data Distilleries did indeed possess substantial data mining capability, but its greatest strength lay in the deployment of that capability into various analytical business applications. Following the acquisition, SPSS has been able to incorporate these capabilities into the products that are known as SPSS PredictiveMarketing, SPSS PredictiveCallCenter, SPSS PredictiveWeb and SPSS PredictiveClaims. In other words, the company has been able to move rapidly from being a provider of tools into a leader in the provision of analytical applications.

Parallel to this move into the provision of analytic solutions, SPSS has also been concentrating on its own infrastructure so that it can now offer a complete Enterprise Platform for Predictive Analytics (see Fig. 3B.8.1) that integrates all of its data/text/Web mining, predictive analytics, statistics, survey, and reporting capabilities into a single platform.

In total, what this has meant for Clementine is that it is now the basis for many powerful analytical applications that can be more readily embedded and used within frontline business operations in areas such as call centers, as well as in generic applications. Overall, it is now apparent that SPSS has a far more comprehensive and coherent strategic view of its role in the intelligent and predictive enterprise, and most significantly of all, it now possesses the means to deliver that vision.

3B.3 Product Availability

As noted, Clementine was first introduced to the market in 1994. The latest version of the product is 9.0, which was released at the end of 2004. Clients are Microsoft Windows based, while the server may run under Windows, AIX, Sun Solaris (including 64-bit support on Solaris 9 and 10), HP-UX, or OS/400.

There are a number of optional products associated with Clementine, most notably Text Mining for Clementine, which leverages the facilities of the company's LexiQuest text mining technology; Web mining for Clementine, which does the same thing with respect to the NetGenesis product, providing analysis of Web-based behavioral data; Cleo, which provides a Web deployment environment for interactive predictive models; and, new in this release, there is optional integration with SPSS Model Manager. A 30-day free trial of Model Manager, which provides repository capabilities, is included with Clementine 9.0.

In addition to these, the SPSS analytic applications derived from the Data Distilleries stable, PredictiveMarketing, PredictiveCallCenter, PredictiveWeb and PredictiveClaims (which is used to determine the validity of insurance claims), can all leverage the facilities provided by Clementine.

3B.4 Software Description

Over the last few years the market for data mining tools has changed dramatically. In particular, we have seen the emergence of analytic applications as a distinct discipline. Previously, the sort of things that analytics now provides belonged within the domain of data mining. For example, market-basket analysis used to be regarded as a pure data mining technique. Today it is packaged and delivered within a variety of analytic software products.

In effect, what has happened is that the heuristic approach provided by data mining tools has led to the identification of algorithms that can be applied to a range of environments such as market basket analysis. Algorithms are also much easier to package for deployment by the casual user.

However, analytic applications (as supplied by nondata mining vendors) are limited by the fact that they are intrinsically about analyzing what has already happened: we are left to make our own extrapolations as to how that might be reflected in the future. Data mining, however, has this predictive ability as well, in addition to the sort of base analytic capabilities that are offered by CRM and other vendors. In fact, SPSS is delivering what we might call genuine analytic applications (in that they provide predictive capabilities), for a number of different environments.

Nevertheless, it is clear that "analytics" has become a popular buzzword and, in order to capitalize on this, SPSS and other vendors have now started to refer to their products' capabilities as providing "predictive analytics."

Most recently, another significant change in the market is that the use of analytics is no longer seen as a minority capability. Leaders in this segment (which includes SPSS) are providing capabilities based on both historical analysis and future prediction to every seat within a business environment that can utilize the benefits of that analysis. So, while a few years ago the sale of a data mining tool would have been to, say, 5 seats, now the analytical application, (which is the data mining tool together with all of its associated deployment elements) is being made available to hundreds if not thousands of seats. As the demand within business is for faster and better decision making closer to the point of contact, be it a production line or a customer, analytical applications are one of the main tools that can be used to empower workers to fulfill that management goal.

Finally, the other thing to note about data mining (or predictive analytics) is that there are very few major players (of which SPSS is one) left in the market. A few years ago there were a plethora of such vendors but most of these have either fallen by the wayside, been taken over, or have diverted into niche markets. That Clementine is one of the market leading products in this area is a testament, not just to its technology, but also to its user-friendly nature.

3B.5 Architecture

Historically, Clementine was a client-only product in which all processing took place on the client itself. While external databases could be accessed, they were treated by the software as local files; this approach necessarily put a heavy burden on the client platform.

Clementine was originally implemented on X-Windows platforms rather than Microsoft Windows. However, with the release of version 5.1 of Clementine in 1999, SPSS introduced a server-based version of the product, with middleware running on a middle-tier server taking the load off the client and using the superior performance of back-end databases to support in situ data mining. Features introduced at that time included SQL optimization, storage of intermediate results and, of course, much reduced network track. More recently, the company has also implemented in-database sampling and automatic operation re-ordering for in situ mining, both of which improve performance and scalability.

Nevertheless, despite the move to a three-tier architecture, SPSS persisted with X-Windows until release 7.0, in which it adopted a Microsoft Windows front end. According to SPSS this meant the largest engineering feat that the company had ever had to face, to build in all the sorts of features that users expect at this level. In fact, the product included a raft of usability enhancements in that release, such as full cut and paste support, the ability to have multiple streams (process models) open at the same time, a Java front-end, automated node connection (with a process model), a preview pane, and an improved script editor that includes syntax checking capabilities amongst others. With the latest releases of Clementine (8 and 9) the integration of the product at both the back end (into the enterprise infrastructure) and the front end (into business applications) has been transformed. Clementine is now an integrated part of the corporate IT arsenal and should not be viewed as a standalone element.

One of the biggest changes has been the move to put the code for data mining functions into the kernel of the database, a move started by Oracle and since followed by IBM with DB2 and, increasingly, Microsoft with SQL Server. Whilst a great idea to enable data mining to be invoked within the kernel, using SQL, the issue that faced early users with Oracle after its initial release was how to actually access the kernel without having to write the relevant SQL. As has been mentioned already, the drive in data mining has been to take that process away from the isolated analytics expert and to empower business users to employ it for themselves. Writing SQL was obviously a retrogressive step in that process. Clementine can now be used as the visual workbench to generate the necessary SQL (integrating directly with Oracle Data Mining, IBM DB2 Intelligent Miner and Microsoft OLAP Services, and enabling we to build and score models directly within these databases) without the user having to have any knowledge of how and what is happening at a technical level.

3B.6 Methodology

SPSS (and ISL before it) has always espoused the use of a formal methodology for data mining and it has been a member of the CRISP-DM (Cross Industry Standard Process for Data Mining) group since its foundation in 1996. This methodology defines a six-stage process for data mining projects, whose steps are detailed below. In addition, Clementine includes a project tool, and an associated help system, that specifically supports the CRISP-DM method. The six steps in that method are illustrated in Table 3B1.1, where generic tasks are shown in bold and outputs are in plain type:

We will consider the facilities offered by Clementine under each of these headings. In addition to methodology support, in the latest release there is also workgroup support, which is provided through the SPSS Model Manager add-on. This makes it possible for a workgroup to share its models, streams, and results in a common repository. The repository, which provides version control, check-in and out, permission support, and search capabilities amongst other features, has its own administrative interface.

3B.6.1 Business Understanding

In practice, of course, there is little that Clementine can do to assist in business understanding, at least in any a priori sense. However, SPSS does offer a number of Clementine Application Templates (CATs), available as add-on modules, which encapsulate best practice and provide a lot of out-of-the-box functionality to gain immediate business advantage. Examples of CATs include:

Table 3B1.1. Methods supported by Clementine

Business Understanding	Data Understanding	Data Preparation	Modeling	Evaluation	Deployment
Determine Business	Collect Initial Data	Data set	Select Modeling Techniques	Evaluate Results	Plan Deployment
Objectives	Initial data collection report	Data set description	Modeling technique	Assessment of Data Mining results w.r.t. Business Success Criteria	Deployment plan
Background Business objectives			Modeling assumptions		
	Describe Data	Select Data			Plan Monitoring and Maintenance
Business success criteria	Data description report	Rationale for inclusion/ exclusion	Generate Text Design	Approved models	
			Text design		Monitoring and maintenance plan
Assess Situation	Explore Data			Review Process	
Inventory of resources	Data exploration report	Clean Data	Build Model	Review of process	
Requirements, assumptions and constraints		Data cleansing report	Parameter settings	Produce Final Report	
	Verify Data Quality		Models	Determine Next Steps	Final report
Risks and contingencies	Data quality report	Construct Data	Model description	List of possible actions	Final presentation
Terminology		Derived attributes		Decision	
Costs and benefits		Generated records	Assess Model	Review Project	
			Model assessment		Experience documentation
Determine Data Mining Goals		Integrate Data	Revised parameter settings		
Data Mining goals		Merged data			
Data Mining success criteria		Format data			
		Reformatted data			
Produce Project Plan					
Project Plan					
Initial assignment of tools and techniques					

Web mining CAT, which analyzes on-line behavior and includes facilities for:

- Web usage statistics.
- Site search optimization.
- User and visit segmentation.
- Activity sequence analysis.
- Responses prediction.
- Campaign performance measurement.
- Telco CAT, which targets improvements in retention and cross selling for the telco vertical with prebuilt modules for churn and cross-selling.
- CRM CAT, which includes prebuilt modules for customer value, RFM response modeling, and customer segment migration and attrition.
- Fraud CAT, for the prediction and detection of fraud instances in financial transactions.
- Microarray CAT for accelerated life sciences work.

Each of these CATs consists of five major components: a realistic set of sample data, a number of prebuilt streams (models) that are based on industry best practices, documentation and help structured according to the CRISP-DM methodology, annotative help, and mapping tools that have been designed to assist in the process of customizing the CAT to our own environment.

3B.6.2 Data Understanding

There are really two aspects to data understanding: the extraction of data and the examination of that data for its utility in the proposed data mining operation. In the latter case, many of the techniques used for examination are essentially the same as those used for visualization purposes during the evaluation phase. Thus, while SPSS lists 3D, panel and animation graphs, for example, as data understanding features, these features might equally well be used at other stages within the data mining process.

As far as data extraction is concerned, Clementine supports a number of methods for accessing miscellaneous data sources. At the simplest level, the product supports both ODBC access to database sources as well as import facilities from file-based data sources. Alternatively, users can submit SQL queries to relational databases. Some aspects of preprocessing, such as table joins, are delegated by Clementine's SQL optimization to the database itself, which should provide superior performance.

A major feature that was introduced in Clementine 7.0 was its integration with LexiQuest text mining. Text mining has emerged recently as one of the hot spots in analytical applications, and SPSS has a proven and mature ordering. This text mining technology uses Natural Language Processing to work at the level of concepts rather than keywords. Thus "customer satisfaction," for example, is treated by the software as a single concept rather than as two undifferentiated words. The product works by automatically generating concept-based indexes against each document and record in the database, and then by classifying these concepts into separate categories and by performing frequency counts against concepts. As far as the classification of concepts is concerned, some of these are predefined, such as people, organizations, products, and locations while we can also add our own categories if we wish. Text mining techniques can be used directly from within Clementine, and concepts, categories, and frequency can be inspected independently or linked into structured data mining. We can also retrieve relevant documents directly from the Clementine user interface. Text mining in Clementine has been advanced in this release with the separation of the analysis from scoring, in order to refine and speed up the process. Within the analysis phase key concepts are identified, those concepts are ranked and they can then be included or excluded from the scoring process, which enables the scoring to be more readily deployed and more rapidly executed.

3B.6.3 Data Preparation

Data preparation (and retrieval) is performed through Clementine's "data manipulation" operations, of which an important part is the CLEM expression language. This can be used in conjunction with operations such as record selection and attribute derivation and it provides a wide range of mathematical and data handling functions, including selection, sampling, merging, sorting, and the aggregation of records. It is also used in data cleansing operations and provides data balancing functionality such as reducing any "skew" in a set of values, as well as dealing with missing values in records.

Historically, CLEM has largely been the domain of experienced users. However, with version 6.5 SPSS introduced its CLEM expression builder. An example illustrating the expression builder is shown in Fig. 3B.8.2; but, to put it simply, the expression builder is to CLEM what 4GLs were to conventional

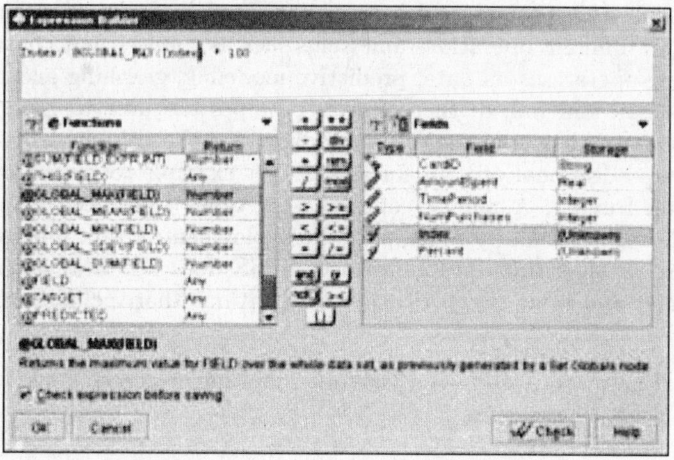

Fig. 3B.8.2. CLEM expression builder

programming languages such as COBOL, i.e., much, much easier to use, especially for inexperienced users.

A major feature of the latest release is a facility called the "partition mode." This is an automated (single step) function that manages the task of splitting data into training and test data sets; or training, test, and evaluation sets. Historically, this was a time-consuming process that added little value to the business; so this represents a significant step forward. It will be especially beneficial as it allows expensive analysis time to be re-distributed to areas of activity with direct business benefit.

3B.6.4 Modeling

Modeling (and visualization) is at the heart of Clementine. Clementine is a fully graphical end-user tool based on a simple paradigm of selecting and connecting icons from a palette to form what SPSS calls a "stream." This differs from a "process model" because a stream may consist of any number of different attempts to analyze the data under consideration and the "model" may therefore include both unwanted and disconnected branches (though the software will prevent us from defining impossibilities). In effect, the environment is more like that of a canvas upon which the user tries out ideas (stream of consciousness), with the current path through the model being highlighted in colors. It should also be noted that streams encompass other parts of the CRISP-DM method, data preparation for instance, that go beyond just the modeling phase.

Streams support the idea of a process model in that proven processes and models can be stored for reuse. Moreover, we can have multiple streams or processes open at the same time. This means that we can switch between proven models and ideas that we are trying out, without the latter interfering with the former. An example of this canvas/process model approach is illustrated in Fig. 3B.8.3.

As illustrated, process nodes are represented by icons of several types, catering for different operations and items such as connecting to data sources, performing operations on data, predictive modeling, graphing and reporting. Once selected and linked, the software prompts the user to input relevant parameters and then automatically paradigm uses the ubiquitous double-click to bring up additional options for a node, for example to invoke the CLEM Expression Builder. A number of new node types have been added in the latest release, particularly to support the deployment of text mining.

As far as algorithms are concerned, SPSS does not believe in a single silver bullet and it supports a range of algorithms that include, but are not limited to:

- Neural networks – Kohonen networks, multilayer perceptrons, and radial basis function networks.

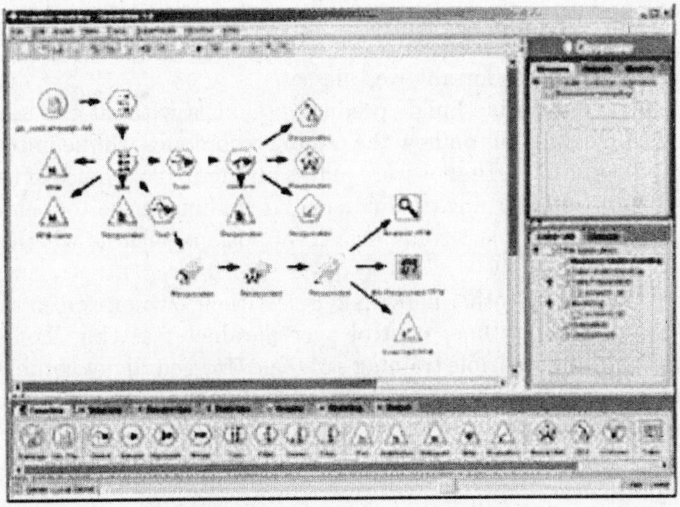

Fig. 3B.8.3. Clementine's canvas/process model approach

- Rule induction – C5.0 and C&RT, both of which are decision tree algorithms. There is also a graphical decision tree browser so that we can select the most intuitive way to view decision trees. In the latest release, there are new decision tree algorithms supported, Quest and Chaid (both standard and exhaustive versions), which are more sophisticated algorithms for the analysis of complex multivariate situations, with less danger of overfitting. There is also support for interactive trees, which provides a means of developing decision trees in a stepwise fashion enabling business insight to guide decisions on how to split the branches of the tree with an evaluation of each step being provided by a gains chart to ensure that the model remains predictive and of value.
- Regression modeling – which uses linear regression to generate an equation and a plot. Logistic regression is also available.
- Clustering – K-means and two-step clustering (the latter provides good performance and can suggest the optimal number of clusters for a dataset). These, together with the clustering-related Kohonen network models, provide a graphical browser that allows us to understand the meaning of the clusters we have discovered.
- Association rule discovery – apriori association detection and generalized rule induction find links between related attributes (and are often used in basket analysis).
- Sequence Detection – this is an extension to association in the sense that it discovers associations that exist over time.

It is also important to appreciate that we can combine algorithms within a single model. This is particularly useful since no algorithm is perfect and this allows us to compensate for any weaknesses.

By default, Clementine builds predictive models without the user having to specify technical details on how the mining techniques will be implemented. Based on the input data, the software makes an intelligent choice of the appropriate algorithm and learning parameters and configures the tool accordingly. However all users can select some high-level options such as whether to aim for generality or accuracy.

Expert users, on the other hand, may select their own on-screen tab, which will give the user much finer control over parameter setting. For example, Clementine supports multiple training strategies for neural networking and the expert user can select which one to use, and modify the technical parameters of each. While on this topic, it is worth mentioning that training can be stopped automatically, and Clementine is able to detect overtraining and revert to the best state obtained.

3B.6.5 Evaluation

Evaluation is all about visualizing the results of the data mining process and, as we have previously noted, this is also very important in understanding the data. As one might expect from a company that has historically specialized in statistical manipulation and the graphical representation thereof, this is an area of particular strength in Clementine.

A brief list of generic visualization capabilities includes tables, distribution displays, plots and multiplots, histograms, webs (in which different line thicknesses show the strength of a connection, with additional facilities for weak connections and automatic layout capabilities) and matrices, as well as the panel and animation graphs mentioned previously. In the latest release, SPSS has added further capabilities such as scatterplot matrices, parallel coordinate maps, heat maps, linkage analysis plots, and so on. In addition, specific evaluation capabilities include gains, lift, response, profit and ROI charts, and we have already mentioned the use of graphical decision tree browsing. Additional detail can be displayed by clicking on relevant displays. For example, with a web diagram we can click on a connection to see a description of its strength and the number of records involved.

Some visualization techniques are available in 3D as well as 2D, and we can overlay multiple attributes onto a single chart in order to try to get more information. For example, we might have a 3D scatter diagram and then add color, shape, transparency, or animation options to make patterns more easily discernible.

In addition to individual graphs, Clementine provides a "data audit" node, which allows us to get an overview of a whole dataset in one report, including thumbnail graphs, statistics, and target overlays for each field.

Perhaps even more significant is the degree of interaction that Clementine provides to the user, which is available for all types of graphs and displays. For example, we can select a region on a plot by dragging with a mouse, which will automatically select the records corresponding to the defined region.

3B.6.6 Deployment

Output data, together with relevant predictions, can be written to files or exported to ODBC compliant databases as new tables or appended to existing ones. It is also possible to use batch mode for clientless operations.

Perhaps the most commonly used deployment options are via the Clementine Solution Publisher. This is used when we want to redeploy a whole scoring stream, including data access definitions, data preparation requirements, and the model combinations that are then to be used. By using a Publisher node within the process model we can automatically capture this information for subsequent deployment either on another platform or as an embedded facility within a customized application. In either case, for example in a call center application, we call the Clementine Runtime to process the resulting files.

Generalized Web-based support is available, with the ability to present results in HTML format so that they can be viewed across the enterprise via a browser. In addition, Cleo is a specialized tool that provides the means to build simple web front-ends that we can use to deploy analytical applications. Since the acquisition of Data Distilleries SPSS has implemented Predictive Application Wizards that take streams and deploys them directly into the various packaged analytic applications.

In some instances, a model-only solution is suficient and in this case models can be exported in PMML 2.0 (Predictive Modeling Mark-up Language, an XML-based industry standard) and embedded in customized applications. SPSS has long supported deployment of PMML models via its SmartScore component, and has now branched out to support deployment through third-party products such as IBM DB2 Intelligent Miner Scoring.

3B.7 Clementine Server

Clementine measures data mining productivity by time to value: the time required for users to get results that deliver value in an organization. To shorten this time to value, Clementine introduced the concept of a visual workflow interface. With Clementine's visual workflow, users are able to move interactively through all steps in the data mining process, applying their business knowledge to guide them to meaningful results more quickly. Other approaches to data mining have stressed raw processing power rather than productivity throughout the entire data mining process. Optimizing raw processing power comes at a price: models, for example, can be built quickly, but the models may not deliver results that offer any value to an organization. We get better performance with a data mining approach that minimizes time to value. In the past, Clementine's approach worked best with sampled data. With the release of Clementine Server, Clementine's interactive data mining approach can be used on much larger datasets. This is because Clementine Server scales the entire data mining process. For example, visualization techniques are scaled for data understanding. Data preparation steps such as field and record operations also see significant gains, as do modeling processes that include preprocessing steps. Finally, model evaluation and deployment can be performed more efficiently.

3B.8 How Clementine Server Improves Performance on Large Datasets

Clementine Server improves performance while mining large datasets by leveraging our investment in a database management system (DBMS). It does this by maximizing in-database mining, delegating as many operations as possible to the DBMS, therefore, taking advantage of database indexing and optimized data operations. The architecture has three tiers (as shown in Fig. 3B.9): the DBMS, or database tier, the application server tier and the client tier. The client tier includes Clementine's visual workflow interface with data mining "streams" that show all steps in our data mining processes. In previous versions, when a Clementine stream was executed, the client's processing was used to perform data analysis.

With Clementine Server, however, the stream processing is pushed back onto the DBMS via SQL queries. Any operation that cannot be represented as

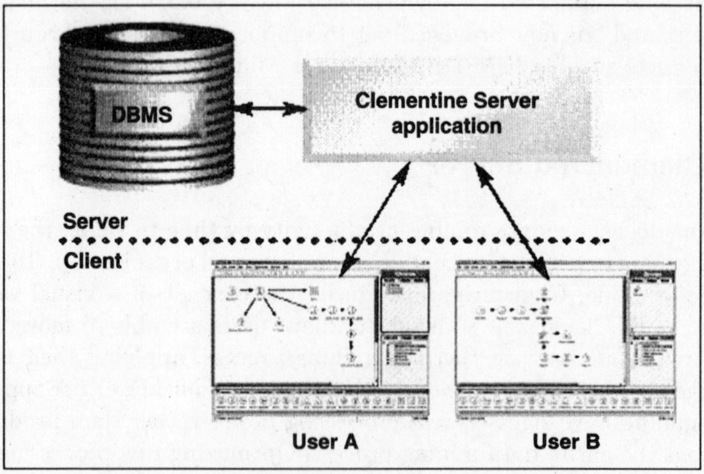

Fig. 3B.9. Clementine three-tier distributed architecture

SQL queries is performed in a more powerful application server tier. Only relevant results are passed back to the client tier. This approach takes advantage of optimized operations in a DBMS and increased processing power found at the application server tier to deliver predictable, scalable performance against large datasets. Clementine gives us feedback when in-database mining is activated. During execution, nodes turn purple if the operations represented by the node are executed in-database. At the point of the last purple node in the stream, the remaining data is extracted and performed at the application server tier. Since in-database mining is almost always faster than application server processing, the more nodes that are pushed back to the database, the better. Clementine maximizes in-database mining by using rules of thumb to order operations automatically. We do not have to worry about the mechanics of stream building, because operation reordering is automatic. Instead, we can focus on the business problem at hand. Operations will not be reordered if the reordering would change the results of the stream.

3B.8.1 Benchmark Testing Results: Data Processing

The average increase in time required to process one million records is consistent at around 69 seconds, demonstrating linear scalability. This stream is an example of the Data Preparation Phase of the CRISPDM process. It shows several common preparation steps (as in Fig. 3B.9.1), including aggregation, merging and deriving new fields that are necessary to prepare data for modeling. The data preparation takes 75% to 90% of the time required for an entire data mining project.

3B.8.2 Benchmark Testing Results: Modeling

The average increase in time required to process one million records increases slightly as millions of records are added, which means model building scales well. This stream is an example of the modeling phase of the CRISP-DM process. It shows deriving a field and generating a C&RT model. The dataset includes eight fields, with a mix of continuous and categorical data. To get the highest performing models we need to try many models quickly. To accomplish this, we use samples to try many models before building models on larger datasets. In addition to increased performance in model building due to in-database processing of data preparation steps in model-building streams (in Fig. 3B.9.2), Clementine Server also delivers improved performance for model building itself. This performance is not accomplished with in-database mining. Instead, the increased processing power and efficiency of the application server delivers better performance.

Performance in training models depends on a number of factors. First, different types of models inherently consume more processing power than other types of models. Neural networks, for example, are more processor intensive than regression models. The number of records and fields in our dataset and

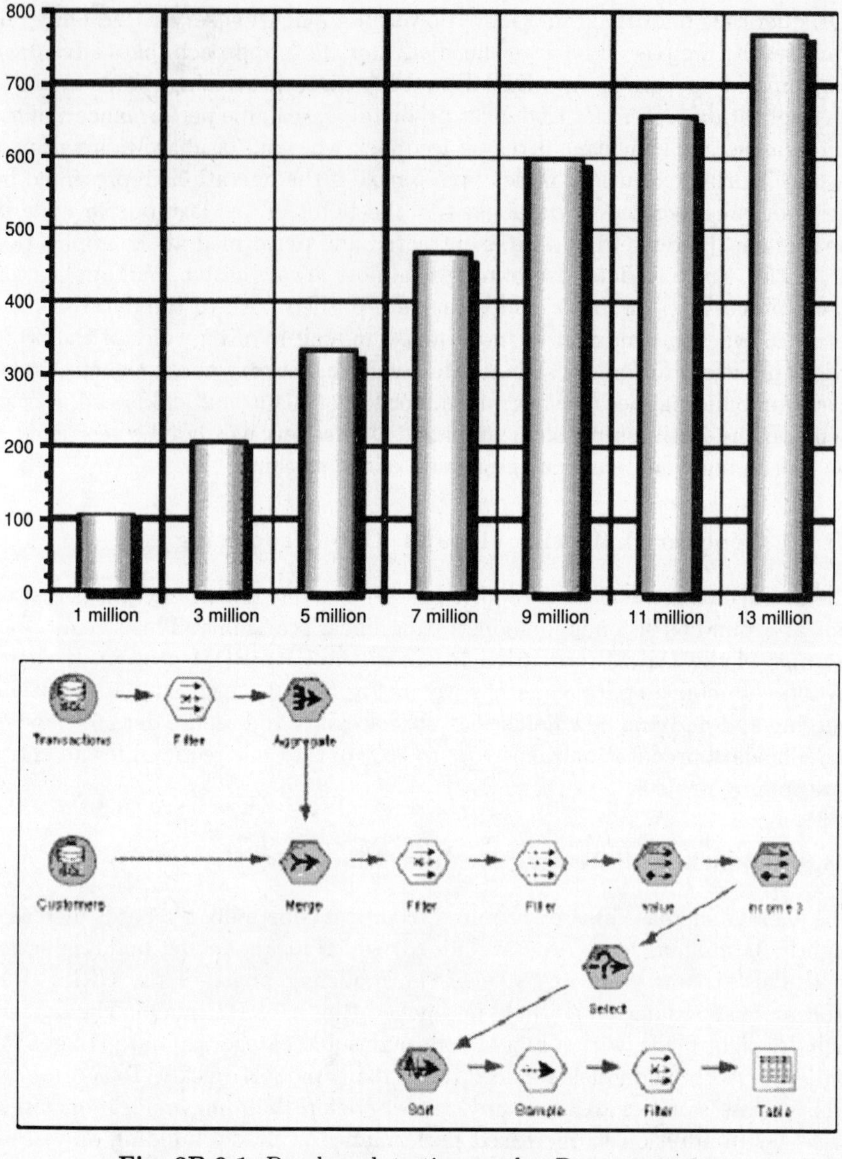

Fig. 3B.9.1. Benchmark testing results: Data processing

our computing system architecture, hardware, and software can have an effect on model building speed.

We must keep in mind that models built quickly are not necessarily good models. We can speed up model building in Clementine by adjusting the model training default settings. Often, accuracy is a tradeoff for speed. When

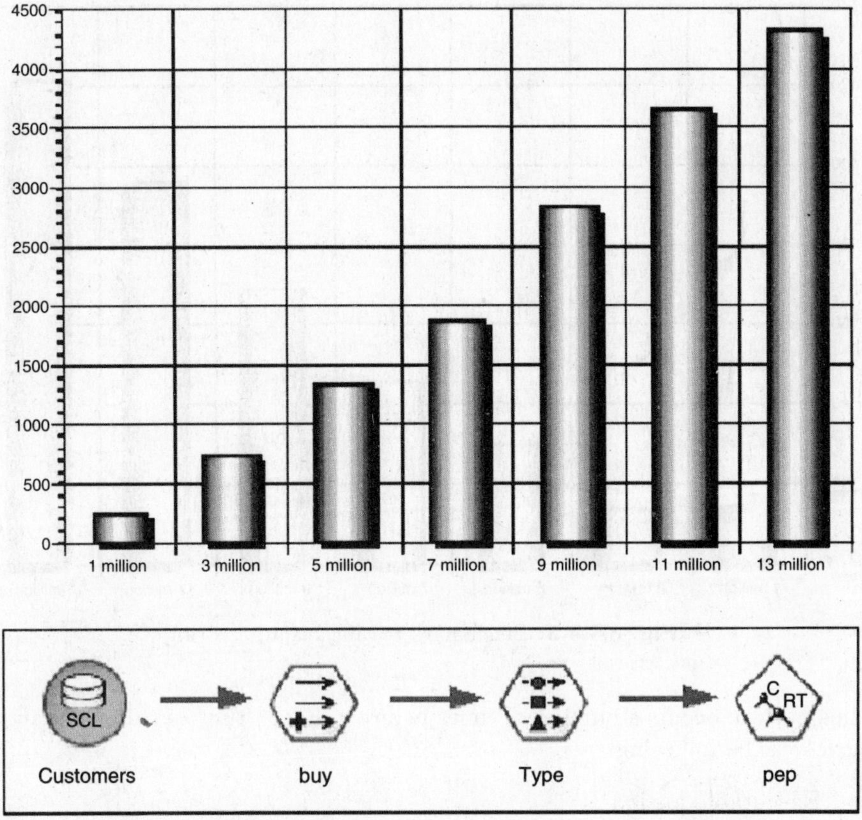

Fig. 3B.9.2. Benchmark testing results: modeling

working with large datasets, it may be best to try models for speed first and then opt for more accurate settings as we determine which models are most appropriate for the task at hand.

3B.8.3 Benchmark Testing Results: Scoring

The average increase in time required to process one million records increases slightly as millions of records are added, which means scoring scales well. This stream is an example of the deployment phase of the CRISP-DM process. It shows data access, applying generated model scores and confidence values, and writing back to a new database table. The second set of tests are designed to show interactive scoring results (as in Fig. 3B.9.3) — single case scoring with a complex stream and several iterations of multiple concurrent tests of the same stream. This type of test is important for applications like Web personalization, when we have many concurrent hits at the same time and

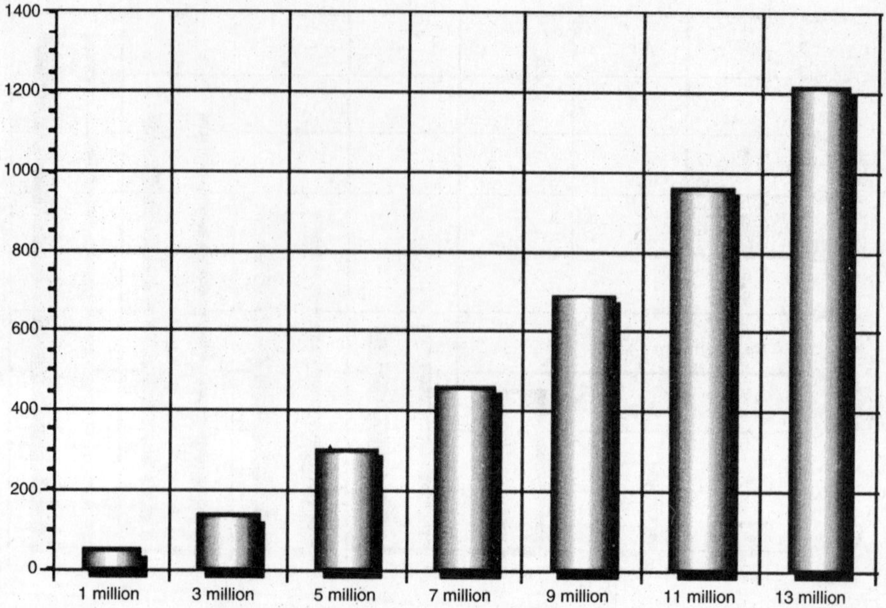

Fig. 3B.9.3. Benchmark testing results: scoring

this pattern occurs a number of times in a row. Each "run" of this application involves the following:

- Executable loads
- Reads data from file of 50+ "special offer" products
- Reads single case (data that's been entered through form) from file
- Combines case/offer data
- Applies model to each case/offer combination
- Applies other logic needed to create list of "best 10" offers for this customer
- Writes "best 10" list to file
- Executable exits

The scoring was run 1000 times with five concurrent processes. The test was performed on a small laptop with 128MB memory. Each run took 0.22 seconds with an average of 255 runs per minute.

3B.9 Conclusion

As the market for advanced analytics continues to grow, with all vendors in this segment recording strong double-digit growth in the last few years, with Clementine, SPSS is continuing to justify its position as one of the leaders in the market.

Clementine, in its latest guise, provides key features of use to business today, notably Web analysis and text mining. The direction that SPSS has for the product – enhancing its integration with the enterprise BI infrastructure, as epitomized by the facility to enable in-database data mining with Oracle, IBM and Microsoft – is in building strong partnerships with key vested interests, and this serves to differentiate Clementine from its key competitor SAS, with Enterprise Miner.

The acquisition of Data Distilleries, that has done so much to reinvigorate the capability of SPSS in the enterprise space, is clearly highly beneficial in enabling the deployment of models from Clementine within the predictive enterprise. Finally, enhancements to the product are being made in all of the key areas of productivity, functionality, and management. All of which leads to the inevitable conclusion that Clementine is a product worthy of serious consideration: all current users can continue to rely upon it with assurance and all potential users should include it on any evaluation shortlist.

The ever-growing amount of data created by organizations presents opportunities and challenges for data mining. Growing data warehouses that integrate all information about customer interactions present new possibilities for delivering personalization, resulting in increased profits and better service delivery. The challenge lies in making this vision a reality. Scaling the entire data mining process with Clementine Server makes mining large datasets more efficient, shortening the time needed to turn data into better customer relationships.

Appendix 3C

Crisp

AbstractThe information presented in this appendix is based on the CRISP-DM process model.

3C.1 Hierarchical Breakdown

The CRISP-DM data mining methodology is described in terms of a hierarchical process model, consisting of sets of tasks described at four levels of abstraction (from general to specific): phase, generic task, specialized task, and process instance (see Fig. 3C.1).

At the top level, the data mining process is organized into a number of phases; each phase consists of several second-level generic tasks. This second level is called generic, because it is intended to be general enough to cover all possible data mining situations. The generic tasks are intended to be as complete and stable as possible. Complete means covering both the whole process of data mining and all possible data mining applications. Stable means that the model should be valid for yet unforeseen developments like new modeling techniques.

The third level, the specialized task level, is the place to describe how actions in the generic tasks should be carried out in certain specific situations. For example, at the second level there might be a generic task called *clean data*. The third level would describe how this task differed in different situations, such as cleaning numeric values versus cleaning categorical values, or whether the problem type is clustering or predictive modeling.

The description of phases and tasks as discrete steps performed in a specific order represents an idealized sequence of events. In practice, many of the tasks can be performed in a different order and it will often be necessary to repeatedly backtrack to previous tasks and repeat certain actions. Our process model does not attempt to capture all of these possible routes through the data mining process because this would require an overly complex process model.

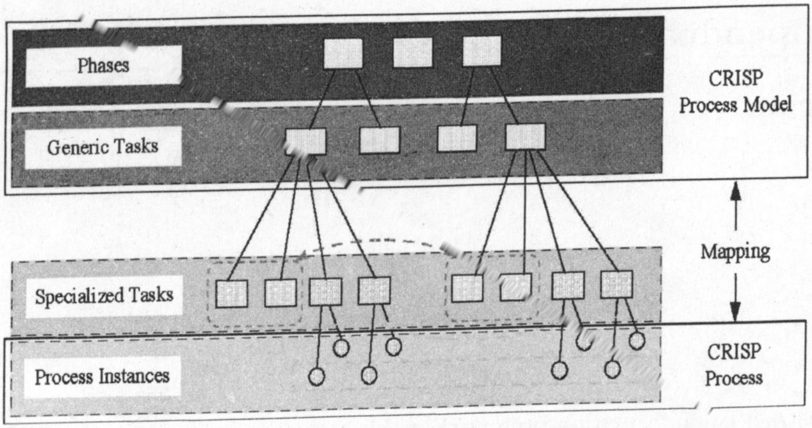

Fig. 3C.1. Four Level Breakdown of the CRISP-DM Methodology

The fourth level, the process instance, is a record of the actions, decisions, and results of an actual data mining engagement. A process instance is organized according to the tasks defined at the higher levels, and represents what actually happened in a particular engagement, rather than what happens in general.

3C.2 Mapping Generic Models to Specialized Models

3C.2.1 Data Mining Context

Mapping between the generic and the specialized level in CRISP-DM is driven by the data mining context. Currently, we distinguish between four different dimensions of data mining contexts:

1. The application domain is the specific area in which the data mining project takes place.
2. The data mining problem type describes the specific class(es) of objective(s), which the data mining project deals with.
3. The technical aspect covers specific issues in data mining, which describe different (technical) challenges that usually occur during data mining.
3. The tool and technique specifies, which data mining tool(s) and/or techniques are applied during the data mining project.

Table 3C.1 below summarizes these dimensions of data mining contexts and shows specific examples for each dimension.

A specific data mining context is a concrete value for one or more of these dimensions. For example, a data mining project dealing with a classification problem in churn prediction constitutes one specific context. The more values for different context dimensions are fixed, the more concrete is the data mining context.

Table 3C.1. Dimensions of Data Mining Contexts and Examples

	Data Mining Context			
Dimension	*Application Domain*	*Data Mining Problem Type*	*Technical Aspect*	*Tool and Technique*
Examples	Response Modeling	Description and Summarization	Missing Values	Clementine
	Churn Prediction	Segmentation	Outliers	MineSet
	...	Concept Description	...	Decision Tree
		Classification		...
		Prediction		
		Dependency Analysis		

3C.2.2 Mappings with Contexts

We distinguish between two different types of mapping between generic and specialized level in CRISP-DM:

1. "Mapping for the Presence": If we only apply the generic process model to perform a single data mining project and attempt to map generic tasks and their descriptions to the specific project as required, we talk about a single mapping for (probably) only one usage.
2. "Mapping for the Future": If we systematically specialize the generic process model according to a predefined context (or similarly systematically analyze and consolidate experiences of a single project toward a specialized process model for future usage in comparable contexts), we talk about explicitly writing up a specialized process model in terms of CRISP-DM. Which type of mapping is appropriate for our own purposes depends on our specific data mining context and the needs of our organization.

3C.3 The CRISP-DM Reference Model

The current process model for data mining provides an overview of the life cycle of a data mining project (in Fig. 3C.2). It contains the corresponding phases of a project, their respective tasks, and relationships between these tasks. At this description level, it is not possible to identify all relationships. Essentially, there possibly exist relationships between all data mining tasks depending on the goals, the background, and interest of the user, and most importantly on the data.

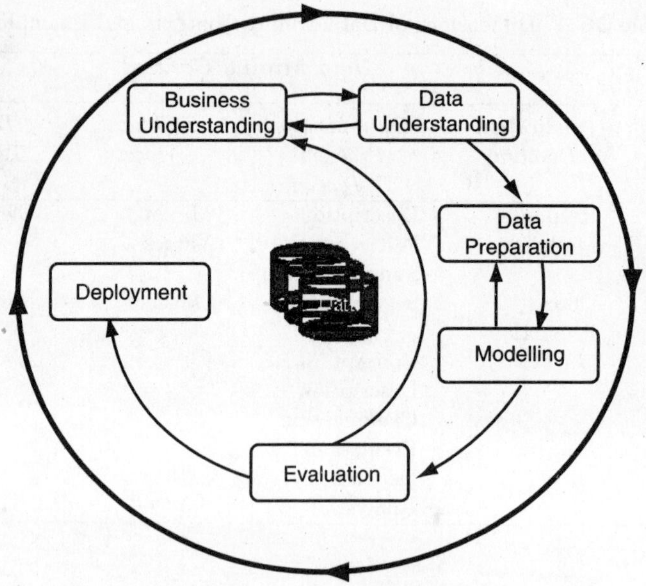

Fig. 3C.2. Phases of the CRISP-DM Reference Model

The life cycle of a data mining project consists of six phases. Figure 3C.2 shows the phases of a data mining process. The sequence of the phases is not strict. Moving back and forth between different phases is always required. It depends on the outcome of each phase which phase, or which particular task of a phase, has to be performed next. The arrows indicate the most important and frequent dependencies between phases.

The outer circle in Fig. 3C.2 symbolizes the cyclic nature of data mining itself. Data mining is not over once a solution is deployed. The lessons learned during the process and from the deployed solution can trigger new, often more focused business questions. Subsequent data mining processes will benefit from the experiences of previous ones.

In the following, we outline each phase briefly:

- **Business Understanding**

This initial phase focuses on understanding the project objectives and requirements from a business perspective, and then converting this knowledge into a data mining problem definition, and a preliminary plan designed to achieve the objectives.

- **Data Understanding**

The data understanding phase starts with an initial data collection and proceeds with activities in order to get familiar with the data, to identify data quality problems, to discover first insights into the data, or to detect interesting subsets to form hypotheses for hidden information.

- **Data Preparation**

The data preparation phase covers all activities to construct the final dataset (data that will be fed into the modeling tool(s)) from the initial raw data. Data preparation tasks are likely to be performed multiple times, and not in any prescribed order. Tasks include table, record, and attribute selection as well as transformation and cleaning of data for modeling tools.

- **Modeling**

In this phase, various modeling techniques are selected and applied, and their parameters are calibrated to optimal values. Typically, there are several techniques for the same data mining problem type. Some techniques have specific requirements on the form of data. Therefore, stepping back to the data preparation phase is often needed.

- **Evaluation**

At this stage in the project we have built a model (or models) that appears to have high quality, from a data analysis perspective. Before proceeding to final deployment of the model, it is important to more thoroughly evaluate the model, and review the steps executed to construct the model, to be certain it properly achieves the business objectives. A key objective is to determine if there is some important business issue that has not been sufficiently considered. At the end of this phase, a decision on the use of the data mining results should be reached.

- **Deployment**

Creation of the model is generally not the end of the project. Even if the purpose of the model is to increase knowledge of the data, the knowledge gained will need to be organized and presented in a way that the customer can use it. Depending on the requirements, the deployment phase can be as simple as generating a report or as complex as implementing a repeatable data mining process. In many cases it will be the customer, not the data analyst, who will carry out the deployment steps. However, even if the analyst will not carry out the deployment effort it is important for the customer to understand upfront what actions will need to be carried out in order to actually make use of the created models. Fig. 3C.3 presents an outline of phases accompanied by generic tasks (bold) and outputs (italic). In the following sections, we describe each generic task and its outputs in more detail.

3C.3.1 Business Understanding

Determine Business Objectives

The first objective of the data analyst is to thoroughly understand, from a business perspective as in Fig. 3C.4, what the client really wants to accomplish.

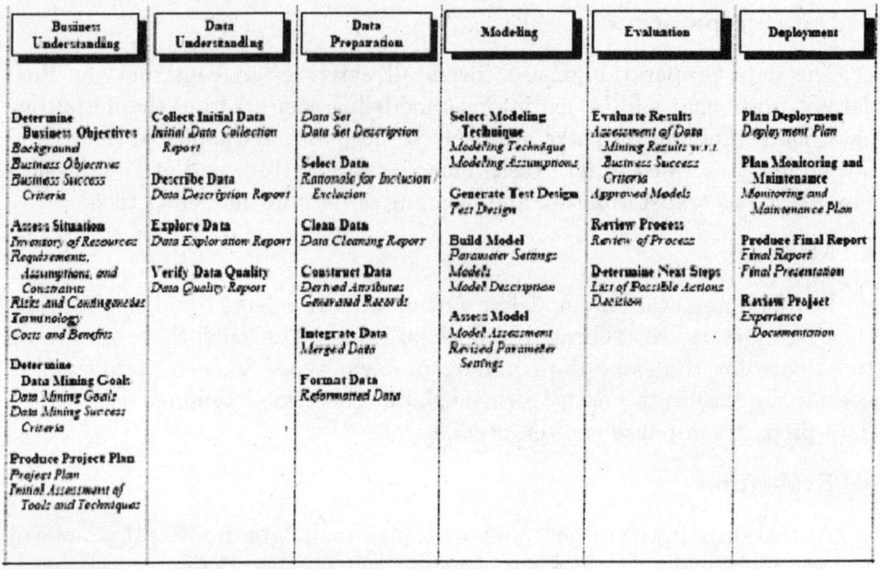

Fig. 3C.3. Generic Tasks (bold) and Outputs (italic) of the CRISP-DM Reference Model

Often the client will have many competing objectives and constraints that must be properly balanced. The analyst's goal is to uncover important factors, at the beginning, that can influence the outcome of the project. A possible consequence of neglecting this step is to expend a great deal of effort producing the right answers to the wrong questions.

Describe the customer's primary objective, from a business perspective. In addition to the primary business objective, there are typically a large number of related business questions that the customer would like to address. For example, the primary business goal might be to keep current customers by predicting when there are prone to move to a competitor. Examples of related business questions are "How does the primary channel (e.g., ATM, visit branch, internet) a bank customer uses affect whether they will stay or go?" or "Will lower ATM fees significantly reduce the number of high-value customers who leave?"

Business Success Criteria

Describe the criteria for a successful or useful outcome to the project from the business point of view. This might be quite specific, such as reduction of customer churn to a certain level, or general and subjective such as "give useful insights into the relationships." In the latter case it should be indicated who will make the subjective judgment.

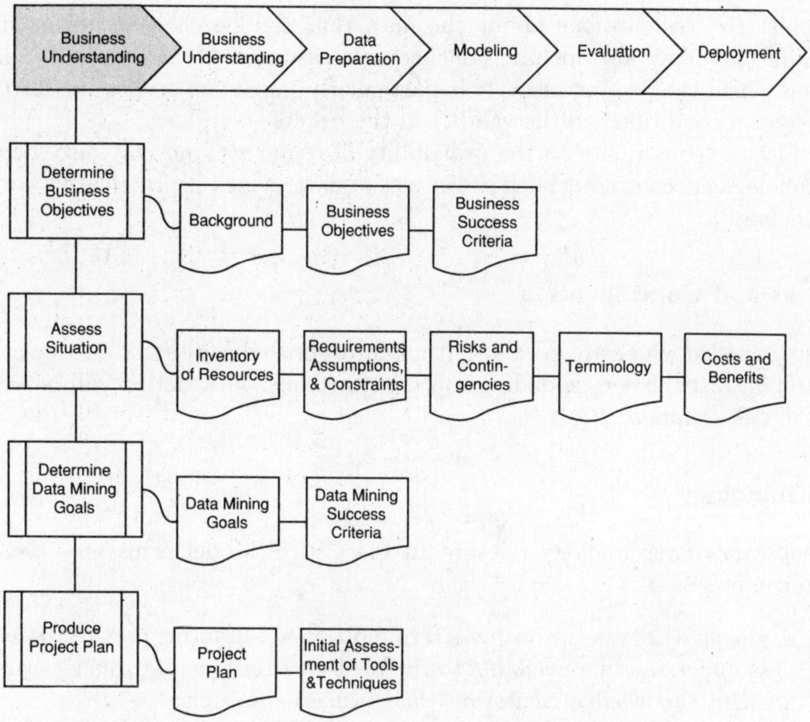

Fig. 3C.4. Business Understanding

Assess Situation

This task involves more detailed fact finding about all of the resources, constraints, assumptions, and other factors that should be considered in determining the data analysis goal and project plan. In the previous task, our objective is to quickly get to the crux of the situation.

Outputs Inventory of Resources

List the resources available to the project, including: personnel (business experts, data experts, technical support, data mining personnel), data (fixed extracts, access to live warehoused or operational data), computing resources (hardware platforms), software (data mining tools, other relevant software).

Requirements, Assumptions, and Constraints

List all requirements, including schedule of completion, comprehensibility and quality of results, and security as well as legal issues.

List the assumptions about the data that can be checked during data mining, but may also include noncheckable assumptions about the business upon which the project rests. It is particularly important to list the latter if they form conditions on the validity of the results.

List the constraints on the availability of resources, and may also include technological constraints such as the size of data, which is practical to use for modeling.

Risks and Contingencies

List the risks, which are events that might occur to delay the project or cause it to fail. List the corresponding contingency plans; what action will be taken if the risks happen.

Terminology

A glossary of terminology relevant to the CRISP-Model. This may include two components:

(1) A glossary of relevant business terminology, which forms part of the business understanding available to the project. Constructing this glossary is a useful "knowledge elicitation" and education exercise.
(2) A glossary of data mining terminology, illustrated with examples relevant to the business problem in question.

Costs and Benefits

A cost-benefit analysis, compare the costs of the project with the potential benefit to the business if it is successful. The comparison should be as specific as possible, for example using monetary measures in a commercial situation.

Determine Data Mining Goals

A business goal states objectives in business terminology. A data mining goal states project objectives in technical terms. For example, the business goal might be "Increase catalog sales to existing customers" while a data mining goal might be "Predict how many widgets a customer will buy, given their purchases over the past three years, demographic information (age, salary, city, etc.), and the price of the item".

Outputs Data Mining Goals

Describe the intended outputs of the project, which will enable the achievement of the business objectives.

Data Mining Success Criteria

Define the criteria for a successful outcome to the project in technical terms, for example a certain level of predictive accuracy, or a propensity to purchase profile with a given degree of "lift." As with business success criteria, it may be necessary to describe these in subjective terms, in which case the person or persons making the subjective judgment should be identified.

Produce Project Plan

Describe the intended plan for achieving the data mining goals, and thereby achieving the business goals. The plan should specify the anticipated set of steps to be performed during the rest of the project, including an initial selection of tools and techniques.

Outputs Project Plan

List the stages to be executed in the project, together with duration, resources required, inputs, outputs, and dependencies. Wherever possible, make the large-scale iterations in the data mining process explicit, for example repetitions of the modeling and evaluation phases.

As part of the project plan, it is also important to analyze dependencies between time schedule and risks. Mark results of these analyses explicitly in the project plan, ideally with actions and recommendations if the risks appear. Note the project plan contains detailed plans for each phase. For example, decide at this point which evaluation strategy will be used in the evaluation phase.

The project plan is a dynamic document in the sense that at the end of each phase a review of progress and achievements is necessary and an update of the project plan accordingly is recommended. Specific review points for these reviews are part of the project plan, too.

Initial Assessment of Tools and Techniques

At the end of the first phase, the project also performs an initial assessment of tools and techniques. Here, we select a data mining tool that supports various methods for different stages of the process, for example. It is important to assess tools and techniques early in the process since the selection of tools and techniques possibly influences the entire project.

3C.4 Data Understanding

3C.4.1 Collect Initial Data

Acquire within the project the data (or access to the data) listed in the project resources. This initial collection includes data loading if necessary for data

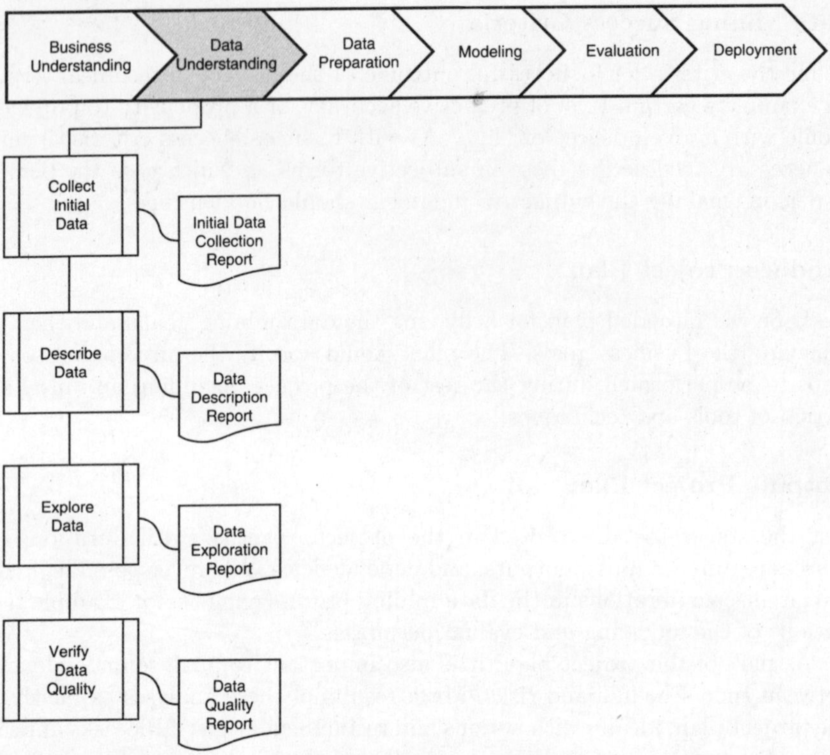

Fig. 3C.5. Data Understanding

understanding. For example, if we apply a specific tool for data understanding, it makes perfect sense to load our data into this tool. This effort possibly leads to initial data preparation steps (as in Fig. 3C.5). Note, if we acquire multiple data sources, integration is an additional issue, either here or in data preparation later.

3C.4.2 Output Initial Data Collection Report

List the dataset (or datasets) acquired, together with their locations within the project, the methods used to acquire them and any problems encountered. (Problems encountered, and any solutions achieved, are listed to aid with future replication of this project or with the execution of similar future projects.)

3C.4.3 Describe Data

Examine the "gross" or "surface" properties of the acquired data and report on the results. Describe the data that has been acquired, including the identifiers of the fields, and any other surface features of the data that have been discovered. Does the data acquired satisfy the relevant requirements?

3C.4.4 Explore Data

This task tackles the data mining questions that can be addressed using querying, visualization, and reporting. These include: distribution of key attributes, for example the target attribute of a prediction task; relations between pairs or small numbers of attributes; results of simple aggregations; properties of significant subpopulations; simple statistical analyses. These analyses may address directly the data mining goals; they may also contribute to or refine the data description and quality reports, and feed into the transformation and other data preparation needed for further analysis.

3C.4.5 Output Data Exploration Report

Describes results of this task, including first findings or initial hypothesis and their impact on the remainder of the project. The report possibly also covers graphs and plots that indicate data characteristics or lead to interesting data subsets for further examination.

3C.4.6 Verify Data Quality

Examine the quality of the data, addressing questions such as: Is the data complete (does it cover all the cases required)? Is it correct or does it contain errors, and if there are errors how common are they? Are there missing values in the data? If so how are they represented, where do they occur and how common are they?

List the results of the data quality verification; if quality problems exist, list possible solutions. Solutions to data quality problems will generally depend heavily of both data and business knowledge.

3C.5 Data Preparation

This is the dataset (or datasets) produced by the data preparation phase, which will be used for modeling or the major analysis work of the project (as in Fig. 3C.6). Describe the dataset (or datasets), which will be used for the modeling or the major analysis work of the project.

3C.5.1 Select Data

Decide on the data to be used for analysis. Criteria include relevance to the data mining goals, quality, and technical constraints such as limits on data volume or data types. Note that data selection covers selection of attributes (columns) as well as selection of records (rows) in a table. List the data to be included/excluded and the reasons for these decisions.

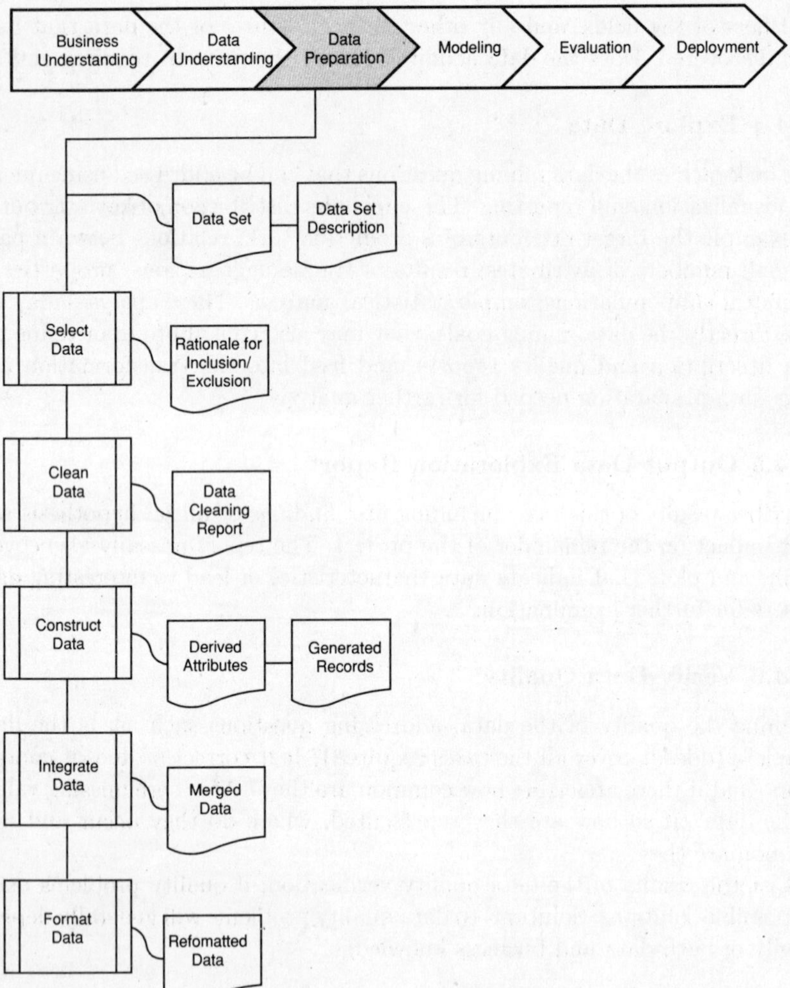

Fig. 3C.6. Data preparation

3C.5.2 Clean Data

Raise the data quality to the level required by the selected analysis techniques. This may involve selection of clean subsets of the data, the insertion of suitable defaults or more ambitious techniques such as the estimation of missing data by modeling.

This set of reports describes what decisions and actions were taken to address the data quality problems reported during the verify data quality task of the data understanding phase. Transformations of the data for cleaning purposes and the possible impact on the analysis results should be considered.

3C.5.3 Construct Data

This task includes constructive data preparation operations such as the production of derived attributes, entire new records, or transformed values for existing attributes. Derived attributes are new attributes that are constructed from one or more existing attributes in the same record. Examples: area=length * width.

3C.5.4 Generated Records

Describe the creation of completely new records. Examples: Create records for customers who made no purchase during the past year. There was no reason to have such records in the raw data, but for modeling purposes it might make sense to explicitly represent the fact that certain customers made zero purchases.

3C.5.5 Integrate Data

These are methods whereby information is combined from multiple tables or records to create new records or values.

3C.5.6 Output Merged Data

Merging tables refers to joining together two or more tables that have different information about the same objects. Example: A retail chain has one table with information about each store's general characteristics (e.g., floor space, type of mall), another table with summarized sales data (e.g., profit, percent change in sales from previous year), and another with information about the demographics of the surrounding area. Each of these tables contains one record for each store. These tables can be merged together into a new table with one record for each store, combining fields from the source tables.

Merged data also covers aggregations. Aggregation refers to operations where new values are computed by summarizing together information from multiple records and/or tables. For example, converting a table of customer purchases where there is one record for each purchase into a new table where there is one record for each customer, with fields such as number of purchases, average purchase amount, percentage of orders charged to credit card, percentage of items under promotion, etc.

3C.5.7 Format Data

Formatting transformations refer to primarily syntactic modifications made to the data that do not change its meaning, but might be required by the modeling tool.

3C.5.8 Reformatted Data

Some tools have requirements on the order of the attributes, such as the first field being a unique identifier for each record or the last field being the outcome field the model is to predict.

It might be important to change the order of the records in the dataset. Perhaps the modeling tool requires that the records be sorted according to the value of the outcome attribute. Most common situation is that the records of the dataset are initially ordered in some way but the modeling algorithm needs them to be in a fairly random order. For example, when using neural networks it is generally best for the records to be presented in a random order although some tools will hand this automatically without explicit user intervention.

Additionally, there are purely syntactic changes made to satisfy the requirements of the specific modeling tool. Examples: removing commas from within text fields in comma-delimited data files, trimming all values to a maximum of 32 characters.

3C.6 Modeling

3C.6.1 Select Modeling Technique

As the first step in modeling, select the actual modeling technique, which is used now. Whereas we possibly already selected a tool in business understanding, this task refers to the specific modeling technique as shown in Fig. 3C.7, e.g., decision tree building with C4.5 or neural network generation with back propagation. If multiple techniques are applied, perform this task for each technique separately.

3C.6.2 Outputs Modeling Technique

This output refers to the actual modeling technique, which is used.

3C.6.3 Modeling Assumptions

Many modeling techniques make specific assumptions on the data, e.g., all attributes have uniform distributions, no missing values allowed, class attribute must be symbolic etc.

3C.6.4 Generate Test Design

Before we actually build a model, we need to generate a procedure or mechanism how to test the model's quality and validity. For example, in supervised data mining tasks such as classification, it is common to use error rates as

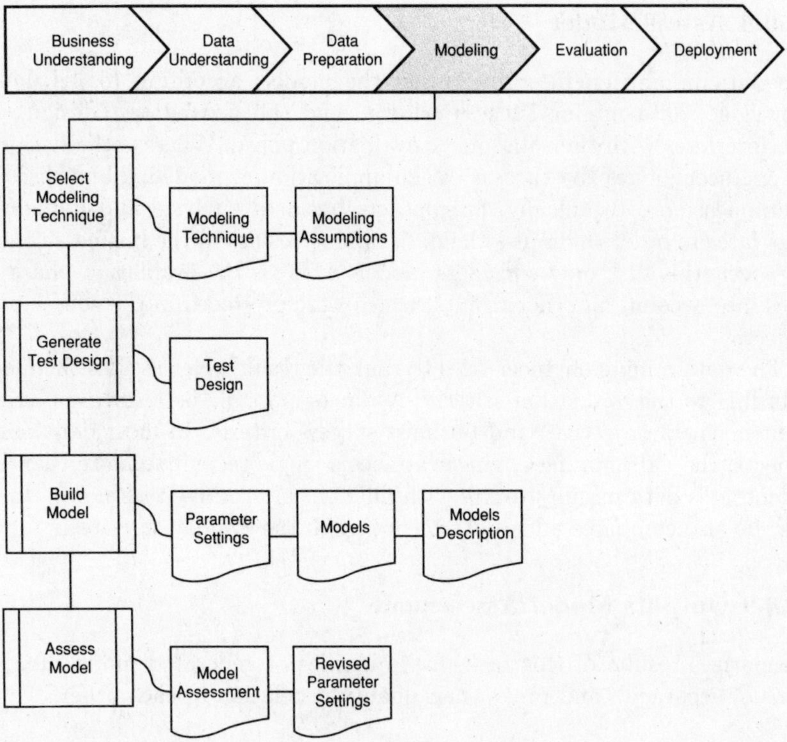

Fig. 3C.7. Modeling

quality measures for data mining models. Therefore, we typically separated the dataset into train and test set, build the model on the train set, and estimate its quality on the separate test set.

3C.6.5 Output Test Design

This deliverable describes the intended plan for training, testing, and evaluating the models. A primary component of the plan is to decide how to divide the available dataset into training data, test data, and validation datasets.

3C.6.6 Build Model

Run the modeling tool on the prepared dataset to create one or more models.

3C.6.7 Outputs Parameter Settings

With any modeling tool, there are often a large number of parameters that can be adjusted. This report lists the parameters and their chosen value, along with rationale for the choice of parameter settings.

3C.6.8 Assess Model

The data mining engineer interprets the models according to his domain knowledge, data mining success criteria, and the desired test design. This task interferes with the subsequent evaluation phase. Whereas the data mining engineer judges the success of the application of modeling and discovery techniques more technically, he contacts business analysts and domain experts later in order to discuss the data mining results in the business context. Moreover, this task only considers models whereas the evaluation phase also takes into account all other results, which were produced in the course of the project.

The data mining engineer tries to rank the results. He assesses the models according to the evaluation criteria. As far as possible he also takes into account business objectives and business success criteria. In most data mining projects, the data mining engineer applies a single technique more than once or generates data mining results with different alternative techniques. In this task, he also compares all results according to the evaluation criteria.

3C.6.9 Outputs Model Assessment

Summarizes results of this task, lists qualities of generated models (e.g., in terms of accuracy), and ranks their quality in relation to each other.

3C.6.10 Revised Parameter Settings

According to the model assessment, revise parameter settings and tune them for the next run in task build model. Iterate model building and assessment until we strongly believe that we found the best model(s).

3C.7 Evaluation

3C.7.1 Evaluate Results

Previous evaluation steps dealt with factors such as the accuracy and generality of the model. This step assesses the degree to which the model meets the business objectives and seeks to determine if there is some business reason why this model is deficient. Another option of evaluation is to test the model(s) on test applications in the real application if time and budget constraints permit.

Moreover, evaluation (as in Fig. 3C.8) also assesses other data mining results generated. Data mining results cover models that are necessarily related to the original business objectives, and all other findings that are not necessarily related to the original business objectives but might also unveil additional challenges, information, or hints for future directions.

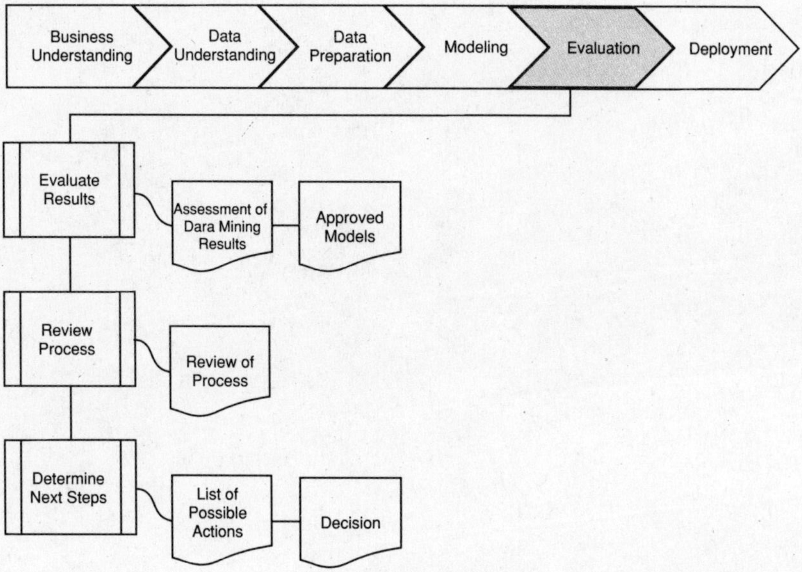

Fig. 3C.8. Evaluation

3C.8 Conclusion

In practice, many of the tasks can be performed in a different order and it will often be necessary to repeatedly backtrack to previous tasks and repeat certain actions. Our process model does not attempt to capture all of these possible routes through the data mining process because this would require an overly complex process model.

The fourth level, the process instance, is a record of the actions, decisions, and results of an actual data mining engagement. A process instance is organized according to the tasks defined at the higher levels, and represents what actually happened in a particular engagement, rather than what happens in general.

Appendix 3D

Mineset

3D.1 Introduction

The information presented in this appendix is based on MineSet 2.6 Silicon Graphics Inc.

MineSet (Silicon Graphics 1998, Brunk, Kelly & Kohavi 1997) is a general purpose data analysis tool that provides database access, analytical data mining, and data visualization in a highly integrated environment that supports the knowledge discovery process (Fayyad, Piatetsky-Shapiro & Smyth 1996). In addition, MineSet is a platform for developing vertical applications that require analytical data mining and visualization. MineSet is an evolving product.

We begin with an overview of the system architecture, then describe the analytical algorithms, the visualization techniques, and the support for KDD process management. We conclude with a brief history of the project and commercial uses.

3D.2 Architecture

MineSet employs a three-tier architecture (Fig. 3D.1). The first tier is the client, which includes Tool Manager and the visualization tools. Tool Manager is the graphical user interface through which the user interacts with MineSet. The visualization tools are used to display data and models of data generated by the mining algorithms. After invoking a visual tool with Tool Manager, the user interacts directly with that tool and sends information from it to other tools via Tool Manager. The second tier is the server, which includes Data Mover and the analytical mining engine. Data Mover is the database access and data transformation component of MineSet. It extracts data from the source, transforms it, and orchestrates moving it from one MineSet component to another. The mining tools are used to generate models of the transformed data, which can be applied to new data or visualized. The third tier is the

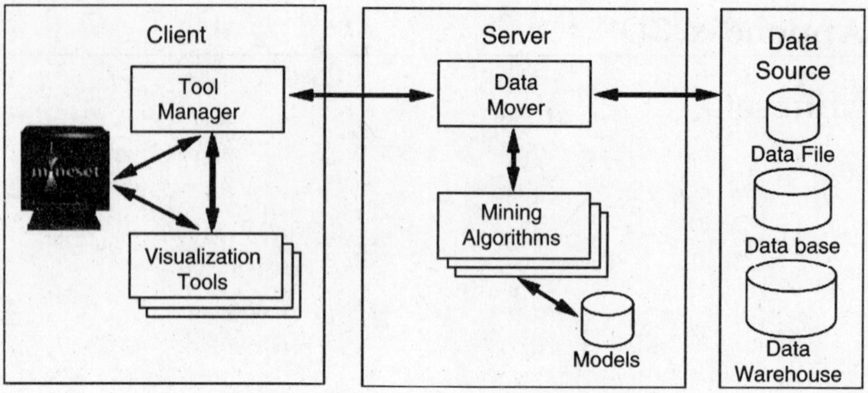

Fig. 3D.1. MineSet's three-tier architecture

data source, which includes the storage subsystem that maintains the user's data. It can be either a file or a commercial database. The tiers are not tied to specific machines: all three can reside on a single hardware platform or three separate platforms. This architecture provides the flexibility needed to scale with the size of the problem. It allows large mining tasks to be performed on a powerful server machine, while smaller pilot projects can be performed on a desktop machine.

Knowledge discovery is a time-consuming and iterative process involving modeling data, then understanding, and validating the model. Useful tools facilitate this process by generating models of the data quickly and allowing the user to interact with and understand those models. Because speed is of primary importance, MineSet's analytical mining algorithms operated on data in core memory and key components have been parallelized to further reduce execution time on multi processor machines.

Although limited to core memory MineSet supports 64-bit addressing allowing access to large amounts of memory.

3D.3 MineSet Tools for Data Mining Tasks

If we have data mining problems requiring classification, regression, and clustering, we will find these MineSet tools useful:

Decision Tree Inducer and Classifier—Induces a classifier resulting in a decision tree visualization.

Option Tree Inducer and Classifier—Induces a classifier similar to a decision tree inducer and classifier. However, it builds alternative options and averages them during classification, usually leading to improved accuracy.

Evidence Inducer and Classifier—Creates its own classifier and produces a visualization to display evidence based on the data provided.

Decision Table Inducer and Classifier—Creates a hierarchical visualization displaying pairs of dimensions at every level. We can drill up and drill down quickly, while maintaining context.

Clustering Algorithm—Groups data according to similarity of characteristics, then displays it as a series of box plots and histograms, similar to the Statistics Visualizer.

The clustering algorithm displays results using the Cluster Visualizer by default, but other visual tools may be used as an alternative.

Regression Tree—Induces a regressor that predicts a real value, that is, results with gradations of value rather than specific predetermined limits.

Column Importance—Determines the importance of specific columns in discriminating one label value from another. Used to observe the varying effects of changing variables, or to suggest columns to map to the axes of the Scatter and Splat Visualizers.

MineSet contains additional tools to aid the knowledge discovery process:

Statistics Visualizer—Data is displayed in the form of box plots and histograms, one per column. Continuous columns are shown as box plots, discrete columns are shown as histograms.

Histogram Visualizer—Data is displayed in the form of histograms. Continuous columns are binned (broken down into ranges).

Record Viewer—The original data is displayed as a spreadsheet.

3D.4 About the Raw Data

The *churn* dataset deals with telecommunications customers—people who use the phone regularly. Customers have a choice of carriers, or companies providing them with telephone service. When these customers change carriers they are said to "churn," which results in a loss of revenue for the previous carrier. A telecommunications company is likely to have a database of call records containing call information (source, destination, date, duration), a billing database, a customer database, and a customer service database.

Relevant information about the customer appears in all these databases. This information, when combined, yields a set of customer signatures. The churn dataset provided with MineSet is such a set; the step of identifying the data and creating customer signatures into records has already been done.

3D.5 Analytical Algorithms

MineSet uses MLC++ as its analytical engine. The naive Bayes, decision tree, option tree, k-means clustering, regression tree, decision table, association rule generation, and feature selection algorithms in MLC++ have been made accessible through MineSet's Tool Manager. The emphasis has been on selecting algorithms that generate interpretable models that facilitate data understanding. Algorithms that create"\ black box" models, like neural networks, provide little insight into the data and have not yet been included in MineSet. A plug-in API provides support for algorithms developed outside the MLC++ framework. For instance, Ultimode has released a MineSet add-on plug-in called ACPro for clustering based on AutoClass. This is extremely important because it is unrealistic to expect a single off-the-shelf tool to provide all the algorithms needed to analyze data in every problem domain. Instead MineSet provides the infrastructure common to the discovery process and allows the user to extend the tool as needed by plug-in algorithms specific to their task.

3D.6 Visualization

MineSet provides a rich set of visualization tools that enable users to interactively explore data and quickly discover new patterns, trends, and relationships. These 2D and 3D visualization capabilities allow direct data visualization for exploratory analysis, including tools for displaying high-dimensional data taking advantage of geographical and hierarchical information. In addition the visualization techniques have been specialized for displaying the models generated by the analytical mining algorithms. The algorithms help the user identify potentially interesting models of the data. The visual tools help make these models more understandable and allow the user to interact with the models to gain more insight into the model and the underlying data.

The human perception system can identify anomalies and patterns much faster in a representative landscape than in a spreadsheet. The visual tools utilize 3D landscapes that take advantage of a person's ability to navigate in space, track movement, and compare objects of different sizes, colors, and shapes. In addition to visualization and navigation, the tools contain filtering and search facilities that allow users to quickly reduce the landscape to items of interest. MineSet includes eight visualization tools. The Statistics Visualizer display basic statistics in histograms and box plots. The Cluster Visualizer extends the Statistics Visualizer to show the attribute-by-attribute differences between clusters identified by the clustering algorithms. The Tree Visualizer displays data hierarchically. Users can determine the hierarchy and map attributes to a histogram at each node.

The Map Visualizer (Fig. 3D.6.2 left) displays data with a spatial component. A polygonal map must be provided and two attributes can be mapped to the polygon's height and color. Multiple maps may be linked together to show different attributes for the same spatial geography. The Scatter Visualizer displays scatter plots with up to eight dimensions: three axes, entity color, entity size, entity rotation, and two independent attributes shown through animation. It is also used to visualize the confidence and support of one-to-one association rules. The Splat Visualizer (Becker 1997) (Fig. 3D.1) extends the

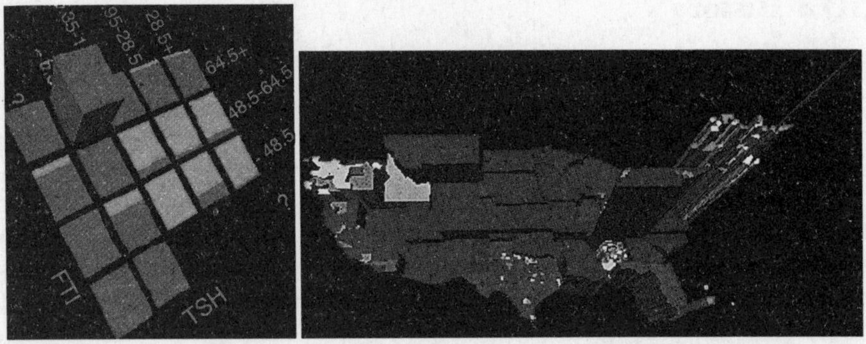

Fig. 3D.6.2. Visualization of a decision table for the hypothyroid database (left).

scatter plots when there are more than tens of thousands of records. It blurs the points using Gaussian smoothing. The Decision Table Visualizer (Kohavi & Sommerfield 1998) (Fig. 3D.6.2 right) shows the break down of class label according to attribute value. Initially, the two most predictive attributes are shown, but the user can show additionally informative attributes by clicking on the cakes of interest and drilling down. This provides visual OLAP (On-Line Analytical Processing) capability. The Evidence Visualizer (Becker, Kohavi & Sommerfield 2001) shows a graphic representation of the naive Bayes model and allows the user to interact with the model by selecting known values, providing what-if analysis.

Additional capabilities shared by most visual tools include: mapping attributes to graphical attributes (color, height, shape); manipulating the scene using thumbwheels and mouse movements for rotation, translation, and zoom; data slicing and animating by manipulating sliders for two additional independent variables as shown in left; searching and filtering of data; drilling-down by pointing to elements in the scene; and sending records associated with selected visual elements to the Tool Manager for further analysis.

As with the analytical algorithms, emphasis has been placed on selecting visualization techniques that are relatively simple to interpret. Techniques, which require extensive training to understand like parallel coordinates and grand tours, have yet to be included in MineSet.

3D.7 KDD Process Management

MineSet is more than an ensemble of data access, transformation, analytical mining, and visualization techniques connected by a common user interface. In addition to providing a consistent interface to all the tools, MineSet's Tool Manager provides a history mechanism that allows users to review and edit the steps performed in the current analysis, and change data sources and performs the same analysis on different data sets. Once satisfied with an analysis sequence it can be stored permanently and applied automatically to future data, or updated to account for changing future conditions.

3D.8 History

MineSet was first released in early 1996 primarily as a visualization product. The importance of connecting to commercial databases was recognized early, with a native connection to Oracle in MineSet 1.0, followed by connections to Sybase, Informix, and flat (ASCII) files in MineSet 1.02.

MineSet 1.1 integrated machine learning algorithms from MLC++ (Kohavi et al. 1997), including decision trees, naive Bayes (evidence), column importance, and automatic (entropy-based) discretization. Support for unknown (null values) was added, as well as support for session management (save/restore), a batch mode, integration with the desktop environment (icon launching), and the ability to define new columns

The plot shows how gross income is affected by age, occupation, and the number of hours worked per week. The density of each splat represents the number of people, its color represents the average gross income. The selected cube represents people over 30, who work over 55 hours per week in a professional specialty or an executive managerial position.

The figure shows the top-level view with two attributes: FTI and TSH. Users can see that several intersections are empty: high TSH values imply unknown FTI (probably not measured), and that most data is in the low range of TSH (below 6.3). High values for FTI (above 64.5) are negative hypothyroid with high probability (dark gray). The interesting intersections are for low FTI and high TSH. MineSet's map visualizer (right) showing refinancing costs, mapped to height, for every US country based on FIPS codes. Deviations from each state's average are colored from blue (zero deviation) to yellow (0.005) to red (0.01).

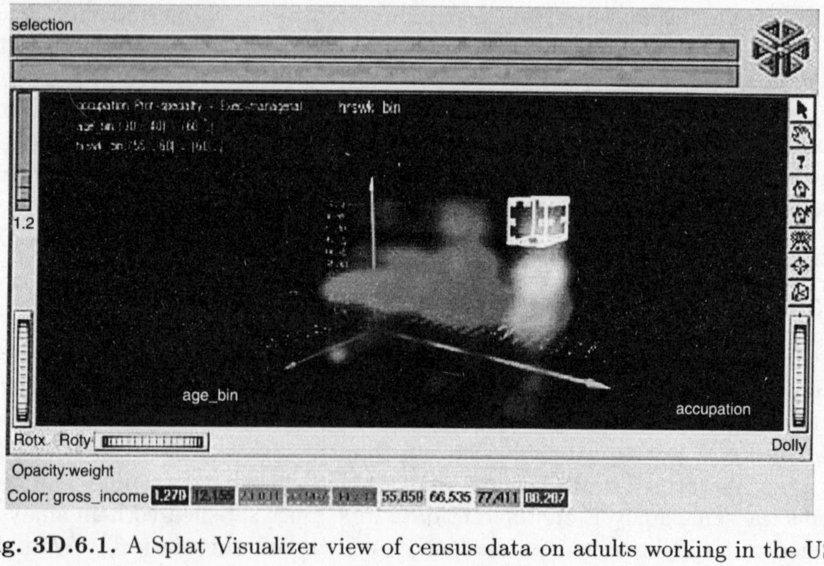

Fig. 3D.6.1. A Splat Visualizer view of census data on adults working in the US.

MineSet 1.2 added Web launching capabilities from machines that have MineSet installed. MineSet 2.0 added drill-through, the Splat Visualizer, Statistics Visualizer, Record Viewer, binary file format, sampling, Option Trees, loss matrices, learning curves, probability estimates from classifiers, and backfitting of data.

MineSet 2.5 added boosting of classifiers, parallelization, clustering, regression trees, and decision tables. Support for multibyte characters for internationalization and 64-bit support for large memory models was added in MineSet 2.6. Also added was a Java-based record viewer and a plug-in architecture for adding new data mining tools.

As of summer 1998, the engineering effort in product development is estimated at over 55 person years, with the engineering team consisting of 18 people.

3D.9 Commercial Uses

MineSet has been used commercially since it was released in 1996, but most customers are reluctant to publicize the exact uses as they perceive data mining as a competitive advantage. As of summer 1998, there are several hundred commercial sites using MineSet and close to a thousand universities. In this section we mention a very restricted set of commercial uses. Incyte Pharmaceuticals (www.incyte.com) provides genomic technologies to aid in the understanding of the molecular basis of diseases. Incyte created the LifeSeq 3D software to give scientists powerful visualization tools for sifting through the vast amounts of genomic data in the LifeSeq database (Incyte Pharmaceuticals 1997, Incyte Pharmaceuticals 1998). LifeSeq 3D is based on MineSet, displaying genomic information as interactive, multidimensional graphics, enabling scientists to easily navigate large data sets and uncover hidden relationships and important trends in gene expression.

Risk Monitors conducts statistical analyses of loan and mortgage data nationwide, building the models that mortgage services and banks rely on to calculate their underwriting risks. It analyzes 11 million loans nationwide dating back to 1989, and applies up to 200 variables to them during statistical analyses. Typical analysis was previously done in group, or cohorts. With MineSet, Risk Monitors was able to work with loans at the individual level.

Procter & Gamble Co.'s health care division uses MineSet for clinical trials and efficiency tests of over-the-counter drugs. On average, the total cost to bring a pharmaceutical drug to market is approximately $500 million, making it important to effectively run clinical trials and process the data effectively. Using data mining, scientists are trying to find an active molecule in a drug, making sure it works, testing it for harmful side effects, and eventually testing it on humans.

3D.10 Conclusion

MineSet provides a set of scalable analytical mining algorithms for identifying interesting patterns in data. MineSet also provides a rich selection of visualization techniques that help make these patterns understandable. But the primary feature that differentiates MineSet from other KDD tools is the integrated environment in which these algorithms and techniques are combined.

Appendix 3E

Enterprise Miner

Abstract. The information presented in this appendix is based on Enterprise Miner, Release 4.1 from SAS Institute Inc.

Data mining is a process, not just a series of statistical analyses. Implementing a successful data mining process requires a solution that gives users:

- Advanced, yet easy-to-use, statistical analyses and reporting techniques.
- A guiding, yet flexible, methodology.
- Client/server enablement.

Enterprise Miner software from SAS *is* that solution. It synthesizes the world-renowned statistical analysis and reporting systems of SAS software with an easy-to-use graphical user interface (GUI) that can be understood and used by business analysts as well as quantitative analysts.

3E.1 Tools For Data Mining Process

A broad set of tools supports the complete data mining process. Regardless of our data mining needs, SAS provides flexible software that supports all steps necessary to address the complex problems at hand in a single integrated solution. Going from raw data to accurate, business-driven data mining models becomes a seamless process, enabling statistical modeling group, business managers, and the IT department to collaborate more efficiently.

An easy-to-use GUI helps both business analysts and statisticians build more models, faster. SAS Enterprise Miner's process flow diagram environment eliminates the need for manual coding and dramatically shortens model development time for both business analysts and statisticians. The process flow diagram also serve as self-documenting templates that can be updated easily or applied to new problems without starting over from scratch. Users can tailor their experience with SAS Enterprise Miner 5.2 via the flexible and interactive display settings. Diagrams can be shared easily with other analysts throughout the enterprise.

SAS Enterprise Miner offer numerous assessment features for comparing results from different modeling techniques. Both statistical and business users share a single, easy-to-interpret view. Model results are shared quickly across the enterprise with the unique model repository system, which links metadata and model management capabilities together in an integrated framework.

Scoring is the process of applying a model to new data and is the end result of many data mining endeavors. SAS Enterprise Miner automates the tedious scoring process and supplies complete scoring code for all stages of model development in SAS, C, Java, and PMML. The scoring code can be deployed in a variety of real-time or batch environments within SAS, on the Web or directly in relational databases. The result is faster implementation of data mining results.

3E.2 Why Enterprise Miner

SAS Enterprise Miner streamlines the entire data mining process from data access to model assignment. It supports all necessary tasks within a single integrated solution while providing the flexibility for efficient collaborations.

SAS provides the most powerful, complete data mining solution on the market with unparalleled model development and deployment opportunities. Delivered as a distributed client-server system, it is especially well suited for data mining in large organizations. SAS Enterprise Miner is designed for data miners, marketing analysts, database marketers, risk analysts, fraud investigators, business managers, engineers and scientists who play strategic roles in identifying and solving critical business or research issues.

To gain an edge in today's competitive market, powerful advanced analytic solutions are required to extract knowledge from the vast stores of data. The emerging field of data mining incorporates the process of selecting, exploring, and modeling. Discovering previously unknown patterns can deliver actionable strategies for decision makers across our enterprise. For those who choose to implement data mining, the payoffs can be huge.

The data preparation step is compounded further as the complexities of business problems increase. As a result, quantitative experts spend considerable time accessing and manipulating disparate data sources before beginning to apply their expertise to building the models required to solve business problems.

The demand for actionable analytical information is growing in every industry, putting increased pressure on data miners to produce more and better models in less time. Today's organizations require enterprise wide collaboration on data mining projects and call for powerful, multipurpose solutions that can be tailored to meet different needs.

SAS Enterprise Miner provides an optimized architecture for mining large quantities of data to provide data miners with more time to create highly accurate predictive and descriptive models. Results of the data mining process can be shared easily throughout an organization to deliver actionable analytical information and incorporate models into business processes.

3E.3 Product Overview

SAS Enterprise Miner 5.2 is delivered as a modern, distributed client-server system. To streamline the data mining process, this software is designed to work seamlessly with SAS data integration, advanced analytics, and business intelligence capabilities. It also provides proven model deployment architecture.

With SAS Enterprise Miner, we get an easy-to-use process flow diagram approach that eliminates the need for manual coding and supports collaborative model development. With the proven, self-guiding (SEMMA) data mining process, both experienced statisticians and less seasoned business analysts can develop more and better predictive analytical models. SEMMA provides a flexible framework for conducting the core tasks of data mining, encompassing five primary steps – sampling, exploration, modification, modeling, and assessment. It is driven by process flow diagrams that can be modified, saved, and shared.

SEMMA makes it easy to apply exploratory statistical and visualization techniques, select and transform the most significant variables, create models with those variables to predict outcomes, validate accuracy, and deploy these rich insights as decision models for the operational day-to-day business environments.

The innovative Java client/SAS server architecture provides unprecedented flexibility for configuring an efficient installation that scales from a single-user system to very large enterprise solutions. Powerful servers may be dedicated to computing, while end users move office to home to remote sites without losing access to mining projects or services. Many process-intensive server tasks such as data sorting, summarization, variable selection, and regression modeling are multithreaded and processes can be run in parallel for distribution across a grid of servers or scheduled for batch processing.

SAS Enterprise Miner provides superior analytical depth with an unmatched suite of advanced predictive and descriptive modeling algorithms, including decision trees, neural networks, memory-based reasoning, clustering, linear and logistic regression, associations, time series, and more. Critical preprocessing tasks such as merging data files, addressing missing values, clustering, dropping variables and filtering for outliners are all handled within SAS Enterprise Miner.

Preparing data for mining usually is the most time-consuming aspect of data-mining endeavors. SAS Enterprise Miner includes several tools that make data preparation a fully integrated and efficient part of the data mining process by providing interactive capabilities to explore and transform data for optimal model training. Extensive descriptive summarization features as well as advanced visualization tools enable users to examine quickly and easily large amounts of data in dynamically linked multi-dimensional plots that support interactive tasks. Quality data mining results that are uniquely and optimally suited individual business problems.

Assessment features for comparing models in terms of lift curves and overall profitability allow analysts to easily share and discuss essential results with business users. The models generated from different modeling algorithms can be consistently evaluated across a highly visual user interface. Business domain experts and statisticians alike can compare data mining from a common framework. The model results packages that contain all relevant information of a data mining process flow provide easy model reporting and management. These model result packages are centrally managed through the SAS Metadata Server and can be viewed and queried by data miners, business managers and data managers via a Web-based model repository viewer – the industry's only Web-based system for effectively managing and distributing large model portfolios throughout the organization.

Model deployment is the final and most important phase in which the ROI from the entire mining process is realized. This can be a tedious process and often entails manually writing or converting scoring code, which can delay model implementation and introduce potentially costly mistakes. Scoring code needs to mirror the entire process that led to the final model, including every data preprocessing step. Often, organizations have different environments for model building and model deployment and scoring code must be provided in different languages. SAS Enterprise Miner automatically generates the score code for the entire process flow and supplies complete scoring code in SAS, C, Java, and PMML. Production data can be scored within SAS Enterprise Miner or on any other machine, and the scoring code can be deployed in batch, real time on the Web, or directly in relational databases.

The customizable modeling environment of SAS Enterprise Miner provides the ability to add tools and include personalized SAS code. A Java API is available to embed data mining algorithms into operational business production systems in a fast and easy manner. Default selection lists can be extended with custom-developed tools written with SAS code or XML logic, which opens the entire world of SAS to data miners.

3E.4 SAS Enterprise Miner 5.2 Key Features

3E.4.1 Multiple Interfaces

- Easy-to-use GUI for building process flow diagrams:
- Build more and better models faster.
- Web deliverable.
- Access to SAS programming environment.

- XML diagram exchange.
- Reuse diagrams as templates for other projects or users.
- Batch processing:
- Encapsulates all features of the GUI.
- SAS macro-based.
- Experimental Java API.
- Web-based model repository:
- Manage large model portfolios.
- Query models by algorithm, rating, target, etc.
- Distribute results such as lift charts, tree diagrams and score code to business and data managers.

3E.4.2 Scalable Processing

- Server-based processing – asynchronous model training. Stop processing cleanly.
- Grid computing:
- Distribute mining process across a cluster.
- Schedule training and scoring tasks.
- Load balancing and resource allocation.
- Parallel processing – run multiple tools and diagrams concurrently.
- Multithreaded predictive algorithms.
- All storage located on servers.

3E.4.3 Accessing data

- Access to more than 50 different file structures.
- Integrated with SAS ETL Studio through SAS Metadata Server:
- Use SAS ETL Studio to define training tables for mining in SAS Enterprise Miner.
- Use SAS ETL Studio to retrieve and deploy SAS Enterprise Miner scoring code.

3E.4.4 Sampling

- Simple random.
- Stratified.
- Weighted.
- Cluster.
- Systematic.
- First N.
- Rare event sampling.

- Distribution plots.
- Statistics breakdown for each level of the class target.
- Bivariate statistics and plots:
- Ordered Pearson and Spearman correlation plot.
- Ordered chi-square plot with option for binning continuous inputs into $n$bins.
- Coefficient of variation plot.
- Variable selection by logworth.
- Other interactive plots:
- Variable worth plot ranking inputs based on their worth with the target.
- Class variable distributions across the target and/or the segment variable.
- Scaled mean deviation plots.

3E.4.10 Graphs/Visualization

- Batch and interactive plots: scatter plots, scatter plot matrix plots, lattice plots, 3D charts, density plots, histograms, multidimensional plots, pie charts, and area bar charts.
- Segment profile plots:
- Interactively profile segments of data created by clustering and modeling tools.
- Easily identify variables that determine the profiles and the differences between groups.
- Easy-to-use graphics wizard:
- Titles and footnotes.
- Apply a WHERE clause.
- Choose from several color schemes.

3E.5 Enterprise Miner Software

Enterprise Miner provides powerful data mining capabilities that meet the needs of a wide variety of users, including information technology staff, business analysts, and quantitative analysts. With Enterprise Miner, users can:

- Identify the most profitable customers and the underlying reasons for their loyalty.
- Determine why customers switch to competitors.
- Analyze clickstream data to improve e-commerce strategies.
- Increase customer profitability and reduce risk exposure through more accurate credit scoring.
- Determine the combination of products customers are likely to purchase and when.
- Set more profitable rates for insurance premiums.

- Save on downtime by applying predictive maintenance to manufacturing sites.
- Detect and deter fraudulent behavior.

Enterprise Miner provides the most complete set of data mining tools for data preparation and visualization, predictive modeling, clustering, association discovery, model management, model assessment, and reporting. A key advantage of using Enterprise Miner is that it generates complete scoring code for all stages of model development. Scoring code can be applied immediately to new data on any SAS system, which results in significant time savings. Scoring code is also available in the C and Java languages to enable users to deploy results outside of the SAS environment. Because it is a software solution for business from SAS, Enterprise Miner can integrate with other SAS software, including the award-winning SAS/Warehouse Administrator software, the SAS solution for online analytical processing (OLAP), and SAS/IntrNet software.

3E.5.1 The Graphical User Interface

Enterprise Miner uses a single GUI to provide the functionality that users need to uncover valuable information hidden in their data. With one point-and-click interface, users can perform the entire data mining process, from sampling data, through exploration and modification, to modeling, assessment, and scoring of new data for use in making business decisions.

The GUI is designed with two groups of users in mind. Business analysts, who might have limited statistical expertise, can use the GUI to quickly and easily navigate through the data mining process. Quantitative analysts, who might want to explore the details, can use the GUI to access and finetune the underlying analytical processes.

3E.5.2 The GUI Components

The primary components of the GUI, as shown in Fig. 3E.6.1, are:

- **Project Navigator.** Use to manage projects and diagrams; add tools, such as analysis nodes to the diagram workspace; and view HTML reports that are created by the Reporter node. The Tools subtab of the Project Navigator displays the data mining nodes that are available for constructing process flow diagrams (PFDs). An example is shown in Fig. 3E.6.2.
- **Diagram Workspace.** Use for building, editing, and running PFDs.
- **Tools Palette.** Contains a subset of the Enterprise Miner tools that are commonly used to build PFDs in the diagram workspace. Users can add to and delete from the Tools palette.
- **Nodes.** A collection of tools that enable users to perform tasks in the data mining process, such as data access, data analysis, and reporting. Nodes have a uniform look-and-feel that makes them easy to use.

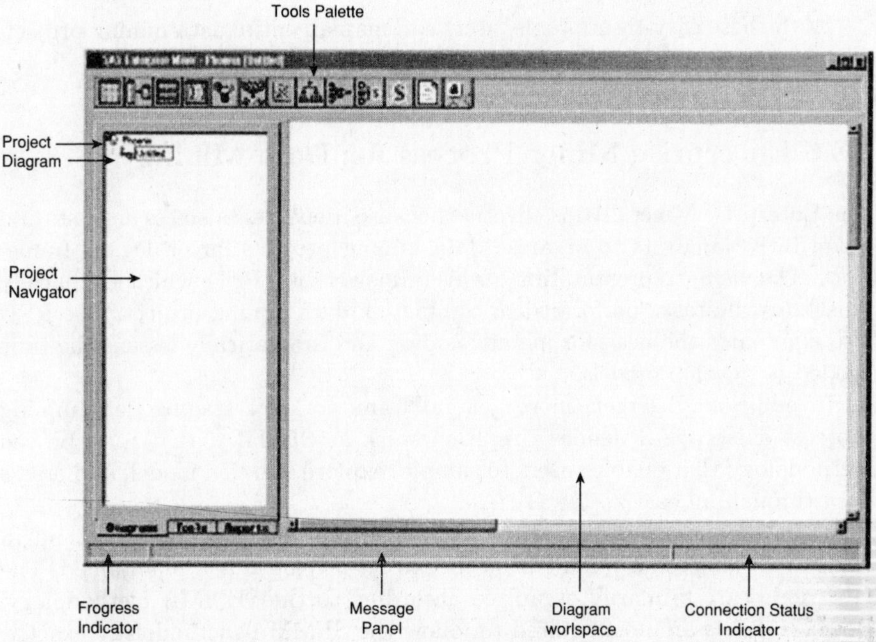

Fig. 3E.6.1. Primary components of GUI

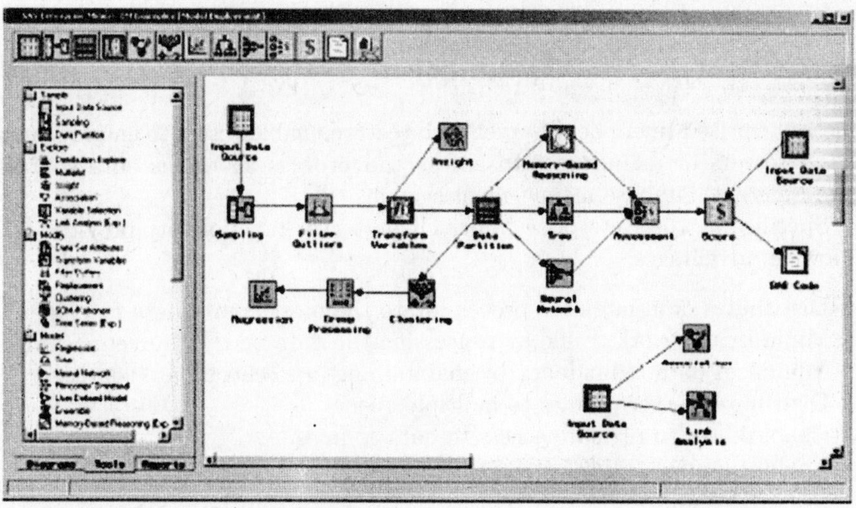

Fig. 3E.6.2. Process flow diagram

- **Message Panel**, **Progress Indicator,** and **Connection Status Indicator.** Use to display information and task completion estimates in response to user actions. These components appear across the bottom of the workspace window.

With these easy-to-use tools, users can map an entire data mining project, execute individual functions and modify PFDs.

3E.6 Enterprise Miner Process for Data Mining

The Enterprise Miner GUI is effective because it enables business analysts and quantitative analysts to organize data mining projects into a logical framework. The visible representation of this framework is a PFD, which graphically illustrates the tasks performed to complete a data mining project. The PFD also eliminates the need for manual coding and dramatically lessens the time needed to develop models.

In addition, a larger, more general framework for staging data mining projects exists. SAS defines this framework as SEMMA, that is, a proven methodology that enables users to sample, explore, modify, model, and assess data through an easy-to-use GUI.

SEMMA provides users with a powerful and comprehensive method in which individual data mining projects can be developed and maintained. The GUI nodes are primarily organized according to the SEMMA methodology. However, users are not required to follow the SEMMA methodology exactly. The relationship between the nodes and the methodology is flexible, which allows for the repetition or the reordering of SEMMA tasks.

3E.7 Client/Server Capabilities

The Enterprise Miner client/server architecture enables users to employ large UNIX servers or mainframes to access and process enormous data sources from different database management systems.

Specifically, the Enterprise Miner client/server functionality provides the following advantages:

- Distributes data-intensive processing to the most appropriate machine.
- Minimizes network traffic by processing the data on the source machine.
- Minimizes data redundancy by maintaining one central data source.
- Distributes server profiles to multiple clients.
- Enables user to regulate access to data sources.
- Enables user to toggle between remote and local processing.

3E.8 Client/Server Requirements

Table 3E.9.1 lists the client and server options available in Release 4.1 of Enterprise Miner.

Base SAS and SAS/STAT are required on the same server or mainframe on which the Enterprise Miner server component resides and processes data.

3E.4.5 Data Partitioning

- Create training, validation, and test data sets.
- Ensure good generalization of your models through use of holdout data.
- Default stratification by the class target.
- Balanced partitioning by any class variable.

3E.4.6 Filtering Outliers

- Apply various distributional thresholds to eliminate extreme interval values.
- Combine class values with fewer than n occurrences.
- Interactively filter class and numeric values.

3E.4.7 Transformations

- Simple: log, square root, inverse, square, exponential, standardized.
- Binning: bucketed, quantile, optimal binning for relationship to target.
- Best power: maximize normality, maximize correlation with target, equalize spread with target levels.
- Interactions editor: define polynomial and nth degree interaction effects.
- Interactively define transformations:
- Define customized transformations using the expression builder.
- Compare the distribution of the new variable with the original variable.

3E.4.8 Data Replacement

- Measures of centrality.
- Distribution-based.
- Tree imputation with surrogates.
- Mid-medium spacing.
- Robust M-estimators.
- Default constant.
- Replacement Editor:
- Specify new values for class variables.
- Assign replacement values for unknown values.

3E.4.9 Descriptive Statistics

- Univariate statistics and plots:
- Interval variables – n, mean, median, min, max, standard deviation, scaled deviation, and percent missing.
- Class variables – number of categories, counts, mode, percent mode, percent missing.

Table 3E.9.1. Client and Server options available in Release 4.1 of Enterprise Miner.

Client Options	Server Options
Microsoft Windows 98	ABI+ for Intel
Microsoft Windows NT	AIX (32, 64 bit)
Microsoft Windows 2000	Compaq Tru64 UNIX (32, 64 bit)
	HP-UX (32, 64 bit)
	Intel Linux
	MVS/ESA (or prior releases) including all releases of OS/390
	Solaris (32, 64 bit)
	Microsoft Windows NT and 2000

3E.9 Conclusion

Enterprise Miner can uncover valuable information hidden in large amounts of data. This software fully integrates all steps of the data mining process beginning with the sampling of data, through sophisticated data analyses and modeling, to the dissemination of the resulting information. The functionality of Enterprise Miner is provided through a flexible GUI that enables users, who have different degrees of statistical expertise, to plan, implement, and refine their data mining projects.

References

M. Holsheimer and A. Siebes. *Data mining: The search for knowledge in databases.* In CWI Report CSR9406, Amsterdam, The Netherlands, 1994.

G. H. John. *Enhancements to the data mining process.* In PhD. Thesis, Computer Science Dept., Stanford Univeristy, 1997.

R. Agrawal, T. Imielinski, A. Swami. *Database mining: A performance perspective.* IEEE Trans. Knowledge and Data Engineering, 5(6) 914–925, December 1993.

P. Bradley, U.M. Fayyad O.L. Mangasarian. *Mathematical Programming for Data Mining: Formulations and Challenges.* Technical Report 98-01, Computer Sciences Department, University of Wisconsin, Madison, WI, January 1998.

U. M. Fayyad, G. PiatetskyShapiro, P. Smyth. *Knowledge discovery and data mining: towards a unifying framework.*

J. Han. *Data mining techniques.* In Tutorial Notes, 1996 ACMSIGMOD Int. Conf. Management of Data, pages 1–71, Montreal, Canada, June 1996.

W. H. Inmon. *Building the Data Warehouse.* John Wiley, 1996.

M. S. Chen, J. Han, and P. S. Yu. *Data mining: An overview from a database perspective.* IEEE Trans. Knowledge and Data Engineering, 8:866–883, 1996.

U. M. Fayyad, G. Piatetsky-Shapiro, P. Smyth, and R. Uthurusamy. *Advances in Knowledge Discovery and Data Mining.* AAAI/MIT Press, 1995.

S. Hasty. Mining Databases. Apparel industry Magazine, 57(5), 1996.

U. M. Fayyad, G. Piatetsky-Shapiro, P. Smyth. *The KDD process for extracting useful knowledge from volumes of data.* Communication of the ACM, 39(11): 27–34, Nov 1996.

C. Glymour, D. Madigan, D. Pregibon, P. Smyth. *Statistical inference and data mining.* Communication of the ACM, 39(11):27–34, Nov 1996.

B. Linoff. *Data mining techniques.* Wieley, 1997.

H. Mannila, H. Toivonen, A. Inkeri Verkamo. *Efficient algorithms for discovering association rules.* In KDD-94: AAAI Workshop on Knowledge Discovery in Databases, 181–192, Seattle, Washington, July 1994.

R. Agrawal, T. Imielinski, and A. Swami. *Mining association rules between sets of items in large databases.* In Proc. 1993 ACM-SIGMOD Int. Conf. Management of Data, pages 207–216, Washington, D.C., May 1993.

S. Brin, R. Motwani, and C. Silverstein. *Beyond market basket: Generalizing association rules to correlations.* In Proc. 1997 ACM-SIGMOD Int. Conf. Management of Data, pages 265–276, Tucson, Arizona, May 1997.

U. M. Fayyad, G. Piatetsky-Shapiro, P. Smyth, and R. Uthurusamy. *Advances in Knowledge Discovery and Data Mining.* AAAI/MIT Press, 1995.

R. Agrawal and R. Srikant. *Fast algorithms for mining association rules.* In Proc. 1994 Int. Conf. Very Large Data Bases, pages 487–499, Santiago, Chile, September 1994.

R. Srikant and R. Agrawal. *Mining sequential patterns: generalizations and performance improvements.* Research report RJ 9994, IBM Almaden Research Center, San Jose, California, December 1995.

R. Agrawal and R. Srikant. *Mining sequential patterns.* In Proc. 1995 Int. Conf. Data Engineering, pages 3–14, Taipei, Taiwan, March 1995.

H. Mannila, H. Toivonen, and A. I. Verkamo. *Discovering frequent episodes in sequences.* In Proc. of the Int. Conference on Knowledge Discovery in Databases and Data Mining (KDD-95), Montreal, Canada, August 1995.

R. Agrawal, S. Gosh, T. Imielinski, B. Iyer, A. Swami. *An interval classifier for database mining applications.* In Proc. Of the VLDB Conf., Vancouver, British Coloumbia, Canada, August 1992.

D. J. Fifield. *Distributed tree construction from large data-sets.* Bachelor's Honours Thesis, Australian National University, 1992.

J. R. Quinla, R. L. Rivest. *Inferring decision trees using minimum decision length principle.* Information and computation, 1989.

J. Ross Quinlan. C4.5: *Programs for Machine Learning.* Morgan Kaufman, 1993.

R. T. Ng and J. Han. *Efficient and Effective Clustering Methods for Spatial Data Mining.* In Proceedings of the 20th VLDB Conference, pages 144–155, Santiago, Chile, 1994.

L. Kaufman and P. J. Rousseeuw. *Finding Groups in Data: an Introduction to Cluster Analysis*, John Wiley & Sons, 1990.

T. Zhang, R. Ramakrishnan, M. Livny. *BIRCH: An Efficient Data Clustering Method for Very Large Databases.* In Proceedings of the 1996 ACM SIGMOD International Conference on Management of Data, pages 103–114, Montreal, Canada, 1996.

S. Guha, R. Rastogi, K. Shim. *Cure: An efficient clustering algorithm for large databases.* In Proceedings of the ACM SIGMOD conference on Management of Data, pages 73–84, Seattle, WA, 1998.

M. Ester, H. Kriegel, J. Sander, and X. Xu. *A DensityBased Algorithm for Discovering Clusters in Large Spatial Databases with Noise.* In Proceedings of 2nd International Conference on KDD, 1996.

M. Ester, H. Kriegel, J. Sander, M. Wimmer, and X. Xu. *Incremental clustering for mining in a data warehousing environment.* In Proc. of the 24th VLDB Conf., NewYork, USA, 1998.

R. Agrawal, J. Gehrke, D. Gunopulos, P. Raghavan. *Automatic subspace clustering of high dimensional data for data mining applications.* In Proceedings of the ACM SIGMOD Conference on Management of Data, pages 94–105, Seattle, WA, 1998.

Agrawal R, Mehta M., Shafer J., Srikant R., Arning (1996) A *The Quest Data Mining System*, Proceedings of The Second International Conference on Knowledge discovery and Data Mining, pp. 244–249, August 1996.

Chaudhuri, S. Dayal, U. (1997) An Overview of Data Warehousing and OLAP Technology. SIGMOD Record 26(1) 65–74.

M. Chen, J. Han, and P. Yu. (1996) Data mining: An overview from database perspective. IEEE Transactions on Knowledge and Data Eng., 8(6) December, pp. 866–883.

Fayyad, U., Piatetsky-Shapiro, G. and Smyth, P. (1996) *Knowledge Discovery and Data Mining: Towards a Unifying Framework.* Proceedings of the Second International Conference on Knowledge Discovery and Data Mining, Portland, OR, pp. 82–88.

Frawley W., Gregory Piatetsky-Shapiro, Matheus, C. (1992) Knowledge Discovery in Databases: An Overview. *AI Magazine* 13(3):57–70.

Advances in Knowledge Discovery and Data Mining (1996) Usama M. Fayyad, Gregory, Piatetsky-Shapiro, Padhraic Smyth, Ramasamy Uthurusamy (Eds.)

Pawlak, Z. 1982 Rough sets. International Journal of Computer Sciences, 11:341–356.

Skowron, A. 1990 The rough sets theory and evidence theory. Fundamental Informitacae, 13:245–262.

Skownron, A. 1995 Extracting laws from decision tables. Computational Intelligence, 11/2:371–388.

Tsumoto, S. and Tanaka, H. 1993 PRIMEROSE: Probabilistic rule induction method based on rough set theory. Rough Sets, Fuzzy Sets and Knowledge Discovery (RSKD '93), Ziarko E. (ed.), 274–281.

Lefley, M., and Austin, V. (1997) Match'em: Using fuzzy logic to profile criminals. Proceedings of 6th IEEE conference on fuzzy logic '97 July 1997, Barcelona, Spain.

New Directions in Rough Sets, Data Mining, and Granular-Soft Computing, Proceedings of The Seventh International Workshop on Rough Sets, Fuzzy Sets, Data Mining, and Granular-Soft Computing, Lecture Notes in Artificial Intelligence 1711, Yamaguchi, Japan, November 9–11, 1999, Zhong, Skowron, A. and Ohsuga, S. (Eds) Dubois, D et al. (eds) 1993 *Readings in Fuzzy Sets for Intelligent Systems.*

Specht, D. F., Probabilistic neural networks for classification, mapping, or associative memory', IEEE Conference on Neural Networks, Vol. I, San

Diego, July 1988, pp525–532. *Readings in Machine Learning* (1990) Shavlik & Dietterich (Eds.)

David E. Goldberg (1989) *Genetic Algorithms in Search, Optimization, and Machine Learning.*

Cios et al., (1998) K. Cios, W. Pedrycz, and R. Swiniarski. Data Mining Methods for Knowledge Discovery, Kluwer, 1998.

Létourneau, Famili, and Matwin (1999) Data Mining to Predict Aircraft Component Replacement, IEEE *Intelligent Systems*, Vol. 14, No. 6, November/December 1999, pp. 59–66.

Rogers (1992) SPIE Proceedings Vol. 1709 *Applications of Artificial Neural Networks III*, Steven K. Rogers (ed.)

Bradley, P.S.; Fayyad, U.; Reina, C (1998) Proceedings Fourth International Conference on Knowledge Discovery and Data Mining, Agrawal, R., Stolorz, P. (Eds), pp. 9–15.

D. A. Bell, S. S. Anand, and C. M. Shapcott. Database Mining in Spatial Databases. *International Workshop on Spatio-Temporal Databases*, 1994.

Cromp, R. F. and Cambell, W. J. Data Mining of Multidimensional Remotely Sensed Images In *Proceedings of 2nd International Conference on Information and Knowledge Management* CIKM 93. (Nov. 1–5, Arlington, VA), ACM, New York, USA, pp. 471–480.

S. Dutta. Qualitative Spatial Reasoning: A Semi-quantitative Approach Using Fuzzy Logic. In *Proc 1st Symp. SSD '89*, pp. 345–364, Santa Barbara, CA, July 1989.

S. Dutta. Topological Constraints: A Representational Framework for Approximate Spatial and Temporal Reasoning. In *Proc. 2nd Symp. SSD'91*, pp. 161–182, Zurich, Switzerland, August 1991.

Falousos, C. and King-Ip, L. FastMap: A Fast Algorithm for Indexing, Datamining and Visualization of traditional and Multimedia datasets In *SIGMOD Record* 24, 2 (June 1995), 163–174.

J. Guan and D. Bell. *Evidence Theory and its Applications, vol. I.* North-Holland, 1991.

J. Han, S. Nishio, and H. Kawano. *Knowledge Discovery in Object-Oriented and Active Databases. In F. Fuchi and T. Yokoi (eds), Knowledge Building and Knowledge Sharing, Ohmsha/IOS Press*, pp. 221–230, 1994.

Keim, D. and Kriegel, H.-P: Visualization Techniques for Mining Large Databases: A Comparison *IEEE Trans. Knowledge and Data Engineering* 8,6 (Dec. 1996), 923–938.

R. Laurini and D. Thompson: *Fundamentals of Spatial Information Systems.* Academic Press, London, 1995.

L. Mohan and R. L. Kashyap: An Object-Oriented Knowledge Representation for Spatial Information. In *IEEE Transactions on Software Engineering*, 5:675–681, May 1988.

P. Van Oosterom and J. van den Bos: An Object-oriented Approach to the Design of Geographic Information System. In *Proc. 1st Symp. SSD'89*, pp. 255–269, Santa Barbara, CA, July 1989.

Andersen, Bjorn 1990. Methodological Errors in Medical Research: an incomplete catalogue. Blackwell Scientific Publications, Oxford.

Bartholomew, D. J. 1995. What is statistics? Journal of the Royal Statistical Society A 158: 1–20.

Begg, C., Cho, M., Eastwood, S., Horton, R., Moher, D., Olkin, I., Pitkin, R.,Rennie, D., Schulz, K. F., Simel, D., and Stroup, D. F. 1996. Improving the Quality of Reporting of Randomised Controlled Trials: the CONSORT Statement. Journal of the American Medical Association 276: 637–639.

Bryan-Jones, J. and Finney, D. J. 1983. On an error in "Instructions to Authors". HortScience 18: 279–282.

Bussell, W. T., Maindonald, J. H. and Morton, J. R. 1997. What is a correct plant density for transplanted green asparagus? New Zealand Journal of Crop & Horticultural Science 25: 359–368.

Chalmers, I. and Grant, A. 1996. Salutory lessons from the collaborative eclampsia trial. Evidence-Based Medicine 1: 39. Available at http://www.acponline.org/journals/ebm/janfeb96/notebook.htm

Cheng, B. and Titterington, D. M. 1994. Neural networks: A review from a statistical perspective. Statistical Science 9: 2–54. Cleveland, W. S. 1993. Visualizing Data. Hobart, Summit, New Jersey.

Cochrane Injuries Group Albumin Reviewers 1998. Human albumin administration in critically ill patients: systematic review of randomised controlled trials. British Medical Journal 317: 235–240.

Comuzzie, A. G., Williams, J. T., and Blangero, J. 1999. The effect of ascertainment correction on linkage results in the COGA data set: A cautionary note. Genetic Epidemiology, to appear.

DeBaun, M. R. and Sox, H. C. 1991. Setting the optimal erythrocyte protoporphyrin screening decision threshold for lead poisoning: A decision analytic approach. Pediatrics 88: 121–131.

Draper, D; Gaver, D P; Goel, P K; Greenhouse, J B; Hedges L V; Morris, C N; Tucker, J R; Waternaux, C M 1992. Combining Information. Statistical Issues and Opportunities for Research. National Academy Press, Washington D.C.

Edelstein. H. 1998. Quoted in Wilson, R. Beware of fool's gold in the data mine. Canberra Times, Tuesday Nov. 10 1998, p. 55.

Elder, J. F. & Abbott, D. W. 1998. A comparison of leading data mining tools. Fourth International Conference on Knowledge Discovery & Data Mining. [Available from http://www.datamininglab.com]

Elder, J. F. and Pregibon, D. 1996. A statistical perspective on knowledge discovery in databases. In Fayyad, U. M., Piatetsky-Shapiro, G., Smyth, P. and Uthurusamy, R.: Advances in Knowledge Discovery and Data Mining, pp. 83–113. AAAI Press/MIT Press, Cambridge, Massachusetts.

Fayyad, U. M., Piatetsky-Shapiro, G. and Smyth, P. 1996. From data mining to knowledge discovery: An overview. In Fayyad, U. M., Piatetsky-Shapiro, G., Smyth, P. and Uthurusamy, R.: Advances in Knowledge Discovery and Data Mining, pp. 1–34. AAAI Press/MIT Press, Cambridge, Massachusetts.

Feinstein, A. 1984. The role of observational databases in the evaluation of therapy. Statistics in Medicine 3: 341–345.

Freedman, D. A. 1991. Statistical models and shoe leather, with discussion by R. Berk, H. M. Blalock and W. Mason. In Marsden, P., ed., Sociological Methodology 1991, pp. 291–358.

Friedman, J. H. 1997. Data Mining and Statistics. What's the Connection? Proc. of the 29th Symposium on the Interface: Computing Science and Statistics, May 1997, Houston, Texas.

Friedman, J. H. 1998. Statistics 315B: Statistical Aspects of Data Mining (Winter 1998). Available from http://www.stanford.edu/~jhf/Stat315B.html

Friedman, J. H. and Fisher, N. I. 1998. Bump hunting in high dimensional data. Available from http://www.stat.stanford.edu/reports/friedman/SuperGEM/prim.ps.Z

Gale, W. A. and Pregibon, D. 1984. REX: an Expert System for Regression Analysis, Proc. Compstat 84, Prague, pp. 242–248.

Gartland, J. 1988. Orthopaedic clinical research. Deficiencies in experimental design and in determinations of outcome. Journal of Bone and Joint Surgery 70: 1357–1364.

Green, S. B. and Byar, D. P. 1984. Using observational data from registries to compare treatments: the fallacy of omnimetrics. Statistics in Medicine 3: 361–370.

Hampel, F. 1998. Is statistics too difficult? Canadian Journal of Statistics 26: 497–513. House Committee on Science 1998. Unlocking Our Future. Toward a New National Science Policy. Available from http://www.house.gov/science/science_policy_report.htm

Kolsky, J. 1998. Statistics and data mining in the analysis of large data sets. Available from http://www.infosense.com/news/article/article1.html

Kuhn, T., 2nd edn, 1970. The Structure of Scientific Revolutions. University of Chicago Press, Chicago.

Lim, T.-S., Loh, W.-Y. and Shih, Y.-S. 1997, revised 1998. An empirical comparison of decision trees and other classification methods. Technical report 979, Department of Statistics, University of Wisconsin. [Available from http://www.stat.wisc.edu/p/stat/ftp/pub/loh/treeprogs/quest1.7/techrep.zip]

Maindonald J. H. 1992. Statistical design, analysis and presentation issues. New Zealand Journal of Agricultural Research 35: 121–141.

Maindonald J. H. and Cox, N. R. 1984. Use of statistical evidence in some recent issues of DSIR agricultural journals. New Zealand Journal of Agricultural Research 27: 597–610.

Mishra, B. 1998. Special Topics in Math Biology. Computational Genomics: G63.2856.002/G22.3033.006 [Course Summary]. Available from http://cs.nyu.edu/cs/faculty/mishra/COURSES/COBIO/cobio.html

Mitchell, T. 1997. Machine Learning. McGraw-Hill, New York.

Morrison, R. 1998. Communicators count the costs. (An interview with Robyn Williams.) Available from http://www.abc.net.au/rn/science/ockham/stories/s17313.htm Moynihan, R. 1998. Too Much Medicine. Australian Broadcasting Corporation.

Ripley, B. D. 1996. Pattern Recognition and Neural Networks. Cambridge University Press, Cambridge U. K.

Sackett, D. L., Richardson, W. S., Rosenberg, W. M. C. and Haynes, R. B. 1997. Evidence-Based Medicine. Churchill Livingstone, New York.

Sackett, D. L., Rosenberg, W. M. C., Gray, J. A. M., Haynes, R. B. and Richardson, W. S. 1998. Evidence-Based Medicine: What it is and what it isn't. Available from http://cebm.jr2.ox.ac.uk/ebmisisnt.html

Sagan, C. 1979. Broca's Brain. Random House, New York. ScHARR (School of Health and Related Research, University of Sheffield). 1998. Netting the Evidence. A ScHARR Introduction to Evidence Based Practice on the Internet. Available at http://www.shef.ac.uk/~scharr/ir/netting.html

Simoudis, E. 1996. Reality check for data mining. IEEE Expert 11:26–33.

Skoufos, E. 1998. Bioinformatics. Scientific discipline or support field. In HMS Beagle: The BioMedNet Magazine (http://hmsbeagle.com/hmsbeagle/1997/01/resnews/meeting.htm), Issue 43 (Nov. 27).

Smith, A. F. M. 1996. Mad cows and ecstasy: chance and choice in an evidence-based society. Journal of the Royal Statistical Society A 159: 367–383.

Transborder 1997. Bits of Power. Issues in Global Access to Scientific Data/Committee on Issues in the Transborder Flow of Scientific Data, U.S. National Committee for CODATA, Commission on Physical Sciences, Mathematics, and Applications, National Research Council. National Academy Press, Washington D. C. Available from http://www.nap.edu/readingroom/books/BitsOfPower/index.html/

Wegman, E. J. 1995. Huge data sets and the frontiers of computational feasibility. Journal of Computational and Graphical Statistics 4: 281–295.

R. Agrawal, H. Mannila, R. Srikant, H. Toivonen, and A. I. Verkamo. Fast discovery of association rules. In U. Fayyad,

G. Piatetsky-Shapiro, P. Smyth, and R. Uthurusamy, editors, *Advances in Knowledge Discovery and Data Mining, AAAI Press/The MIT Press*, pages 307–328, 1996.

H. Ahonen, O. Heinonen, M. Klemettinen, and A. I. Verkamo. Applying data mining techniques in text analysis. In *Report C-1997-23, Dept. of Computer Science, University of Helsinki*, 1997.

G. Barish, C. A. Knoblock, Y.-S. Chen, S. Minton, A. Philpot, and C. Shahabi. Theaterloc: A case study in information integration. In *IJCAI Workshop on Intelligent Information Integration*, Stockholm, Sweden, 1999.

C. Borgelt. apriori. http://fuzzy.cs.Uni-Magdeburg.de/ ? borgelt/.

M. Craven, D. DiPasquo, D. Freitag, A. McCallum, T. Mitchell, K. Nigam, and S. Slattery. Learning to extract symbolic knowledge from the World

Wide Web. In *Proceedings of the Fifteenth National Conference on Artificial Intellligence (AAAI-98)*, pages 509–516, 1998.

M. Craven, D. DiPasquo, D. Freitag, A. McCallum, T. Mitchell, K. Nigam, and S. Slattery. Learning to construct knowledge bases from the world wide web. *Artificial Intelligence*, 2000.

J. S. Deogun, V. V. Raghavan, A. Sarkar, and H. Sever. Rough sets and data mining: Analysis of imprecise data. In T. Y. Lin and N. Cercone, editors, *Data mining: Trends in research and development*, pages 9–46. Kluwer Academic, 1996.

D. Freitag. *Machine Learning for Information Extraction in Informal Domains.* PhD thesis, Carnegie Mellon University, 1999.

M. Hearst. Untangling text data mining. In *Proceedings of ACL '99: the 37th Annual Meeting of the Association for Computational Linguistics*, 1999.

C. Knoblock, S. Minton, J. L. Ambite, N. Ashish, P. Modi, I. Muslea, A. G. Philpot, and S. Tejada. Modeling web sources for information integration. In *AAAI-98*, 1998.

N. Kushmerick. *Wrapper Induction for Information Extraction.* PhD thesis, University of Washington, 1997.

J. R. Quinlan. Learning logical definitions from relations. *Machine Learning*, 5:239–2666, 1990.

E. Riloff and R. Jones. Learning Dictionaries for Information Extraction Using Multi-level Boot-strapping. In *Proceedings of the Sixteenth National Conference on Artificial Intelligence (AAAI-99)*, pages 1044–1049. The AAAI Press/MIT Press, 1999.

S. Slattery and M. Craven. Combining statistical and relational methods for learning in hypertext domains. In *Proceedings of the 8th International Conference on Inductive Logic Programming (ILP-98)*, 1998.

S. Soderland, D. Fisher, and W. Lehnert. Automatically learned vs. hand-crafted text analysis rules. Technical Report TC-44, University of Massachusetts, Amherst, CIIR, 1997.

S. Soderland and W. Lehnert. Wrap-up: A trainable discourse module for information extraction. *Journal of Artificial Intelligence Research (JAIR)*, 2:131–158, 1994.

SPSS. Clementine. http://www.spss.com/clementine/.

Hinke, T.H., J. Rushing, H. Ranganath and S. J. Graves, "Techniques and Experience in Mining Remotely Sensed Satellite Data," Artificial Intelligence Review (AIRE, S4): Issues on the Application of Data Mining, pp. 503–531, 2001.

Keiser, K., J. Rushing, H. Conover, and S. J. Graves, "Data Mining System Toolkit for Earth Science Data," Earth Observation (EO) & Geo-Spatial (GEO) Web and Internet Workshop, Washington, D.C., February 1999.

Hinke, T. H., J. Rushing, S. Kansal, S. J. Graves, H. Ranganath, E. Criswell, "Eureka Phenomena Discovery and Phenomena Mining System," AMS 13th Int'l Conference on Interactive Information and Processing Systems (IIPS) for Meteorology, Oceanography and Hydrology, 1997.

XML: Extensible Markup Language, http://www.w3.org/XML

Fu, Yongjian, "Distributed Data Mining: An Overview", 8th IEEE International Conference on Network Protocols, November 2000.

Ramachandran, R., M. Alshayeb, B. Beaumont, H. Conover, S. J. Graves, N. Hanish, X. Li, S. Movva, A. McDowell, and M. Smith, "Earth Science Markup Language," 17th Conference on Interactive Information and Processing Systems for Meteorology, Oceanography, and Hydrology, 81st American Meteorological Society (AMS) Annual Meeting, Albuquerque, NM, January, 2001.

Hinke, Thomas, J. Novotny, "Data Mining on NASA's Information Power Grid," Proceedings of the Ninth IEEE International Symposium on High Performance Distributed Computing, Pittsburgh, Pennsylvania, August 1–4, 2000.

Steve Tanner, Ken Keiser, Helen Conover, Danny Hardin, Sara Graves, Kathryn Regner, and Matt Smith, EVE: An Environment for On-Orbit Data Mining, IJCAI Workshop on Knowledge Discovery from Distributed, Dynamic, Heterogeneous, Autonomous Data and Knowledge Sources, Seattle, Washington, August 4–10, 2001.

Ramachandran, R., H. Conover, S. J. Graves, K. Keiser, "Algorithm Development and Mining (ADaM) System for Earth Science Applications," Second Conference on Artificial Intelligence, 80th AMS Annual Meeting, Long Beach, CA, January, 2000.

Rushing, J., H. Ranganath, T. Hinke, S. Graves, "Using Association Rules as Texture Features", IEEE Transactions on Pattern Analysis and Machine Intelligence, 2001.

Allen, J., Alan Christie, Willima Fithen, John McHugh, Jed Pickel, Ed Stoner. 2000. *State of the Practice of Intrusion Detection Technologies.* CMU/SEI-99-TR-028. Carnegie Mellon Software Engineering Institute. (Downloaded from http://sei.cmu.edu/publications/documents/99.reports/99tr028abstract.html).

Lunt, T. 1993. Detecting intruders in computer systems. In *Proceedings of 1993 conference on auditing and computer technology.* (Downloaded from http://www2.csl.sri.com/nides/index5.html on 3 February 1999.)

Teng, H., K. Chen, and S. Lu. 1990. Adaptive real-time anomaly detection using inductively generated sequential patterns. In *Proceedings of 1990 IEEE computer society symposium on research in security and privacy held in Oakland, California, May 7–9, 1990*, by IEEE Computer Society, 278–84. Los Alamitos, CA: IEEE Computer Society Press.

Debar, H., M. Becker, and D. Siboni. 1992. A neural network component for an intrusion detection system. In *Proceedings of 1992 IEEE computer society symposium on research in security and privacy held in Oakland, California, May 4–6, 1992,* by IEEE Computer Society, 240–50. Los Alamitos, CA: IEEE Computer Society Press.

Lee, W., S. Stolfo, and K. Mok. 1998. Mining audit data to build intrusion detection models. In *Proceedings of the fourth international conference on knowledge discovery and data mining held in New York, New York, August 27–31, 1998*, edited by Rakesh Agrawal, and Paul Stolorz, 66–72. New York, NY: AAAI Press.

Ilgun, K., and A. Kemmerer. 1995. State transition analysis: A rule-based intrusion detection approach. *IEEE Transaction on Software Engineering* 21(3): 181–99.

Orchard, R. 1995. *FuzzyCLIPS version 6.04 user's guide.* Knowledge System Laboratory, National Research Council Canada.

Agrawal, R., and R. Srikant. 1994. Fast algorithms for mining association rules. In *Proceedings of the 20th international conference on very large databases held in Santiago, Chile, September 12–15, 1994*, 487–99. San Francisco, CA: Morgan Kaufmann. (Downloaded from http://www.almaden.ibm.com/cs/people/ragrawal/papers/vldb94_rj.ps on February 1999.)

Kuok, C., A. Fu, and M. Wong. 1998. Mining fuzzy association rules in databases. *SIGMOD Record* 17(1): 41–6. (Downloaded from http://www.acm.org/sigs/sigmod/record/ issues/9803 on 1 March 1999). Presented at the National Information Systems Security Conference (NISSC), October 16–19, 2000, Baltimore, MD.

Luo, J. 1999. *Integrating fuzzy logic with data mining methods for intrusion detection.* M.S. Thesis, Mississippi State University.

Mannila, H., and H. Toivonen. 1996. Discovering generalized episodes using minimal occurrences. In *Proceedings of the second international conference on knowledge discovery and data mining held in Portland, Oregon, August, 1996*, by AAAI Press, 146–51. (Downloaded from http://www.cs.Helsinki.FI/research/fdk/ datamining/pubs on 19 February 1999.)

Porras, P., and A. Valdes. 1998. Live traffic analysis of TCP/IP gateways. In *Proceedings of the 1998 ISOC symposium on network and distributed systems security held in March, 1998.* (downloaded from http://www2.csl.sri.com/emerald/downloads.html on 1 March 1999.)

Wang, W., and S. Bridges. 1999. Genetic algorithm optimization of membership functions for mining fuzzy association rules. Submitted for publication to the 7th International Conference on Fuzzy Theory and Technology (FT&T 2000).

Shi, Fajun, Susan M. Bridges, Rayford B. Vaughn 2000. The Application of Genetic Algorithms for Feature Selection in Intrusion Detection. In preparation.

Mukkamala, R., J. Gagnon, and S. Jajodia. 2000. Integrating data mining techniques with intrusion detection methods. In *Research Advances in Database and Information Systems Security*, Vijay Atluri and John Hale, editors, Kluwer Publishers, Boston, MA. 33–46.

Zadeh, L. A. 1973. Outline of a new approach to the analysis of complex systems and decision processes. *IEEE Transactions on Systems, Man, and Cybernetics*, SMC-3.

Freier, A.O., Karlton, P., Kocher, P.C., Transport Layer Security Working Group. Netscape Communications, November 18, 1996. http://wp.netscape.com/eng/ssl3/draft302.txt

R. Housley Cryptographic Message Syntax (CMS)Algorithms http://www.ietf.org/rfc/rfc3370.txt

E. Rescorla, A. Schiffman The Secure HyperText Transfer Protocol http://www.terisa.com/shttp/1.2.1.txt

Box, D., Ehnebuske, D., Kakivaya, G., Layman, A., Mendelsohn, N., Nielsen, H.F., Thatte, S., Winer, D., Simple Object Access Protocol (SOAP) 1.1,W3C Note 08 May 2000 http://www.w3.org/TR/SOAP.

Hallam-Baker, P., XML Key Management Specifications http://www.w3.org/TR/xkms2/

Bartel, M., Boyer, J., Fox, B., LaMacchia, B., Simon, E., XML Signature Syntax and Processing W3C Recommendation 12 February 2002 http://www.w3.org/TR/xmldsig-core/

Imamura, T., Dillaway, B., Simon, E., XML Encryption Syntax and Processing, W3C Proposed Recommendation 03 October 2002. http://www.w3.org/TR/xmlenc-core/

DES NIST FIPS 46–3: Data Encryption Standard(DES). October1999. http://csrc.nist.gov/publications/fips/fips46–3/fips46–3.pdf

HTTP RFC 2616: Hypertext Transfer Protocol – HTTP/1.1..J. Gettys, J. Mogul, H. Frystyk, L. Masinter, P. Leach, and T. Berners-Lee. Standards Track, June 1999. http://www.ietf.org/rfc/rfc2616.txt

N. R. Adam and J. C. Wortman. Security-control methods for statistical databases. *ACM Computing Surveys*, 21(4):515– 556, Dec. 1989.

D. Agrawal and C. C. Aggarwal. On the Design and Quantification of Privacy Preserving Data Mining Algorithms.*PODS* 2003, June 912, 2003, San Diego, CA. Copyright 2003 ACM1581136706/ 03/06. . . $5.00. *Proc. of the 20th ACM Symposium on Principles of Database Systems*, pages 247–255, Santa Barbara, California, May 2001.

R. Agrawal, A. Evfimievski, and R. Srikant. Information sharing across private databases. In *Proc. of the 2003 ACM SIGMOD Int'l Conf. on Management of Data*, San Diego, CA, June 2003.

R. Agrawal and J. Kiernan. Watermarking relational databases. In *28th Int'l Conference on Very Large Databases*, Hong Kong, China, August 2002.

R. Agrawal, J. Kiernan, R. Srikant, and Y. Xu. Hippocratic databases. In *Proc. of the 28th Int'l Conference on Very Large Databases*, Hong Kong, China, August 2002.

R. Agrawal and R. Srikant. Privacy preserving data mining. In *ACM SIGMOD Conference on Management of Data*, pages 439–450, Dallas, Texas, May 2000.

Evfimievski, R. Srikant, R. Agrawal, and J. Gehrke. Privacy preserving mining of association rules. In *Proc. of the 8th ACM SIGKDD Int'l Conference on Knowledge Discovery and Data Mining*, Edmonton, Canada, July 2002.

M. Kantarcioglu and C. Clifton. Privacy-preserving distributed mining of association rules on horizontally partitioned data. In *ACM SIGMOD Workshop on Research Issues on Data Mining and Knowledge Discovery*, June 2002.

Y. Lindell and B. Pinkas. Privacy preserving data mining. *Journal of Cryptology*, 15(3):177–206, 2002.

S. J. Rizvi and J. R. Haritsa. Privacy-preserving assooiation rule mining. In *Proc. of the 28th Int'l Conference on Very Large Databases*, August 2002.

J. Vaidya and C. W. Clifton. Privacy preserving association rule mining in vertically partitioned data. In *Proc. of the 8th ACM SIGKDD Int'l Conferenceon Knowledge Discoveryand Data Mining*, Edmonton, Canada, July 2002.

L. Willenborg and T. deWaal. *Elements of Statistical Disclosure Control.* Springer Verlag Lecture Notes in Statistics 155, 2000.

C. Yao. How to generate and exchange secrets. In *Proc. of the 27th Annual Symposium on Foundations of Computer Science*, pages 162–167, Toronto, Canada, October 1986.

Technology Policy and Assessment Center (TPAC) at Georgia Institute of Technology. http://www.tpac.gatech.edu Porter AL. Text Mining For Technology Foresight. http://www.tpac.gatech.edu/~darius/papers/foresight-outline.html, 2000.

Office of Naval Research (ONR) Science & Technology.
http://www.onr.navy.mil/sci_tech/special/technowatch/default.htm

Kostoff RN. Information Extraction From Scientific Literature with Text Mining, 2001.

A Roadmap to Text Mining and Web Mining.
http://www.cs.utexas.edu/users/pebronia/text-mining/

Van Gemert J. Text Mining Tools on the Internet. *Intelligent Sensory Information Systems (ISIS) technical report series*, September, 2000.

Watts RJ, and Porter AL. Innovation Forecasting.
http://www.tpac.gatech.edu/toa/inov.shtml

Porter AL and Detampel MJ. Technology Opportunities Analysis. *Technological Forecasting & Social Change*, Vol. 49, 237–255, 1995.

Suhail Ansari, Ron Kohavi, Llew Mason, and Zijian Zheng, Integrating E-Commerce and Data Mining, *Blue Martini Software Technical Report*, 2001. Available from the articles section of http://developer.bluemartini.com.

Eric Schmitt, Harley Manning, Yolanda Paul, and Joyce Tong, Measuring Web Success, *Forrester Report*, November 1999.

Gregory Piatetsky-Shapiro, Ron Brachman, Tom Khabaza, Willi Kloesgen, and Evangelos Simoudis, An Overview of Issues in Developing Industrial Data Mining and Knowledge Discovery Applications, *Proceeding of the second international conference on Knowledge Discovery and Data Mining*, 1996.

Ralph Kimball, *The Data Warehouse Toolkit: Practical Techniques for Building Dimensional Data Warehouses*, John Wiley & Sons, 1996.

Ralph Kimball, Laura Reeves, Margy Ross, Warren Thornthwaite, *The Data Warehouse Lifecycle Toolkit: Expert Methods for Designing, Developing, and Deploying Data Warehouses*, John Wiley & Sons, 1998.

Ron Kohavi, Mining e-commerce data: The good, the bad, and the ugly (invited industrial track talk). In Foster Provost and Ramakrishnan Srikant, editors, *Proceedings of the Seventh ACM SIGKDD International Conference on Knowledge Discovery and Data Mining*, August 2001. http://robotics.Stanford.EDU/users/ronnyk/goodBadUglyK DDItrack.pdf

Robert Cooley, Bamshad Mobashar, and Jaideep Shrivastava, Data Preparation for Mining World Wide Web Browsing Patterns, *Knowledge and Information Systems*, 1, 1999.

Bettina Berendt, Bamshad Mobasher, Myra Spiliopoulou, and Jim Wiltshire, Measuring the Accuracy of Sessionizers for Web Usage Analysis, *Workshop on Web Mining at the First SIAM International Conference on Data Mining*, 2001.

J. Pitkow, In search of reliable usage data on the WWW, *Sixth International World Wide Web Conference*, 1997.

Shahana Sen, Balaji Padmanabhan, Alexander Tuzhilin, Norman H. White, and Roger Stein, The identification and satisfaction of consumer analysis-driven information needs of marketers on the WWW, *European Journal of Marketing*, Vol. 32 No. 7/8 1998.

Osmar R. Zaiane, Man Xin, and Jiawei Han, Discovering Web Access Patterns and Trends by Applying OLAP and Data Mining Technology on Web Logs, *Proceedings of Advances in Digital Libraries Conference (ADL'98)*, Santa Barbara, CA, 1998.

B. Mobasher, H. Dai, T. Luo, M. Nakagawa, Y. Sun, and J. Wiltshire, Discovery of Aggregate Usage Profiles for Web Personalization, *Proceedings of KDD '2000 Workshop on Web Mining for E-Commerce - Challenges and Opportunities (WEBKDD '2000)*, p. 1–11, 2000.

Jim Nail, Chris Charron, Tim Grimsditch, and Susan Shindler, The Email Marketing Dialogue, *Forrester Report*, January 2000.

Barry Becker, Ron Kohavi, and Dan Sommerfield. Visualizing the Simple Bayesian Classifier, Chapter 18, pages 237–249, *Information Visualization in Data Mining and Knowledge Discovery*, Morgan Kaufmann, Publishers, San Francisco, 2001.

Saharon Rosset, Uzi Murad, Einat Neumann, Yizhak Idan, and Gadi Pinkas, Discovery of Fraud Rules for Telecommunications: Challenges and Solutions, *Proceedings of the Fifth ACM SIGKDD International Conference on Knowledge Discovery and Data Mining*, 1999.

Hussein Almuallim, Yasuhiro Akiba, and Shigeo Kaneda, On Handling Tree-Structured Attributes, *Proceedings of the Twelfth International Conference on Machine Learning*, p. 12–20, 1995.

Abu-Mostafa: Learning from hints in neural networks. *Journal of complexity* 6: 192–198, 1990.

Bergadano, F., Giordana, A., & Ponsero, S. (1989). Deduction in top-down inductive learning. Proceedings of the Sixth International Workshop on Machine Learning (pp. 23–25). Ithaca, NY: Morgan Kaufmann.

Bratko, I., Muggleton, S., Varvsek, A. Learning qualitative models of dynamic systems. In *Inductive Logic Programming*, S. Muggleton, Ed. Academic Press, London, 1992.

Bratko, I. Innovative design as learning from examples. In *Proceedings of the International Conference on Design to Manufacture in Modern Industries*, Bled, Slovenia, June 1993.

Bratko I, Muggleton S (1995): Applications of inductive logic programming. *Communications of ACM* 38 (11):65–70. Carnap, R., Logical foundations of probability, Chicago, University of Chicago Press, 1962.

Danyluk, A. (1989). Finding new rules for incomplete theories: Explicit biases for induction with contextual information. Proceedings of the Sixth International Workshop on Machine Learning (pp. 34–36). Ithaca, NY: Morgan Kaufmann.

Dzeroski, S., DeHaspe, L., Ruck, B.M., and Walley, W.J. Classification of river water quality data using machine learning. In: ENVIROSOFT '94, 1994.

Dzeroski S (1996): Inductive Logic Programming and Knowledge Discovery in Databases. In: Advances in Knowledge Discovery and Data Mining, Eds. U. Fayad, G., Piatetsky-Shapiro, P. Smyth, R. Uthurusamy. AAAI, MIT Press, pp. 117–152.

Flach, P., Giraud-Carrier C., and Lloyd J.W. (1998). Strongly Typed Inductive Concept Learning. In Proceedings of the Eighth International Conference on Inductive Logic Programming (ILP '98), 185–194.

Flann, N., & Dietterich, T. (1989). A study of explanation-based methods for inductive learning. Machine Learning, 4, 187–226.

Halpern J. Y, An analysis of first-order logic of probability. *Artificial Intelligence* 46: 311–350, 1990.

Hirsh, H. (1989). Combining empirical and analytical learning with version spaces.

Hyafil L, Rivest RL, Constructing optimal binary decision trees is NP-Complete. *Information Processing Letters* 5 (1), 15–17, 1976.

Kendall M.G., Stuart A. (1977) *The advanced theory of statistics*, 4th ed., v.1.Charles Griffin & Co LTD, London.

D. Koller, A. Pfeffer, Learning probabilities for noisy first-order rules, In: Proc. of the 15th Int. Joint Conf. on Artificial Intelligence, Nagoya, Japan, 1997.

Kovalerchuk B (1973): Classification invariant to coding of objects. *Computational systems.* 55:90–97, Novosibirsk. (in Russian).

Kovalerchuk, B. (1976), Coordinating methods for decision rules and training data in pattern recognition. Ph. D. Diss., Institute of Mathematics, USSR Academy of Science, Novosibirsk, 146 p. (in Russian).

Kovalerchuk B, Vityaev E, Ruiz JF. (1997). Design of consistent system for radiologists to support breast cancer diagnosis. *Joint Conf. of Information Sciences*, Duke University, NC, 2: 118–121, 1997.

Kovalerchuk B, Vityaev E (1998): Discovering Lawlike Regularities in Financial Time Series. *Journal of Computational Intelligence in Finance* 6 (3): 12–26.

Kovalerchuk B., Vityaev E. Data Mining in Finance: Advances in Relational and Hybrid methods. (Kluwer international series in engineering and computer science; SECS 547), Kluwer Academic Publishers, 2000, p. 308.

Lebowitz, M. (1986). Integrated learning: Controlling explanation. Cognitive Science, 10. Mitchell (1997): Machine Learning, Prentice Hall.

Mitchell, T., Keller, R., & Kedar-Cabelli, S. (1986). Explanation-based learning: A unifying view. Machine Learning, 1, 47–80.

Mooney, R., & Ourston, D. (1989). Induction over the unexplained: Integrated learning of concepts with both explainable and conventional aspects. Proc. 6th International Workshop on Machine Learning (pp. 5–7). Ithaca, NY: Morgan Kaufmann.

Muggleton S. Bayesian inductive logic programming. In *Proceedings of the Eleventh International Conference on Machine Learning* W. Cohen and H. Hirsh, Eds. (1994), pp. 371–379.

Muggleton S (1999): Scientific Knowledge Discovery Using Inductive Logic Programming, Communications of ACM, vol. 42, N11, pp. 43–46.

Muggleton, S., & Buntine, W. (1988). Machine invention of first-order predicates by inverting resolution. Proceedings of the Fifth International Workshop on Machine Learning (pp. 339–352). Ann Arbor, MI: Morgan Kaufmann.

Pazzani, M. (1989). Explanation-based learning with weak domain theories. Proceedings of the Sixth International Workshop on Machine Learning (pp. 72–74). Ithaca, NY: Morgan Kaufmann.

Pazzani, M. J. (1990). Creating a memory of causal relationships: An integration of empirical and explanation-based learning methods. Hillsdale, NJ: Lawrence Erlbaum Associates.

Pazzani, M., Kibler, D. (1992). The utility of prior knowledge in inductive learning. Machine Learning, 9, 54–97.

Pazzani, M., (1997), Comprehensible Knowledge Discovery: Gaining Insight from Data. First Federal Data Mining Conference and Exposition, pp. 73–82. Washington, DC

Pazzani, M., Brunk, C. (1990), Detecting and correcting errors in rule-based expert systems: An integration of empirical and explanation-based learning. Proceedings of the Workshop on Knowledge Acquisition for Knowledge-Based System. Banff, Canada.

Pfanzagl J. (1971). Theory of measurement (in cooperation with V. Baumann, H. Huber) 2nd ed. Physica-Verlag.

Quinlan, J. R. (1990). Learning logical definitions from relations. Machine Learning, 5, 239–266.

Russel S, Norvig P (1995): Artificial Intelligence. A Modern Approach, Prentice Hall.

Samokhvalov, K., (1973). On theory of empirical prediction, (Comp. Syst., #55), 3–35. (In Russian)

Shavlik, J., & Towell, G. (1989). Combining explanation-based learning and artificial neural networks. Proceedings of the Sixth International Workshop on Machine Learning,pp. 90–93. Ithaca, NY: Morgan Kaufmann.

Ahlert, H.: Enterprise Customer Management: Integrating Corporate and Customer Information. In: Henning-Thurau, T., Hansen, U. (eds.): Relationship Marketing, Springer, Berlin Heidelberg New York (2000) 255–264.

Arndt, D.; Gersten, W.: External Data Selection for Data Mining in Direct Marketing. In: International Conference on Information Quality 2001, MIT, Boston (2001) (to appear)

Arndt, D.; Gersten, W.; Nakhaeizadeh, G.; Wirth, R.: eCustomers – How the Internet affects our relationship to our customers. In: DaimlerChrysler (ed.): eMagine: Journey into the eFuture, Stuttgart (2000), 23–27.

Berry, L.L.: Relationship Marketing. In: Berry, L.L., Shostack; G.D., Upah, G.D. (eds.): Emerging Perspectives in Service Marketing 1983, Chicago (1983) 25–28.

Berry, M.J.A., Linoff, G.S.: Mastering Data Mining. Wiley, New York (2000).

Chojnacki, K.: Relationship Marketing at Volkswagen. In: Henning-Thurau, T., Hansen, U. (eds.): Relationship Marketing, Springer. Berlin Heidelberg New York (2000) 49–58.

Diller, H.: Vahlens großes Marketinglexikon. Deutscher Taschenbuchverlag, München (1994).

ECCS: CRM Defining customer relationship marketing and management. In: http://www.eccs.uk.com/crm/crmdefinitions/define.asp, 27.01.1999 (1999).

English, L.P.: Improving Data Warehouse and Business Information Quality. Wiley, New York (1999).

Gersten, W., Wirth, R., Arndt, D.: Predictive Modeling in Automotive Direct Marketing: Tools, Experiences and Open Issues. In: Proceedings of the 6th International Conference on Machine Learning. ACM, New York (2000) 398–406.

Muther, A.: Electronic Commerce. Springer, Berlin Heidelberg New York (1999).

Pyle, D.: Data preparation for data mining. Morgan Kaufmann Publishers, San Francisco (1999).

Schmidt, R.E.: Mit Customer Relationship Management zum Prozessportal. In: Bach, V., Österle, H.: Customer Relationship Management in der Praxis. XY, Berlin et al. (2000).

Schweiger, A., Wilde, K.D.: Database Marketing – Aufbau und Management. in: Hilke, W.: Direct Marketing. Gabler, Wiesbaden (1993) 89–125.

Gallant, S., Piatesky-Shapiro, G. and Pyle, D. (2000): Successful customer relationship management in financial applications. Tutorial PM-1. KDD-2000, ACM SIGKDD 7th annual conference on Data Mining and Knowledge Discovery.

Ripley, B.D. (1996): Pattern Recognition and Neural Networks. Cambridge University Press, Cambridge UK.

Saarenvirta, G. (1998): Mining customer data. A step-by-step look at powerful clustering and segmentation methodology. DB2 magazine. http://www.db2mag.com/db_area/archives/1998/q3/98fsaar.shtml

Thearling, K. (2000). Data mining and customer relationships. http://www3.shore.net/~kht/text/whexcerpt/whexcerpt.htm. Excerpted from *Building Data Mining Applications for CRM* by Alex Berson, Stephen Smith, Kurt Thearling (McGraw Hill, 2000).

Barry, M.J.A. and G. Linoff (1997), *Data Mining Techniques: For Marketing, Sales, and Customer Support*, John Wiley, New York. Breiman, L., J.H. Friedman, R.A. Olshen, and P.J. Stone (1984), *Classification and Regression* Trees, Wadworth International Group, Belmont, CA.

Carlett, J. (1991), Megainduction: Machine Learning on Very Large Databases, Ph.D. Thesis, Department of Computer Science, University of Sydney, Australia.

Cattral, R., F. Oppacher, and D. Deugo (2001), Supervised and unsupervised data mining with an evolutionary algorithm, *Proceedings of the 2001 Congress on Evolutionary Computation*, IEEE Press, Piscataway, NJ, pp. 767–776.

Cios, K., W. Pedrycz, and R. Swiniarski (1998), *Data Mining: Methods for Knowledge Discovery*, Kluwer, Boston, MA.

Dugherty, D., R. Kohavi, and M. Sahami (1995), Supervised and unsupervised discretization of continuous features, *Proceedings of the 12th International Machine Learning Conference*, pp. 194–202.

Duda, R.O. and P.E. Hart (1973), *Pattern Recognition and Scene Analysis*, John Wiley, New York.

Fayyad, U.M. and K.B. Irani (1993), Multi-interval discretization of continuously-valued attributes for classification learning, *Proceedings of the 13th International Joint Conference on Artificial Intelligence*, pp. 1022–1027.

Fukunaga, K. (1990), *Introduction to Statistical Pattern Analysis*, Academic Press, San Diego, CA.

Han, J. and M. Kamber (2001), *Data Mining: Concepts and Techniques*, Morgan Kaufmann, San Diego, CA.

John, G., R. Kohavi, and K. Pfleger (1994), Irrelevant features and the subset selection problem, *Proceedings of the 11th International Conference on Machine Learning, ICLM.94*, Morgan Kaufmann, San Diego, CA, pp. 121–127.

Kovacs, T. (2001), What should a classifier system learn, *Proceedings of the 2001 Congress on Evolutionary Computation*, IEEE Press, Piscataway, NJ, pp. 775–782.

Kusiak, A. (2000), Decomposition in data mining: an industrial case study, *IEEE Transactions on Electronics Packaging Manufacturing*, Vol. 23, No. 4, pp. 345–353.

Kusiak, A. (2001a), Rough set theory: a data mining tool for semiconductor manufacturing, *IEEE Transactions on Electronics Packaging Manufacturing*, Vol. 24, No. 1, pp. 44–50.

Kusiak, A., J.A. Kern, K.H. Kernstine, and T.L. Tseng (2000), Autonomous decision-making: A data mining approach, *IEEE Transactions on Information Technology in Biomedicine*, Vol. 4, No. 4, pp. 274–284.

Pawlak Z. (1982), Rough sets, *International Journal of Information and Computer Science*, Vol. 11, No. 5, pp. 341–356.

Pawlak, Z. (1991), *Rough Sets: Theoretical Aspects of Reasoning About Data*, Kluwer, Boston, MA.

Quinlan, J.R. (1986), Induction of decision trees, *Machine Learning*, Vol. 1, No 1, pp. 81–106.

Ragel, A. and B. Cremilleux (1998), Treatment of missing values for association rules, *Proceedings of the Second Pacific Asia Conference, PAKDD.98*, Melbourne, Australia.

Slowinski, R. (1993), Rough set learning of preferential attitude in multi-criteria decision making, in Komorowski, J. and Ras, Z. (Eds), *Methodologies for Intelligent Systems*, Springer-Verlag, Berlin, Germany, pp. 642–651.

Tou, J.T. and R.C. Gonzalez (1974), *Pattern Recognition Principles*, Addison Wesley, New York.

Venables, W.N. and B.D. Ripley (1998), *Modern Statistics with S-PLUS*, Springer-Verlag, New York.

Wickens, G., S.E. Gordon, and Y. Liu (1998), *An Introduction to Human Factors Engineering*, Harper Collins, New York.

Wilson, S.W. (1995), Classifier fitness based on accuracy, *Evolutionary Computation*, Vol. 3, No. 2, pp. 149–175.

Bianco, A., Garcia Ben, M., Martinez, E. and Yohai, V. (1996). "Robust procedure for regression models with ARIMA errors", *COMPSTAT '96 Proceedings of Computational Statistics, 27–38*. Physica-Verlag.

Bruce, A. G. and Martin, R. D. (1993). "Tree based robust Bayesian estimation of time series structural models", in *Proceedings of the 25th Symposium on the Interface.*

Denby, L. and Martin, R. D. (1979). "Robust estimation of the auto regression parameter", *Journal of the American Statistical Assoc.*, **74**, 140–146.

Fox, A. J. (1972). "Outliers in time series", *Jour. of the Royal Statist. Soc.*, **B**, **34**, 350–363.

Chang, I., Tiao, G.C., and Chen, C. (1988). "Estimation of time series in the presence of outliers", *Technometrics*, **30**, No. 2, 193–204.

Martin, R. D., Samarov, A. and Vandaele, W. (1983). "Robust methods for ARIMA models", in Applied Time Series Analysis of Economic Data, edited by E. Zellner.

Martin, R. D. and Yohai, V. J. (1985). "Robustness in time series and estimating ARMA models", invited paper for *Handbook of Statistics, Volume V: Time Series in the Time Domain*, edited by Hannan, Krishnaiah and Rao, North-Holland.

Martin, R. D. and Yohai, V. J. (1996). "Highly robust estimation of autoregression integrated time series models", *Publicaciones Previas No. 89, Facultad de Ciencias Exactas y Naturales*, Universidad de Buenos Aires.

Muler, N. and Yohai, V. J. (2001). "Robust estimates for ARCH processes", to appear in *Jour. of Time Series Analysis.*

S+ChangeDetection User's-PLUS 2000, preprint, Insightful Corporation.

Tsay, R. S. (1988). "Outliers, level shifts and variance changes in time series", *Jour. Of Forecasting*, **7**, 1–20.

West, M. and Harrison, J. (1989). *Bayesian Forecasting and Dynamic Models*, Springer-Verlag.

C. Bettini. Mining temportal relationships with multiple granularities in time sequences. *IEEE Transactions on Data & Knowledge Engineering*, 1998.

G. Das, K. Lin, H. Mannila, G.Renganathan, and P. Smyth. Rule discovery from time series. In *Proceedings of the international conference on KDD and Data Mining(KDD-98)*, 1998.

G. Das, D. Gunopulos, and H. Mannila. Finding similar time seies. In *Principles of Knowledge Discovery and Data Mining '97*, 1997.

Cen Li and Gautam Biswas. Temporal pattern generation using hidden markov model based unsuperised classifcation. In *Proc. of IDA-99*, pages 245–256, 1999.

Wei Q. Lin and Mehmet A. Orgun. Applied hidden periodicity analysis for mining discretevalued time series. In *Proceedings of ISLIP-99*, pages 56–68, Demokritos Institute, Athens, Greece, 1999.

Wei Q. Lin and Mehmet A. Orgun. Temporal data mining using hidden periodicity analysis. In *Proceedings of ISMIS 2000*, University of North Carolina, USA, 2000.

Wei Q. Lin, Mehmet A. Orgun, and Graham Williams. Temporal data mining using multilevel-local polynomial models. In *Proceedings of IDEAL2000*, The Chinese University of Hongkong, Hong Kong, 2000.

S. Jajodia and S. Sripada O. Etzion, editor. *Temporal databases: Research and Practice.* Springer-Verlag,LNCS1399, 1998.

P. Baldi and S. Brunak. *Bioinformatics & The Machine Learning Approach.* The MIT Press, 1999.

Z. Huang. Clustering large data set with mixed numeric and categorical values. In *1st Pacific-Asia Conference on Knowledge Discovery and Data Mining*, 1997.

R. Agrawal, T. Imielinski, and A. N. Swami. Mining association rules between sets of items in large databases. In *Proc. of the 1993 Int'l Conf. on Management of Data*, pages 207–216, 1993.

R. Agrawal and R. Srikant. Fast algorithms for mining association rules in large databases. In *Proc. of the 1994 Int'l Conf. on Very Large Data Bases*, pages 487–499, 1994.

C. Bettini, S. Jajodia, and X.S. Wang. *Time granularities in databases, data mining, and temporal reasoning.* Springer-Verlag, 2000.

R. Kohavi and C. Brodley. 2000 knowledge discovery and data mining cup. Data for the Cup was provided by Blue Martini Software and Gazelle.com, 2000. http://www.ecn.purdue.edu/KDDCUP/.

B. Leban, D. McDonald, and D. Foster. A representation for collections of temporal intervals. In *Proc. Of AAAI-1986 5th Int'l Conf. on Artifical Intelligence*, pages 367–371, 1986.

Y. Li, P. Ning, X. S. Wang, and S. Jajodia. Discovering calendar-based temporal association rules. In *Proc. of the 8th Int'l Symposium on Temporal Representation and Reasoning*, 2001.

S. Ramaswamy, S. Mahajan, and A. Silberschatz. On the discovery of interesting patterns in association rules. In *Proc. of the 1998 Int'l Conf. on Very Large Data Bases*, pages 368–379, 1998.

R. Agrawal, T. Imielinski, and A. Swami, "Mining Association Rules between Sets of Items in Large Databases," in *Proc. of the ACM SIGMOD Int'l Conf. on Management of Data*, Washington D.C., 1993, pp. 207–216.

R. Agrawal and G. Psaila, "Active Data Mining," in *Proc. of the 1st Int'l Conf. on Knowledge Discovery and Data Mining*, Montreal, Canada, 1995.

R. Agrawal and R. Srikant, "Fast Algorithms for Mining Association Rules," in *Proc. of the 20th Int'l Conf. on Very Large Data Bases*, Santiago, Chile, 1994, pp. 487–499.

W.-H. Au and K.C.C. Chan, "An Effective Algorithm for Discovering Fuzzy Rules in Relational Databases," in *Proc. of the 7th IEEE Int'l Conf. on Fuzzy Systems*, Anchorage, Alaska, 1998, pp. 1314–1319.

W.-H. Au and K.C.C. Chan, "FARM: A Data Mining System for Discovering Fuzzy Association Rules," in *Proc. of the 8th IEEE Int'l Conf. on Fuzzy Systems*, Seoul, Korea, 1999, pp. 1217–1222.

W.-H. Au and K.C.C. Chan, "Classification with Degree of Membership: A Fuzzy Approach," in *Proc. of the 1st IEEE Int'l Conf. on Data Mining*, San Jose, CA, 2001.

K.C.C. Chan and W.-H. Au, "Mining Fuzzy Association Rules," in *Proc. of the 6th Int'l Conf. on Information and Knowledge Management*, Las Vegas, Nevada, 1997, pp. 209–215.

K.C.C. Chan and W.-H. Au, "Mining Fuzzy Association Rules in a Database Containing Relational and Transactional Data," in A. Kandel, M. Last, and H. Bunke (Eds.), *Data Mining and Computational Intelligence*, New York, NY: Physica-Verlag, 2001, pp. 95–114.

D.W. Cheung, J. Han, V.T. Ng, and C.Y. Wong, "Maintenance of Discovered Association Rules in Large Databases: An Incremental Updating Technique," in *Proc. of the 12th Int'l Conf. on Data Engineering*, New Orleans, Louisiana, 1996, pp. 106–114.

J.Y. Ching, A.K.C. Wong, and K.C.C. Chan, "Class-Dependent Discretization for Inductive Learning from Continuous and Mixed-Mode Data," *IEEE Trans. on Pattern Analysis and Machine Intelligence*, vol. 17, no. 6, pp. 1–11, 1995.

V. Ganti, J. Gehrke, R. Ramakrishnan, and W.-Y. Loh, "A Framework for Measuring Changes in Data Characteristics," in *Proc. of the 18thACM SIGMOD-SIGACT-SIGART Symp. on Principles of Database Systems*, Philadelphia, PA, 1999, pp. 126–137.

B. Liu, W. Hsu, and Y. Ma, "Integrating Classification and Association Rule Mining," in *Proc. of the 4th Int'l Conf. on Knowledge Discovery and Data Mining*, New York, NY, 1998.

B. Liu, Y. Ma, and R. Lee, "Analyzing the Interestingness of Association Rules from the Temporal Dimension," in *Proc. of the 1stIEEE Int'l Conf. on Data Mining*, San Jose, CA, 2001.

H. Mannila, H. Toivonen, and A.I. Verkamo, "Efficient Algorithms for Discovering Association Rules," in *Proc. of the AAAI Workshop on Knowledge Discovery in Databases*, Seattle, Washington, 1994, pp. 181–192.

R. Srikant and R. Agrawal, "Mining Quantitative Association Rules in Large Relational Tables," in *Proc. of the ACM SIGMOD Int'l Conf. On Management of Data*, Montreal, Canada, 1996, pp. 1–12.

Agrawal, R., T. Imielinski, et al. "Mining Association Rules between Sets of Items in Large Database", in *Proc. SIGMOD '93.*

Agrawal R., R. Srikant R. "Mining Sequential Patterns", In *Proceedings of the 11th International Conference on Data Engineering (ICDE '95)*, Tapei, Taiwan, March 1995.

Blake C. L. and Merz C. J., UCI Irvine ML Database Repository: http://www.ics.uci.edu/~mlearn/MLRepository.html, University of California, Irvine, Dept. Of Information and Computer, 1998.

Brin S., Motwani R., Ullman J. D., Tsur S., "Dynamic Itemset Counting and Implication Rules for Market Basket Data", in *Proc. SIGMOD '97.*

N. Lavrac, P. Flach, and B. Zupan "Rule Evaluation Measures: A Unifying View", in *Ninth International Workshop on Inductive Logic Programming (ILP '99)*, Vol. 1634 of Lecture Notes in Artificial Intelligence, pages 174–185. Springer-Verlag, June 1999.

R. Agrawal, T. Imielinski, and A. Swami, "Mining Association Rules between Sets of Items in Large Databases," in *Proc. of the ACM SIGMOD Int'l Conf. On Management of Data*, Washington D.C., 1993, pp. 207–216.

R. Agrawal and R. Srikant, "Fast Algorithms for Mining Association Rules," in *Proc. of the 20th Int'l Conf. on Very Large Data Bases*, Santiago, Chile, 1994, pp. 487–499.

W.-H. Au and K.C.C. Chan, "An Effective Algorithm for Discovering Fuzzy Rules in Relational Databases," in *Proc. of the 7th IEEE Int'l Conf. on Fuzzy Systems*, Anchorage, Alaska, 1998, pp. 1314–1319.

W.-H. Au and K.C.C. Chan, "FARM: A Data Mining System for Discovering Fuzzy Association Rules," in *Proc. of the 8th IEEE Int'l Conf. on Fuzzy Systems*, Seoul, Korea, 1999, pp. 1217–1222.

W.-H. Au and K.C.C. Chan, "Classification with Degree of Membership: A Fuzzy Approach," in *Proc. of the 1st IEEE Int'l Conf. on Data Mining*, San Jose, CA, 2001, pp. 35–42.

K.C.C. Chan and W.-H. Au, "Mining Fuzzy Association Rules," in *Proc. of the 6th Int'l Conf. on Information and Knowledge Management*, Las Vegas, Nevada, 1997, pp. 209–215.

K.C.C. Chan and W.-H. Au, "Mining Fuzzy Association Rules in a Database Containing Relational and Transactional Data," in A. Kandel, M. Last, and H. Bunke (Eds.), *Data Mining and Computational Intelligence*, New York, NY: Physica-Verlag, 2001, pp. 95–114.

J. Han and M. Kamber, *Data Mining: Concepts and Techniques*, San Francisco, CA: Morgan Kaufmann, 2001.

D. Hand, H. Mannila, and P. Smyth, *Principles of Data Mining*, Cambridge, MA: The MIT Press, 2001.

K. Hirota and W. Pedrycz, "Fuzzy Computing for Data Mining," *Proc. of the IEEE*, vol. 87, no. 9, pp. 1575–1600, 1999.

D.H. Lee and M.H. Kim, "Database Summarization Using Fuzzy ISA Hierarchies," *IEEE Trans. on Systems, Man, and Cybernetics – Part B: Cybernetics*, vol. 27, no. 4, pp. 671–680, 1997.

B. Liu, W. Hsu, and Y. Man, "Integrating Classification and Association Rule Mining," in *Proc. of the 4th Int'l Conf. on Knowledge Discovery and Data Mining*, New York, NY, 1998.

O. Maimon, A. Kandel, and M. Last, "Information-Theoretic Fuzzy Approach to Data Reliability and Data Mining," *Fuzzy Sets and Systems*, vol. 117, pp. 183–194, 2001.

H. Mannila, H. Toivonen, and A.I. Verkamo, "Efficient Algorithms for Discovering Association Rules," in *Proc. Of the AAAI Workshop on Knowledge Discovery in Databases*, Seattle, Washington, 1994, pp. 181–192.

J.R. Quinlan, *C4.5: Programs for Machine Learning*, San Mateo, CA: Morgan Kaufmann, 1993.

R. Srikant and R. Agrawal, "Mining Quantitative Association Rules in Large Relational Tables," in *Proc. Of the ACM SIGMOD Int'l Conf. on Management of Data*, Montreal, Canada, 1996, pp. 1–12.

R.R. Yager, "On Linguistic Summaries of Data," in G. Piatetsky-Shapiro and W.J. Frawley (Eds.), *Knowledge Discovery in Databases*, Menlo Park, CA: AAAI/MIT Press, 1991, pp. 347–363.

L. Zadeh, "Fuzzy Sets," *Inform. Control*, vol. 8, pp. 338–353, 1965.

Mary E. Califf, Relational Learning Techniques for Natural Language IE, *Ph.D. thesis*, Univ. Texas, Austin, www.cs.utexas.edu/users/mecaliff

Claire Cardie, 'Empirical methods in information extraction', *AI Journal*, 18(4), 65–79, 1997.

Fabio Ciravegna, Alberto Lavelli, and Giorgio Satta, 'Bringing information extraction out of the labs: the Pinocchio Environment', in *ECAI 2000, Proc. of the 14th European Conference on Artificial Intelligence*, ed., W. Horn, Amsterdam, 2000. IOS Press.

Fabio Ciravegna, 'Learning to Tag for Information Extraction from Text' in F. Ciravegna, R. Basili, R. Gaizauskas (eds.) *ECAI Workshop on*

Machine Learning for Information Extraction, Berlin, August 2000. (www.dcs.shef.ac.uk/~fabio/ecai-workshop.html)

Aaron Douthat, 'The message understanding conference scoring software user's manual', in *the 7th Message Understanding Conf.*, www.muc.saic.com

Dayne Freitag, 'Information Extraction from HTML: Application of a general learning approach', *Proc. of the 15th National Conference on Artificial Intelligence (AAAI-98)*, 1998.

Dayne Freitag and Andrew McCallum: 'Information Extraction with HMMs and Shrinkage', *AAAI-99 Workshop on Machine Learning for Information Extraction*, Orlando, FL, 1999, www.isi.edu/~muslea/RISE/ML4IE/

Dayne Freitag and Nicholas Kushmerick, 'Boosted wrapper induction', in F. Ciravegna, R. Basili, R. Gaizauskas (eds.) *ECAI 2000 Workshop on Machine Learning for Information Extraction*, Berlin, 2000, (www.dcs.shef.ac.uk/~fabio/ecai-workshop.html)

Ralph Grishman, 'Information Extraction: Techniques and Challenges. In *Information Extraction: A Multidisciplinary Approach to an Emerging Information Technology*, in M.T. Pazienza, (ed.), Springer, 97.

N. Kushmerick, D. Weld, and R. Doorenbos, 'Wrapper induction for information extraction', *Proc. of 15th International Conference on Artificial Intelligence, IJCAI-97*, 1997.

I. Muslea, S. Minton, and C. Knoblock, 'Wrapper induction for semi-structured, web-based information sources', in *Proc. of the Conference on Autonomous Learning and Discovery CONALD-98*, 1998.

Steven Soderland, 'Learning information extraction rules for semi-structured and free text', *Machine Learning*, (1), 1–44, 1999.

Roman Yangarber, Ralph Grishman, Pasi Tapanainen and Silja Huttunen: "Automatic Acquisition of Domain Knowledge for Information Extraction" In *Proc. of COLING 2000, 18th Intern. Conference on Computational Linguistics*, Saarbrücken, 2000.

R. Agrawal, H. Mannila, R. Srikant, H. Toivonen, and A. I. Verkamo. Fast discovery of association rules. In U.M. Fayyad, G. Piatetsky-Shapiro, P. Smyth, and R. Uthurusamy, editors, *Advances in Knowledge Discovery and Data Mining*, pages 307–328. AAAI Press, Menlo Park, CA., 1996.

Liu Bing, Wynne Hsu, and Yiming Ma. Integrating classification and association rule mining. In *Proceeding of the Fourth International Conference on Knowledge Discovery and Data Mining*, pages 80–96, New York, NY, 1998. AAAI Press.

Sergey Brin, Rajeev Motwani, and Craig Silverstein. Beyond market baskets: Generalizing association rules to correlations. In *Data Mining and Knowledge Discovery*, volume 2, pages 39–68, 1999.

J. Catlett. On changing continuous attributes into ordered discrete attributes. In *European Workshop on Machine Learning*, pages 164–178. Springer-Verlag, 1991.

David Jensen and Paul Cohen. Multiple comparisons in induction algorithms. In *Machine Learning (in press)*. Boston, MA: Kluwer, 1999.

R. Kohavi. A study of cross validation and boostrap for accuracy estimation and model selection.

In *Proceedings of the Fourteenth International Joint Conference on Artificial Intelligence*, pages 1137–1143. Morgan Kaufmann, 1995.

J. Ross Quinlan. *C4.5: Programs for Machine Learning.* Morgan Kaufmann Publishers, Inc., 1994.

R. Rymon. An SE-tree based characterization of the induction problem. In *Proceedings of the Tenth International Conference on Machine Learning*, pages 268–275. San Francisco: Morgan Kaufmann, 1993.

R. Vilalta, G. Blix, and L. A. Rendell. Global data analysis and the fragmentation problem in decision tree induction. In *9th European Conference on Machine Learning*, pages 312–326. Lecture Notes in Artificial Intelligence, Vol. XXX, Springer-Verlag, Heidelberg, Available: http://www.research.ibm.com/people/v/vilalta, 1997.

Ricardo Vilalta and Daniel Oblinger. A quantification of distance-bias between evaluation metrics in classification. In *Proceedings of the 17th International Conference on Machine Learning*, pages 1087–1094. Morgan Kaufman, 2000. [VO00]

Geoffrey I. Webb. Systematic search for categorical attribute-value data-driven machine learning. In N. Foo and C. Rowles, editors, *Proceedings of the Sixth Australian Joint Artificial Intelligence Conference*, pages 342–347, Singapore, 1993. World Scientific.

G. I.Webb. Opus: An efficient admissible algorithm for unordered search. *Journal of Artificial Intelligence Research*, 3:431–435, 1995.

Geoffrey I. Webb. Efficient search for association rules. In *Proceedings of the Sixth ACM SIGKDD International Conference on Knowledge Discovery and Data Mining*, pages 99–107, 2000.

A.P. White and W.Z. Liu. Bias in information-based measures in decision tree induction. *Machine Learning*, 15:321–329, 1994.

D. Wolpert. The lack of a priori distinctions between learning algorithms and the existence of a priori distinctions between learning algorithms. *Neural Computation*, 8:1341–142, 1996.

http://rock.cs.ndsu.nodak.edu/smiley/

http://midas.cs.ndsu.nodak.edu/~smiley/

Smiley: a web-based remote sensing data mining system

Dr. William Perrizo, Longjun Chen, Dennis Amundson Computer Science Department North Dakota State University

Genetic Algorithms in Search Optimization, and Machine Learning, Goldberg David E., First Edition, Addison Wesley Publishing Inc, 1989. Goldman.

http://www.cs.ndsu.NoDak.edu/~perrizo/classes/765/pct.html

Knowledge Discovery and Data Mining, Institute of Electrical Engineers, London 1999.

A Genetic Algorithm-Based Approach to Data Mining, Flockhart Ian W., Radcliffe Nicholas J., Department of Mathematics and Statistics University of Edinburgh, London, 1995.

Fundamentals of Database Systems, Third edition, Elmasiri R., Navathe S.B., Addison Wesley, 2000.

D. P. Benjamin, editor. *Change of Representation and Inductive Bias.* Kluwer Academic Publishers, Boston, 1990.

L. B. Booker, D. E. Goldberg, and J. H. Holland. Classifier Systems and Genetic Algorithms. *Artificial Intelligence*, 40:235–282, 1989.

J. Cheng and M. J. Druzdzel. AIS-BN: An adaptive importance sampling algorithm for evidential reasoning in large Bayesian networks. Journal of Artificial Intelligence Research (JAIR), 13:155–188, 2000.

G. F. Cooper and E. Herskovits. A Bayesian Method for the Induction of Probabilistic Networks from Data. Machine Learning, **9**(4):309–347, 1992.

G. F. Cooper. The computational complexity of probabilistic inference using bayesian belief networks. *Artificial Intelligence*, 42(2–3):393–405. Elsevier, 1990.

K. J. Cherkauer and J. W. Shavlik. Growing Simpler Decision Trees to Facilitiate Knowledge Discovery. In *Proceedings of the Second International Conference of Knowledge Discovery and Data Mining (KDD-96)*, Portland, OR, August, 1996.

K. A. DeJong, W. M. Spears, and D. F. Gordon. Using genetic algorithms for concept learning. *Machine Learning*, 13:161–188, Kluwer Academic Publishers, 1993.

G. Elidan and N. Friedman. Learning the Dimensionality of Hidden Variables. In *Proceedings of the Seventeenth Conference on Uncertainty in Artificial Intelligence (UAI-2001)*, Morgan-Kaufmann, 2001.

N. Friedman and M. Goldszmidt. *Learning Bayesian Networks From Data.* Tutorial, American National Conference on Artificial Intelligence (AAAI-98), Madison, WI. AAAI Press, San Mateo, CA, 1998.

N. Friedman, M. Linial, I. Nachman, and D. Pe'er, Using Bayesian networks to analyze expression data. In *Proceedings of the Fourth Annual International Conference on Computational Molecular Biology (RECOMB 2000)*, ACM-SIGACT, April 2000.

D. E. Goldberg. *Genetic Algorithms in Search, Optimization, and Machine Learning.* Addison-Wesley, Reading, MA, 1989.

C. Guerra-Salcedo and D. Whitley. Genetic Approach to Feature Selection for Ensemble Creation. In *Proceedings of the 1999 International Conference on Genetic and Evolutionary Computation (GECCO-99).* Morgan-Kaufmann, San Mateo, CA, 1999.

D. Heckerman, D. Geiger, and D. Chickering, Learning Bayesian networks: The combination of knowledge and statistical data. *Machine Learning*, 20(3):197–243, Kluwer, 1995.

R. L. Haupt and S. E. Haupt. *Practical Genetic Algorithms.* Wiley-Interscience, New York, NY, 1998.

G. Harik and F. Lobo. *A parameter-less genetic algorithm.* Illinois Genetic Algorithms Laboratory technical report 99009, 1999.

W. H. Hsu, M. Welge, T. Redman, and D. Clutter. Genetic Wrappers for Constructive Induction in High-Performance Data Mining. In *Proceedings of the Genetic and Evolutionary Computation Conference (GECCO-2000)*, Las Vegas, NV. Morgan-Kaufmann, San Mateo, CA, 2000.

W. H. Hsu, M. Welge, T. Redman, and D. Clutter. Constructive Induction Wrappers in High-Performance Commercial Data Mining and Decision Support Systems. *Knowledge Discovery and Data Mining*, Kluwer, 2002.

R. Kohavi and G. H. John. Wrappers for Feature Subset Selection. *Artificial Intelligence, Special Issue on Relevance, 97*(1–2):273–324, 1997.

M. Faupel. GAJIT genetic algorithm package.
URL: http://www.angelfire.com/ca/Amnesiac/gajit.html, 2000.

T. M. Mitchell. *Machine Learning.* McGraw-Hill, New York, NY, 1997.

S. L. Lauritzen and D. J. Spiegelhalter. Local computations with probabilities on graphical structures and their application to expert systems. *Journal of the Royal Statistical Society, Series B* 50, 1988.

R. E. Neapolitan. Probabilistic Reasoning in Expert Systems: Theory and Applications. Wiley-Interscience, New York, NY, 1990.

R. M. Neal. Probabilistic Inference Using Markov Chain Monte Carlo Methods. Technical Report CRG-TR-93–1, Department of Computer Science, University of Toronto, 1993.

J. Pearl and T. S. Verma, A theory of inferred causation. In *Principles of Knowledge Representation and Reasoning: Proceedings of the Second International Conference.* Morgan Kaufmann, San Mateo, CA, 1991.

M. Raymer, W. Punch, E. Goodman, P. Sanschagrin, and L. Kuhn, Simultaneous Feature Extraction and Selection using a Masking Genetic Algorithm, In *Proceedings of the* 7th *International Conference on Genetic Algorithms*, pp. 561–567, San Francisco, CA, July, 1997.

R. D. Schacter and M. A. Peot. Simulation approaches to general probabilistic inference on belief networks. In *Uncertainty in Artificial Intelligence 5*, p. 221–231, Elsevier Science Publishing Company, New York, NY, 1989.

R. L. Welch. Real-Time Estimation of Bayesian Networks. In *Proceedings of UAI-96*, Morgan-Kaufmann, 1996.

Aksoy, A. And T.B. Culver, Effect of sorption assumptions on aquifer remediation designs, *Groundwater*, 38(2), 200–208, 2000.

Albert L.A., and Goldberg, D.E., Efficient Evaluation Genetic Algorithms under Fitness Functions, IlliGAL Report No. 2001024, July 2001.

Cantu'-Paz, E., A survey of Parallel Genetic Algorithms, *Calculateurs Paralleles, Reseaux et Systems Repartis*, Vol. 10, No. 2, pp. 141–171, Paris: Hermes, 1998.

Cantu'-Paz, E., Designing efficient and accurate parallel genetic algorithms, PhD thesis, 1999.

Clement, T. P., RT3D - A modular computer code for simulating reactive multi-species transport in 3-Dimensional groundwater aquifers, *Battelle Pacific Northwest National Laboratory Research Report*, PNNLSA-28967. (http://bioprocesses.pnl.gov/rt3d.htm.), 1997.

Clement, T. P, Sun, Y., Hooker, B. S., and Petersen, J. N., Modeling multi-species reactive transport in groundwater, *Ground Water Monitoring and Remediation*, 18(2), 79–92, 1998.

Clement, T. P., Johnson, C. D., Sun, Y., Klecka, G. M., and Bartlett, C., Natural attenuation of chlorinated solvent compounds: Model development and field-scale application, *Journal of Contaminant Hydrology*, 42, 113–140, 2000.

Harik G. R., Cantu-Paz E., Goldberg D. E., and Miller B. L., The gambler's ruin problem, genetic algorithms and the sizing of populations, In *Proceedings of the 1997 IEEE Conference on Evolutionary Computation*, pp. 7–12, IEEE press, New York, NY, 1997.

Gopalakrishnan G., Minsker B., and Goldberg D.E., Optimal sampling in a Noisy Genetic Algorithm for Risk-Based Remediation Design, *Journal of Hydroinformatics*, in press, 2002.

Grefenstette J.J. and Fitzpatrick J.M., Genetic search with approximate function evaluations, In Grefenstette, J.J. (Ed.), *Proceedings of an International Conference on Genetic Algorithms and their Applications*, pp. 112–120, Hillsdale, NJ, 1985.

Hogg, R., and Craig, A., *Introduction to Mathematical Statistics.* Macmillan Publishing Co., Inc., New York, 1978.

Liu, Y., and B. S. Minsker, "Efficient multiscale methods for optimal in situ bioremediation design." *Journal of Water Resources and Planning Management*, in press, 2001.

McDonald, M.G., and Harbaugh, A.W. (1988). "A modular three-dimensional finite-difference ground-water flow model." Techniques of Water Resources Investigations 06-A1, United States Geological Survey.

Reed P., Minsker B. S., and Goldberg D. E., Designing a competent simple genetic algorithm for search and optimization, *Water Resources Research*, 36(12), 3757–3761, 2000.

Reed, P. Striking the Balance: Long-Term Groundwater Monitoring Design for Multiple Conflicting Objectives, Ph. D. Thesis, University of Illinois, 2002.

Ritzel, B.J., J.W. Eheart, and S. Ranjithan, Using genetic algorithms to solve a multiple objective groundwater pollution containment problem, *Water Resources Research*, 30(5), 1589–1603, 1994.

Smalley J. B., Minsker B. S., and Goldberg D. E., Riskbased In Situ bioremeditation design using a noisy genetic algorithm, *Water Resources Research*, 36(20), 3043-3052, 2000.

Wang, Q.J., The genetic algorithm and its application to calibrating conceptual runoff models, *Water Resource Research*, 27(9), 2467–2471, 1991.

Wang, M. And C. Zheng, Optimal remediation policy selection under general conditions, *Groundwater*, 35(5), 757–764, 1997.

John R. Koza, *Genetic Programming: On the Programming of Computers by Means of Natural Selection*, MIT Press, Cambridge, MA, USA, 1992.

Sean Luke, "Genetic programming produced competitive soccer softbot teams for robocup97," in *Genetic Programming 1998: Proceedings of the Third Annual Conference*, John R. Koza, Wolfgang Banzhaf, Kumar Chellapilla, Kalyanmoy Deb, Marco Dorigo, David B. Fogel, Max H. Garzon, David E. Goldberg, Hitoshi Iba, and Rick Riolo, Eds., University of Wisconsin, Madison, Wisconsin, USA, 22–25 July 1998, pp. 214–222, Morgan Kaufmann.

Reiko Tanese, "Parallel genetic algorithms for a hypercube," in *Proceedings of the Second International Conference on Genetic Algorithms*, John J. Grefenstette, Ed. 1987, Lawrence Erlbaum Associates, Publishers.

T. C. Fogarty and R. Huang, "Implementing the genetic algorithm on transputer based parallel processing systems," in *Parallel Problem Solving from Nature*, Berlin, Germany, 1991, pp. 145–149, Springer Verlag.

Reiko Tanese, *Distributed Genetic Algorithms for Function Optimization*, Ph.D. thesis, University of Michigan, 1989, Computer Science and Engineering.

David E. Goldberg, "Sizing populations for serial and parallel genetic algorithms," in *Proceedings of the Third International Conference on Genetic Algorithms*, J. D. Schaffer, Ed., San Mateo, CA, 1989, Morgan Kaufman.

Erick Cant'u-Paz, "Designing efficient master-slave parallel genetic algorithms," IllGAL Report 97004, The University of Illinois, 1997, Available on-line at: ftp://ftp-illigal.ge.uiuc.edu/pub/ papers/IlliGALs/97004.ps.Z.

Erick Cant'u-Paz, "Designing scalable multi-population parallel genetic algorithms," IllGAL Report 98009, The University of Illinois, 1998, Available on-line at: ftp://ftp-illigal.ge.uiuc.edu/pub/ papers/IlliGALs/98009.ps.Z.

Ricardo Bianchini and Christopher Brown, "Parallel genetic algorithms on distributed-memory architectures," Technical Report 436, The University of Rochester, The University of Rochester, Computer Science Department, Rochester, New York 14627, May 1993.

Shyh-Chang Lin, William F. Punch, and Erik D. Goodman, "Coarse-grain parallel genetic algorithms: Categorization and new approach," in *Proceedings of the Sixth IEEE Symposium on Parallel and Distributed Processing*, 1994, pp. 28–37.

Tsutomu Maruyama, Tetsuya Hirose, and Akihito Konagaya, "A fine-grained parallel genetic algorithm for distributed parallel systems," in *Proceedings of the Fifth International Conference on Genetic Algorithms*, Stephanie Forrest, Ed., San Mateo, CA, 1993, pp. 184–190, Morgan Kaufman.

John J. Grefenstette, Michael R. Leuze, and Chrisila B. Pettey, "A parallel genetic algorithm," in *Proceedings of the Second International Conference on Genetic Algorithms*, John J. Grefenstette, Ed. 1987, pp. 155–161, Lawrence Erlbaum Associates, Publishers (Hillsdale, NJ).

T. C. Belding, "The distributed genetic algorithm revisited," in *Proceedings of the Sixth International Conference on Genetic Algorithms*, L. Eschelman, Ed. 1995, pp. 114– 121, Morgan Kaufmann (San Francisco, CA).

David E. Goldberg, Kerry Zakrzewski, Brad Sutton, Ross Gadient, Cecilia Chang, Pillar Gallego, Brad Miller, and Eric Cant'u-Paz, "Genetic algorithms: A bibliography," IlliGAL Report 97011, Illinois Genetic Algorithms Lab. University of Illinois at Urbana-Champaign, December 1997.

Erick Cant'u-Paz, "A survey of parallel genetic algorithms," IllGAL Report 97003, The University of Illinois, 1997, Available on-line at: ftp://ftp-illigal.ge.uiuc.edu/pub/papers/ IlliGALs/97003.ps.Z.

Mariusz Nowostawski, "Parallel genetic algorithms in geometry atomic cluster optimisation and other applications," M.S. thesis, School of Computer Science, The University of Birmingham, UK, September 1998, http://studentweb.cs.bham..ac.uk/~mxn/gzipped/mpga-v0.1.ps.gz.

Shumeet Baluja, "A massively distributed parallel genetic algorithm (mdpga)," Technical Report CMU-CS-92-196R, Carnagie Mellon University, Carnagie Mellon University, Pittsburg, PA, 1992.

Shumeet Baluja, "The evolution of genetic algorithms: Towards massive parallelism," in *Proceedings of the Tenth International Conference on Machine Learning*, San Mateo, CA, 1993, pp. 1–8, Morgan Kaufmann.

David Goldberg, Kalyanmoy Deb, and Bradley Korb, "Messy genetic algorithms: Motiation, analysis, and first results," *Complex Systems*, vol. 3, pp. 493–530, 1989.

Bianchini, R., C. Brown. Parallel Genetic Algorithms on Distributed-Memory Architectures. Technical Report 436, Computer Science Department. University of Rochester, Rochester NY, August 1992.

Carriero, N., and D. Gelernter. How to Write Parallel Programs: A First Course. Massachusetts: MIT Press, 1991.

J. G. Elias, "Genetic generation of connection patterns for a dynamic artificial neural network," in *COGANN-92, Combinations of Genetic Algorithms and Neural Networks*, eds. L. D. Whitley and J. D. Schaffer, IEEE Computer Society Press, Los Alamitos, CA, pp. 38–54, 1992.

Brunk, C., Kelly, J. & Kohavi, R. (1997), MineSet: an integrated system for data mining, in D. Heckerman, H. Mannila, D. Pregibon & R. Uthurusamy, eds, 'Proceedings of the third international conference on Knowledge Discovery and Data Mining', AAAI Press, pp. 135 to 138. http://mineset.sgi.com

Fayyad, U. M., Piatetsky-Shapiro, G. & Smyth, P. (1996), 'The KDD process for extracting useful knowledge from volumes of data', Communications of the ACM 39(11), 27 to 34.

Kohavi, R. & Kunz, C. (1997), Option decision trees with majority votes, in D. Fisher, ed., 'Machine Learning: Proceedings of the Fourteenth International Conference', Morgan Kaufmann Publishers, Inc., pp. 161 to 169. http://robotics.stanford.edu/users/ronnyk.

Kohavi, R. & Sommerfield, D. (1998), Targeting business users with decision table classifiers, in R. Agrawal, P. Stolorz & G. Piatetsky-Shapiro, eds, 'Proceedings of the Fourth International Conference on Knowledge Discovery and Data Mining', AAAI Press, pp. 249 to 253.

Kohavi, R., Sommerfield, D. & Dougherty, J. (1997), 'Data mining using MLC++: A machine learning library in C++', International Journal on

Artificial Intelligence Tools 6(4), 537 to 566. http://www.sgi.com/Technology/mlc.

Quinlan, J. R. (1993), C4.5: Programs for Machine Learning, Morgan Kaufmann, San Mateo, California.

Silicon Graphics (1998), MineSet User's Guide, Silicon Graphics, Inc. http://mineset.sgi.com.

Java Specification Request 73: Java Data Mining (JDM), Version 1.0, Final Review.

XML for Analysis Specification version 1.0.

Predictive Model Markup Language, Version 2.1.2, http://www.dmg.org.

OLE DB for Data Mining Specification, Version 1.0.

SOAP Version 1.2, http://www.w3.org/TR/soap/.

WS-Security, http://www-106.ibm.com/developerworks/webservices/library/ws-secure/.

WS-Resource Framework, http://www.globus.org/wsrf/.

XML Specification, http://www.w3.org/TR/2000/REC-xml-20001006.

The Data-Mining Industry Coming Of Age Gregory Piatetsky-Shapiro, Knowledge Stream Partnerswww.kdnuggets.com/ gpspubs/ieee-intelligent-dec-1999-x6032.pdf

Current issues in modeling Data Mining processes and results Panos Xeros [pxeros@cti.gr]& Yannis Theodoridis [ytheod@cti.gr] PANDA informal meeting, Athens,19 June 2002 dke.cti.gr/panda/tasks/meetings/2002–06-Athens-informal/CTIpresentation-Athens-19June02.ppt

The Physiology of the Grid: An Open Grid Services Architecture for Distributed Systems Integration. I. Foster, C. Kesselman, J. Nick, S. Tuecke, Open Grid Service Infrastructure WG, Global Grid Forum, June 22, 2002.

Java Specification Request 73: Java Data Mining (JDM)–JDM Public review Draft 2003/11/25: JSR-73 Expert Group